The armouries of the Spanish nobility

Roberto González Ramos /
Jesús María Ruiz Carrasco /
Carlos Ramírez Raya (eds.)

The armouries of the Spanish nobility

Studies on a powerful signal of social distinction and inequality

PETER LANG

Berlin - Bruxelles - Chennai - Lausanne - New York - Oxford

Library of Congress Cataloging-in-Publication
A record in the CIP catalog has been requested for this book
of the Library of Congress.

Bibliographic information published by the Deutsche Nationalbibliothek
The Deutsche Nationalbibliothek lists this publication in the Deutsche Nationalbibliografie; detailed bibliographic data is available on the Internet at http://dnb.d-nb.de.

The publication of this book was supporte by the European Union:

European Union (Fondo Europeo de Desarrollo Regional); Agencia Estatal de Investigación, Ministerio de Ciencia, Innovación y Universidades, Gobierno de España. Program: Programa Estatal de Generación de Conocimiento y Fortalecimiento Científico y Tecnológico del Sistema de I+D+i y en el marco del Programa Estatal de I+D+i Orientada a los Retos de la Sociedad, del Plan Estatal de Investigación Científica y Técnica y de Innovación 2017-2020.

Cover illustration © Source : Allegory of the Sense of Touch (detail). 1618.
Peter P. Rubens and Jan Brueghel the Elder. Madrid, Museo del Prado

ISBN 978-3-631-91557-8 (Print)
E-ISBN 978-3-631-91558-5 (E-PDF)
E-ISBN 978-3-631-91559-2 (E-PUB)
DOI.10.3726/b21950

Published by Peter Lang GmbH, Berlin, Germany

info@peterlang.com - www.peterlang.com

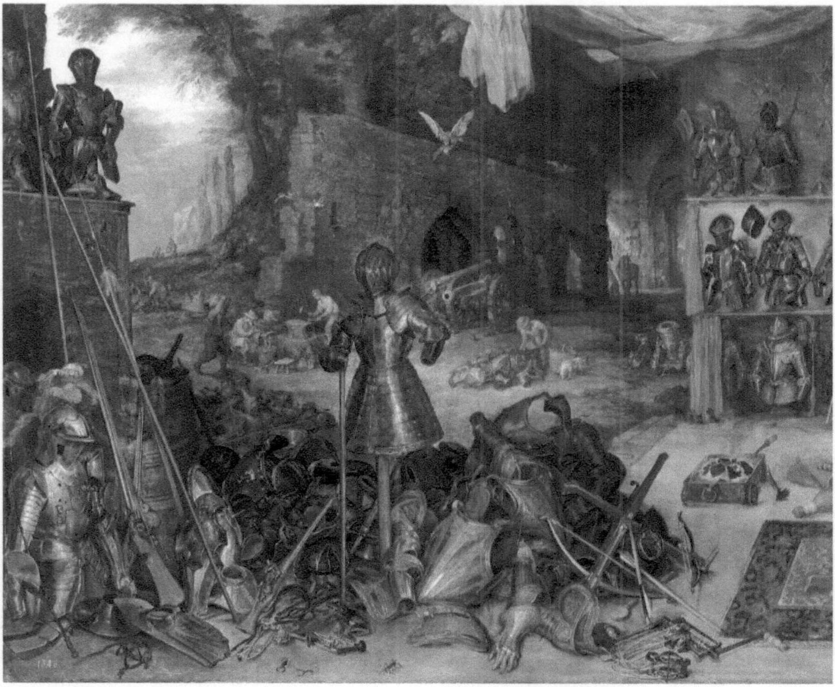

The Sense of Touch (Detail). 1618. Peter P. Rubens and Jan Brueghel the Elder. Madrid, Museo Nacional del Prado. Foto: © Museo Nacional del Prado

The present book is the result of the labours of the investigative team of the I+D+i project entitled "Las armerías de la nobleza española. Identidad y cultura aristocrática en la Edad Moderna [The Amouries of the Spanish Nobility: Aristocratic Identity and Culture in the Modern Age]" (ref. PID2019-104250GB-100), subsidised by the Spanish government's Ministerio de Economía y Competitividad-Agencia Estatal de Investigación, assigned to the Universidad de Córdoba (Spain). This project has developed investigations centred on a theme as novel as it is important for understanding the development of Spanish noble culture as a fundamental part of the History of Collecting, all within the framework of Social History and Cultural History.

The noble armouries were one of the main symbolic visual manifestations of their owner's power and social position. They were sets that seemed to emerge as symbolic gatherings of objects in the second half of the sixteenth century and appear to have developed especially around 1600. Effectively, one of the greatest indicative elements of the nobility's elevated social position and the justification for inequality, along with chapels, palaces, cabinet rooms and libraries, among others, were arms, given the aristocracy's traditional military function. Looking beyond the military aspect, considering that private armies no longer had the meaning they had at the end of the Middle Ages, these armouries overall served to demonstrate the lineage's importance and prestige, in that arms, alongside letters, were the fundamental basis of the social category known as nobility. As such, the most noteworthy noble houses formed important sets of arms with implications on social order and status, to demonstrate their lineage's antiquity, prestige, importance and worth by presenting the family's antique weapons with an undeniable expository meaning. These weapons were obtained during important events in the service of the Monarchy, even as gifts and exchanges, from exotic sources, etc., that they had been able to bring together in what is one of the most meaningful elements of the symbolic use of the gathering of objects as part of a cultural vision that structured social networks.

The study of armouries, although not new in absolute terms, has not been developed significantly in the past. Museums, conservators and collectors related to arms and armour certainly have provided relevant studies focused on pieces, typologies and even more or less specific sets, especially within Europe outside Spain. Nonetheless, a study of the formation, origin, chronology, evolution and use of noble armouries has never been addressed in its entirety until now, either in the Spanish case or apparently in any other context. Furthermore, the most important aspect in any historiographical work about the subject, i.e, the analysis of the armouries' cultural importance, their meaning and their social symbolism, has not been approached in a way that covers it from a broad perspective, with the study of the arms collections from a general point of view, starting with their description, to reach the sphere of interpretation.

Here, the principal investigator and other members of the research team present some of the main results of their inquiries, analysis and interpretations, a result of the work carried out over several years in which the research project has been developed. Along with them, other specialists, whom we should thank for their great collaboration, have contributed their chapters to bring about a broad and comprehensive vision of the examined phenomenon, helping the reader understanding all its broad dimension the subject whose study is initiated in these pages. In fact, the texts that make up this volume bring together the most important results of the investigative work developed, though other important studies will doubtlessly appear in other publications of these authors, and we trust that going forward other results and studies will appear with increasing force. We trust that the catalyst given by the project and this publication will definitively open the study of armouries and arms collecting to the interest of historiography, as it recognises their importance and their value for interpreting the past's society and culture. There is no doubt that the relevance and extension of the phenomenon means that the research performed, although intense and oriented towards the broadest possible vision of the issue, cannot bring the issue to a close. We believe that we have offered a rather complete perspective but, above all, we hope that we have inaugurated a field of study that doubtlessly must be much more fruitful and full of nuance in the future.

The study of armouries developed by the research team, along with the collaboration of specialist researchers who have generously shared their knowledge and collaboration, has been carried out in various manners and activities, including scientific meetings in which materials, reflections, interpretations and points of view have been shared. All of these researchers have provided important keys to situate the issue conveniently in its context and to advance the understanding of the examined phenomenon, ensuring a broad vision that has included the contacts of the Hispanic world with its European environment, which in the case of armouries is entirely indispensable. The works have included the basic scope of primary source information, both archival and literary, as well as the contextualisation and necessary explanation, from the most diverse perspectives possible.

The book has been structured as to offer a selection of the results of the most relevant investigations and the broadest panorama possible of the phenomenon that is Spanish noble armouries. It has been organised, taking great care in the coordination between the different authors, according to the different subjects, chronologies and spectrums of study. In this manner, the various chapters written by each one of the participating researchers have been structured into three sections. Firstly, three chapters focus on the global analysis of the phenomenon of armouries, taking into account their characteristics and in a panoramic manner, their typologies and their distribution, headed by a specific chapter the global explanatory analysis of the meaning, chronology, functions and extension of the armouries, without overlooking a general review of the examples that have been identified, described and analysed. They continue with the analysis of aspects related to the physical, typological and organisational framework of the different sets.

Secondly, we find a section of six chapters, which focus on specific cases, selected from among the most important and least studied to date, either showing specific examples or making comparative views between some of them. They are preceded by the necessary texts of contextualisation of the phenomenon of armouries from a European perspective, and regarding the parallels of the Hispanic examples in the Iberian Peninsula. The section includes the study of some of the most important cases of Hispanic noble armouries, and ends with an analysis of the novel and unexpected aspects in relation to formats and geographical latitudes, although always in connection with the Hispanic world. The third section addresses the no less relevant issue of the dispersion of these sets, the formation of arms collections and the expansion of the collection of pieces from the old Spanish noble armouries, both in the private and museum spheres. Finally, we did not want to miss the opportunity to offer a select, yet very complete, repertoire of transcribed sources, which are collected in an extensive documentary appendix.

In the first chapter, Roberto González Ramos (Universidad de Córdoba), offers us a text of global pretentions, a general interpretive study, in which the cultural, symbolic, chronological, gendered, functional bases, among many other details that explain Spanish noble armouries are analysed. To do this, his chapter displays the information collected throughout the research work on all the armouries that have been possible to uncover, both from large noble houses and other medium or smaller ones. This information is analysed in the second part of this chapter alongside the documentary transcription of a wide selection of uncovered archival documentation, which we mentioned before, and which is added to this volume as a documentary appendix.

The second chapter, by the researcher and curator of reference in the study of the universe of Spanish arms and armouries, Álvaro Soler del Campo (Real Armería, Madrid), focuses its gaze on various general explanatory issues regarding noble armouries, stopping both at the definition of typological categories and at the study of the organisation and exhibition uses of the different armouries known and studied within the framework of the research project. His contribution is fundamental in offering this particular analysis with the data that has been collected in the course of the research. The chapter that Jesús M. Ruiz Carrasco (Universidad de Córdoba) offers focuses on the armouries' architecture. That is, it analyses the buildings that housed the noble armouries and attempts to explain what meaning, and under what architectural conditions and distribution of space, they had, among other issues related to the containers of at least the most important and significant armouries.

Once these questions are dealt with, the book is focused on specific cases of Spanish noble armouries and its contextualisation.The next chapter, written by the renowned specialist Pierre Terjanian (Metropolitan Museum, New York), offers an initial contextualising approach to the Spanish armouries within the European context, influencing the general vision of European noble armouries, which are the indelible framework for the creation of the Hispanic ones in their origin and importance, perhaps even feeding back to them, exemplified through a specific Central European case. The book continues with another chapter that focuses on a case that is close and very connected to the Spanish one, that of the armouries of Portugal, which António Oliveira (Universidade de Coimbra), develops together with the analysis of the magnificent example of the armoury of the 5th Duke of Braganza in Vila Viçosa.

The following chapters are dedicated to the monographic study of several cases of Spanish armouries, especially those in Castile. They consist of a selection of some of those that, being lesser known, had been able to be studied and analysed thanks to the works carried out within the framework of the aforementioned research project. Of great interest is the study of the armoury of the Counts of Benavente that Juan Francisco Torres Cubero (Universidad de Córdoba) contributes. He has unravelled the history of this collection that was probably the most important of the Castilian nobility, perhaps with the exception of that of the Dukes of the Infantado, examining the cultural framework that made it possible and analysing its formation, locations, content and evolution as one of the most outstanding complexes with symbolic meaning that existed, until its destruction at the beginning of the nineteenth century. Next, Eduardo Muñoz Baudot (Universidad Autónoma de Madrid) is the author of a complete study of the armoury of the Dukes of Alba, addressing all the relevant aspects of a particularly scenographic ensemble and manifestation of the glories and exploits of the house's title holders, but especially of Fernando Álvarez de Toledo. No less noteworthy is the chapter that another specialist in the world of arms, Germán Dueñas Beráiz (Museo del Ejército, Toledo), offers us. He is the author of the chapter focused on the dynastic armoury of the Constables of Castile in

Burgos. It consists of a study on a complex that is really very significant given how and why it was created, because of its location, its deployment, content and symbolic value for one of the most important aristocratic houses in the kingdom. This central body of specific studies ends with the chapter written by Roberto González Ramos (Universidad de Córdoba), who offers a comparative study of two armouries of the highest magnitude. It jointly discusses the armoury of the Dukes of Medinaceli in Cogolludo and the armoury of the Duke of Lerma in Ampudia, two very different establishments, around 1600. The former was a product of a noble house of great power since the Middle Ages, that in the sixteenth century founded a dynastic armoury, albeit not a very large one, but one closely linked to the glory of the noble house because it was built with objects, and in a palace, of great dynastic symbolism. In contrast to this armoury, we have the armoury of the Duke of Lerma, who, thanks to his political position, could bring together an enormous set of arms and suits of armour in a recently acquired castle. It is a set in which the value of quantity and quality prevailed over the value of lineage, acquiring a symbolic significance that was no less than, but different from, that of the Medinaceli family. Other themes merit being treated in this edited volume as the project's research into Spanish noble armouries has brought us equally interesting discoveries. In this way, the chapter written by Manuel Gámez Casado (Universidad de Sevilla), refers us to the reflection of the phenomenon of armouries in Hispanic America, analysing the transport of not only military but also symbolic weapons to the "Indies", the formation of armouries especially in the world of viceregal courts and in specific buildings in the American viceregal capital cities.

Finally, to close this study of Spanish noble armouries, two chapters address the diffusion and ultimate fate of these armouries throughout the nineteenth and twentieth centuries, a point touched upon in previous chapters. The first such chapter comes from Jonathan Tavares (Art Institute, Chicago), who offers us a detailed study about how many pieces, and even entire armouries, travelled to the United States of America, focusing on the specific case of the acquisition of the armoury of the Marquises of Dos Aguas. From a similar perspective, Francisco Javier López Martín gives us a timely account of the formation of the Kernoozer's Club in the United Kingdom and how the British passion for collecting antique arms had strong connections to Spain, especially in the figures of the Count of Valencia de Don Juan and his son-in-law Guillermo de Osma.

All the research work that has led to these results, and that will undoubtedly produce more in the coming years, could not have been carried out without the institutional support that comes with having a research project like the one that has been supporting this team. But it is no less important that the researchers have been able to rely on the help, support and collaboration of various people and institutions, who have greatly facilitated the arduous labours. In this sense, we must thank the good work, advice and support of the staff of various Spanish public archives, such as the Archivo Histórico de la Nobleza de Toledo, especially the always attentive Miguel F. Gómez Vozmediano; the Archivo Histórico de Protocolos de Madrid; the Archivo Histórico Provincial de Guadalajara; the Archivo Histórico Provincial de Burgos, especially Antonio Amigo González; the Archivo Histórico Nacional (Madrid); the Archivo Histórico Provincial de Valladolid; the Archivo de Indias of Seville; etc. At least as important have been the private archives, some of them noble archives of primary importance, to whose staff we also owe heartfelt gratitude. Specifically, we speak of the Instituto Valencia de Don Juan, namely María Ángeles Santos Quer; the Biblioteca y Archivo Francisco de Zabálburu; the Archivo de la Casa Ducal de Medinaceli, especially Rosalía Marqués de los Ríos; the Archivo Ducal de Alba, especially José Manuel Calderón Ortega; and the Archivo Ducal de Alburquerque, most notably Lucía Velasco Herrero. Many other archive and library professionals should have our thanks here, and we hope they can forgive us for not mentioning them explicitly. Likewise, we must thank the welcoming and exquisite treatment of the staff of the Metropolitan Museum of New York Arms and Armor Department (Stuart W. Pyhrr, Donald LaRocca, Pierre Terjanian, Markus Sesko, John Byck, Stephen Bluto, George Sferra, Edward Hunter, Sean Belair and Catherine Chesney), the archive and library of the Victoria & Albert Museum in London, and that of the British Library, where research visits directly related to the project were performed. Other people who have been interested in the research and the work carried out, and to whom we owe our gratitude are Stefan Krause and Enrique Soria Mesa, who participated enthusiastically in the scientific meeting that took place in Cordoba in November 2022 within the framework of the research project. We also owe our recognition to the Casa Árabe de Córdoba (Ministerio de Asuntos Exteriores de España) and the great host that is its manager, Javier Rosón, who welcomed us at said meeting. Thanks also to the kindness of Desiderio Vaquerizo. Appreciation is also offered to the Universidad de Córdoba, in whose institutional setting the project has been developed, and especially to its rector (president), Manuel Torralbo, who had the courtesy of opening the scientific meeting, as well as the dean of the Facultad de Filosofía y Letras, Ricardo Córdoba de la Llave. Mark Lewis, who provided us with important information about his studies on some Bejar armoury arms. We cannot forget to show our appreciation for their interest in the project and in this publication to Alfredo J. Morales, Jesús Cantera Montenegro, Begoña Alonso Ruiz, Pedro Luengo, Edward Cooper (RIP), Jesús F. Pascual Molina, Ángel Campos-Perales and James C. Knarr.

TABLE OF CONTENTS

Roberto González Ramos

Universidad de Córdoba

HISPANIC NOBLE ARMOURIES: FORMS, MEANINGS AND OVERVIEW

Covarrubias's 1611 dictionary, the *Tesoro de la lengua castellana*, does not include the word "armoury", strangely or perhaps significantly, but the *Diccionario de autoridades* in the eighteenth century (1726) offers us a very clear definition of the concept, as

> *la casa o sitio donde se colocan y ponen en custodia varias especies y suertes de armas: que por lo regular suele ser por ostentación, memoria y grandeza del príncipe o señor que la tiene*[1] [The house or place where various species and types of arms are placed and kept in custody: which is usually due to ostentation, memory and greatness of the prince or lord who has it].

Although it is undoubtedly a partial definition, it is very useful to begin to frame the issue that brings us here. This definition highlights some uses and functions of these ensembles of weapons and, although it focuses on the premises, points out the ostentation of the privileged social situation of the aristocracy, and its mission to preserve and manifest the memory of the house and greatness of the lineage: arms as "*eslabones de una ensoñada memoria caballeresca* [links to a dreamy chivalric memory]" in the words of Bouza[2].

The arms appear in noble literature as a reflection of the noble house's history and as an incentive to increase the lineage's glory, as seen in texts such as those by Luis Zapata (1566) and Juan Benito Guardiola (1591)[3]. The arms signified heroic deeds of the noble knights, which were expected to be emulated and even surpassed by their descendants. Moreno de Vargas, in 1636, when speaking of the ancients, says that "*no solo ponían en los zaguanes insignias y estatuas de sus familias, mas los despojos ganados a sus enemigos* [not only did they put their families' insignias and statues in the hallways, but also the spoils they won from their enemies]", that is, coats of arms and images of ancestors, and trophies, adding "*ansí mismo las banderas, armas y despojos* [likewise, the flags, weapons,

1 See: Real Academia española, 1726: I, 395.

2 Bouza, 2003: 158.

3 Zapata, 1566: 141. Guardiola, 1591: 35 and 64.

and spoils of war]"[4], making all of them "*recuerdo y representación de la nobleza, virtud, valor y hazañas de los difuntos, honrándolos [...] y excitando a los vivos [...] procurasen hazer hechos valerosos* [a memory and representation of nobility, virtue, courage and exploits of the deceased, honoring them [...] and exciting the living [...] to try to do brave deeds]"[5]. It is the memory of the ancestors that gives lustre to the house and justifies its privileges, transforming the arms and trophies into the palpable evidence of their value and the family's prestige, and, in so doing, encouraging their descendants to continue their feats.

The Countess of Aranda, who criticised the nobles' lukewarmness and immorality in 1619, alludes to the inefficiency of the armouries in their function to encourage honour and acts of arms when she says:

> *Poco importa guardar en su Armería la espada y armas con que el abuelo hizo insignes hazañas el que no sabe vestir armas ni sacar la espada en las ocasiones de honra. Aquellos despojos son fiscales mudos, que tácitamente reprenden la ociosidad del nieto de tales abuelos, y los retratos de ellos son espejos, en que verán sus propias faltas*[6],
> [It matters little to keep in his Armoury the sword and arms with which the grandfather made famous feats he who does not know how to wear arms or draw the sword on occasions of honour. Those spoils are silent prosecutors, who tacitly reprimand the idleness of the grandson of these grandparents, and their portraits are those mirrors, in which they will see their own faults],

but again she points out the functions attributed to the armouries. Saavedra Fajardo, in 1640, also alluded to the meaning of these ensembles of arms, criticising them especially when the nobleman is not educated and neglects letters, "*habiéndose vuelto los museos en armerías, las garnachas en petos y espaldares, y las plumas en espadas* [the museums having become armouries, robes into breastplates and backplates, and feathers into swords]"[7].

The noble armouries were thus constituted as manifestations of the family's history and importance, of its relations and influence in the circles of power, as well as a fundamental element to demonstrate the noble house's antiquity and prestige, something essential for the lineage's recognition. Often, they are also linked to other elements that build an image of the family and the lineage, in addition to being full of a symbolic, historical, political and genealogical significance. Of course, they were a fundamental element in the discourse of

4 Moreno de Vargas, 1636: 120.
5 Moreno de Vargas, 1636: 90.
6 Padilla, 1619: 6.
7 Saavedra Fajardo, 1640: 203. Morán/Checa, 1985: 203.

justification of inequality and privilege[8]. Noble armouries are, in this sense, one of the most significant elements of the symbolic use of a set of objects as part of a visual culture that structured social cleavages[9].

The issue of noble armouries, and specifically the Hispanic ones, has been scarcely discussed, although in recent times some progress has been made in this field, and historiography today offers some data and interpretations. The search for information for a process of analysis and explanation of the phenomenon has indicated several things to us. As happens on other occasions when studying any aspect of the past, we only have very partial information about what must have existed. Furthermore, during the development of the investigations within the framework of the project on the armouries, we have found ourselves quite limited, because we depend heavily on the degree of conservation and access to the basic sources. For example, by seeing if the noble houses have kept archives, and in some cases this has happened very partially, or not at all. When we have sources, for example for primary descriptive information (inventories), we depend on their characteristics. When they are notarial documents, which constitute the majority of sources about these armouries, the information they offer is very generic, and they hardly give us specific references about the origin or meaning of the objects. When it comes to records or documents made by and for the use of the noble house itself, a minority of available sources, sometimes they reflect origins, antiquity, meanings, uses, etc., and they provide us with an appreciable amount of information about the pieces, although sometimes that information is manipulated by the legend or by the desire to stand out. When it comes to other documents and sources, these are so brief that they hardly provide us with data of interest. Fortunately, today we have an important mass of information on different Spanish houses and armouries, which allow us to have a fairly accurate view of the matter.

First, it is verified that the armouries were something that concerned only men. That is, they were a specifically masculine field and the documents reflect this fact well. In the sources there are hardly any cases of arms linked to women. This is logical based on the social role of both genders, although the ladies of the houses sometimes in their role as heads of the lineage, were involved in linking the armouries to the "*mayorazgo*" (entailed estate), and even in their reorganisation and conservation, as we can see in the cases of the 6th Duchess of the Infantado, related to the management, organisation and connection to the

8 Soria Mesa, 2011: 5–10. Cruz/Franganillo/Sanz, 2021.

9 Morán/Checa, 1985: 118 and 204. See: Appadurai, 1996.

entailed estate of the armoury of her house, that of the Countess of Benavente Manuela López de Zúñiga, or in that of María Manrique in the case of the funeral armoury of the "Gran Capitán" Gonzalo Fernández de Córdoba.

As for the issue of whether the arms really belong to a collecting phenomenon, we find various categories of nobles involved and various types of sets. One category is the gathering of groups of arms with prestigious roles and appreciation for certain specimens, and another is the possession of items for different utilitarian purposes, and even for current use in the performance of noble activities. Everything indicates that the different types and categories of nobles had their arms and armour, usually as accessories for their activities and social function, among which was hunting, tournaments and "*alardes* [displays]"[10] and, in general, as part of the clothing and class image[11]. Some noblemen, although not in a significant number, seem to have collected specimens approaching the modern sense of what we understand by collecting: gathering unique weapons for pleasure and knowledge. Especially in the world of swords, a true expert and knowledgeable field can be appreciated, as in the case of the 4th Duke of the Infantado, who had a very special relationship with his swords, and an erudite knowledge of them, having noted the names of the swordsmiths, and the linkage of mastership and discipleship among several of them[12]. In many other cases, it is clearly seen that certain swords were owned based on the prestige of their manufacturer or place of manufacture. Also, certain firearms and other objects of war may be included in the very select range of items collected for special reasons[13]. We can find in records of private chambers, true sets of select specimens, in a behaviour linked to expertise, which became an important part of noble culture. One of those cases is the one that Carducho cites about the knight of the Order of Santiago Gerónimo Funes. Speaking of the paintings, he comments:

> *Porque tuvo el Príncipe de Gales noticia dellos, fue a verlos a su casa, al qual le presentó ocho, los que el Príncipe escogió con algunas espadas, montantes, ballestas, arcabuzes de lo mejor que labraron los mas primorosos Artifices destas armas, asi en España, como fuera della, porque siempre se preció de lo mas escogido de todo aquello que fuere de un virtuoso cauallero*[14].
>
> [Because the Prince of Wales had news of them, he went to see them at his house, to whom he presented eight, which the Prince chose with some of the best swords,

10 Carrillo/Pereda, 2000: 183–198. Frieder, 2008. Fallows, 2010.
11 Bernis, 1962. Civil, 1990: 307–319. Burke, 1990: 142–173. Cruz/Perry, 1992.
12 González Ramos, 2013: 337.
13 Puddu, 1984. Verrier, 1997.
14 Carducho, 1633: 150.

> longswords, crossbows, arquebuses made by the most exquisite Artificers of these arms, both in Spain, as well as outside of it, because he always prized himself of the most select of all that was of a virtuous gentleman].

In the case of the armouries, a vision related to the exaltation of power and privilege predominates. Sometimes defining features of the phenomenon of collecting can be found in them, but their meaning went beyond this area. In addition, the possibility of using the pieces was never abandoned, which could be hunting, related to parades and tournaments, even for military purposes or for lending to the king. We found loans to relatives, some who even left receipts. Some noblemen who went to war took weapons from the family armoury, as the Duke of the Infantado did in the mid-seventeenth century[15]. It is recorded that the kings requested elements from armouries to arm the troops in times of military difficulties, as occurred in the emergency of the siege of Fuenterrabía by the French, the war in Catalonia or Portugal, during the reign of Philip IV, but also in the eighteenth century. That is why the formation of armouries, although it can be placed within the study of collecting, goes beyond what it implies. The armoury of the lineage, of the noble house, represented something more than mere collecting. It was a manifestation of something more transcendent, and that surpassed the individual to become part of the noble title itself.

The highest nobles, with the oldest and most prestigious houses, assembled armouries in the strict sense of the term, as large and important as possible based on their aristocratic level and resources. Other, less important, nobles or those with less prominent families or possibilities, carried away by the rise of the phenomenon, joined the trend and created medium- or small-sized armouries. There were others who knew how to take advantage of a status of importance for political reasons, to join the group of creators of armouries of a certain level. It is noticeable that there is a desire for differentiation, that the creation and possession of an armoury manifested the category of the house, especially when the need to distance itself from the "*nobleza nueva* [new nobility]" arose.

Armouries were ensembles of weapons, situated in places of special significance, with their own entity, and named as such in documents and records. A specific sense of arrangement and deployment is noted in them, normally with the harnesses and suits of armour placed on shelves or supporting structures, or in cabinets, etc., as was the case with other types of arms. Sometimes these were displayed hanging on walls and ceilings, sometimes forming panoplies, although we also found them stored in chests. We even find more elaborate

15 González Ramos, 2013: 343.

elements, sometimes even mannequins (of people or horses)[16]. These armouries were displayed in specific halls and, if possible, as part of the "*escenografías palaciegas* [palace scenographies]" alongside paintings, tapestries, furniture, silver and gold tableware and other elements. They demonstrated the prestige of the lineage and its position in the noble ranks, showing its importance, antiquity and fame, and on many occasions they are linked to the family library[17], as in the case of Constable Juan Fernández de Velasco (1613) and in the exemplary case of the Infantado, or are combined with the gallery of portraits of the family and ancestors, and sometimes even with the noble house's archive[18].

The large armouries were usually located in palaces or castles of the villages or towns of noble property and jurisdiction or of traditional entailment with the lineage, the so-called "*casas solariegas*", or manor houses[19]. They were installed in halls of buildings that were representative of the family and important to the lineage, and this is undoubtedly a characteristic of these sets. However, on a few occasions we can find similar groups of arms in the residence in the capital city, in what are called courtly armouries, specific sets of weapons with a similar meaning to the dynastic ones, but limited in their significance. Normally it was the main castle-palace, although sometimes an alternative one was used specifically for the deployment of the armoury exclusively or combined with other elements of a symbolic order, as in the case of the Infantado. Therefore, the location in familiar buildings endowed the armoury with an increase of meaning linked to the lineage. Nonetheless, it also meant that the armouries were far from the places of political operation, such as the court, which partially deprived them of an expository use of great pretensions and direct propaganda capacity, which court sets did have. It gives the impression that the armouries were used to be seen above all as something purely symbolic that was known especially by word of mouth, although it could be shown to distinguished travellers and visitors, by the lord himself if he lived on his lands and not at court, or by his administrators if he no longer inhabited the building. We have some references related to these visits, but they are not very abundant, and when the court became the obligatory residence of the nobles who stood out politically and socially, these are testimonials[20]. We know that the 5th Duke of the Infantado personally showed

16 González Ramos, 2017: 302.
17 Morán/Checa, 1985: 155. González Ramos, 2017: 302.
18 González Ramos, 2017: 302.
19 González Ramos, 2017: 301.
20 González Ramos, 2017: 303.

his armoury to the historian Esteban de Garibay[21], and on another occasion to the court on the way to Zaragoza in 1585, and that Rodrigo Méndez Silva visited the Benavente armoury in 1645[22]. Perhaps part of the explanation is that, being far away, they were safe from creditors and bankruptcy proceedings. In the eighteenth century, in some cases, the transfer of the family armoury to the court became effective.

These armouries, as in the cases of the Infantado, Béjar and the Constable of Castile, relied on a staff specifically hired to work in them, at least until the middle of the seventeenth century. The so-called "*armeros* [armourers]" were the primary components of these staffs; they were in charge of fixing and maintaining the weapons and keeping them in good condition[23]. These were normally specialists, and had tools specific to their work in the armoury itself. After this time period, one most commonly found other hired personnel, such as custodians and guards, and a greater carelessness and, sometimes, negligence is generally noted.

On occasion, the armouries ended up being linked as a group to the entailed estate or "*mayorazgo*", which made them, like land and other properties, inalienable as well as symbolically an inherent part of the noble title itself. We found several documented cases, although surely there were more. The 6th Duchess of the Infantado, Ana de Mendoza, connected her family's armoury to the entailed estate between 1624 and 1633[24]; the 1st Duke of Lerma did the same in his 1617 will and the Constable of Castile linked his armoury and his library together in 1613; the Dukes of Béjar did the same late in 1718[25]. There are more examples, such as that of the Marquis of Leganés[26]. In 1550, the Count of Monterrey connected several "*cosas de guerra* [items of war]" such as a gilt Moorish sword, rich "*jaeces* [harness]" for horses, forty-five harnesses "*de seguir* [for war]", various tents, two harnesses "*de justa real* [for royal jousting]", four more "*de justa de guerra* [for war jousting]", 110 infantry corselets with their morions, 600 pikes, etc. to the "*mayorazgo*"[27]. Although there is no mention of an armoury *per se*, it is a very eloquent bringing together of a set. At times, we have been able to verify links to the entailed estate of a part of the armouries

21 González Ramos, 2014: 165.
22 Simal, 2002: 103.
23 González Ramos, 2017: 302.
24 González Ramos, 2013: 343.
25 González Ramos, 2020: 26.
26 See later.
27 Archivo Ducal de Alba (hereafter ADA), Casa de Alba, C. 158, nº. 15.

or of various arms with a special value. Known cases of specific arms linked to the "*mayorazgo*" are those of the sword of the 2nd Count of Tendilla, a papal rapier linked in 1515[28], or the arms and clothing of the King of Granada Boabdil, imprisoned in Lucena in 1483, trophies linked to the Marquises of Comares "*mayorazgo*"[29]. We also know of the case of the grandfather's sword that linked to the entailed estate the 7th Duke of Medina Sidonia[30], among other examples. In 1575 the Duke of Alba did the same with two swords and the artillery that formed part of the Alba de Tormes armoury[31].

Regarding typologies, although in Álvaro Soler's chapter this subject appears much more developed, we find, first of all, armouries properly said. These were the large ones, belonging to high aristocratic houses, sometimes of an enormous size, a fact that shows the existence of a certain competition between them. We also found medium-sized armouries, belonging to noble houses that, for various reasons, could not or did not want to have such large armouries, a fact that, however, did not affect their symbolic meaning, quality or selection of their content. There are also small ones, usually belonging to nobles who sought to have that symbol of prestige of their lineage, but who did not have the economic means or sociopolitical influence comparable to the great houses. Many of them were also located in lineage buildings. Among all of them we can also distinguish those that are of ancient formation, which could have arms from the ancestors in a good number, normally belonging to powerful houses since the Middle Ages, and those of recent formation, gathered at a certain moment, for social and political reasons, and perhaps thanks to the fact that one of the members of the lineage had reached relevant positions due to his military performance, or for being "*valido* [favourite of the king]" or belonging to the government then in power. This category did not generally have old family weapons, but the creators of the ensemble used to make large, bulk, purchases and had important specimens obtained by the influence of their positions and by political and power relationships, as evidenced by the cases of the Duke of Lerma or the Marquis of Leganés.

Some of the medium and small armouries were brought together because their owner had played a role in an outstanding military event, allowing him to gather a good quantity, or better, quality, of prestigious pieces, and that were

28 Urquízar: 106.
29 Urquízar, 2007: 75, 106.
30 Urquízar, 2007: 103.
31 ADA, Alba, C. 158, nº. 15.

often trophies won in battle. According to the criteria of the most important, these could be installed in palaces in the rural possessions of the lineage, such as that of the Marquis of Santa Cruz in his El Viso palace, or displayed in the court, like the collections of prestigious arms, although contrary to these, more than in private rooms, they seem to have been deployed in halls on purpose for their exhibition and eventual shows to illustrious visitors. One example is that of the Count of Priego in 1579 that we will see later.

Regarding their chronology and evolution[32], armouries as these symbolic groups of arms, seem to have been born and formed in the second half of the sixteenth century, mostly since 1570, although the general trend seems to have intensified around 1600, with exceptions. Although we found large ensembles from before the middle of the sixteenth century, everything seems to indicate that these were sets of arms for personal use or arsenals to equip private retinues and armies. It would be about military armouries, whose *raison d'être* was limited by the political action of a Monarchy that no longer viewed the direct warlike power of the nobility positively, especially during Philip II's reign[33]. Perhaps the aforementioned linking of arms to the entailed estate of the Count of Monterrey in 1550 points to the first formation of an armoury of a symbolic nature, although we cannot be sure. Perhaps the ensemble of arms, very abundant and of quality that the Duke of Calabria, of the royal house of Naples/Aragon, bequeathed to the convent of San Miguel de los Reyes in Valencia in the same year, formed an armoury of the house, although the name "*armería*" was not used to refer to it in the descriptive source. Surely these are first examples and it seems that they are exceptional cases. Surely one of the precursors was the Count of Benavente, very aware of what was done in the Habsburg context. He seems to have started what would later become an armoury in the strict sense, around 1540.

Afterwards, the armouries, now with that name, and with the characteristics of the symbolic order that we have been analysing, appear to have been established beginning around 1570, as is the case with that of the Duke of the Infantado or that of Benavente. Of course, some of the most famous ones do not appear until the end of the century, such as that of Medinaceli, or until at the beginning of the seventeenth century, such as the Constable of Castile's or Duke of Lerma's. In earlier ensembles, arms are usually registered with other objects and groups of goods, and they are not cited as "*armerías*" until the second half of the sixteenth century. The data seems to indicate a clear coincidence with

32 González Ramos, 2017: 302.
33 García Hernán, 2000: 285–300.

the tacit ban of private noble armies and with the fact that Philip II definitively organised Madrid's royal armoury in its own building and thus providing new meaning to the royal collection in 1562. This fact seems to have motivated the emulation of the nobility and the competition between the houses[34], although some nobles may have predated it in some ways, surely based on what they saw in Europe, and what Charles V himself manifested with his attitudes.

The phenomenon of armouries seems to have peaked around 1600 and to have remained important until the central years of the seventeenth century, although we can find as early as 1626, references as Fernández Navarrete's, who noted:

> *No ha muchos años que en todas las casas de los nobles se acostumbraua a tener cantidad de arneses, picas y arcabuzes, con que en ellos, y en sus hijos se despertauan los espíritus militares heredados de sus passados. Ya todo este varonil aparato ha cessado con las costosas alhajas de que se adornan; o por mejor decir se afean las casas*[35].
> [Not many years ago it was customary in all the houses of the nobles to have a quantity of harnesses, pikes and harquebuses, with which the military spirits inherited from their ancestors were awakened in them and in their children. All this manly apparatus has now ceased with the expensive objects with which they adorn them; or rather to say the houses become ugly].

Some sets were gathered during that relatively late period, such as the Duke of Pastrana's armoury or Marquis of Leganés's, which existed as such since at least 1642, created for special needs due to his military and political elevation, in the shadow of the Count-Duke of Olivares. However, in many of the best-known cases studied, there is a progressive abandonment of the armouries beginning in the mid-seventeenth century, although some continued to grow with new contributions or by marriages between houses. Many inventories and related documents from the second half of the seventeenth century describe carelessness, loss of arms, their disarrangement and even disassembly, often due to negligence on the part of the administrators or those responsible for their custody. In the case of the Infantado armoury, there was even a lawsuit against the guards, and finally a fire in 1702, although what was saved from destruction, quite a large amount of items, was taken to the main palace in Guadalajara[36]. In Cogolludo, we found various complaints from the Duke of Medinaceli about the disappearance of pieces and lack of care of his armoury[37]. In Béjar there was a general neglect and,

---

34 Soler, 1998: 25–37.

35 Fernández Navarrete, 1626: 246. I thank Juan F. Torres Cubero for the reference of this text.

36 González Ramos, 2013: 344–45.

37 Archivo Ducal de Medinaceli (hereafter ADM), Cogolludo, Leg. 3, D. 44, fol. 17 r.

in Benavente something similar that led to the rearrangement and inventory of the armoury in 1702[38]. The reason for this apparently general situation is surely found in a lack of interest in such symbolic elements of aristocracy, which had been losing a good part of their meaning and significance.

In the eighteenth century, the armouries in certain cases clearly once again interested their owners. Surely, this interest came about because the old houses once again needed to justify their position and to demonstrate their status through this type of prestige signalling, given the increasingly expansive and, in a certain sense, upstart new nobility. In 1718, the Béjar armoury was rebuilt and rearranged, and linked to the ducal house's entailed estate[39]. In the middle of the century, the Duke of Medinaceli collected his Cogolludo armoury and other sets of the lineage, and took them to Madrid, where he displayed them alongside the family library[40].

The objects that made up the noble armouries were very varied. The most important items was the different types of arms, but they were not the only goods that could be found in them. The most common arms, and surely the most valued and sought after, were harnesses and suits of armour, of all kinds and origins. There were ancient and modern ones, those from German and Central European workshops, these often from Milan, although there are Hispanic specimens. We find them for jousting, parade, tournament, ceremonial or declared military use. Some were very rich and ornamented, others less elaborate, with all their accessories, including helmets, morions, and armour for horses ("bards") and saddles with their steels. No less important were the swords, of all kinds and conditions, among which we can also find rapiers, tucks, longswords, arming swords, scimitars, falchions, cutlasses, with all kinds of sword fittings and hilts, with their daggers, scabbards, belts, waistbands, etc. There were even loose hilts and loose blades, sometimes with inscriptions. Other common elements were the morions and the rondaches ("*rodelas*"), sometimes of great wealth or with figurative decorations, in relief. Polearms were not lacking either. These included lances, halberds, partisans, "*chuzos*" and other similar pieces, including axes, hammers, and maces "*de armas*". Other specimens were exotic, among which we found weapons and various accessories from Eastern Europe, Africa, Turkey or the Granada War against the Moors and particularly from America[41] ("*las*

38 Simal, 2002: 109.

39 González Ramos, 2020: 26.

40 ADM, Archivo Histórico, Leg. 127, 2, nº. 51, fol. 249 r. Biblioteca Nacional, Madrid, mss/12024, V.1, fol. 63 v.

41 Yaya, 2008: 173–188. Urquízar, 2011: 205–222.

Indias"), and even on occasions from Japan or China. On many occasions we find firearms, including artillery, arquebuses, shotguns and pistols, sometimes in large quantities and/or in examples of special richness or technical characteristics, both Hispanic and from other origins, often gifts of various kinds. Some of these had to do with a primordial activity in noble culture such as hunting, as well as crossbows, hunting knives or falconry accessories that also often appear. In some cases we even found pavilions, bell tents and camping tents.

The arms had different origins, all of them with a greater or lesser degree of symbolic importance. Those who belonged to the lineage and their feats stood out, which were the ones that held the most symbolism and importance. We also find military trophies, especially from deeds of people related to the house by blood or marriage. A part of the arms was the result of acquisitions of different items, in Spain itself, sometimes made in foreign territories, sometimes in auctions from other noble houses. There was no shortage of rare and exotic weapons, only within the reach of a few, such as weapons brought from varied geographies that demonstrated the exercise of viceroyalties, military or political positions of family members. Others were gifts from relatives, and even from members of the same party or political faction[42]. When the documents provide us with detailed information, and the nobleman in question was of a certain category, it is discovered that they were gifts, sometimes diplomatic, from great princes, nobles, and even from the king himself (and indeed, a few from foreign monarchs), gifts that showed the social and political, local, regional and international importance of the lineage, and its influence and contacts at the most varied and high levels, present, and above all, past. An exemplary case, in the opposite direction, is that of Ferdinand, Archduke of Tyrol, who in 1583 requested the portrait of the Grand Duke of Alba and the suit of armour that he had worn against the Schmalkaldic League. For his part, Emperor Rudolf II also requested the portrait and the suit of armour of the Marquis of Santa Cruz[43]. But, in our case, it is not strange to find harnesses, rondaches, morions, swords, pistols and other elements that had been gifts from European princes or from noblemen of different categories and geographical origins. In fact, as is well known, the phenomenon of armouries in the rest of Europe was very important, such as the alluded to, and extremely famous, of Ferdinand of Tyrol in Ambras, whose content was published in 1601 by Jakob Schrenck von Notzing, thus

42 Davis, 2000. Scott, 2006. Heal, 2014.

43 Bouza, 2003: 108–109.

contributing greatly to the expansion of the phenomenon[44]. Everything seems to indicate that the European models of this type of ensemble would influence Spaniards, although surely some of the characteristics were their own.

Another interesting issue is that sometimes we find huge masses of corselets, spears, swords, morions, firearms and other elements, which more than armies, provided retinues of "*hombres de armas* [men of arms]", or the noble retinues. But on other occasions we discover individual objects or forming small groups, of great wealth due to their materials and workmanship, undoubtedly select pieces. The latter usually changed locations, and often we can find them in cabinets and chambers rather than in the armoury.

When choosing, keeping and displaying weapons, the existence of defined criteria is verified, especially in particularly symbolic and significant items. It was very important that they were family arms, but in general the pieces antiquity mattered, true or false. Another relevant aspect was the quality and richness of the specimens. Likewise, its origin and rarity were important, and the criteria of exoticism could not be ignored. In this latter aspect the armouries have an undoubted contact with the cabinets of curiosities, given that they share some of their characteristics, especially in the selection criteria of the objects and, in addition, the armouries themselves sometimes had non-military elements that were properly objects of "wunderkammer". In short, these criteria were fundamental as signs of the house's prestige, which the armouries provided and reaffirmed, because the pieces thus selected were manifestations in themselves of how far and with what importance some lineage was related and exalted, in addition to the fact that in occasions they were directly a manifestation of the military glories of the family nucleus itself, past or more recent.

The family origin of the arms was, as we have seen, a fundamental element. Ancestral arms were used to demonstrate the genealogical line of military performance, but sometimes dynastic armouries were completed with arms from contemporary relatives. In the formation of many of the armouries, special care is noted in the acquisition, conservation and care of family arms, used by members of the lineage or obtained by them, especially in "*hechos de armas* [deeds of war]". The criterion of antiquity is also seen in the possession of harnesses and swords of the ancestors, but also of illustrious characters of the past such as El Cid, even in what we can call "archaeological" pieces, such as swords that are said to have belonged to Julius Caesar, the Visigothic King

44 Schlosser, 1988: 64–68. Gamber, 1977: 24–59. Morán/Checa, 1985: 65. Montgomery-Campbell, 2005.

Reccared, the first Count of Castile Fernán González, or Suero de Quiñones the protagonist of the famed "Paso honroso", for example. In the case of swords, the reference to their antiquity is very recurrent, but it also appears in the description of other arms, which demonstrates the importance of this criterion. The criterion of quality and richness is reflected in the manufacturing techniques and extraordinary materials, as well as the production area or the authorship of many of the specimens. It is verified in the engraved and gilded harnesses, from centres such as Augsburg, Milan, Flanders, Peñacerrada or Fuentidueña, or from armourers such as Desiderius Helmschmid or Pompeo della Cesa; in the arquebuses with engraved and gilded barrels or butts with mother-of-pearl inlay or details and inlays of gold, silver or ivory on the cases; on the rondaches with matching morion with scenes in relief and/or gilded; in swords and other pieces from prestigious production centres such as Toledo, Valencia, Calatayud, Bilbao, San Clemente de la Mancha, Fuentidueña, etc. The criteria of origin and rarity can be seen in swords that were said to belong to the King of France or to Turks; harnesses with the Bourbon coat of arms; pieces of Milanese, German, English, Flemish manufacture; weapons and trophies taken from the enemy; oddly shaped items such as convertible swords, oddly shaped crossbows, ornate bows, three- or four-barreled firearms, or with novelty locks and loadouts. The criterion of exoticism is reflected above all in trophies, but not exclusively. The best armouries, although these objects were also in cabinets and chambers, had Japanese harnesses, katanas and other arms, which arrived in Spain from the Japanese embassies to the courts of Philip II and Philip III, although they may have arrived by other means[45]. In general, the noble armouries had swords, bows, turbans and scimitars taken from the Turkish enemy, highlighting the trophies from the battle of Lepanto. African specimens also stand out, such as the rondaches made with elephant skin, but above all the Islamic Berbers, although the Granada specimens such as the spears or swords of the "Rey Chico", among others, are undoubtedly the most outstanding of the arms of Muslim origin. We can also find weapons and other objects from Central Europe, Poland, Hungary and Protestants in general. And of course, warlike or non-warlike objects from "the Indies" (the Americas) such as feather head-dresses, feather rondaches, stone maces (of obsidian?), bows, hammocks, cotton breastplates and more.

Another important element that we find is artillery. The large armouries had a good number of cannons and other artillery pieces of various types, sometimes

45 See especially, Ruiz de Alarcón, 1973: 22–28. Torreiro, 1992: 37–44. Soler, 2003: 59–67. Gil Fernández, 1991. Sola, 1999.

emblazoned. We did not find elements of this type in all of them, but we did find them in a good number. This artillery reflects the importance of modern warfare, but some of the pieces were old and in general we do not know if they were in use or if they were pieces for pure exhibition, as in the case of Béjar or Fernán Núñez. The armouries had pieces of good size, but it is very common to find miniature pieces. In any case, the presence of artillery pieces is more noticeable in military noblemen who had fought or exercised their command, such as the Duke of Alba, who we saw linked the artillery of his armoury to the entailed estate, or the Count of Fernán Núñez, who probably displayed artillery in his palace as a type of armoury, showing the achievements of his house by this way[46]. The case of the castle of San Leonardo is especially significant, because its owner, the son of the 2nd Duke of Nájera, the captain general of the artillery Juan Manrique de Lara, had it built as an artillery bulwark[47].

THE NOBLE ARMOURIES

The armouries that we have identified can be listed and described below based on their size and characteristics. Starting with the largest ones, we have the armoury of the Dukes of the Infantado, created as such a symbolic group by the 5th Duke, Íñigo López de Mendoza (Fig. 1), which was probably the best of the Hispanic nobility, and probably the most varied, with more than 160 harnesses and almost 6,000 objects[48]. We know that it already existed by 1585, when the court visited it and was described on the occasion in his memoirs by Enrique Cock, one of the king's Flemish archers, but it is certain that it was established before, albeit after the 1566 death of the 4th Duke. Quite a model in quality, variety and quantity of pieces, it was located not in the Palace of the Infantado, but instead in the former palace of the "Gran Cardenal [Great Cardinal]" Pedro González de Mendoza in Guadalajara, where it occupied several halls. It is an exemplary representation of the importance that the combination of "*las armas y las letras* [arms and letters]" had for the Mendozas, as the armoury was located together with the library of the Marquis of Santillana, linked to the entailed estate by the 1st Duke of the Infantado. The set was very careful and had above all the arms of the family, its clientele and the political environment of John of Austria, whose political party included both the Duke Íñigo and his brother and heir, Rodrigo de Mendoza (Fig. 2)[49].

46 Espejo Gutiérrez, 2020: 182.
47 Cobos/Castro, 1998: 266.
48 González Ramos, 2014: 153–198.
49 González Ramos, 2013: 338–341.

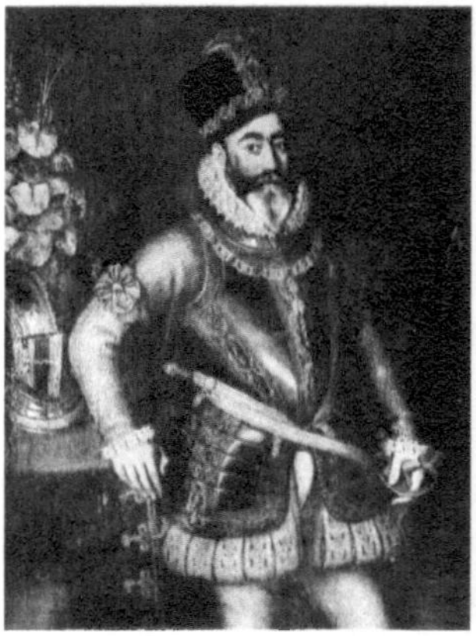

Fig. 1. *Portrait of Íñigo López de Mendoza, 5th Duke of the Infantado*. ca. 1580. Anonymous painter. Casa del Infantado. (Taken from:Arteaga,Cristina de (1940), *La Casa del Infantado, Cabeza de los Mendoza*. Madrid, pl. 50).

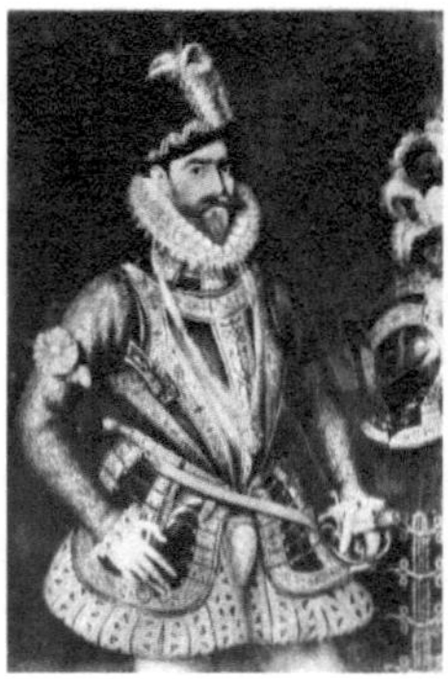

Fig. 2. *Portrait of Rodrigo de Mendoza, Count of Saldaña*. ca. 1580. Anonymous painter. Casa del Infantado. (Taken from: Arteaga,Cristina de (1940), *La Casa del Infantado, Cabeza de los Mendoza*. Madrid, pl. 56).

The Infantado's armoury continued to be expanded in the time of the 6th Duchess and the Duke consort, Juan Hurtado de Mendoza (+ 1624). He obtained a good number of quality pieces, including the arms of his brothers, the Admiral of Aragon (Fig. 3) and the general of the galleys of the Order of Saint John (the "Bailío de Lora"), and would continue to increase later, although the ensemble also lost pieces, sometimes at the request of the king. Halfway through the seventeenth century, it was in a lamentable state, although it was able to be reconstructed, only to partially disappear in the fire that decimated the old palace of the "*Gran Cardenal* [Great Cardinal]" Mendoza in 1702, shortly after having it incorporated the pieces from the armoury of the Dukes of Pastrana, given the connection between the two houses. Although the damage must have been considerable, many of the arms from the Infantado armoury were saved and transferred to the family's main palace, and in the nineteenth century, they went to Osuna's house.

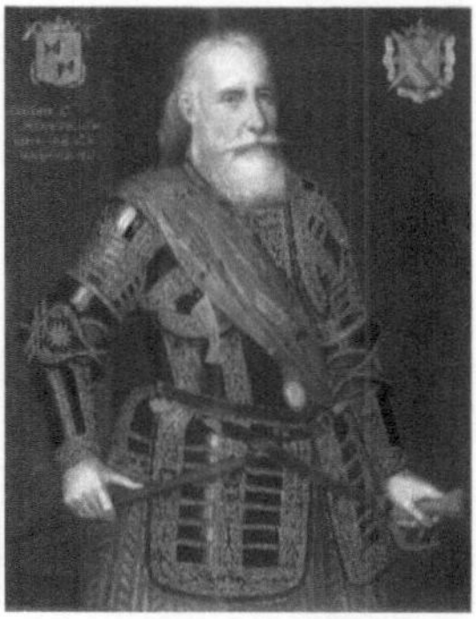

Fig. 3. *Portrait of Francisco de Mendoza, Admiral of Aragón*. 1600–1602. Anonymous painter. Amsterdam, Rijksmuseum.

The Counts of Benavente's armoury is described by travellers and chroniclers as the largest of the Castilian nobility, and was at least as stocked as the previous one (maybe more). Its existence as such is recorded in literary sources as early as 1577. It was first established in the Benavente hospital founded by the Counts, where arms of great importance are recorded between 1541 and 1574. Later it was reorganised and moved to the Benavente fortress in the same town, which is when it was established as an armoury in the strict sense[50]. We are not going to go too far into the description and analysis, since this volume contains a monographic study of this armoury, except to say that, like others, it had its best period around 1600, and that since then it had been losing importance at the same time as pieces. In 1702 it was rearranged together with the archive and the library, to which the armoury was linked as symbols of lineage. In its final

50 Cooper, 1991: 472. Cobos/Castro, 1998: 143.

location, it was deployed first in three, and then in five, rooms and, although it was related to the library of the house, it did not show as manifest a sense as that of the Infantado, since it shared other spaces and models of palatial scenery, and therefore it had a much broader and more diffuse meaning related to lineage. In the Napoleonic wars it endured a fire that almost completely destroyed the castle-palace, and practically the entire armoury[51].

Another of the famous great armouries of Castile was that of the Dukes of Alburquerque in Cuéllar, which we will study in depth elsewhere. Already in 1526 the Duke had an astonishing number of arms, some of them in the castle of that town, where the family had finished the construction of an impressive fortress (Fig. 4)[52]. Likewise, in 1560 there was a good group of arms in the possessions of the Duke, but the sources do not speak of armoury itself[53]. However, in 1584, thanks to a record book, we know that a spectacular armoury already existed in the castle of Cuéllar that had a huge number of arms and armour, which are arranged by category in the inventory book[54]. In the section referring to "*Arneses de seguir dorados y grabados, blancos y de colores* [gilded and engraved harnesses of war, metal-coloured and coloured]" no less than 108 items are specified in detail, including harnesses and corselets, some of them belonging to members of the household or relatives, and gifts. Those of the Count of Lemos (son-in-law of the 3rd Duke of Alburquerque) were incorporated in 1594 adding another ten harnesses, and another twenty-two arrived later, sent from Zaragoza by the Duke in 1595, which totaled 140 suits of armour. But, in addition, we found 171 morions (some of quality and 150 plain); chain mail pieces; strong breastplates; plumes; rondaches; halberds and partisans; 349 arquebuses; musketry; pistols; drums; spears and lances; swords; rapiers and scimitars; banners and flags; war saddles; shaffrons and horse tack; and nineteen artillery pieces.

51 Simal, 2002: 114.

52 Cobos/Castro, 1998: 136.

53 Rodríguez Villa, 1883: 68.

54 Archivo Histórico de la Casa Ducal de Alburquerque (hereafter AHCDA), 161, nº. 2.

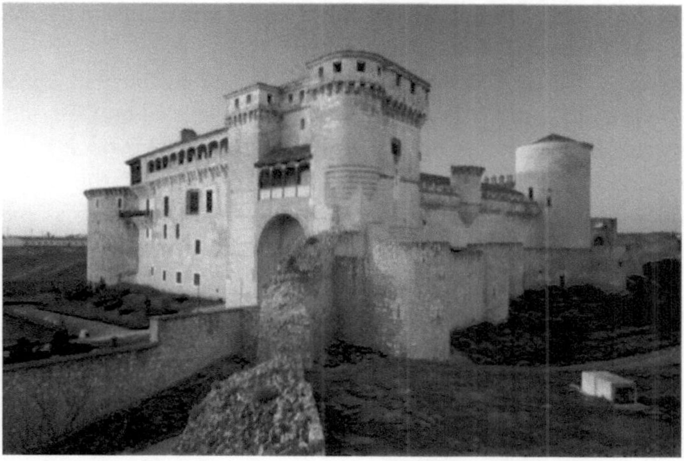

Fig. 4. *Castle of the Dukes of Alburquerque*. Cuéllar, Spain (Wikimedia Commons).

The armoury continued to exist in its fullness into the seventeenth century, when its contents were recorded numerous times, especially in the second half of the century. Like that of the Infantado and others, it was used to supply the armies of Philip IV, especially for the siege of Fuenterrabía, which involved the extraction of forty-nine harnesses, as well as shaffrons and some other elements[55], and the same would happen due to the war with Portugal[56]. In 1685 there were still thirty-five full suits of armour, sixty-eight "*medios cuerpos* [half bodies]", as well as different usual elements: polearms (among them the lance "*de ristre* [for tournaments]" of the 1st Duke), rondaches, swords and cutlasses, morions, banners and flags, some firearms, saddles, etc., as well as artillery pieces. In the eighteenth century, there were two armouries in the Cuéllar fortress, one small and the other large, although they were already quite depleted. The last record is from 1786, which indicates that the large armoury housed thirty suits of armour, sixty-nine "*medios cuerpos* [half bodies]", 317 swords, as well as large amounts of polearms and morions, flags, saddles, and various pieces in old trunks, as well as artillery pieces. The little armoury ("*armería chica*") mostly brought together select pieces and cabinet objects[57]. Everything points to the fact that the armoury

55 AHCDA, 160, nº. 41.
56 AHCDA, 160, nº. 36.
57 AHCDA, 161, nº. 18.

of the Duke of Alburquerque was looted by Napoleonic troops in the Peninsular War, so nothing remains of the ensemble[58].

Another of the great Castilian armouries was the armoury of the Dukes of Béjar in their main castle-palace (Fig. 5). It already existed by 1591, although it was surely created earlier. Like other ensembles, it had a "*maestro armero* [master armourer]" hired for the purpose of caring for and keeping the arms in good condition. This armoury collected a large number of arms such as spears and other polearms; tournament swords; artillery pieces; shaffrons; hilts and guards of swords; and eighty-one complete harnesses on their wooden stands, and many pieces of loose armour. The sum was around 3,000 objects[59].

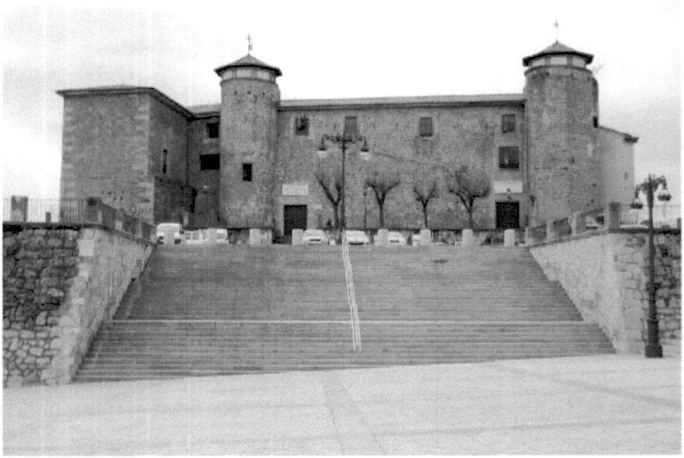

Fig. 5. *Castle-palace of the Dukes of Béjar*. Béjar, Spain (Wikimedia Commons).

The armoury continued to exist in the seventeenth century with acceptable quantity and quality, although losing some pieces whilst adding others. Its contents were inventoried in 1637, when it housed sixty harnesses "*de seguir* [for war]" and nineteen for jousting, including that of the Duke himself and that of his son, and more than forty swords, among other pieces. In 1671 another record placed the arms in two rooms completely occupied by the items, the layout of which is described in detail. The eleven artillery pieces were distributed on the

58 Cobos/Castro, 1998: 137. Velasco Bayón, 1981. Montes González, 2016.

59 González Ramos, 2020: 16–29.

castle wall, in the entrance area. A sword was special, antique, and belonged to the mythical founder of the house of Zúñiga-Béjar, Íñigo Arista, the first king of Pamplona (ninth century). This sword was included in the "*mayorazgo* [entailed estate]". With regard to the suits of armour, one was placed on horseback at the front of the main hall. In 1682, a gilded harness "*que vino de Madrid* [that came from Madrid]" was added to the ensemble. It was a gift from the king (apparently, it belonged to Philip II). In 1686 the Duke of Béjar died in the siege of Buda (Hungary) and various Turkish weapons and other items were added to the armoury in memory of his heroic death in combat. In 1702 several pieces of artillery were handed over to the king, and in 1718 the entire armoury was linked to the "*mayorazgo* [entailed estate]", when it was rearranged and its walls were decorated with red hangings. Some swords had inscriptions, including the one "*del mayorazgo* [of the entailed estate]" ("Timebunt Gentes"), or the one with the arms of the Mendozas and the inscription "Marchio Rodrigo de Vibar" (the first Marquis of Zenete), which is now preserved in the British Royal collections[60]. In addition, there were two katanas and a Japanese suit of armour, as well as firearms (one with "Gothic" lettering), and the Turkish weapons from the Buda siege. The armoury passed to the house of Osuna before 1827, when it was already very depleted, probably due to the Peninsular War, since the mentioned sword was acquired in England in 1810[61].

The armoury of the Constables of Castile was founded as such a symbolic group by Juan Fernández de Velasco, 5th Duke of Frías, 6th Constable. It was located in Burgos and, in his will of 1613, it was linked to the entailed estate of the house together with the library, forming a noble symbolic set very similar to that of the Dukes of the Infantado, expressly pairing arms and letters[62]. It was located in the ancestral palace of the house in Burgos, the Casa del Cordón (Fig. 6)[63], to which a wing or addition to the main building was added, which is very reminiscent of the initiative of Philip II when erecting a building in front of the Alcázar in Madrid to locate the royal armoury. We will not go into the detail of its content here either, since in this same book there is a specific study of the

60 Information provided by Mark Lewis, Saint Lawrence College, Canada. Lewis, 2024. See: Laking, 1904: 12.

61 González Ramos, 2017: 289–297. González Ramos, 2020: 16–29. See the arms in Frías and in Medina de Pomar: Cadiñanos, 1997: 57–61.

62 Morán/Checa, 1985: 203. Carlos, 2005: 225. Montero/González-Sánchez/Rueda/Alonso, 2014: 172.

63 Cámara Fernández, 1994: 407–412. Zalama/ Andrés, 2002: 124.

ensemble. We will simply emphasise, to assess its importance, that in 1613 it had slightly less than 1,200 items, including around eighty harnesses.

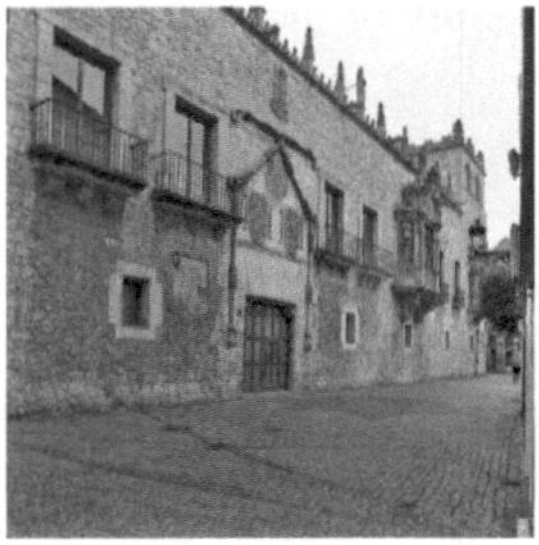

Fig. 6. *Palace of the Constables of Castile, called Casa del Cordón*. Burgos, Spain (Wikimedia Commons).

Another great symbolic ensemble was that of the Admirals of Castile, in their castle-palace in Medina de Rioseco. However, although we have some information with which we must continue working, including inventories and other documents to locate and study, we have only brief references, which tell us about a military arsenal rather than a symbolic armoury. In the year 1540 an inventory of the arms of the fortress of Medina de Rioseco was made, which includes sixty different artillery pieces. Distributed in several rooms, we find almost 350 corselets, some with helmets and others with morions; in addition to crossbows; horse covers; 111 shotguns from Flanders; 338 from Biscay; sixteen harnesses and corselets for cavalry; in addition to sixty-eight "*serones* [esparto baskets]" of dismounted harnesses; thirty heavy old halberds; and another 700 ash and beech polearms. It also includes a tent or "*tienda Real que era del Rey Católico* [royal tent that belonged to the Catholic Monarch]" —Ferdinand II of Aragon—and other various tents and pavilions[64]. Surely this group of arms, which in principle was not yet an armoury of the lineage, must have been the basis of the one that would be formed later and that the admirals had in Medina de Rioseco, which we suspect contained up to 200 harnesses, which would make it one of the greatest among those of the high Castilian nobility. Unfortunately, the fall from grace of Juan Tomás Enríquez de Cabrera, 11th and last Admiral of Castile and 7th Duke of Medina de Rioseco, due to his support for the Austrian cause in

64 ADA, Alba, C. 79, nº. 19.

the War of Spanish Succession[65], has deprived us of a documentary corpus that allows us to address this great ensemble in more detail. The documents handled by María L. González Mezquita and other researchers do not provide us with relevant information on the size and quality of that armoury[66]. We only have some data of the admirals' arms at the end of the seventeenth century, but they are generally from specific properties in Madrid[67].

The Marquis of Poza Francisco de Rojas had a very important armoury. In 1544 an inventory tells us about an important set of arms, but not an armoury[68]. However, when the armoury appears as such an ensemble, it was located not in his estates, but instead in his Madrid palace. In 1605 an inventory of his goods was made[69], and in it we find, under the heading "Armoury", no less than 100 harnesses, some of them engraved and gilded, from Germany, from Fuentidueña, and from Milan, in addition to 115 wooden display stands for harnesses and a multitude of loose pieces of armour, "*aceros* [steels]" for saddles, and a steel horse's armour. It was an important set in which there was no lack of maces, stirrups, spurs, halberds, spears, partisans, pikes, axes, spears and "*lanzas de sortija* [lances for tilting at the ring]" from Biscay; engraved and plain morions from Milan; fifty-four swords from Bilbao; others from Cuenca, Galicia, Germany or Toledo, as well as Moorish Berber and Turkish cutlasses; daggers; an engraved rondache from Milan; a large number of arquebuses from Aragon, Biscay or Milan; pistols; and three bronze artillery pieces mounted on their carriages. In addition, in the private rooms we find six select swords.

The Count of Monterrey must have had an armoury, but we know very little about it. However, the good quantity of harnesses (fifty-one) and corselets (110) and other objects linked to the entailed estate in 1550 points to the origin of a possible armoury of great importance[70]. In 1560, various war items were collected in the Count's chamber, but nothing referred to an armoury, and the publications

---

65 Agüero Carnerero, 2017: 132–153.

66 González Mezquita, 2007. Fernández Duro, 1902.

67 Archivo Histórico de Protocolos de Madrid (hereafter AHPM), T. 6233, fol. 258 r, 323 r. Archivo Histórico de la Nobleza (hereafter AHNOB), Osuna, C. 498, D. 2, fols. 390r-390v. See: Fernández Duro, 1902.

68 Biblioteca Francisco de Zabálburu (hereafter BFZ), Altamira, 32, G. D. 3, D. 12.

69 AHPM, T. 2176, fol. 234 r. Cited by Morán/Checa, 1985: 204. See also: Martínez Hernández, 2004. Martínez/Fernández, 2005.

70 ADA, Alba, C. 158, nº. 15. See: Rivas Albaladejo, 2015, p. 41. Diéguez Delgado, 2015.

on the tapestries, paintings and sculptures of the 6th Count, the viceroy of Naples, indicate nothing about our matter, which will have to be further studied[71].

The Marquises of Los Vélez must have also had an important armoury, based on the many weapons that the 3rd Marquis had in Madrid in 1579[72], but we have not been able to find out about it either, not even where it could be, because the documents of the house, preserved in the ducal archive of Medina Sidonia, do not offer much information in this regard. The latest comprehensive works on the lineage do not give us specific information either[73].

Other houses had good armouries, but we can classify them as medium or intermediate, somewhat less large and assorted. The most prominent of this class was surely that of the Dukes of Medinaceli, not too large given the house's importance, which at the time of its possible creation was at a low point economically and politically. Due to chronology, deployment and location, this ensemble follows the line of the most famous and well-known armouries, since it was deployed in a palace of symbolic importance for the lineage, that of Cogolludo. We do not know exactly when it was formed as an armoury as such, but we have previous documents that inform us of the arms that the De la Cerda family possessed beginning in the middle of the sixteenth century, details that we deal with in a specific chapter in this book. What is clear is that the Cogolludo armoury already existed in 1594, when its contents were recorded for the first time, according to the documents we have. It had around 1,000 items, including twenty-one suits of armour and a large number of loose pieces with which to assemble more. During the seventeenth century it continued to grow, especially with pavilions, tents and artillery. In this period it came to have thirty-three harnesses, undergoing, like other armouries, a striking abandonment in the second half of the century, which was partially resolved. In the first half of the eighteenth century it apparently recovered its importance for the holders of the Medinaceli duchy. Halfway through the century, it was transferred to Madrid, where it became a court armoury, next to the family library, and would become one of the most famous and largely preserved, adding weapons from other family branches. The Duke of Medinaceli, for example, had another armoury in Segorbe that that house had bequeathed to him when it joined that of Medinaceli, to which we will allude later.

---

71 Pérez Sánchez, 1977: 417–425. Burke/Cherry, 1997: 510–520. Simal, 2013: 345–365. Ramírez Ruiz, 2015.

72 AHPM, T. 992, fol. 1 r. Cited by Morán/Checa, 1985: 164–165.

73 Díaz López, 2007. Rodríguez Pérez, 2011.

The Count of Luna, Claudio Fernández de Quiñones, a character closely related to the court and active as ambassador to the Empire and in Trento, also had his armoury, according to his post-mortem inventory of 1563, carried out in León, which would make it an early example related to his courtly and international activities[74]. There, in addition to pieces of mail, horse harness, a saddle, a rapier and two adargas in the private rooms, followed by paintings and sculptures brought from Flanders, we find horse saddles, as well as paintings and curious objects, including some leaves of paper with drawings of "armas" (we do not know if coats of arms or armour). In the family palace there was also a room called the "gymnasium", divided into two rooms. One was filled with portraits of the Emperor, Philip II and Philip the Handsome, the lord of the house, and a multitude of high-ranking figures from Spain and other places, among other religious and mythological paintings. The second room of the "gymnasium" housed the library, which included portraits of Spanish and foreign ladies, astrolabes, and bronze and clay figures, as well as an interesting set of books. Surely linked to this, we then find the armoury in its own room. In it there were eighty-seven corselets "hung above", probably from the walls, and another five lower down, as well as numerous pieces such as sixty-three left vambraces, and 177 morions hanging between the corselets. Hanging from nails were ten muskets with their flasks, 112 harquebuses with their large flasks, as well as moulds for making bullets. The thirty-eight harnesses recorded below stand out, apparently some complete, and others composed of various pieces, as well as three cuirasses and thirteen shaffrons, in addition to twenty jousting saddles, only one with steels. We have to assume that one of those harnesess would be the one commissioned by the Count to the German armourer Anton Pefferhauser. That harness would be finished by Jörg Sorg, an engraver who decorated it and who anoted it in the famous book in which he wrote down his works, acompanied by drawings (Fig. 7). Perhaps another of the harnesses was the one decorated by the same craftsman, on the armour made by Matheus Frawenbreis, which is recorded as being made in 1549 for Gaspar de Quiñones[75]. As seen in the chapter on Benavente's, the house of Luna entered the former in the beginning of seventeenth century, and probably added this armoury.

---

74 Casado Quintanilla, 2015: 822–858.
75 Becher/Gamber/Irtenkauf, 1980: 48–51 and 90–91.

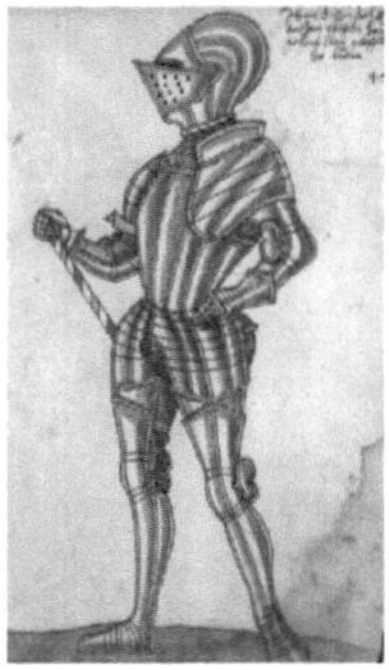

Fig. 7. *Harness for the Count of Luna.* 1563. Anton Pefferhauser and Jörg Sorg. (Taken from: Fallows, Nigel (2010): *Jousting in medieval and Renaissance Iberia.* Woodbridge, Boydell Press).

In addition to adargas, four pieces of artillery and a field tent, in the armoury there was a pile of black stone brought from Genoa and other similar sculptural elements of what could have been a fountain. But, in addition, in a room called "cuadra de las armas [arms room]", we find sixteen partisans and halberds, four spears from Germany, two rondaches, a large mirror, six black cuirasses and six morions, also black and one white. In some boxes there were also three jousting breastplates with their tassets, six jousting buffes, some other pieces, and five left vambraces, a jousting "sobrebarriga [stomach protector]" with its tassets, two jousting gauntlets, and other similar pieces.

This display is a model of what we understand by a palatial scenography, which reveals the importance of a house by linking the portraits of important people, the library and the weapons in a very complete way, although it does not seem excessively like a dynastic complex, but linked to the person of Count Claudio himself.

Another good armoury was that of the Count of Oropesa, which was located in the castle of the same name (Fig. 8). Thanks to the investigation ordered by Philip II about the weapons that were in different localities of the kingdom, we found that in 1573 there were in the Oropesa armoury thirty harnesses "*con toda armadura para hombre y caballo* [with all the armour for man and horse]", one more from Germany, several "*coracinas* [brigandines]", in addition to artillery and other things, such as arquebuses, some pistols and shotguns, thirty crossbows, a spear, a halberd, four rondaches, and "*hay cantidad de coracinas y coseletes enteros* [there are a number of brigandines and full corselets]". In addition, in the arms register of the rest of the town, it is stated that in the house of the Count's uncle, Francisco Álvarez de Toledo, "vicerey del Perú, [hay] dos çeladas de ynfante y otra çelada de seguir la guerra y una testera de caballo y una maça e una silla alemana de caballo e unas grebas y dos pares de quixotes e unos boçetes (sic) de mangas de malla e una testera de arnés [viceroy of Peru, there are two infantry sallets and another war sallet and a shaffron for a horse and a mace and a German saddle and some greaves and two pairs of cuisses and some mail sleeve holes and a harness shaffron]"[76].

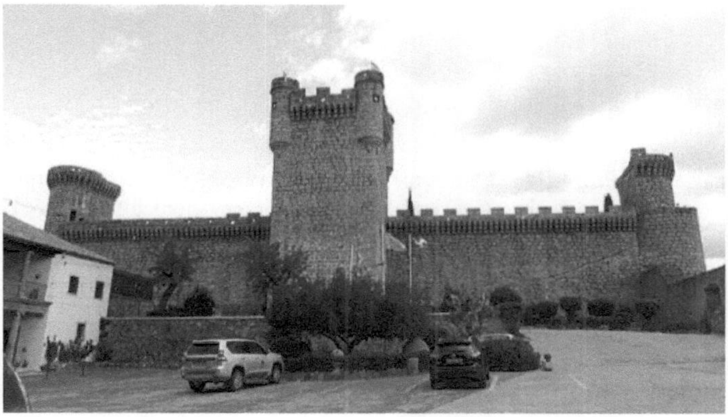

Fig. 8. *Oropesa castle*. Oropesa, Spain (Wikimedia Commons).

For his part, the Marquis of las Navas had an interesting armoury, which we only know about thanks to the report on the weapons he owned, also carried out by order of Philip II in 1572. According to the document, he had 750 arms in his house, of which 330 arquebuses distributed in other places, which offers a total of almost 310 items. But it was still an interesting armoury, although it is not called by that name. There three harnesses from Germany are described; "*jacos* [chain mail shirt]", "*coracinas* [brigandines]" and "*almillas* [loricas]" from Milan; six rondaches; helmets and "*sombreros fuertes* [strong head coverings]"; twenty harnesses for "*caballos ligeros* [light cavalry]" with their "*celadas borgoñonas* [Burgundian sallets or burgonets]"; ten corselets with their morions; chain mail pieces; fifty-six infantry corselets; 110 morions; thirty-four polearms; firearms; and fifty-two miscellaneous artillery pieces[77]. We know that the Marquis, Pedro

76 AHNOB, Frías, C. 1277, Doc. 11. See: Gutiérrez/Moreno/Hernández, 1985. Franco Silva, 2006: 85–224.

77 García Hernán, 1998: 341. Pérez Mínguez, 1930: 745–816. Martínez Hernández, 2004.

Dávila, commissioned a harness from the German armourer Lutzenberger, which the decorator Jörg Sorg engraved in 1551 (Fig. 9)[78].

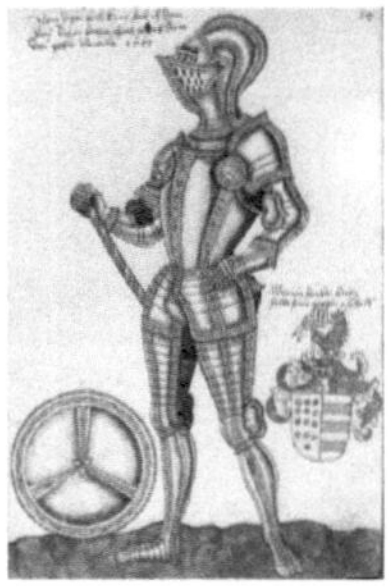

Fig. 9. *Harness for Pedro Dávila, Marquis of las Navas.* 1551. Hans Lutzenberger and Jörg Sorg. (Taken from: Becher, Charlotte/Gamber, Ortwin/Irtenkauf, Walter (1980): Das Stuttgarter Harnish-Musterbuch, 1548–1563. Viena, Anton Schroll & Co., p.61).

Another relatively well-known armoury, but apparently also of intermediate importance due to its size, was that of the Duke of Alba in Alba de Tormes, with a building converted to house the set of arms. Of particular interest is the "*torre de la armería* [armoury tower]" of the ducal palace, which not only housed arms and trophies of the Grand Duke of Alba, Fernando Álvarez de Toledo (Fig. 10), but also a pictorial programme extolling his life as a soldier and politician, especially the battle of Mülhberg, and a whole classic display of allegories[79]. It is a case that is studied in another section of this book and, although the family must have supplied it with good number of items given the importance of its arsenals in the late Middle Ages and first half of the sixteenth century[80], it is apparently the armoury assembled by a specific character, based on his military exploits. It could have been formed as early as 1559, according to documents on the arrangement of arms[81]. The highlight, however, was surely the artillery, which the Duke linked to his entailed estate in 1575 along with two swords[82]. In 1637 the artillery consisted of ninety guns of various types. He had his hired "*armero* [armourer]" in charge of the objects, and a good number of harnesses,

78 Becher/Gamber/Irtenkauf, 1980: 60–61.

79 Martínez de Irujo, 1962: 33. García Cueto, 2006: 217–218. González Zymla, 2014: 65–80.

80 ADA, Alba, C. 23, nº. 29. Calderón Ortega, 2005: 384–385.

81 ADA, Alba, C. 164, nº. 23.

82 ADA, Alba, C. 158, nº. 15, fol. 15 r.

in addition to the armour of the Duke of Saxony from the battle of Mühlberg and those of Fernando Álvarez de Toledo himself[83]. In 1705, forty of the artillery pieces were taken by Philip V for his military campaigns[84]. The castle-palace was burnt down in the Napoleonic wars, leaving only the "*torre de la armería* [armoury tower]" with part of its pictorial and artistic program.

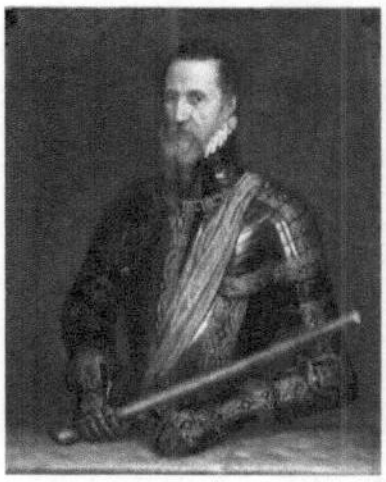

Fig. 10. *Fernando Álvarez de Toledo, Duke of Alba.* 1549. Titian (copy after). Liria Palace, Madrid, Spain (Wikimedia Commons).

A similar case is the armoury formed by the 1st Marquis of Santa Cruz, in his Renaissance palace of El Viso, built there in the Italian style and decorated with mural paintings, in a cycle in which the naval actions of the Marquis stood out[85]. From the data we have, it was basically a set of trophies and peculiarities of military action, in this case naval, of the warrior lord. In fact, in 1584 Álvaro de Bazán linked a large number of arms to the "*mayorazgo* [entailed estate]", which we will discuss elsewhere[86], including the gilded German harness with which he went to war, the trophies taken from the French and Portuguese on Terceira Island of the Azores, and various lanterns from galleys taken from the enemy, as well as hundreds of arquebuses, shotguns and pikes, and some drums, arms chosen in battle, and "*una espada gineta antigua que era de mi abuelo guarnecida de esmaltes de oro con que sirvió a los reyes Cathólicos en la guerra de Granada* [a

83 ADA, Alba, C. 157, nº. 43. It is quite possible that the complete horse bard made by the German armourer Desiderius Helmschmid and engraved by Jörg Sorg in 1551, for the Duke, was found in his armory. Becher/Gamber/Irtenkauf, 1980: 62–63. Surely, the armour of the Duke of Saxony is the one kept in the Royal Armory of Madrid. Valencia de Don Juan, 1898: 366.

84 ADA, Cartas, C. 256, nº. 25.

85 López Torrijos, 2009, with further references.

86 See: Pérez Pastor, 1895, pp. 389–393.

horseman's sword that belonged to my grandfather decorated with gold enamel with which he served the Catholic Monarchs in the Granada War]"[87]. Located in a room in the lower body of the palace[88], the armoury apparently had at least three pieces of artillery, fourteen harnesses with their morions, one of them belonging to the marquis himself, and loose pieces of armour; iron rondaches; helmets; twenty-six adargas, six of them Moorish, according to subsequent inventories (1748). Although the highlight is the display of the naval glories of Álvaro de Bazán, among which stood out the galley lanterns that were trophies of the victories he achieved, located around the palace courtyard, with their explanatory inscriptions (that of the marquis himself; of the Turkish captain's galley in Lepanto; of "Mahamet Bey hijo de Hacán Bey rey de Argel [Muhammad Bey, son of Hasan Bey, king of Algiers]"; that of "*Francia* [France]", of a French galley; and that of "Portugal", won in the war against the Prior of Crato)[89]. The set was completed, in addition to the wall paintings, with several canvases portraits of generals, and four large and small paintings of "the naval battle"[90]. The Bazans, moreover, had linked the artillery they owned in Gibraltar, consisting of fifty-two pieces of various shapes, sizes and origins, to the entailed estate since 1537[91].

Very singular was that of the Duke of Sessa, Gonzalo Fernández de Córdoba, about whose content we have very little information, albeit several arms and armour said to have belonged to him were distributed in several armouries. Formed in an extraordinary way as a display of trophies of a great soldier at the service of the Monarchy, the "Gran Capitán", around his tomb in the main chapel of the church of the monastery of San Jerónimo in Granada, is a very special case, which included arms and flags[92]. The fact that the family glories were concentrated in a funerary space surely prevented the formation of an armoury in the strict sense, although the Fernández de Córdoba family owned select collections of weapons during their stay in Rome (1526)[93], and of swords in Madrid in the mid-seventeenth century[94].

87 AHNOB, Santa Cruz, C. 32, D. 1, fols. 83 r-84 r.
88 López Torrijos, 2017: 136.
89 All this pieces were moved to the Royal Armoury in Madrid: Valencia de Don Juan, 1898: 380.
90 AHNOB, Santa Cruz, C. 35, D. 9–18.
91 AHNOB, Santa Cruz, C. 32, D. 1.
92 Bermúdez de Pedraza, 1608: 114. Prescott, 1848: 290. Fernández de Córdoba, Francisco, 2021: 217–229.
93 BFZ, Altamira, 29, D. 10.
94 AHPM, T. 6932, fol. 1117 v. AHPM, T. 7152, fol. 313 r.

Another example of a little-known armoury, but in any case not very large, is that of the Dukes of Pastrana. We know that already in 1596, according to an inventory of the family palace, although not called an armoury, there was a relatively modest set of weapons that, over time, would become a full-fledged armoury. In 1596 there were seven rondaches, some from "*India*"; some quivers; arquebus powder flasks; two Turkish bows with their quivers and arrows; a cutlass; two pistols; two arquebuses; five swords, three "*adargas de Berbería* [Barbary adargas]", etc. Counting on the weapons that the Duke had in Madrid that year, it seems very small group of arms[95]. However, after his stay in Rome as ambassador in 1627, the Duke received and acquired some arms, such as four arquebuses "*que presentó el Duque de Florencia a su Exª* [that the Duke of Florence presented to His Excellency]", the small pistols "*que trajo su Exª de Nápoles* [that His Excellency brought from Naples]", the pistols that "*presentó don Luis de Portugal* [Don Luis of Portugal presented]", several bows "*de indios* [from the Indies]", an "*alfanje de la China* [cutlass from China]" –surely a Japanese katana-, the eighteen arquebus barrels that "*trujo de Génova su Exª del Sr Duque* [His Excellency, the Duke, carried from Genoa]" and a good quantity of swords and other elements, which would serve surely to nurture the incipient ensemble[96]. The armoury existed as such when, with the union of the Silva house with that of the Mendoza-Infantado, it was transferred to the armoury of this great house from Pastrana between 1690 and 1698, including particularly luxurious objects, among them forty-five complete suits of armour with morions; various tournament pieces; forty-five rondaches; a child's corselet; firearms; cutlasses; as well as wooden display stands; and about forty pieces of artillery, some of which were placed before one of the entrances to the Guadalajara armoury building[97].

Two very interesting court armouries due to their character, location, and content are the very personal ones of Hernando Carrillo de Mendoza, Count of Priego, and Juan de Borja, Count of Ficallo. Both were modest and very special because they were not armouries of the lineage, but a depository for the arms and trophies of two figures who, due to their military career, managed to collect a good number of arms and other objects of a special nature. The Count of Priego, who participated actively and decisively in the battle of Lepanto together with John of Austria, to whom he was "*mayordomo* [steward]", created in his

---

95 AHPM, T. 1617, fol. 1840 r. The year 1580 he had arms in common use: AHNOB, Osuna, C. 1839, D. 2.

96 AHNOB, Osuna, C. 1838, D. 5.

97 González Ramos, 2013: 344. See: Sancho/Zabalburu, 1882: 524–531.

Madrid residence, as stated in his inventory of 1579, a "*camarín de despojos de la batalla de la Liga* [cabinet room for the spoils of war for the Battle of the League]"[98], which is what that naval confrontation was called. All the content reconstructed and exposed the development of the battle through the exhibition of maps, paintings of the battle, portraits of the main participants, Christians and Turks, and a large number of trophies[99]. The set of objects did not continue in the Priego house, as the king moved most of them to El Escorial in 1583[100]. The case of the Count of Ficallo Juan de Borja, known to us from a record from 1600, the "*camarín de las armas* [chamber of arms]" of his house focuses on the activities of a character who developed his military career in Central Europe at the service of the Habsburgs. Among the pieces that were displayed in his court residence we found Hungarian and German objects, "*rodelas de la China* [Chinese rondaches]", and an enormous portrait gallery[101].

In the "Crown" of Aragon there were good armouries, but we do not have much information. The one belonging to the Duke of Calabria, from the Italian branch of the Royal house of Aragon, is well known. It contained a good set of arms, abundant in both quality and quantity, which the Duke bequeathed in 1550 to the convent of San Miguel de los Reyes in Valencia that he had founded. Although it is not called an armoury, and is related to his general collection and its great library[102]. However, it was a very well endowed and organised set, with eighteen flags and "*banderillas* [small flags]", among them banners with the arms of Pope Alexander VI, those of Florence, and Aragon; rondaches; eleven tents and pavilions; six harnesses; fifteen corselets; helmets; breastplates and jousting pieces; as well as chain mail pieces; shaffrons; jousting saddles; lances; etc. Also are recorded almost twenty Valencian, German, Hungarian and Moorish swords; as well as rapiers and scimitars, daggers, shotguns, crossbows, bows and arrows, etc[103].

Other examples are more evolved and specifically named armouries, such as those of the Duke of Gandía, the one owned by the Duke of Villahermosa in his Pedrola palace, and the one of the Duke of Segorbe. We know that already by 1588, in the time of Carlos of Borja, 5th Duke of Gandía, there was an armoury in the town of the noble title, which contained more than 800 arms. In the inventory

98 Morán/Checa, 1985: 164.

99 AHPM, T. 759, fol. 385 r. AHNOB, Priego, C. 2, D. 30. Quintanilla Raso, 1992: 381-402.

100 Martin, 2012: 52.

101 AHPM, T. 933, fol. 379 r. Morán/Checa, 1985: 158 and 163.

102 Morán/Checa, 1985: 45.

103 Vignau, 1871: 12–15, 28–29, 92–93, 108–110.

made that year we found that in Castellón there were many more, apparently the most noteworthy being in Gandía[104]. There were more than twenty-six pieces of artillery; large quantities of arquebuses (156); twenty-four corselets that had been brought from Milan in 1566; an engraved harness; morions; war saddles; drums; swords and rapiers; polearms of all kinds (partisans, spears, "*lanzones* [lances]", pikes, "*lanzas ginetas* [calvary lances]", jousting and tournament lances); swords and rondaches; tents and pavilions; and numerous "*aderezos de caballos* [horse reins]" in the stable.

The Duke of Villahermosa possessed particularly curious objects, in addition to an armoury itself, of which we do not know much. The katana "*del tártaro* [of the Tartars]" that he gave to Lastanosa came from his armoury. In 1627, Gaspar Galcerán de Castro, Count of Guimerá among his antiquities, had ancient arms, including a piece of polearm and a helmet or "*galeo*", surely from ancient Roman times; a rondache of the infant Don Alonso of Aragon and Castile, 1st Duke of Villahermosa and Count of Ribagorza; a chess piece that Miramamolín (the Almohad caliph, Muhammad al-Nasir) played with just before the battle of Las Navas de Tolosa; or the dagger of King Peter IV of Aragon[105]. We do not know much more about armouries in the Crown of Aragon, except for the case of the very famous armoury of Vicencio J. de Lastanosa in Huesca, which is surrounded by legend and which Álvaro Soler del Campo has studied[106]. Although tradition had given it the reputation of being a huge armoury[107], in reality it was not that big, but it was nonetheless of high quality and well equipped. We found some boots of the King Henry IV of France; the dagger of King Peter IV of Aragon; ancient and curious arms; arquebuses; musketry; pistols; halberds; crossbows; quivers; bows; a horn from Japan; breastplates; curiasses; rondaches and a good amount of horse reins[108]. The Segorbe armoury is only known to us from a late document. The Segorbe duchy was one of those that belonged to the "*Grandes de España* [Grandees of Spain]" since the creation of this category by Charles V. It ended up being united by matrimonial links to that of Medinaceli in the person of Luis Francisco de La Cerda y Aragón, and when his descendant, the 11th Duke of Medinaceli, wanted to gather in Madrid the armouries of his lineages in 1759, an inventory was drawn up indicating that in the castle of Segorbe there were, in

104 AHNOB, Osuna, C. 568, D. 127.
105 Biblioteca Nacional, Madrid (hereafter BNE), Mss. 6423. Morán/Checa, 1985: 190. Morán Turina, 2009: 247 and 267.
106 Soler, 2007: 117–123.
107 Arco y Garay, 1934: 219–225.
108 Arco y Garay, 1934: 219–220.

addition to artillery pieces and some more firearms, eighty-one breastplates and "*espaldares* [backplates]" with pieces for arms and legs, and twenty iron helmets, all in bad condition[109]. Exceptional for having been preserved in the family's Valencian palace in the eighteenth century, it was the armoury of the Marquises of Dos Aguas, whose formation and fate are discussed in another section of this book. It was not a very large armoury, and certainly late, with family weapons such as a "*petral y pistolera* [rein and holster]" that a family member carried when King Philip III entered Valencia; the crossbows of the "*moriscos* [Muslim converts to Christianity]" who rebelled against the expulsion in 1609; the Count of Albatera's suit of armour and another that the Count of Peralada took to Valencia in 1689; a shaffron of the 1st Marquis of Dos Aguas in 1691, etc[110].

Another category in the world of armouries is, as we have seen, those formed *ex novo* by nobles who, without having a family background, nor a prominent origin in the universe of the Castilian nobility, came to form very personal armouries based on their political fortune. The most impressive, probably because it was formed when the phenomenon was at its peak, was the armoury of the Duke of Lerma, which is also studied in more detail in another chapter in this volume. Despite coming from a lineage that had suffered difficulties and had not been able to rise to prominence, court action allowed the 1st Duke of Lerma to rise up as a favourite of King Philip III. The springs of power, and the economic situation achieved thanks to his position, allowed Lerma to hoard not only a good amount of rich, ancient and exotic arms, but especially to create in one of his recently acquired properties, Ampudia, an armoury that for its size was surely the most surprising in Castile of its time. The armoury of the castle of Ampudia, however, stood out more for quantity than for quality, given that the Duke could only rely on limited historical and family arms, although he succeeded in a certain way. It is more an ensemble built on the basis of large acquisitions of Milanese arms, than an armoury of an ancient lineage.

The armoury of the Marquis of Leganés in Morata de Tajuña is also that of a man of not very high origin who, thanks to his kinship and the political support of the Count-Duke of Olivares, managed to rise up and reach important positions[111]. His new position, and his political and military activity, allowed him not only to put together one of the largest and best artistic collections of the time[112], but also to combine part of it with a medium but very interesting

109 ADM, Archivo Histórico, Leg. 82, nºs. 155–158.

110 García Martínez, 2015: 176. See the inventory of 1843: AHNOB, Osuna, CT. 538, D. 42.

111 Elliott, 1991. Arroyo Martín, 2002: 145–185. Arroyo Martín, 2012.

112 Pérez Preciado, 2010.

armoury, set up at a time when these sets had lost much of their *raison d'être*, but it was manifested as a symbolic element of the importance of its owner. Even though it was a late ensemble and we believe that it was incipiently anachronistic, it was made keeping many of the characteristics of the most powerful armouries, being located in a palace in the possessions of the marquisate, combining it with galleries of portraits and other aristocratic symbolic elements, and exhibiting pieces of importance in cabinets and with a very prominent exhibition sense.

The Leganés armoury in his Morata de Tajuña palace had a perfectly ordered group at his death in 1655. The first news dates from the inventory of his assets from 1642, which records a large number of select arms and trophies, as well as of the incipient armoury, registered in its own section of the document[113]. Vicente Carducho even refers to the artillery that Leganés had in his house in Madrid, when he points out about the Marquis: "*Vimosle en un patio entre culebrinas, y cañones de todos géneros, […] con algunos Artilleros, y Fundidores* [We saw him on a patio amongst culverins and cannons of all types . . . with some artillerists and founders]"[114]. But when it was fully formed, the armoury was located in the aforementioned palace in a series of seven wooden cabinets with metal mesh doors that exhibited the contents. In these cabinets we find eleven complete harnesses and "*medios cuerpos* [half bodies]", as well as breastplates; rondaches; crossbows; scimitars; "*machetes*"; firearms and other items. In addition, there were twenty-six small artillery pieces[115]. The marquis, as is found on other occasions in the noble documentation, linked his armoury to the entailed estate or "*mayorazgo*" of his house[116], and had left various arms to relatives in his will[117]. It was combined with the gallery of paintings, an analysis that would be convenient to carry out. The Morata de Tajuña armoury appears registered in 1753, pointing out the habitual negligence and carelessness, when it is indicated that the cabinets and weapons were "*muy maltratados* [very poorly treated]", and their content would not be inventoried in the smallest detail[118].

We have also located a good number of armouries of a minor category due to their importance, size and deployment. These are surely sets that began to be made as a manifestation of the aspirations of some nobles who were not first class,

113 AHPM, T. 5993, fol. 218 r, 361 r and 408 r. Pérez Preciado, 2010: 575–576.
114 Carducho, 1633: 148. Morán y Checa, 1985: 192.
115 AHPM. T. 6267, fol. 473 r.
116 Instituto Valencia de Don Juan (hereafter IVDJ), Manuscritos, 26-II-20, fol. 99 r.
117 AHPM, T. 6265, fol. 363 r.
118 AHPM, T. 30083, fol. 348 r. Pérez Preciado, 2010: 912. See also: Agulló y Cobo, 1994. Arroyo Martín, 2012: 698.

but who were somehow dragged along by the trend of gathering sets of arms to indicate a certain aristocratic level and claims of promotion. Many of them seem to have tried to follow the characteristics of the most important, gathering arms and armour, if possible from the lineage, in castles and family palaces.

The armoury owned by the Count of Fuensalida is scarcely known[119]. In 1662 the owner of the house, Don Bernardino López de Ayala Velasco y Cárdenas, had a small collection of select arms in his chamber in his Madrid residence, especially swords, but as usual, the armoury is not recorded as such in the document[120]. However, in 1593 the castle of Guadamur, owned by the Count, was endowed with arms and supplies, in the "*sala de las armas* [room of arms]", which included ten pieces of artillery; a pair of harnesses; more than fifty sets of greaves, "*brazaletes* [vambraces]" and other pieces, and various breastplates that would be used for several more suits of armour; about fifteen shotguns and almost the same number of crossbows; various pavises; drums and other items, with a portrait of the Count[121]. However, we know that existed in Guadamur Castle (Fig. 11) an armoury as such, and quite equipped, especially because Philip IV requested arms from that ensemble in 1637 for the war with France, on the occasion of the siege of Fuenterrabía[122]. The corresponding receipt indicates that thirty "*querpos de cosoletes* (sic) *de infantes* [bodies of infantry corselets]" were delivered[123]. The same would happen in 1639[124], and, later, in 1641, when the king's letter indicates that the request was motivated "*porque se ha entendido que tenéis cantidad de armas* [because it has been understood that you have a quantity of arms]"[125]. Unfortunately, in these last two petitions there is no record of the arms delivered or of those that remained. The armoury continued to exist, we do not know under what conditions, and in the nineteenth century it was recovered, after the sale of the building to the Count of Asalto, at the time of the restoration of the castle of Guadamur, but without relation to the original group of arms[126].

---

119 Fuente Escribano, 2019: 231–263.

120 AHPM, T. 9801, fol. 27 v.

121 AHNOB, Frías, C. 24, D. 113. See Cadiñanos, 1997: 61.

122 AHNOB, Frías, C. 26, D. 23

123 AHNOB, Frías, C. 26, D. 31.

124 AHNOB, Frías, C. 26, D. 25.

125 AHNOB, Frías, C. 26, D. 24.

126 Fuente Escribano, 2019: 231–263, with further references.

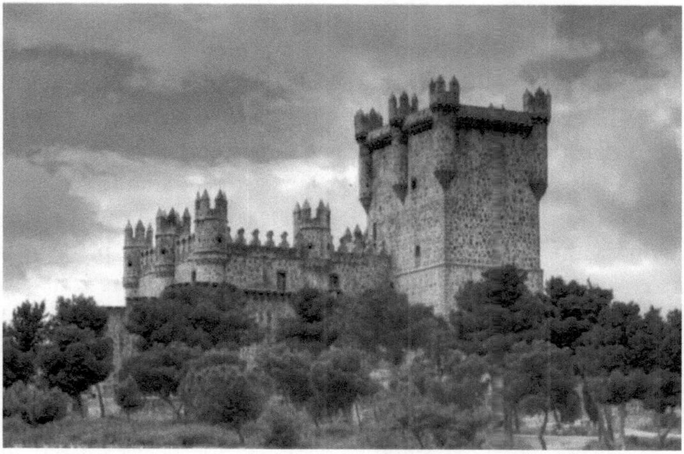

Fig. 11. *Guadamur Castle.* Guadamur, Spain (Wikimedia Commons).

One of the most curious and interesting armouries of this type was the one assembled by the Marquis de Velada in his castle of Villatoro, a building that was modified according to the Marquis' own plans[127]. This estate was incorporated into the house of Velada in 1607[128], therefore it does not appear in the 1596 inventory[129]. In the inventory carried out in 1616, upon the death of the 2nd Marquis Gómez Dávila de Toledo, the creator of the armoury, the arms that he had are described, showing a good display of quality and selection, including more than seventeen harnesses (one of which had belonged to Duke of Medinaceli)[130]. Some reports even indicate that he had the sword of the "*repoblador* [settler]" of Ávila, Nalvillos Blázquez, which Alonso de Aragón would have given him, and that it had previously belonged to Alfonso VI of Castile and, although this information is not reflected in the documents we handle, it could be the "*espada ancha a lo antiguo con los seis roeles que se halló en Ávila en un sepulcro* [antique broadsword with six heraldic rondels that was

127 Martínez Hernández, 2003: 62.
128 Martínez Hernández, 2003: 69.
129 AHPM, T. 1810, fol. 1291 r.
130 BFZ, Altamira, 198, D. 1, fol. 71.

found in a tomb in Ávila]"[131]. In Villatoro, there were mainly pieces of harness that totaled 127 "*brazaletes* [vambraces]", legs and other elements, but not breastplates or "*espaldares* [backplates]"[132]. With the house of Velada attached to the Marquisate of Astorga[133], we find a full-fledged armoury in the palace of the town of Velada itself, which was located in one of the towers of the building in 1678. In addition to select arms in the cabinet, we find in the armoury several morions, some carved; twenty-six breastplates with their "*espaldares* [back plates]", along with "*brazaletes*[vambraces]" and other loose pieces of armour; shaffrons; crossbows; "*montantes* [longswords]"; firearms; a knife "*de Japón* [from Japan]"; chain mail parts, etc[134]. Shortly after, in 1698, the armoury was already quite abandoned, when it is described as located in an attic with the arms "*tan pasadas y mohosas del tiempo que no pueden servir de nada* [so old and mouldy with time that they cannot be of any use]", although there were around 150 pieces[135]. In 1737, another record of the palace's assets indicates, under the heading "*armería* [armoury]", that the arms were still in an attic, "*todo muy oriniento y maltratado por el trascurso del tiempo* [all very rusty and poorly treated by the passage of time]", which indicates that, as on other occasions, they were sets that had lost their meaning and importance[136].

---

131 Ariz, 1607. Martín Carramolino, 1872. Ballesteros, 1896: 119. On the sword, see, Sandoval, 1792: 21. The coat of arms of Blázquez, in addition to a bend, shows indeed, six rondels.

132 BFZ, Altamira, 198, D. 1, fol. 120. On the armoury of Villatoro see: Martínez Hernández, 2005. Martínez Hernández, 2004.

133 There are drawings of a harness made in 1551 for Alonso de Osorio, Marquis of Astorga (Fig. 12), in the book of Jörg Sorg: Becher/Gamber/Irtenkauf, 1980: 17 v-18 r.

134 IVDJ, Manuscritos, 25-V- 13, fol. 124 r.

135 IVDJ, Manuscritos, 25-V- 14, fol. 304 r.

136 IVDJ, Manuscritos, 25-V- 14, fol. 340 v.

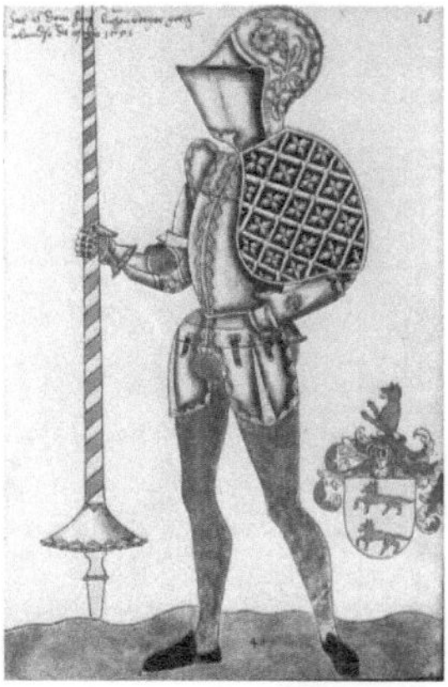

Fig. 12. *Suit of armour for Alonso de Osorio, Marquis of Astorga*. 1551. Jörg Sorg. (Taken from: Becher, Charlotte/Gamber, Ortwin/Irtenkauf, Walter (1980): Das Stuttgarter Harnish-Musterbuch, 1548–1563. Viena, Anton Schroll & Co., p.65).

From the Silva lineage, there is another interesting armoury, small in size, but with good pieces. It is about the ensemble that the 6th Count of Cifuentes Fernando de Silva had in his castle of Cifuentes already in 1593, upon his death[137]. This group of arms, which had obviously been formed before that date, although we do not know how much, had 122 arms according to the inventory[138]. We find seven rich harnesses, some gilt and engraved, and others black; artillery; and a good number of interesting, rich, exotic and ancient arms. Among them, various engraved and gilded rondaches and morions, a set of them with "*una historia de relieve* [an embossed history]"; in addition to pistols, arquebuses and other arms also of great value. Like other armouries, there was a small artillery piece there; seven rich Turkish cutlasses; over twenty select ancient swords and sword blades; and even Turkish bows. The Count of Cifuentes also had arms in the castle of Barcience (Fig. 13), totaling sixty "*medios arneses* [half harnesses]"; twenty-one artillery pieces, some with the Count's coat of arms, and their ammunition; and other arms, but in this case it is surely not a symbolic armoury, but rather the manning of the fortress.

137 Ávila Seoane, 2006: 395–438. Layna, 1955.

138 AHPM, T. 10801, fols. 173 r-178 r, 211 r-219 r.

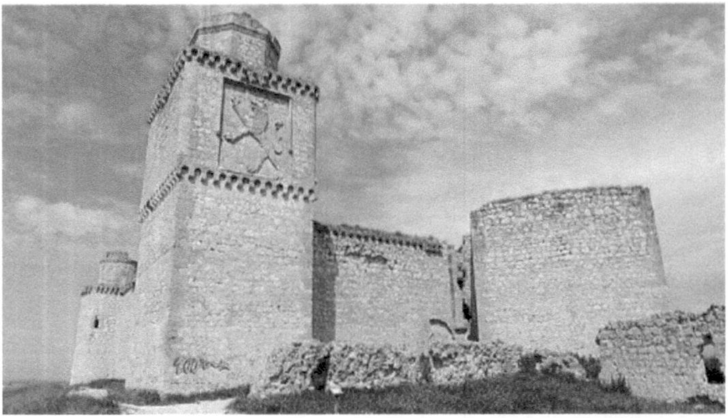

Fig. 13. *Castle of the Counts of Cifuentes.* Barcience, Spain (Wikimedia Commons).

On the other hand, the Marquis of Auñón put together an interesting armoury in his Madrid palace, so we once again have a court armoury, perhaps in the absence of other locations. In 1604, twenty-two harnesses were recorded there, one of them rich, and many loose pieces of armour, among other arms[139].

Although we have no information to confirm it, the armoury of the Marquises of Villafranca was probably of this same category, and it was in the family castle, as not very extensive, so the sources indicate (Fig. 14). Certainly, in 1650, the Villafrancas had a good collection of swords, rapiers and pistols in Madrid, which due to their number and quality were not just everyday objects[140]. On the other hand, we know that in 1802, in the Villafranca del Bierzo fortress[141], in the south-west tower, the second floor "*solo sirve para tener en él una gran porción de morriones, casquetes, pedazos de cotas de malla, cañones, lanzas, muchas piezas de armería todo muy maltratado* [only serves to have in it a grand portion of morions, helmets, pieces of mail, barrels, lances, and many pieces of armour, all very poorly treated]"[142], which indicates the existence of the armoury, some of its contents and its location, as well as its status in disrepair. It is quite possible that the complete multi-piece armour made by the German armourer Anton

139 AHPM, T. 1990, fol. 227 r, 238 r.

140 AHPM, T. 4499, fol. 97 r.

141 Cooper, 1991: 385–386. Cobos/Castro, 1998: 258.

142 Bosch Ballbona, 2013–2014: 89. See also: Halcón, 2021: 249–274, where there is no reference to the armoury.

Pfeffenhauser and engraved by Jörg Sorg in 1551, for García de Toledo, was among those pieces[143].

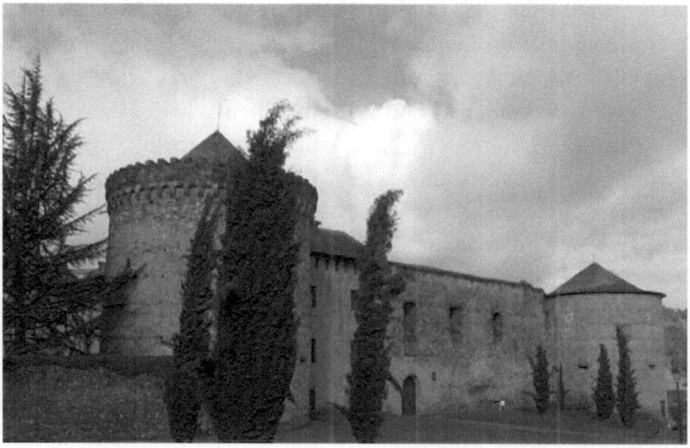

Fig. 14. *Castle of Marquises of Villafranca.* Villafranca del Bierzo, Spain (Wikimedia Commons).

Another interesting armoury must have been that of the Dukes of Nájera. Unfortunately we have only located one document that indicates the arms they had, but they were in his Madrid residence. It seems that the Nájera fortress was abandoned at the end of the sixteenth century[144]. In 1600, Juan Esteban Manrique de Lara, 4th Duke of Nájera, died, leaving his daughter Luisa as his heir. In Madrid we found only a couple of harnesses, some chain mail pieces, cutlasses, and a few "*montantes* [longswords]", several rapiers, some swords and a handful of "*pistoletes* [pocket pistols]" and arquebuses, as well as two metal rondaches, a morion and a helmet[145]. However, in 1604 an inventory was made of the assets of Luisa Manrique de Lara, Duchess of Nájera and her husband Bernardino de Cárdenas, Duke of Maqueda (who died in 1601), in which we found a true court armoury. It was a set of objects placed in seven large boxes, including an engraved and gilt harness from Germany; five morions from Calatayud; no less than fifty very select swords, among which there were old ones, rich "*de oro* [of gold]", from

143 Becher/Gamber/Irtenkauf, 1980: 64–67.

144 Ceniceros Herreros, 2004: 519–530.

145 AHPM, T. 2606, fol. 987 v.

Bilbao, Sahagún, Valencia, "moriscas [Moorish]", from renowned swordsmiths such as the "*el moro de Zaragoza* [the Moor of Zaragoza]", Sebastián Hernández, Tomás de Ayala, Maese Domingo, Francisco Ruiz "*el Viejo* [the elder]", of Juanes de la Orta, or Pedro de Toro, and of the "perrillo"; various rapiers; cutlasses; fifty-four arquebuses from Germany, from Biscay and from "*Bresa* [Brescia]"; nineteen pistols from Germany and Italy; forty-three crossbows and crossbow prods of important craftsmen as Luis Moreno, Puebla, Juan Blanco, Santa María, el Sordo de Zamora, Grajeras, etc[146]; in addition to thirty-five Turkish bows and quivers, "*viras* [darts]", bolts and other things, such as a "machete y cuchillo judíos [Jewish machete and knife]"[147]. In 1616, the possessions were distributed among the heirs, in whose documentary record we have an impressive number of arms[148].

An armoury that existed in 1630 was that of the Duke of Arcos in Marchena[149]. Located in the family palace, in the main floor near to the chapel, library and principal rooms, we have no previous registers, but several documents (around 1579–1581) that points out to the formation of an interesting set that had at least 330 corselets, twenty sallets, 178 "*barbotes* [bevors]", eighteen morions[150], halberds, partisans, swords, and so on, some of which were used in 1580 in the war of annexion of Portugal, when is mentioned the Duke's "cámara de armas". In the 1630 inventory of objects incorporated to the "mayorazgo [entailed state]" we have various arms registered in the lord's chambers, among which we find two "*rodelas negras* [black rondaches]"; many longswords; "*espadas anchas de a caballo* [cavalry broadswords]", one of them rich; sword blades; several Turkish cutlasses; arquebus barrels; elements for horses, etc. A cutlass had historical significance, since it is said that it belonged to King "Avenavo", whose name refers to Abén Aboo (Diego López) a leader of the revolt of the "Moriscos" of Granada in the time of Philip II, successor of the "*rey de los andaluces* [king of the Andalusians]", Aben Humeya[151]. This piece was a war trophy obtained in 1571 by Luis Cristóbal Ponce de León, Duke of Arcos (1530–1573) himself by means of his participation in that conflic and the dead of the revolt leader. But in the armoury itself we found registered (linked to the "*mayorazgo*"), in addition to various "*sillas de armas* [saddles of arms]"; five rondaches, one

146 On Spanish crossbow craftmen: Martínez de Espinar, 1644: 12-13.
147 AHPM, T. 2619, fol. 1337 r.
148 AHNOB, Frías, C. 890, D. 12.
149 Salvago, 1957: 47-84.
150 AHNOB, Osuna, CT. 550, D. 57.
151 Mármol, 1797: XV.

of them "*veneciana dorada* [gilded Venetian]"; and five harnesses, one gilt with its helmet, another of the Duke itself and several made of "pasta" (from Colmenar, from Milan, from "Agusta de un menino", probably a child's suit of arms from Augsburg). Also are described several rich breastplates forming a set with morions; jousting pieces; "*tarjetas* [Stechtarschen]" and shaffrons with the Duke's coat of arms; tournament helmets; "*petos y espaldares a prueba* [bullet-proof breastplates and backplates]", etc[152]. We do not have a complete and detailed inventory of the armoury, but this data gives information about a complete set, and the value for the family of arms linked to the "*mayorazgo*". In the family palace of Marchena in 1658, although we find an inventory of a large number of carbines and pistols from Brescia and other places, as well as swords, daggers and some cutlasses, we do not find direct allusions to the set cited in 1630, nor to the harnesses or corselets[153], but in the 1673 inventory[154], a total of forty-eight weapons were recorded and, finally, breastplates and backplates, and "bracelets, legs and feet and hands and loose pieces" in the chamber. According to the report, the breastplates and backplates weighed eighty-seven "arrobas", that is, 1,000 kg., and the rest of the armour pieces reached fifty-six "arrobas" (644 kg.), which amounted to 1,644 kg, a total weight that far exceeds that of the harnesses of 1630, and which suggests a fairly large set of suits of armour.

A small and yet famous and very selective armoury was that of the Duke of Alcalá de los Gazules, which was connected to the family artistic collections in the Casa de Pilatos in Seville[155]. According to the inventory of 1637, in the armoury hall, in addition to "*un lienzo copia de las armas que estaban en el comedor* [a painted copy of the arms that were in the dining room]"[156], two "*medios cuerpos* [half bodies]" of engraved and gilded suits of armour were kept; another suit armour; a suit of armour with a "*sobrepeto*"; two engraved and gilt shaffrons[157]; three breastplates and "*espaldares* [backplates]" with "*faldones a prueb*a [bullet-proof skirts or tassets]" made in Barcelona; adargas; a bow; Turkish muskets; a shotgun; a blued and gilded steel rondache; sword hilts; dagger

152 AHNOB, Osuna, C. 125, D. 176.

153 AHNOB, Osuna, C. 1591, D. 49.

154 AHNOB, Osuna, C. 1625, D. 236.

155 Morán/Checa, 1985: 233–234. Lleó, 2017: 167 and 185. Urquízar, 2007: 75 and 126. See also: *I exposición*, 1971. Godoy, 1997: 139–148.

156 BFZ, Altamira, 215, D. 1, unfoliated (nº. 38). See: Brown/Kagan 1987: 231–255, where the armoury is not studied.

157 There is a shaffron in the Victoria & Albert Museum supposedly from the Casa de Pilatos. See Hayward, 1948: 253. Hayward, 1965: cat.12. Patterson, 2009: 41.

blades; nine select swords, almost all of them old, some with inscriptions on the blade, one of which "*dizen fue de Diego Gómez de Ribera, 2º adelantado mayor de la frontera* [it is said belonged to Diego Gómez de Ribera, 2nd *adelantado mayor* of the borderlands]"; a rapier with a golden inscription of Pope Paul IV, with verses; cutlasses, some with inscriptions, others Turkish arms; two pairs of stirrups that apparently had belonged to the Marquis of Tarifa Fadrique de Ribera; and a small bronze artillery piece. The armoury hall, therefore, contained antique arms, especially those related to the family and important events of the lineage, such as the papal rapier that is to be related to the 1st Duke, Perafán de Ribera; the sword of the "*adelantado*" who died in 1434; the Marquis' stirrups, and the arms made in Barcelona, connecting the armoury with Fernando Afán de Ribera, who among other positions was viceroy of Catalonia. In its location in the palace, it was combined with the library, modelled on some of the great armouries.

The Count of Chinchón in 1617 had thirty-eight pieces of artillery in the castle of Chinchón. A solitary document informs us that on an undetermined date the Count ordered three black harnesses, one with gilded decoration, to be brought to one of his possessions (perhaps Chinchón)[158]. However, in his Villaviciosa de Odón fortress there was a "room of armoury", which in 1665 contained a small group of arms, in which there was only an engraved harness, some firearms and various crossbows[159]. We must assume that earlier this set was more important.

Also Luis Carrillo de Toledo, Marquis of Caracena, and since 1624 Count of Pinto, owned an armoury in the castle of Pinto. We know that in 1622 this armoury displayed fourteen harnesses, forty-four muskets and other weapons, and a portrait gallery, in another modest attempt to create a symbolic scenography of the lineage[160].

As we have seen, the Count of Fernán Núñez also possessed, above all, artillery, which he placed in the family palace. He had various pieces ensconced in the building, which Diego Gutiérrez de los Ríos y Guzmán had taken from the English admiral Robert Blake in a naval battle off the coast of Cádiz in 1657[161], and which would be linked to the "*mayorazgo* [entailed estate]"[162]. He also had some memorabilia from the wars in Granada in the Middle Ages, such as a "*estribo y*

158 AHNOB, Frías, C. 1436, D. 25.
159 AHNOB, Frías, C. 1435, D. 4.
160 AHPM, T. 2679, fols. 1248 r and ss.
161 Espejo Gutiérrez, 2020: 182.
162 Espejo Gutiérrez, 2020: 287.

pieza del jaez del rey de Granada [stirrup and harness piece belonging to the king of Granada]" and pictures of mediaeval and modern events in the house (such as the painting of Count Diego when he defeated the English "Armada de Cromuel (sic) [Cromwell's Navy]"), but not an armoury *strictu sensu*[163].

We can also cite here the case of the Duke of Feria, about whose armoury we know almost nothing, except that at least a part of it passed to that of Medinaceli. However, it is documented that there were weapons in the castle of Zafra. An inventory from the Medinaceli archive refers to the adargas, lances, partisans, spears, arquebuses, helmets, hammers and axes "*de armas* [of arms]", maces, stirrups, reins and other elements that were in that castle in 1552, upon the death of Pedro of Córdoba and Figueroa, Count of Feria. In 1624, there were suits of armour that were stored in iron boxes, as well as the house's books and portraits, such as those of Gómez Suárez and Lorenzo Suárez de Figueroa[164]. By 1571, in Madrid, there were hardly any arms in the Duke's possession[165].

Like other nobles, the Count of Salinas de Añana, Duke of Híjar by marriage, had in Madrid in 1631 various "*cosas de caballeriza, guarda arnés y armería* [items of cavalry, horse harnesses, and armoury]". In addition to saddles and other equestrian elements, there was an "*arnés liso* [plain harness]", another "*arnés de justa* [jousting harness]" engraved and gilded, and one more of "Charaum"; as well as adargas, "*rodelas de la India* [rondaches from India]", some pistols and shotguns; quivers; crossbows; spears; swords and tents, in what is a small court armoury[166]. Like other ensembles that we have been able to locate, actually they were sets of select arms, like the one owned by the Marquis of Villanueva del Río in 1599, abundant in swords[167], or that of the Marquis del Valle de Cerrato, who in 1616 owned a good set—in addition to swords and other arms—of arquebuses, crossbows and knives[168], which he left in his 1610 will to his son and heir so that he could practise hunting[169]. The Marquis of Salinas del Río Pisuerga and president of the Council of the Indies, twice viceroy of New Spain and viceroy of Peru, Luis de Velasco, in 1617 had some arms, among which four harnesses

163 AHNOB, Fernán Núñez, C. 494, D. 4. AHNOB, Fernán Núñez, C. 1417, D. 2–1.
164 Both references in: ADM, Archivo Histórico, Leg. 82, D. 65.
165 AHPM, T. 742, fol. 1 r. See: Godoy, 1997: 139–148. Valencia Rodríguez, 2000.
166 AHNOB, Osuna, C. 3910, D. 149.
167 AHPM, T. 1418, fol. 1663 r.
168 AHPM, T. 2661, fol. 1.
169 AHPM, T. 2656, fol. 826 v.

stand out, and six more "de la China [from China]" (perhaps Japan), as well as a multitude of American memorabilia[170].

Surely there were more armouries than we have been able to locate[171], and it is probable that more groups of arms such as those described here appear, but the search and analysis work carried out up to now allows us to offer a broad overview of a phenomenon that will have to be expanded and qualified in the future.

BIBLIOGRAPHY

Agüero Carnerero, Cristina (2017): "El ocaso de los Enríquez de Cabrera. La confiscación de sus propiedades y la supresión del almirantazgo de Castilla". In: *Tiempos Modernos*, 33, 2, pp. 132–153.

Agulló y Cobo, Mercedes (1994): *Documentos para la Historia de la Pintura Española I*. Madrid, Museo del Prado.

Appadurai, Arjun (ed.)(1996): *The Social Life of Things. Commodities in Cultural Perspective*. Cambridge, Cambridge University Press.

Arco y Garay, Ricardo del (1934): *La erudición aragonesa del siglo XVII*. Madrid, Cuerpo Facultativo de Archiveros, Bibliotecarios y Arqueólogos.

Ariz, Luis (1607): *Historia de las grandezas de la ciudad de Ávila*. Alcalá de Henares, Luis Martínez Grande.

Arroyo Martín, Francisco (2002): "El marqués de Leganés. Apuntes biográficos". In: *Espacio, tiempo y forma. Serie IV. Historia Moderna*, 15, pp. 145–185.

Arroyo Martín, Francisco (2012): *Poder y nobleza en la primera mitad del siglo XVII: El I marqués de Leganés*. Getafe, PhDr. Dissertation, Universidad Carlos III.

Ávila Seoane, Nicolás (2006): "El señorío de los Silva de Cifuentes en los concejos de Atienza y Medinaceli (1431–1779)". In: *Revista de Historia Moderna*, 24, pp. 395–438.

Ballesteros, Enrique (1896): *Estudios histórico de Ávila y su territorio*. Ávila, Manuel Sarachaga.

170 AHPM, T. 2320, fol. 1.

171 Because we know, for example, many commissions for armour decorated by the engraver Jörg Sorg (who collected his works in his famous book): Bernardino Manrique de Lara, 1549; Rodrigo de Benavides, 1550; Andrés de Ribera, 1550; Francés de Álava y Beaumont, 1550; Alonso de Osorio, 1551; Alfonso de Bustos, Marquis of Corvera, 1551; Álvaro de Sande –governor of de Milán-, 1563. Becher/Gamber/Irtenkauf, 1980.

Becher, Charlotte/Gamber, Ortwin/Irtenkauf, Walter (1980): *Das Stuttgarter Harnish-Musterbuch, 1548–1563.* Vienna, Anton Schroll & Co.

Bermúdez de Pedraza, Francisco (1608): *Antigüedad y excelencias de Granada.* Madrid, Luis Sánchez.

Bernis, Carmen (1962): *Indumentaria española en los tiempos de Carlos V.* Madrid, CSIC.

Bosch Ballbona, Joan (2013–2014): "La fortaleza que quiso ser palacio. Noticia de Camillo Camiliani en España (1604)". In: *Locus Amoenus*, 12, pp. 79–106.

Bouza, Fernando (2003): *Palabra e imagen en la corte. Cultura oral y visual de la nobleza en el Siglo de Oro*. Madrid, Adaba.

Brown, Jonathan/Kagan, Richard (1987): "The Duke of Alcalá: His Collection and Its Evolution". In: *The Art Bulletin*, 69, 2, pp. 231–255.

Burke, Marcus B./Cherry, Peter (1997): *Collections of Paintings in Madrid, 1601–1755.* Los Angeles, Provenance Index of the Getty Information Institute.

Burke, Peter (1990): "L'homme de Cour". In: Garin, Eugenio (dir.): *L'homme de la Renaissance.* Paris, Editions du Seuil: pp. 142–173.

Cadiñanos Bardeci, Inocencio (1997): "Noticias de arquitectura fortificada en España". In: *Castillos de España*, 108, pp. 55–62.

Calderón Ortega, José M. (2005): *El ducado de Alba. La evolución histórica, el gobierno y la hacienda de una estado señorial (siglos XIV-XVI).* Madrid, Dykinson.

Cámara Fernández, Carmen (1994): "Aportación al estudio del coleccionismo en el siglo XVII. La armería del Condestable de Castilla en su palacio de Burgos". In: Nieto Alcaide, Víctor (ed.): *Los Clasicismos en el Arte Español. Actas del X Congreso del Comité Español de Historia del Arte.* Madrid, UNED: pp. 407–412.

Carducho, Vicente (1633): *Diálogos de la pintura.* Madrid, Francisco Martínez.

Carlos, María Cruz de (2005): " 'Al modo de los antiguos'. Las colecciones artísticas de Juan Fernández de Velasco, VI Condestable de Castilla". In: Alonso, Begoña/De Carlos, María C./Pereda, Felipe (eds.): *Patronos y coleccionistas. Los condestables de Castilla y el arte (siglos XV-XVII).* Valladolid, Universidad de Valladolid: pp. 207–277.

Carrillo, Jesús/Pereda, Felipe (2000): "El caballero: identidad e imagen en la España Imperial". In: *Carlos V: las armas y las letras. Sociedad Estatal para la conmemoración de los Centenarios de Carlos V y Felipe II.* Madrid, Sociedad Estatal para la Conmemoración de los Centenarios de Carlos V y Felipe II: pp. 183–198.

Casado Quintanilla, Blas (2015): *Claudio Fernández de Quiñones, Conde de Luna, embajador de Felipe II en el Imperio y en el Concilio de Trento.* Madrid, PhDr. Dissertation (1984), Universidad Complutense.

Ceniceros Herreros, Javier (2004): "Alcázar de Nájera: primeras investigaciones arqueológicas". In: Iglesia, José I. (coord.): *Conflictos sociales, políticos e intelectuales en la España de los siglos XIV y XV*. Nájera, Instituto de Estudios Riojanos: pp. 519–530.

Civil, Pierre (1990): "Corps, vêtement et société: le costume aristocratique espagnol dans la deuxiéme moitié du XVIe siècle". In: Redondo, A. (ed.): *Le corps dans la société espagnole des XVIe et XVIIe siécle.* Paris, Publications de la Sorbonne: pp. 307–319.

Cobos, Fernando/Castro, José J. (1998): *Castillos y fortalezas. Castilla y León.* León, Edilesa.

Cooper, Edward (1991): *Castillos señoriales en la Corona de Castilla.* Salamanca, Universidad de Salamanca.

Cruz, Anne J./Franganillo, Alejandra/Sanz, Carmen (eds.)(2021): *La nobleza española y sus espacios de poder (1480–1715).*Madrid, Sanz y Torres.

Cruz, Anne J./Perry, Mary E. (eds.)(1992): *Culture and Control in Counter-Reformation Spain.* Minneapolis-Oxford, University of Minneapolis.

Davis, Natalie Z. (2000): *The Gift in Sixteenth-century France.* Madison, University of Wisconsin Press.

Díaz López, Julián P. (et al.) (2007): *El marquesado de los Vélez: Señorío y poder en los reinos de Granada y Murcia.* Murcia, Consejería de Cultura, Región de Murcia-Junta de Andalucía.

Diéguez Delgado, Agustín (2015): *La Casa de Monterrey: de señorío gallego a grandeza de España.* Ourense, Diputación de Ourense.

Elliott, John H. (1991): *El conde duque de Olivares.* Barcelona, Crítica.

Espejo Gutiérrez, Francisco M. (2020): *El patrimonio cultural de la Casa de Fernán Núñez en Córdoba y su trasmisión.* Córdoba, PhDr. Dissertation, Universidad de Córdoba.

Fallows, Nigel (2010): *Jousting in medieval and Renaissance Iberia.* Woodbridge, Boydell Press.

Fernández de Córdoba, Francisco (2021): "Monasterio de San Jerónimo (Granada), Mausoleo funerario de los duques de Sessa y vicisitudes de sus espadas". In: Peláez, Manuel/Vázquez, Hugo/Murcia, Domingo/Toro, Francisco (eds.): *Los Fernández de Córdoba:Nobleza, hegemonía y fama: Segundo congreso: Homenaje a Miguel Ángel Ladero Quesada.* Jaén, Ayuntamiento de Alcalá la Real: pp. 217–229.

Fernández Duro, Cesáreo (1902): *El último Almirante de Castilla Don Juan Tomás Enríquez de Cabrera*. Madrid, Tipografía de la viuda e hijos de M. Tello.

Fernández Navarrete, Pedro (1626): *Conservacion de monarquias y discursos politicos sobre la gran consulta que el Consejo hizo al Señor Rey Don Felipe Tercero*. Madrid, Imprenta Real.

Franco Silva, Alfonso (2006): "El condado de Oropesa". In: *Cuadernos Abulenses*, 35, pp. 85–224.

Frieder, Branden (2008): *Chivalry and the Perfect Prince. Tournaments, Art, and Armor at the Spanish Habsburg Court*. Kirksville, Truman State University Press.

Fuente Escribano, Alejandro de la (2019): "La restauración del castillo de Guadamur en el siglo XIX como expresión del romanticismo en España". In: *Espacio, Tiempo y Forma. Serie VII Historia del Arte*, 7, pp. 231–263.

Gamber, Ortwin (1977): "Erzherzog Ferdinand und die Ambraser Rüstkammer". In: Elisabeth Scheicher/Gamber, Ortwin/Wegerer Kurt/Auer, Alfred (eds.): *Kunsthistorisches Museum, Sammlung Schloss Ambras. Die Kunstkammer*. Vienna, Kunsthistorisches Museum: pp. 24–59.

García Cueto, David (2006): *Seicento boloñés y Siglo de Oro español*. Madrid, Centro de Estudios Europa Hispánica.

García Hernán, David (1998): "Felipe II y el levantamiento de tropas señoriales". In: Martínez, José (dir.): *Felipe II (1527–1598): Europa y la Monarquía Católica. Congreso Internacional*. Madrid, Parteluz: II, pp. 333–344.

García Hernán, David (2000): "La función militar de la nobleza en los orígenes de la España Moderna". In: *Gladius*, 20, pp. 285–300.

García Martínez, José L. (2015): "El legado de los Rabassa de Perellós, fundadores de la Casa de Dos Aguas". In: *e-rph. Revista electrónica de patrimonio histórico*, 17, pp. 170–191.

Gil Fernández, Juan (1991): *Hidalgos y Samuráis. España y Japón en los siglos XVI y XVII*. Madrid, Alianza.

Godoy, José A. (1997): "Las armaduras de la Casa Ducal de Medinaceli, colección del Museo del Ejército". In: *Militaria. Revista de Cultura Militar*, 9, pp. 139–148.

González Mezquita, María L. (2007): *Oposición y Disidencia en la Guerra de Sucesión Española. El Almirante de Castilla*. Valladolid, Junta de Castilla y León.

González Ramos, Roberto (2013): "The armoury of the Dukes of the Infantado. Collecting, prestige and meaning". In: *Journal of the History of Collections*, 25, 3, pp. 335–350.

González Ramos, Roberto (2014): "La colección de armas de Íñigo López de Mendoza, V duque del Infantado". In: *Gladius*, 34, pp. 153–198.

González Ramos, Roberto (2017): "Las armerías nobiliarias castellanas del siglo XVII como manifestación de identidad cultural". In: Carrasco, Adolfo (ed.): *La nobleza y los reinos. Anatomía del poder en la Monarquía de España (siglos XVI-XVII)*. Madrid, Iberoamericana-Vervuert: pp. 269–312.

González Ramos, Roberto (2020): "La armería de los duques de Béjar". In: *Goya. Revista de arte*, 370, pp. 16–29.

González Zymla, Herbert (2014): "El castillo palacio de Alba de Tormes, simbolismos clásicos en un edificio medieval". In: De Maria, Sandro/Parada, Manuel (eds.): *De Trajano a Carlos V. Clasicismo y poder en el arte español*. Bologna, Bononia University Press: pp. 65–80.

Guardiola, Juan B. (1591): *Tratado de nobleza, y de los títulos y ditados que oy dia tienen los varones claros y grandes de España*. Madrid, Viuda de Alonso Gómez.

Gutiérrez José M./Moreno, Ángel/Hernández, José M. (1985): *Oropesa y los Álvarez de Toledo*. Toledo, Diputación Provincial.

Halcón, Fátima (2021): "La colección artística de los marqueses de Villafranca del Bierzo en el siglo XVIII". In: *Laboratorio de Arte*, 33, pp. 249–274.

Hayward, John F. (1948): "Some Augsburg Armours Etched by Jörg Sorg". In: *The Burlington Magazine*, 90, pp. 253–257.

Hayward, John F. (1965): *European Armour*. London, V & A. Museum.

Heal, Felicity (2014): *The Power of Gifts: Gift-exchange in Early Modern England*. Oxford, Oxford University Press.

I exposición (1971): *I exposición histórico-militar. Casa de Pilatos*. Seville, Gráficas del Sur.

Laking, Guy F. (1904): *The Armoury of Windsor Castle*. London, Bradbury, Agnew & Co.

Layna Serrano, Francisco (1955): *Historia de la villa condal de Cifuentes*. Madrid, Compañía Bibliográfica española.

Lewis, Mark (2024): " 'Major Monsoon' and the Tizona of Béjar: the provenance of a late medieval sword in the British Royal Collection". In: *Arms & Armour*. (Forthcoming).

Lleó, Vicente (2017): *La Casa de Pilatos. Biografía de un palacio sevillano*. Seville, Universidad de Sevilla.

López Torrijos, Rosa (2009): *Entre España y Génova. El palacio de Don Álvaro de Bazán en el Viso*. Madrid, Ministerio de Defensa.

López Torrijos, Rosa (2017): "La vida en un palacio: imágenes y datos sobre espacios masculinos y femeninos en el siglo XVI". In: Birriel, Margarita M.

(ed.): *La(s) casa(s) en la Edad Moderna*. Zaragoza, Inst. Fernando el Católico-Diputación: pp. 125–154.

Mármol, Luis del (1797): *Historia del rebelión y castigo de los moriscos del Reyno de Granada*. Madrid, Sancha.

Martín Carramolino, Juan (1872): *Historia de Ávila, su provincia y obispado*. Madrid, Juan Aguado.

Martin, Therese (2012): *Reassessing the Roles of Women as 'Makers' of Medieval Art and Architecture*. Leiden, Brill.

Martínez de Espinar, Alonso (1644): *Arte de ballesteria y montería*. Madrid, Imprenta Real.

Martínez Hernández, Santiago (2003): " 'Obras… que hazer para entretenerse'. La arquitectura en la cultura nobiliario-cortesana del Siglo de Oro. A propósito del marqués de Velada y Francisco de Mora". In: *Anuario del Departamento de Historia y Teoría del Arte*, 15, pp. 59–77.

Martínez Hernández, Santiago (2004): *El Marqués de Velada y la corte en los reinados de Felipe II y Felipe III. Nobleza cortesana y cultura política en la España del Siglo de Oro*. Salamanca, Junta de Castilla y León.

Martínez Hernández, Santiago (2005): *Don Gómez Dávila y Toledo, II Marqués de Velada, y la Corte en los reinados de Felipe II y Felipe III (1553–1616)*. Madrid, PhDr. Dissertation, Universidad Complutense.

Martínez José/Fernández, Santiago (dirs) (2005): *La Monarquía de Felipe II: la casa del rey*. Madrid, Mapfre.

Montero, Juan/González-Sánchez, Carlos A./Rueda, Pedro J./Alonso, Roberto (2014): *De todos los ingenios los mejores. El Condestable Juan Fernández de Velasco y Tovar, V Duque de Frías (c. 1550–1613)*. Seville, Real Maestranza de Caballería de Sevilla.

Montes González, Francisco (2016): *Mecenazgo virreinal y patrocinio artístico. El ducado de Alburquerque en la Nueva España*. Seville, Real Maestranza de Caballería.

Montgomery-Campbell, M. (2005): *The Armour of Schloss Ambras*. Godmanchester, Ken Trotma Publishing.

Morán Turina, José M. (2009): *La memoria de las piedras. Anticuarios, arqueólogos y coleccionistas de antigüedades en la España de los Austrias*. Madrid, Centro de Estudios Europa Hispánica.

Morán, José M./Checa, Fernando (1985): *El coleccionismo en España. De la cámara de maravillas a la galería de pinturas*. Madrid, Cátedra.

Moreno de Vargas, Bartolomé (1636): *Discursos de la nobleza de España*. Madrid, María de Quiñones.

Padilla, Luisa M. de, condesa de Aranda (1619): *Idea de nobles y sus desempeños en aforismos. De nobleza virtuosa.* Zaragoza, Hospital Real.

Patterson, Angus (2009): *Fashion and Armour in Renaissance Europe: Proud Lookes and Brave Attire.* London, V & A Museum.

Pérez Mínguez, Fidel (1930): "El castillo de los Marqueses de Las Navas". In: *Boletín de la Real Academia de la Historia*, 97, pp. 745–816.

Pérez Pastor, Cristóbal (1895): "Los trofeos de D. Álvaro de Bazán". In: *Boletín de la Real Academia de la Historia,* 26, pp. 389–393.

Pérez Preciado, Juan J. (2010): *El marqués de Leganés y las artes.* Madrid, PhDr. Dissertation, Universidad Complutense.

Pérez Sánchez, Alfonso E. (1977): "Las colecciones de pintura del conde de Monterrey (1653)". In: *Boletín de la Real Academia de la Historia,* 174, pp. 417–425.

Prescott, William H. (1848): *Historia del reinado de los Reyes Católicos D. Fernando y Dª Isabel.* Madrid, Imprenta de la Biblioteca del Siglo.

Puddu, Raffaele (1984): *El soldado gentilhombre.* Barcelona, Argos Vergara.

Quintanilla Raso, María C. (1992): "El condado de Priego de Cuenca. Un ejemplo de estrategia señorial en la Baja Edad Media castellana". In: *Historia. Instituciones. Documentos,* 19, pp. 381–402.

Ramírez Ruiz, Victoria (2015): "La colección de tapices de los condes de Monterrey". In: *Libros de la Corte,* 10, pp. 30–59.

Real Academia española (1726): *Diccionario de la lengua castellana, en que se explica el verdadero sentido de las voces.* Madrid, Francisco del Hierro.

Rivas Albaladejo, Ángel (2015): *Entre Madrid, Roma y Nápoles. El VI Conde de Monterrey y el gobierno de la Monarquía Hispánica (1621–1653).* Barcelona, PhDr. Dissertation, Universidad de Barcelona.

Rodríguez Pérez, Raimundo A (2011): *El camino hacia la corte. Los marqueses de los Vélez en el siglo XVI.* Madrid, Sílex.

Rodríguez Villa, Antonio (1883): *Inventario del moviliario, alhajas, ropas, armería y otros efectos del Excmo. Sr. D. Beltrán de la Cueva, tercer Duque de Alburquerque. Hecho en el año 1560.* Madrid, Imprenta de D. G. Hernando.

Rueda, T. de (1973): "Armaduras japonesas en la Real Armería". In: *Reales Sitios*, 38, pp. 22–28.

Saavedra Fajardo, Diego (1640), *Idea de un príncipe político cristiano.* Munich, Nicolao Enrico.

Salvago, José (1957): "La Casa Ducal de Arcos en la Historia de Marchena". In: *Archivo Hispalense,* 26, 81, pp. 47–84.

Sancho, José/Zabalburu, Francisco de (eds)(1882): "Armería del Duque del Infantado en Guadalajara". In: *Colección de Documentos Inéditos para la Historia de España, 79*. Madrid, Viuda de Calero.

Sandoval, Prudencio de (1792): *Historia de los reyes de Castilla y León. II*. Madrid, Benito Cano.

Schlosser, Julius von (1988): *Las cámaras artísticas y maravillosas del renacimiento tardío*. Madrid, Akal.

Scott, Alison V (2006): *Selfish Gifts: The Politics of Exchange and English Courtly Literature, 1580–1628*. Madison, University Press Copublishing Division-Fairleigh Dickinson University Press.

Simal López, Mecedes (2013): "Antes y después de Nápoles. Iniciativas artísticas del VI conde de Monterrey durante el virreinato partenopeo, y fortuna de sus colecciones a su regreso a España". In: Denunzio, Antonio E./Birra, Ciro (eds.): *Dimore Signorili a Napoli. Palazzo Zevallos Stigliano il mecenatismo aristocrático a Napoli dal XVI al XX secolo*. Naples, Sanpaolo.

Simal López, Mercedes (2002): *Los Condes-Duques de Benavente en el siglo XVII. Patronos y coleccionistas en su villa solariega*. Benavente, Centro de estudios Benaventanos "Ledo del Pozo".

Sola, Emilio (1999): *Historia de un desencuentro. España y Japón (1580–1614)*. Alcalá de Henares, Fugaz.

Soler del Campo, Álvaro (1998): "La Armería de Felipe II". In: *Reales Sitios*, 135, pp. 25–37.

Soler del Campo, Álvaro (2003): "Embajadas japonesas en la Real Armería". In: Alfonso, Marina/Martínez, Carlos/Arias, Matilde R./Fisac, Taciana/Viguera, María J. (eds.): *Oriente en Palacio. Tesoros asiáticos en las Colecciones Reales Españolas*. Madrid, Patrimonio Nacional: pp. 58–67.

Soler del Campo, Álvaro (2007): "La armería de Vicencio Juan de Lastanosa". In: Morte, Carmen/Garcés, Carlos (eds.): *Vicencio Juan de Lastanosa (1607–1681): la pasión del saber*. Huesca, Instituto de Estudios Aragoneses-Diputación de Huesca: pp. 117–123.

Soria Mesa, Enrique (2011): "La imagen del poder. Un acercamiento a las prácticas de visualización del poder en la España Moderna". In: *Historia y Genealogía*, I, pp. 5–10.

Torreiro Luengo, Carlos (1992): "Armas y armaduras japonesas en la Real Armería de Madrid". In: *Reales Sitios*, 114, pp. 37–44.

Urquízar Herrera, Antonio (2007): *Coleccionismo y nobleza. Signos de distinción social en la Andalucía del Renacimiento*. Madrid, Marcial Pons.

Urquízar Herrera, Antonio (2011): "Imaginando América: objetos indígenas en las casas nobles del renacimiento Andaluz". In: *Historia y Genealogía*, 1, pp. 205–222.

Valencia de Don Juan, Conde viudo de. (1898): *Catálogo Histórico-descriptivo de la Real Armería de Madrid.* Madrid, Hauser y Menet.

Valencia Rodríguez, J. M.(2000): *Señores de la tierra. Patrimonio y rentas de la Casa de Feria (siglos XVI y XVII).*Mérida, Ed. Regional Extremadura.

Velasco Bayón, Balbino (1981): *Historia de Cuéllar.* Segovia, Diputación Provincial.

Verrier, Frédéric (1997): *Les armes de Minerve. L'Humanisme militaire dans l'Italie du XVIe siècle.* Paris, Université Paris-Sorbonne.

Vignau, Vicente (1871): "Armas y efectos de guerra del duque de Calabria". In: *Revista de Archivos, Bibliotecas y Museos*, 1, pp. 12–15, 28–29, 92–93, 108–110.

Yaya, Isabel (2008): "Wonders of America: The curiosity cabinet as a site of representation and knowledge". In: *Journal of the History of Collections*, 20, 2, pp. 173–188.

Zalama, Miguel A./Andrés, Patricia (2002): *La colección artística de los Condestables de Castilla en su palacio burgalés de la Casa del Cordón.* Burgos, Caja de Burgos.

Zapata, Luis de (1566): *Carlo Famoso.* Valencia, Juan Mey.

Álvaro Soler del Campo

Real Armería, Madrid

SPANISH NOBLE ARMOURIES IN THE MODERN AGE: TYPOLOGIES AND LEGACIES

The historical image of Western armouries is due in large part to the interventions they experienced in the second half of the nineteenth century, which has in certain circumstances remained until the present. If we roll back time, the Renaissance and Baroque iconography allows us to have a limited set of often partial and scattered images, generally referring to specific details about the forms of exhibition and the storage of these armouries' collections. To understand how they might have been between the sixteenth and eighteenth centuries, we must include the deficiencies of the iconography from the small number of studies carried out that have attempted to reconstruct their appearance and the criteria of classification, an aspect of the history of collecting to be treated in depth below[1]. As such, this work proposes an initial approach that seeks to advance as far as possible in these aspects by analysing a set of inventories that reflect the different types of armouries, the main types of weapons present in them and their manner of organisation, both from a typological and a topographical viewpoint[2]. Ultimately, it is worthwhile to inquire which elements survived from

---

1 As a starting point on this topic, see: Cámara, 1994. Franco/Soler, 1994. González Ramos, 2013, 2014, 2017, 2020. Romero Medina, 2020. Simal López, 2002. Soler del Campo, 1998.

2 For space reasons, a sufficiently meaningful documental sampling has been made without the intention of it being exhaustive. To avoid continual notations, they are listed below, grouped chronologically, alluding to the titles of nobility to which they refer to make locating them easier: 1531, 2nd Duke of Alba, Madrid, Archivo Ducal de Alba (hereafter ADA), Alba, C.23, nº. 29. 1536, 2nd Duke of Béjar, Toledo, Archivo Histórico de la Nobleza (hereafter AHNOB), Osuna, C. 222, doc. 49, fol. 20r-22r. 1537, Mayorazgo de Álvaro de Bazán, 1st Marquis of Santa Cruz de Mudela, AHNOB, Santa Cruz, C. 32, doc. 1., fols. 17v-19v. 1540, 1st Duke of Medina de Rioseco, ADA, Casa de Alba, C.79, nº. 19. 1541, 3rd Duque de Benavente, AHNOB, Osuna, C. 424, D. 32–34. 1544, 3rd Duke of Sessa, Madrid, Biblioteca Francisco de Zabálburu (hereafter BFZ), Altamira, 32, grupo documental 3, documento 12. 1571, 1st Duke of Feria, Madrid,

the Renaissance and Baroque armouries, if they influenced the conception of nineteenth-century armouries or not, and up to which point had they conditioned

Archivo Histórico de Protocolos de Madrid (hereafter AHPM), T. 742, fols. 1–2, 19r-19v, 45. 1573, 4th Count of Oropesa, AHNOB, Frías, C. 1277, D. 11. 1579, 7th Count of Priego, AHPM, T.759, fols. 385–425. 1580, 4th Count of Salinas, Archivo Histórico Provincial de Valladolid (hereafter AHPV), P. 386, fols. 587–588v, 605v-614v. 1584, 6th Duke of Alburquerque, Cuéllar, Archivo Histórico de la Casa Ducal de Alburquerque (hereafter AHCDA), Leg.161, nº. 2. 1584, 1st Marquis of Santa Cruz, AHNOB, Santa Cruz, C. 32, D. 1607. 1588, 5th Duke of Gandia, AHNOB, Osuna, C. 568, D. 127. 1593, 6th Count of Cifuentes, AHPM, T. 10801, fols. 173–178, 211–218v. 1593, 5th Count of Benavente, AHNOB, C.427, D.11, fols.1–12. 1593, 4th Count of Fuensalida, AHNOB, Frías, C.24, D. 113, fols. 4v-5v. 1594, 5th Duke of Medinaceli, AHPM, T.628, Cogollудо, fols. 647–653v. 1600, 4th Duke of Nájera, AHPM, T. 2606, fols. 987v-988v. 1601, 5th Duke of the Infantado, AHNOB, Osuna, C.1948, Doc.3. 1605, 3rd Marquis of Poza, AHPM, T. 2176, fols. 234–245. 1607, 2nd Count of Olivares, AHPM, T.1824, fols.789v, 831v-863v. 1607 and 1611, 1st Duke of Lerma, Archivo Ducal de Medinaceli (hereafter ADM), Sección Archivo Histórico, Leg. 262. 1613, 6th Duque de Alburquerque, AHPM, T.2649, 819–827v. 1610, 1st Marqués del Valle de Cerrato, AHPM, T. 2656, fol.823. 1614, 1st Duke of Lerma, ADM, Secc. Lerma, Leg.55, nº. 16, fols.1–5. 1616, 2nd Marquis of Velada, BFZ, Altamira, 198, D. 1. 1617, 1st Marquis of Salinas de Río Pisuerga, AHPM, T. 2320, 1, fols. 10v-12, 57v, 80v, 88v-122. 1620, 2nd Marquis of Salinas de Rio Pisuerga, AHPM, T.2320, fol.125, 156v-174v . 1627, 3rd Duke of Pastrana, AHPM, T. 3361, fols. 647v, 671v-673v. 1628, Constable of Castile, 6th Duke of Frías and 8th Count of Haro, Burgos, Archivo Histórico Provincial de Burgos (hereafter AHPBU), P. 6191, fols. 1467–1491v. 1630, 3rd Duke of Arcos, AHNOB, Osuna, C. 125, D. 176, fols. 24v and following.1631, 8th Count of Salinas, AHNOB, Osuna, C. 3910, D. 149, fol. 3r. 1637, 6th Marquise of Villanueva del Fresno, AHPM, T. 5201, fols. 1, 62–71, 89–92. 1637, 7th Duke of Béjar, AHNOB, Osuna, C. 244, D. 43, fols. 2, 136–137. 1642, 6th Duke of Sessa, AHPM, T. 6932, fol. 1117v-1122v. 1647, Admiral of Castile, 5th Duke of Medina de Rioseco, AHPM, T. 6233, fols. 323–325v, 363–364v. 1650, 7th Marquis of Villafranca del Bierzo, AHPM, T. 4499, fols. 97–101, 116–118. 1655, 1st Marquis of Leganés, AHPM, T. 6267, fols. 472v-473, 757–758v. 1659, Duke of Sessa, AHPM, T. 7152, fols. 313–328. 1665, 8th Duke of Benavente, AHNOB, Osuna, C. 427, D. 96. 1667, 7th Duke of Medinaceli, ADM, Sección Cogolludo, Leg.3, doc.44. 1671, 10th Duke of Béjar, AHNOB, Osuna, C.253, D.101. 1675, 3rd Count of Molina de Herrera, AHPM, T. 12006, fol. 480v. 1672, 7th Duke of Montalto, AHPM, T. 7152, fol. 378. 1678, 4th Marquis of Velada and Astorga, Instituto Valencia de Don Juan (hereafter IVDJ), 25.5.13, fol. 122–124. 1681, 3rd Marquis of Castel Rodrigo, AHPM, T. 9861, fols. 606v, 677, 681, 794v. 1682, 10th Duke of Béjar, AHNOB, Osuna, C.255, D. 143. 1702, 9th Duke of Benavente, AHNOB, Osuna, C. 436, D. 69–72. 1746, 11th Duke of Medinaceli, ADM, Sección Cogolludo, Leg.5, nº. 38. 1746, 11th Duke of Medinaceli, ADM, Sección Cogolludo, Leg.5, nº. 38.

our conception of historical armouries. To achieve this end, different examples will be presented that will allow us to approximate the physical appearance of the noble armouries as an essential part of the Renaissance and Baroque court culture, as well as their possible imprint up to the present. In addition to these effects, the analysis of the inventories presents some determining factors by their characteristics or nature.

First, we might presume that an inventory reflects a topographical order, but it does not necessarily happen in this manner, which leads us to an initial division into two blocks: those that react to an expressly designed topographical organisation and those that stick to recording the weaponry present in an armoury typologically. A clear example of this latter typology is the inventory of the court armoury of the 8th Count of Salinas from 1631, in which it is specified that it does not follow a typographical order "*porque cada cosa dél cuando se buscare se halle con más presteza* [because each item is found more promptly at the moment it is sought]". In contrast to that example, the topographic inventories in most cases allow us to confirm that the armouries' arrangement did not respond to the formation of strictly unitary sets. Even in the best organised armouries, there are always isolated elements inserted in groups of arms of a different typology, such as crossbows between the horse harnesses, or white arms intermixed with firearms. Nonetheless, the grouping of the weapons was carried out according to their nature or typology, because in the overall analysis, in some cases we can cast aside the one-off intrusion of other types of arms as incidental.

Second, we must be conscious of the fact that the organisation of an armoury was not fixed, because it evolved over time for various reasons such as its growth, a change of location, the reunification of dispersed groups, the adaptation to an exceptional event that requires a new orientation or a partial rearrangement of spaces, as in the case of the Mühlberg and Lepanto campaigns. Thus, in their evolution, we run the risk of not detecting transcendental changes due to the possible loss of intermediary documents between key dates, such as in the example of the inventory of 7th Count of Priego from 1579 in which a room is described exclusively dedicated to the celebration of the battle of Lepanto laid out as a triumphant scenography.

At other times, for the purpose of assessing a development of the calculation of their collections, we must take into account that its number can vary from one inventory to another in the same armoury merely due to its rearrangement and without the need to take into consideration the collection's expansions or retractions. This circumstance is especially applicable to the number of suits of armour and loose pieces existing, given that between two different moments, the simple fact of assembling the loose pieces for a rearrangement or a new staging might change the aspect and the vision of the whole. These practices are common and there is a clear knowledge of their incidence, as is stated in 1628 in the armoury of the Constable of Castile, in which it is noted "*que como se han ido juntando y componiendo de lo viejo no se acabaron de juntar todas las piezas y sólo tienen las que aquí están desglosadas* [that as they have been bringing together and forming from the old, not all the pieces have been put together and they only have those that are itemised here]".

For these reasons, the analysis of each armoury must be timely, specific to a given moment, given that it can be an object even of a radical change. Therefore, the inventories on which we rely offer us only a snapshot concerning the nature and characteristics of noble armouries because we do not have a complete documentary series that covers the history of each and every one of them available.

Finally, it is important to state that a small number of documents also provides fixed images of the elements that composed their staging. These help us to be able to imagine how they should have been and to go about forming a repertoire of display elements that might be combined to offer us possible images for the reconstruction of their appearance.

The project entitled "Las armerías de la nobleza española. Identidad y cultura aristocrática en la Edad Moderna [The Amouries of the Spanish Nobility: Aristocratic Identity and Culture in the Modern Age]", to which this work belongs, is ground-breaking in that it opens a new line of research that should be continued in the future, not only in clearing out Spanish archives, but also in the analysis and reconstruction of noble armouries[3]. This work only claims to introduce some aspects about the composition and presentation of these armouries, a line of work that should be amplified and detailed in the future. Although the analysis of the whole suggests different readings and topics of inquiry, we focus here on three specific aspects: their classification according to the nature and the magnitude of their collections, according to their topographical organisation, and, lastly, in the partial reconstruction of their appearance through their display elements and the containers used in their installations.

---

3 I thank Roberto González Ramos for his invitation to participatein it and the documentation provided for the drafting of this work.

TYPES OF ARMOURIES ACCORDING TO THE NATURE AND MAGNITUDE OF THEIR COLLECTIONS

In general, the inventories allow us to establish at least four clearly defined categories according to the nature and size of their items. These categories are not exclusive, in that some noble houses might have scattered their armoury in more than one location, each one with its own character within these typologies.

Noble armouries at court

The inventories focused on the Madrid court describe mostly small noble armouries with weapons and horse harnesses intended to cover the needs of court attire and etiquette. In them, only a small number of pieces to meet the commitments implied in their owners' social status are stored, generally related to hunting or some chivalrous game. They were not, however, the only typology present in the capital, because some houses might have a seignorial armoury of a very superior magnitude at court. Those small collections were generally kept in the chamber and were composed of no more than three suits of armour if there were any, some crossbows, a limited number of firearms and, above all, dress swords for daily use and cavalry harnesses, each one of different quality. This typology is found from the first half of the sixteenth century until the eighteenth century. A variant of these armouries is one that furthermore incorporates some element of prestige, mainly suits of armour or weapons from Turkey or the Indies, creating medium-sized collections that meet personal needs and those of the image of power associated with a figure or a family[4].

4 See the armouries of the 2nd Duke of Béjar in 1536, that of the 1st Duke of Feria from 1571, those of the Count of Salinas in 1580 and 1621, that of the 4th Duke of Nájera in 1600, that of the 2nd Count of Olivares from 1607, that of the 6th Duke of Alburquerque in 1613, that of the 2nd Marquis of Salinas del Río Pisuerga in 1620, that of the 6th Marquise of Villanueva del Fresno in 1637, that of the Admiral of Castile and the 5th Duke of Medina de Rioseco in 1647, that of the 7th Marquis of Villafranca del Bierzo in 1650, that of the Duke of Sessa in 1659, that of the 3rd Count of Molina de Herrera in 1675, that of the 7th Duke of Montalto in 1672, that of the 3rd Marquis of Castel Rodrigo in 1675 and that of the 4th Marquis of Velada in 1678. In this context, "the Indies" most likely refers to the Americas, but could also include some pieces from Asia.

Armouries of dynastic families

The second group are the great armouries of dynastic families, the armouries *par excellence* as they are currently known. They are prestigious collections located in the family houses and in the buildings most representative of a specific family, well away from court or in manor houses or fortresses. In them, there is a large variety of categories, including luxury and exotic weapons, trophies, ancient arms in poor condition, Turkish, Moorish (Nasrid), and even American and Japanese collections, following the connection between armouries and the so-called cabinets of curiosities.

In this second category, the number of arms, the typology to which they answer and the information about their materials, their realisation and decoration are key to determine their importance. Clearly, an armoury filled entirely with military corslets is not the same as another with a large number of armoured harnesses for war, jousting, cavalry or infantry with their corresponding reinforcement pieces.

These armouries ended up being family depositories that were transmitted over time with a clear character as a family "*Memento*", as was spelled out in 1702 in the inventory of Benavente, in which it was specified that this was its only purpose, other than being decorative. This character might be due to the mere effect of generational accumulation, but also a declared willingness to do so. Such might be the case in 1584 of the connection to the right of primogeniture of the Marquisate of Santa Cruz on the part of its first holder, Álvaro de Bazán y Guzmán, of his own military trophies, his father's, and his grandfather's to the glory and prestige of the family, so that it became part of the armoury.

Among the large armouries, Count of Benavente's armoury in 1593, with 152 suits of armour, eighteen brigandines [*coracinas*], forty shaffrons and forty-two saddles, sixty-five horse harnesses, fifty-nine "*rodelas* [rondaches]", and more than 200 polearms and swords, to which more than 300 different weapons and harnesses were added, stands out. The armoury of the 1st Duke of Lerma in Ampudia with 168 suits of armour in 1607, more than 2,000 polearms and with more than a 100 white arms and firearms, in addition to other equally important ensembles[5], as well as the Constable of Castile's armoury in 1628 with ninety-seven suits of armour, also are noteworthy.

5 ADM, Sección Archivo Histórico, Leg. 262, fols. 265 r-270 v.

A particularly paradigmatic example is the large armoury that Beltrán III de la Cueva y Castilla, 6th Duke of Alburquerque maintained in his Castillo de Cuéllar in 1584. There, he stored 140 suits of armour, nineteen coats of mail, 225 polearms, 168 firearms, forty-seven saddles and 569 different arms, in addition to eleven artillery pieces.

As an example of a great armoury, albeit one located at court, that of the 3rd Marquis of Poza stands out. It contained 100 suits of armour, almost 100 replacement pieces, a bard and 672 weapons of various types, in addition to the harnesses and equestrian items. Its character as a prestigious element is doubtless as it contained 115 "*perchas* [wooden display stands]". for the mounting of at least the 100 described suits of armour, which allows us to imagine a staging for this purpose.

Finally, it should be noted that, although the dynastic armouries mainly preserve the weaponry of the greatest prestige, we should not forget either that sometimes there are more simplistic venues with large collections that fulfil this representative role, such as what takes place with the ducal armoury of Medinaceli, inventoried in Cogolludo in 1594, 1667 and 1746, as we will see.

Storerooms in fortifications or palaces

Generally, the storerooms of fortifications are usually secondary to or dependent on a dynastic armoury. They tend to be very irregular in their items, although some sets like artillery stand out in them. Artillery pieces, except for small pieces, are not generally present in other types of armouries given their complex transport or storage in buildings unsuitable for them[6]. A clear example is the armoury of Santa Cruz in Gibraltar from 1537 with fifty-two pieces of artillery. Along with the artillery, important sets of "*armas viejas* [old weapons]", probably late mediaeval, are noteworthy in these types of armouries. In specific instances, familial armouries are included in this typology, understanding both as the same set. Such is the case of the 1593 testament of the 6th Count of Cifuentes, in which his armouries of Barcience are connected to a clearly military armoury in which artillery and likely outdated weapons predominate, and that of the Castle of Cifuentes, clearly seignorial for the quality, diversity and origin of its weapons and suits of armour.

6 See the armouries of the 4th Count of Oropesa in 1573, that of the Count of Fuensalida in Guadamur in 1593 and that of the 6th Count of Cifuentes in Barcience.

Specialised or monographic armouries

Specialised or monographic armouries are so characterised because their collections revolve around a specific origin or historical event of great relevance. The most representative examples are the armoury of the Count of Priego from 1579 dedicated to Lepanto and that of the 1st Marquis of Salinas de Río Pisuerga from 1617 in which exotic objects stand out[7]. Alongside them is Alba de Tormes's well-known tower and armoury room dedicated to the battle of Mühlberg, adorned with Cristophoro Passini frescoes.

The Madrid armoury that the 7th Count of Priego had in 1579 constitutes a unique case, divided into two rooms on different floors. The first room was dedicated to the battle of Lepanto through eighty-two commemorative objects and trophies that included portraits of the main protagonists including Turks, paintings, sculptures, weapons, pontifical vestments and Turkish arms along with various other objects of Turkish origin. In addition to this room, there was another in which a small personal armoury was combined with a select collection of exotic objects, among which were also Turkish items, in the style of a cabinet of curiosities. This armoury would have, thus, a double monograph character, due to the origins and the nature of its collections.

The example of the 1st Marquis of Salinas de Río Pisuerga's armoury is doubtlessly unique for various reasons. First, it is necessary to state that the Marquis had been the viceroy of New Spain (1590–1595) and Peru (1596–1604), and that he had married María de Ircio y Mendoza in 1564, who was the first *criolla* vicereine given that she was the daughter of Martín de Ircio, a captain under Hernán Cortés, and Mariana de Mendoza, illegitimate daughter of Iñigo López de Mendoza, 2nd Count of Tendilla and Marquis of Mondéjar. His armoury should be understood in this context. It was small in size and consistent with the rest of his belongings and ornamentation in his house in terms of its items' Japanese and American origin, denoting a certain "*orientalizante* [Orientalist]" style. It contained two katanas, six Japanese suits of armour, an armour with its reinforcement pieces, three bullet-proof suits of armour, two shotguns, two "*rodelas* [rondaches]" and two shaffrons. The number of Japanese suits of armour therefore exceeds the usual proportions independent of the total number of collections in an armoury and underlines that this armoury, along with the rest of its items, might have had an exotic character in its own time similar to the cabinet of curiosities. It should be noted, however, that three years

7 AHPM, T. 2320, 1, fols.88v-92r.

later, upon the death of his son, the 2nd Marquis, this paternal armoury is not mentioned, so it could have been dispersed or the will with which we are familiar intentionally did not mention it. Be that as it may, this example suggests that, except in the case of the largest dynastic armouries, transmission does not always take place, although occasionally this transmission of the armoury was expressly ordered in the inventory, as in the case of the 1610 will of the 1st Marquis of Valle de Cerrato.

Armouries without an apparent location

Along with the aforementioned categories, we should also consider that in some instances there was no space for the armoury. Rather, it was kept in containers within an undefined space, as the inventory of the 6th Duke of Sessa in 1642 suggests, or it was intermixed among other objects within the building's interior. Given the dearth of the number of weapons, it could be considered that it did not constitute a proper armoury, but it represented the minimal amount of equipment for personal use. In Sessa's case, the arms were stored in two chests. In one, twenty-five swords, fifteen blades, five pistols, a carbine, six pairs of spurs and seven saddles were kept. The second contained ten coats of arms. This typology and its interpretation also might be due to the fact that the documentation offers us a distorted image if there is no type of topographic indication.

Armouries of a unique nature divided between different locations

Logically, all the noble houses must have had their armouries divided into different locations, which might be reconstructed at different temporal moments. Only on rare occasions can we confirm the grouping of more than two locations in a single document that offers us a fixed image of the whole, for which it has been considered a special category. The dispersion in more than two settings also permits us to appreciate that in these cases some of them can respond to the aforementioned classifications according to the function that was assigned to them.

A paradigmatic example is the armoury of the 1st Duke of Lerma in 1607, divided into four locations that represent three different typologies: that of Burgos at La Ventosilla palace, that of Palencia at Ampudia, the ducal one at Lerma and the court one in Madrid. The first might belong to the abandoned or the antiquated armouries of the fortifications, with only twenty-four polearms, two crossbows, three bows and an arquebus. The Ampudia castle reflects the opposite model, with a large seignorial dynastic armoury for which we lack topographical indications, but which undoubtedly must have been a very notable example due to its large number of items. In it, 2,188 polearms, 168 suits of armour, including one "*del rey de Francia* [of the king of France]," two bards, 149 white arms, 119 firearms, 139 defensive elements and 170 horse shaffrons among other smaller groups, are noteworthy. As for the court armoury in Madrid, it consisted of, as we have already noted, a small personal armoury enlarged for reasons of prestige with a number of suits of armour. Considering the three taken together, the result is an armoury of considerable dimensions composed of 2,205 polearms, 178 suits of armour, 290 defensive elements, 432 horse harness pieces, 129 firearms, 175 white arms, 120 maces, fifteen crossbows and arrows, seventeen artillery pieces and two metallic horse bards.

The ducal arms only became the object of an inventory four years later, in 1611. Ampudia seems to have maintained his armoury without large changes, like in Madrid, but the same did not happen in the locations of Lerma and Ventosilla, which experienced an important increase of collections among which a likely set of eight Japanese suits of armour. We do not know if this set came to Spain with the first Japanese embassies or if it is related to the 1611 arrival of a Spanish embassy that had went to Mexico in 1610 with the governor of the Philippines, a trip that later inspired one by Date Masamune that Hasekura Tsunenaga would lead between 1613 and 1620.

In the instance of the Ampudia location, a new inventory has been uncovered from 1614, three years after the previous one. Its armoury continued to have the seignorial character of previous years without having undergone major changes, but some differences of a certain character are seen, such as the presence of twelve more suits of armour than in 1611, as well as thirty-five head pieces, the lack of twenty-one horse shaffrons and fourteen polearms. These changes must be due to a lack of strict control over the armoury and its subsequent disarray. At the beginning, the document specifies that until this inventory, nobody had control of the castle of Ampudia since the death of its most recent warden, after which it was found abandoned and closed, in poor condition and with the roof collapsing. As a result, the 1st Duke of Lerma ordered an inspection "*viendo por vistas de ojos cada pieza de por sí, la reparación de todo lo que fuera necesario y que la casa debía habitarse y de encender lumbre en ella y para esto conviene que haya casero en ella* [viewing each piece by eye sight individually, the repair of everything that was necessary and that the house ought to be inhabited and a fire lit in it and, for this, it is convenient that there be a caretaker in it]".

Similarly, the armoury of Fernando de Silva, 6th Count of Cifuentes, in the castles of Barcience and Cifuentes, inventoried in 1593 and 1590, is also a clear example. In Barcience, artillery along with long firearms devoid of decoration, incomplete suits of armour without head pieces, some "*quebradas* [broken]", and various loose parts are collected together. On the contrary, Cifuentes held suits of armour and engraved pieces, among which the following examples stand out: a parade set composed of a morion and a rondels with "*una historia de relieve* [an embossed history]", a rare saddle to carry the standard, most probably with an iron bracket, a combined spear and pistol weapon, Turkish cutlases and bows, engraved and gilded spears and double-lock pistols with ivory stocks and other luxury weapons, a set typical of a seignorial armoury rather than the antique one in Barcience.

In the case of the 2nd Marquis of Velada, his armoury was found dispersed in three locations in 1616: Velada, Villatoro and Madrid, with weapons intermixed or inventoried with other items, presumably scattered around the property. Velada appears to be a small armoury of residual use with some firearms or crossbows for hunting and old pieces of armour. Villatoro was a disordered warehouse of loose parts and Madrid was properly the seignorial collection. Its weapons, however, are found scattered throughout the building, which would have included seventeen suits of armour, thirty-nine loose pieces and another twenty-four arms of various types, had they been collected in an armoury room. It was not a cared for collection because some of the pieces were "*tomadas de herrumbe* [taken of rust]".

TYPES OF ARMOURIES ACCORDING TO THEIR SPATIAL DISTRIBUTION

One of the most complex aspects when it comes to discerning what the Renaissance and Baroque armouries were like rests in being able to document what type of rooms they were in, how many they occupied and how arms were distributed within them. In this sense, only a small number of documents provide us with sufficient details, but even with these we can propose an initial spatial classification. In this regard, it is important to consider that the weapons and pieces of suits of armour could nonetheless be found in any room of the house, in addition to those specifically dedicated to them. The inventory of the 3rd Duke of Pastrana from 1627 includes one of these cases. Here, we find "*dos arcabuces de rueda que estaban en el retrete* [two wheelock arquebuses that were in the private room]", a term that at this time alluded to a room of retirement devoid of modern connotations. Depending on the number of rooms and of the exhibitors and containers, the following categories can be considered.

One single room with or without cabinets

The inventories expressly indicate the existence of armouries that occupied one room, like the royal armoury itself. These single rooms were clearly differentiated from the other rooms. Their arrangement is defined, as in the case of the 3rd Duke of Sessa's 1544 inventory, but their content, given that in them it is specified that the armoury collection is composed of "*todo los que está dentro de la armería* [everything that is within the armoury]", a definition that even Philip II adopted in his testamentary regulations to define the content of the royal armoury[8].

At times, we have news of the possible appearance of these single rooms, as in the case of the armoury of the 2nd Duke of Alba, who in 1531 distributed his pieces on mounts or supports, in chests, in trunks or panniers and a large number of pieces on the walls. The 1541 inventory of the 2nd Duke of Benavente, which we analyse below, also reflected this type, as did the ducal armoury of Lerma, made familiar through three consecutive inventories, of 1607, 1611 and 1614. Of those, only the 1614 example provides information about its assembly. This armoury was on a level of the castle that was accessed by a staircase that led to a single room where the entire collection was concentrated. Its floor was completely covered with an esparto mat, despite the large dimensions it must have had, given that in 1607 it contained 182 suits of armour, 170 shaffrons and more than a hundred pieces of mail, firearms, maces and white arms along with 300 defensive elements of items including two horse bards. The access staircase was decorated with twenty-three native American feather head-dresses, in a staging of great exociticism as an element of prestige complementary to the armoury itself.

Armouries of two rooms without cabinets

The armouries of two rooms without cabinets, with the arms or other items visible, were designed to create a staging clearly representative of personal character or familial prestige. A paradigmatic example is the armoury of the 7th Count of Priego in 1579, *mayordomo mayor* to John of Austria and participant at Lepanto. As has already been noted, it was divided on two floors, with a first room that overlooked the house's patio that was split, in turn, into two milieus. The first milieu was the Duke's study whilst the second was dedicated to the battle of Lepanto, referred to as the "*camarín de los despojos de la batalla de la Liga* [alcove of the loot from the battle of the League]". It was a piece of great

8 For everything about the Royal Armoury in Philip II's era, see: Soler, 1998.

importance due to the great number and origin of its items, of a monographic character, in which the arms were complemented with various objects related directly or indirectly with the battle. Among them the protagonists' portraits, including those of Turks, papal vestments of Pius V, maps, sculptures, rosaries, costumes, Turkish white arms and archery weapons, harnesses, sculptures and "*un dosel para el Agnus Dei* [a canopy for the Lamb of God]" stood out. This entire collection brought together a total of eighty-two objects.

On the upper floor was located the Duke's chamber, in which the weapons for personal use, especially pieces of armour and hunting weapons coexisted with thirty-three valuable or exotic objects, mostly American, in the manner of a cabinet of curiosities. The totality of both was clearly aimed at providing the palace with an element of prestige in the social and political context of the Madrid court. It also stood out for its size, given that it contained a total of 528 objects that represented these categories.

Another case study highlighted is the armoury of Béjar, of which we have preserved various inventories dated to 1637, 1671[9] and 1682, belonging to the 7th and to the 10th Dukes of Béjar. Although all are equally accurate, they do allow us to determine its appearance in some periods and the changes experienced in a short time. In the 1637 inventory, topographical arrangements are not specified, but it does record the existence of more than one room dedicated to the armoury by stating that there were some "*sillas que están en una sala de la armería* [saddles that are in an armoury room]". At that time, it listed polearms, eighty-one suits of armour, pieces for war, shaffrons and white arms as the most emphasised groupings. In 1671, the inventory of the 10th Duke of Béjar is more accurate and notes a reduction in the number of suits of armour by almost half, with weapons displayed on the walls and suits of armour presumably on *perchas* [wooden display stands], encircling the perimeter walls. At this late date, the armoury was endowed with a certain staging character typical of a moment in which its content no longer made sense or had a practical application, either for court or military life. The entrance, which had over it a panoply with steel saddle plates and a hackamore, along with the arrangement of the harness located next to the entrance, to the left, are very meaningful. The wall opposite the entrance is dominated by a mannequin of a horse with a man-at-arms flanked, in turn, by three suits of armour on each side whilst, on the wall, presumably in the centre, is placed another suit of armour surrounded by horse harnesses and other loose pieces. Continuing clockwise, there was a large wall without windows on which

9 See: González Ramos, 2020.

twenty men-at-arms were arranged, whilst, on the opposing wall, in the sections interrupted by their doors and two windows, three groups of five, four, and thirteen men-at-arms were placed, that is, a total number of twenty-two suits of armour, analogous to the opposite wall. On the wall, forming "*colgajos* [flaps]," hypothetically panoplies, there was a certain number of horse harnesses, loose pieces and polearms.

In the middle of the second room, three benches or trestles accommodated sixty-one saddles, so we must consider that each trestles must have been approximately twelve metres in length, which leads us to conceive of a room of at least fourteen metres on its axis. Along with the trestles, there was a drawer with loose parts. To the room's right, there were twenty-one armed men and sixteen to the left, whilst on the walls, loose or in panoplies, there were harnesses, polearms and even parts of white arms. The sensation of mass was further accentuated by the presence of loose elements of armour and elements of horse harnesses hanging from the ceiling's four beams in an arrangement reminiscent of arsenals such as the Austrian one in Graz.

An important feature of this armoury is that the men-at-arms were placed facing the walls, which presupposes their display on complete mannequins or mannequins with stands. The latter are the most probable, as we already know through other armouries that used them, such as the Royal Armoury. This does not cease to be a feature of modernity and of the use of armoury as an element of prestige, distinct from those in which the weapons were kept in chests and therefore were not visible to the naked eye.

The 1682 inventory features a drastic change only eleven years later. The suits of armour are reduced in number to sixteen accompanied by 117 loose items and ninety-five shaffrons as the most emphasised elements. Even so, there is mention of a shipment from Madrid between 1671 and 1682 that increased the number of the suits of armour to seventy men-at-arms, thus recouping part of the existing discrepancy from the 1637 inventory. Perhaps these adjustments and divergences are due precisely to remodelling activities at the court armoury in Madrid. In any case, these activities and staging changes that we are able to observe here reinforce the idea that the images that the inventories transmitted are exclusive images of a specific moment in their history through which neither their image nor their composition of a specific armoury can be fixed given. That is, these armouries are in a constant process of alteration[10].

10 Within this typology, also belongs the armoury of the 2nd Duke of Arcos from 1630 and probably that of the Marquis of Poza, inventoried upon his 1605 death in Madrid. Given their augmented number of items, however, they might have relied on an additional room.

Armouries of three rooms, occasionally divided by typologies

Some armouries clearly divided their content into spaces that accommodated one or more typologies of weapons, avoiding mixing one with the other, as was very common. To this group would belong the armoury of the 1st Duke of Medina de Rioseco, Admiral of Castile, in Medina de Rioseco in 1540. Its first room brought together 221 infantry suits of armour. In the second, there were seventy-one more, along with crossbows, firearms and harnesses. The third organised its contents in panniers or baskets along with some polearms. Finally, the fourth held 700 polearms and eighteen shotguns barrels.

The armoury of the 7th Duke of Medinaceli in the Cogolludo palace inventoried in 1667 might also belong to this typology. Although the inventory does include a topographic division, it does not provide more information on how they were exhibited, so we have no express mention of their placement on the walls, a defining feature of the next type of armoury that we will analyse. Its inventory expressly cites the existence of an armoury divided among three rooms, calling the first "*la primera pieza de dicho palacio* [the premier room of this palace]", which could indicate that the armoury was located on the ground floor near the entrance, although it makes no express mention on the matter. The most noteworthy element are the forty-one suits of armour it holds, but these articulated the scenery of its three spaces being divided among them equally; thirteen in the first two and fifteen in the third. All of them intentionally lacked gauntlets, which were stored away but not mounted.

The relationship of objects that each room held suggests that the first two were smaller than the third, but in each of them a specific relationship was established between the rest of the collection and each room's set of suits of armour, which presumably might have been located in the centre judging by the types of pieces with which they were associated. The inventory specifies that the suits of armour were found "*en estantes* [on shelving]," a confusing term because it may have a double meaning, either as a shelf, a meaning recognised by Sebastián de Covarrubias Orozco in his 1611 *Tesoro de la Lengua Castellana o Española* [*Treasury of the Castilian or Spanish Language*], or as wooden display stands on which they were mounted. This latter possibility does not find support in Covarrubias, so we must lean towards its meaning as a shelf, in which the other weapons would also be placed, although we cannot provide any more details. The first room contained polearms and bows. The second held polearms and loose pieces of armour. Therefore, an initial typological differentiation was made that would be widely surpassed in the third room that doubtlessly, from lowest to highest, constituted the heart of the armoury after the first two, which were merely introductory in nature. In the third room, the knightly weapons clearly related to an important seignorial armoury were clearly stored, compared to the less-defined character of the arms in the first two rooms. A horse bard, a youth's suit of armour, a parade set with a battle scene, brigandines, shaffrons, horse armoury pieces, fourteen small artillery pieces and 106 armour pieces kept in two chests are noteworthy. Despite all of this, if we compare it with other armouries studied, it appears clearly to be a noble armoury, but plainly not the family's best. It fulfilled its roles as an element of prestige, but it does not include decorative details or references to the creators of the weapons that allow us to consider it an armoury of the first class in terms of quality.

This armoury at Cogolludo had previously been inventoried in 1594 when it belonged to the 5th Duke of Medinaceli, in which no reference was made to the topographic distribution but which does offer other interesting information. In 1594, it contained twenty-two fewer suits of armour, but it did have 574 loose pieces that do not appear in 1667, which might explain the difference in the number of suits of armour if these had been mounted in the interim, as happened with the horse bard, which in 1594 probably was conserved in them dismounted. Overall, the most important difference in the seventy-three years that separated both inventories resides in the fact that in 1594 the weapons were stored in "*cofres* [trunks]", "*arcas* [chests]" and "*cajas* [boxes]", not on shelving as in 1667. Although these terms are related to portable containers, their interpretation is not entirely clear if we look at their large capacities, given they could manage to store ninety-eight loose pieces and twelve corsets or two saddles, twenty-four horse pieces, and eleven loose pieces in only two of them. This suggests that we cannot rule out the possibility that this armoury had large wooden cabinets, as did Philip II's Royal Armoury, where they were called "*cajones* [drawers]". Be that as it may, it does seem clear that a change was made in the installation of weapons, which went from containers to shelves to have a more scenic character without there being significant changes, maintaining its condition as a secondary armoury. In the 1746 inventory carried out on the orders of the 11th Duke of Medinaceli, it is proven that the set was maintained, but by then it was already undergoing a new deterioration by abandonment given that part of the suits of armour "*estaban por el suelo* [were on the ground]".

Armouries divided into three or more rooms with different typologies of arms, but always with the presence of elements of suits of armour on the wall or harnesses in rooms as elements on which the others revolve

This typology would be conceptually the immediate antecedent of nineteenth-century armouries. The 1655 inventory of the fortress and palace of the 8th Duke of Benavente is the clearest example. It was divided into five rooms. In the first, alongside the entrance to the fortress, a small number of weapons are brought together, accompanied by "*paredes llenas de pedazos de armas* [walls full of pieces of weapons]". The second contained a number of pieces slightly superior to those of the first, in addition to four suits of armour, but it maintained that the walls were "*llenas de retazos de armas* [full of scraps of weapons]". The third had a window overlooking the moat and held twenty-three suits of armour, a small artillery piece and 149 partisans. These first three rooms were introductory to the fourth and fifth rooms, clearly the main ones given their assembly and content, located on the corridor above the moat. It contained eighteen partisans, forty-five "*rodelas* [rondaches]" and adargas, eighteen half suits of armour, three mounted *jinetes*[11] with bards, riding saddles, sixty-nine weapons of various types and "*las paredes colgadas de pedazos de armas en número indeterminado* [the walls hung with an indeterminate number of pieces of arms and armour]". In the fifth, forty-nine suits of armour, sixty-four pavises, cannonballs and other weapons stand out.

In 1702, the 9th Duke performed a new inventory that indicates a renovation of the five rooms to balance the distributions of weapons and suits of armour. They maintained the criteria of having pieces on the walls, but although the common thread is the armour, pieces of armour and shields, each room will be characterised by a unique set. In the first room, a heroic entrance with two horses armed with bards and their respective *jinetes* surrounded by fifty-two shields with the ducal coat of arms or in the Moorish style, in addition to thirty-eight half-suits of armour, a banner with the image of the Virgin and the ducal arms and four *atabales* [drums] stand out. In this room, there are no pieces on the walls, but there are in the next ones. In the second room, there were twelve artillery pieces and three mannequins of horses with *jinetes* also armed, accompanied by ten half-suits of armour. The walls were covered with fifty-two shields, among them twenty rondaches with the ducal arms, along with ten adargas, forty-two morions and other pieces "*que pasan de doscientas* [that exceed two hundred]", white arms and harnesses. This room was without a doubt the main one because it was dominated by a half-body portrait of King Philip, without specifying if it was Philip II or Philip III. In the third room, there were twenty-two suits of armour, polearms and thirty-three adargas, some of them Moorish (Nasrid) and others with the ducal arms, a set whose purpose is clearly explained when it is noted "*que solo sirven de*

11 Horse riders.

memento y adorno [that they only serve as a memento and adornment]", all accompanied by about twenty pieces of armour on the walls. Compared with the previous rooms, the fourth room was of rather reduced dimensions, which suggests that it was a passageway between rooms. In it, the inventory points out that "*todo es de poco provecho* [everything is of little use]", which is why a workbench for the armourer was set up and it remained configured as a small, cluttered workshop in the second half of the armoury. Finally, the fifth room was fundamentally dedicated to firearms surrounded by walls full of pieces of armour and bards that merely served as decoration. This distribution clearly illustrates the transformation of armouries into collections of social prestige, but it is striking that it coincides with a dynastic change that, in turn, meant a change of era and paradigms generally.

Armouries established in their own edifice, isolated from the palace or castle, where, in turn, there might be a small armoury for personal use

The armoury that responds to this typology is the one that the 3rd Duke of Benavente had in 1541 in the Hospital de Nuestra Señora de la Piedad [Hospital of Our Lady of Piety], in Benavente. The entire armoury was contained in only one room, but in the inventory's description, two large sets appear that are expressly cited in its introduction: the seigniorial armoury, "*las armas que su señoría tiene así de su persona* [the weapons that his lordship has on his person]" and a military armoury, alluded to as "*de la gente de su casa* [of the people of the house]". These two basic blocks are divided, in turn, into generic sets according to the different arms typologies. In the seigniorial armoury, twelve garnitures for jousting and warfare with their respective sets of replacement pieces, along with another fourteen suits of armour, twenty-nine brigandines and sixty-six loose pieces without counting the replacement harnesses, in addition to three bards and saddles of an indeterminate number, polearms and shields, white arms and horse coverings are highlighted. For its part, the military armoury reproduces the seigniorial armoury's distribution by blocks, but only in its first three sets: suits of armour, horse harnesses and polearms. This sequence of items in the inventory description suggests that the centre of the room would hypothetically be occupied by suits of armour, saddles and four bards. The wall opposite the entry for the polearms and the other loose pieces, shields, white arms must have hung on the walls as there were no references to chests, esparto baskets or cabinets, in accordance with a form of storage known in other contemporary armouries.

In 1601, the armoury of the 5th Duke of the Infantado, Iñigo López de Mendoza, is a special case because it was divided into at least two clearly differentiated spaces: one small armoury in his palace with 100 weapons of various types, accompanied by 102 accessories and an edifice in town expressly used as an armoury, where military weapons and noble arms, which might also include military items, are differentiated by floors[12]. This is the true armoury, because we should consider the arms of the palace for immediate use. This building is divided on the ground floor in two rooms that contained a large quantity of military firearms "*que están ordenados* [that are organised]", from which it can be deduced that they were mounted on structures that allowed their arrangement without us being able to know what kind of supports they used. On the upper floor, the seigniorial armoury was found, the main element of which were 137 suits of armour that were exhibited on 128 "*estantes*", which we should interpret as shelves upon which the suits of armour was placed, probably wooden display stands in small cabinets as, in the classical Castilian meaning, which defines the *estante* as an cabinet without doors. In Infantado's case, nothing is said that allows us to discern from either of these possibilities, but the armoury's 1628 inventory by the Constable describes *estantes* with four doors to protect the suits of armour. In 1655, in the armoury of the 1st Marquis of Leganés, the *estantes* are described as of walnut, with doors fitted with a metal underlay, similar to that of libraries, and of an interior curtain. The presence of a door indicates a cabinet, according to the description of Leganés, which opens the possibility that the *estantes* were built as shelves but could also have had curtains, as shown clearly in the painting entitled *El Tacto* [The Sense of Touch] by Rubens and Brueghel[13]. In any case, we must assume that it was a room of considerable size, if we take into account that each suit of armour would need at least a metre of width to accommodate the shaffrons and saddles as well, which would mean around 122 linear metres of wall, which would require a room of fifty metres of length, similar to that of the Royal Armoury. Nothing is said of the other weapons, so we cannot rule out that Infantado's armoury consisted of a central line of double-sided cabinets to leave the walls free for the rest. In addition to the suits of armour, 718 weapons of various types, an indeterminate number of flags, and four pieces of artillery were inventoried. Given that there are no references to boxes, trunks or panniers in

12 About this armoury, see: González Ramos, 2013, 2014.

13 For suits of armour on shelves above a hanger, see the painting *El Tacto* [The Sense of Touch] by Peter Paul Rubens and Jan Brueghel the Elder, dated to 1618. Museo Nacional del Prado, inv. P001398.

which to store them, we must hypothetically assume that the scheme of a central line of cabinets with suits of armour and the rest on the walls is the most probable, given the difficulty of displaying this number of pieces would entail. The room also would have had a small workshop area for maintenance of the collection and the servicing of the weapons themselves, judging by the presence of pinchers, scissors, "*bastones para limpiar las armas* [sticks for cleaning the arms]", two anvils, and two mallets to break and grind emery, a clear allusion to the cleaning material to inhibit the pieces' oxidation and also to sharpen the weapons.

The third space was a room exclusively dedicated to storing thirty-seven boxes that contained different types of arms and thirty-nine "*arpilleras*", which we must identify as the burlap bags used to transport part of the armoury's pieces including waxed canvases that presumably wrapped the pieces. Other waxed canvases covered the lower windows of rooms of firearms and part of the main room. There is no indication as to their purpose, if to keep out the light, act as insulation or in place of windows, but this last purpose would be strange for evident reasons of conservation.

The fourth space was an attic, but it is unclear if that was in the same edifice or another one. In it, without logic or order, apparently 122 loose pieces of different types were stored.

Armouries divided in two places of different location, but inventoried as one

A similar case to that of Infantado in 1601 is the 1588 inventory of the 5th Duke of Gandia, which differs in that his seats of Gandia and Castellón are considered as one and there is no clear division of spaces in which the weapons are stored. They would therefore be complementary and appear divided into two clear groupings, although they are not exclusive between themselves, sharing the same type of object. In Gandia, the seigniorial arms are basically found, consisting of suits of armour, saddles and white arms. In contrast, in Castellón, a much larger number of arms are stored, among which polearms with a set of 400 Milanese pikes, 156 arquebuses, thirty-six artillery pieces and a large quantity of gunpowder and components.

THE FURNITURE. DISPLAY ELEMENTS AND CONTAINERS

One of the most interesting aspects of the inventories of the noble armouries is the information that they provide about their physical aspect, which allows us partially to reconstruct how they appeared. The documentation provides isolated information fundamentally about their containers and mounts that provides us a set of variables to reconstruct hypothetically the possible interiors in which they might have been found. Broadly speaking, these elements could be divided into two sets: display cases and containers.

Among the first category, cabinets stand out, such as the one in the armoury of the 3rd Duke of Sessa from 1544, the capacity of which allowed for the storage of three sets of armour, ninety-two weapons of different types, four saddles, a plume crest and a shaffron. The inventory does not provide any further details, but it is possible that it was similar to the ones at the Royal Armoury described in 1594. The armoury of the 5th Duke of the Infantado in Guadalajara, from 1601, constitutes a special case, wherein 128 cabinets, presumably in the centre of the room, are noted.

A second element is the "*estantes*", a word with a confusing interpretation because, as has been pointed out, it can have a triple meaning as "*estanterías* [shelving]", the meaning recognised by Covarrubias in 1611, or as wooden display stands on which suits of armour are mounted. This shelving system is documented in the cited painting of Bruegel, in which the suits of armour are arranged over the shelves mounted on hangers, but there is no allusion to the latter associated to the *estantes* in the documentation consulted. However, the inventory of the armoury of the Constables from 1628 includes this same term associated with the presence of doors, which at least in that instance, we must consider the *estantes* as cabinets that would already be enclosed. In other examples, such as the armouries of the 5th Duke of the Infantado in Guadalajara from 1601 and that of the Duke of Medinaceli de Cogolludo from 1667, the arms and bards might have been arranged on *estantes* because there is no information to suppose otherwise.

A third display case of great importance are the very walls of the armouries, which lend a great staging character to their rooms. It is important to emphasise that this system would be normalised in the modern armouries of the nineteenth and twentieth century, thus it would not be a novel museographic element in contemporary times. The use of walls is documented in the armoury of Philip II, although it is likely that it was already used in its first installation in 1567. Among the noble houses, it is present in the armoury of the Dukes of Alba in Alba de Tormes in 1531 and in Medina de Rioseco in 1540, the first with twenty-one corslets, thirty-three breastplates and three sallets, whilst in the second there would be 221 "*colgados* [hung]" corslets. The system repeats itself in Benavente's Hospital de la Piedad [Hospital of Piety] in 1541, where it persists in 1655 and 1702, and in the armoury of the Duke of the Infantado de Guadalajara at least in 1601. To reach them, stairs were used, included in the armoury of the Constables from 1628 and that of Benavente from 1655.

The wooden display stand for suits of armour, understood as a support equipped with a base and some arms in a cross, is a system known at least since the sixteenth century and documented graphically by nineteenth-century painting and photography of the Royal Armoury[14]. The inventories consulted also include their use and provide some details. The 3rd Marquis of Poza's armoury in Madrid on his 1605 death had 115 wooden display stands "*de pino de Cuenca* [of Cuenca pine]", and that of the Constables of Castile in Burgos contained in 1628 "*otras doce cruces de pino nuevas que se han hecho para poner armas* [some twelve new pine crosses that have been made to place weapons upon]". In the examples of the armouries of the 10th Duke of Béjar from 1671 and that of the Counts of Benavente in 1628, there must have been a similar system, in that in the first instance the men-at-arms were lined up on the walls and in the latter, the large number of suits of armour allows us to assume so.

The presence of horse mannequins sculpted in round shape is also documented, not only in the royal armoury, which in the era of Philip II had six, but also in the armoury of the Duke of Lerma in Ampudia for two bards, in the Benavente armoury, which has three armoured horses in 1655 and later in 1702, by which time its number of equestrian figures increased to five.

For saddles, trestles or benches were used, as is noted in the Duke of Sessa's armoury in 1544. In it, three benches held sixty-one saddles, which permits us to conclude that the length of each piece would be close to twelve metres. Similarly, the armoury of the Constables in 1628 had saddles on "*sus caballetes de madera de pino* [their pinewood trestles]".

Regarding the containers, overall the presence of boxes, chests, and trunks stand out to store various types of weapons. Although this term might lead us to think about "*cajones* [drawers]", i.e., cabinets, in the styles of the Royal Armoury, in the majority of the inventories, the term "*caja* [box]" refers to the containers of a small size, judging by their small number of contents[15].

---

14 Calvert, 1907: plates 34, 37, 45, 47ª, 49, 51–54.

15 In 1531, they were included in the inventory of Alba de Tormes containing all types of weapons, in 1540 in Medina de Rioseco, they were used for equestrian coverings, in 1540 those of the son of the Marquis of Poza were covered in leather, in 1580 the personal armoury of the Count of Salinas was stored in two chests, in 1584 those of the 6th Duke of Alburquerque had a label specifying that they were chambered weapons, in 1585 those of Antonio Pérez were ofwalnut covered by red leather, in 1588 those of the 5th Duke of Gandía were painted white and contained Milanese arms, in 1604 those of the Count of Aguilar were of leather-lined pine, in 1630 those of the 3rd Duke of Arcos were stamped, probably with the ducal arms and in 1642 the Duke of Sessa

Within the boxes, trunks or chests, the arms might have been protected within red or yellow fabric cases, a system occasionally followed by the 6th Duke of Alburquerque and the 2nd Marquis of Velada in 1584 and 1616, respectively. In 1628 the Constable of Castile had chests in Burgos to store the cases, which at that time did not contain weapons, as well as "*dos cortinas grandes que parecen de campana de lienzo muy largas que tienen diferentes figuras forradas de fustán que la una sirve de guardapolvo en la armería y la otra en la sillería de palacio* [two large curtains that look like very large canvas bells that have different figures lined with fustian, one that serves as a dust cover in the armoury and the other in the palace stalls]" and "*cuatro cortinas de bocací colorado que toman desde lo alto a lo bajo con que se cubren las armas doradas de su excelencia* [four red bocasine curtains that hang from top to bottom with which the golden arms of His Excellency are covered]". Another similar system was the covers made with burlap sacks and waxed covers for the transport of weapons, as noted in the 5th Duke of the Infantado's armoury in 1601.

A cruder container than boxes were the panniers or baskets in which items could be piled. Their presence in Pedraza and Torremormojón[16] in 1506 and 1512 allows us to assume that they were a well-known system at the end of the Middle Ages, but they survived, as evidenced in the armouries of Alba de Tormes in 1531 and that of Medina de Rioseco in 1540. The latter was endowed with sixty-eight panniers for as many suits of armour.

Finally, an important area within them was the space dedicated to the armourers tasked with the collection's custody and maintenance, where they had their workshop and tools such as cleaning sticks, anvils, mallets, scissors, pincers, files, every manner of clearing material and specific furniture for them. At times, they were very well equipped spaces, as in the armoury of the Constable in 1628, or that of the 10th Duke of Béjar in 1682.

CONCLUSION

The study of noble armouries from the sixteenth and seventeenth centuries on these and other aspects still requires an in-depth analysis. Here, they are only outlined in these aspects superficially. Nevertheless, we yet can come to initial concussions that should be examined further in the future. First, we should be conscious that every inventory simply reflects a moment and thus

stored his entire personal armoury for the court in two trunks, as did the Count of Salinas previously.

16 Franco/Soler, 1994.

is only representative of that moment. In instances where there is more than one inventory, changes can be seen as a result of multiple factors, such as the movement of pieces between venues, redecoration or simply loss of control of the collections because they sometimes had been abandoned, as in the case of the castle of Ampudia in 1614. They do not include fixed patterns when it comes to organising an armoury topographically, nor are there any in the order of the blocks by typologies, which logically indicates that adapting the container to the edifice according to the nature of the content prevails. For the installation of an armoury, there does not appear to be a reference model, not even the Royal Armoury, which, in fact, differs in many ways from noble armouries. In noble armouries, the systems of custody and exhibition used most of the time are trunks and the use of the walls, reserving wooden display stands and complete mannequins for exceptional instances. All this leads us to reflect upon the continuity of the solutions of these installations in the contemporary era until the end of the nineteenth century and the beginnings of the twentieth century. The romantic images of European and American museums of that period would be very conditioned by the very past of the Renaissance and Baroque armouries, with pieces on the walls, wooden display stands for the suits of armour, mannequins of wood for the horses and equestrian figures and large sets of suits of armour concentrated in the centre or along the room's walls. These stagings should not be considered as an innovative product of romanticism, but as a direct recreation of the old noble armouries.

BIBLIOGRAPHY

Calvert, Albert F. (1907): *Spanish Arms and Armour.* London-New York, Ed. John Lane.

Cámara Fernández, Carmen (1994): "Aportación al estudio del coleccionismo en el siglo XVII. La armería del Condestable de Castilla en su palacio de Burgos". In: *Actas del X Congreso del CEHA. Los Clasicismos en el Arte Español.* Madrid, UNED: pp. 407–412.

Franco Silva, Alfonso. Soler del Campo, Álvaro (1994): "Los arsenales de dos fortalezas castellanas: inventarios de Torremormojón (1505) y Pedraza de la Sierra (1512)". In: *Historia. Instituciones. Documentos*, 21, pp. 309–343.

González Ramos, Roberto (2013): "The armoury of the Dukes of the Infantado. Collecting, prestige and meaning". In: *Journal of the History of Collections*, 25, 3, pp. 335–350.

González Ramos, Roberto (2014): "La colección de armas de Íñigo López de Mendoza, V Duque del Infantado". In: *Gladius*, XXXIV, pp. 153–198.

González Ramos, Roberto (2017): "Las armerías nobiliarias castellanas en el siglo XVII como manifestación de identidad cultural". In: Carrasco Martínez, Adolfo (ed): *La nobleza y los reinos. Anatomía del poder en la monarquía de España (Siglos XVI-XVII)*. Madrid, Iberoamericana-Vervuert: pp. 269–314.

González Ramos, Roberto (2020): "La armería de los duques de Béjar". In: *Goya*, 370, pp. 16–29.

Romero Medina, Raúl (2020): "Como la espada, así la vaina. La colección de armas blancas de Don Juan de la Cerda y Silva, IV Duque de Medinaceli (1552–1575)". In: *Potestas*, 16, pp. 81–101.

Simal López, Mercedes (2002): *Los Condes-Duques de Benavente en el siglo XVII. Patronos y coleccionistas en su villa solariega*. Benavente, Centro de Estudios Benaventanos "Ledo del Pozo".

Soler del Campo, Álvaro (1998): "La armería de Felipe II". In: *Reales Sitios,* 135, pp. 24–37.

Jesús M. Ruiz Carrasco

Universidad de Córdoba

THE ARCHITECTURE FOR THE ARMOURIES IN SPAIN IN THE EARLY MODERN AGE

The content of this monograph focuses mainly on how those historicist compilations known as armouries can be analysed and even typified through the study of their origin, belonging, use, significance, history, composition, classification or exhibition. However, another of the most relevant aspects to examine in order to deepen our knowledge of Spanish armouries is that concerning their original location. In this sense, complementing the content of the chapter dedicated to the arrangement of the armouries, we should not only take into account the specific spaces where the nobiliary war heritage was kept, but we should also proceed to analyse both the location of the rooms and the architectural characteristics of the building to which they belonged. On the one hand, the study of these sites and buildings would help to determine or conjecture about the category, use, classification, display or meaning given to each armoury by its owners. On the other hand, this architectural analysis would serve to establish a series of conceptual relationships between the constituent spaces of the stately building according to their function, content and location; as well as between this interior arrangement and the exterior image of the complex. In this way, the armoury as an architectural entity (independent or integrated into a whole) must acquire a meaning through the relationship between the pieces that compose it and the space that houses them, but it must also be defined by the private exhibition of a power that has previously been staged through the monumentality of the public building. Therefore, this is the first specific study on the architecture of noble armouries in the Modern Age, dedicated to presenting a global investigation of this heterogeneous group of constructions and to clarifying the particularities of the most significant examples concerned.

Firstly, in view of the complexity of the subject, it is necessary to specify the essential principles of this study and to explain the reasoning behind the selection of cases, as well as to set out a series of specific factors that will facilitate the understanding of the general argumentation, the consequent typological classification, the distinction of singular exceptions and the formulation of conclusive ideas. Firstly, it should be noted that the content of this text is not based on the desire to present a compendium of noble architecture according to the aesthetic relevance of the cases, but rather on the relationship of the most relevant and unique armouries of which there is evidence with their building context. Therefore, the argumentative development is based on a series of paradigmatic examples of different types, the choice of which derives from the characteristics that define the armoury they house, as well as the particularities of the link between content and continent. Thus, the sense of this section of the research presented in this publication is not understood as an independent complement that uses the subject to talk about architectural aesthetics, as it is born from the study of the armouries to provide another main aspect that completes the knowledge about them. However, once the principles of this chapter have been established and before going deeper into the development of its content, it is worth revealing a series of specific factors that will determine the development of the subject to be dealt with.

The architectural diversity of the examples to be exhibited allows us to present a typological classification based on the spatial conception, the location within the palace complex, the relationship with other rooms and the singularity of certain distinguished cases. In this sense, it should be noted that not all the cases correspond to the same model, not only because the distribution and composition of the buildings are not homogeneous, but also because of the importance given to the armoury according to its location, as well as its distinction with respect to the palace complex. At the same time, a number of aspects must also be taken into account that determine the significance given to the armoury, since, in addition to the situational value with regard to the architectural layout of the manor building, its relationship to the adjacent spaces and their contents must also be emphasised. In addition to this, both the size of the rooms and the number of rooms used for the display and conservation of the noble arms, as well as their morphological characteristics and ornamentation, are also considered to be relevant. Therefore, we proceed to establish a typological classification that will be duly divided and exemplified, according to the principles set out above and taking into account the aforementioned principles that have just been presented.

Firstly, it is worth highlighting the case of those armouries that present an absolute relationship with the architecture, dividing the location of the war pieces along different rooms of a medieval fortification destined to vindicate the dynastic value of its owner. This is the case of the castle of Ampudia, a medieval building acquired by the Duke of Lerma in 1597 and intended to stage his power as Secretary of State to the King, as well as his brand new status into noble cathegory[1] (Fig. 1). This castle, which was the casual seat of the Spanish royal court, housed a dynastic armoury in several rooms of the complex, thus demonstrating the direct relationship between the exterior architectural language and the interior content, intended to magnify the figure of a new noble power that had the necessary resources but not a lineage commensurate with its pretensions.

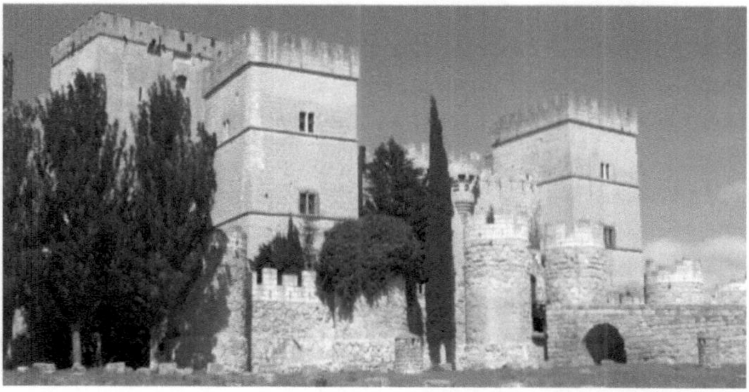

Fig. 1. *Castle of Ampudia*. Ampudia, Spain (Wikimedia Commons).

Secondly, we must consider the most common group of cases as far as the architecture of the armouries and their location are concerned: those located in building complexes of medieval aesthetics with certain later additions or (even) completely modernised and distributed around a central courtyard. Normally, these fabrics tended to have a fortified appearance, alluding both to their original defensive character and to the later nobiliary significance that their castellated appearance entailed[2]. In this sense, from its exterior visualisation, the continent presented a staging of the authority linked to the lordship of the owners of the building, so that the visitor or spectator could predict in advance the ostentation that would be presented in the architectural interior[3]. Thus, an anticipated relationship was established between the external presence of the building and the spaces that made up the castle or palace, where the location of the corresponding rooms and their contents were to be organised through a

1 Izquierdo, 2011: 24.
2 Cooper, 1991: 29–154. Varela, 1999: 41–61.
3 Molero, 2022: 141–169.

personalised discourse that set out the characteristics of the aristocratic dynasty that housed it[4]. Each room was to house a series of elements which, complemented by their appropriate decoration, would transmit a specific message whose background was determined by the basic idea of representing noble power. Thus, given their military idiosyncrasy, the armouries were mostly located in a symbolic place in the building, but also in initial and accessible spaces, so that they could be seen after entering the architectural complex and thus continue the discourse of power established from the outside. Therefore, beyond the architectural relationship between the exterior and the interior, a relationship must be established between the preceding visual staging of noble power and its logical discursive continuation through the location of the armouries near the entrance to the building complex.

In this respect, it is worth pointing out some representative examples. The first of these corresponds to the house of Benavente, located in the town of the same name and structurally dated to the fifteenth century with later additions[5]. Beyond the external presence of this castle, which was elevated due to its geographical location and the decoration inspired by the "Moorish" aesthetic[6], the large armoury must have been representative of the power of the Pimentel family dynasty. Specifically, after studying the inventory of the armoury dated 1655 and its corresponding allusions (to the moat, the entrance and the staircase), it can be established that the five rooms dedicated to the armoury must have been located near the entrance to the castle[7]. These spaces, probably adapted from the original configuration of the complex, were a compendium of complex rooms that preceded the central courtyard and which had to be crossed before reaching the main rooms, as can be deduced from the plan drawn by Blein dated 1809[8]. This is the most relevant case of all those that form part of this model, given the multiplicity of the rooms and the evident exhibition of the armoury through the spatial structuring, which obliged the visitor to contemplate partially or totally the representativeness of power through the pieces of a war-like nature duly exhibited.

Within this same typological group we can add the two examples that belonged to the Count of Cifuentes, located in the town of the same name and

---

4 For more information on this issue, see Bisso, 1870.

5 Beceiro, 1997: 185–204.

6 González/Regueras/Martín, 1998: 71.

7 Archivo Histórico de la Nobleza, Toledo (AHNOB),Osuna, C. 427, D. 108, fol. 5v.

8 Regueras, 2019: 194–201.

in Barcience, which housed two parts of the house's armoury[9]. In both cases the fortified and clearly medieval architecture is organised around a large irregular courtyard with large towers at its corners, but with some later additions near the main entrances. This would suggest that these were modifications intended to house the armoury, especially in the case of the castle of Cifuentes, where the main pieces were moved and where the display of weapons must have been much more representative.

There are also two known cases belonging to this group where the construction was conceived following the new Renaissance aesthetic coming from Italy, leaving the medieval fortification model, but with an analogous situation of the armoury. These are the palaces of Cogolludo, belonging to the house of Medinaceli and completed in the 1490s[10]; and the palace of the Marquis of Santa Cruz (in Viso del Marqués), built between the 1560s and 1580s[11] (Fig. 2). In the first case, an inventory of 1667 specifies that the three rooms where the armoury was located were situated in "the first room of the said palace", so it can be reliably deduced that they were located on the ground floor just after the main entrance of the palace[12]. On the other hand, in the second case, it is known that in 1646 "the smallest and closest room to the entrance" of the two main rooms that made up the eastern side of the palace was the one dedicated to the armoury, which could be accessed from the courtyard just after crossing the vestibule of the complex[13]. These two examples show how, beyond the stylistic renovation of Spanish architecture, the location of the armouries was prepared to stage the stately power and to vindicate the dynastic value of their owners, even adapting the architectural distribution to the exhibition of war pieces or selecting very specific structural segments for this purpose.

9 Archivo Histórico de Protocolos de Madrid, Madrid (AHPM), 10801, fol. 173r-178r, 211r-218v.

10 Pérez/Pérez, 2012: 41–86.

11 López Torrijos, 2004: 129–138.

12 Archivo Ducal de Medinaceli, Toledo (ADM), Cogolludo, Leg.3, exp .44.

13 López Torrijos, 2009: 172–177.

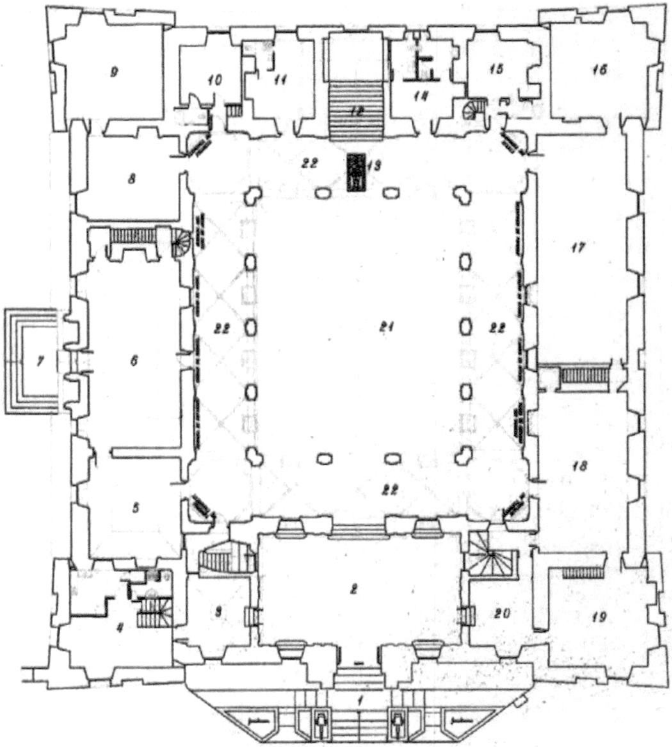

Fig. 2. *Ground floor of the Palace of the Marquis of Santa Cruz.* El Viso del Marqués, Spain. 1564–1586 (Wikimedia Commons). The room where the armoury was located is marked with the number 18.

A variant of this model is the medieval castle of Cuéllar, owned by the Duchy of Alburquerque, where an extension was carried out between 1530 and 1550 that consisted of the incorporation of a new architectural complex on the south side, where the entrance to the stables, a series of rooms and a porticoed gallery open to the main courtyard whose arcades were intended to correspond to the new Roman aesthetic (Fig. 3). It is known that this new architectural addition housed the dining room, the library and the armoury on the first floor[14], which must have been located next to the entrance to the stables. The specification of

14 Montes, 2016: 51–52.

the type of entrance is due to the fact that the castle of Cuéllar had two different entrances, the one to the town and the one to the stables, next to which the armoury was located. In this sense, it could easily be interpreted that the weapons were located near the aforementioned entrance so that they could be accessed in case their use was required, which is also one of the aspects that determined the location of the armouries. However, it should be borne in mind that the entrance to the stables was the most prominent entrance to the palace and the one through which the guests of the lord who lived there entered, so its location also corresponds to the desire to represent the power of the noble dynasty. Even so, it is worth highlighting this variation with respect to the common cases described above, given that it is a location that specifies the proximity of the armoury to the library. This binomial composed of the staging of primordial military power and noble virtue through knowledge was one of the signs of identity of the most important dynasties from the sixteenth century onwards, when armouries began to be located next to libraries in different spatial arrangements, which will be analysed below.

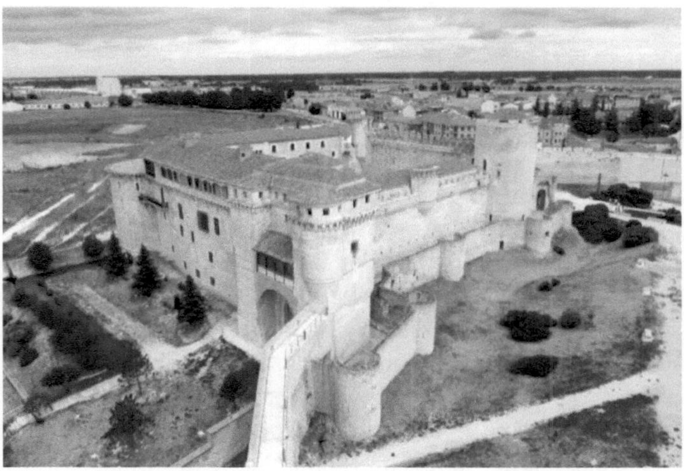

Fig. 3. *Castle of the Dukes of Alburquerque*. Cuéllar, Spain (Wikimedia Commons).

In relation to the previous case, the third of the groups presented is the one in which the armouries are expressly located next to spaces dedicated to knowledge and culture, combining dynastic representativeness with the virtue that the noble status was to acquire. This was the case of the armoury located in the urban palace popularly known as "Casa de Pilatos", belonging to the lordship of Alcalá de los Gazules, where the 3rd Duke of that dynasty ordered the construction in 1604 of upper rooms to house both the library and the armoury[15] (Fig. 4). Unlike the cases described above, the armoury was not located near the entrance to the palace, but on the upper floor of a very specific wing of the complex, with sole access through the library. Furthermore, it is worth noting the Duke's famed humanistic concerns and the consequent importance given to the library, decorated with frescoes by Francisco Pacheco and a place for intellectual meetings between the nobleman and his closest circle. It is therefore a specific case, differentiated from the previous ones by the theme of the adjacent rooms and the relationship established between noble virtue and lineage.

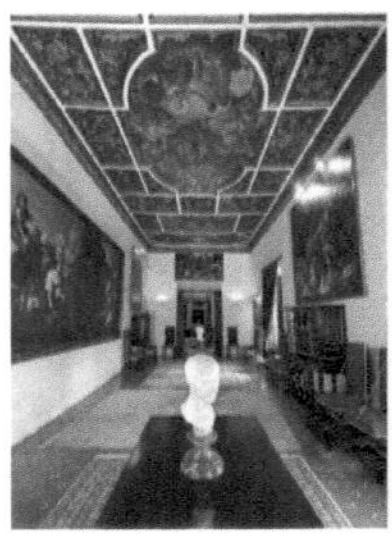

Fig. 4. *Casa de Pilatos*. Seville, Spain (Wikimedia Commons).

Along the same lines, but with a significant variation, is the case of the armoury of the house of the Infantado, which in 1601 is recorded as being located in a palace separate from the manor house where the library of the family (linked to the entailed estate) was also located[16]. This disappeared building, which had been commissioned by Pedro González de Mendoza in the fifteenth century[17], once again combined a content that summed up culture and noble power, symbolising the virtue of the dynasty beyond the power it wielded. However, unlike the previous case, this was an express decision to locate the armoury in a separate building outside the Duke's residential palace, an ensemble that also symbolised the cultural value of the Mendoza family and its primordial role in both cultural and artistic patronage. Therefore, its relevance lay not only in combining the

15 Beltrán, 2020: 262.
16 González Ramos, 2014: 153–198.
17 Romero, 2010: 1–21.

idiosyncratic meanings of the pen and the sword, but also in its relationship with a building that in itself represented the dynastic culture of the Mendoza family on the outside.

Finally, and fourthly, two examples of armouries located in spaces separated from the rest of the architectural ensemble or adjacent to it for practical or representative reasons are worth mentioning. The first of the two cases is related to the previous one because of its apparently artistic content, but it is not a work intended to combine letters and arms with such zeal: the tower of the armoury of the ducal palace of Alba de Tormes, a late medieval construction owned by the house of Alba. The tower, the only fully preserved remains of the ruined castle, was intended to house the arms of the Duchy of Alba, which were accompanied by a sixteenth-century pictorial mural programme depicting great military exploits of the dynasty[18]. Thus, beyond the artistic analysis that can currently be given to these paintings, it is an exhibition space where the weapons were combined with the visual representation of their use and meaning. It is therefore a visual manifestation of the staging of war power and the relevance of a rising lineage. On the other hand, the last example is that of the Constable of Castile in his palace in Burgos, to which an adjacent space was added in the second half of the sixteenth century to house the stately armoury[19]. This was a space independent of the palace, intended for the proper storage of all the war pieces and, supposedly, to arrange them in a more clear-sighted and representative exhibition sense. However, it is not known to have been combined with any other type of content, but rather had a purely practical purpose, probably following directly the example of the Royal Armoury building ordered by Phillip II[20].

By means of this tour of the sites where several of the most important armouries on record were located, complementing other texts in this publication and in line with them, the aim has been to provide an initial analysis of the relationship between architecture and armorial compendia. The aim has not only been to highlight a link between the representativeness of architectural power and the military power granted by the possession of arms, but also to establish a classification according to their location within the buildings, their adaptation to the spatial structure, their modification due to the armouries, their individualisation and the relationship with other messages equally relevant to the noble lineages. It can be established that the armouries have had a direct link

18 González Zymla, 2014: 66–80.

19 Cámara, 1994: 407–411.

20 Soler, 1998: 24–37.

with an architecture, both because of the need to display them in certain rooms and because of the importance of their location and the ideological message that they wished to convey as a result. In short, it is a link between container and content that has largely determined the manifestation of power in many ways, as well as the very essence of noble values, the dynastic sense of each family and the knowledge about them that has endured to the present day.

BIBLIOGRAPHY

Beceiro Pita, Isabel (1997): "La fortaleza de Benavente en el siglo XV". In: *Brigecio: revista de estudios de Benavente y sus tierras*, 7, pp. 185–204.

Beltrán Fortes, José (2020): "El III Duque de Alcalá y sus intereses epigráficos. Notas sobre su colección lapidaria en Sevilla (Siglo XVII)". In: *Spal*, 29, 2, pp. 259–279.

Bisso, José (1870): *Castillos y tradiciones feudales de la Península Ibérica*. T. 1. Madrid, Achille Ronchi.

Cámara Fernández, Carmen (1994): "Aportación al estudio del coleccionismo en el siglo XVII. La armería del Condestable de Castilla en su palacio de Burgos". In: *Los clasicismos en el arte español. Actas del X Congreso del CEHA*. Madrid, Universidad Complutense: pp. 407–412.

Cooper, Edward (1991): *Castillos señoriales en la Corona de Castilla*. Vol. 1. Salamanca, Junta de Castilla y León.

González Ramos, Roberto (2014): "La colección de armas de Íñigo López de Mendoza, V duque del Infantado". In: *Gladius*, 34, pp. 153–198.

González Rodríguez, Rafael/ Regueras Grande, Fernando/ Martín Benito, Jose Ignacio (1998): *El castillo de Benavente*. Benavente, Centro de Estudios Benaventanos "Ledo del Pozo".

González Zymla, Herbert (2014): "El castillo palacio de Alba de Tormes, simbolismos clásicos en un edificio medieval". In: Sandro De Maria/ Parada López de Corselas, Manuel (eds.): *El Imperio y las Hispanias de Trajano a Carlos V. Clasicismo y poder en el arte español*. Bologna, Bononia University Press.

Izquierdo Misiego, Jose Ignacio (2011): "El duque de Lerma y la villa de Ampudia". In: *La Corredera, Revista Cultural de Ampudia*, 1, pp. 24–48.

López Torrijos, Rosa (2004): "Arte e historia común en el palacio del Viso". In: Boccardo, Piero/Colomer,José L./ Di Fabio Clario (eds.): *España y Génova. Obras, artistas y coleccionistas*. Madrid, Fernando de Villaverde Ediciones: pp. 129–138.

López Torrijos, Rosa (2009): *Entre España y Génova. El Palacio de Don Álvaro de Bazán en el Viso*. Madrid, Ministerio de Defensa.

Molero García, Jesús (2022): "El castillo medieval en la Península Ibérica: ensayo de conceptualización y evolución tipológico-funcional". In: *Vínculos de Historia*, 11, pp. 141–169.

Montes González, Francisco (2016): *Mecenazgo virreinal y patrocinio artístico: El ducado de Alburquerque en la Nueva España*. Seville, Fundación Real Maestranza de Caballería de Sevilla.

Pérez Arribas, Juan Luis/Pérez Fernández, Javier (2012): *El palacio de Cogolludo*. Guadalajara, Gea Patrimonio.

Regueras Grande, Fernando (2019): *Elogio de un fantasma. El Castillo de Benavente*. Benavente, Centro de Estudios Benaventanos "Ledo del Pozo".

Romero Medina, Raúl (2010): "La casa del Cardenal Mendoza en Guadalajara. Una traza del arquitecto Lorenzo Vázquez con la colaboración de canteros tardogóticos valencianos y maestros moros aragoneses". In: *XII Encuentro de Historiadores del Valle del Henares. Alcalá de Henares*, Ayuntamiento de Alcalá de Henares: pp. 1–25.

Soler del Campo, Álvaro (1998): "La armería de Felipe II". In: *Reales Sitios,* 135, pp. 24–37.

Varela Agüi, Enrique (1999): "Fortificación medieval y simbolismo. Algunas consideraciones metodológicas". In: *Medievalismo*, 9, pp. 41–61.

Pierre Terjanian

Metropolitan Museum, New York

CONTEXTUALISATION OF NOBLE SPANISH ARMOURIES FROM A BROADER EUROPEAN PERSPECTIVE: THE CASE OF ARMOURY OF THE DUKES OF PFALZ-NEUBURG AT NEUBURG CASTLE ON THE DANUBE

While the term "armoury" often designates a dedicated repository for storing the personal armors, weapons and related objects of a nobleman and his immediate entourage, in the Renaissance the term was also often used more conceptually, to refer to such items as a distinct category of material possessions, for example when writing one's last will and testament. An armoury was quite different from an arsenal. The term "arsenal" was typically used to refer to a space, generally a warehouse, for housing armors, weapons and related objects for the use of anonymous soldiers. Those items belonged not to the soldiers but to their employers—noble territorial rulers, provincial and municipal governments.

Textual sources are invaluable for the study of noble European armouries. Registers of payments for their upkeep, observations made by visitors and above all inventories of their contents provide invaluable information that the objects themselves or the physical facilities in which they were kept are unable to yield. Although relevant records —especially inventories—generally abound, there is no readily accessible census of those sources and the vast majority of the relevant documents have not been published or even studied. In view of the current state of knowledge, it thus seems premature to seek patterns among noble armories across Europe. By contrast, by concentrating on individual noble armouries and by publishing complete transcriptions of inventories and other relevant primary sources where possible, it is possible to make a valuable contribution to expanding the foundations of our understanding.

Over time, as the documentation expands and is made more accessible, it may become possible to single out features —whether goals, contents or organisational principles— that noble armouries in different parts of Europe shared, to monitor how given armouries evolved over times, and even develop typologies of armouries.

The following seeks to contextualise noble Spanish armouries by shedding light on a noble German armory; that of the Dukes of Pfalz-Neuburg at Neuburg on the Danube, Germany. The main purpose of this study is to illustrate the kinds of information that an inventory of that armoury recorded and invite a reflection on what a noble armoury can represent.

Before examining the Neuburg armoury, it may be helpful to introduce a few facts regarding the duchy of Pfalz-Neuburg and its successive rulers. The duchy did not exist prior to 1505, when it was created that year as the result of an arbitration of Maximilian I (1459–1519), then King of the Romans, following the deaths of Rupert of Palatinate (1483–1504) and his wife, Elisabeth of Bavaria-Landshut (1478–1504). Created in favour of their two sons, Otto Henry (1502–1559) and Philip (1503–1548), the principality was a small compensation for the territories posthumously lost by the parents as result of the Landshut War of Succession (1502–1505), which ended with the defeat of their party.

Fredrick II of the Palatinate (1482–1556) administered the duchy on behalf of Otto Henry and Philip, his nephews, until 1522, when the two brothers were declared old enough to rule personally. They ruled jointly until 1535, when Otto Henry became sole ruler of the duchy. In 1557 Otto Henry resigned his title of Duke of Pfalz-Neuburg after he had become prince elector the Palatinate, in favour of his cousin, Wolfgang, (1526–1569), Duke of Pfalz-Zweibrücken. Following Wolfgang's death, Wolfgang's eldest son Philip Ludwig (1547–1614) became Duke of Pfalz-Neuburg. As a result of his marriage to Anna of Jülich-Kleve-Berg (1552–1632), their son Wolfgang Wilhelm (1578–1643), became Duke of Jülich-Berg in 1614 when he also succeeded his father as Duke of Pfalz-Neuburg. The next Dukes of Pfalz-Neuburg were all direct descendants until 1742, when Karl III Philip (1661–1742) died without a male issue, and his cousin, Karl Theodor (1724–1799), Duke of Pfalz-Sulzbach, succeeded him as Duke of Pfalz-Neuburg. As in turn, Karl Theodor died without a legitimate issue, the principality was inherited by his cousin, Maximilian I Joseph (1756–1825), Duke of Pfalz-Zweibrücken, who became prince elector of Bavaria in 1799 and king of Bavaria in 1805. The duchy of Pfalz-Neuburg was declared a Bavarian province and in 1808 it was formally abolished, and its territories were incorporated into the Upper Danube Circle of the recently created kingdom.

The town of Neuburg was the administrative capital of the duchy of Pfalz-Neuburg. The old-fashioned medieval castle there, which Otto Henry and Wolfgang transformed into a Renaissance palace, served as the Dukes of Pfalz-Neuburg's chief residence until Wolfgang Wilhelm's rule. Because Wolfgang

Wilhelm was also Duke of Jülich-Berg, he had a competing residence in Düsseldorf, which was the administrative capital of the duchy of Jülich-Berg. Over the course of his reign, Düsseldorf eventually became his preferred place of residence. Subsequent Dukes of Pfalz-Neuburg spent occasional time in Neuburg, but as many of them ruled over additional, sometimes more prestigious lands, their chief residences generally were elsewhere: Düsseldorf, Heidelberg, Mannheim and Munich.

Otto Henry, who played such an important part in turning the Castle of Neuburg into a suitable residence, likely initiated the practice of keeping his personal armours and weapons there. Even though he became prince elector the Palatinate in 1556 and moved his chief residence to Heidelberg as a result, the fact that many armours and weapons that had personally belonged to him were kept in Neuburg Castle in 1628 suggests that he had left his armoury in place in Neuburg when Wolfgang succeeded him in 1557. The available evidence, which lists armours that had personally belonged to Wolfgang, Philip Ludwig and Wolfgang Wilhelm in Neuburg Castle in 1628 suggests that Wolfgang and the next two rulers of Pfalz-Neuburg similarly kept their armouries in there and that they had been successively incorporated into one another[1]. In the late eighteenth century, the castle's armoury, which was sometimes referred to as a cabinet of antiques (*antiquitaeten Cammer*), was reportedly located in the west wing[2]. It is not clear whether it was precisely located. The contents of the castle armoury were seized by the French when they occupied Neuburg in 1800. What they did not take was removed to Munich the following year and for the most part it was either gifted or sold shortly thereafter[3].

The Neuburg Castle armoury provides a prime case study of a noble European armoury. Except when they also ruled over Bavaria, the resources of the Dukes of Pfalz-Neuburg paled in comparison with those of the Empire's prince electors and other powerful territorial rulers such as the Archdukes of Tyrol. In that regard, they belonged to a larger circle of distinguished but not nearly as wealthy territorial rulers, which included the various branches of the landgraves of Hessen (such as Hessen-Darmstadt, Hessen-Cassel), of the Dukes of Saxony (such as Saxony-Coburg, Saxony-Weimar), and of the margraves of Baden (such as Baden-Baden, Baden-Durlach), to name a few. The Neuburg armoury is also

1 The evidence is provided by an inventory of the Neuburg armory that was compiled in 1628.

2 Riedel, 1796: 16. Rieder, 1886: 213.

3 Reitzenstein, 1940: 41–43.

of interest because a documentation for studying it has survived in the form of three inventories and additional materials; because its contents were dispersed, which was the fate of most other noble armouries; because objects that were part of it survive and may be identified; and because the Neuburg armory has already attracted a fair amount of scholarly attention, and for that reason it is reasonably well understood.

Although important visitors and guests of the Dukes of Pfalz-Neuburg were sometimes shown the castle's armoury, few have left written accounts and those that are known are laconic[4]. No pictorial depictions of the armoury's interior appear to exist, and no registers of accounts recording payments for the upkeep of the armoury and its contents are known either. The three known inventories of the Neuburg armoury are thus the chief documentary sources. They date from 1628, 1654 and 1750, respectively. Two of them have been published[5]. However, as none was transcribed in full, it would be highly desirable for all three inventories to be published in their entirety and that the same transcription and editorial standards be used to facilitate comparisons among them. For the time being, the documentary basis of the present inquiry is provided by the partial published transcription of the 1628 inventory, the earliest of the three.

The 1628 inventory is kept in the State Archives of Bavaria in Munich. There can be little doubt that earlier inventories must have existed, but if they have survived their whereabouts are uncertain. The reasons for compiling an inventory of the Neuburg armoury in 1628 are not clear. As there are no significant events in the life of Wolfgang Wilhelm that year that would have required it, it is possible that a mundane development, such as the appointment of a new keeper of the armoury, prompted the document's compilation.

---

4 For example, see the entry in the diary of Christian the Younger (1599–1656), Prince of Anhalt, following his visit to Neuburg, in 1622, in Krause, 1858: 72: "Die Rüstkammer so in etliche Zimmer verteilet ist wol versehen, unter anderen ist Pfalzgraff Otto Heinrichs und Pfalzgraff Philips (welcher die Wien Stadt wieder den Solymanum so wol defendiert) rüstungen, zu notiren."

5 The 1628 inventory is kept in the Bayerisches Hauptstaatsarchiv München, Dreissigjähriger Krieg XXIV Fasc. 222 and was partly published in anonymous 1823 and Reitzenstein 1973a. The 1654 inventory is kept in the same archive, I. Abteilung, Pfalz-Neuburg Akten Nr. 128, and unpublished. The 1750 inventory is kept in the same archive, III. Abteilung, Geheimes Hausarchiv Nr. 2347 III, and was partly published in anonymous 1873. A partial copy of it is kept in the Staatsarchiv Neuburg, AB 1909 Nr. 14.

As published, the inventory appears to be organised into thirty-one sections. The headers of those sections are as follows (author's translation of the published German text)[6]:

(1) *Allerley Feldküris und Trabharnisch* [All kinds of field armours and light cavalry armours]
(2) *Gemeine Trabharnische* [Ordinary light cavalry armours]
(3) *Nun folgen allerlei Stech und Turnier Kürіß* [Now follow all kinds of jousting and tournament armours]
(4) *Harnisch So weiland herz. Philippsen Pfalgr geweßen, und zu Neuburg gefunden worden* [Armour that once was Duke Philip's, and found in Neuburg]
(5) *Harnisch So weilandt herz. Ottheinrichs gewesen* [Armour that once was Duke Ottheinrich's]
(6) *Nun folgen die Stechzeug* [Now follow the jousting armours]
(7) *Rennzeug und anders, so alles an der Wand gegen den Schloß Traid Kasten hangt* [Armours for the joust of war and else, which are all hung on the wall towards the castle's granary]
(8) *Nun folgen auch schwarze Harnisch* [Now also follow the black armours]
(9) *Noch ist vorhanden, so uf der neuen Rüstkammer anejzo auch hangt* [Still remaining, presently also hanging in the new armoury]
(10) *Sättel, so zu Neuburg gefunden worden* [Saddles found in Neuburg]
(11) *Sporn und Rennbis* [Spurs and horse bits]
(12) *Seittenwöhren* [Side arms]
(13) *Uf der alten Rüstkammer ist noch vorhanden* [Still remaining in the old armoury]
(14) *Banzerwerk so zu Neuburg gefunden* [Works of mail armour, found in Neuburg]
(15) *Banzerwerk, so auch in diesem Kämmerlein* [Works of mail armour, also in this small room]
(16) *Banzerwerk zu den Sieben Trabharnischen* [Mail armour belonging to the seven light cavalry armours]
(17) *Sattel, so an der Wand gegen dem Schloß in diesem Kämmerlein hangen* [Saddles hung in this small room on the wall towards the castle]
(18) *Sturmhauben* [Burgonets]

6 I am indebted to Fabian Brenker, Curator in the Imperial Armoury of the Kunsthistorisches Museum Vienna, for his thoughts and advice on the proposed translations.

(19) *Feldzeichen und Fahnen* [Standards and flags]
(20) *Fausthämmer* [War hammers]
(21) *Wöhren, so aus Heidlberg kammen* [Weapons that came from Heidelberg]
(22) *Wöhern, so zu Neuburg gefunden* [Weapons found in Neuburg]
(23) *Wöhren und Schwerdter und Dolchen, so Pfalzgraf Wolfg. vor den Zug in Frankreich hat machen lassen* [Weapons, swords and daggers made at the orders of Pfalzgraf Wolfgang before the campaign in France]
(24) *Wöhren, so herz. Philipp Ludwig aufs Rüsthaus geliefert hat* [Weapons delivered to the armoury by Duke Philip]
(25) *Scheflin oder Spieß, so vor diesem die Spießjungen geführt* [Javelins or spears carried by the spear-pages in front of Duke Philip]
(26) *Nun folgen knöbel, Schwein und Federspies* [Now follow the hunting spears]
(27) *Sattel, so auch uf diesem Boden hangen* [Saddles, which are also hung in this room]
(28) *Nachfolgende Köcher sind an den Brettern gegen der Thonau* [The following quivers are mounted on the boards towards the Danube]
(29) *Verschiedene Stück zu den alten Stechzeugen gehörig* [Various items that belong to the jousting armours]
(30) *Feuerspieß oder Schürßer* [Incendiary spears or Schürtzer]
(31) *Noch ist vorhanden, so oben an den Brettern dieses Bodens liegen oder hangen* [Still remaining, lying on or hung above the boards of this floor].

The published transcription lists a total of 172 entries under those various headers.

The entries provide summary descriptions of the inventoried objects. Modern museum cataloguing standards typically require the record of an object to include a name for the object, information about the materials the object is made of, its maker(s), place of manufacture, date of manufacture, physical dimensions, a unique registration number (such as inventory or accession number) and an account of how it entered the collection (credit line). In the field of armour and arms it is common for the object name to give information about the object's function (such as field, tournament, ceremonial, sporting, etc.) and to name the person for whom the object was made or to whom it was presented whenever the information is available. It is generally also customary for a modern object record to list all of the components that make up the object (such as the various elements of an armour), and to indicate where the object is stored or exhibited, by whom it was owned prior to its acquisition, where it was exhibited and where it was published. In the 1628 inventory, some of this information is also recorded, however, not in a consistent fashion. Instead, some entries give more weight to certain types of information. The location of the objects, for example, is sometimes carefully recorded and at times not given at all.

Rather than following consistent descriptive standards, the 1628 inventory relies on a variety of features, some of which are not visible, to identify the objects. The features that were deemed significant for the purpose include the colour of metals and textile elements:

> *Ein eisenfarber Feldküris zu Fuß mit vergolten Geschmeidt und Nägl* [An iron-color field armor [for use] on foot[7] with gilt fittings and rivets].
> *Ein ganz eisenfarber Feldküriß mit vergoltem Geschmaid und schwarzen Sammet eingefaßt* [A complete iron-color field armor with gilt fittings and lined with black velvet].

Singularities of form, for example the design of a helmet's face defense, are sometimes noted:

> *Ein glatter Planker Küriß mit weis Neglen und Geschmeid, das Viesier am Helmlin mit einem Löwenkopf, auf dem Helmlin zween Flügel mit schiefen getrieben* [A smooth bright full armor with white rivets and fittings, the visor on the helmet with a lion's head, on the helmet two wings embossed with slants][Fig. 1].

---

7 The German term "Küriß" is generally translated into "cavalry armour" in English; however, as the inventory mentions a "Feldküris zu Fuß", it seems that the term was used in the inventory to designate a full armour (one that encloses the body from head to toes) rather than a cavalry. Therefore, in this essay the term will be translated to "full armour" instead. "Feldküriß" will be translated into "field armour," an English term that always designate a full armour.

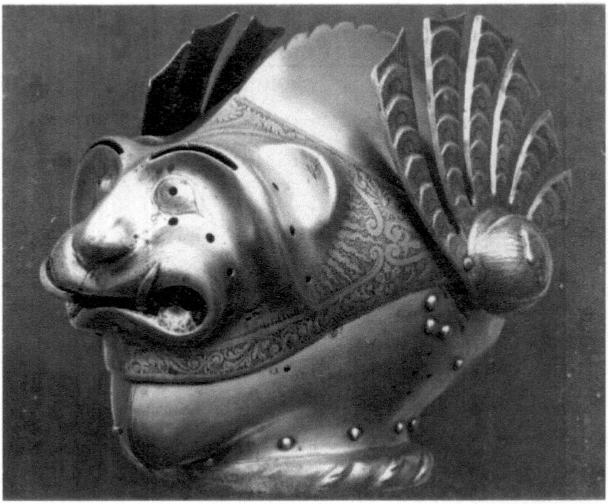

Fig. 1. *The armet of an armour of Duke Wolfgang of Pfalz-Zweibrücken and of Pfalz-Neuburg*. ca. 1530–1540. Musée de l'Armée, Paris (Archives of the Department of Arms and Armor, The Metropolitan Museum of Art).

Not just colour and form, but also the decoration of the objects sometimes receives special attention. The entries sometimes call out techniques that were used to execute the ornamentation:

Ein weißer stehlner glatter Küriß samt Zugehör auf den Orthen getrieben und geezt, desgleichen auf den Spanneröhl, Achseln, Armen und Kniebuggeln getrieben [A white steel smooth full armor with its accessories, embossed and etched at the edges, the spaulders, pauldrons, vambraces and poleyns likewise embossed].

The entries can also rely on the type of ornament or the presence of distinctive emblems to facilitate the identification of the objects:

Ein blau illuminirter [Feldküriss] mit vergolten Blumen [A blue illuminated [cavalry armor] with gilt flowers]
Ein blauer, Spanischer von Gold allerlei Gesprengwerk vornen mit einem spanischen Kreuz, Kronen und Szepter geezter Küriß [A blued Spanish etched full armor [decorated] with all sorts of gold trusswork, in the front with a Spanish cross, crown, and scepter].

Inscriptions, including abbreviated mottos, are equally relied upon:

Ein eiserner Fausthammer, vornen mit einer Schneiden, diese Buchtstaben M. D. Z. darauf, seindt vergolt und geezt [An iron hand hammer with a cutting [head] at the front, the initials M. D. Z. on it, which are gilded and etched].

The inventory's entries do not solely provide information on the intrinsic qualities of the objects. Many include information on how they are mounted and exhibited.

> *Ein weißer glatter stehlener Kürißer […] uf ein hölzenen Roß mit aller seiner zugehören. Ein violfarben sammeten Schurz mit ein geblümten silbern Stuck verbrembt hat der Küriser an* [A white smooth steel armored figure […] on a wood horse, with all its accessories. The armored cavalier has a skirt of violet-colored velvet trimmed with a floral silver piece [i.e., ornament]].

Some descriptions rely on specific accessories held or carried by armoured figures to facilitate their identification:

> *Ein kleiner planker Küriß mit meßenen Geschmaidt, so Landgraf Wilhelm zu Hessen, als mein gstr Fürst u. Herr Pfalzgr Philipp Ludwig uf der Hochzeit im Niederland gewesen, meinem gstn Fürsten und Herrn Pfalzgr Wolfgang Wilhelm verehrt. Hat in der Hand einen Spieß mit gebüekhten Schaft, und mit schwarzen Sammet und seidnen Franzen oben gefaßt. An der Seiten ein Rappier mit sameter Schaiden und silbern Ohrband, nebst einem Dölchen; die Schaiden mit silbern Dräthen umwunden* [A small full armor for the tilt with brass fittings, which Landgrave Wilhelm of Hesse presented to my severe Prince and Lord, Count Palatine Wolfgang Wilhelm, as my severe Prince and Lord, Count Palatine Philip Ludwig, attended the wedding in the Netherlands. [The figure] has a spear with a stooped shaft and [which is] covered with black velvet and silk fringes. At the side [the figure carries] a rapier with a velvet scabbard and silver chape, in addition to a dagger, the scabbards wound with silver wire].

The descriptions sometimes extend beyond the characteristics of individual objects and focus on their relationship to other objects that are kept separately instead. The exchange and reinforcing elements of an armour garniture that are not assembled with the others, but stored and shown separately, like extra helmets or reinforcing breastplates, for example, are identified as belonging to the core elements:

> *10 weiß gereifte Harnisch […] Unter den Zugehör zu diesen Harnischen ist 1) eine weiße Brust, auf den Orten geezt und vergolt in der Mitte U. L. Frauen mit dem Kindlein. 2) Eine schwarze Brust mit dem nämlichen Bild von getriebner Arbeit. 3) Ein dergleichen gleichfalls mit unser L. Frauen Bildniß versehen* [10 white fluted armors […] Among the accessories of these armors there is a white breastplate, etched and gilt in places with our Dear Lady with the infant in the middle; a black breastplate with the previously named image worked in embossing; a likewise [breastplate that is] similarly furnished with the portrait of our Dear Lady].

Some descriptions include information on the location of object to facilitate their identification. The locations can be an area of the armoury, for example, the middle of it, the walls or even the floor:

Nachfolgende Köcher sind an den Brettern gegen der Thonau [The following quivers are on the boards on the side of the Danube [river]].

The locations can also refer to how a given object fits into an arrangement of like objects, for example, its order in a sequence of objects presented in a row:

Ott Heinrichs Küriß zu Neuburg gefunden worden, der achte in der Raien [The full armor of Otto Henry was found, the eighth in the row [or line]].

In addition, the inventory entries often provide information on objects by naming the persons for whom they were acquired made, to whom they were presented as gifts or who deposited them into the armory. For example, the inventory lists several armours personally belonged to the successive Dukes of Pfalz-Neuburg; fourteen armors of Otto Henry, three of Philip, two of Wolfgang, three by Philip Ludwig and nine by Wolfgang Wilhelm. The inventory mentions additional historical owners, including "Duke Friedrich" (possibly Count Palatine Friedrich II (1482–1556)) and "Duke Georg" (i.e., George of Bavaria-Landshut (1455–1503)).

The entries often specify what functions the objects were intended to serve. They list armours for use in the battlefield:

Ein ganz blauer Feldküriß, hinden und vorn beschossen [A fully blued field armor, shot in the front and back].

They similarly point out that some armours were designed for specific forms of the tournament:

Ein eisenfarber glatter Küriß sambt aller Zugehörung [...] Dieser Küriß ist zum Thurnier und Balgen gemacht [An iron-color smooth full armor with all accessories [...] This full armor is made for the tourney and the tilt].

In addition to objects for battle and the tournament, some entries mention weapons for the hunt:

Ein Sauschwerdt mit vergoldten meswerk [A sow-sword with gilt brass work].

The origin of the objects is another aspect that is given significant attention. Their respective mode of acquisition —including purchases, gifts, trophies and simply found in situ— are often stated. Some were made to order:

Pfalzgr. Philip Ludwigs eisenfarber glatter Küriß. Den haben Se. fürstl. Gnaden Ao. 1567 auf Dero Leib zu Lauingen von neuen schlagen lassen, und ist daß uf diesen Raien der 5te [Iron-color smooth full armor of Count Palatine Philip Ludwig, which his princely grace had made from scratch in Lauingen to [the measure of] his body in the year 1567, and in this row [or line] it is the fifth one].

Other objects were purchased:

> *6 blaue Feldkürifß von Christ. Bolzen in Augsburg erkauft* [6 blue field armors bought from Christ. Bolzen in Augsburg].

Gifts also loom large among the various inventoried objects:

> *Ein klein Harnisch, so Pflzgr Johannes meinem gnädigsten Fürsten Wolfgang Wilhelm verehrt.* [A small armor, which Count Palatinate Johann presented to my most gracious prince Wolfgang Wilhelm].

The entries indicate that some objects were probably acquired as trophies of war as the result of military campaigns:

> *Ein Janitscharen Heeckhlein, so Wolf Dietrich aus Ungarn gebracht, welches Ihr frst. Dchlt Seufridt von Gollnitsch verehrt* [A janissary's small hammer (?), which Wolf Dietrich brought back from Hungary, [and] which was presented to his Serene Highness by Seufridt von Gollnitsch].

The entries go as far as to record the places where commissioned armors were made (Augsburg, Lauingen (Bavaria), Neuburg, Nuremberg, Onolzbach [i.e., Ansbach], Stuttgart and "Vislingen" [probably Vlissingen, Flushing in English, in Zeeland, unless it is Villingen-Schwenningen in Baden-Württemberg]) as well as the names of armorers who crafted them (Anton Peffenhauser, Anthoni Müller, Hans Roth and Zacharias Bitterlein, Melchior Schwab and Hans Schultheiss). Similarly, the entries often mention the places in which purchased armours were acquired (Augsburg, Cologne and Stuttgart) as well as information about the sellers (such as "an Italian"; Vice-chancellor (*Vizkanzler*) Dr. Bernhard Mosmüller (d. after 1615); the wife of Friedrich Georg Schrott (d. after 1610), clerk and administrator (*Landgerichtsschreiber undLandrichteramts-Verwalter*) of the provincial tribunal at Monheim on the Rhine (North Rhine-Westphalia); and the armourers Anthoni Müller, Baltaß Visch[er] and Hans Schultheiss.)

The entries are similarly specific about armours and other objects that were presented as gifts. Recorded donors of armours include territorial rulers such as Duke Wilhelm of Jülich, Cleve and Berg (1516–1592); Margrave Carl II von Baden-Durlach (1529–1577); Landgrave Wilhelm IV of Hesse-Cassel (1532–1592) and Duke Johann Friedrich von Pfalz-Neuburg-Hilpoltstein (1587–1644). Other listed donors include a number of fief-holders. It appears to have been a tradition upon the confirmation or renewal of a fief for the beneficiary to mark the occasion by presenting the liege lord (i.e., the ruling Duke of Pfalz-Neuburg and Duke of Jülich-Berg) with the gift of an armour. The recorded donors include the abbot of the Abbey of Keisheim (now Markt Keisheim, Bavaria); the treasurer

(*Pfennigmeister*) of Pfalz-Neuburg Georg von Zöschlin (d. after 1642); the von Strahlenfels family; the Stieber von Büllenach family; the representative of the provincial estates and military commander (Obrist) of the city of Nuremberg Johann von Leubelfing zu Gansheim (1578–1648); Baptist Großschedl (probably the noble alchemist and esoteric author Johann Baptist Großschedl von Aicha (1577–1630)) and the provincial administrator and judge of Parkstein and Weiden (both Bavaria), Georg Peter von Satzenhoven, who was also the military commander of the fortress Parkstein. Examples of armours presented as gifts by fief-holders include:

> *1622 liefert Lehenprobst einen schwarzen Feldküriß mit schwarzem Sammet eingefaßt, welchen Georg Zöschlein Pfenningmeister wegen Naderholz zu Lehen gab.* [In 1622 the fief provost [i.e., a Pfalz-Neuburg official] delivered a black field armor lined with black velvet, which the treasurer Georg Zöschlein donated in relation to the fief of Naderholz]. *Ein mit verzinntem Geschmeid versehener Feldküriß, welchen Sebast. v. Röttenhann der Vest und Guet Luzmannstein zum Lehen gegeben den 2. Jul. 1624.* [A field armor furnished with tinned fittings, which Sebastian von Röttenhann has donated on July 2, 1624 [in relation to] the Lutzmannstein fortress and property].

The 1628 inventory thus provides historical information about the objects. Such information is not limited to the circumstances for their acquisition, but it also encompasses information about their past use with references to people, places, dates and specific events.

Military campaigns for which objects were made or used are mentioned. They include Duke Philip's defense of the duchy of Württemberg, of which he was the imperial administrator, against the joint forces of Landgrave Philip I of Hesse-Kassel (1504–1567) and Duke Ulrich of Württemberg (1487–1550) at the battle near Lauffen in 1534; Duke Wolfgang and Duke Philip Ludwig's military campaign against the Ottoman Turks in Hungary in 1566 and Duke Wolfgang's military campaign in France in support of the Huguenot party there in 1568 and 1569.

Examples of inventoried objects relating to the 1566 military campaign in Hungary include:

> *Pfalzgr. Philip Ludwigs eisenfarber glatter Küriß […] Der Gaul, worauf der Kürißer gestelt ist, hat einen ganz schwarz sammethen Sattel […] Ueber diesen Gaul ein gelb wullen Deck von gutem Tuch mit einem burgundischen rothen Kreuz, und an Orthen mit schwarzen Sahmet eingefaßt, so Ao. 1566 in Ungarn gemacht worden* [An iron-color smooth full armor of Count Palatine Philip Ludwig […] the horse onto which the armored rider is installed has a completely black velvet[-covered] saddle […] On this horse [there is] a yellow woolen trapper of good cloth with a red Burgundian cross [i.e., the raguly cross of Burgundy used by imperial forces] and [which is] bordered in places with black velvet; which was made in Hungary in the year 1566].

7 dergleichen [i.e., Trabharnische], so Philipp Ludwig in Ungarn gehabt [7 comparable light cavalry armors, which Philip Ludwig had [by him] in Hungary].
Ein Eisenfarber Trabharnisch, von Herz. Philipp Ludwig gebraucht 1566 [An iron-color light cavalry armor, which Duke Philip Ludwig used in 1566].
Ein Banzer schurz, so Philipp Ludwig in Ungarn geführt Ao. 66. [A mail skirt, which Philipp Ludwig wore in Hungary in [15]66].
Als 7 Schurz und Ermel. NB. Diese hat Phil. Ludw. in Ungarn Ao. 66 gekauft [7 mail skirts and arm sleeves. N.B. Philip Ludwig bought these in Hungary in the year [15]66].
Ein Schwerdt [...] Ist Pfgr Wolfgang Ao. 66 in Ungarn geschenkt worden [A sword [...] which was presented as a gift to Count Palatine Wolfgang in Hungary in [15]66].
2 Reit schwerdter [...] habens Ao. 66 die Knecht in Ungarn geführt [2 riding swords [...] [which] were carried by the soldiers in Hungary in [15]66].
Ein schöner länger Landsknecht Spieß, den Pfgf Philipp in Ungarn gebraucht [An attractive long Landsknecht pike, which Count Palatine Philip used in Hungary].
Ein Reutter Fahnen, blau und weis geweckt gleichfalls von Pfgr Philipp in Ungarn gebraucht [A cavalry flag, with blue and white rhombuses, likewise used by Count Palatine Philip in Hungary].

Examples of objects relating to the military campaign in France in 1568 and 1569 include:

2 dergleichen [Trabharnische], so Herz. Wolfg. Wilh. in Frankreich geführt [2 like light cavalry armors, which Duke Wolfgang Wilhelm wore in France].
Ein blau französische Rennfahne mit weißem Kreuz [A blue French guidon flag featuring a white cross].

In addition to providing information on the acquisition of objects, the inventory sometimes records their subsequent movement. Specifically, it identifies a total of 108 armours that came from the Netherlands (i.e., the duchy of Jülich-Berg), and a total of 155 armours no longer kept in the Neuburg armoury, which were sent to the Netherlands (i.e., Jülich-Berg). As mentioned before, following the death of Duke Philip Ludwig in 1614, his eldest son, Wolfgang Wilhelm became Duke of both Pfalz-Neuburg and of Jülich-Berg. As a result, Wolfgang Wilhelm had two chief residences, one in Neuburg and the other in Düsseldorf. Wolfgang Wilhelm spent his time between both places and over the latter part of his reign mostly in the latter. Some armours in Düsseldorf followed him to Neuburg; others were sent from Neuburg to Düsseldorf for similar reasons. In Southern Germany the areas of the German-speaking lands that were adjacent to the Netherlands (such as the duchy of Jülich-Berg) were commonly referred to as being part same region[8].

---

8 See Reitzenstein, 1940: 45.

In addition to preserving the memory of physical transfers, the inventory provides information about the upkeep of some objects. For example, it lists some armours that were supplemented or restored with new pieces:

> *12 weiße glatte [Küriße] mit geetzen Strichen […] Zu diesen Kürißen hat herz. Philip Ludwig 1575 manches machen lassen* [12 white smooth full armors with etched bands […] To these full armors Duke Philip Ludwig had some [work] done in 1575].
> *Ein weißer altfränkischer Küriß […] An diesen Küriß hat Pfalzgraf Wolfgang 1563 einige Stück machen lassen in Augsburg* [A white 'old-Franconian' [i.e., antique, probably late-Gothic-style] cavalry armor […] Of this armor Count Palatine Wolfgang had a few elements made in Augsburg in 1563].

If the 1628 inventory was originally compiled to keep track of what was being kept in the Neuburg armoury, and probably hold its custodians accountable for its contents, it was subsequently put to other uses. In 1823, it was relied upon to preserve the memory of the armoury's former existence and contents after it was emptied out and dispersed in 1800. The inventory also provided the documentary basis for rediscovering objects that were part of it, a task that was a focus of arms and armour scholars such as Hans Stöcklein (1874–1936), who was chief conservator of the Bavarian Army Museum from 1920 until 1925, when he became its director, and Alexander von Reitzenstein (1904–1986), who similarly was chief conservator of the Bavarian Army Museum from 1936 until 1965, and then its director[9].

Beyond the help that it provided to recognise objects that had once been in the Neuburg armory, the 1628 inventory was put to use to better understand their significance.As was noted previously, the inventory documents original features and alterations of objects; provides a basis for attributing and dating them; rediscovering their historical ownership, functions, and uses; and their place among larger groups objects and narratives.Beyond its value for scientific purposes, the inventory provided a moral if not legal foundation to establish title and reclaiming objects. It was the basis for Reitzenstein to have objects kept in the Musée de l'Armée, Paris seized during World War II, and shipped to Germany (most were seized by allied forces at the end of the war and brought back to Paris).

The inventory is also an invaluable source for studying the history of collecting antique arms and armour in the modern era, from the art market and

---

9 See, for example, Reitzenstein, 1940; 1964; 1973a, esp. n. 2; 1973b. I am grateful to Stuart W. Pyhrr, Curator Emeritus, former member of The Metropolitan Museum of Art's Arms and Armor staff, for bringing several of those publications to my attention.

collectors to the institutional repositories in which some items from Neuburg found a home. This group includes the Musée de l'Armée, Paris; the Bayerisches Nationalmuseum, Munich; the Armeria Reale, Turin; the Wallace Collection, London; the British Museum, London; the State Hermitage Museum, Saint Petersburg; the Metropolitan Museum of Art, New York and the Musée royal de l'Armée et d'Histoire militaire, Brussels.

The inventory can additionally be a resource to better understand the history, organisation, management and function of other historical armouries such as the armoury of the Dukes of Jülich-Berg and Pfalz-Neuburg in Düsseldorf since some objects kept in it were transferred to Neuburg and vice versa; the arsenals of Munich and Strasbourg, to which some objects from the Neuburg armory were moved in the early 1800s and other noble German and European armouries, including of course noble Spanish armouries, by providing a basis for comparisons.

In conclusion, this overview of the armoury of the Dukes of Pfalz-Neuburg, as documented by the inventory of 1628, provides an example of the many things that noble armouries can be and represent. A noble armoury can be a property. It can be a physical space or structure as well, intended to house specific types of items (in this instance first and foremost armour, edged and shafted weapons, equestrian equipment; and not firearms which long belonged in a gun cabinet (*Gewehrkammer*), a different kind of repository). It can represent an administrative unit of a court structure (usually a branch of the stables, which are under the responsibility of the master of the stables (*Stallmeister*)). A noble armoury can be as a venue for the purposes of self-representation. It can help legitimise its noble owners by providing material proof of the antiquity or distinction of their house, and by preserving the memory of their respective physical appearance, names and deeds, and their pedigree and matrimonial alliances. It can also fulfill the ambitions of others, by legitimising the donors of objects, whose associations with the armoury's owners is manifested by their gifts. A noble armoury can be turned into a source of objects that victors take as trophies; a source of objects that fuel the art market and collecting (and by the same token it can become a provenance that lend luster to collectors' possessions)[10]; a stake in the historical renegotiation of power (for example as a symbol of the successful invasion of France by the Nazis, and as symbol of their ultimate demise when the Allies brought the objects that they had removed back to France); and a resource for the scholarly study of arms and armour, by providing useful information on

10 For example, see Reitzenstein, 1973a: n. 13.

points such as terminology, authorship, typology and style. It is hoped that as the study of other noble European armouries progresses some of those themes will resonate with the patterns that will emerge.

BIBLIOGRAPHY

Anonymous (1823): "Auszuege aus dem Inventarium ueber das fuerstliche Zeug- und Ruesthaus zu Neuburg, vom Jahre 1618". In: *Wochen-Blatt der koeniglich Baierischen Stadt Neuburg* 27 (July 5, 1823): p. 107; 29 (July 19, 1823), *Beilage*: unpaginated; 30 (July 19, 1823): p. 119; 36 (September 6, 1823): pp. 143–144; 45 (November 8, 1823): pp. 179–180; 46 (November 15, 1823): p. 183; 47 (November 22, 1823): pp. 187–188; 48 (November 29, 1823): pp. 191–192; 49 (December 6, 1823): pp. 194–195.

Anonymous (1873): "Inventarium über das Churfürstl. Zeug- und Rüsthaus, so nach Absterben Weil. Ihrer chrfüsrtl. Drctl. Carl Philipp Pfalzgrafens anno 1750 verfaßtworden". In: *Collektaneen-Blatt für die Geschichte Bayerns, insbesondere für die Geschichte der Stadt Neuburg a. d. D., und des ehemaligen Herzogthums Bayerns* 37, pp. 80–110.

Krause, G. (ed.) (1858): *Tagebuch Christians des Jüngeren, Fürst zu Anhalt.* Leipzig, Verlag der Dyk'schen Buchhandlung.

Reitzenstein, Alexander von (1940): "Die Harnische der Neuburger Rüstkammer". In: *Zeitschrift für historische Waffen- und Kostümkunde* n.s. 7: 2, pp. 41–51.

Reitzenstein, Alexander von (1964): "Der Ringlersche Harnisch des Pfalzgrafen Ottheinrich". In: *Anzeiger des germanischen Nationalmuseums*, pp. 44–56.

Reitzenstein, Alexander von (1965): "Eine Harnischbrust des Hans Ringler im Britischen Museum". In: *Waffen- und Kostümkunde*, pp. 122–123.

Reitzenstein, Alexander von (1973a): "Die Harnischkammer des Neuburger Schlosses im Jahre 1628". In: *Waffen- und Kostümkunde*, pp. 146–58.

Reitzenstein, Alexander von (1973b): "Ein Harnisch Valentin Siebenbürgers in französischem Museumsbesitz". In: *Waffen- und Kostümkunde*, pp. 99–108.

Riedel, Adrien von (1796): *Reise Atlas von Bajern oder Geographisch-geometrische Darstellung aller bajrischen Haupt- und Landstraßen mit den daranliegenden Ortschaften und gegenden nebst kurzen Beschreibungen alles dessen, was auf und an einer jeden der gezeichneten Strassen für den reisenden merkwürdig seyn kann.* Munich.

Rieder, Otto (1886): "Das kgl. Kreisarchiv Neuburg an der Donau und seine Vorläufer seit 1785". In: *Archivalische Zeitschrift* 11, pp. 199–225.

António Conduto Oliveira[1]

Universidade de Coimbra

THE ARMOURY OF TEODÓSIO I, 5TH DUKE OF BRAGANÇA

ARMOURIES IN LATE MEDIEVAL AND EARLY MODERN PORTUGAL: AN OVERVIEW

Portugal is not known for its collections of arms and armour. Whilst the reasons for this are manifold, most can be traced back to the kingdom's military organisation in the late Middle Ages and up to the late sixteenth century.

A substantial portion of late medieval and early modern armouries are the result of stockpiling due to the necessities of war and chivalric pursuits, such as jousting, tournaments, or parades. These accumulations of weapons usually occurred in systems in which the Crown heavily relied on a number of its noblemen to provide their own hosts in times of war, in exchange for lands, power and revenue. The ability to own weapons and armour, not just for oneself but for one's own men, was inextricably tied to social standing (and one's defence of it, when it came to internal conflict within the upper echelons of society[2]). In these systems, both the Crown and the nobility were required to amass vast quantities of weaponry, some of which would be passed down and added to across generations, and go on to form the arsenals of the state (in the first case) or the initial core of early modern armouries (in the second). Several Iberian collections of arms and armour, including many examples in the present volume, began precisely thus.

The situation was somewhat different in Portugal. On the one hand, the kingdom was ever too small to sustain several large noble houses with their own private hosts, a fact which severely curtailed the Portuguese nobility in terms

---

1 PhD candidate at the Faculty of Arts and Humanities of the University of Coimbra, member of the Centre for the History of Society and Culture (CHSC). This paper was funded by National Funds under an individual PhD grant from the Portuguese Foundation for Science and Technology (2021.04826.BD).

2 As it happened with some frequency in some regions of Europe, such as Castile and Aragon (Gallastegi/Rojas, 2021), England (see, for example, Hyams, 2003) or France (for example, Firnhaber-Baker, 2010: 37–76).

of individual agency compared to their European counterparts. On the other hand, it was also a kingdom which municipalities played a substantial role in expanding and consolidating by providing bodies of men and equipment for military operations. Indeed, the military system in late medieval Portugal was heavily reliant on men of the Third Estate, from its cities and towns, recruited and organised by agents of the Crown into a royal host, instead of armies largely patched together from the hosts of individual seigneurs[3]. These circumstances made it easier for the Crown to concentrate all of the power in itself instead of dispersing it across lesser polities[4], a process which further drew military power and importance away from the nobility and no doubt helped prevent large-scale seigneurial or private conflicts[5]. Additionally, the arms industry in fifteenth-century Portugal could be classed as an almost literal cottage industry, with a low-to-medium quality output and little high-quality armour, most of which had to be imported from abroad[6] at considerable expense. Lessened revenue streams, a somewhat reduced military relevance compared to other European kingdoms, and Portugal's location at the tail-end of commercial circuits all meant a diminished ability to hoard weapons, even for wealthier individuals, throughout the entire Late Middle Ages and up to the sixteenth century.

Taken together, these factors meant that most of the weapons in Portugal were either in the hands of its urban-dwelling population or stored in the Crown's arsenals[7], the most important of which was located in Lisbon. In the middle of the sixteenth century, Lisbon's royal arsenal is purported to have stored enough arms and armour in it to equip at least 40,000 men of foot and 3,000 of horse[8], making

---

3 On this topic, see for example, Gomes Martins, 2014: 23–138, Gouveia, 1998: 27–79, 90–98 and Gouveia, 2003: 192–200 for an overview of recruitment in Portugal from the thirteenth to the fifteenth century.

4 These factors, amongst many others, explain why the beginnings of the modern state in Portugal are usually drawn back to the thirteenth century. See, in this regard, a reasonable summary in Gonçalves de Freitas, 2011.

5 To the point where no substantial armed conflicts between seigneurs are recorded in Portugal for the entirety of the fourteenth, fifteenth and sixteenth century. One exception is the conflict between the first Duke of Bragança's and Prince D. Pedro in the events leading up to the battle of Alfarrobeira in 1449, but even this stemmed from a royal order and could therefore be argued to be an extension of the Crown's monopoly of violence (Baquero Moreno, 1973: 349–357, specifically).

6 Sousa Viterbo, 1920: 158. Oliveira Marques, 1987: 340.

7 Oliveira Marques, 1987: 351.

8 Numbers given for 1552, according to Cristóvão Rodrigues de Oliveira (Rodrigues de Oliveira, 1987: 171); the same exact figures given by Damião de Góis in 1554 (Góis,

it one of the largest arsenals in Europe at the time. There were also additional arsenals and armouries in the mainland, such as Porto, Barcarena or Santarém[9], as well as in the overseas territories of the nascent empire, such as Goa[10].

That is not to say that there were no private stores of arms and armour in Portugal. We know of certain small assemblies of pieces which could be considered forerunners to collections, or proto-collections[11], and certain individuals were expected to field a number of their own men equipped from their own stores[12], as indeed were the military orders[13]. These arsenals, in any case, don't seem to have formed the basis of any curated collection as such. The system—partly by accident, partly by design—greatly aided the Crown in ensuring its monopoly on sovereign authority, but dissuaded the Portuguese nobility from amassing large caches of weapons[14]. Its one singular exception was the ducal armoury of the Braganças, the object of the present study[15].

2002: 175). Substantial numbers, mostly supported by the available documentation: a 1568 quittance issued to Marco Dias, *almoxarife* (overseer or comptroller) of the arsenal, confirms his receipt of, amongst other things, 20.336 harquebuses, 95.974 javelins (of two different kinds), 10.539 spears (of at least four different types), 20.050 unhafted spearheads, and 12.901 pikes and demi-pikes—all precisely between 1552 and 1554 (Viterbo, 1908: [n. p.]).

9 On Barcarena, see Cardoso, João/Gomes, José.2005: pp. 24–27. The armouries of Porto and Santarém (Sousa Viterbo, 1920: 160–161) have yet to be researched, though in the case of Santarém there is ample documentation to be consulted (partly transcribed by Anselmo Braancamp Freire in several volumes of the *Archivo Historico Portuguez*).

10 See Lee de Jesus, 2006.

11 A good example of which was the collection of Gonçalo Gil Barbosa, who throughout his voyages amassed an assortment of European, "Moorish" and Indian pieces of arms, armour, and military accoutrements—at least 41, at the time of his death. See Pereira dos Reis, 2022: 208–2010.

12 Gouveia, 2001: 28.

13 Gouveia, 2001: 29.

14 An example of this is a muster of arms from the *fidalgos* of the town of Tavira, in southern Portugal, dating from around 1500 (Arquivo Nacional—Torre do Tombo, Lisbon (ANTT), Colecção de cartas, Núcleo Antigo 880, n.º 23; transcribed and under revision). In this list, most of the *fidalgos*, who belonged to well-respected and affluent families such as the Corte-Real, declare to have only enough pieces to equip themselves and between one and four attendants—with most of them only being able to equip two.

15 A brief overview of the armoury and its contents, titled "A Armaria na Casa de Bragança e a sua organização military: a resposta de D. Teodósio e seu pai aos desafios impostos pela revolução da pólvora", written by Vitor Rodrigues, was published in 2018 as part of the volume "De Todas as Partes do Mundo" (Hallet/Senos, 2018: 269–280), which

THE HOUSE OF BRAGANÇA

The history of the armoury of the Braganças is intimately tied to the history and fortunes of the ducal house itself. The duchy of Bragança was created in 1442 in favour of Afonso of Portugal, natural son of King João I. It was the third ducal title created in Portugal, and the only one, up to 1475, which hadn't been created for a prince of Portugal. Afonso of Portugal's holdings were already substantial before his investiture as Duke, with most of his patrimony made up of the estates and lands which Nuno Álvares Pereira, Constable of Portugal and the seventh Count of Barcelos, had endowed his daughter Beatriz, Afonso's wife[16]. Afonso's ducal title was, among other factors, a recognition of his power and influence in the kingdom, on par with that of any Portuguese prince. Afonso's elevation inaugurated a steady rise in power for the ducal house, which was sustained by his son and grandson, the Dukes Fernando I[17] and Fernando II[18]. By the time of Duke Fernando II, the ducal house of Bragança was by far and away the most powerful house in the kingdom.

However, this ascendancy was cut short in 1483, with Fernando II's execution on charges of treason against king João II. Though there seems to have been a confirmed plot to assassinate the king, and Fernando II might have been one of its ringleaders, it is likely João II seized the opportunity to have the Duke killed because of the clear threat the Braganças presented to the Crown[19]. All of the ducal lands and assets were seized and Fernando's son, Jaime, fled to Castile for safety.

Fernando II's death closed the first chapter in the family's history. After João II's death, his successor, Manuel I, perhaps deciding that the house of Bragança was simply too important to the stability of the kingdom, not only pardoned Jaime but reinstated him as a Duke, with all his attending lands and other titles, as well as designating him as his own heir, should he die childless[20]. Thus, the second chapter—one of renewal—began.

---

examines the post-mortem inventories of Teodósio's possessions in great detail. This paper is, in a sense, an expansion of that article.

16 Soares da Cunha, 1990: 79–97.

17 A summary of Fernando I's life can be read in Melo, 1944: 96–112. For his patrimonial acquisitions, see Soares da Cunha, 1990: 97–99.

18 Melo, 1944: 113–117. Soares da Cunha, 1990: 99–102.

19 Moreira, 2020: n.p.

20 Góis, 2014: 26–27. Soares da Cunha, 2014: 155.

Part of Jaime's process of reinvention of his house was its literal reconstruction in stone and mortar: a new and magnificent palace in the duchy's seat of Vila Viçosa, which began being built on the occasion of his wedding[21]—a space in which the armoury, or at least part of it, would come to be installed. Jaime became instrumental in the Crown's policy of conquests in Africa, which culminated in his nomination to lead the conquest of the Moroccan city of Azemmour in 1513[22] with a great host of men, many of them his own—arrayed in the new-fangled *companhias de ordenança*[23]—and outfitted with armour bought specifically for that campaign[24].

The Azemmour campaign could be said to have been Jaime's crowning achievement. Through his deeds, the house of Bragança had managed to overcome every limitation imposed on it by the late medieval socioeconomic, political and military system or structures of the kingdom of Portugal. As one of the armed agents of the Crown, it was allowed, even tasked with, amassing an arsenal of its own to act. And it was those arms and armour, both Jaime's and his men's, that would go on to form the core of the Bragança's armoury at Vila Viçosa. Sizeable though this initial arsenal might have been, however, it was but a fragment of what we know was stored in the armoury in the mid-sixteenth century. To glimpse the armoury at its peak, we must look to the extensive inventories made in 1563 after the death of Teodósio, Jaime's son and the 5th Duke of Bragança[25].

As the head of the ducal house of Bragança post-Azemmour, Teodósio was fabulously wealthy. His assets, holdings and trade monopolies, as well as the ducal house's importance within the Portuguese kingdom and empire, ensured that Teodósio could stand shoulder to shoulder with any of the grandees of Spain.

---

21 Góis, 2014: 550–551.

22 Góis, 2014: 551.

23 On the structure and functions of the Manueline *companhias de ordenança* (ordinances), see Rodrigues, 2004: pp. 185–196. However, the entire system of ordinances and its early sixteenth century origins is currently being researched and entirely reconsidered by Paulo M. Dias (NOVA FCSH).

24 Melo, 1944: 122.

25 Hallet/Senos, 2018a: 15–16. The entire inventory—in full, *Inventário dos bens do 5º duque de Bragança, D. Teodósio I*; simply *Inventário* for short—can be found in a digital edition in Hallet/Senos (2018b). "De Todas as Partes do Mundo. O património do 5º duque de Bragança, D. Teodósio I. Volume 2 – Documentos". In: <https://run.unl.pt/handle/10362/66157> [05 May 2023]. For ease and convenience of reading, every entry in the digital *Inventário* will be given as "*Inventário*, X", where X is its corresponding entry number.

The inventories—the so-called *Inventário de D. Teodósio I*, or simply *Inventário*—written in the wake of harsh disputes between his heirs[26], are a testament to that wealth: an extensive set of documents with 657 folios, over 1,300 pages and 6,303 entries[27], detailing the 5th Duke's great collections of clothing, jewellery, textiles, furniture, books and weaponry at Vila Viçosa, with pieces acquired everywhere from Spain and Germany and Flanders to India, Ceylon, China and Brazil. The *Inventário* lists not only each piece or set of pieces in great detail, but also their location within the compound of the ducal palace, so that the appraisers who wrote the inventories knew exactly where to find them; as well as valuations, which are key in assessing the prices of these pieces and the overall economy of the arms trade in Portugal during the sixteenth century. It is within this great mass of possessions that we can find a precise list of the contents of the armoury.

THE DUCAL ARMOURY

Definition and Location

It is important to define *a priori* what exactly we mean when we refer to the ducal armoury of Bragança. Teodósio's collections (plural) of arms and armour were spread across several different spaces. *Exempli gratia*: João Correia and João Gomes Vieira, tasked with the custody of the Duke's *guarda-roupa* [wardrobe], list as part of their charges dozens of precious swords with their scabbards and belts; daggers, *adargas* and rotellas, spears and maces and horse trappings[28], objects with not only intrinsic value but cherished possessions of the deceased Duke. They include too a *casinha das armas* [a "weapons room"] in which a small portion of ornate weaponry was stored, including crossbows, firearms, plain and decorated powder flasks, and a few odd pieces of armour, such as an assortment of helmets in a chest[29] or a *"rodela E hum morriam dourado que foi do Duque de Cesa Laurado de Romano* [rondache and gilded morion that belonged to the Duke of Cesa, decorated in the Roman fashion]"[30]. Yet the *armaria* (lit. "armoury"), overseen by Bento Esteves, was a special collection unto itself, a wholly independent part of the *Inventário* and, apparently, of the ducal palace.

26 See Mafalda Soares da Cunha's chapter "Uma história quase interminável? O processo das partilhas por morte de D. Teodósio" in Hallet/-Senos, 2018: 23–37.

27 Hallet/Senos, 2018: p. 16.

28 *Inventário*, 1139–1142, 1487–1508, 1551–1608, including pieces 2089–2106 and 2284–2316 which were stored in the *casinha das armas.*

29 *Inventário*, 1820–1895.

30 *Inventário*, 1894.

The *armaria* alone held Teodósio's personal harnesses and those of his father, as well as the majority of all weapons, armour and firearms in the *Inventário*, along with the tools needed to keep all of these pieces in good repair. Because of its self-contained nature, then, the focus of this analysis will fall solely on this group of pieces, and will disregard Teodósio's other collections of arms and armour described in the *Inventário*.

That this collection was seen as such in period is clear from the documentation; its location however is somewhat controversial. According to the late seventeenth century genealogist António Caetano de Sousa, the armoury had its own dedicated building "*no Paço, a casa das Armas, huma das singulares, que se conhecem* [and in the palace, the armoury, one of the most exquisite there are]"[31], that Teodósio I himself had ordered to be built. In his 1683 description of the town of Vila Viçosa and its palace, António de Oliveira de Cadornega, a seventeenth century historian born in the town in 1623, described that "*entrando por baixo dos arcos [...] estavam, por baixo de Palácio, as açoteias [...]. Por diante das açoteas, iam correndo outras logeas* [As one entered under the arches [...] there were, under the Palace, the terraces [...]. In front of the terraces, there were other loggias]"[32] fitted with "*ginelas de grades de ferro pera fora que o curioso via tudo quanto dentro havia [...]*[all with windows barred with iron from the outside, so that the curious visitor could see everything within]"[33]. In these were stored many "*bestas de ferro [...] saias de malha, maças de ferro, espadas ao antigo, muitas armas de remesso, pelotas, e muitas outras cousas com que se maravilhavam muito em ver as armas com que antigamente se peleijava* [iron crossbows [...] mail hauberks, iron maces, ancient swords, many missile weapons, projectiles, and many other things they wondered at in seeing the weapons with which war had once been fought]"[34]. This would point to the armoury being located inside the palace grounds, in its own space. Additionally, the *Inventário* itself seems to appertain only to the contents of the palace compound and the goods therein—the literal objects of contention between D. Teodósio's heirs.

However, Vítor Rodrigues[35], as José Teixeira before him[36], argued that the ducal armoury was not located in the palace but on the nearby fortification of Vila Viçosa, built in the early sixteenth century as a modern complement to (and

31 Sousa, 1739: 95.
32 Cadornega, 1982: 87.
33 Cadornega, 1982: 87.
34 Cadornega, 1982: 87.
35 Hallet/Senos, 2018: 270–271.
36 Teixeira, 1983: 39. Teixeira indicates corroborating evidence in folio 77v of Francisco de Moraes Sardinha's *Antiquissimo Parnaso novamente achado e descuberto em Vila*

partly as a replacement of) the old medieval castle overlooking the town[37]. They based this assertion on references to arms and armour stored in the castle, such as the account of Cardinal Alessandrino Michele Bonelo's passage through Vila Viçosa in 1571[38], and the sheer number of pieces listed in the *Inventário*, which in their view would require a dedicated space away from the ducal compound[39]. This could've been the series of rooms in the castle described by Cadornega as:

> *as salas e as paredes delas tudo de cavides de alto abaixo, de deversidade de armas de fogo todas mui limpas, e cantidade de espadas e traçados. Havia nestas salas duas carreiras de colunas, ou postes, todas de alto abaixo cobertas de armas brancas reluzentes como espelhos, peitos, e espaldares, morrões e capacetes, manoplas, saias de malha. E, entre poste e poste, um cavaleiro armado sobre madeira, com rodas pera se moverem, todos armados, e até os mesmos cavalos, com seus piques nas mãos, tudo muito reluzente, trabalho de seis homens armeiros, que continuamente estavam alimpando e sacalando tudo [...] Os almazéns por baixo alojavam cantidade de picaria, levantada do chão sobre postes piquenos e vigas que atravessavam por cima, perservado do rigor do tempo que tudo consome*[40]. [The rooms and their walls covered in hangers from top to bottom, a great diversity of firearms all very clean, and a quantity of swords and *terçados*. There were in these rooms two rows of pillars, or posts, all covered with cold steel weapons as glittering as mirrors, breastplates, and backplates, morions and helmets, gauntlets, and shirts of mail. And, between each post, a knight armed on wood [supports], with wheels to move around, all fully armed, and the horses the same, with their pikes in their hands, all very shiny, the work of six armourers who were continually cleaning and scrubbing everything [...] The storerooms below housed a lot of pikes, raised from the ground on small posts and beams that crossed over them, preserved from the rigours of time that consumes everything].

Viçosa from 1618 (BNP, Reservados, cod. 107), which I have not be able to consult; nor was it possible to acquire a copy of Cristopher C. Lund's critical edition of said text (Lund, Cristopher (2003). *O parnaso de Vila Viçosa: leitura do texto, introdução, notas e indices*. Rio de Janeiro, H.P. Comunicação Editora).

37 Hallet/Senos, 2018a: 270–271.

38 On this, the cardinal's secretary Giovanni Battista Venturino di Fabriano wrote that "*nell'intrar della terra Il Duca u'há una fortalezza et bona doue ce tiene soldati et tutta la sua armaria et altre cose* [as one enters the land [of Vila Viçosa], the Duke has a good fortress there, where he has soldiers and all his arms and other things]. Brandão, 1926: 411.

39 In this, Rodrigues conflated the contents of the *armaria* with the list of "*artilheria* [artillery]" under the supervision of Vicente Fernandes, who was also tasked with the "*guarda-reposta* [the pantry]". Hallet/Senos, 2018: 270–271. As with the contents of the armoury, the *Inventário* never makes clear where Teodósio's artillery was stored.

40 Cadornega, 1982: 110.

This detailed description reminds us of the design of other early modern armouries, such as Graz's *Landeszeughaus*, and would perfectly accommodate all of the pieces listed in the 1563 inventories.

It is clear both sources cannot be simultaneously correct. To be sure, there is a temporal gap to consider—an entire century between the *Inventário* and Sousa's and Cadornega's writings, in which the armoury could have been transferred from one location to another. However, given our sources, and on the balance of probability, there seemed to have been important collection of arms and armour of some description in the palace, which might have corresponded to the collection of the *casinha das armas*; yet most, if not all of the armoury's contents were most likely stored in the nearby castle. Though there are many examples of large armouries seamlessly integrated within the fabric of sixteenth century palaces, there are just as many such examples of armouries assembled in nearby fortresses, which would more easily allow them to equip any necessary forces.

The Contents of the Armoury

In total, the collection of the ducal armoury of Vila Viçosa was comprised of between 7,000 and 7,500 items which can be classed as armaments—either whole pieces or components thereof[41]— as well as dozens of additional uncounted items and accessories, such as a group of old and useless pavises[42]. This would rank it amongst the largest armouries in the Peninsula in the sixteenth century—for comparison—in 1601, the great armoury of the house of the Infantado held in it about 6,000 items which could be classed as armaments[43]. It is impossible to analyse every piece in depth in this paper but, by separating the contents into thematic areas, we can sketch an overview of the armoury's contents.

Body Armour

Defences for the body are undoubtedly the most emblematic component of any armoury. In Vila Viçosa, these included twenty-eight harnesses, from garnitures to full harnesses to incomplete ensembles. Of these, twelve sets in total, including

41 The *Inventário* lists every individual piece in an ensemble, including parts such as faulds, tassets or plackarts, and even missing pieces in a set, which allows counting almost every piece available.

42 *Inventário*, 3489.

43 González Ramos, 2013: 4.

four sourced from Flanders[44] and one from Biscay[45], have several exchange pieces for the field and for jousting.

The entire repertoire of sixteenth century decorative techniques is present in the *Inventário*, including blueing, engraving[46], gilding, a combination of these or employed on their own, alongside plain burnished surfaces. Along with Teodósio's harnesses there was one older set belonging to Duke Jaime[47], as well as the corselet with which Jaime had entered Azenmour in 1513[48]. An interesting feature of the armoury are its sets of armour for children: along with a complete child's harness[49], there were also two sets of corselets[50] with missing pieces—perhaps not surprising if we take into account Teodósio's two sons[51] and their formative years as noblemen and knights[52].

This category also included 186 corselets. The *Inventário* is a prime example of how polysemic the word "corselet" truly was in the sixteenth century, with these corselets ranging from simple setups, such as the forty corselets from Biscay with nothing more that breastplates with faulds, backplates and sallets, to ensembles practically indistinguishable from full harnesses in coverage, number of pieces and decoration. The presence of so many corselets, along with the pikes and arquebuses we'll see below, expresses the armoury's most pragmatic facet as a source for the equipment of the ducal forces, and not simply storage for the Duke's more high-end pieces.

Some other highlights in this category include 324 breastplates, sixty-two morions, a large number of sallets, war hats and assorted bevors (mostly from

44 *Inventário*, 3385.

45 *Inventário*, 3390.

46 At least two of the Duke's harnesses (*Inventário*, 3474, 3475) are described as "pretos [black]", engraved and gilded, a description which is highly evocative of contemporary harnesses such as the famous "Blue & Golden" harness made for Maximilian II in Augsburg around 1557 (Kunsthistorisches Museum Wien, Hofjagd—und Rüstkammer, A578).

47 *Inventário*, 3386.

48 *Inventário*, 3392.

49 *Inventário*, 3389.

50 *Inventário*, 3492.

51 Jaime, who would die at nineteen in the battle of Ksar-el-Kebir in 1578, and João, who would succeed Teodósio as Duke. Melo, 1944: 134–135.

52 Elsewhere, the Inventário also includes riding accoutrements for children (*Inventário*, 5275).

Biscay), seventeen *couraças* [brigandines[53]], four *almilhas*[54] and six *rodelas*, both decorated and plain. There were also an astounding number of assorted complementary mail pieces—sleeves, half-sleeves, skirts, standards, hauberks and two of nine mail "*bragas* [brayettes]"[55] in Teodósio's possession, the only such pieces known thus far in Portuguese sources.

Equestrian Equipment

With these many harnesses, it was only natural for the armoury to store an equally large number of riding accoutrements. Of bardings, old and new, there were fifty-six, mostly of leather [probably *cuir bouilli*], along with twenty-nine saddles and thirty-five shaffrons. Most of the shaffrons are paired with—and in fact seemed to have acquired with—specific harnesses or corselets, as it frequently happened with sixteenth century garnitures[56]. A large number of the saddles are paired with a specific bard and shaffron, indicating that some of the pieces were stored and subsequently counted as set pieces on display.

Tourneys and jousting seemed to have been one of the favourite ducal pursuits. Apart from the many exchange pieces for the joust listed for most of the harnesses, and apart from the many saddles, bardings and shaffrons which might've been used both for war and for jousting, the armoury included at least

---

53 *Inventário*, 3422. Until the late sixteenth century, "*couraça*" generally seems to have meant a brigandine (Oliveira, 2022: 337–340), and the seventeen items in this entry conform to that meaning. However, several of the harnesses and corselets are described as having "*couraças*", sometimes with their "*volante* [plackart]" and some of these "*couraças*" (for example, in *Inventário*, 3391) are specified as being composed of a "*peito* [breastplate]" and "*espaldar* [backplate]". The *Inventário* is likely one of the earliest witnesses to the semantic shift of the word "*couraça*", which from the eighteenth century onwards would come to mean a solid steel piece of protection for the torso (Oliveira, 2022: 340, note 101), as it does in many other European languages.

54 *Inventário*, 3406 and 3513. Enrique de Leguina rightly argued the similarity between "*almilhas*" and brigandines (Leguina, 1912: 58). Judging from the *Inventários* entries, "*almilhas*" likely correspond to a specific type of very tight-fitting mid-to-late sixteenth century brigandine, which was complemented by pieces for the limbs made in the same style as the body. Extant pieces such as these can be found, among others, in the Metropolitan Museum of Art (Acc. no. 14.25.1531 and 14.25.1532).

55 *Inventário*, 3430 and 3478. The other seven pairs (*Inventário*, 5198), were stored with equestrian equipment for the *gineta*, under the supervision of stable master ("*estribeiro*") Francisco do Carvalhal.

56 A good example is the harness called "*das granadas*" or garnets (*Inventário*, 3473) and its bard, two shaffrons and two saddles (*Inventário*, 3486).

twenty-seven shaffrons specifically for the joust, 150 jousting lances, twenty-seven "worn" swords for the tourney and ten pairs of jousting spurs. Additional equestrian equipment was stored elsewhere the palace, including several pieces (*Inventário*, entries 1391 to 1486) maintained by João Correia and João Gomes Vieira, and many others scattered in other lists, such as silver spurs and bridle fittings counted as part of the Duke's silver.

Weapons

The mainstays of the ducal armoury's weapons were spears and pikes, meant to outfit the ducal *companhias de ordenança*[57]. An exact 1,000 pikes, in two lots of 500 each, were apparently no longer fit for use when the inventory was written; they were to be replaced by 485 pikes from Biscay and 470 pikes from other sources, though most of the new pieces were as yet unhafted in 1563. Old spears for the "*gineta*" abounded, with heads in the shape of olive-tree leaves, along with 130 old spears from Naples.

Other hafted weapons included halberds—187 of them, eighty-seven of which with black hafts and gilded heads—seven *chuços* [a short type of spear], guisarmes and spontoons. There were also a few swords, estocs and maces. As mentioned earlier, most of the Duke's better swords were kept in his wardrobe, along with their attending daggers, belts and baldrics, or in the *casinha das armas*.

Firearms

Last but certainly not least, the armoury included a great number of firearms—581 in total—of several different types. Modern "*arcabuzes* [harquebuses]", sourced from Biscay (198), from the Holy Roman Empire (102) and elsewhere, coexisted with older typologies, such as thirty-nine "*espingardas*", and 173 "*espingardas*" or "*mosquetes* [muskets]". The inventories show that Teodósio kept a close eye on the latest developments in firearms, as reflected in a "*mejo mosquete*", or demi-musket[58], which had been acquired as a working sample [*amostra*]. This preoccupation which is also reflected in a number of more special firearms included in the *casinha das armas*: five harquebuses "*de murão*"[59], one harquebus "*turquesquo* [Turkish or in Turkish fashion]", one flintlock [probably a snaplock]

---

57 For Vitor Rodrigues' analysis of the composition of the ducal forces, see Hallet/Senos, 2018a: 272–275.

58 *Inventário*, 3454.

59 *Inventário*, 1834, 1836, 1839, 2288. Possibly meaning siege arquebuses.

harquebus and several "*pistoletes*" or handguns[60]. The firearms in the armoury are often paired with powder flasks, some perfectly (and literally) non-descript, others described as being richly decorated or made of fine materials, such as ivory. The armoury also included eighty-one crossbows, proving their longevity well into the early decades of the age of gunpowder.

The impact of these firearms in the development of body armour is also attested to in the *Inventário*, with no fewer than six different pieces[61] described as being "*de proua* [of proof]" against harquebus fire, including one rondache in the *armaria* that had actually been made in Vila Viçosa[62].

Preferred Markets and Sources

Reading the inventory of the armoury is to see the embodiment and ramifications of an extensive European trade network of arms and armour which connected Portugal with the best centres of arms and armour production in Europe. Few of the more specialised pieces were produced locally; in 1573, in fact, there was but one single plate armourer living in Lisbon—a Byscaian, by the name of Martim Perez[63]. Though some pieces were acquired closer to home, in local fairs[64], most of the Duke's harnesses were instead acquired from well-reputed armour-making centres, such as the four harnesses "brought from Flanders"[65], the harness from Milan[66] or the harness "*das granadas* [of the pomegranates]" with its tonlet, ordered from distant Germany[67]. Biscay seems to have been a preferential source of arms and armour for equipping the ducal infantry, providing everything from pikes to harquebuses to large numbers of corselets and individual pieces[68]. In fact, Biscay was the main supplier of armaments for Portugal in the sixteenth century, a relationship which might be explained as

60 For example, *Inventário*, 2295 to 2302.

61 *Inventário*, 3469, 3498, 3509, 3516.

62 *Inventário*, 3516.

63 According to a letter written by king Sebastião to the councilmen of Lisbon. Viterbo, 1920: 159–160.

64 Such as the eight helmets in *Inventário*, 3436.

65 *Inventário*, 3385.

66 *Inventário*, 3393.

67 *Inventário*, 3473.

68 Hallet/Senos, 2018a: 278–279.

a combination of geographical proximity and special treatment (with reduced prices) shown to Portuguese agents[69].

The *Inventário* shows a clear distinction between these centres and their output: whereas Flanders provided Teodósio with high-quality cavalry harnesses and cuirasses, as well as a substantial amount of artillery, Biscay produced what may be deemed munitions—grade pieces, such as hundreds of pikes, harquebuses, corselets and complementary pieces of armour (bevors, helmets, cuisses). Unlike most similar documents from the time, the *Inventário* also records a number of individual armourers and craftsmen[70] who supplied pieces to the armoury.

As correctly stated by Vítor Rodrigues, it is likely most of these pieces arrived via ship in Lisbon, before being sent on their way to Vila Viçosa[71]. A 1564 inventory of the contents of the Bragança's palace in Lisbon[72] includes an assortment of arms and armour, including 500 spears and 500 pikes (once again from Biscay) which were on their way to Vila Viçosa when the Duke died[73]. Pieces such as these could arrive pre-hafted or in their separate component parts, to be assembled in Vila Viçosa by the Duke's own craftsmen.

A Dynastic Armoury?

These more than 7,000 pieces were but a portion of the ducal stores, since every castle under the ducal house included its own independent arsenal, and other possessions of the Braganças, such as their urban palace in Lisbon, stored their fair share of weapons as well[74]. But Vila Viçosa's armoury was set apart from all other caches of weapons owned by the ducal house. Was it just a question of size, location or did this collection have any special significance to the ducal house?

---

69 Hinted at in a letter from Francisco de Mimença to the king (either João III or Sebastião), asking for an official letter which he could present to "*ser fauorecido em bizcaya na compra negoçiaçam e embarcamento* [be favoured in Biscay in his acquisition, negotiation and shipment']" of the arms and armour which he was charged with acquiring there for Lisbon's arsenal (Arquivo Nacional – Torre do Tombo, Lisbon (ANTT), Colecção de São Vicente, Livro 10, fol. 429 r). The rather formulaic way the request is made seems to suggest a habitual practice.

70 Such as João Fortes, Diogo de Rupol and Agostinho de Rupol, the latter of which was in charge of the armoury's upkeep and maintenance (Hallet/Senos, 2018a: 292).

71 Hallet/Senos, 2018a: 278.

72 Transcribed and analysed in Saraiva, 1948. A more recent transcription is included in Hallet/Senos, 2018b: 359–368.

73 Saraiva, 1948: 21. Hallet/Senos, 2018b: 363.

74 See previous note.

To put the question in broader terms: when do private armouries move beyond being mere stockpiles of weapons, signalling a person's capacity to wage war, and become not only visual displays of wealth but also, and most importantly, places of memory, places which narrate a family history in steel, leather and wood?

There are many examples of that conceptual construction in the present volume, but for the ducal armoury at Vila Viçosa that transition seems to take place during Teodósio's lifetime. The collection in the armoury didn't include the trophies and curiosities we see elsewhere, such as those in the armoury of Alba; even though Teodósio had accrued his fair share of special items, most of them were kept elsewhere in the palace, as stated above. Yet the armoury did house the armours of Duke Jaime, including the one the Duke had worn to Azemmour in 1513[75]. Teodósio was keen on displaying the memory of that conquest wherever he could: on the palace's grand stairway, he ordered three life-sized murals of the campaign to be painted, a dazzling portrayal of the fourth Duke and the Bragança forces[76]. Caring for and displaying Jaime's armour rings as another manifestation of Teodósio's preoccupation with perpetuating his father's memory, and consequently the family's history and glory. In this regard, it is telling that none of the pieces which belonged to Jaime are described as being poorly maintained. Elsewhere, when pieces are old, rusted or otherwise rendered unusable, the inventory makes careful note of the fact. Examples of that are a number of worn leather bards which had seen service in Azemmour[77] and a batch of 500 "ancient" pikes, which could very well have been acquired for the 1513 campaign but which, fifty years later, were unfit for purpose[78]. It is true that Duke Jaime's former armours aren't worth much, compared to other contemporary armours in the inventory, but that is due to their age[79], and not because they were in disrepair. Teodósio's armoury, like his library[80] and particularly the palace itself[81], was very much a part of a greater programme of self-promotion we see in other Iberian noble houses at the time, one consummated aspect of the principle of self-aggrandisement via "*las armas y las letras* [arms and letters]".

75 *Inventário*, 3392.

76 Teixeira, 1983: 43–51.

77 *Inventário*, 3397.

78 *Inventário*, 3456.

79 "*Por serem velhas* [for being old]", Cadornega, in 1982: 88.

80 See "Aspectos da livraria de D. Teodósio: uma grande biblioteca do Renascimento" in Hallet/Senos, 2018a: 281–299.

81 See "A ampliação do paço de Vila Viçosa" in Hallet/Senos 2018a: 109–134. Teixeira, 1983: 34–37.

CONCLUSION

If this facet of glorification through ownership of arms and armour is still somewhat open to discussion in 1563–1566, by the late seventeenth century it had morphed into a certainty, and the armoury, or at least part of it, existed specifically as a collection of treasures to awe and impress. As stated, in the loggias of the palace grounds there were many "*antiguidades de armas* [antiquities of arms]"[82] kept there "*por grandeza* [for grandeur]"[83] and display. These may have been pieces from Teodósio's *casinha das armas*, select pieces of the *armaria*, or both. The bulk of the armoury, however, remained in the castle, to where esteemed visitors such as Cardinal Alessandrino were also led.

Cadornega's seventeenth century description is sadly the last recorded instance of Teodósio's armoury, and of early modern arms and armour in Vila Viçosa. Unlike the Lisbon arsenal, whose contents were mostly dispersed during the Duke of Alba's 1580 incursion in Portugal[84] and then successively replenished over the years[85], we have no further news of weapons at Vila Viçosa after the beginning of the eighteenth century. Sant'Anna Dionísio claims, without presenting any sources, that most of the sixteenth century arms and armour had been lost in the disastrous African campaign of 1578[86], and that Dukes Teodósio II and João II had subsequently replenished the armoury in anticipation of the Portuguese uprising against Habsburg rule in 1640[87]. Whatever the truth, upon João II's ascension to the throne of Portugal in 1640, as João IV, Vila Viçosa was no longer a ducal palace but one of many royal residences, and its holdings were prone to being scattered amongst other royal collections. It is also possible that a portion of the pieces might have been stolen during the French Invasions[88]. By the nineteenth century, there was nothing left of Teodósio's collection. The small armoury we see in the palace today is an amalgamation of the pieces collected by the last kings of the Bragança dynasty, permanent loans from other museums, such as the National Museum of Ancient Art, and pieces acquired at

82 "Ancient arms" Cadornega, 1982: 88.

83 Cadornega, 1982: 88.

84 According to the Duke, on his arrival to Lisbon, there wasn't a single piece of equipment left in the city, since the pretender to the throne, D. António, had taken everything to equip his army and armada (Miraflores/Pidal/Salvá, 1842: 557).

85 Considering that the Lisbon arsenal outfitted some of the ships of the Invincible Armada of 1588, it necessarily had to be replenished with arms and armour since 1580.

86 A good summary of the campaign can be read in Costa e Sousa, 2009.

87 Dionísio, 1947: 155.

88 Dionísio, 1947: 120. Again, no sources are given that support this assertion.

auctions during the twentieth century[89]; the castle holds nothing but a collection of hunting rifles and implements.

In a way, the current armoury is still a place of memory, but of a different kind: much like Lisbon's Military Museum evokes the great arsenals of the state, the armoury at Vila Viçosa reminds us of one of the richest and greatest collections of arms and armour in Iberia during the sixteenth century, a collection whose memory rests within the yellowed pages of the *Inventário*.

BIBLIOGRAPHY

Baquero Moreno, Humberto (1973): *A batalha de Alfarrobeira: antecedentes e significado histórico. Volume I.* Coimbra, Biblioteca Geral da Universidade de Coimbra.

Brandão de Buarcos, João (1990): *Grandeza e Abastança de Lisboa em 1552.* Lisbon, Livros Horizonte.

Brandão, Mário (1926): "Uma carta àcêrca da viagem do cardeal Alexandrino a Portugal". In: *Biblos*, 2, pp. 401–415.

Cadornega, António (1982): *Descrição de Vila Viçosa.* Lisbon, Imprensa Nacional-Casa da Moeda.

Cardoso, João Luís / Gomes, José Luís (2005): "As "Ferrarias Del Rey" em Barcarena: Subsídios para a sua História". In: *Estudos Arqueológicos de Oeiras*, 13, pp. 9–194.

Costa e Sousa, Luís (2009): *Alcácer Quibir 1578. Visão ou delírio de um rei?* Lisbon, Tribuna da História.

Dionísio, Sant'Anna (1947): *Museu-Biblioteca de Vila Viçosa.* Lisbon, Fundação da Casa de Bragança.

Firnhaber-Baker, Justine (2010): "Seigneurial War and Royal Power in Later Medieval Southern France". In: *Past & Present*, 208 - 1, pp. 37–76.

Gallastegi, Ekaitz/Rojas, Jon Andoni (coords.) (2021): *Edad Media. Las coronas de Castilla y Aragón (siglos XIV y XV).* Saragossa, Prensas de la Universidad de Zaragoza.

Góis, Damião (2002): *Elogio da Cidade de Lisboa-Urbis Olisiponensis Descriptio.* Lisbon, Guimarães Editora.

Góis, Damião (2014): *Crónica do Felicíssimo Rei D. Manuel.* Braga, Edições Vercial.

89 Loureiro de Figueiredo, 2001.

Gomes Martins, Miguel (2014): *A arte da guerra em Portugal: 1245 a 1367*. Coimbra, Imprensa da Universidade de Coimbra.

Gonçalves de Freitas, Judite (2011): *O Estado em Portugal (séculos XII -XVI)*. Lisbon, Alétheia Editores.

González Ramos, Roberto (2013): "The armoury of the Dukes of the Infantado. Collecting, prestige and meaning". In: *Journal of the History of Collections*, 25–3, pp. 335–350.

González Ramos, Roberto (2014): "La colección de armas de Íñigo López de Mendoza, V Duque del Infantado". In: *Gladius. Estudios sobre armas antiguas, arte militar y vida cultural en oriente y occidente*, 34, pp. 153–198.

Gouveia Monteiro, João (1998): *A Guerra em Portugal nos finais da Idade Média*. Lisbon, Editorial Notícias.

Gouveia Monteiro, João (2001): *Armeiros e Armazéns nos finais da Idade Média*. Viseu, Palimage Editores.

Gouveia Monteiro, João (2003): "Organização e Formação Militares". In: Severiano Teixeira, Nuno/Themudo Barata, Manuel (coords.): *Nova História Militar de Portugal (Vol. I)*. Rio de Mouro, Círculo de Leitores.

Hallet, Jessia/Senos, Nuno (coords.) (2018a): *De Todas as Partes do Mundo. O património do 5º duque de Bragança, D. Teodósio I. Volume 1-Estudos*. Lisbon, Tinta da China.

Hallet, Jessia/ Senos, Nuno (coords.) (2018b): *De Todas as Partes do Mundo. O património do 5º duque de Bragança, D. Teodósio I. Volume 2 – Documentos*. Lisbon, CHAM – Centro de Humanidades / Tinta-da-China <https://run.unl.pt/handle/10362/66157> [05 May 2023].

Hyams, Paul (2003): *Rancor and Reconciliation in Medieval England*. Ithaca, Cornell University Press.

Lee de Jesus, Roger (2006): "Abastecer a guerra noutro oceano: O armazém das armas de Goa em 1545–1546". In: Rogado, Francisco/Domingues, Contente (coords.): *Actas do XXIV Colóquio de História Militar "Nos 600 Anos da Conquista de Ceuta-Portugal e a Criação do Primeiro Sistema Mundial"*, pp. 169–220.

Leguina, Enrique (1912): *Glosario de Voces de Armería*. Madrid, Librería Felipe Rodríguez.

Loureiro de Figueiredo, João (2001): *Armaria do Paço Ducal de Vila Viçosa*. Caxias, Fundação Casa de Bragança.

Salvá, Miguel (1842): *Colección de documentos inéditos para la historia de España. Vol. 33*. Madrid, Imprenta de la Viuda de Calero.

Melo, Francisco (1944): *D. Teodósio II. Duque de Bragança*. Porto, Civilização.

Moreira, Filipe (2020): "Morte, espetáculo e encenação de poderes em relatos de execuções 'políticas'. Fontes e modelos comparativos". In: Miranda García, Fermín/Guereño Sanz, María Teresa (dirs.): *La Muerte de los Príncipes en la Edad Media. Balance y Perspectivas Historiográficas*. Madrid, Collection de la Casa de Velázquez <https://books.openedition.org/cvz/22782> [05 May 2023].

Oliveira Marques, António (1987): *Portugal Na Crise Dos Séculos XIV e XV*. Lisbon, Editorial Presença.

Oliveira, António (2022): "'Uma peça muito preciosa de significado ainda desconhecido': Decifrar o termo gibanete". In: *Medievalista*, 31, pp. 317–354 <https://medievalista.iem.fcsh.unl.pt/index.php/medievalista/article/view/467> [10 May 2023].

Pereira dos Reis, Pedro (2022): *Depois da Índia. Regresso, morte e posteridade de Gonçalo Gil Barbosa* [Master's dissertation]. Lisbon, Faculdade de Ciências Sociais e Humanas da Universidade Nova de Lisboa.

Rodrigues, Vítor Luís Gaspar (2004): "As companhias de ordenanças em Marrocos nos reinados de D. Manuel e D. João III". In: Teodoro de Matos, Artur/Carneiro, Roberto (coords.): *D. João III e o Império. Actas do Congresso Internacional comemorativo do seu nascimento*. Lisbon, CHAM / CEPCEP: pp. 185–196.

Rodrigues de Oliveira, Cristovão (1987): *Lisboa em 1551. Sumário em que brevemente se contém algumas coisas assim eclesiásticas como seculares que há na cidade de Lisboa*. Lisbon, Livros Horizonte.

Saraiva, José (1948): *Memoria sobre um inventário artístico do ano de 1564 do Paço dos Duques de Bragança em Lisboa*. Lisbon, Arquivo Histórico do Ministério das Finanças.

Soares da Cunha, Mafalda (1990): *Linhagem, parentesco e poder. A Casa de Bragança (1384–1483)*. Lisbon, Fundação Casa de Bragança.

Soares da Cunha, Mafalda (2000): *A Casa de Bragança (1560–1640). Práticas Senhoriais e Redes Clientelares*. Lisbon, Editorial Estampa.

Soares da Cunha, (2014): "A herança de D. Jaime". In: Correia Guedes, Natália (coord.): *Homenagem a João Gonçalo do Amaral Cabral*, Vila Viçosa, Fundação da Casa de Bragança.

Sousa Viterbo, Francisco (1908): *A Armaria em Portugal*. Lisbon, Academia Real das Sciencias.

Sousa Viterbo, Francisco (1920): *Artes e artistas em Portugal: contribuições para a história das artes e industrias portuguezas*. Lisbon, Livraria Ferin.

Teixeira, José (1983): *O Paço Ducal de Vila Viçosa. Sua Arquitectura E Suas Colecções*. Lisbon, Fundação Casa de Bragança.

Juan Francisco Torres Cubero

Universidad de Córdoba

THE ARMOURY OF THE COUNTS OF BENAVENTE

In 1808, General Sir John Moore commanded the British army deployed on the northern section of the Iberian Peninsula in the campaign against Napoleon. With him travelled the Scottish artist Robert Ker Porter, who had the opportunity to admire the fortress of Benavente and to leave us his impressions during the soldiers' residence in the area. One of the elements that caused him the greatest fascination was the armoury, in which he could study the large number of "coats of mail, barbed steeds, rondaches, helmets, crossbows and weapons of every description" which formed that "ancient treasury of heroes"[1]. The Benavente armoury was one of the most important and recognised of the Spanish noble houses, according to the limited information we have. Simal López[2] analysed it generally in her study about the family's patronage and collecting proclivities, and González Ramos[3] echoed the importance of the collection with some more detail. Regueras Grande also referred to the relevance of the collection in his monograph about the Benavente castle[4]. Nonetheless, an extended study that considers the available sources and analyses the house of Pimentel armoury's content, meaning, importance and circumstances has not been made, an effort that is intended to be carried out in the space provided below.

THE ARMOURY'S ORIGIN

Little is known about the arms that the Benaventes possessed in the Middle Ages and the beginnings of the Modern Age. The first information that we have are related to the figure of Don Alonso Pimentel (†1530)[5]. In the fortress, he undertook a series of renovations to accommodate it for residential use and

1 Robert Ker Porter in one of his letters, cited in González Ramos, 2017: 284.

2 Simal López, 2002.

3 González Ramos, 2017: 269–312.

4 Regueras Grande, 2019: 40–43.

5 Simal López, 2002: 24. Ledo del Pozo, 1853: 300. Sandoval, 1847: 555.

erected the "Torre del Caracol [Snail Tower]", the only vestige that remains today. In Benavente, he and his wife, Doña Ana de Velasco, focused on the construction of the Hospital de Nuestra Señora de la Piedad [Hospital of our Lady of Piety], founded in 1517. Moreover, he ordered the construction of a familial palace in Valladolid (1515–1516). During the Revolt of the Comuneros, Don Alonso strongly backed the emperor, contributing a noticeable amount of human and material resources[6]. According to his post-mortem inventory, he possessed some "*arneses de justa real* [harnesses for royal jousting]" and "*armas de la gineta* [arms for riding]"[7], jousting lances, short spears, swords and daggers that were scattered throughout the inventory[8]. This undoubtedly represents the nobleman's personal endowment, but it tells us nothing of an armoury in the strict sense.

Don Antonio Alonso Pimentel (1514–1575), son of Don Alonso and his wife, Doña Ana de Velasco, succeeded to the title in 1530. His first years as head of the house of Pimentel were characterised by the European wars of Charles V, whom he accompanied to Germany beginning in 1535 and on the journey to Tunis. Three years later, the monarch entrusted him with the imperial standard during the French campaign[9]. Additionally, he participated in the different festive occasions that were offered upon the emperor's passage through Italy. First, after the victory in Tunis, he entered Naples and Rome, occasions in which he rode with the sovereign[10]. His position also allowed him to consolidate his relationship with other important figures of the house of Habsburg. The main ones were Prince Maximilian, later the second of that name as imperial sovereign, and his father, Ferdinand of Austria[11], still King of Hungary and Bohemia, with whom he struck up a friendly relationship that, as we will see, was reflected in his weaponry. In the 1540s, he became close with Prince Philip and his entourage, hosting them in 1554 when he was travelling to wed Queen Mary in England, which led Don Antonio Alonso to pay for some extraordinary celebrations in Philip's honour, according to Andrés Muñoz, the heir's servant. At the end of the decade, now with Philip II on the throne, Don Antonio participated in the

6 Kamen, 1995: 49. Berdum de Espinosa, 1753: 95v. Sandoval, 1847: III, 242.
7 Archivo Histórico de la Nobleza (hereafter AHNOB), Osuna, C. 423, D. 52, fol. 69 r and v.
8 AHNOB, Osuna, C. 423, D. 52, fol. 16 r, 71 v.
9 *Origen de los condes duques de Benavente y su apellido Pimentel* by Domingo de Ascargorta, Biblioteca Nacional, Madrid (BNE), Mss. 11569, fol. 159 r. Simal López, 2002: 25–26.
10 Morales Folguera, 2015: 99. Santa Cruz, 1922: 324.
11 Simal López, 2002: 26.

festivities organised to celebrate the marriage between the king and Elisabeth of Valois in Toledo[12]. His political career would culminate in his naming as viceroy of Valencia, an office he occupied between 1567 and 1572. He died in Benavente on 1575[13].

Don Antonio Alonso's participation in those grand events that contributed to the forging of the Monarchy's fate was cloaked in a constant and magnificent symbolic display appropriate to his status. This began in the palace-castle of Benavente, in the seignorial setting that the Count embellished with the presence and works of artists, writers, actors and musicians that formed his humanistic court[14]. In this way, the 6th Count of Benavente's residence became an earthly artistic enclave where the gifts of Apollo prospered, but it also dedicated a space to project the virtues of powerful Mars through weaponry. Indeed, it was from Don Antonio Alonso's hand that the Pimentel armoury arose as a Virgilian hymn to the family's military glories, its history and its chivalrous greatness. Throughout its existence, the collection won the praise of the many people who visited the fortress and left their written impressions when studying what, according to Simal López, was the most attractive part of the edifice[15]. However, it is curious that the first documentary information about the armoury places it outside the castle, specifically in the Hospital de Nuestra Señora de la Piedad, patronised by the Count.

The hospital's erection came relatively quickly after its 1517 founding. The projected edifice had a square plan, organised around a central patio surrounded by semicircular arches and a second level with windows. The atrium provided access to the chapel, in line with the entrance, the room for pilgrims, and the other rooms in the hospital. In 1520, it was already built and six years later, it received the ordinances that made its assistance exclusively for the "*pobres y peregrinos que pasan por la villa de Benavente en romería a Santiago* [poor and pilgrims that pass through the town of Benavente on pilgrimage to Santiago de Compostela]"[16]. Its administrators, initially a prior and later a servant of the house of Pimentel, remained subject to the patrons' authority, which gave the complex a glint more of splendour. Various members of the family were interred in its chapel and, moreover, endowed the hospital with its rich domestic

12 Muñoz, 1877: 47. Gómez de Castro, 1561: 90–91.

13 Ferrer Valls, 2007: 198.

14 Simal López, 2002: 27.

15 Simal López, 2002: 106.

16 González Rodríguez, 1998: 173–174.

furnishings[17] that, along with the service staff, aroused the sensation in Villalba y Estaña (1577) of finding himself in a "*palacio* [palace]"[18].

A vital element in this expression of grandeur was the collection of arms that the Count brought together in the hospital, according to various account books unpublished until now. The documents, which include charges made in different years, refer to pieces located in the "*sala grande y cuadra del dicho hospital* [large room of this hospital]" that belonged both to the Count and to "*la gente de su casa* [the people of his house]"[19]. The first recorded charge was found on 21 May, 1541, to Pedro de Grijalba, the Count's servant, and it will be the basis of the analysis. After this, we find other deliveries signed for by the different managers, who also write down particularities and circumstances of the collection with respect to the previous charges: Juan de Valdivieso (1550) and the armourers Pedro de Rivas (1571), Francisco de Canseco (1573) and Juan Rodríguez de Mercado (1574). The weapons were entrusted to Rodríguez de Mercado until 1581, the date of his death, when the task passed to his brother, friar Francisco, the hospital's prior, who was tasked with giving an account of their condition before the chief armourer, Antonio de Salazar. The size of the collection was extraordinary and this is reflected in the document, which, at times, groups the pieces according to their type but, at other instances, places them in disparate points from similar weapons. Thus, to present a categorisation of armaments that is both effective and illustrative, we will analyse the weapons according to their criteria of typology and function, largely following the document's organisation.

In 1541, harnesses formed the first great set of pieces with a quantity that surpassed 100. In the list, they are ordered according to their purpose, location, origin and morphological similarities, although the variety is a transversal constant, given that a large part of the pieces have distinctive features relative to the others. Those fabricated for jousting, royal or for war, number eight; seventy five for engaging in war, all of which are recorded as "*arneses de guerra complidos* [complete war harnesses]"[20], except four that are described in more detail. Next comes the sets, the function of which is not determined or those that have double pieces. About twenty of these were for riding carelessly. Along with these, we find morions, pauldrons, arm guards, gauntlets, plackarts and cuisses, among a large number of loose pieces that precede the "*cotas y cosas de malla* [coats and

17 Pérez Álvarez, 2016: 350–351. Simal López, 2002: 163–164.

18 Villalba y Estaña, 1886: 365.

19 AHNOB, Osuna, C. 424, D. 32–34.

20 AHNOB, Osuna, C. 424, D. 32, fol. 62 r.

things of mail]": three coats, five besagews and three collars. Under "*armas de la gineta* [arms of the horse rider]" some helmets, sallets and *capacetes*[21], cuirasses and *coracinas*[22] with different ornamentation and other pieces for arms and legs are grouped together. One of the most noteworthy sets of suits of armour was the gold and crimsoned-trimmed one belonging to "*conde don Alonso que haya gloria* [Count Don Alonso, let there be glory]"[23], which had been the one that the 5th Count of Benavente wore upon entering Valladolid in 1521. With it were harnesses of other members of the family such as Alonso Pimentel Maldonado and his brother Don Juan Pimentel, Philip II's gentleman of the bedchamber and knight of Calatrava[24].

Also are found other pieces acquired from the monarch by various nobles, such as Don Íñigo de Guevara (*c.* 1500), Lord of Salinillas; Don Claudio Fernández de Quiñones, the 4th Count of Luna and Don Alonso Manrique, Lord of Las Grañeras and Don Pedro Sarmiento[25]. Certain purchases were made from workshops in Milan, others from German merchants such as the Fugger family and a certain "*Enrique alemán* [German named Henry]", who must have been an armourer.

As was common in the era, the Count also received a number of gifts from people with whom he had a relationship, whether personal or political[26]. In the first place, the chiselled white harness that the emperor gifted him stands out. We also have the gilded one and the two corslets that Don Juan Esteban Manrique de Lara, 3rd Duke of Nájera, a man very close to the Count in the military events of the 1530s, sent[27]. Moving on to foreigners, the black, gilded and engraved damascened corslet that "*il Moro*", Alessandro de Medici (1512–1537), Duke

---

21 A late Middle Ages helmet produced in Spain that preceded the morion.

22 A type of cuirass that protects the internal part of the chest, often used by pikemen.

23 AHNOB, Osuna, C. 424, D. 33, fol. 10 r.

24 López de Haro, 1622: 134.

25 On the family history of Guevaras and Manriques, see Salazar y Castro, 1795: 77, 155. It fails to note the identity of Don Pedro Sarmiento, who, almost assuredly, would belong to the Count of Salinas branch, given that its holder is mentioned repeatedly throughout the account books.

26 Davis, 2000.

27 The Dukedom of Nájera is a constant in the inventories of the Hospital de la Piedad. It occurs in two fundamental ways: the owner and the armouries that Manrique had at his disposal, of which the name of "maese Hernando vizcaíno" and that of Albaina as a centre of production stand out. Regarding Don Juan Esteban, it is worthwhile to consult the chronicler Alonso de Santa Cruz, who cites him a number of times alongside Pimentel, for example, in Nice (1538). See Santa Cruz, 1992: 507.

of Florence, gifted him is noteworthy[28]. Moreover, the opulent engraved white harness that a gentleman of Ferdinand of Austria brought him merits special attention. The scribe transliterated this gentleman's name, who had to have been Czech or German, into Castilian as "Pérez Coz" and "Pez Coz"[29].

For its part, the mounting team brought together a notable quantity of coverings and protections (at time inventoried along with the accompanying harnesses), among which cruppers, breastplates, collars, frenteras and saddles stand out. Given their connection, vamplates and other elements of jousting harnesses were also incorporated[30]. The first pieces of horse vestments were inherited by the Count from his father, made in brocade "*con franjas de oro y seda colorada con los hierros y hebillas doradas* [with gold stripes and coloured silk irons and gilded buckles]", surely a pair of its harness[31]. With this, a dozen steel covers that the Count purchased from Desiderius Helmschmied, the prestigious Augsburg armourer, of which nine were "*historiadas de cosas antiguas* [storied of ancient things]", decorated with yellow stripes and woollen tassels. Another covering, this one of black velvet with stripes and almost three-hundred gold-thread tassels, was manufactured for the 1535 entrance into Rome. Normally, these harnesses corresponded in appearance to the bridles and, of course, the saddles, which the Count had in good numbers at the Hospital, some of them acquired from the same Desiderius and others from Milan. Among the most opulent is a saddle with the gilded edges and swords "*guarnecida de terciopelo morado recamado de oro y franjas de oro y seda morada cubierta de paño blanco* [trimmed with purple velvet embroidered with gold and golden stripes and purple silk covered with white cloth]". We will highlight another saddle with the gilded and engraved swords[32].

The other main category consists of hand weapons, beginning with the pointed ones, in which thirty-one partisans, sixty-seven halberds, twenty-two "roncas"[33], twelve pokers, thirty-seven loose irons as well as a total of thirty-one spears and lances are recorded. Second, fifty-seven rondaches with a large ornamental diversity are noted. Although the majority of them are steel and gilded with twisted silk trim, they were painted with stripes, heraldic motifs, specific colours, animal and human figures as well as mythological themes like

28 Pingaro, 2016: 87.
29 AHNOB, Osuna, C. 424, D. 32, fol. 27 v-31 v.
30 AHNOB, Osuna, C. 424, D. 32, fol. 3 r-43 v.
31 AHNOB, Osuna, C. 424, D. 32, fol. 53 r.
32 AHNOB, Osuna, C. 424, D. 32, fol. 38 r-38 v.
33 A weapon similar to a partisan.

the story of Gaius Mucius Scaevola and the kidnapping of Ganymede. The most noteworthy were the two rondaches "*rodelas* [rondaches]" of feathers "*de Indias* [from America]", one of them decorated with jaguar skin around it[34]. After the attraction the feathering caused, although the following do not belong to Aztec art, the feathers protagonise another element, in this case linked to the helmet's decoration. In 1563, a charge for the remaining "*plumas de armas* [feathers of arms]" from the tournament celebrated by the Count in 1560 to celebrate the wedding of Philip II to Elisabeth of Valois was recorded, which he took out for the Toledo celebrations as well. They are followed by the feathers given in 1562 by Hernando Pimentel, *cacique* of Texcoco, who not only maintained a good relationship with the Count but also received his distinguished surname through the Count's intercession with the emperor[35].

The section continues with "*espadas y estoques y dagas con talabarte* [swords, rapiers and daggers with belts]", which is one of the most interesting due to its significance and the material quality of its arms. In this segment, the most numerous and varied group consists of swords, the knights' essential steel, which total twenty-six in number and whose typological variation helps one to get an idea about the function for which they were intended and the importance the Count placed on them. In the armoury, there was a pair of swords for hunting; a hand-and-a-half sword, seven for tournaments, four greatswords, six rapiers and an unspecified trio. Of note is the greatsword[36] with an undulating blade. The falchion gifted by Don Alfonso Dávalos (1502–1546), 2nd Marquis of Vasto[37], likewise stands out. It is a piece with a silver guard and grip that is stored in a purple velvet scabbard with a bocal, ferrule and silver thread around it. Also included in it are two scimitars adorned with gold that had a head of a Turk and a bird on the pommels as well as three damascened scimitars, one of Tunisian origin with gold and silver decoration and the pommel with a bird's head, and two exceptional scimitars of Turkish origin gifted by Ferdinand of Austria. The monarch also rewarded him with a bow equipped with a quiver with nineteen

34 AHNOB, Osuna, C. 424, D. 32, fol. 46 r-46 v. AHNOB, Osuna, C. 424, D. 33, fol. 105 r.

35 AHNOB, Osuna, C. 424, D. 32, fol. 107 r and v. Don Antonio Alonso personally requested that the monarch grant his surname and arms to the American, as recorded in AGI, México, 1089, Leg. 4, fol. 422 v.

36 Also known as a Zweihänder or, in Scotland, a claymore.

37 A good part of his military renown was gained in the Italian wars, where he fought alongside Don Antonio Alonso in various events according to the imperial chronicler, one of them being the 1536 entrance into Lucca. See Santa Cruz, 1922: 365.

arrows along with another thirty in a wooden box[38]. Among the firearms, we have three arquebuses, of which two marched with the Marquis of Villena. Harnesses, girths, and an interesting group of hackles, antecedents to the most recent registry, which corresponded to the armament of the "*hombres de armas* [men at arms]" charged separately on 16 June, 1541[39].

In 1550, there already were more than 2,000 pieces and, according to what we know, the collection was separated into two locations. Juan de Valdivieso himself signed a plan that reveals the existence of a space and a catalogue of pieces called an "*armería* [armoury]" in the castle, which is specified in connection with one of the harnesses. This detail indicates that the entirety of the arms the Count possessed was divided between the Hospital de la Piedad and that fortress, which already housed part of the armaments of Don Alonso Pimentel in 1530. Likewise, the fact that a space with this function is indicated invites us to consider that the hospital complex must have acted as a complementary location, surely on a temporary basis, until the necessary rooms were prepared. Obviously, the patrimonial fortifications of the great noble families accumulated armaments since the Middle Ages, but not all of them could have accommodated the armouries that were being formed in the sixteenth century within their walls.

If we follow the account books, the functionality of the hospital as a storehouse of arms lasted about forty years. The weapons were definitively moved to the fortress under the supervision of Juan Rodríguez de Mercado (1574). The same pieces that had been recorded up until then, except for some losses and gifts, such as the "*arneses dobles y sencillos que parece que hay en la armería del conde mi señor* [double and single harnesses that seem to be in the armoury of my Lord, the Count]", appear on 3 April of that year[40]. In 1581, there is already clear talk of the "*armas de la armería de su señoría* [arms of his lordship's armoury]" in the position of the senior armourer, Antonio de Salazar[41].

THE FLEETING PLEASURE OF THE BENAVENTE ARMOURY

The Pimentel armoury was clearly constituted in 1573, the year in which the Count was able to communicate to the king that he possessed a collection of

38 AHNOB, Osuna, C. 424, D. 32, fol. 48 r-49 v. The bow in AHNOB, Osuna, C. 424, D. 33, fol. 105 r.

39 AHNOB, Osuna, C. 424, D. 32, fol. 44 r-51 v, and fol. 67 r.

40 AHNOB, Osuna, C. 424, D. 33, fol. 111 r

41 AHNOB, Osuna, C. 424, D. 33, fol. 5 r.

arms capable of equipping a good number of people on foot and on horse[42], a description that Villalba y Estaña would confirm in his *Pelegrino Curioso* [*Curious Pilgrim*] four years later:

> *quitando la del Rey, es la mejor cosa que hay en España. Habrá en ella más de dos mil coseletes todos con el aderezo necesario y unos espejos que os podéis mirar en ellos, y otras muchas invenciones y géneros de armas, de que nuestro pelegrino quedó muy pagado*[43]. [apart from the king's, it is the best thing that there is in Spain. In it, there are more than two-thousand corslets, all with the necessary dressing, and some mirrors in which you can see yourself, and many other inventions and types of arms for which our pilgrim was very well repaid].

By 1574, the fort of Benavente had become the sole site of the Count's armoury. Elevated on a hill on the outskirts of town, it stood as a symbol of the lineage and witness to the passage of time, the roots of which were sunk in the fourteenth century, developing fully beginning in the following century. From then on, the renovations undertaken by Don Rodrigo Alfonso Pimentel, 4th Count of Benavente and his son, in addition to the works already mentioned, finished preparing its grandeur. For instance, they incorporated a brand new garden and the "*Bosque* [Forest]", illustrative of its hunting function[44]. To approach the fortress, we have a multitude of literary and graphic sources, although the fundamental document remains the plan outlined on 6 January, 1809, but the French engineer in the context of the War of Independence[45].

The fortress had an irregular plan and was flanked by towers on each of its corners and by bastions that strengthened its defence. The moat on the northern facade and the western side that ended in the Órbigo River also contributed to its protection. It was accessed through the "*callejón de Santiago* [Santiago alleyway]", and later via the bridges that crossed the moat and introduced the traveller into a courtyard, the castle's initial interior area. In the centre was placed the main atrium that Münzer (1494) mentions. It was surrounded by pavilions and other rooms richly decorated in the Mudejar style with marble and jasper columns and gilded coffered ceilings. Nearby was the "*Torre del homenaje* [Donjon Tower]" and to the south was the "*Torre del Caracol* [Snail Tower]" (Fig. 1). In the interior, it is necessary to highlight the Lineage room, "*con escudos de armas de las casas principales de España* [with coats of arms of Spain's principal noble houses]" and

42 García Hernán, 1998: 337.
43 Villalba y Estaña, 1886: 365. Simal López, 2002: 87.
44 Beceiro Pita, 1997: 185–186. Simal López, 2002: 74–75, 80. Martínez Muñoz, 2018: 117.
45 Regueras Grande, 2019: 194–201.

the "*cámara rica* [rich camera]", in which an archway opened to the Órbigo River and exposed the Leonese countryside[46].

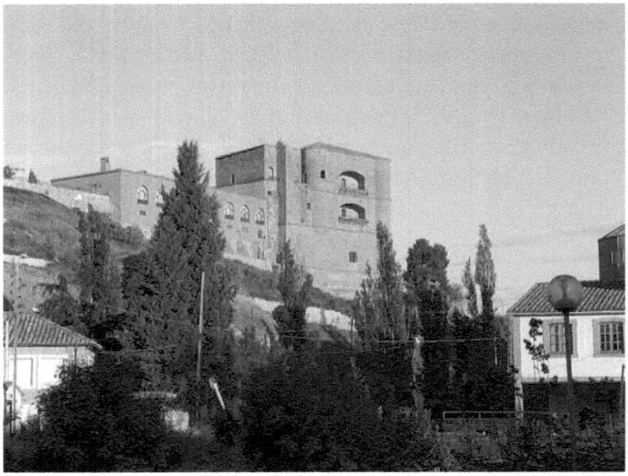

Fig. 1. *Torre del Caracol*. Benavente, Spain (Wikimedia Commons).

With the armoury complete between the palatial walls, the next, central stage began, the circumstances in the development of which can be seen more clearly. Don Antonio Alonso was able to act as the guardian of the complex for about four years. When he died, the responsibility fell to his son, Don Luis Alonso (†1576), who died without issue only a few months after inheriting the estate. It would be Don Antonio Alonso's second son, perhaps the most famous head of the house of Pimentel, who would direct the fate of the armoury for years to come.

Don Juan Alfonso Pimentel (1553–1621) was the 8th Count of Benavente and Count consort of Luna under the terms of his marriage to Doña Catalina Vigil de Quiñones, the "*hija única y heredera de la ilustrísima casa de Luna* [only daughter and heir of the illustrious house of Luna]"[47], on whose death the title entered into the Pimentel lineage. At the end of the century, he was honoured as viceroy of Valencia for four years (1598–1602)[48], during which time he hosted

46 González Rodríguez/Regueras Grande/Martín Benito, 1998: 64. Regueras Grande, 2019: 33–35.

47 BNE, VE/141/84/7, fol. 11 r.

48 Mateu Ibars, 1963: 186.

two important weddings: that of the infanta, Isabel Clara Eugenia, to Albert, Archduke of Austria and the marriage of Philip III to Margaret of Austria. Likewise, in 1600 he hosted the monarch for some time in his palace in Valladolid and in 1602 he received one of the positions most desired by any nobleman of the period: viceroy of Naples[49]. His stay in the Italian kingdom piqued the Count's cultural interests, which already stood out for his refined courtly behaviour, so much so that Lope de Vega himself gave him tenth place among the Muses[50]. In Italy, he became involved in the artworks and antiquities trade and was able to acquire a large quantity of objects of extraordinary wealth and symbolic value for the house of Pimentel, which returned with him to Benavente and increased the collection's quality there. In 1607, in Naples, the Count received some gilded weapons from Don José de Acuña in the name of his son, Don Juan Pimentel y Zúñiga, and he acquired a series of swords, crossbows and suits of armour that he transported to the Benavente armoury[51]. Later, he occupied the role of the queen's *mayordomo mayor* [high steward] and the presidency of the Council of Italy. There is no doubt that Don Juan Alfonso was an active patron of the arts and he possessed deep intellectual interests that he did not hesitate to apply with plenty of intensity in the mechanisms of noble representation, in addition to being a knight of valour, gallantry and courage, as Miguel Daza notes[52]. During his term as Count, the house of Pimentel entered into one of the most noteworthy periods in its history, although the economic and political circumstances ended up signalling significant difficulties[53].

The first information we have about the armoury in the period of the 8th Count of Benavente comes from the 1581 charge to Antonio de Salazar. In it, we find some novelties, such as the gold-banded harness that was purchased from Gabriel Zapata in Madrid as well as references to other suits of armour that lost pieces or had been given away. For example, the Enrique Alemán harness was lent to Don Diego Sarmiento, son of the Count of Salinas, and a small harness was gifted to the first-born son of the Marquis of Viana del Bollo, Don Juan

49 Ledo del Pozo, 1853: 290. Berdum de Espinosa, 1753: 67. BNE, Mss. 11569, fol. 159 v-161 r. Simal López, 2002: 34–36. *Libro donde se trata de los Virreyes, Lugarestenientes de este Reyno y de las cosas tocantes a su grandeça,* edited by Jusepe Raneo, BNE, Mss. 2979, fol. 21 r-25 v. As viceroy of Naples, he was an important actor in Lerma's foreign policy. See García García, 1996.

50 Regueras Grande, 2019: 43.

51 Simal López, 2002: 180–182. González Ramos, 2017: 281.

52 Martínez Muñoz, 2018: 127.

53 Simal López, 2002: 87–88; 2005: 32–37. Yun Casalilla, 2002: 16, 113, 151, 211.

Pimentel Manrique, as was another to Don Bernardino Pimentel, son of the Marquis of Távara[54].

Later, on 4 November 1593, the "*armas y otros bienes* [arms and other goods]," property of his lordship, were recorded[55]. The document lists 161 corslets, of which forty-six (including two small ones) were complete, three double corslets, five for jousting "*de planchón* [of iron]", three without detail, and thirty-one remnants "*de munición* [munition-quality ones]". The incomplete ones number forty-one of this last type, which perhaps could have been completed with the numerous related loose pieces, and seventy-seven sets of breastplates, backplates and morions. A large part, as happened in Don Antonio Alonso's era, were gilded, chiselled and blued, although there were various white and black ones, perhaps some of which were from Alessandro de Medici. Others stood out for being engraved and gilded with three or five stripes and having fleurons on different pieces. An exceptional case is the blued and gilded harness with the image of "*Nuestra Señora de la Anunciación en los pechos* [Our Lady of the Annunciation on the breasts]". Equally, there were some linked to certain figures, such as what happened with the small harness with red outlines and five engraved and gilded stripes that belonged to a young Don Luis Fajardo, 4th Marquis of Vélez, who was only seventeen years of age[56]. Three jousting harnesses and a blued war harness belonged to Don Alonso Pimentel, knight of Calatrava and Don Juan Alfonso's uncle[57]. In the area of purchases, we know that Don Juan Alfonso acquired a harness from the celebrated master armourer Pompeo della Cesa in the second half of the sixteenth century[58] along with various other ones "*del vizcaíno* [from the Biscayan]"[59]. Also, cuirasses of different materials and padded fabric tunics are found.

---

54 AHNOB, Osuna, C. 424, D. 33, fol. 4 r, 11 v, 13 r. Both marquisates had been granted to members of the Pimentel family. Soler Salcedo, 2020: 142–143.

55 AHNOB, Osuna, C. 427, D. 111.

56 AHNOB, Osuna, C. 427, D. 111, fol. 4 r, 5 r and v, 9 v.The 8th Count of Benavente married a second time to Doña Mencía de Zúñiga, who had Don Luis before being widowed by the 3rd Marquis of los Vélez. This political connection was reinforced by blood when Fajardo married his half-sister, Doña María Pimentel y Quiñones. See Soler Salcedo, 2020: 145.

57 AHNOB, Osuna, C. 427, D. 108, fol. 1 r and v. López de Haro, 1622: 134–135.

58 Simal López, 2002: 107.

59 This denomination brings us closer to the Duchy of Nájera and the sets acquired from the master Hernando by Don Antonio Alonso. Perhaps they were still preserved in the family armoury, AHNOB, Osuna, C. 427, D. 108, fol. 1 r.

With regard to horse riding, although stirrups, frenteras and groups of trimmings (which everything indicates would be for the men-at-arms) are distributed among the corslets, a set of forty-one saddles is noted. Of these, nineteen were weapons, four were Moorish, ten were for jousting and the others for unspecified reasons. Next, we find a heading entitled "*cubiertas de caballo de terciopelo bordadas* [embroidered velvet horse coverings]", where, in addition to skirts (six), breastplates (three), adornments and more loose pieces, a dozen covers and another dozen smocks and short tunics are noted. The most detailed of these is the one embroidered of gold, white, and yellow cloth, "*con unas hojas y troncos* [with some leaves and trunks]" that had a neck, breastplate and haunches with different tassels, all trimmed with a fringe of gold and purple silk[60].

Next, we have a total of fifty-nine rondaches, of which forty-five were Milanese, four were "*a prueba de arcabuz* [arquebus-proof]" and another "*de pluma con las armas Pimentel* [of feathers with the Pimentel coat of arms]". Four Stechtarschen[61] and a certain number of *pavises* accompany them. Among the pointed weapons, we find thirty-one guisarmes and Milanese partisans, twenty-five cavalry lances, fifty infantry lances, six short lances and a spear "*de palo alemán nudoso* [of knotty German wood]". Fifty-nine halberds come before the forty-eight irons without shafts that complete the group. We also find four maces and a war hammer. The swords total seventy nine, thus surpassing the number of positions in the hospital but the document offers a minor typological and functional detail: forty-nine hold twenty-nine daggers, they being sheathed swords, and thirty-two waistbands; five are old and without scabbards; the same number of tournament ones and six rapiers and four scimitars. Other weapons that are recorded were two bows, twenty-one crossbows, eleven muskets and twenty arquebuses with their large and small powder flasks. Finally, some pieces that were moved to Leon are discussed[62]. Similarly, we find two Turkish "*atabalillos* [small kettledrums]" and a total of four mirrors[63].

The next reference to the Benavente armoury is literary and involves Carlos de Requesens, who in 1595 referred to the collection as a "*gran armada de diferentes armas, y llena de pinturas y cosas curiosas está* [grand armada of different weapons and full of paintings and curious items]"[64]. Another documentary

60 AHNOB, Osuna, C. 427, D. 111, fol. 4 r-9 v.

61 A small shield of German origin laced to a jouster's left shoulder.

62 AHNOB, Osuna, D. 111, fol. 1 r and v, 5v, 10 r and v.

63 AHNOB, Osuna, C. 427, D. 111, fol. 2 v, 8 v.

64 González Ramos, 2017: 281.

mention comes from 1600, the year in which the Count was acting as viceroy in Valencia. It consists of an inventory under the supervision of the wardrobe attendant Gerónimo Ruiz of the arms and suits of armour that Don Juan Alfonso brought together in the viceregal capital either by acquisition or by shipment from the fortress[65]. It is worth noting that in 1599 the wedding between Philip III and Margaret of Austria was celebrated. Thus, these pieces were probably needed to march in the parades and compete in the associated tournaments. Even if it does not strictly refer to the armoury, it does provide information of specific pieces that came from it. Furthermore, we know that they returned with new pieces through some marginal notations. The defensive armaments are not very numerous and they appear organised into three sets: one consisting of a rondache—"*rodela*"—and a morion, another containing a corslet without protection for the lower train and a third consisting of a breastplate with a morion and a strong rondache. Among the hand weapons, we find a pair of very detailed cavalry swords, one of which is known as "*la rica* [the rich one]" the hilt, grip, pommel and contera of which were silver plated with gold lace. It is accompanied by a silk and gold baldric with lace and yellow tassels, as well as two silver capes, a clear opulence that must have pleased the Marquis of Vélez when he later received it as a gift. Furthermore, the inventory included a sword with dagger, two large hunting knives and four crossbows, of which three belonged to Don Juan Alfonso[66].

Three years later, in 1603, the armourer in charge of the armoury, Juan de Castro, drafted another inventory in which the armoury's pieces were inscribed[67]. This document is the first that provides certain data about the organisation of the collection and the describes the space in which it was located: three rooms or quarters whose walls "*están cubiertas de despojos de armas que son brazales y manoplas y quijotes y guardabrazos y grebas y morriones, y arandelas y otras cosas* [are covered with spoils of war including brassards, gauntlets, cuisses, arm guards, greaves, morions, vamplates and other items]"[68]. Additionally, there were six chests and two of them were full of the same items. The listed weapons totaled almost 4,500 in number, of which 3,360 are "*cuadrillos grandes y pequeños* [large and small arrows]", a group that is now included. We sense that

---

65 Simal López, 2002: 175 includes the information of the same task with another documentary reference. Ours, although different, refers to the same data in AHNOB, Osuna, C. 429, D. 116.

66 AHNOB, Osuna, C. 429, D. 116, fol. 27 r and v.

67 AHNOB, Osuna, C. 427, D. 108.

68 AHNOB, Osuna, C. 427, D. 108, fol. 5 v.

those that are missing among the remaining 1,100 and 2,000 mentioned above would be on the walls and in the chests. The variety seen previously continues, but interesting pieces appear. For instance, among the eight crossbows that the armoury contained, one was of "*marfil de Brasil contrahecho con la verga de acero y la nuez de marfil* [imitation Brazilian ivory with the prod of steel and the catch of ivory]" whilst another was "*de ataujía labrada de hueso de marfil con la verga de nervios sin nuez* [of damascene carved of bone of ivory with the prod of cartilage without a catch]"[69]. We also have novelties among the firearms, which include nine muskets, eleven cannons "*de munición* [munition-quality ones]" and a total of eleven old arquebuses, inventoried alongside "*una lanza que dicen ser de Bernardo del Carpio* [a lance that they say was from Bernardo del Carpio]", the most noteworthy of the pointed weapons and one surely quite appreciated by the Count[70].

That same year, the Count of Benavente rose to become viceroy in Naples and, as such, he could participate much more easily in the Italian market, so much so that the collection ended up being strengthened in many aspects[71]. In this sense, the 1607 documents are central, by which we know that Don Juan Alfonso added rich pieces to his wardrobe, such as some harnesses gilded with the heraldry of the "*Gran Capitán*", Gonzalo Fernández de Córdoba (1453–1515). Moreover, he received a group of "*armas doradas* [gilded arms]" that José de Acuña sent to Don Juan Pimentel y Zúñiga, the Count's son[72].

On his return to the Iberian Peninsula, the house of Pimentel's situation was poor given the increased debt that the Count had contracted in Italy, the satisfaction of which became, at the very least, complex. As a result of these needs, an auction was held in Valladolid in 1614[73], whose relationship is interesting for the Benavente family's weapons, as the sale of some of those that were in the Count's palace wardrobe in this city is reported. This and two additional pieces of documentary evidence relating to the free goods of the Counts and the chamber

69 AHNOB, Osuna, C. 427, D. 108, fol. 2 v.

70 Although there are those who questioned some of his most imaginative feats, the figure of Bernardo del Carpio was highly revered in the Modern Age, so much so that any object connected to him ended up sharing that respect. For more on the hero, see Rubio García, 2000: 7–30.

71 Simal López, 2002: 89.

72 Simal López, 2002: 180–182. González Ramos, 2017: 281. We identify the recipient as the 1st Marquis of Villar de Grajanejos (†1626), eldest son by the second marriage of Don Juan Alfonso. Soler Salcedo, 2020: 145.

73 AHNOB, Osuna, C. 430, D. 95.

of his lordship[74] are those that allow us to know more about his possessions in these years. They included weapons that, as in other cases, the nobleman had at his direct disposal, not from the armoury, although in many occasions the pieces might have come from the symbolic set or ended up becoming part of it. The inventory of the free goods that the Count had in Benavente in February 1614[75] shows us an extensive list of weapons in which twenty-seven firearms stand out. Of these, twenty-two are pistols, some of which are first-rate. To them is added a rich sword with a gilded and red dagger "*con el hábito de Santiago por guarnición tiros y pretina de terciopelo carmesí* [with the insignia of Santiago as decoration, straps and crimson velvet waistband]"; and different black swords[76]. This document also includes six Turkish muskets "*en sus cajas y con sus vaquetas doradas y grabadas* [in their containers and with their gilded and engraved ramrods]"[77].

In 1618, "*todas las cosas de armería de su excelencia* [all of the items of His Excellency's armoury]" were again recorded, this time under the supervision of jewelkeeper Juan de Vivero and armourer Alonso de Alvarado. The volume is practically the same as the 1603 inventory and the pieces are identical, which means that during these years the collection experienced a period of stability in its composition. Even so, it is necessary to highlight three tournament swords "*de los señores pequeños* [of the young lords]" with two daggers and the most distinctive pointed arms: an "*africano* [African]" iron lance, an English pike, spears on German shafts and irons for larger spears and bull lances[78].

Beginning in 1619, furniture and other objects were transferred from the fortress to the Valladolid palace and to the Madrid residence due to the political responsibilities that required the Count's presence at court[79]. The pieces that were moved were those of daily use (swords, books, daggers, crossbows, suits of armour), given that large quantity of the collection remained in the fortress's armoury, the indisputable bulwark of the family's symbolic power[80]. A memoir of the things he kept in the Valladolid chamber, drawn up after the Count's death, helps to reaffirm this idea, as it reveals a majority of opulent pieces at the owner's disposal. These include a pair of black swords and a "*espada y daga y espuelas*

74 AHNOB, Osuna, C. 429, D. 114, fol. 14 v.
75 AHNOB, Osuna, C. 3916, D. 1.
76 AHNOB, Osuna, C. 3916, D. 1, fol. 219 r-220 r.
77 AHNOB, Osuna, C. 3916, D. 1, fol. 218 v-219 r.
78 AHNOB, Osuna, C. 427, D. 107.
79 Simal López, 2002: 103.
80 González Ramos, 2017: 281.

todo guarnecido con sangre de dragón [sword, dagger and spurs all decorated with dragon's blood]", which followed the "*guarnecido de oro* [gold decorated]" staff of the queen's *mayordomo mayor* [high steward][81].

The relocation continued during the incumbency of his grandson, Don Juan Francisco Pimentel, 10th Count of Benavente, who died at the end of December 1652 (Fig. 2). A few days later, an inventory was made of the objects that were in the chamber and the Valladolid palace's "*aposentico de las armas* [chamber of weapons]"[82], that includes the existence of eleven swords, seventeen firearms, four crossbows, three bows and some pieces of armour. The document referring to the "*cosas que son antiguas* [things that are ancient]" of the house's estate[83], which includes different interesting arms, must be from this time as well. The "*alfanje damasquino con su caja de terciopelo carmesí y piedra turquesa que se dice fue del gran Solimán* [damascene cutlass with its crimson velvet case and turquoise stone said to have belonged to the great Suleiman]" and the broadsword "*que trajo el señor san Isidoro de León* [that Saint Isidore of Leon, carried]", stored in a black box "*forrada con toda reverencia* [upholstered with complete reverence]" are striking[84]. As González Ramos states, its authenticity remains in doubt, but this does not take away from the fact that they had special meaning for the family and, as such, it would be able to return to the armoury. To this same group must be included the sword found amongst the pieces of the Count's estate belonging to Suero de Quiñones, famed for his "*pas d'armes*"[85].

81 Archivo Histórico de Protocolos de Madrid, P. 4429, fol. 738 v. Teijeiro, 2014: 364.

82 AHNOB, Osuna, C. 427, D. 100.

83 AHNOB, Osuna, C. 433, D. 82.

84 AHNOB, Osuna, C. 427, D. 100, fol. 59 r.

85 González Ramos, 2017: 281–282. According to Soler del Campo, during the seventeenth century, it was common to find swords attributed to Suero de Quiñones among the goods of Spain's noble families. The knight of the *pas d'armes* belonged to the House of Luna, whose titles and estates entered the lineage of the House of Pimentel at the end of the sixteenth century.

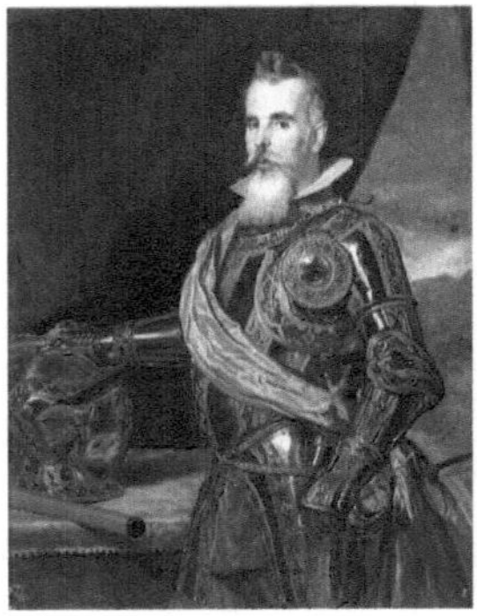

Fig. 2. *Portrait of Juan Francisco Pimentel, Count of Benavente*. 1648. Diego Velázquez. Madrid, Museo del Prado (Wikimedia Commons).

The fortress's inventory written up in April 1655, distributes the armoury across five rooms, two more than the 1603 document, which means that the number of rooms to hold the collection were increased. The information that this registry offers us and its comparison to the 1809 French plan, moves it to the area near the fortress's entrance, the moat, and the staircase, so that it was undoubtedly located in the rooms of the northern facade[86]. This is how it must have been since Don Antonio Alonso ordered the weapons transfer from the hospital at the beginning of the 1570s.

In 1655, the eighty-nine medium and full armoured bodies, the fifty-five breastplates and back plates, four corslets, seventy-one morions, sixty-four *pavises*, among other items, stood out. There were twenty-nine firearms, and they included three shotguns, eighteen pistols, seven wall muskets and a bullet. We also found eleven crossbows. There still remained eighty-four partisans, twenty-three maces and three loose greatsword blades, which are related to campaign saddles, *atabales* [kettle drums], a pair of mirrors of armour, a cannon and a general's sceptre. Although we highlight the "*cuna dorada con las armas de la Casa* [golden cradle with the house's arms]"[87] as an interesting piece, there would be other important arms amongst the odds and ends that covered the walls. In fact, it appears that they continued to be functional because a wooden ladder was listed to reach them. Although a certain neglect is perceived towards the weapons collection, its magnificence nonetheless continued to pervade the castle

86 AHNOB, Osuna, C. 427, D. 108, fol. 5 v.
87 AHNOB, Osuna, C. 427, D. 108, fol. 1 r.

with the timeless essence of family prestige, so much so that the edifice's main attraction resided in the armoury[88].

THE SWAN SONG

During the rest of the seventeenth century, there would be some news of the fortress, but not specifically of the armoury. One must wait until 1702 to find documentary descriptions of its contents, already under very different circumstances. This year an inventory was carried out on the order of the Countess Doña Manuela López de Zúñiga, wife of Don Francisco Antonio Casimiro Pimentel, the 12th Count of Benavente[89]. By then, the king had gone on the Italian campaign and the Count of Benavente, in his capacity as *Sumiller de Corps*, accompanied him. According to Simal López, this would explain the desire to reorganise the collection. It must be stated that the fortress also preserved the family archive, inventoried in 1746, and the library, although it had lost a good number of its items as a result of transfers. Despite these changes, it continued being the symbolic bastion of the family's history and power[90].

The disposition of the collection continued to occupy five rooms in which the large part of the pieces were "*muy antiguas y del tiempo de los moros* [very old and from the time of the Moors]", highlighting some harnesses that had been assembled on mannequins and wooden frames, undoubtedly of the most attractive and excellent craftsmanship. In the first room, one found "*dos caballos fingidos y armados sobre madera y los homb^s^ de cuerpo entero* [two armoured fake horses on wood and the men in full body]" and another three armoured horses with their "*jinetes*" in armour, among other items in the second[91]. Also were inventoried forty-five half bodies with armour, forty-two morions, thirty-two bucklers, 106 adargas, twenty-six swords and an abundance of "*pedazos de petos, espaldares, brazos, piernas y guantes todo de hierro templado* [pieces of breastplates, back plates, arm guards, leg plates, and gloves, all of tempered iron]" that hung from the walls, distributed in the first four rooms. Additionally, the many muskets and arquebuses are placed alongside their small powder flasks, the fifth room being full of them. Located on the walls, two portraits of the Marquises of Vasto, another of "*la señora doña Catalina Fajardo* [the Lady, Doña Catalina Fajardo]" at the age of four years, and one of "*Felipe Rey de España*

88 Simal López, 2002: 103–108. González Ramos, 2017: 282.
89 AHNOB, Osuna, C. 436, D. 69.
90 González Ramos, 2017: 284.
91 Simal López, 2002: 109–110. Previously published by Cadiñanos Bardeci, 1997: 61–62.

vestido a lo antiguo [Philip, King of Spain, dressed in the old style]" could be seen. After inventorying the fifth room, the gunpowder store was opened, where there was a large pot with twenty *arrobas*[92] of bullets, two chests with about five *arrobas* of gunpowder and eighty-four loose iron shot, as well as an artillery ramrod.

Moreover, along with these pieces and the accessories needed for their care and maintenance, objects resulting from the fascination with the *mirabilia*, which were linked with the Mannerist style of the sixteenth and early-seventeenth centuries, continued to be preserved[93]. The aforementioned great rod was kept, which is said to have come from India and belonged to a giant who fished with it, along with some from bones from an elephant "*cuyo pellejo está a la entrada de la fortaleza* [whose hide is at the fortress's entrance]"[94]. In any case, times had changed and, barring a few exceptions, the armoury's antiquarian status was undeniable: the majority of the arms and suits of armour could not serve as "*más que de memoria y adorno* [little more than a memory and a decoration]"[95]. This organisation had been the swan song, the final meaningful expression of the collection, which would not return to the documented history until the end of the century and the beginning of the nineteenth century, a preamble to its ill-fated destiny.

The fortress was occupied in 1808 by the English, "*que demostraron ser los más destructivos huéspedes que jamás había albergado […] se encendieron hogueras al lado de las finas paredes; se apilaron cuadros de valor desconocido como simple combustible* [who proved to be the most destructive guests who had ever taken shelter in it […] bonfires were lit next to the thin walls; paintings of unknown value were piled up as simple fuel]"[96]. At year's end, the French arrived, who followed up the devastation during the previous period to the great fire that devoured the fortress that Christmas. After the disaster, an attempt was made to salvage something in 1814, but with little success, and we know nothing more of the armoury until the 1837–1840 period. At that time, Don Pedro de Alcántara Téllez-Girón y Beaufort Spontin, 11th Duke of Osuna and 16th Count of Benavente (heir of his grandmother, Doña María Josefa Pimentel, the end of the lineage in the male line), was interested in recovering whatever arms and

92 A historical measure of Spanish weight equal to about eleven and a half kilograms today.

93 Simal López, 2002: 108–110. González Ramos, 2017: 283–284.

94 In 1518, Don Alonso Pimentel brought an elephant to the fortress to show it to the emperor, who visited Benavente that year. See García Mercadal, 1962: 712.

95 AHNOB, Osuna, C. 436, D. 69, fol. 3 v-5 r.

96 Regueras Grande, 2019: 65.

suits of armour he could from the old castle. Excavations were made and locals were questioned but the result was not at all satisfactory. In total, twenty-eight *golines*[97], twelve frenteras, eleven helmets, 152 arm guards and fifty-nine pieces for hands were recovered. Some more pieces were also found in a cart but they were unable to "*formar armadura ninguna* [form any suit of armour]". Thus, with the Duke's hopes dashed, his aspirations vanished and the Pimentel armoury, maintained over the centuries, went along with them[98].

BIBLIOGRAPHY

Beceiro Pita, Isabel (1997): "La fortaleza de Benavente en el siglo XV". In: *Brigecio: revista de estudios de Benavente y sus tierras*, 7, pp. 185–204.

Berdum de Espinosa, Ignacio (1753): *Derechos de los Condes de Benavente a la Grandeza de primera clase*. Madrid, Lorenzo Francisco Mojados.

Cadiñanos Bardeci, Inocencio (1997): "Noticias de arquitectura fortificada en España". In: *Castillos de España*, 108, pp. 55–62.

Davis, Natalie Z. (2000): *The Gift in Sixteenth-century France*. Oxford, Oxford University Press.

Ferrer Valls, Teresa (1993): "Torneo organizado por el conde de Benavente en su villa de Benavente para festejar a Felipe II (1554)". In: *Nobleza y espectáculo teatral (1535–1622)*. Valencia, Universitat: pp. 177–180.

García García, Bernardo José (1996): *La pax hispánica: política exterior del Duque de Lerma*. Leuven, Leuven University Press.

García Hernán, David (1998): "Felipe II y el levantamiento de tropas señoriales". In: Martínez Millán, José (ed.): *Congreso Internacional Felipe II (1598–1998), Europa dividida, la monarquía católica de Felipe II* (20–23 abril 1998, Universidad Autónoma de Madrid). Madrid, Parteluz: pp. 333–344.

García Mercadal, José (1962): *Viajes de extranjeros por España y Portugal*. Madrid, Aguilar, Vol. 1 .

Gómez de Castro, Álvaro (1561): *Recebimiento que la imperial ciudad de Toledo hizo a la magestad de la reina nuestra señora doña Isabel, hija del rey Enrique II de Francia, cuando nuevamente entró en ella a celebrar las fiestas de sus felicíssimas bodas con el rey don Filipe nuestro señor, II deste nombre* . Toledo.

González Ramos, Roberto (2017): "Las armerías nobiliarias castellanas del siglo XVII como manifestación de identidad cultural". In: Carrasco Martínez,

97 Considered to be a smaller version of a gorget.

98 Simal López, 2002: 114, 280–283. González Ramos, 2017: 284–285.

Adolfo (ed.): *La nobleza y los reinos: Anatomía del poder en la Monarquía de España*. Madrid, Iberoamericana-Vervuert: pp. 269–312.

González Rodríguez, Rafael (1998): "Escrituras fundamentales del Hospital de la Piedad de Benavente". In: *Brigecio: revista de estudios de Benavente y sus tierras*, 8, pp. 169–192.

González Rodríguez, Rafael/Regueras Grande, Fernando/Martín Benito, José Ignacio (1998): *El Castillo de Benavente*. Benavente, Centro de Estudios Benaventanos "Ledo del Pozo".

Kamen, Henry (1995): *Una sociedad conflictiva: España, 1469–1714*. Madrid, Alianza.

Ledo del Pozo, José (1853): *Historia de la nobilísima villa de Benavente*. Zamora, Imprenta Vicente Vallecillo.

López de Haro, Alonso (1622): *Nobiliario genealógico de los reyes y títulos de España*. Madrid, Imprenta Real.

Martínez Muñoz, Ana (2018): "Imagen nobiliaria y ficción renacentista: la memoria linajística de la Casa de Benavente en un libro de caballerías manuscrito". In: *Cuadernos de Historia Moderna*, 43, pp. 111–131.

Mateu Ibars, Josefina (1963): *Los virreyes de Valencia: fuentes para su estudio*. Valencia: Ayuntamiento de Valencia.

Morales Folguera, José Miguel (2015): "El viaje triunfal de Carlos V por Sicilia tras la victoria de Túnez". In: *Imago: revista de emblemática y cultura visual*, 7, pp. 97–111.

Morán Turina, José Miguel/Checa Cremades, Fernando (1985): *El coleccionismo en España: de la cámara de las maravillas a la galería de pinturas*. Madrid, Cátedra.

Muñoz, Andrés (1877): *Viaje de Felipe segundo a Inglaterra*. Madrid, Imprenta de Rivadeneyra.

Pérez Álvarez, María José (2016): "Caridad y proyección social de la nobleza en la Edad Moderna: el Hospital de la Piedad de Benavente". In: *Chronica Nova*, 42, pp. 343–366.

Pingaro, Claudia (2016): "'Suntuose & splendide noze'. L'importanza del matrimonio di Cosimo I con Eleonora di Toledo per la costruzione dello Stato mediceo". In: *Pedralbes*, 38, pp. 85–111.

Regueras Grande, Fernando (2019): *Elogio de un fantasma. El Castillo de Benavente*. Benavente, Centro de Estudios Benaventanos "Ledo del Pozo".

Rubio García, Luis (2000): "Historia y poesía: Bernardo del Carpio". In: *Estudios románicos*, 12, pp. 7–30.

Salazar y Castro, Luis de (1795): *Árboles de costados de gran parte de las primeras Casas de estos reynos, cuyos dueños vivían en el año de 1683*. Madrid, Imprenta de Antonio Cruzado.

Sandoval, Prudencio de (1846): *Historia del emperador Carlos V, rey de España*. Madrid, Establecimiento tipográfico de P. Madoz y L. Sagasti, Vol. III.

Santa Cruz, Alonso de (1922): *Crónica del emperador Carlos V*. Madrid, Patronato de Huérfanos de Intendencia e Intervenciones Militares.

Simal López, Mercedes (2002): *Los condes-duques de Benavente en el siglo XVII. Patronos y coleccionistas en su villa solariega*. Benavente, Centro de Estudios Benaventanos "Ledo del Pozo" (CECEL-CSIC).

Simal López, Mercedes (2005): "Don Juan Alfonso Pimentel, VIII Conde-Duque de Benavente, y el coleccionismo de antigüedades: inquietudes de un Virrey de Nápoles (1063–1610)". In: *Reales Sitios*, 164, pp. 30–49.

Soler Salcedo, Juan Miguel (2020): *Nobleza española: grandezas inmemoriales*. Madrid, Visión.

Villalba y Estaña, Bartolomé (1886): *El pelegrino curioso y grandezas de España*. Madrid, Sociedad de Bibliófilos Españoles.

Yun Casalilla, Bartolomé (2002): *La gestión del poder: corona y economías aristocráticas en Castilla (siglos XVI-XVIII)*. Madrid, Akal.

Eduardo M. Baudot

Universidad Autónoma de Madrid.

ARMS IN THE HOUSE OF ALBA: A BRIEF STORY OF FUNCTIONALISM

> … Para que los dichos sucesores la tengan perpetuamente para solo servicio de los Reyes de castilla que siempre fueren y defensa de sus Reynos y republica y conservación de los estados de la dicha casa e mayoradgo de Alva [… So that said successors may have (the artillery) perpetually for the sole service of the Kings of Castile and the defense of their Kingdoms and republic and the preservation of the states of the said house and mayoradgo (entailed state) of Alba].

This quote serves as a precise compendium of the meaning of arms, artillery and other similar objects in the house of Alba between the fifteenth and seventeenth centuries. Nevertheless, analysing these objects as mere instruments of war removes complex layers of understanding and comprehension. Through these pages we will try to solve that gap. To achieve this, however, we must accept certain generalisations due to the broad timeline we are focusing on, for instance, the fifteenth and seventeenth centuries have little in common regarding certain social attitudes, and etiquette, regardless of the "foundation" of the duty, ambitions and intentions of the nobiliary houses, at least, in Castile.

The house of Alba was one of the more influential and powerful houses in Castile through those centuries. This status was achieved through marriage alliances, courtiers' strategies and on the battlefield. Undoubtedly, they excelled in this area, at least, until the 3rd Duke of Alba. Their constant appearance in battles validated them as noblemen serving the king, Castile, or God, but also forced them to buy and maintain arms and armour[1]. Likewise, their victories enlarged their capability to amass spoils of war. This, as we shall see below, is important when considering one nobiliary house or another: the battles won (and the enemies defeated) were more than mere memoirs or souvenirs.

Their historical records have been studied elsewhere[2]. Therefore, we will focus on the diverse periods where historical records related to the armoury have survived. Still, it is important to note that several documents in relation

1 Archivo Ducal de Alba (hereafter ADA), C. 164, nº. 23. For payments for maintaining and cleaning arms [*Pagos por la limpieza y arreglos de armas del duque de Alba*] in 1558/1559.

2 See Martínez Llorente, 2008: 103–118.

to arms and armour prior to the 2nd Duke of Alba are preserved in the archive of the Liria Palace, Madrid. For example, several archival documents provide a unique overview of the different offices and professions linked to the castle and the armoury: "*Alféreces de peones, lombarderos, espingarderos* [second lieutenant of peons, bombard and espingarda operators]". Besides, the so called "*Libro maestro* [main book]" gathers accounts, payments and jobs directly attached to the armoury as well as to other interventions in the castle of Alba de Tormes[3].

As noted, the documentation related to the 2nd Duke of Alba, Fadrique Álvarez de Toledo, reveals fascinating information associated to the armoury. The most relevant testimony is the post-mortem inventory, dated on 1531. Inside this varied and rich account of items (it contains books, paintings, personal belongings…), the contents of the armoury are described. It is noteworthy that in Alba de Tormes there were two different locations intended to preserve or display arms: the prominent tower that still exists, and another which no longer survives [Fig. 1]. This division of spaces is especially relevant due to their apparently different functions.

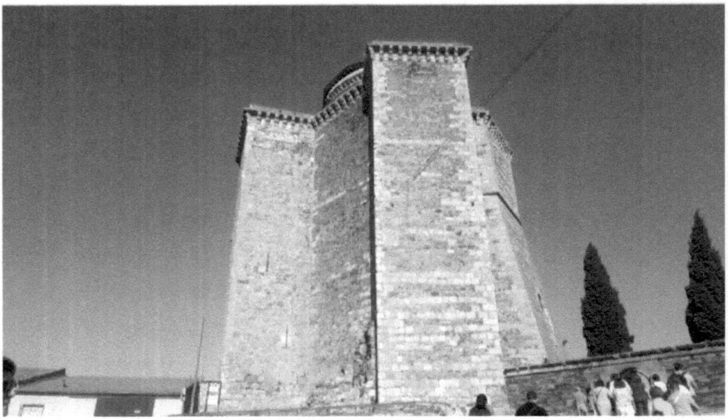

Fig. 1. *Torre de la Armería*. Alba de Tormes, Spain (Wikimedia Commons).

3 The *Libro maestro* is preserved in ADA, C. 301. A summary of the contents of this book has been published in Calderón Ortega, 2005: 409. See also Castro Santamaría, 2002: 39, 70, 218–225.

First, we have no evidence of a premeditated display of arms and similar items in Alba de Tormes previous the 2nd Duke of Alba. Therefore, functional motives must prevail when analysing these spaces, such as those related to defensive purposes. The three-story construction that is the tower of the castle is not actually "practical" in terms of moving, transferring or carrying artillery objects, so besides its balking purpose, other objects must have been preserved inside. Unfortunately, the document does not specify the location of these items, although it indicates the way they were stored[4]:

En un baúl viejo están las cosas siguientes
Un faldaje de armas dorado los bordes fecho a manera de faldaje de sayo con su funda
dos quixotes con hebillas doradas y guarnición de terciopelo negro con sus fundas
otros dos quijotes de la mesma manera con sus fundas
dos grebas con sus escarpes y correas de carmesí raso con su funda
un quixote con su guarnición de cuero blanco con su funda
otro quijote con su guarnición de cuero blanco con su funda
Otro baúl
otra barrata con su babera con su funda de acero
otro barbote con su funda
dos braçales con una guarnición de raso morado en sus fundas
dos alças guarneçidas con una clavazón dorada en sus fundas
una çelada blanca en su funda con su barbote.
[In an old crate there are the following things.
A gilded arms skirt with edges made like a tunic skirt with its sheath.
two cuisses with golden buckles and black velvet lining with their sheaths
two other cuisses in the same way with their sheaths.
two greaves with their sabatons and straps of crimson satin with their sheath
a cuisse with its white leather garniture and sheath
another cuisse with its white leather garniture with its sheath
another crate
another "barrata" with its bevor with his steel sheath
another bevor with its sheath
two vambraces with a purple satin trim in their sheaths
two "alças" adorned with gilded nailing in their sheaths
a white sallet in its sheath with its bevor].

Inside these crates [*baúles*] we distinguish an (apparently) disorganised accumulation of military objects. Besides the *baúles*, there were items stored in arks [*arcas*], different typologies of boxes [*caxas*, *caxas viejas*] and baskets [*serones*]. At first glimpse, this inventory shows no evidence of a deliberated

4 The inventory of the II Duke of Alba is located in ADA, C. 23, nº. 29.

organisation of these items. In Álvaro Soler del Campo's research on this book, he distinguishes certain patterns of arrangement for arms, armour and artillery. The varied way of preserving these objects shows the lack of an ideal model or reference, being the sense of functionalism the essential characteristic. Also, this supposed disarrangement (that is common to other armouries) should make us reconsider when and how do we use the word "collection". In other words, the arrangement of "true" "collections" is more restrictive and complex than the mere assemblage of certain objects. Nevertheless, the inventory of the 2nd Duke of Alba offers an interesting record of what an utilitarian armoury would have looked like. Still, certain items add layers of interpretation to this armoury, like the flags and trophies. As mentioned before, this aspect reinforces the status of the house of Alba as a family constantly serving their king, not only at court, but also on the battlefield.

Despite the fact that only scant information has been preserved of this matter in relation to the house of Alba, some primary sources after the 2nd Duke of Alba offer an interesting overview of this topic in the late sixteenth century.

Very much of this panorama comes from the 3rd Duke of Alba, Fernando Álvarez de Toledo, commonly known as the Great Duke of Alba. Fortunately, both primary sources (textual and visual) as well as secondary ones remain about this historical character that allow us to theorise over certain aspects we will discuss below, such as the meaning of the armoury, the value of its contents and how the visual arts interact with arms, armour and artillery.

To do so, it is imperative to outline a couple of events to comprehend the following assumptions. First, García Álvarez de Toledo died prematurely on the battlefield (battle of the Gelves, Tunisia, 1510), and so the heir of the 2nd Duke of Alba left young Fernando as the inheritor of the duchy. The succession took place in 1531, as we have seen above. By then, Fernando was 24 years old, and he had proven himself apt in matters of war and strategy. His value on the battlefield as well as in courtly matters allowed him to travel throughout Europe, experiencing firsthand the visual culture that predominated outside of Spain. Also, this gave him access to the best armourers, such as Helmschmied's workshop.This experience was probably decisive at the time the castle of Alba the Tormes was reformed. Nonetheless, the refurbishment of the castle cannot be only ascribed to the Duke's recently acquired visual culture. Tradition, connotations, and values attached to certain architectural forms were also decisive factors. In other words, the "conciliation" (or reception) of classical trends exemplifies the way the Duke wished to exteriorise his values and legacy, and, consequently, those of his house.

Thus, the 3rd Duke of Alba continued the reforms projected by his grandfather, the 2nd Duke[5]. This renovation followed the incipient classical trend spread from Italy (profuse vegetal decoration, *candelieri*, *putti*, etc.) in most of the castle except the tower of the armoury. This space was conceived very differently from other armouries studied in this book. The inside of the tower is cylindrical, therefore the walls of the second floor are rounded. In order to access into this location there is a spiral staircase located approximately in the centre of the room. Thereupon, the first thing visitors see once they access this floor is a colourful and vivid representation of classical architectures and characters and, most importantly, three scenes that depict the Battle of Mühlberg, 1547.

Two accounts from the nineteenth century offer unique testimonies of this space: Antonio Ponz, an erudite that wrote a description of Spain during his *Viage* [tour], wrote that "Tambien es cosa digna de verse la Armería, así por sus armas, y armaduras, como por las pinturas que adornan las paredes [...]. Una de ellas es (las batallas que decoran la armería), en la que quedó prisionero el Duque Mauricio de Saxonia, de quien se guarda allí mismo un busto de mármol [The Armoury is also something worth seeing, both for its arms and armour, and for the paintings that adorn the walls [...]. One of them is (the battles that decorate the armoury), in which Duke Maurice of Saxony was captured, of whom a marble bust is kept there.]"[6].

The next account comes from a student from the University of Salamanca, who visited the Palace of Alba de Tormes:

> *Alua è una terra di fuochi.*
> *Tiene 10 Parrochie, 2 Conuenti, uno di San Geronimo, altro di San Francesco fuori della Città, 3 Monasterij, Carmelitas Descalças, Benitas, come di Santa Isabella, chiamano il di dentro, il di fuora.*
> *Il Palazzo del Duca in un alto che guarda il Ponte San Girolamo uerso Salamanca. Tiene 60 Pezzi d'artigleria in circa in su carri fuori del Palazzo, però mal trattati, 2 grandi di ferro fuori dell'Armeria detta.*
> *In una Torre stan dipinte 3 Istorie, la una si conosce, che è quando presenta il Duca Giouanni Federigo all'Imperatore Don Carlos, l'altre due deuono essere la medesima guerra.*
> *Vi son dipinti certi Ciclopi con la fucina che battono fulmini.*
> *8 o 10 teste con i busti di marmo, et bronzo.*
> *Vn tauolino di pietra nel mezzo*[7].

5 See Martínez de Irujo, 1962. Mulcahy, 2013: 154–156. Pérez de Tudela, 2013 : 179–180. Sánchez Jiménez, 2014.

6 Ponz, 1788, XII: 291.

7 Text extracted from Haley, 1977: 169.

> [Alua is a land of fire.
> It has 10 Parishes, 2 Conuents, one of San Geronimo, another of San Francesco outside the City, 3 Monasteries, Carmelitas Descalças, Benitas, as of Santa Isabella, one inside the walls, the other, outside.
> The Duke's Palace ovelooks from above the San Girolamo Bridge towards Salamanca. It holds approximately 60 artillery pieces on wagons outside the Palace, but poorly treated, 2 large iron ones outside the aforementioned Armoury.
> In a Tower 3 stories are painted, one is known, when Duke Giouanni Federigo is presented to the Emperor Don Carlos, the other two must be of the same war.
> There are painted certain Cyclopes with the forge, striking lightning.
> 8 or 10 heads with marble and bronze busts.
> There is a small stone pit in the middle.]

These two travellers agree in some parts but differ in others. The divergences are noticeable on the contents of the armoury. The former does not mention any of the busts but Maurice of Saxony's. The latter speaks of "*8 o 10 teste con I busti di marmo, et bronzo* [8 or 10 heads with marble and bronze busts]". One of them, was most likely the one of the Elector of Saxony. Again, the second traveller remembers having seen a "*tauolino* [a table]" in the middle of the room. On the other hand, the similarities are, of course, the mural paintings depicting the battle of Mühlberg and their accounts on armours and artillery. Sadly, they do not specify how and where those objects where located, but we can assume a selection of these items was displayed inside the tower[8] [Fig. 2].

8 It is noteworthy the fact that the monogram 'F' [Fernando, the Duke] and 'M' [Maria, the duchess] was displayed on the walls of the armoury. This is also detectable on the top of the tapestries that depict the War against the Schmalkaldic League (The Liria Palace, Madrid). These tapestries were attached to the *Mayorazgo* along to other objects that belonged to both. This apparently strong bond is evident in their joint testament (ADA, C. 344 n.16). Still, why would the name of the duchess of Alba (the monogram of her name, at least) would appear in a strictly masculine space? Is this recurrent in other armouries?

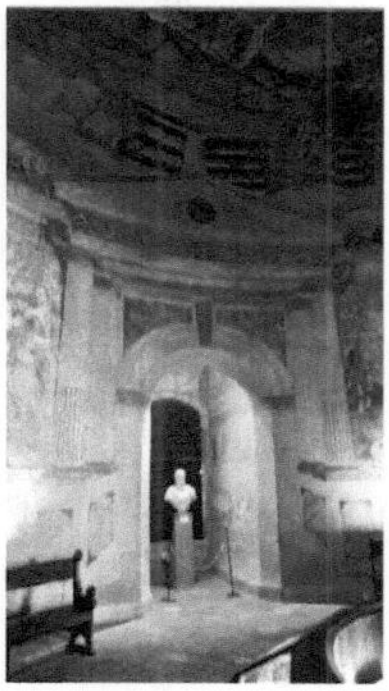

Fig. 2. *Interior view of the Torre de la Armería.* Alba de Tormes, Spain (Wikimedia Commons).

The idea of selecting and singling out certain items to be placed inside the tower is highlighted in another document dated in the seventeenth century, during the time of the 5th Duke of Alba. This primary source is a list of arms, armour and similar objects sent to reinforce the equipment of the Spanish troops after the French attack on Fuenterrabía (the Basque Countrie, Spain), in 1637. The house of Alba had equipment for battle at hand (something that can be appreciated in certain payments for armour upkeeping), but the 5th Duke banned specific items to be sent there:

> *Francisco de Aranguren ha venido aquí con comisión de Su Magestad en que me manda le entregue las armas q hubiere de provecho para llevallas a Fuenterrabía, haréis que les den todas las q fueren de provecho dexando las que no lo son como es las del duq de Saxonia las de la persona de mi abuelo y si hay algunos a prueba de arcabuz tomando del su carta de pago nro Señor os guarde en la Abadía 30 de Abril 1637*[9].
> [Francisco de Aranguren has come here with a commission from His Majesty in which he orders me to give him weapons that could be useful to take them to Fuenterrabía, you will ensure that they give them all those which could be suitable, leaving those that are not, such as those of the Duke of Saxony those of my grandfather and if there are some harquebus-proof, taking from his payment letter our Lord preserve you in La Abadía April 30, 1637]

Therefore, there were, at least, two armours specially reserved as "mementos". The rest, which were sent to batter, were reserved for utilitarian purposes. Apart from these findings, the absence of other armour which the Duke did not allow to

9 ADA, C. 164, nº. 23.

leave the premises of the castle must be considered. Were there only preserved in the ducal palace pieces of armour that belonged to one Duke? If true, this should be analysed in further studies, since perhaps there were other objects of greater significance in terms of prestige, lineage or other personal purposes and goals. As other scholars have noted, swords were very valuable (both economically and sentimentally) as the artillery, expressly those won in battle.

The document related to the objects attached to their "*mayorazgo* [entailed estate]" gives relevant information on this matter, specifically about the utilitarian purposes of artillery (and its consequent reputation) in terms of the ducal objectives:

> *Iten metemos para el dicho augmento de maioradgo toda la artillería que yo el dicho Duque gane en las jornadas de Alemania, Italia y Flandes, con todos los pertrechos y municiones según que todo ellos lo truximos con grande costa y trabajo nuestro, y lo pusimos y tenemos puesto la mayor parte dello en la nuestra fortaleza de Alva y la resta en otras partes [...] y la procuraremos dexar, para que los dichos sucesores la tengan perpetuamente para solo servicio de los Reyes de castilla que siempre fueren y defensa de sus Reynos y republica y conservación de los estados de la dicha casa e mayoradgo de Alva*[10].
> *Otrosi condición e additamento que exceptuamos siempre de la venta el almoneda que se ha de hazer de todos los dichos nuestros bienes muebles esemovientes toda la dicha nuestra artillería con sus pertrechos y municiones que tenemos en la dicha nuestra fortaleza de Alva y en otras partes, por quanto la dicha artillería se ha de conservar y estar en pie perpetuamente, como y para los efectos, que de suso queda proveydo y dispuesto.*
> [We put into the said increase of the Entailed Estate all the artillery that I, the Duke, gained in the battles of Germany, Italy and Flanders, with all the supplies and ammunition that we brought with a lot of effort and work, and we put them inside our fortress of Alva and what remains in other places [...] and we will try to leave it so that said successors may have it perpetually for the sole service of the Kings of Castile and for the defense of their Kingdoms and republic and the preservation of the states of the said house of Alba.
> In addition, we always exclude them from the auction that must be made of all our personal properties, cattle and all our artillery with the equipment and ammunition that we have in our fortress of Alba and in other places, and the aforementioned Artillery must be conserved and be always ready, for the purpose its use will be provided for].

About swords, there are clear references to "*dos espadas de mí el dicho Duque, la una dellas con una guarnición llana dorada, la otra con una guarnición negra de vueltas y la hoja con tres canales en los recazos que no pasan abaxo* [two swords that belong to me the said Duke, one of them with a plain gilded hilt, the other

10 ADA, Salón Flamenco, Vitrina 1, nº. 6, Ff. XXXIII r. – XXXV.

with a black hilt "*de vueltas*" and the blade with three grooves in the ricasso that do not pass below]".

It is important to mark that these items were attached to the *mayorazgo* and therefore prohibited to be sold or given away. The decision made by the Duke is therefore deliberate. Why would he select from his possessions artillery and not armours? Are there missing documents that could shed light on this issue?

This attitude towards armours is rare if we compare it to the Duke's contemporaries, like Ferdinand II and his "Armoury of heroes" or Philip II of Spain and his newly conceived "Royal Armoury"[11]. The notion of "memento" looks crucial in order to preserve these items. Why would a warrior like Fernando Álvarez de Toledo neglect them? Again, it is possible to assume that this was stipulated somewhere, but the document about the *mayorazgo* is wholly preserved, and armours were never mentioned[12].

What about other Castilian noble houses? As Roberto González Ramos has argued, the house of the Infantado had some consideration of their arms and armour[13]. Can we talk here about collecting? Is therefore the armoury of Ambras a "collection"? These interrogations cannot be answered here, but it is possible to assume that, at least with the Alba's, collecting weapons was not in their agenda.

With the 3rd Duke of Alba ends the trend in this family of serving the Crown fearlessly on the battlefield, so there is no more artillery attached to the *mayorazgo* and, apparently, no other kind of weapons. This conjecture comes from the testament of the 9th Duke of Alba, where it states that the document of the entailed estate is followed "*de manera irregular* [unreliably]".

Therefore, the epitome of the house of Alba was the Great Duke, and going back to the tower of the armoury, he clearly stated that. The tower was an apotheosis of the battle of Mühlberg, and only about that. This could be confusing due to the assumed importance of the ancestors and inheritance. Also, other members of the house of Alba fought in numerous battles, so the resolution of depicting just Mühlberg manifests a clear individualism.

The reason of this choice is unknown, apart from the pursue and search of the ideal of fame. This goal also raises another issue: the trophies. All over the mentioned vault of the tower in Alba de Tormes, there are depicted plentiful of

11 Soler del Campo, 2010: 27 and 41.

12 This does not mean a dispersion of armours after the passing of a Duke of Alba in the Early Modern era. The MET Museum, NY, have an armour attributed to Luccio Piccinino (inv. Num. 14.25.714a–h, j) that was sold by the house of Alba in 1861. Currently, the House of Alba preserves pieces of armour yet to catalogue and analyse.

13 González Ramos, 2014: 153–198.

flags that, according to a vexillologist study, belonged to the banner's foes the Duke defeated in Germany[14]. As mentioned, artillery were spoils of war, and we might consider flags as well. But how? Did the Duke have also these flags? If so, certain assumptions related to the "original" object back in the sixteenth century should be reconsidered.

Another episode can shed light on this matter. When the 3rd Duke of Alba ended his governorship in Flanders, he sent his belongings back to Spain. Inside the diverse trunks, boxes and other containers, the following is registered:

- *Catorce banderas de las que ganó el señor don Fadrique junto en Mans.*
- *Más va en dicho cofre diez banderas de la que ganó el señor don Fadrique junto a Mons de Senao*
- *Más va en dicho cofre dos libros de las banderas que han ganado el duque mi señor y el señor don Fadrique en estos estados.*
- [Fourteen flags won by don Fadrique in Mons
- And in the same coffer there are ten flags won by don Fadrique along Mons de Seano
- And in the same coffer there are two books with the flags won by the Duke my Lord and don Fadrique in these states].

Not only the Duke and his son Fadrique (the future 4th Duke of Alba) captured enemy flags on the battlefield. He commissioned illuminated books registering these spoils of war. Nevertheless, there are no references in the accounts cited here from the nineteenth century about flags. Where they lost, moved to another palace or stored in the functional armoury instead of the tower?

Despite this, it is noteworthy the will to register the triumphs of the noble house. Although this disposition, apparently, was not perpetual (at least there are no traces of these objects being attached to the *mayorazgo*). These kinds of decisions could elucidate the value and valorisation of the visual and material culture back then. For example, the Duke's aforementioned luggage contained paintings representing battles the Duke took part in: "*Cuatro cuadros de la jornada de Frissa del conde Loduvyco* [Four pictures of the Frissa event of Count Ludovico]" (probably scenes depicting the battles of Gronningen and Jemmingen, both in 1568 against Luis (Loduvyco) of Nassau). Besides this case there are more. For example, a canvas depicting his siege of Rome during his conflict with the papacy, in 1556: "*le mandó que sobre un lienço de 8 pies en alto y 12 pies en ancho le pintase la ciudad de Roma con el cerco que su Ex.*[cia] *le tenía*

14 Gómez Herrera, 2019: 50–102.

puesto en el año de 1556 [He ordered him to paint the city of Rome on a canvas 8 feet high and 12 feet wide with the siege that his Excellency had placed on it in the year 1556]"[15].

However, neither of these representations were included in the *mayorazgo*. The battle scenes that were attached were the tapestries portraying the German campaign (the same war depicted on the walls in Alba de Tormes). Was the material (the tapestries) and its richness the only motive to include them in the entailed estate?

There are so many questions left to answer, and this book offers a unique overview to begin with. Comparison is a useful tool in art history, but one must be cautious about what and how do we compare. Continuing with the house of Alba, what should be compared? Are the arms, artillery or armour preserved accurate measurers to solve these problems? Maybe their attitude towards them or the way they used these objects? The possible answers are certainly biased: there are preserved way more documentation from the age of the 3rd Duke of Alba than previous or future Dukes. Nevertheless, some intriguing questions could be resolved.

Their apparent lack of interest in preserving armours is one to study further. We currently know thanks to primary sources such as documents, payments, or portraits, that the 3rd Duke had several armours, one, at least, from the distinguished workshop of the Helmschmied's[16]. The armour made from the famed workshop from Augsburg was sent to the Armoury of heroes, proving that the value was relative compared to other objects. Was this attitude old-fashioned set by the trend of "collecting" or preserving armours?

---

15 Fragment of the letter sent by Matías Gast, brother of the painter Miguel Gast, to the 3rd Duke of Alba. Text extracted from Falcó y Osorio, 1891: 130–131.

16 See the portraits of the 3rd Duke preserved in the Liria Palace, Madrid. The armour designed by the Helmschmied's is the one kept in the Kunst Historisches Museum (inv. Num. Hofjagd- und Rüstkammer, A 420), Vienna. This is relatable to the portraits painted by Anthonis Mor (Hispanic Society, NY. Inv. Num. A105) and the model from the Royal Armouries (inv. Num. I.54), although the latter is not blued, a decision yet to interpret. This commissioned is registered in a payment, ADA, C. 211.

Information relates to other armourers like "master Willem" was published in Pérez de Tudela, 2013: 185. Notes in p.414. The 3rd Duke also had an armourer to clean and repair his arms, information in ADA C. 164–23. A horse saddle made for the 3rd Duke of Alba was published by Pierre Terjanian in his analysis of the "Thun sketchbooks". See Terjanian, 2016: 151–292.

By studying the house of Alba it is immediately apparent the constant strive between tradition and modernity. Until the 2nd Duke, it looks like arms function as utilitarian items. With the 3rd Duke this attitude does not change, but with the arrival of new ways of promoting the self-image (specially with the classical and Italian trends that focus on heroic representations) and the new manner of making war, generates the tension mentioned above. His legacy for his descendants (the *Mayorazgo*) does not include armours but swords and artillery, as we have noted. The first, an item alluding to a knightly past and, more importantly, the ideal of their past, present and future. The latter, a pragmatic legacy to defend their king and lands as well as a signal of power and strength.

From the time of the 4th Duke of Alba, the luster of the house would fade, and the way of promoting and becoming "powerful" varied drastically. The artillery became an annoying legacy, as well as the rest of the arms and armour.

So it is intriguing how sixteenth-century Castilian noblemen changed their mind in the way they used arms and armour. By examining the archival documentation compiled in this research project and the correlatives studies we offer a clarification of one of the most intriguing and complex discriminations about possessing objects. That is, differentiating which were "held" and which were "had"[17].

BIBLIOGRAPHY

Blom, Philipp (2003): *To Have and to Hold: An Intimate History of Collectors and Collecting*. London, Penguin.

Calderón Orterga, José Manuel (2005): *El ducado de Alba. La evolución histórica, el gobierno y la hacienda de un estado señorial (siglos XIV-XVI)*. Madrid, Dykinson.

Castro Santamaría, Ana (2002): *Juan de Álava. Arquitecto del Renacimiento*. Salamanca, Caja Duero.

Falcó y Osorio, María del Rosario (1891): *Documentos escogidos del archivo de la casa de Alba*. Madrid, M. Tello impresores.

Gómez Herrera, Rafael Luis (2019): *Banderas en los murales sobre la batalla de Mühlberg*. Madrid, Sociedad española de vexicología.

González Ramos, Roberto (2014): "La colección de armas de Íñigo López de Mendoza, V duque del Infantado". In: *Gladius*, 34, pp. 153–198.

17 For this essential question, see Blom 2003. Also, see note 1 in Urquízar, 2015: 386.

Haley, George (ed.) (1977): *Diario de un estudiante de Salamanca. La crónica inédita de Girolamo de Sommaia.* Salamanca, Ediciones Universidad de Salamanca.

Martínez de Irujo y Artázcoz, Luis (1962): *La batalla de Mühlberg en las pinturas murales de Alba de Tormes.* Madrid, Sucesores de Ribadeneyra.

Martínez Llorente, Félix Javier (2008): "*De Stirpe Gothorum* los Alvarez de Toledo y su linaje hasta don Fernando Alvarez de Toledo". In: Ser Quijano Gregorio del (coord.): *Congreso V Centenario del Nacimiento del III Duque de Alba, Fernando Alvarez de Toledo. Actas.* Diputación de Ávila. Institución Gran Duque de Alba, Salamanca, Diputación Provincial de Salamanca: pp. 103–118.

Mulcahy, Rosemarie (2013): "The Manifestation of His Magnificence: The Third Duke of Alba and the Arts". In: Ebben, Maurits/ Lacy-Bruijn, Margriet/van Hövell tot Westerflier, Rolof (eds.): *Alba. General and servant to the crown.* Rotterdam, Karwansaray: pp. 137–167.

Pérez de Tudela, Almudena (2013): "The Third Duke of Alba: Collector and patron of the arts". In: Ebben, Maurits/ Lacy-Bruijn, Margriet/ van Hövell tot Westerflier, Rolof (eds.): *Alba. General and servant to the Crown.* Rotterdam, Karwansaray: pp. 168–191.

Ponz, Antonio (1788): *Viage de España, ó Cartas, en que se da noticia de las cosas mas apreciables, y dignas de saberse que hay en ella. Vol. XII.* Madrid, Joaquín Ibarra impresor.

Sánchez Jiménez, Antonio (2014): "Furor, mecenazgo y enárgeia en la Arcadia (1598): Lope de Vega y los frescos de Cristoforo Passini para el palacio del Gran Duque de Alba". In: *Etiópicas. Revista de Letras Renacentistas*, 10, pp. 55–110.

Soler del Campo, Álvaro (coord.) (2010): *El arte del poder. La Real Armería y el retrato de corte.* Madrid, Museo Nacional del Prado.

Terjanian, Pierre (2016): "The art of the armourer in late medieval and Renaissance Augsburg: The rediscovery of the *Thun sketchbooks* (part II)". In: *Jahrbuch des Kunsthistorischen Museums Wien*, Band 17/18, pp. 151–292.

Urquízar Herrera, Antonio (2015): "'Making invisible things visible and palpable': visual marks of nobility in Early Modern French social theory and the embodiment of social estates in collections, 1550–1650". In: *Word & Image*, 31, 3, pp. 386–397.

Germán Dueñas Beraiz

Museo del Ejército, Toledo.

THE ARMOURY OF THE CONSTABLES OF CASTILE: A REPRESENTATIVE EXAMPLE OF A SEVENTEENTH-CENTURY NOBLE CASTILIAN ARMOURY

INTRODUCTION

The present work's objective is to analyse and to explain an important collection of Spanish noble armour that for various reasons has gone almost entirely unnoticed, both in the general historiography as well as in the the specialised historiography of historical armoury. When it came to agreeing to participate in this research project about noble armouries, there were numerous possibilities for work. Some were specific, i.e., on a particular noble house, whilst others were comprehensive, i.e., approaching the research from the field of historical armour. But when I examined several of the sources on the Constables of Castile's armoury, and its wealth, in number, in nuance and in its descriptive richness, it turned out to be the most interesting and abundant in data.

The Constables of Castile's armoury was founded by the 6th Constable of Castile upon his 1613 death. It was located, along with the library, in the palace called Casa del Cordón in Burgos. This collection's existence is hardly discussed, either in the general or the specialised literature. Valencia de Don Juan mentions its existence when trying to connect a rondache (Real Armería de Madrid, A 109), located in Burgos, and acquired by Alfonso XIII[1]. Also, an *Inventario*

1 Marchesi, 1849. The Count of Valencia de Don Juan (Valencia, 1898: 39), notes this piece, decorated with a saltire or St. Andrew's cross and some "*eslabones* [firesteels—in the shape of the letter B—]", with the royal armoury's harness A 108–A 111, so-called "*eslabones*" for having some elements of the Order of the Golden Fleece decorating part of the suit of armour, attributed to the German armourer Colman Helmschmid. This suit of armour appears included in both the "Relación de Valladolid" and in the Illuminated Inventory, the latter pointing out that there appeared among its decoration a figure of Our Lady on the breastplate and Saint Barbara on the back plate, since lost. This is when this armoury located in Burgos is referenced, saying that it was disbanded and disappeared, according to him, during the War of Independence

manuscrito de la armería del Condestable de 1705 [*Manuscript inventory of the Constable's armoury from 1705*] is mentioned in that work, without providing any further details about the document's location or content[2].

Even if the noble house's importance and its relationship with artistic patronage and collecting have already been discussed by some authors, the armoury's collections have not been examined thoroughly[3]. Passing references to its existence have been made when discussing the subject of certain Constables or the noble houses to which they belonged[4]. Likewise, it has also been discussed in the context of the relationship between Don Juan Fernández de Velasco and the intellectual world of engineers and poliorcetics[5]. It was even alluded to in documentary appendices and with short reviews when talking about the collecting aspect of this noble house[6].

But it was not until the Professor Cámara's monographic study, after locating an inventory of the armoury in Burgos Archivo de Protocolos, that its actual content is analysed in greater depth[7]. This article provides a tour of the collection, highlighting some of the most interesting aspects to the author. More recently, on the topic of the nobility armoury upon which we focus, the armoury has also been mentioned and identified as one of the most potentially important in Castile between the sixteenth and seventeenth centuries[8].

We will also attempt to focus precisely on the Constables within the historical and political panorama, both due to the variety of different noble houses that composed it, as well as the important political and military roles that their members had in sixteenth- and seventeenth-century Spain. These factors could explain or frame the formation and composition of all or some pieces of its magnificent armoury. We also intend to examine this armoury on its own

---

or Peninsular War. This shield was probably located in that armoury as a gift from the emperor to the Constables.

2 The reference that Valencia de Don Juan provides is the following: *Inv. del C. de C. Inventario de las armas del Condestable de Castilla. Burgos, 1705.* It might be the inventory that Professor Payo Hernanz (1992: 767–784) references. Archivo Histórico Provincial de Burgos (hereafter AHPB), Juan Francisco de Gandía 6878. This inventory has also been examined by us, albeit not transcribed in its entirety, verifying what Professor Cámara's data relates: that its composition does not vary substantially.

3 Carlos, 2005. Zalama/Andrés, 2002.

4 Bouza, 1983. Payo Hernanz, 1992.

5 Vázquez Manassero, 2019: 83–100.

6 Zalama/Andrés, 2002.

7 Cámara Fernández, 1994: 407–411.

8 González Ramos, 2017: 269–314.

merits and not as a military arsenal or other grouping or accumulation of arms, groups that should not be denominated an armoury or a dynastic armoury. This explains, moreover, why we do not include here the inventory of the fortress in Berlanga[9], given that it is more of an arsenal of a stronghold than a collection of representational or dynastic arms.

Its conception as a sixteenth-century armoury comes from the fact that, although its creation as such and the inventories we deal with are all from the seventeenth century, except one from the beginning of the eighteenth century, it was formed throughout the sixteenth century. Therefore, the arms collected and displayed in it mostly belong to that century, despite the importance of exceptional elements, such as short firearms, which could have been collected throughout the initial years of the seventeenth century due to the presence of the Constable having command in the Duchy of Milan.

We consider it a dynastic armoury because it establishes an estate over the armoury and the library, providing staff and money for its maintenance. In the words of the 1613 will of Juan Fernández de Velasco himself:

> *Item porque los sucesores en mi casa tengan memoria y se acuerden que así como las dos cosas principales con que se adquiere la nobleza y se conserva son las armas y las letras, quiero y mando en señal de esto que esté siempre vinculada en el mayorazgo de mi casa la Armería que dexo y también toda la librería que tengo […]*
> *Y también mando que se dé salario competente al armero para que tenga cuidado de tener limpias y lucidas las armas y que así mismo se le pague de mis bienes*[10].
> [Because the successors in my house have memory and recall that the two main things with which nobility is acquired and preserved are arms and letters, I desire and command as a signal of this that the Armoury that I leave and also all the library that I have, always be linked in my house's estate, […]
> And I also order that a competent salary be given to the armourer so that he takes care to keep the arms clean and splendid and that he also be paid from my assets].

The will, at the same time, creates a staff and a specific location, the Palacio del Cordón in Burgos, for the armoury's establishment so that it maybe looked after and visited in an orderly and continuous manner as time passes.

Attempting to classify this armoury within the framework of the different noble armouries existing in Spain is difficult due to the fact that, until very recently, this topic has not permitted systematic and profound studies to

9 Inventory of arms in the Berlanga fortress: Archivo Histórico de la Nobleza (hereafter AHNOB), Frías, C. 336, D. 16–21.

10 Archivo Histórico de Protocolos de Madrid (hereafter AHPM), T. 1830, fols. 289 v-290 r.

be to carried out about this phenomenon[11]. Although, we can say that it is a magnificent collection, with high-quality pieces, balanced in its composition and with all the very aspects of its nature. It is true that some aspects stand out, as we will see later, but the collection does not leave any typology unrepresented in any more or less meaningful way. Once again, we believe that many of the pieces that form a noble armoury come shaped by the history and duties that its owners performed and that these factors shaped their armouries' composition decisively. At the same time, the personal tastes of the noble members themselves determine the possession of specific pieces. Additionally, the armoury possesses other enriching elements when studying the topic of noble armoury collections in general and armour in particular, such as its distribution or museography; the origin and composition; its evolution as if the pieces were living organisms or if they were watertight or closed compartments due to their connection with the entailed estates; and other issues such as its day-to-day operations with the existence of personnel and materials for its maintenance and rearrangement.

Finally, we must state that we are are going to focus on the Constables's noble armoury, not on other groupings of arms that the different houses possessed for different reasons, such as the arsenal in Berlanga, or the personal arms that Juan Fernández de Velasco had in his house upon his death. The arms located in Pedraza, already partially discussed[12], and ones that, as we will see below, became part of the collection located in the Palace of Burgos, pointing out their origin, perhaps because they were part of the Tovar family estate, will be a separate case[13]. This fact also allows for a comparative analysis between a military armoury from the beginning of the sixteenth century (1512) and the one that arrived in Burgos in the first quarter of the seventeenth century.

METHODOLOGY AND SOURCES

The present work does not seek to be an exhaustive study of the Constables's armoury nor of its evolution and history, but instead it offers a fixed image of its composition, structure and importance at different specific moments: 1608, 1628, 1633 and 1705. These dates came about through the analysis of different inventories dated to the aforementioned moments. We must differentiate the inventories in terms of their purpose and, therefore, their method of preparation. Some are made by external experts with the intention of appraising the pieces

11 González Ramos, 2017: 269–314.
12 Franco/Soler, 1994: 309–344.
13 Franco/Soler, 1994: 309–344.

economically. While in other cases, such as those of 1628 and 1705, the inventories were carried in the context of transfer of responsibility between the armourers that had custody over them. In some instances, we believe that they were copied directly from one of the others, as in the cases of the 1628 and 1633 inventories, wherein a gilded rondache with an inscription that alludes to a Constable is said to be decorated with a "*nudo de Salomón* [Solomon's knot]" in 1628 whilst the 1633 version says "*niño de Salomón* [Solomon's child]"[14]. This proves a poor transcription from one inventory to the other, without taking into account the element represented on the rondache.

All the inventories, as we will see below, have an internal structure, framed, we believe, by their topographical disposition, which allows us to traverse and to understand the internal organisation of a noble Castilian armoury at the beginning of the seventeenth century. In addition to listing the arms within the armoury, these inventories also described other elements that made it up, such as furniture, tools, accessories, etc. Items typically found in stables, such as saddles, stirrups, bits, bridals, etc., which also formed part of the armoury's structure, likewise stand out.

We will focus, more or less, on the composition of the arms and the identification of typologies and characteristics of the arms that appear in the inventories, on the verification of their geographical origins and manufacturers, in the rare instances in which they are identified, as well as on the mode of entry into the armoury. That is, we will discern if they are dynastic arms, "*heroicas* [heroic]" arms, gifts, purchases, etc. Some of the inventories, especially those of the specialists, that is, the custodial armourers, are very lavish in their descriptions, the number of pieces and their characteristics, which allows us to familiarise ourselves with or at least speculate about the typologies of the arms described, especially in the cases of harnesses and the firearms. It is also possible to know the most abundant tastes or typologies in this armoury. Although defensive arms, i.e., suits of armour, stand out in number and type, as is normal in these type of collections, other sets such as shields, especially rondaches and bucklers, and above all firearms, especially pistols and arquebuses of the hunting type, are also noteworthy. In the case of white arms and crossbows, as they are mostly objects for daily use, they are more connected to the Constable's chamber for his personal use and not to the dynastic armoury.

The 1608 inventory[15].

14 AHPB, P. 6332, fol. 53 v.

15 AHPM, Jerónimo Díaz del Real, T. 24580, fols. 531 r-572 r.

This first inventory was made at the request of the Constable Juan Fernández de Velasco on 9 July 1608, as a result of the death of his first wife, María de Girón, daughter of the Dukes of Osuna, in February of that year. He wanted an inventory to be carried out of all the Constable's assets and it is significant that the Constable owned paintings, arms and other goods in Burgos that he wanted to be inventoried and appraised. As such, it is not an armoury inventory *per se*, because it was not constituted as such, but there was an important collection of armaments in the Palacio de los Condestables in Burgos with a clear sense and intention of armoury as was later established. The inventory must have been presented to the Constable's accountant, Francisco Cuéllar, and the magistrate of Burgos was requested to carry it out using experts in each field. The armourer Pedro de Lomana was named to appraise the armoury on 7 January 1609.

It consists of an appraisal inventory, i.e., a valuation is placed on the side of each piece's identification. It is rather comprehensive, including aspects such as the most significant decorative motifs, techniques and textile elements. In the case of garnitures, which are always referred to using the generic term "*armas* [arms]", Lomana identifies each one of the pieces that form the set from his point of view. He even indicates when he sees that they may not be part of a set. Simultaneously, he indicates on the margin if they were the Constable's property and even their other traits, such as from where they came. In some cases, identifying the armourer is possible, as in the case of Pompeo della Cesa. It becomes significant that on many of the garnitures he placed marginal notes such as "*traídas de Milán* [brought from Milan]"; "*fechas en Milán* [made in Milan]" or simply "*de Milán* [from Milan]"[16], perhaps to distinguish those that arrived with the Constable from his stay in the Duchy of Milan after having exercised his mandate there from those that the armourer considered fabricated there. With regard to the other places of fabrication, he mentions a suit of armour "*hecha en Flandes color de agua* [made in Flanders, water-coloured]"; and a bullet-proof breastplate "*de Alemania* [from Germany]". There are also various references to Spanish centres of fabrication little known at the time, such as Calatayud, Segovia and Peñacerrada.

In the majority of cases, he identifies the exterior aspect that nature gives to the set, alluding to the decorative techniques and the colours derived from them. Thus, the majority of the arms are gilded and engraved, further identifying other meaningful differentiating elements: Marian images, crosses of Burgundy, etc. On other occasions, they are white, steel-coloured, black, gilded and blued. At

16 Mann, 1962. Boccia/Godoy, 1985. Godoy/Phyrr, 1999.

times, he provides a summary of various aspects, such as "*armas con un poco de oro negras llanas* [plain black armour with a little bit of gold]", or he includes functional aspects: "*armas negras chicas de torneo* [small black tournament armour]". He also distinguishes the defensive arms' function using expressions like "*de torneo* [for tournaments]", "*viexas* [old]", "*de seguir* [for war]", "*de justa antiguo* [ancient for jousting]", "*peto a prueba* [bullet-proof breastplate]", etc.

He would inventory a premier group of harnesses, arms that indicate higher quality and appraisal, composed of a dozen suits of armour with interchangeable parts. Among these were some antique ones and a pair of "*a prueba* [bullet-proof]" corslets, one composed of a breastplate, backplate, large tassets and a morion with blackear flaps with white stripes. The other one was a black breastplate and backplate with gunshot marks, two on the breastplate and another on the back. Next, he goes on to prepare an inventory of the rest of the armoury's collections (firearms, white arms, polearms, etc.) among which are included objects of furniture, the stable, attire, etc., that are found in the same physical space as the armaments.

The 1613 inventory[17].

This inventory reflected the need to make an inventory and appraisal as a result of the death of the Constable Juan Fernández de Velasco on 15 March, 1613. His widow and second wife, Doña Juana de Córdoba y Aragón, requested it as guardian and administrator of his first son, Bernardino de Velasco, who later would be the new Constable, Councillor of State, Viceroy of Aragon and Governor of Milan.

It included the arms that the Constable had in his Madrid residence and except for some specific pieces such as the iron helmet covered in black velvet, it consisted of the personal arms that he used at court: hunting weapons, crossbows, some auxiliary items and the white arms that he commonly carried as part of his daily attire, which were called dress or everyday swords at the time, or, in more recent times, rapier swords, with their corresponding matching daggers[18]. These were located in his room, salon or sleeping quarters. It becomes especially interesting in its description of some crossbows and swords, and especially for the identifications of the firms that fabricated both typologies of arms, even including drawings of the marks that the manufacturers of this type of object made on them, especially in the case of crossbows. It is probable, and we will

17 AHPM, T. 24850, fol. 535 r-556 r.
18 Martínez, 1992.

try to study it, that some of these personal or everyday weapons later became dynastic arms in the armoury itself.

The 1628 inventory[19].

This is an inventory of an armoury already constituted as such, carried out as a result of the transfer of functions between one outgoing armourer, Pedro de Sagabedia, and the incoming one, Diego Pascual. That is, it is an exhaustive inventory between two armourers, wherein the aim is to describe the arms as much as possible to identify them for the benefit of inventory and assumption of responsibilities.

This is the main inventory studied, transcribed in its entirety, due to the quantity and quality of the information provided. Moreover, it is located at a time more or less central, chronologically speaking, with regard to the other inventories. It also allows us to analyse a collection straddling between the arms of first-class sixteenth-century Castilian noblemen, ancestors of Juan Fernández de Velasco, and those who had participated directly in conflicts such as the Granada War, the wars with France, the wars of religion, the campaigns in Flanders, etc., and that of a less militarised nobleman, but one with a high diplomatic and courtly relevance in the Spanish empire, especially in the Duchy of Milan. Furthermore, firearms, crossbows and other types of armoury pieces begin to gain influence compared to the ancestral knightly harnesses in this inventory.

The document almost perfectly repeats the scheme and order of the arms included in the 1608 inventory, except for the absence of the appraisal. A small group of harnesses is incorporated under the heading of arms "*del señor conde de Haro* [of his lordship, the Count of Haro]". We might be able to consider that this entire subset of pieces would constitute a portion belonging to the Count of Haro and that it was incorporated later, given that it does not appear in the 1608 inventory. Since its foundation, the house of Haro was linked to the Velasco family's firstborn son by entailed estate.[20]

Moreover, the collection of shields, bucklers and rondaches increased. Rondaches engraved with heraldic motifs and painted bucklers or those with bosses of gold appear for the first time. An armourer's tools and a group of loose, deteriorated pieces that are being arranged and repaired, probably arriving

19 AHPB, Toribio Diez del Real, P. 6191.

20 It could have included the armoury of the first son of Juan Fernández de Velasco, Don Iñigo, who married a daughter of the Duke of Sessa, who governed the Duchy of Milan in the absence of his father and who predeceased his father, the Constable.

from another of the Constable's possessions and that were incorporated into the armoury, are included as well, as seems logical.

The 1633 inventory[21].

This inventory was carried out by the armourer Diego de Pasqual at the request of the *mayordomo* and *alcaide mayor* of the Constable's accounts, Pedro Juan de Maeda y Salazar, via the *corregidor* of Burgos. It is an inventory without economic appraisal of the value of pieces "*de la armería y pertrechos y demás cosas tocantes a ella* [of the armoury, the equipment, and other things related to it]". Although it repeats the order and typologies of the pieces, it is more concise in terms of its description and attached data, such as an allusion to owners, place of manufacture and decorative motifs. It continues to use the marginal annotation for objects belonging to the Constable, for instance. These range from folio 49 to 65, the inventory itself being on all folios except the first and last. We believe that the information and identifications on many of the objects were included from previous inventories, such as the 1628 inventory, given that there were no meaningful differences between them. The order established in the previous inventory is repeated and it would be, like the previous one, topographical, or it simply would be copied from the previous one.

The 1705 inventory[22].

This new inventory is carried out by the armourer Blas de Domingo de la Torre on 12 June 1697, so that the new armourer, master locksmith Juan Alonso de Biadero, "*que trabajaba en el oficio de armero, seguridad, custodia, aseo y limpieza de las armas y harneses* [who worked in the office of armourer, security, custody, tidying and cleaning of arms and harnesses]" could take charge of the armoury. Although the number and description of the arms are similar to the previous inventories from 1608 and 1628, curious information is provided about some objects, but we do not know if this information is accurate in many cases. It is probable that now began the mythical attributions of some of the arms that, rightly or wrongly, were passed down over time with erroneous or fanciful ascriptions. In some cases, some objects were simply changed.

New pieces are also noted that are incorporated into the collection over time, revealing the open nature of the collection. Thus, in discussing a set of pieces of armour, of iron colour, it is said that they belonged to the person of

21 AHPB, P. 6332, fols. 49 r- 65 v.

22 AHPB, Juan Francisco de Gandía, P. 6878–1, fol. 46 r-67 v.

Don Francisco de Velasco "*conque sirvió en Milán siendo gobernador y capitán general en dicho estado el señor Don Bernardino Fdez de Velasco y Tovar su padre* [who served in Milian with Don Bernardino Fdez de Velasco y Tovar, his father, who was governor and captain general in that state]"[23]. This set, owing to its typology and references to the owner, corresponds to a suit of armour from the mid-seventeenth century, when Bernardino Fernández de Velasco (1609–1652) was governor of the Duchy of Milan (1645–1647). Therefore, it has to have been added to the armoury after the 1633 inventory, demonstrating, as we have said, that it was a living collection of the dynastic type.

On another suit of armour (No. 12), the inventory makes references to it "*está hecha en Flandes de color de agua* [being made in Flanders in the colour of water]", the same reference found in previous inventories, although in this one the annotation "*son armas fuertes hechas a prueba, llenas de trofeos labradas con barniz guarnezidas de ante con espaldarcete de ynfante del uso nuevo* [they are strong arms, made to be bullet-proof, full of trophies carved with lacquer, lined with suede with newly used infantry pauldron]" is added. The copying from previous documents with small changes is evident. At times, it is more precise, as when discussing musket or arquebus-proof arms, the document specifies the areas where the marks of the bullet tests are noted, as in number 15, which states that it has two gunshots on the chest on both sides and a third on the back.

Another set of defensive arms, which are referred to as "*grandes y blancas antiguas de medias lunas* [large and white antiques of half moons]" are attributed to the "*Gran Capitán*". Meanwhile, on other previous inventories, it was only a helmet that was associated with Gonzalo Fernández de Córdoba. This set consisted of a jousting suit of armour that includes a sallet-helmet with the text reading "*Don Juan de Cardona año de 1552* [Don Juan de Cardona, year of 1552]", already mentioned in the other previous inventories without a link to the "*Gran Capitán*", as well as a heterogeneous group of arms from different periods and workmanship, such as small shields "*de jugar* [for playing]", and a white breastplate with an image of the Virgin and a cherub at her feet, these images being removable though the use "*de unos clavillos* [of a few little nails]". The set's heterogeneity and incongruent nature makes its association hardly credible, exemplifying these false or erroneous attributions.

References are made again to various jousting harnesses (No. 17) with reinforcement pieces made in Peñacerrada, without additional information, and subsequent set (No. 18), but only the upper section made Segovia for tournaments,

23 This set made up a breastplate, a backplate, a burgonet and two half-arm guards.

still lacking the backplate. Another tournament suit of armour belonging to the Count of Haro, engraved, "*del color del cielo* [of the colour of the sky]", without mentioning the leg protection, might have been a set for tournament on foot or "*de barrera* [for the barrier]": "*Otros cinco arneses blancos que dizen hicieron en Peñacerrada y el otro era de D^{n} Juan de Velasco* [Another five white harnesses that it is said were made in Peñacerrada and another belonging to Don Juan de Velasco]". No reference is made to the set of arms that came from Pedraza, and they are cited in succession with the others. Therefore, it is noted that the order relating to antique weapons of war, shields and firearms was maintained, later to include white arms, polearms and other typologies without substantially changing the explanations and quantifications of the previous inventories.

A significant number of loose or reinforcement parts, armourers' tools and additional parts from the stables, such as saddles, stirrups, bits, etc. continue to be recorded. Bulky pieces such as pavilions and tents also appear. The total amounts to 398 entries, although that number is misleading, because under the same number different pieces are included that may or may not form a set, or that simply serve to group typologies. Finally, the inventory states that "*se ha cotejado el presente inventario* [the present inventory has been compared]" with the one signed by the previous armourer Blas Domingo, recognising differences, and adding that there were different loose unrelated pieces, such as mails, gauntlets, vambrace pieces, thigh guards and other pieces that could complete the armoury's missing items.

THE HOUSE OF THE CONSTABLES OF CASTILE

The title of Constable has its origins in the Middle Ages and had a direct link with the development of military activity. The titleholder was the highest royal officer in this matter because he had supreme command over the army. Even in the absence of the king, he could consider himself to be his substitute or delegate. The office was created by King John I of Castile to replace the king's "*alférez mayor*". The first Constable was Alfonso of Aragon the Elder, Marquis of Villena, the office being for life rather than hereditary in nature. In 1473, Henry IV of Castile named Pedro Fernández de Velasco as Constable, after which the title became hereditary in his family.

The Duke of Frías

The creation and awarding of the title of Duke of Frías to the house of Velasco occurred in the figure of Constable Bernardino Fernández de Velasco (1492–1512), to which it was conceded based on the his father's achievements in the

Granada War[24]. This figure is the one who married Doña Blanca de Herrera, Lady proprietor of the estate of Pedraza de la Sierra[25]. The castle of Pedraza and its possession by the Constables would be key to understanding the composition of their armoury collections later. Bernardino took part in many of the actions in northern Africa, such as the captures of Mers El Kébir (1505), Peñón de Vélez de la Gomera (1508), Oran (1509) and Tripoli (1510).

Upon his death without male heirs, his noble titles passed to his brother, Iñigo Fernández de Velasco (1466–1529), the 3rd Constable. He was married to María de Tovar, Lady of Berlanga. He had also fought alongside his father and brother in different conflicts, most notably in the Revolt of the Comuneros and the wars against France at the beginning of the sixteenth century. After being named Constable, he received the new king, Charles I, in Becerril de Campos in 1517, and he was awarded the Order of the Golden Fleece. His firstborn son, Pedro Fernández de Velasco y Tovar (1529–1559) succeeded him. He was one of the leaders of the army that defeated the *Comuneros* [commoners] in Villalar (1521) and liberated Logroño from the French siege at the head of an imperial army. Due to the entailed estate of the house of Tovar, his brother, Juan Sánchez de Velasco y Tovar, kept the title Marquis of Berlanga. But, upon Pedro's childless death, the estates, titles and other goods of the house of Velasco, went to his nephew, Íñigo Fernández de Velasco y Tovar, the 4th Duke of Frías, third son of the aforementioned Marquis of Berlanga. The Count of Haro was a title created in 1430 by John II of Castile in reward for the help given by Pedro Fernández de Velasco against John II of Aragon, for the males of the house. Beginning with Iñigo Fernández de Velasco, the firstborn son of the Dukedom of Frías would receive this title and it had, thus, a life-long and hereditary character.

Íñigo Fernández de Velasco y Tovar (1520–August 22, 1585) was, therefore, the 5th Constable of Castile, the 4th Duke of Frías, the 6th Count of Haro, the 2nd Marquis of Berlanga, the 14th Lord of Velasco and the 12th Lord of Tovar. He had a more courtly and political profile as ambassador in Rome, councillor of state, president of the Council of Aragon and of the Council of Italy and Philip II's great chamberlain. He was succeeded by his son, Juan Fernández de Velasco y Tovar (c. 1550–Madrid, March 15, 1613), the 6th Constable of Castile, the 5th

24 Royal Decree promulgated in Granada on 20 March 1492. "En remuneración de los grandes servicios y lealtad, con que Don Bernardino Fernández de Velasco, Condestable de Castilla, había engrandecido el reino peleando contra los enemigos de la fe, hasta la entrada de los Reyes en Granada". Del Arco, 1899: 67–69.

25 This was his first wife, the second being Juana de Aragón, illegitimate daughter of King Ferdinand.

Duke of Frías, the 4th Marquis of Berlanga and the 7th Count of Haro. He held important positions in the Spanish court, such as ambassador in Rome (1585), governor of the Duchy of Milan, councillor of state and president of the Council of Italy, among many others. He was a key figure in the formation of the armoury, given that he established an entailed estate over it and the library upon his 1613 death[26]. His son, Bernardino Fernández de Velasco y Tovar, the 7th Constable, who was, like his father, a diplomat, a councillor of state and governor of the Duchy of Milan in the middle of the seventeenth century, succeeded him.

THE CONSTABLES' ARMOURY: ITS QUALITATIVE AND TYPOLOGICAL COMPOSITION

Taking into account what we have stated about the inventories carried out in the first half of the seventeenth century, and especially those richest in descriptive detail about the objects, we have arrived at a series of conclusions. In principle, the armoury reflects a Renaissance armoury's classic configuration, following the conventions set forth by Madrid's Royal Armoury and other counterparts, such as the one in Vienna. The suits of armour or harnesses are the armoury's most important and significant pieces, in their three configurations: parade, joust or tournament and war. In the first two cases, the harnesses are connected to the heads of the noble house that owned them as a demonstration of their economic and social power as pre-eminent members of the Castilian nobility. The military suits of armour also appear, although they are less distinguishable, forming part of larger sets or as interchangeable elements of the most ample garnitures to form different suits of armour[27].

There are only a few references that are made regarding their origin or fabrication in the sources, though those that we find are very meaningful. There are significantly few that are identified as being carried out in the first half of the sixteenth century's main production nucleus, Germany, and its different centres, such as Augsburg, Nuremberg, Landshut, etc[28]. There is also a reference to the Flemish hub, as a piece is identified "*de Flandes* [from Flanders]"[29].

We will use as an important reference the first inventory with which we are familiar, that of 1608, as a foundation for the armoury's order and composition. Additionally, in being the oldest, we might be able to conceive of it as the most

26 Alonso/Carlos/Pereda, 2005. Montero/González/Rueda/Alonso, 2014.

27 Soler, 2010: 25–39.

28 Terjanian, 2013. Terjanian, 2016.

29 Soler, 2006.

precise with regard to the information and references about the pieces' origin, fabrication and ownership. Although, the 1628 inventory, by being something later, complete and professional, i.e., an inventory between armourers, is the richest and thus the main source for this study.

The armoury, in addition to the arms themselves, contained other objects, supplies and items connected to this type of collection. First, there were objects related to hunting activities, such as bird calls "*para cazar al açor* [for hunting with goshawk]", or an ivory horn with the coat of arms of Portugal "*para la brama* [deer hunting in bellowing season]". It is also worthwhile to emphasise the large quantity of firearms, arquebuses and wheel-lock muskets, of different calibres and barrel lengths, which have a clear hunting function. Similarly, the large quantity of crossbows and their accessories stand out, produced by the world's most prestigious craftspeople at the time, the main function of which was also hunting.

The objects of saddlery are rather present, in both quantity and quality, i.e., every type of saddle and accessory to ride a horse, both "*a la brida*" and in the "*jineta* [jennet]" style. Campaign equipment also appears, such as tents and pavilions, flags and standards, etc. Even so, we focus on the very arms themselves, given they are the main object of our field of study.

By being a dynastic armoury, it is normal that arms belonging to direct ancestors or indirect ancestors who held other noble titles appear and that, for various reasons, became part of the main branch. The clearest case is a garniture that belonged to the Count of Haro. Taking into account that the title of Count of Haro was automatically awarded to the firstborn son of the Fernández de Velasco family, both for life and hereditarily, they might be arms belonging to Iñigo as Juan's firstborn son, who was not the firstborn, and that his son inherited the positions when his brother Pedro died without issue. Although, it is also possible that these were the arms belonging to the first Count of Haro[30]. The fact that it is not stated categorically, given that it uses the formula "*que dicen fueron* [that they say they were]" and the fact that they are incomplete tournament arms, could support this hypothesis[31].

---

30 Fernández de Velasco, Pedro. Conde de Haro (I).?, 1390-Medina de Pomar (Burgos), 1470.

31 Another "*cencillada*" sword of silver is noted in the 1628 inventory as belonging to the Count of Haro.

Defensive arms: The suits of armour

Regarding the composition of the armoury, we can say that it reflected the blueprints of the other referential armouries, namely the Habsburg and Burgundian armouries, by highlighting the harnesses and the garnitures of suits of armour. We do not want to enter into the relationship between the figure of the Christian knight as the highest exponent of the noble ideal nor of the harnesses' symbolic weight in the mediaeval and Renaissance imagination, which made these arms the highest exponent of propaganda and power of the militarised nobility in Europe and especially in Castile[32].

The composition or description of them, in addition to the consideration of being antiques or "*viexas* [old]", allows us to identify them as belonging to the most belligerent dynastic ancestors. As such, there are twisted defensive arms and those with gilded borders very much in fashion since the end of the fifteenth century and the first decades of the sixteenth century, and that might have belonged to the Velascos who participated in the Revolt of the Comuneros war and in the campaigns against France. The same can be said for helmets, small bevors, breastplates, backplates, etc., that reflect the Castilian typologies and the end of the fifteenth century and the beginning of the sixteenth century.

The qualification as antique is rather relative and it is necessary to understand that it is a concept that is applied with regard to the arms and typologies of the first decade of the seventeenth century. Thus, when referring to a set of lance vamplates, which are called antique, we find that one had the date 1552 engraved on it. That is, it was only eighty years old when the 1628 inventory was made. Another example of this chronological range of the concepts "antique" or "old" applied to some arms or sets of them is found in a group of jousting armour made in Peñacerrada, called "*buenas y antiguas* [good and antique]".

Among the suits of armour, clearly the dynamic ones, that is, those that belonged to the ancestors of different houses that formed the house of the Constables, stand out. Many of them are recognisable by displaying Marian decorative motifs, very popular during the reign of Emperor Charles V. Sporting suits of armour for jousting and tournaments are also important, either because they are so identified as such, or because of their typology and composition, which include reinforcing or oversized elements, such as gauntlets, Stechtarschen[33], breastplates, etc. Some of these harnesses are garnitures consisting of interchangeable pieces to form suits of armour for war, jousting, or

32 Soler, 2010: 25–39.

33 A small shield of German origin laced to a jouster's left shoulder.

tournaments, for infantry, or for equipment. The different pieces that made up each set are identified, listing each one in the sources. In this sense, the existence of easily recognisable identifications of generic pieces is emphasised, such as the Burgundian sallets, breastplate and backplate, lance rests, tassets, gauntlets, thigh guards and greaves. Others are specific, such as the pieces for jousting and tournaments: "*tarjetas y tarjetillas* [Stechtarschen and small Stechtarschen]"; buffes and "*bufetas* [small buffes]", and "*bufeta tarjeta* [buffe Stechtarsche]"[34]; jousting sallets[35]; "*un placarte que es a manera de tarjeta que sube para justar* [a plackart that is the style of a Stechtarschen that rises for jousting]"; some "*sobremanoplas para justar* [gauntlet covers for jousting]", etc… Occasionally, pieces also appear that are new at the time when identifying parts of suits of armour, as in the case of the epaulettes, which are systematically referred to using the descriptive name of "*espaldarcetes* [pauldrons]".

The references to colours, techniques and decorative motifs are numerous, as are the places of origin and the textile accessories that accompany them. Sometimes, those harnesses are complemented, as was customary in the time period, with other matching arms such as shaffrons for the horse, lance vamplates and shields[36]. There are many loose pieces belonging to lost harnesses or partially acquired from every type of piece, for jousting, tournament or war. Within these sets, sometimes harnesses are interspersed that are described as exclusively for tournaments or jousting, and although they may have war elements, those of a sporting nature were prioritised in number and importance.

Furthermore, there was a small but illustrative set of suits of armour for war from the second half of the sixteenth century, apart from some older loose pieces of the dynastic type. Specifically, there were two sets of complete cavalry or jousting armour, along with another for light cavalry. We believe that they would correspond to the Constables's political term of office in that area of Italy around the end of the sixteenth century and the beginning of the seventeenth century due both to their origin, "*traydas de Milan* [brought from Milan]" and their typology. Surely, their collection would reflect the desire to remember a specific war event or the period in which the highest military authority in the region was held.

As we have previously mentioned, the 1628 inventory establishes the existence of a set of pieces coming from Pedraza, that is, the castle of Pedraza,

---

34 AHPM, T. 24850, fol. 536 r.

35 AHPM, T. 24850, fol. 537 r.

36 AHPM, T. 24850, fol. 537 v.

the Constables's property since the sixteenth century[37]. The composition of the armament coming from that place clearly consists of a small and heterogeneous group of antique infantry and cavalry arms of a military type[38]. Its relationship with Pedraza was lost in the 1633 inventory, where the set was integrated with the rest of the elements, even though it was differentiated again in the 1705 inventory.

We believe that the majority of the high-quality suits of armour that appear at the beginning of the inventory are of German origin. We make this statement owing to the informative wealth that appears on them about the decorative motifs and which allows us to relate them to the typologies of German suits of armour from the second- and third quarter of the sixteenth century and their iconographic devices, especially in the different Marian representations. That being said, direct attribution to the German manufacturing centre to any of its most reputable armourers does not exist.

Most interesting are the references to little-known Iberian production centres from the sixteenth century. Although some, such as Calatayud or Segovia, are mentioned more regularly, the most repeated and numerous is Peñacerrada. It is cited in all the inventories, although the number of pieces with this centre of fabrication varies. These consist of pieces of decorated and engraved garnitures for heavy and jousting cavalry of a sporting and military type with specific pieces such as "*un placarte que es a manera de tarxeta suelto que sube para justar* [a plackart that is like a loose Stechtarschen that rises for jousting]"; "*unas sobremanoplas para justar* [some gauntlet covers for jousting]" and "*una sobre guarda que es navaja del braçal* [a guard cover that is a couter of a vambrace]". Pieces appear of a certain complexity and rarity, such as a gupfe: "*una cresta de celada aventanada* [a crest of a helmet with small windows on the visor]". Another set of complete suits of armour is noted later as also fabricated in Peñacerrada. Of this very poorly known armoury centre, we can say that it corresponds with the Alava's province locality of the same name[39]. Since the beginning of the sixteenth century, at least, there is news of its ability to fabricate armour for war

37 Franco/Soler, 1994: 309–344.

38 The 1608 inventory resolves any doubt by considering this set under the heading "Armas viexas que estaban en Pedraza [Old arms that were in Pedraza]". AHPM, T. 24850, fol. 542 r.

39 Peñacerrada-Urizaharra. The Basque name might be translated as "old water", which is related to the existence of foundries moved by water since the end of the Middle Ages. The study of this and other Basque armoury production centres is one of our principal lines of research.

in significant quantities, this, however, being the first time that there is evidence of it having the ability to produce arms of a sporting type.

Firearms

A truly potent and meaningful set within the Constables's collection is the one containing firearms. As these are inventories from the beginning of the seventeenth century, the presence of this type of arms is noted for their capacity as luxury weapons, for hunting, defence or collection, with no recognisable weapons of war existing. It is logical to consider that a figure like Juan Fernández de Velasco, interested in military treatises, artillery, engineering, fortification and mechanical devices, would have a good collection of high-quality pistols with strange and forward-thinking systems and mechanisms. Despite their number and quality, the sources do not provide much information about their origin, except for intermittent references to Germany, relating some pieces with a decorative technique or materials, such as when it is indicated that they were of "*atauxía de hueso de Alemania* [damascene of bone from Germany]", of that they had "*caxas labradas de hueso de Alemani*a [carved butts of bone from Germany]", which bring to mind Puffer-type models.

In the Italian instance, a model with the barrel signed by Lararino stands out. It seems clear that the periods of the term of office in the Duchy of Milan would influence the time of acquiring Brescian and Milanese items so in fashion at that time. There are even models gifted by top-level contemporary figures that reflected the Constable's tastes. Two small wheel-lock pistols with of bone inlay that have the coat of arms of the Savoy family and that were gifted to the Constable, who speaks in the first person in the inventory, as if it brought back a personal memory about their origin, are good examples[40]. In another inventory, the delivery of some pistols to the Constable of Castile from the Constable of France "*en Borgoña* [in Burgundy]" is described[41]. From the Flemish and Dutch centres, not one piece is identified, but there is an item that says that it had a Flemish cartridge below the "*la caxa* [the butt]". Another instance is a wheel-lock pistol made in ivory and the butt finished with two balls, reminiscent of Dutch models. With regard to the ignition systems, the only one mentioned is that of wheel-lock. Regarding the barrels, there are at least two examples that are said

40 "Two other small wheel-lock cocked pistols that the Duke of Savoy gave me with their butts of very fine damascene of white bone with its ramrods and cover".

41 AHPM, T. 24850, fol. 531 r., fol. 543 v.

to be rifled, although we cannot know if they are rifled externally or in the bore. Most were blued, silver or gold and one even had a serpent near the sight.

The set of pistols, also identified as such, with a great descriptive profusion in materials and techniques, even occasionally specifying the number of barrels and the ignition system, stands out, for instance, a "*pistola de dos cañones con rueda* [double-barreled wheel-lock pistol]". Another one says it had two barrels and two locks, albeit without specifying any more about how it functioned, which would allow us to venture that it was a wheel-lock pistol with bone, ivory and mother-of-pearl inlays. Perhaps it was a Puffer-type model or other derivatives, as in the case of the German pistols.

Only one of them appears inventoried giving a functionality to the band, that is, it was meant to be worn hanging from the belt or waistband. The butts' materials are most varied and rich, although the majority are walnut, with a couple of metallic specimens appearing. They are complemented with mother-of-pearl, gold, silver, bronze, brass, ivory and especially bone. On them different motifs appear, especially on the butt, some anthropomorphic ones (Flemish ladies, half-bodied men with hats); zoomorphic ones (squirrels, lions); or vegetables (two white roses). A pistol stands out, the butt of which has a medallion depicting the "*fuerças de Hércules* [strengths of Hercules]", a term synonymous in the era with the Labours of Hercules. Moreover, almost all the examples are accompanied with their covers, accessories and locks, which are typically missing in many instances. The inventories are also very precise in the covers' materials and colours.

The number of arquebuses is also important and repeats the pistols' scheme, that is, the majority are lavish. They are of different sizes and barrel lengths and are profusely decorated with bone, silver, gold, ivory, etc. The locks are wheeled[42], the butts of walnut with bone, ivory and mother-of-pearl inlay. Some of them have inscriptions, such as one of the named "*furioso* [furious]"; another with that of its owner ("*soy de don Pedro Girón* [I belong to Don Pedro Girón]"); or the name of the maker (Juan Salado).

The firearms are completed with loose pieces, among which are wheel-locks, arquebus barrels and accessories, especially high-quality powder flasks and small powder flasks of bone and ivory. In 1705, ten very old and deteriorated powder flasks, of a military nature, of supply, for muskets and arquebuses, were included, as were "*llaves de serpentín* [serpentine locks]" for muskets.

42 In the 1705 inventory, three flintlocks are included without any further detail.

White arms

Here, as we noted above, it is advisable to differentiate between the personal arms of the 6th Constable described in his 1613 post-mortem inventory, where the swords that he possessed for use at the court in Madrid were included, and those that formed part of his armoury located in Burgos since at least 1608, that we know by the inventory after his first wife's death. In this second document, they appeared under the heading "*Estoques, espadas, montantes, alfanxes y báculos y bastones de general* [Rapiers, swords, longswords, scimitars and general staffs and walking sticks]" (fol. 549 v). First, a combined arm is included that, shaped like a staff, hid a sword blade and a wheel-lock pistol with its barrel in its interior. The knob was formed by a gilded brass Turk's head and at the top of the foot was another gilded cap with an iron tip. The piece came in an ebony box of Milanese provenance, we do not know if it was a place of manufacture or origin, but it accompanied the Duke upon his return to Spain. Also, different long swords appear, one to engage in the sport of fencing, of the sporting type, surely without edge or point, which was not valued highly because it was "*muy antiguo* [very antique]". Others are also noted, one "*muy largo sin puño* [very long, without grip]" and another with a gilded blade[43]. Next, the set of general's batons or sceptres, some made of iron and others made of ash wood, painted and decorated with cords is worth highlighting, despite not being arms. An iron one with many gilded numbers and another decorated with flames of fire are worth mentioning.

Afterwards, various Turkish scimitars are mentioned, all considered antiques, and for which the colours and the scabbards are described. Another piece could be, by its description, a katana of Japanese origin.

> *Otro alfanje turquesco con vaina de zapa dorado el puño y guarnición contera y en la vaina hay estremos de plata dorada y en la cruz de la guarnición un xacinto y otro encima de la empuñadura con talabarte rojo.*
> [Another Turkish scimitar with a leather sheath, gilded grip, chape and hilt. On the sheath, there are edges of gilded silver and a hyacinth on the hilt's cross and another on top of the hilt with red sword belt].

The remainder of the swords are short and wide, for horse riding, some rapiers for lance rests and military swords, without additional details about their owners, qualities or origin marks, perhaps because the higher-quality and

43 In the 1613 inventory, they referred to it as "another longsword gilded on the blade with inscription of each part and the hiltcarved in damascene style, with red pommeland without scabbard".

more-important swords were under the Constable's supervision in Madrid for his daily use.

The majority contains the label of swords and rapiers, one of four corners, with references to the scabbards or to the garnitures or pommels. One of the swords with its dagger is described as "*aventanada* [windowed]", and also gilded; and another we find wide, short, and corrugated blade and hilt also "*aventanada* [windowed]". Another had a German hilt with a blade of only one cut. Loose blades also appear, such as the one in which the mark that it carried ("*un globo crucífero* [a globe with a cross]") was reflected, valued at ten *reales*. Larger sets of antique military arms are also cited, such as seventeen horse swords for men-at-arms, with tinned/soldered hilts, accompanied by forty-two daggers. In the 1613 inventory, two more high-quality swords with details in the hilt appear.

With regard to their origins, little is said, except some reference to the type: "*otra espada con guarnición alemana de un corte y vaina baya* [another sword with German hilt of one cut and low scabbard]" (1628), or of Italian origin, such as "*una espada dorada y daga çencillada hecha en Milán con sus talabartes de terziopelo negro atrencillado de oro* [a gilded sword and an engraved dagger made in Milan with its swordbelt of black velvet braided with gold]" valued at ten ducats (1608). It is much more extensive and explicit when talking about the colours and materials of the straps, scabbards and suspension systems of each white arm the Constable owned, than about the shape of handles and hilts, limiting itself in the best of cases to commenting on an element of the blade, use or shape: "*otra espada ancha la guarnición plateada y el puño de seda y plata vaina negra* [another broadsword, silver-plated hilt, the grip of silk and silver, [with a] black scabbard]" (1628).

This lack of reference to manufacturers and places of fabrication changes in the inventory of the objects that Constable left in his Madrid residence after his 1613 death. There, along with other arms, especially crossbows and objects from horse harnesses, such as stirrups, spurs and bits, etc., appear his personal, daily-use swords. These offer us a rich repertoire of types of hilts and blades, especially with reference to manufacturers (Sebastián Hernández, Hortuño, Juanes de la Orta) and production locations (Toledo, San Clemente, etc.), surely alluding to the inscriptions they carried on them[44].

44 Palomares, 1762. Dueñas Beraiz, 1999. Dueñas Beraiz, 2003: 269–306.

Polearms

The set of polearms is not very extensive. These are those formed by a wooden shaft on the end of which is placed a point or a metallic item that, depending on its form and use, as well as the length of the shaft, carries a particular name. Also noteworthy is the presence of throwing weapons of a hunting nature, namely spears. They range from simple hunting weapons to luxurious shafts that are "*retortixadas* [twisted]", warped, or even silk- or velvet-lined with gold studs and colourful tassels.

Lances used by the cavalry appear, referring both to the nineteen suits of armour for light horses as well as to the war harnesses or the thirty-nine white ones that had a "*lança con banderola de tafetán amarillo* [lance with a yellow taffeta banner]" in addition to a rapier and a mace. We could think that the former were shorter and smaller horse-rider lances used by light cavalry, and the latter were tournament spears for heavy cavalry. Thus, in the armoury's 1638 inventory, there already exists a set of arms under the name "*Lanças venablos y armas de asta* [Lances, spears and polearms]", in which the following appear:

> *Cinquenta y cinco lanças xinetas con sus yerros y astas blancas y al pie su punçón*
> *Catorce lanças de ristre las doce con banderolas y las dos sin ellas*
> *Tres medias lanças de ristre quebradas y sin hierros*
> *Mas otras veintidós lanças de ristre algunas de ellas sin hierros y banderolas y todas estas lanças de ristre son de los mismos arneses*
> *Ocho astas de lanças que están pintadas de amarillo y plata.*
> [Fifty-five horse-rider lances with their irons and white shafts and its awl at the base
> Fourteen tournament lances, twelve with banners and two without them
> Three half tournament lances, broken and without irons
> Another twenty-two tournament lances, some of them without irons and banners and all these tournament lances are from the same harnesses
> Eight lance shafts that are painted yellow and silver].

In 1628, other polearm typologies or loose elements of polearms were named. In this group, we would like to highlight the presence of a great war hammer: "*Una asta de la estatura de un hombre con un martillo y al otro lado su punta de pico quadrado con una punta larga que se mete con tornillo* [A shaft the height of a man with a hammer and, on the other side its square-pick point with a long point that is inserted with a screw]", an antique arm represented iconographically and of which not many models have survived. Another noteworthy item is the symbol of infantry captains: "*Una xineta de capitán con borla y cordón de seda negra y plata y asta de pino teñida de negro* [A captain's horse-rider spear with black and silver silk tassel and cord and black-stained pine shaft]".

The presence of two combined arms, those that had elements of different typologies and which were fashionable in the middle of the sixteenth century, is also noteworthy. The first was formed by a German axe with a firearm incorporated, similarly to some existing ones, shaped by a wooden handle with ivory inlays and a wheel-lock on its side and a barrel on the metal end of the axe head. Similar in conception is another combined, or mixed, arm: "*Un venablo guarnecido en terciopelo açul y oro y la cuchilla labrada de atauxía con dos pistolas que se disparan por medio del asta del venablo* [A spear trimmed in blue and gold velvet with the carved blade of damascene with two pistols that are fired from the spear's shaft]". This type of hunting weapon was used to hunt wild boars, providing the hunter, in addition to a strong spear to stab the animal, with two shots to finish it off[45].

Shields

A meaningful number of shields also appears, mainly consisting of two circular types: rondaches and bucklers. The former are larger and more metallic, of a military nature. These are found as matching parts of some garniture, loose as bullet-proof rondaches or with religious representations of a protective nature, relatively common in the second half of the sixteenth century: "*Otra rodela fuerte con un Xto [Cristo] en medio plantado en un Calvario forrada en terciopelo colorado* [Another strong rondache lined with red velvet with a crucified Christ in the centre]" (1628)[46]; or once again Marian: "*Otra rodela con un escudo dorado […] con una cifra de María y una corona* [Another rondache with a gilded coat of arms […] with a monogram of Mary's name and a crown]" (1628)[47]. One of the

45 Various pieces similar to the one mentioned here are found in the Armoury of the Electors of Saxony (Dresden Rustkammer X 0515; X 0514.01), the first being attributed to the renowned Munich arquebus-maker, Peter Pech.

46 It might be the same one that was described in 1608 as "*Una rodela negra fuerte con un Xto de plata y ocho clavos con que están guarnecidas las correas que hacen muestra en la misma rodela forrada de gamuza con pasamanos negro en cinto y cinquenta ducados* [A strong black rondache with a silver Christ and eight fasteners with which the straps are decorated, which are shown in the same chamois-lined shield with black handrails on the belt and fifty ducats]".

47 "*Una rodela blanca dorado el friso y en medio un escudo con la cifra de María esta forrada de en terciopelo Rojo con pasamanillo de oro y plata tasado en ciento y diez ducados* [A white rondache, the frieze gilded, and in the middle a shield with the figure of Mary. It is lined with red velvet with a gold and silver handrail, valued at 110 ducats]" (1608).

most important that appears mentioned is a possible gift by the inscription that it carries and its high valuation:

> *Una rodela dorada que tiene en medio el nudo de Salomón y un letrero alrededor que dice para el Condestable mi señor forrada de terciopelo negro y unas sirenas doradas en ella tasada en cien ducados.*
> [A gilded rondache that has Solomon's knot in the middle and a notice around it that says "for the Constable, my Lord", lined with black velvet and gilded sirens on it, valued at one hundred ducats].

Wooden rondaches painted with various mythological or Renaissance motifs, as are ones with decorate elements embedded in them, are also mentioned: "*Otra rodela de madera pintada y dorada con quatro veneras de latón y una fuente y ninfas del Parnaso y otras labores* [Another painted and gilded wooden rondache with four brass scallop shells, a fountain and nymphs of Parnassus and other works]".

The bucklers, smaller wooden rondaches of a civil or sporting nature, with metal boss, a hook and a handle for holding it, also appear, albeit in fewer numbers. Some are antique: "*Un broquel antiguo de hierro aperdigonado con su empuñadura de madera* [An antique pelleted iron buckler with its wooden grip]" whilst others were high quality: "*Otro broquel pavonado con un mascarón con oro en medio y clavazón dorada alrededor forado en terciopelo carmesí con su manera* [Another blued buckler with a figurehead with gold in the middle and a gilded set of fasteners around it, lined with crimson velvet, with a grip]".

Another significant set (thirty-seven pieces), although with a more decorative character than from collector's desire, are the pavises. They are called antiques of the house and appear encircling the armoury "*por adorno* [for adornment]." Some are found painted on the exterior with heraldic emblems of "*los Velasco* [the Velasco family]", the royal or Leonese coats of arms, or angels. Lastly, it is necessary to mention the possession, as it could not be otherwise, of two adargas for "*correr cañas* [play canes]"[48], one with the Constable's arms and the other plain, the two valued together at 122 reales in 1608.

Arms of propulsion of launching

Crossbows also form an important set of pieces, not only for their quantity, but also for the quality, and the quality of information that they provide about their

48 Referring to the cane game, in which two lines of mounted men, usually nobles, threw canes (in lieu of spears) at each other, repelling them with shields of the type mentioned here, in the city square for popular amusement and spectacle.

materials, their manufacturers and their ownership by one of the Constables, given that they were personalised arms. Mountain ash is the most noted material for the butt's fabrication, although ebony also appears mentioned. In addition, some of the most famous crossbow makers of the sixteenth century are cited as manufacturers of the lathes and prods, among them Puebla el viejo, Juan Blanco and Juan de Azcoitia, even indicating, at times, their marks[49]. Lesser-known Iberian Peninsula production centres are also mentioned, such as Barbastro, in Aragon:

Ballestas. 13 ballestas con 10 gafas tásase por Xtobal de Escobar ballestero en esta manera
Ballesta de don Juan de Velasco
Una ballesta de ébano con una verga de [...] con su gafa tásase en 100 reales digo en 2800 reales
Otra del Excm Agnus Dey
Otra que dice de Juan Blanco con un tablero de serbal y la gafa de Çamora tásase en seis ducados
Otra del condestable don Yñigo
Otra ballesta intitulada el condestable don Yñigo con su gafa tásase en ocho ducados
Otra de Balbastro (sic)
Otra ballesta de Balbastro (sic) Que tiene dos PP con corona en la verga en seis ducados
Otra ballesta de Puebla el viexo el tablero de serbal con sus gafas que tiene en la verga Jesús María en 4 ducados
Otra de serbal.
[Crossbows. 13 crossbows with 10 spanning levers, appraised by Christobal de Escobar, crossbow maker, in this manner
A crossbow of ebony with a prod of [...] with its spanning lever, appraised at 100 *realesI say 2,800reales*
Another one The Excellency of the Lamb of God
Another one that says it is by Juan Blanco with a butt of mountain ash and spanning lever from Zamora, appraised at six ducats
Another one of the Constable Don Iñigo
Another crossbow entitled to the Constable Don Iñigo with its spanning levers, appraised at eight ducats
Another one from Barbastro
Another crossbow from Barbastro, which has two PP with crown on the prod, six ducats
Another crossbow by Puebla el viejo, the butt of mountain ash with its spanning lever that has "Jesús María" on the prod, 4 ducats
Another one of mountain ash].

49 Martínez de Espinar, 1644.

CONCLUSION

The armoury of the Constables of Castile is a good example of a noble collection developed throughout the sixteenth century that acquired its status as such when the 6th Constable provided it a specific location with funds and a staff for its maintenance, as well as a museographic organisation, with installation systems, supports for armour, curtains, delimited spaces, several specialised rooms, etc.

Not all the Constables's arms formed part of the armoury in Burgos, as many personal arms were found located in the Constable's residence, primarily in Madrid, such as his swords for daily use, some firearms and many crossbows, although some pieces could have later formed part of the armoury. The origin of these arms were direct purchases, especially in Italy, Flanders and Germany, the main production centres and where the Constables held various positions. Few, yet noteworthy, are the pieces produced on the Iberian Peninsula. There also appear gifts from other Spanish or foreign noble houses, although they are not very numerous or identified as such. With regard to the composition, we can say that it reflects classical models of the time period, every type of harnesses, garnitures, white arms, polearms and especially the firearms and combined arms standing out to the taste of an intellectual like Juan Fernández de Velasco, with knowledge and interests in mechanics and military systems.

The existence of several different inventories of the same armoury in a chronological framework over more than a century is very interesting when studying its evolution. The comparison with other similar armouries, as well as new data on historical armouries will enrich our knowledge and allow this collection to be contextualised at the national level and thus permit us to learn more about it, although it will still be difficult to try to locate any of its items still existing in modern collections worldwide.

BIBLIOGRAPHY

Alonso, Begoña/De Carlos, María C./Pereda, Felipe (2005): *Patronos y coleccionistas. Los condestables de Castilla y el arte (siglos XV-XVII).* Valladolid, Universidad de Valladolid.

Boccia, L. G./Godoy, José A. (1985): *Museo Poldi-Pezzoli. Catalogo della Armeria.* Milan, Electra editrice.

Bouza, Antonio L. (1983): *Museo de los Condestables de Castilla.* Burgos, Ed. Santiago Rodríguez.

Cámara Fernández, Carmen (1994): "Aportación al estudio del coleccionismo en el siglo XVII. La armería del Condestable de Castilla en su palacio de Burgos".

In: *Los Clasicismos en el Arte Español. Actas X Congreso del Comité Español de Historia del Arte.* Madrid, CEHA, pp. 407–411.

Carlos, María Cruz de (2005): "'Al modo de los antiguos'. Las colecciones artísticas de Juan Fernández de Velasco, VI Condestable de Castilla". In Alonso, Begoña/De Carlos, María C./Pereda, Felipe (eds.): *Patronos y coleccionistas. Los condestables de Castilla y el arte (siglos XV-XVII).* Valladolid, Universidad de Valladolid: pp. 207–277.

Dueñas Beraiz, Germán (1999): *Los gremios espaderos españoles S. XVI-XVII.* Salamanca, U. Degree Universidad de Salamanca.

Dueñas Beraiz, Germán (2003): "Francisco de Santiago y Palomares y los espaderos toledanos". In: *Gladius*, XXIII, pp. 269–306.

Franco Silva, Alfonso/Soler del Campo, Álvaro (1994): "Los arsenales de dos fortalezas castellanas: inventarios de Torremormojón (1506) y Pedraza de la Sierra (1512)". In: *Historia. Instituciones. Documentos*, 21, pp. 309–344.

Godoy, José A./Pyhrr Stuart W. (1999): *Heroic Armor of the Italian Renaissance: Filippo Negroli and His Contemporaries.* New York, Metropolitan Museum of Art.

González Ramos, Roberto (2017): "Las armerías nobiliarias castellanas del siglo XVII como manifestación de identidad cultural". In: Carrasco, Adolfo (ed.): *La nobleza y los reinos. Anatomía del poder en la Monarquía de España (siglos XVI-XVII).* Madrid, Iberoamericana-Vervuert: pp. 269–312.

Mann, James G. (1962): *Wallace Collection Catalogues, European Arms and Armour.* London, Clowes and Sons Ltd.

Marchesi, José M. (1849): *Catálogo de la Real Armería.* Madrid, Aguado impresor de cámara de S.M.

Martínez de Espinar, Alonso (1644): *Arte de ballesteria y monteria* [...]: *dividida en tres libros.* Madrid, Imprenta Real.

Martínez del Peral Forton, Rafael (1992): *Las armas blancas en España e Indias.* Madrid, Editorial Mapfre SA.

Montero Delgado, Juan/González Sánchez, Carlos A./Rueda Ramírez, Pedro J./Alonso Moral, Roberto (2014): *De todos los ingenios los mejores: el condestable Juan Fernández de Velasco, V duque de Frías (1550–1613).* Seville, Real Maestranza de Caballería.

Palomares, Francisco de Santiago (1762): *Marcas de que usaron sus espadas los últimos y famosos armeros de Toledo hasta la extinción de esta fábrica.* Toledo, s. e.

Payo Hernanz, René J. (1992): "Estudio de algunas colecciones de la Nobleza burgalesa a comienzos del Siglo XVI". In: *Hidalguía. La Revista de Genealógia, Nobleza y Armas,* 235, 6, pp. 767–784.

Soler del Campo, Álvaro (2006): "La producción de armas personales, 1500–1700". In: García Hernán, Enrique/ Maffi Davide (eds.): *Guerra y sociedad en la monarquía hispánica: política, estrategia y cultura en la Europa moderna (1500–1700)*, 2. Madrid, Laberinto-CSIC, pp. 843–860.

Soler del Campo, Álvaro (2010): "La consideración de las armaduras como obras de arte e imagen del poder en el contexto de la Real Armería". In: Soler, Álvaro (ed.): *El arte del poder: la Real Armería y el retrato de corte*. Madrid, Museo del Prado-Patrimonio Nacional: pp. 25–39.

Terjanian, Pierre (2013): "The art of the armorer in late medieval and Renaissance Augsburg: The rediscovery of the *Thun sketchbooks*". In: *Jahrbuch des Kunsthistorischen Museums Wien*. (Part 1).

Terjanian, Pierre (2016): "The art of the armorer in late medieval and Renaissance Augsburg: The rediscovery of the *Thun sketchbooks*". In: *Jahrbuch des Kunsthistorischen Museums Wien*. (Part 2).

Valencia de Don Juan, Conde Viudo de (1898): *Catálogo histórico-descriptivo de la Real Armería de Madrid*. Madrid, fototipias de Hauser y Menet.

Vázquez Manassero, Margarita A. (2019): "Juan Fernández de Velasco y los ingenieros. Redes de poder e intercambios científicos entre España e Italia". In: Cámara, Alicia/ Vázquez, Margarita (ed.): *Ser hechura de: ingeniería, fidelidades y redes de poder en los siglos XVI y XVII*. Madrid, Fundación Juanelo Turriano, pp. 83–100.

Zalama, Miguel A. and Andrés, P. (2002): *La colección artística de los Condestables de Castilla en su palacio burgalés de la Casa del Cordón*. Burgos, Caja de Burgos.

Roberto González Ramos

Universidad de Córdoba

AROUND THE YEAR 1600: THE MEDINACELI AND LERMA ARMOURIES

The armouries of the Castilian nobility had differing characteristics, albeit within the common framework related to their social and prestigious intentions. One of the elements that is going to be examined here, and this text's main investigative framework, is to carry out a comparative vision between the two armouries that allows us to contrast these characteristics, in addition to presenting (both) the Medinaceli and Lerma armouries in the Modern Age. Both armouries were important, but they were created for different reasons, by people and noble houses very different in their social position, and through very distinctive systems. We will see that the Medinaceli armoury is surely a dynastic armoury, accumulated, arranged and formed by and for an ancient lineage, which has been gathering suits of armour and other objects over a long period of time and which, at any given moment, could display them to highlight the family's prestige, antiquity and position. The Lerma armoury is more provisioned than its Medinaceli counterpart, but it has very distinctive characteristics because the Lerma family's position in the social fabric of the Castilian aristocracy was very different, although the family itself is ancient. In fact, everything appears to indicate that Lerma house's modest importance prevented it from establishing a considerable amount of arms. Only the 1st Duke of Lerma's very rapid ascension as "*valido* [favourite]" allowed him to create an *ex novo* armoury that would reproduce many of the elements of a dynastic armoury. However, it resulted not from a family's actions rather than from a specific individual's efforts.

THE MEDINACELI ARMOURY

We do not know exactly when this armoury was established, but it seems clear that it was decided to place it in a specific palace. In this case, it was not located in Medinaceli, seat of the family's principal title, but instead in a place of no less symbolic importance: the palace in Cogolludo (Fig. 1). It consisted of an edifice linked to the Marquisate of the same name, one highly representative and of outstanding architecture. One of the first Renaissance palaces in Castile, its construction began under Lorenzo Vázquez in 1492[1]. It exhibits a representative bossage facade of "Quattrocento" inspiration and is articulated around a main courtyard and another second one[2].

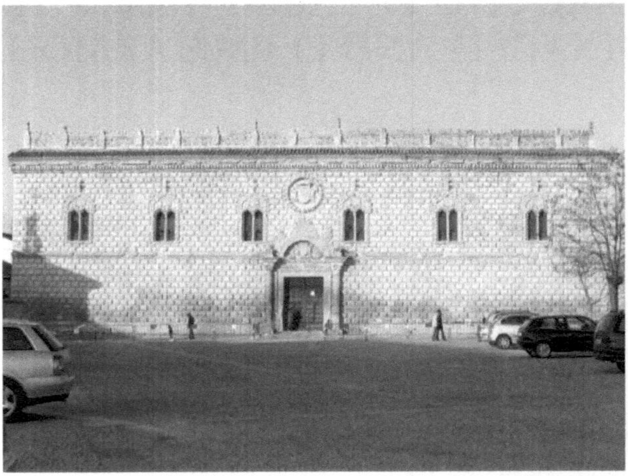

Fig. 1. *Dukes of Medinaceli Palace*. Lorenzo Vázquez. Cogolludo, Spain (Wikimedia Commons).

The first documents about the arms related to the Medinaceli family of which we are aware come from the sixteenth century and consist of an inventory of harnesses and a list of other objects and arms, without a date or location, but that must have dated to 1552, and the post-mortem inventory of Juan de la Cerda y Silva, 4th Duke of Medinaceli, carried out in Madrid in 1575. The first of these sources indicates that it was an inventory of the harnesses of the 3rd Duke Gastón de la Cerda, who died in 1552, that he left to his heir, his brother Juan, as well as a list of various goods taken from his auction, by which we must date it to that year[3]. According to the document, in the deceased Duke's chamber, there were forty-six harnesses with their gorgets, breastplates, backplates, sallets, etc.; a harness with two shots from an arquebus in the breastplate (perhaps a result of a battle); a corslet with pieces in the style of a cuirass, blued, with a sallet and

1 Nieto/Morales/Checa, 1993: 36. Martínez, 1995.
2 Pérez/Pérez, 2008.
3 Archivo Ducal de Medinaceli (hereafter ADM), Medinaceli, legajo 104, doc. 6.

a morion; fourteen lance rests with four that had some harnesses put on them; fifteen steels for saddles, forty-six horse shaffrons; forty-eight sabaton points; bevors and sixteen breastplates that had been brought from Medinaceli, along with various pieces of older arms, which indicated that all this surely was located in Cogolludo, and everything here will possibly be the seed of a future armoury. The goods taken from the auction were crossbows (some of them were given to various people); partisans; a black rondache; a sword made by the swordsmith, the Moor of Zaragoza[4]; a scimitar; a very broad sword; two Turkish bows; a German arquebus; crossbow arrows; crossbow strings; a coat of arms; drums and similar items[5].

The other document is the inventory of goods of the 4th Duke, Juan de la Cerda, in Madrid, where he died in 1575. He was viceroy of Sicily, viceroy of Navarre and governor of the Netherlands, and accompanied Philip II to England[6]. In this source, the majority of the items recorded were weapons, in 150 sections, which totaled around 300 pieces. In addition to some pages with "*trazas de armas* [drawings of arms]", we find six harnesses placed in Flemish chests. One was gilded and engraved, and the other metal-coloured with gilded ends, with his helmets and some morions, and all of them were strong and arequebus-proof. Moreover, we find a large quantity of cutlasses, and blades of others, with their scabbards, many of great wealth, adorned with stones[7]. In addition, we find twenty-four examples of swords and sword blades, some of them "*de las viejas* [of the elders]". We also find irons for spears; engraved irons for pikes; two lances for lance rests; war axes and metal rondaches. Furthermore, it is necessary to add some firearms (arquebuses, pistols), and others for hunting, such as crossbows and quivers, hunting knives and falconry elements. Five Flemish bows with their cases stand out. Finally, two campaign pavilions and two tents, among other items, appear[8].

The existence of this group of arms in Madrid in 1575 does not indicate anything about the existence of a Medinaceli armoury in Cogolludo. However, in all likelihood, some of those arms would end up being moved to Cogolludo at some point in the future. With the information at our disposal, we cannot determine if the armoury already existed with the 3rd Duke or if his successor,

4 Famous swordsmith. Dueñas, 2000.
5 ADM, Medinaceli, legajo 104, doc. 6.
6 Romero Medina, 2017: 350–371.
7 For a partial study of some of the document's aspects, see Romero Medina, 2020: 81–101.
8 Archivo Histórico de Protocolos de Madrid (hereafter AHPM), T. 939, fol. 307 r.

the 4th Duke, or a subsequent Duke, founded it. What is certain is that it already existed upon the 1594 death of the 5th Duke of Medinaceli, Juan Luis de la Cerda, as it appears in the inventory of the palace of Cogolludo made that year. Although the noble house was one of the largest and oldest of the Castilian nobility, it does not have the largest and most stocked armoury in the kingdom—it was among the mid-level ones—perhaps due to the family's poor economic situation. Of course, if its creation was due to the 4th Duke, who died in 1575, it would be one of the earliest.

The 1594 post-mortem inventory brings together a portion of what the 5th Duke had in Madrid, which reveals that there were almost no weapons at court, contrasting with what we have seen in 1575. There was only one pistol, two Spanish swords of high quality, two campaign pavilions, two tents, a crossbow and an arquebus later sold in the auction, i.e., weapons used in the day-to-day life of a nobleman.[9] However, at this time, the registry of goods that were in the palace of Cogolludo was added to the inventory and we specifically find in it weapons and in great quantities. The Cogolludo armoury in 1594 had a total of approximately 340 entries of arms, which totals around 1,000 items. Although the document does not speak of an armoury, it was undoubtedly a display that from all points of view we must qualify as such[10].

In the palace, there were seventeen "*hombres de armas* [men-at-arms]", that is, installed suits of armour, surely on wooden hangers, as was commonplace in these types of sets. There were three more harnesses and the "*arnés cumplido de Su Excelencia* [complete harness of His Excellency]". Moreover, we find coats of mail and various sallets, more diverse blued breastplates and brigandines, as well as around 850 loose pieces of harnesses for men and another seventy one "*de armar caballos* [to arm horses]", among which are those that formed a "*buena* [good]" horse armour, or gilded bard. We also found about fifteen morions, some old, others gilded, with gilded stripes, a gilded Turkish (or Turkish-style) one and another "*grabado de figuras* [engraved with figures]". There was, moreover, seven rapiers, and thirty-two swords with their daggers and hilts, some loose. We find them gilded, blued, broad, etc., but it stands out that twenty-one are characterised as "*viejas* [old]" —for tournaments, wide, gilded, for fencing—which indicates that about two-thirds of these arms were found in Cogolludo

9 AHPM, T. 628, fol. 830 r. A copy of the inventory of cited goods, of the goods from Madrid, exists in the Archivo General de Palacio, caja 322, legajo 11.

10 AHPM, T. 628, fol. 647 r.

according to their age. We also find two old cutlasses, which leads us to question the whereabouts of those who belonged to the Medinacelis in 1575.

The inventory tells us also of fourteen rondaches, among which one made of turtle shell stands out; others bullet-proof; an old painted one; others of wood and of steel, etc. The most notable was "*acerada, de figuras* [strengthened, with figures]". Furthermore, we have nine arquebuses and shotgun butts as well as fifteen saddles, both armoured and leather. The bows stand out, among which we find sixteen English bows with English arrows and strings; thirteen Turkish ones and two steel ones, in addition to arrows, quivers, etc. Finally, we have four drums, six iron maces and polearms. These were no less than seventeen lances "*de sortija* [for tilting at the ring]", as many unspecified, three pikes, five halberds, two partisans and six lances, in addition to a lance with the iron in the shape of a fleur-de-lis, the heraldry of the de la Cerda (Medinaceli) family. Finally, we find a tent, two "*caballos de madera* [wooden carriages]" for artillery pieces and tools to clean and to repair the weapons.

It indicates a very complete set, although the inventory often provides very little descriptive detail, in general. Some of the pieces bring to mind others by their characteristics that we have seen in the 1552 and 1575 documents, such as the harnesses, the brigandines, the rondaches, some of the bows, the lances and others elements such as the campaign pavilions and drums. The rest came from places, and by circumstances and reasons, that we cannot discern, but without a doubt they held symbolic meaning for the set.

Although the documentation speaks of inventories carried out during the seventeenth century, we have only located one partially preserved, from 1654, and those carried out in 1667 and in 1685. We undoubtedly lack references to when this armoury was most important. It was probably in the first decades of the century, for example in 1607, upon the death of the 6th Duke of Medinaceli, who, by the way, only had little more than a few swords and two complete harnesses, a black one for tournaments, in Madrid at a time when it was still customary to solemnise various occasions in tournaments and jousts[11].

The 1654 inventory only includes the final part of the document. Above all, campaign pavilions and tents are recorded, curiously, most of them decorated in blue, white and red stripes, which correspond to the heraldic colours of the house of de la Cerda. We have two pavilions—they would be the ones we have already seen from 1575—and one more and a large quantity of tents, up to eighteen—some with imitation stonework, battlement decorations, and banners;

11 Frieder, 2008. Fallows, 2010. Terjanian, 2010.

all with their cordsand mounting sticks—which tells us that the collection continued receiving items, being a living and evolving entity, both increasing and decreasing, much like other armouries[12].

The 7th Duke, Juan Antonio de la Cerda y Dávila, in 1667 had in the Cogolludo armoury, already named as such and occupying three palace rooms, a total of approximately 350 pieces, and more than 200 items stored in boxes that were pieces of armour for men and horses[13]. The decrease in absolute terms of the number of pieces is evident, but there are also some one-time increases, such as the fourteen artillery pieces that are recorded and that were not previously mentioned. Regarding the usual elements, we find thirty-four harnesses or "*cuerpos armados* [armed bodies]" on their display stands, a growth attributable to the arrival of several of them and/or to the fact that those dismantled had been partly recomposed. Similarly, we find a blued black breastplate and backplate with cuisses and other pieces, fourteen more breastplates—two of them black—plus another two sets of breastplates and backplates with gorget, morion, vambraces and "piernas [legs]" could well have been other such harnesses. There was also a small blued black harness on display stand, which might well have been for a youth or child. Along with those, we find twenty-six suits of armours for horses (bards), perhaps recomposed of the many existing pieces, and one more engraved and gilded on its display stand, probably the one existing in 1594.

Additionally, we have eight gilded and studded crimson cuirasses; three pieces of mail; twenty morions; twenty-five swords without hilts; eleven steel rondaches, one engraved with a battle in relief and its matchching morion, both similar and surely the same set which had already been recorded in 1594[14]; two more rondaches of wood; two adargas; seven muskets; forty horse saddles, one Morish; thirteen bows; another fourteen longbows "*en forma de báculo* [in the shape of a staff]" —these would be English longbows—two good copper drums with the drumheads broken; and the polearms, of which there were fourteen lances "*de ristre* [for tournaments]", ten pikes, five spears, two partisans, three halberds, another unspecified lance and nine lances with vamplates. The three pavilions and the nearly twenty tents were preserved, along with pieces of other unserviceable ones. It is noted that there is a slight loss of pieces in some

12 Archivo Histórico Provincial de Guadalajara, P. 3995, s.f.

13 ADM, Cogolludo, legajo 3, doc. 44, fol. 1 r. Transcribed by Pérez/Pérez, 2008: 175–177.

14 It must have been the set that Ponz saw in the Medinaceli armoury, then in Madrid: "a shield in which a battle is represented […] a morion wherein the victory of David over Goliath and the story of Judith are depicted". Ponz, 1776: 330.

categories, such as swords, firearms, polearms, drums and tents, but the essentials seem to have been preserved rather well. The distribution of the rooms shows that the first contained thirteen harnesses, polearms, muskets and the bows. The second had another thirteen harnesses, more polearms, drums, breastplates and horse shaffrons. The third contained fourteen harnesses and suits of armour, the gilded bard, the rondaches, morions, etc[15].

On the occasion of the 1667 inventory's completion, the Duke maintained a correspondence with his administrators who, in turn, provided him with reports on the armoury and the other assets at Cogolludo. In those documents, there is talk of some losses as well as of a search for solutions to the apparent state of neglect in the weapons' custody[16]. The efforts clearly must have worked, and the set must have been recomposed to a certain extent, which coincides with the interest that is noted in some noble houses to tidy up their family armouries in the second half of the seventeenth century. In fact, in 1685 in the era of the 8th Duke of Medinaceli, another inventory includes forty harnesses, one of them recently arrived, and the harness of the previous Duke. In other items, a slight decrease is noted, as is the case of the rondaches, tents and pavilions. In the case of artillery, we found that the fourteen pieces decreased to eleven.The bows and arrows suffered the most, as only nine unspecified bows remained, compared to those in the previous registry[17].

In the eighteenth century in Cogolludo, the collection remained more or less the same, with specific progressive decreases. In 1716, a description of the building alludes to the armoury being found in the rooms of the edifice's second courtyard, alongside the stables[18]. In 1746, the forty armed men were inventoried and found to be in somewhat of a dilapidated condition, their vambraces lying on the ground. The gilded horse armour is also mentioned, as well as drums and other pieces. The lances (seven), firearms (four arquebuses), the morions (eighteen) declined in number and there is no mention of either tents or pavilions, nor bows and quivers[19]. Perhaps these circumstances, but surely the renewed interest shown in their armouries by some noble houses, especially those of the ancient nobility, must have motivated the 11th Duke of Medinaceli, Luis Fernández de Córdoba, to order the movement of what remained, first to

---

15 See the proposal, not very well founded, but plausible, for the rooms' location in the palace in Pérez/Pérez, 2008: 150–151.

16 ADM, Cogolludo, legajo 3, doc. 44, fol. 17 r. See Pérez/Pérez, 2008: 152.

17 ADM, Cogolludo, legajo 5, doc. 1, fol. 1 r.

18 ADM, Cogolludo, legajo 3, doc. 42, fol. 7 v. Pérez/Pérez, 2008: 94.

19 ADM, Cogolludo, legajo 5, doc. 38. Pérez/Pérez, 2008: 186–193.

his town of Valdeavero and later to Madrid, in 1758. Thus, he was able to verify that the armoury was no longer what it was, although it preserved a part of what we have seen. The harnesses were preserved, albeit very dismembered; one more cannon had mysteriously disappeared, as had muskets and other items[20]. The rest of the Cogolludo armoury was brought to its new location with various difficulties. Indeed, in 1764 in the Madrid palace of the Medinaceli family, there was already an armoury room[21], which would be the origin of the one that became famous shortly thereafter[22].

THE DUKE OF LERMA'S ARMOURY

In contrast to the Medinaceli armoury, which apparently consisted of weapons and elements from the noble house, for the most part, and was a manifestation of the lineage's antiquity, the armoury of Francisco Gómez de Sandoval y Rojas, the "*valido* [favourite]" of Philip III, was an *ex novo* construction and was brought into existence by taking advantages of its creator's powerful resources. That is, the Duke not only accumulated riches, titles and privileges, he also acclaimed his lineage with an armoury of certain importance. One must not forget that the favourite married Catalina de la Cerda, daughter of the 4th Duke of Medinaceli, and that he married his sons and daughters off to the most noteworthy noble houses of his time, such as that of the Infantado, which had important armouries.

The Duke decided that his armoury would be located in the castle of Ampudia (Fig. 2), in accordance with the practice of placing these collections in the palaces and fortresses of symbolic dynastic significance. The town of Ampudia, linked to the Rojas family, was incorporated into the titles of Denia-Lerma following a long dispute between different branches of this family, via a 1597 court ruling, ratified in 1599 and permanently confirmed in 1607[23]. The possession of Ampudia was a manoeuvre that was part of the activities carried out by Lerma to recover his family's prestige and to establish a large patrimony. It is known that the Duke's lineage was in a lowly position ever since Diego Gómez de Sandoval supported the losing side in a noble conspiracy against the King in the fifteenth century, by

---

20 ADM, Archivo Histórico, legajo 127, 2, docs. 47–55. There is a partial inventory of Cogolludo's arms, with a budget for their repair, in ADM, Archivo Histórico, legajo 82, doc. 64, s.f..

21 Biblioteca Nacional, Madrid, mss/12024, V.1, fol. 63 v.

22 Romero Medina, 2023: 273–306.

23 Williams, 2010: 65. Alvar, 2010: 129.

which it had been dispossessed of titles, assets and fiefs[24]. The favourite not only obtained the town of Ampudia, but he was elevated to a Count of the same town in 1599, with confirmation in 1602, and in his will, he linked it to his "*mayorazgo* [entailed estate]".

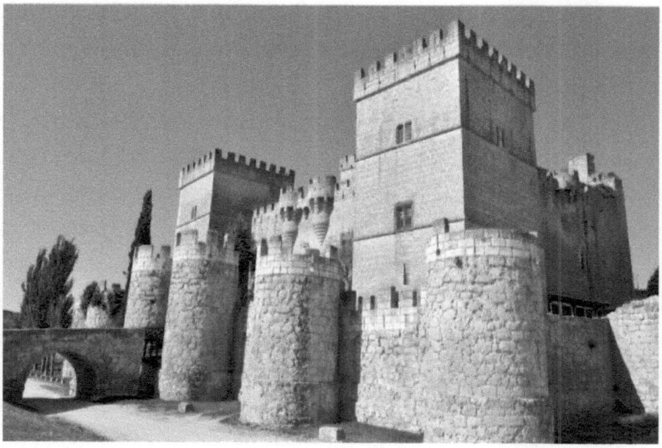

Fig. 2. *Castle of Ampudia.* Ampudia, Spain.

With the court in Valladolid between 1601 and 1606 it is known that Ampudia, so near the city, was a stronghold that allowed Lerma to control the king. The favourite distanced the king from the court there and brought the royal family to the castle-palace on different occasions[25]. It is necessary to take into account that the town of Lerma, the centre of the Duke's construction programme that turned it into the architectural and urban reference of a "*villa ducal* [ducal town]", was not prepared until years later, with the court already back in Madrid. The Ampudia castle was found in good condition, given that it had been improved by the Counts of Salvatierra in the fifteenth century and set up in 1525[26].

The Duke of Lerma, according to what we are seeing, decided that he was going to establish an armoury as a fundamental part of his operations to ensure his prestige and that of his descendants and that this armoury would be located in Ampudia, but we do not know when he made this decision. Of course, there was no Sandoval armoury, nor did the Sandovals have the opportunity to create

24 Feros, 2002: 70–80. Williams, 2010: 15–18. Alvar, 2010: 45–51.

25 Izquierdo, 2011: 29.

26 Cooper, 1991: 228. Cobos/Castro, 1998: 201–203. Izquierdo, 2011: 24.

one, although there was always the possibility of trying to collect weapons from some family branch. In any case, Francisco Gómez y Sandoval y Rojas became favourite in 1598, and some information confirms that the armoury could not have been established as such a complex before 1597. It should be recalled that 1599 brought about the confirmation of the possession of Ampudia, which in 1602 became a County, and that in 1607 the final judicial confirmation was produced. Thus, we believe that the set's official constitution must not predate the 1599 valuation.

In April 1603, the Duke paid the Basque armourer Alejandro de Armaolea to "*limpiar y adereçar las armas que Su Exª tiene en la villa de Ampudia* [clean and arrange the weapons of His Excellency in the town of Ampudia]" over the six months previous[27]. Moreover, in December 1602, when creating the sculptures of the Duke and his wife for their tomb in San Pablo in Valladolid, and after alluding to the portraits and armour that were needed to create the praying figure of the favourite (Fig. 3), the sculptor Juan de Arfe states in a letter that "*Vm. me avise si ha venido de Ampudia Alejandro de Armaolea para que le escriba, que será presto menester* [Let me know if Alejandro de Armaolea has come from Ampudia so that I can write him, which will be necessary soon]"[28]. Later, it would be this Alejandro's brother, Pedro de Armaolea, that is found in the Duke's service in the cleaning of the arms of Ampudia in 1607[29]. Meanwhile, the armoury was completed with sets of arms. In May 1604, 2,000 war pikes were paid for, which another Basque armourer, Juan de Gamarra, delivered along with a pike with gilded decorative nails "*que ha de servir para la propia mano de Su Excelencia* [that must be from His Excellency's own hand]", as well as six "*extraordinarias* [extraordinary]" "*lanzas ginetas* [cavalry lances]" that had the same purpose. They were all delivered to Ampudia, the document also stating that another 1,000 pikes had been delivered in the year 1603[30].

---

27 AHPM, T. 1847, fol. 396 r. Cervera Vera, 1967: 46 and note 758. Cervera Vera, 1996: I, 39. Izquierdo, 2011: 31 and note 30.

28 Uhagón, 1916: 412–413. Viñaza, 1889: II, 32.

29 AHPM, T. 1854, fol. 2123 r. Cervera Vera, 1967: 46 and note 758.

30 ADM, Lerma, legajo 85, doc. 17.

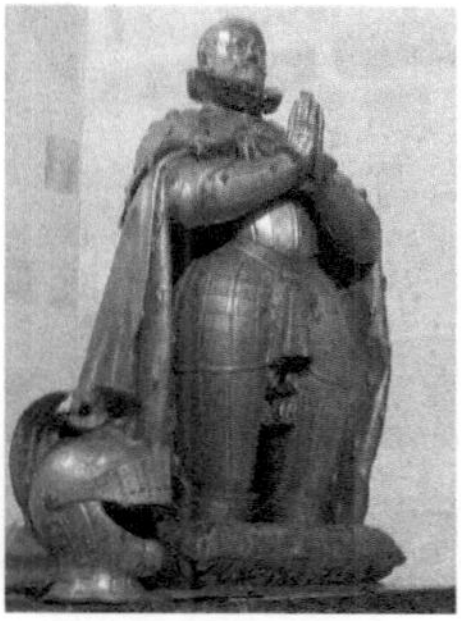

Fig. 3. *Praying figure from the Tomb of the Duke of Lerma.* 1602. Juan de Arfe, Pompeo Leoni and Lesmes Fernández del Moral. Valladolid, Museo Nacional de Escultura (Wikimedia Commons).

We have different inventories that describe for us the context of the Duke of Lerma's armoury in Ampudia, in addition to the other sets of arms that he had in other places. The oldest one we have is from 1607 and it is included in a general inventory of the Duke's belongings. Another one dates from four years later and is included in another general inventory. Finally, we have the one carried out in 1614.

The 1607 inventory indicates the existence in the armoury of Ampudia (named as such) of something more than 3,800 objects, the majority of which were arms and their accessories, in seventy-one entries, certainly an important total[31]. First, we have 168 harnesses. Of these, 100 were of "*caballo ligero y dos de ellos son grabados y dorados con todas sus pieças* [light horsemen and two of these are engraved and gilded in each of their pieces]"; fifty were for infantrymen; thirteen are described as older ones "*de infante de a caballo* [for mounted infantrymen]", except one, which was "*del Rey de Francia, de justa con todas pieças, gravado y blanco* [from the King of France, for jousting, with all its pieces engraved and white]", which is doubtlessly a gift of diplomatic origin. We also find two other harnesses "*que eran del S^r^ marqués de Villamiçar, el uno gravado y dorado [...] y el otro blanco y dorado de color al olio de amarillo y plata [...]de tornearde a pie* [that belonged to the Marquis of Villamiçar, the one engraved and gilded [...] and the other white and gold painted in oil, yellow and silver[...] for jousting on foot]". This reference means that they belonged to a member of Lerma's family, specifically his brother, Juan de Sandoval y Borja (1553–1625), who the king had made the

31 ADM, Lerma, legajo 52, doc. 14, fols. 265 r-270v.

Marquis of the aforementioned title in 1599 through the favourite's influence and who had been the viceroy of Valencia since 1604[32]. Another harness would be inventoried separately, "*a prueba dado de color, al fuego grabado de dibujo, con [...] murrión y rodela* [bullet-proof, coloured using fire technique and engraved with drawings with [...] morion and rondache]", and two more are described as "*grabados acanalados y dorados [...] y una rodela de la misma labor [...], el uno de acaballo y el otro de a pie, y el de a caballo tiene unas armas de caballo de la misma obra y labor con una silla de caballo [...] con sus açeros* [engraved, channelled and gilded [...] and a rondache of the same work [...], the one for horse riding and the other of foot, and the one for horse riding has a horse bard of the same work with a saddle [...] with its steels]"[33]. In addition to eight coats of mail and 121 pairs of mail shoes, we find two black strong steel-proof breastplates, two "*antiguos* [ancient]" strong breastplates "*el uno tiene un Cristo pintado* [one with a painted Christ figure]" and eight corslets "*de ynfante con golas tudescas y murriones* [for a infantry soldier with German gorgets and morions]". For the horses, we find two engraved and white iron coverings (bards) as well as 112 bits, 112 pairs of stirrups, 170 white shaffrons, 120 pairs of spurs "*de munición* [munition-quality ones]", as well as thirty-nine steels of white saddles. We also find another eight saddles without steels, and another twelve of white leather.

It is very interesting to come across two sections that tell us of the method of exhibiting the armour of people and horses in high modernity. One of them indicates the existence of two "*caballos de bulto hechos a propósito en que están puestas las armas de los caballos* [sculpted horses made to to place bards upon]", referring to the two engraved iron ones cited above. The other speaks of fifty-four "*máscaras diferentes una de atrás para adorno de los arneses de infante* [different masks, one for rear of the head, to decorate the infantryman's harness]", elements made of wood, cardboard or modelling clay to place in these suits of armour, which would be mounted standing on their wooden display stands, simulating the faces of those who would wear them, as if mannequins.

We also find 100 white morions, twenty-two rondaches that are also white, two old morions, twelve black helmets with their lining and two "*rodelas extraordinarias* [extraordinary rondaches]", "*la una negra y la otra de colores de la India* [one black and the other of colours from India]". Firearms could not be overlooked, of which there are up to 123 examples, among which we find 100 pistols with their leather holsters, ramrods and moulds for bullet making;

32 Williams, 2010: 94.

33 ADM, Lerma, legajo 52, doc. 14, fol. 268 r.

another eight from Germany, one of them two-barreled, with a wooden butt with ivory outlines, another single-barreled with iron butt and gilded decoration and the rest with butts carved of iron; another two French ones of three *palmos*[34] and wheel locks, with their holsters and another two "*con caxas de hierro grabado y lo mismo los cañones hechos en Aragón* [with butts of engraved iron and the same with the Aragon-made barrels]", also with their holsters. Moreover, we have ten arquebuses, one of which was particularly valuable:

> *Un arcabuz de rueda con el cañón de plata, y las abraçaderas de oro y en la llave cuya chapa es de plata, tiene dos pieças de oro la sobrerueda, y pieça del gatillo que sustenta el tornillo y ansimismo la mira, y el cabo de la vaqueta de oro y está sin punto*[35].
> [A wheel-lock arquebus with a silver barrel, and the clamp of gold and on the matchlock, the lock of which is silver, the wheel-lock has two pieces of gold, and the piece of the trigger that supports the bolt, and so the sight, and the ramrod end of gold as well, and it is without sight on the end of the barrel].

We have another wheel-lock arquebus from Germany with the butt carved of bone; another long flint-lock with a Turkish barrel; an arquebus "*de guerra* [of war]", with the barrel carved of damascene with its munitions store and powder horn; another plain one of war; five other arquebuses described as "*turquescos* [Turkish or Turkish-styled]"; and a wheel-lock musket for "*tirar a grullas* [crane-shooting]". The munition stores and powder horns were not missing. Furthermore, the existence of a small piece of artillery "*de hasta diez libras de metal con su caxa de madera* [of up to ten pounds of metal with its wooden butt]" is also noted.

In the noble armouries, the swords always stand out and the Lerma armoury is no exception. There were 100 swords with white hilts, another eighteen for tournaments and ten "*de hombre de a caballo* [horseman's]" swords. Within this last category, two particular pieces stand out, given that one of these swords was "*la del Rey de Françia* [the King of France's]" and other was "*la de Jullio Çéssar* [Julius Caesar's]"[36]. The first is surely a diplomatic gift related to the aforementioned harness and the other one of the numerous fabrications referring to historical figures. Fabrications, yes, but specimens whose historical provenance was not discussed excessively and that carried a symbolic weight. Also are described twelve great swords[37] with black sheaths, another two gilded,

34 A Spanish measurement of the period. One *palmo* is about 21 modern centimetres.
35 ADM, Lerma, legajo 52, doc. 14, fol. 268 v.
36 ADM, Lerma, legajo 52, doc. 14, fol. 269 r.
37 Also known as Zweihänder or, in Scotland, claymores.

another one that was "*buido de tres esquinas* [three-cornered]" and a damascene knife with gilded garniture and ivory handle. In addition, seven "*turquescos* [Turkish or Turkish-style]" cutlasses are included, one with gold garniture.

Polearms are also included in the armoury in Ampudia, such as the 2,148 pikes that have to do with the acquisitions discussed above; or the two black pikes, one without iron, and the six "*lanzas ginetas* [cavalry lances]" that are described in the document. Along with them, there are twelve golden halberds lined with velvet and another twelve plain ones. Moreover, we find five crossbows, two Turkish bows and a Turkish quiver, and another similar bow "*de dos cuerdas* [of two strings]", twelve small trumpet flags and various objects of American origin—said to be from India or the Indies, as that continent was then known in Castile—in addition to the ones already seen. It consisted of a basket full of plumage, six wooden bows "*de indios con sus flechas de cañas* [of Indians with their reed arrows]", and two wooden Indian (native American) circular rondaches.

It can be seen that the armoury established in Ampudia presents several characteristics that resemble those of other great Castilian noble houses, but also some peculiarities. Like other first-rate armouries, the quantities of arms, in this case especially harnesses, pieces of armour and elements for horses and polearms, are rather noteworthy, those here rivalling the most important collections. In part, they are elements arranged to arm troops and entourages, but it is known that in this period private armies did not have a significant function, except in relation to displays and visual demonstrations (such as the one that Lerma was involved in Valladolid in 1605)[38]. We also find a good representation of elements such as high-quality swords and firearms, in large quantities. As for arms and other family elements, as well as others notable for their historical, exotic and diplomatic origin, without being lacking, they are certainly less notable in number in comparative terms. In general, we note a large number of arms and accessories that are clearly large generic sets, as if purchased *en masse*. These types of weapons are not absent in the most important armouries, such as the Duke of the Infantado's, but in this case they form a much larger set. This fact leads us to consider Lerma's special interest in quantity, rather than in quality or specific symbolism in the armoury sets, and how this gathering policy would fit with the formation of a large armoury in a limited time yet with great resources, albeit without a family foundation comparable to the other examples, nor with a

38 Pinheiro, 1916: 77 and following. Cervera Vera, 1967: 132.

special intensity—where there was not no time to unfurl—regarding elements of symbolic, political, historical or exotic importance.

The other inventories to which we make reference, all done in the Duke of Lerma's lifetime, confirm and at times clarify what we have seen until now. In 1611, we find a section referring to arms that were in Ampudia, and who were under the supervision of the caretaker of the "*casa y castillo* [house and castle]", Juan de Acebedo[39]. The registry demonstrates a large continuity in the content of the armoury, with some occasional losses. The twelve gilded halberds and the twelve white ones had been taken to the ducal property in Ventosilla[40], the decorative masks for the harnesses now only numbered thirty four, some spurs were broken, the bullet-proof harness engraved with a drawing is now said to be "*dado de color açul al fuego* [turned blue by fire]", doubtlessly blued, the arquebus of silver and gold specifies that it "*vino de las Indias* [came from the Indies]" and the plumages from the Indies in the basket were "*en la escalera de la armería* [on the armoury's stairs]"[41].

Upon the death of the caretaker of Ampudia, Juan de Acebedo, by which we discovered that he was a knight of Santiago and brother of the bishop of Burgos, Fernando de Acebedo, one of the favourite's closest collaborators, another inventory of the armoury was made in 1614[42]. This inventory mainly includes the same pieces as the previous inventories and almost all of them described in the same order, including the sculpted horses and the thirty-four masks. However, we find clear differences that we must bring to light, given their interesting nature. First, it is said that the infantry soldiers' morions not only numbered 100, but that there were thirty-six more than had not been counted in "*el inventario antiguo* [the old inventory]"[43]. Similarly, now it says they carried "*las armas reales* [royal coat of arms]," which is surprising and indicative of the privileged situation of the favourite. In place of the 170 white shaffrons for horses, now we only find 149. With regard to the 100 pistols, it says that there are ninety nine, "*de las quales faltaron dos llaves quando por mandado de su Ex^a^ se dieron a don Bernardino de Velasco para la xornada de Valencia* [of which two locks were missing when, by order of His Excellence, they were given to Don Bernardino

39 ADM, Lerma, legajo 52, doc. 15, fols. 383 r-386 v.

40 Williams, 2010: 84–85.

41 Concerning weapons and other American objects in collections, see Yaya, 2008: 173–188. Urquízar, 2011: 105–222.

42 ADM, Lerma, legajo 55, doc. 16. On Fernando de Acebedo, see Williams, 2010: 111–113.

43 ADM, Lerma, legajo 55, doc. 16, fol. 2 r.

de Velasco for the journey to Valencia]". The small trumpet flags and some other loose pieces also went missing.

The most important changes that we find are relative to the harnesses' description. In its individualised section, we have the "*arnés açul fuerte con su rodela y murrión* [strong blue harness with its rondache and morion]", and we also have the entry referring to the fifty infantry harnesses, but now the others are recorded as:

> "*Yten çiento y veinte y seis arneses blancos con el Rey de Francia y los dos de los caballos todos ellos con celadas gola peto espaldar gebras (sic) y bracetes (sic) y manoplas y faldones […]*
> *Yten cinco arneses grabados y dorados y los dos acancelados (sic) con todas sus piezas exceto el uno que le faltan las grebas*
> [126 white harnesses including the King of France's and the two for the horses, all of them with sallets, gorget, breastplate, backplate, greaves, vambraces, gauntlets and skirts (or tassets) […]
> Five engraved and gilded harnesses and two of them channelled with all their pieces, except for the one that lacked the greaves]"[44].

Because the eight infantry soldiers' corslets with German gorgets and morions no longer appear, they would be included among the harnesses, which would give us an increase of six specimens. In any event, we now have 180 harnesses, a clearly spectacular number.

With regards to the exotic objects and those from the Indies, we find in place of plumages twenty-three "*tocados de Yndias de plumas amarillos y colorados* [Indian head dresses with yellow and red feathers"[45], which were on the armoury's staircase. The two rondaches from the Indies continue to exist, but now we find four "*paletas a manera de peines de Yndias* [pallets in the manner of combs from the Indies]", "*un casco de conchas de la mar* [a helmet of shells from the sea", a breastplate of "sea snails", "*una tabla a manera de rabel pintada de diferentes colores* [a wooden panel in the rebec manner painted in different colours]" and "*dos llaves doradas que se dieron a su Exª la primera vez que entró en esta villa* [two golden keys that were given to His Excellency the first time he entered this town". No less important is the inclusion in the armoury of "*un medio tiro pedrero* [stone-throwing cannon]".

---

44 ADM, Lerma, legajo 55, doc. 16, fol. 2 v.
45 ADM, Lerma, legajo 55, doc. 16, fol. 4 v.

The inventory, speaking of the mats that covered the floors, leaves us with a brief description of the different rooms, which is very useful to understand the palace's distribution and the armoury's location:

> *Yten se le entregaron y se hallaron en la dha fortaleça las esteras de zinco salas que están puestas en ellas y más otras dos viexas que están abajo yten más la pieça de la armería que está esterada*
> [The mats for five rooms of this fortress were put under his custody as well. They were put in the rooms and two older ones that were downstairs, in addition to the piece of the armoury that is matted][46].

In his 1617 will, the Duke of Lerma linked to his "*mayorazgo* [entailed estate]" not only the town of Ampudia, but also "*las armas y armería que tengo en la dicha villa* [the arms and the armoury that I have in that town]", which gives us the idea of its symbolic importance[47]. But later, in 1622, now out of power and as a cardinal of the Church, the Duke of Lerma made a public declaration of his assets on a printed text by order of the king. There, the Duke declared that in the castle in Ampudia "*tenemos algunos coseletes, mosquetes, arcabuzes, picas y otras armas, que la suerte principal costó en Milán 1.554U117 maravedís, y por los derechos de los puertos pagamos 203U698 y por el porte dellas 198U951 que todo monta 1.956U766 y algunas pieças dellas avremos dado a diferentes personas, y puesto otras en su lugar* [we have some corslets, muskets, arquebuses, pikes, and other arms, that the principal amount cost 1,554,117 *maravedís*[48] in Milan, and for harbour rights we paid 203,698 and for their port 198,951 which brings the total to 1,956,766 and some of those pieces we have given to different people, and put others in their place]"[49]. In this extract he points out the method employed to bring together such a large quantity of arms in such a short time: making use of his power as favourite. Even so, in another place the same document, he indicates that his houses in Madrid[50] contained other arms, given that the richest and most special ones overall would move around according to need. We are not going to go into great detail, but the document states that in the Madrid palace there were forty-eight halberds, some firearms, eight swords, a few crossbows and up to nine harnesses[51]. We also find cited the artillery that Lerma had in

46 ADM, Lerma, legajo 55, doc. 16., fol. 5 r.
47 Archivo Histórico de la Nobleza (AHNOB), Osuna, C. 1955, D. 4 (1), fol. 29 v.
48 Spanish currency of the Middle and Early Modern Ages.
49 *Descripcion*, 1622.
50 Lopezosa, 1998: 457–486.
51 ADM, Lerma, legajo 52, doc. 14, fols. 271 r-273 v. ADM, Lerma, legajo 52, doc. 15, fols. 386 v-388 v and 389 r.

the castle in Denia, where the fortress had been renovated to include some royal quarters and there were large quantities of arms, a set that consisted of defensive weapons for the stronghold rather than a symbolic armoury[52]. He also states in the 1622 text that in Lerma there were artillery and other weapons. Indeed, there are more artillery pieces listed there since 1607[53], but at least since 1616 there was a curious set of arms included in the ducal cabinet room, which was a set of exotic and curious objects, among which we found arms of special value. There, we locate firearms; hunting knives; rich, Turkish, and antique swords; as well as Turkish and Berber cutlasses of great value; at least two entries of Japanese arms and American pieces, among which four native American hatchets stand out, which have stone in place of iron[54]. By 1628, there were not many remnants from this set, perhaps because it was taken to Ampudia or Madrid[55].

The Ampudia armoury survived the death of the 1st Duke of Lerma and, as with other noble armouries, served as a quarry for the wartime needs of the Spanish Monarchy. In 1637, arms were taken from there for breaking the French siege of Fuenterrabía and in 1641 to deliver to Mérida during the war in Portugal. These consisted mainly of weapons to arm troops, "*reservando las grabadas y que fuesen de las personas de los señores de esta casa* [reserving the engraved ones and those belonging to the persons of the Lords of this house]". The set still had to have been still rather large because 130 engraved morions from Milan, forty-six breastplates with their backplates, gorget and vambraces, twenty cavalry sallets; six rondaches and two strong breastplates were contributed in 1641[56]. In the eighteenth century, the armoury also supplied the royal army with cannons for the war with Portugal (1762).[57] As late as 1798, the armoury still existed, even though Ampudia belonged to the Duke of Berwick, because an envoy with the mission of providing information about the town briefly described the castle and assured the Duke of Berwick that there were still weapons in the armoury, although it had lost much of its contents[58].

---

52 Seser, 2013: 82–83. Campos, 2019: 27–39.

53 ADM, Lerma, legajo 52, doc. 14, fol.175 v and 270 v. ADM, Lerma, legajo 52, doc. 15, fol. 388 v.

54 Archivo Histórico Provincial de Burgos (hereafter AHPB), P. 1195, fols. 435 r and ss. AHPB, P. 1196, f. 665 r. Cervera Vera, 1967: 46–47.

55 ADM, Lerma, legajo 54, doc. 3. AHPB, P. 1368, fols. 538 r-548 r.

56 AHNOB, Osuna, C. 1965, doc. 25, unfoliated.

57 Archivo Ducal de Alba (hereafter ADA), C. 190, D. 16.

58 ADA, C. 190, doc. 16, unfoliated. See: Cobos/Castro, 1998: 201. Izquierdo, 2011: 24.

BIBLIOGRAPHY

Alvar Ezquerra, Alfredo (2010): *El Duque de Lerma. Corrupción y desmoralización en la España del siglo XVII*. Madrid, La Esfera de los Libros.

Campos Perales, Ángel (2019): "Bienes y espacios de privanza en el castillo y palacio del duque de Lerma en Denia". In: *Ars Longa*, 28, pp. 23–141.

Cervera Miralles, Luis (2003): "El Palacio del Duque de Lerma". In: González Pablos, Emilio (ed.): *El Palacio Ducal de Lerma: de ayer a hoy*. Lerma, Asociación de amigos del Palacio Ducal: pp. 35–50.

Cervera Vera, Luis (1967): *Bienes muebles en el palacio ducal de Lerma*. Madrid, Castalia.

Cervera Vera, Luis (1996): *El conjunto palacial de la villa de Lerma*. Lerma, Asociación de amigos del palacio ducal. (Facsimile of the Valencia edition, 1967).

Cobos, Fernando/Castro, José J. (1998): *Castilla y León. Castillos y fortalezas*. León, Edilesa.

Cooper, Edward (1991): *Castillos señoriales en la Corona de Castilla*. Valladolid, Consejería de Cultura y Turismo, Junta de Castilla y León.

Descripción… (1622): *Descripcion, e inuentario de las rentas, bienes, y hazienda del Cardenal Duque de Lerma, en Conformidad de lo que su Magestad, que Dios guarde, manda por su decreto real*. Madrid, s.e.

Dueñas Beraiz, Germán (2000): "Julián del rey. Nuevos datos sobre su figura". In: *Gladius*, 20, pp. 269–284.

Fallows, Nigel (2010): *Jousting in medieval and Renaissance Iberia*. Woodbridge, Boydell Press.

Feros, Antonio (2002): *El duque de Lerma. Realeza y privanza en la España de Felipe III*. Madrid, Marcial Pons.

Frieder, Branden (2008): *Chivalry and the Perfect Prince. Tournaments, Art, and Armor at the Spanish Habsburg Court*. Kirksville, Truman State University Press.

Izquierdo Misiego, José I. (2011): "El duque de Lerma y la villa de Ampudia". In: *La Corredera. Revista cultural de Ampudia*, 1, pp. 23–48.

Lopezosa Aparicio, Concepción (1998): "La residencia del duque de Lerma en el Prado de San Jerónimo, traza de Gómez de Mora". In: *Madrid. Revista de Arte, Geografía e Historia*, 1, pp. 457–468.

Martínez Tercero, Enrique (1995): *La primera arquitectura renacentista fuera de Italia. Lorenzo Vázquez en Guadalajara*. Guadalajara, Colegio Oficial de Arquitectos.

Nieto, Víctor/Morales, Alfredo J./Checa, Fernando (1993): *Arquitectura del Renacimiento en España: 1488–1599*. Madrid, Cátedra.

Pérez, Juan L./Pérez, Javier (2008): *El Palacio de Cogolludo*. Guadalajara, Gea Patrimonio.

Pinheiro da Veiga, Tomé (1916): *Fastiginia o fastos geniales*. Valladolid, Imprenta del Colegio de Santiago.

Ponz, Antonio (1776): *Viage de España, Vol. V.* Madrid, Joaquín Ibarra.

Romero Medina, Raúl (2017): "Don Juan de la Cerda (c.1515–1575), IV duque de Medinaceli. El hombre, el político y el mecenas en la Corte del Rey Prudente". In: *Tiempos Modernos: Revista electrónica de Historia Moderna*, 34, pp. 350–371.

Romero Medina, Raúl (2020): "Como la espada, así la vaina. La colección de armas blancas de Don Juan de la Cerda y Silva, IV duque de Medinaceli (1552–1575)". In: *Potestas*, 16, pp. 81–101.

Romero Medina, Raúl (2023): "La armería de los duques de Medinaceli. Una propuesta de estudio desde la historia del coleccionismo". In: Ramírez, Victoria/Rubio, Abraham (eds.): *El Coleccionismo de artes decorativas en España en el cambio de siglo (finales del S. XIX-Principios del S. XX)*. Madrid, Museo de Artes Decorativas: pp. 273–306.

Seser Pérez, Rosa (2013): *La ciutat de Dénia en temps del Duc de Lerma*. Denia, Ayuntamiento de Denia.

Terjanian, Pierre (2010): "El espectáculo del rey guerrero. Las armaduras reales y pintadas de Felipe III". In: De Jonge, Krista/García, Bernardo J./Esteban, Alicia (eds.): *El legado de Borgoña. Fiesta y ceremonia cortesana en la Europa de los Austrias*. Madrid, Fundación Carlos de Amberes-Marcial Pons: pp. 625–636.

Uhagón, Francisco R. de (1916): *Series de los más importantes documentos del archivo y biblioteca del excelentísimo señor Duque de Medinaceli elegidos por su encargo y publicados á sus expensas por A. Paz y Mélia*. Madrid, s. e.

Urquízar Herrera, Antonio (2011): "Imaginando América: objetos indígenas en las casas nobles del renacimiento Andaluz". In: *Historia y Genealogía*, 1, pp. 205–222.

Viñaza, conde de la (1889): *Adiciones al Diccionario histórico de los más ilustres profesores de las Bellas Artes en España de D. Juan Agustín Ceán Bermúdez*. Madrid, Tipografía de los Huérfanos.

Williams, Patrick (2010): *The great favourite. The Duke of Lerma and the court and government of Philip III of Spain, 1598–1621*. Manchester, Manchester University Press.

Yaya, Isabel (2008): "Wonders of America: The curiosity cabinet as a site of representation and knowledge". In: *Journal of the History of Collections*, 20, 2, pp. 173–188.

Manuel Gámez Casado

Universidad de Sevilla

NOBLE ARMS AND ARMOURIES IN HISPANIC AMERICA: IMAGE, COLLECTION AND SOCIETY

In the examination of the close relationship achieved between noble and military classes in the Modern Age Americas, the use of arms and the structure of armouries stands out as one of the chapters of great interest in historical studies. The need to demonstrate a certain status in the context of the social exigencies in the West Indies resulted in an incipient accumulation of arms on the part of families with the greatest influence at viceregal courts, constituting a phenomenon of interest for the fields of social history and art history[1]. Thus, it must be analysed if the collection of arms became an activity in itself for the elites, who saw a means to materialise their power in front of their colleagues with these objects. For this reason, it is crucial to state that, although some of these weapons had a functional use on an imaginary battlefield, others became an element in the image of high society, in a phenomenon that, although it was consolidated in Europe from the end of the Middle Ages, experienced a notable transfer to the American viceroyalties beginning in the sixteenth century.

This subject should be mutually exclusive from any local approach because it connects with other mechanisms linked to the process of globalisation initiated in the sixteenth century and consolidated in the subsequent centuries, creating a series of cultural relationships that shook the social roots of the Americas and reinforced the interconnection between both shores of the Atlantic. The resources that the new comparative history, defender of transnational currents from a prevailing methodology, offers can be applied to the analysis of the history of weaponry in the Americas, understanding that they were necessary tools in the shaping of a new viceregal society that emphasised gradual *criollo* power[2]. Simultaneously, it will be interesting to examine if these weapons had a true practical function or if, on the contrary, they were only used as objects

1 On this matter, see Dacosta, 2021: 31–49.
2 Gruzinski, 2004: 30–75.

for collecting within armouries. This would make evident a representative use transferred to distinct iconographic types, which incorporated certain pieces within the arts as an attribute for a new image of power.

The methodologies that the historiography has proposed in recent decades allow for the broadening of the focus towards arms and armouries in the modern era, overcoming the mere documentary contribution of inventories or lists to bring about a more complex approach. Thus, this approach to this topic has its foundation in a concern about knowing the functioning of the relationships between social groups that shared space and time, whose division was accentuated by the set of community migrations and the transfer of objects, amongst which arms are found. In fact, the inclusion of weapons as a form of cultural transfer has remained excluded from previous studies, which are more concerned with analysing the reception and adaptation of traditional artistic objects[3]. As such, this paper seeks to bridge that gap, including the phenomenon of arms transfer as a tool necessary in the articulation of a global history, involving different agents concerned with collecting them, conserving them and using them.

The analysis of the leading agents of the phenomenon of globalisation is crucial to lay the foundations of the issue here addressed, because arms conveyance was framed within the networks of technological transfer typical of cultural transfer between Europe and the Americas. This cultural transfer was fostered by the navigation routes of ships and galleons that participated in an endless series of voyages back and forth in which they moved objects that enriched the cultural panorama of both continents. Arms had to have been added to the well-known objects like paintings, sculptures and luxury objects in these shipments, fostering an artistic miscegenation resulting from the new realities and customs that emerged in the Americas. In sum, the combination of themes presented constitutes the foundation of the present text, in which, in addition to showing some examples of arms within the iconographic types of the West Indies, their examination will be linked with the concept of the global circulation of objects and military knowledge, establishing a new definition of the image of the Spanish empire on the other side of the Atlantic[4].

3 Werner/Zimmermann, 2004: 15–52.
4 Headrick, 2010: 21–67.

THE SAME SPACE, AN UNEQUAL INTERPRETATION AND TWO DIFFERENT IMAGES

The research in different national archives in which documents relative to the history of the Americas are preserved has allowed for the locating of numerous new pieces of information about inventories, in which lists of weapons belonging to viceroys and governors are collected. These new documents, which will be analysed with greater accuracy than has been done in the past, demonstrated the importance of those objects in the set of American collections, a circumstance that explains their inclusion in a certain iconographic type alongside pictorial portraits, which contributes to their consolidation of imperial Spanish power on the other side of the Atlantic. Thus, the arms were used within the artistic creation as a tool to forge a specific image of the Spaniard, which fulfilled the ideal of a soldier who had crossed the ocean to implement the emerging policies of the Hispanic empire. In reality, the objective was none other than to separate himself from the local or indigenous social prototype, using arms as an easily recognisable conventional resource. In this way, the visual arts were forging a specific image of the *conquistador*, making these original characters into authentic heroes through the use of certain attributes, highlighting the use of arms, armour and other elements of a military character.

This function served, in the same way, to develop a certain profile of the Spanish soldier and nobleman had moved to America to make his fortune, so he had to display himself in the most sophisticated manner as possible, bringing together his best weapons to demonstrate his origin and his interest in the most advanced weaponry of the time. In this way, likewise, he also distinguished himself from the native American, who was represented carrying antiquated weapons made of local materials, showing a lack of power in the face of the new Spanish militancy. Evidence of this phenomenon are the models used by the new Spanish painters Juan and Miguel González for their twenty-four *tablas enconchadas* about the conquest of Mexico made in 1698 and based on different chronicles written by mid-sixteenth-century travellers and expeditionaries (Fig. 1)[5]. In these texts, the arms that were used during the taking of New Spain were described clearly, highlighting the use of crossbows, long pikes and rondaches. These weapons were indispensable for occupying the region quickly, and became, over time, part of the first viceregal armouries, putting into

5 Of the chronicles that described the taking of Mexico by the army commanded by Hernán Cortés, the one written by López de Gómara in 1552 stands out.

practice the courtly costumes defined by the literature of the sixteenth century. Amongst other authors, Baltasar de Castiglione, writer and diplomat linked to the court of Emperor Charles, already defined the good courtier based on several requirements, among which were speaking prudently, cautiously, and in a cultured manner as well as using weapons properly[6]. This resulted in an entire iconographic body of work that would fundamentally develop in painting, forging an image that allows for the delimiting of specific historical categories such as space, interpretation and image.

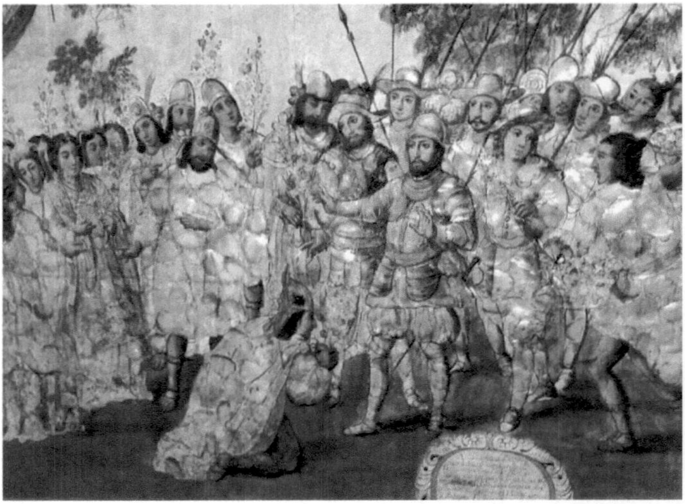

Fig. 1. *La Conquista de México* (detail). 1698. Juan and Miguel González. Madrid, Museo de América (Wikimedia Commons).

The use of certain weapons in Hispanic America is related to the concept of cultural interpenetration, through which an attempt is made to explain the mutual interests that arose between two different societies due to their respective cultures, traditions and lifestyles[7]. This idea is based on the concern demonstrated by Spanish soldiers to obtain weapons belonging to native Americans, such as arrows, feather rondaches or straw armour, which would be reported on their initial return trips to the Iberian Peninsula as authentic

6 See Torres, 2006: 93–94.

7 Kamen, 2004: 12–36.

treasures. In this case, it must be remembered that these objects, initially devoid of material value, formed part of the first chattels transported from the Americas to Spain, alongside hammocks, birds' feathers, and flowers[8]. Among all these objects, perhaps Montezuma's head-dress is the most well known for its symbolic importance within the context of the conquest of Mexico, it being considered a referential object for the demonstration of the new power established on the other side of the Atlantic. This provides proof of the participation of indigenous weapons and armour within the process of interculturality created between Europe and the Americas during the modern age, these objects being included in the main artistic collections linked to Spanish noble houses.

A defining moment for the transport of arms and armour from Europe to the Americas with the intention of participating in the new image of power would be the year 1681. It was this year that the *Recopilación de las leyes de los reinos de Indias sancionada por Carlos II* was published. Herein was included a series of rules about protocol recompiled by public officials Antonio de León and Juan de Solórzano. Among other topics, they referred to the celebrations carried out by the viceroys, such as triumphal entries or funerals, sanctioning them to carry their own weapons or jewels. In this manner, the new viceroys would be permitted to bring their own armoury to the Americas, although it was exclusively for the protection and defence of their families and never for profit from its sale. Prior to this time, i.e., in 1614, King Philip III promulgated a law that would be sanctioned by his successor, which specified the list of weapons that the viceroys destined for New Spain had to carry, containing twelve halberds, twelve partisans, twelve swords, twelve daggers, twelve arquebuses and twelve rodaches. The transport of gold and silver weapons was prohibited to avoid commerce in contraband, which might corrupt the viceroy during his term of office. It is possible that this is one of the reasons that justifies the lack of great armouries in Spanish America, as the export of pieces of interest from one side of the Atlantic to the other had been limited[9].

It becomes evident that weapons formed part of celebrations and viceregal ceremonies and thus played a decisive role in the creation of the imperial image projected to American society. The use of heraldic arms was decisive to such an extent that the Casa de la Contratación [House of Trade] was tasked with introducing them within the baggage that would accompany the viceroy on his shipboard transport to America, including both that would act for his defence

8 Urquízar, 2011: 205–221.

9 Chiva, 2012: 81.

as well as those that would be part of his personal collection. This allowed the viceroy to enter the city within a heavily armed procession, ensuring his status and having an impact on the power established by the crown on the other side of the Atlantic[10]. The inclusion of those elements in some paintings is good evidence of this heraldic use, for example in the folding screen of the entrance of the viceroy dated to 1780 and preserved in Madrid's Museo de América. In this piece, it is seen how the viceroy makes his entrance to the city accompanied by a large horse-mounted court, carrying belongings, flags and heraldic arms as a demonstration of his power. These resources were brought together as a useful means to transmit a certain discourse that assured imperial power in America, supplanting not only its political and military leaders, but also the objects that had to have participated in that message in the form of weapons, triumphal decorations and ceremonies. A similar case can be seen in the painting that depicts the triumphal arch erected in Puebla's cathedral for the entrance of the viceroy, the Marquis de las Amarillas, in 1756, where, in addition to the ample display itself of ephemeral architecture, the procession richly equipped with weapons and regalia alluding to the new viceregal power can be seen in the lower section (Fig. 2). This example constitutes a new case wherein the use of these objects in specific ceremonies linked to American political power is shown.

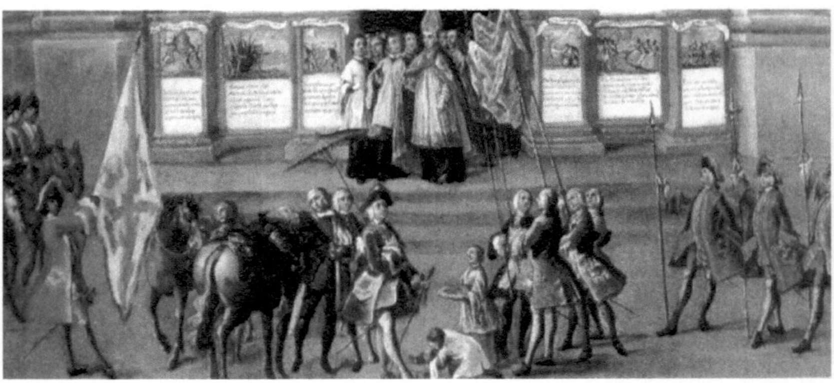

Fig. 2. *Entrada del virrey marqués de las Amarillas* (detail). 1756. José Joaquín Magón. México, Private collection (Wikimedia Commons).

10 Chiva, 2012: 81–88.

In short, this first debate opens new analytical possibilities around the use of weapons, understanding that they were not only considered objects in a particular collection, but they also had a specific use in the projection of a certain discourse of power that delved into the development of the visuals arts. In reality, the Spanish viceregal hierarchy used these resources intelligently since the sixteenth century, falling back on a long weaponry tradition that dates back to ancient times and made them into a fundamental agent for intercultural relations with American society. It was all related to the need to reorganise the vast American territory through political strategies that would counter the problems created by the defence and the control of the new continent's territory.

FROM ONE SHORE TO ANOTHER: TRANSFERS, TASTES AND OBJECTS

The inclusion of the study of weaponry in the Americas within a global discourse allows us to question if these objects truly formed part of the dialogue established between the European and American cultures, as other material goods, specific oral discourses and numerous writings did during the Modern Age. Thus, the concept of transfer of tastes and material goods is important to understand the consolidation of an intercultural dialogue between societies that should understand their customs, traditions and values[11]. Specifically, weapons, as material goods of interest, should be considered as objects inherent to these practices because they formed part of this international circuit of cultural exchanges that were to be received by other societies. This discourse, which connects to ideas proposed by Gruzinski, must be considered within a global perspective because it provided a certain homogeneity in the taste for and use of these weapons both on and off the battlefield[12].

11 On this topic, consult Appadurai, 1988.

12 Gruzinski, 2004: 35–70. The case of the transfer of weapons to the Americas that we know about due to the mandatory authorisations is very interesting, and would deserve a specific study. Some examples: in 1578 Pedro de Alvarado moved to Las Charcas two double harnesses, twelve swords, six coats of mail, roundaches and two harquebuses. In 1558 Hernán Sánchez Morillo took to Peru two harnesses, four coats of mail and cuirasses with their helmets, a dozen harquebuses, twelve partisans and halberds, swords and daggers. In 1580, the Alguacil Mayor of Mexico City was authorised to bring from Spain two corselets, two harnesses, four coats of mail, six harquebuses, six halberds and six swords with their daggers. Obviously, the viceroy carried more things, and in 1580 the Count of Coruña was authorised to take to New Spain six harnesses with their war saddles, twenty-four coats of mail, twenty-four head armours, eight

The trans-Atlantic transfer of arms should be understood as a phenomenon intrinsically linked to the process of globalisation developed during the Modern Era, considering that weapons formed part of the set of material goods necessary to exercise imperial Spanish power in America. As such, it is necessary to establish a clear difference between utilitarian arms, that is, those used in a military encounter, and others, which might be termed heraldic arms, likely those of more interest for art history because they have a clear aesthetic or representative aspect. Precisely these later weapons, when they were transported to America, had already had a similar use among *peninsular* elites, creating a taste for their collecting that was partly transferred to the American continent with the idea of reinforcing the *peninsular* presence in the growing American society. For this, it should be remembered that the Spanish treasure fleet had a clear human dimension, as we have numerous data that allows us to understand the influence exercised by the material goods transported from the Americas to Europe and vice versa, consolidating a phenomenon of acculturation in which weaponry also participated. Therefore, it is worth questioning what influence the products exchanged in the Spanish treasure fleet exercised over lifestyle and societal tastes, weapons being included within this phenomenon. These products infiltrated international markets, creating new tastes through their reception[13]. Along with fabrics, foodstuffs, books and other products, heraldic arms were considered entertainment goods for a certain segment of society interested in their collection as a means of escape from the typical routine of court and diplomacy. This explains why swords and hilts, knives, pikes, helmets, pistols, muskets and arquebuses were exported to the Americas in the galleons that cruised the Atlantic beginning in the sixteenth century. For instance, Rahn Phillips already highlighted that in 1706 a fleet departed for the Spanish Main exclusively to transport shotguns, gunpowder, ramrods, daggers and sword fittings, objects that would be used as heraldic elements in ceremonies and festivals, of which we hardly have any examples as they were melted down at the beginning of the nineteenth century in the context of the beginning of the American independence movements[14].

---

cuirasses, twenty-four harquebuses, three dozen swords with their daggers and two dozen irons for spears. See Archivo General de Indias, Seville (hereafter AGI), Charcas, 415, L. 1, fol. 13 r; Lima, 567, L. 8, fol. 361 v; Mexico, 1091, L. 9, fol. 184 v; Mexico, 1091, L. 9, fol. 190 v-191 r.

13 Phillips, 2003: 188.

14 Phillips, 1991: 80–140.

The introduction of certain arms and armour into the portraits of viceroys demonstrates the construction of a particular prototype, in which the weapons have a notable presence, confirming a model perfectly comparable to the one created by the Spanish monarchs in their court portraits. Thus, a series of resources were brought together on both shores of the Atlantic, framed within a process of the globalisation of the image of power, which were applied locally by the painters tasked with depicting the armour-clad viceroys as symbols of established authority. The Spanish crown used this image to diffuse its presence among local society, establishing an imposed hierarchy that surpassed the increasing *criollo* power, in a phenomenon aligned with the construction of fortifications, viceregal palaces and public buildings decorated with coat of arms and other heraldic elements[15]. As an example, it is worth noting the portrait that the painter Nicolás Rodríguez Juárez made of the viceroy Francisco Fernández de la Cueva, the 10th Duke of Alburquerque, depicting him dressed in his own attire of the Spanish style of the early eighteenth century and equipped with a sword as a symbol of his power in charge of the Viceroyalty of New Spain (Fig. 3). It should also be noted that the shaping of a local taste influenced by the transfer of material goods was based on themes unique to the American world, such as the harquebusier angels, which in the use of weapons have some recurring attributes and were enormously widespread among American society.

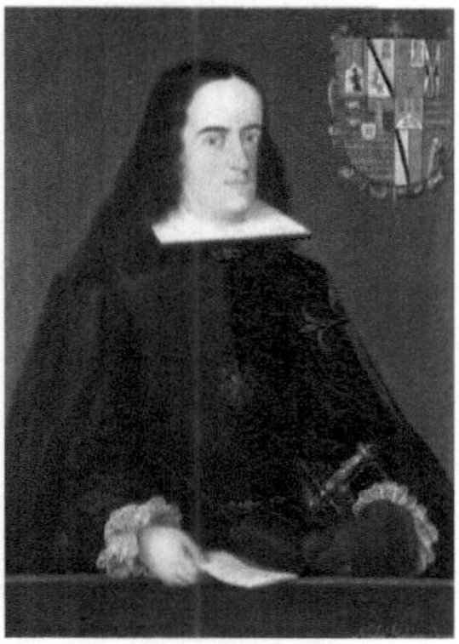

Fig. 3. *Portrait of the viceroy Francisco Fernández de la Cueva, 10th Duke of Alburquerque. ca.* 1708. Anonymous painter. México, Museo Nacional de Historia (Wikimedia Commons).

15 Gámez, 2019: 765– 796.

BETWEEN ARMOURIES: ARCHITECTURES, SPACES AND NOVELTIES ABOUT SOME AMERICAN ENSEMBLES

A very common problem at the present of measuring the importance of weapons in Spanish America resides in our insistence in comparing the documented cases in the overseas territory with those already known in the Iberian Peninsula. In this sense, it is erroneous to accept that the majority of the objects that our museums possess travelled the world with the sole function of responding to the demands of a clientele that sought in them a certain functionality, because this utilitarian character disappears upon their transformance into a museum piece. At the same time, we are inclined to consider that the majority of the arms that are registered in the inventories of goods of the deceased constitute true artwork collections, they being acquired for reasons of taste, when in reality they had been purchased with the more routine purpose of decorating a house, fulfilling a hobby or creating a specific image of power. This phenomenon confirms the quantitative differences between the Spanish and American arms collections in the past, because the exigencies of the societies were quite distinct. Thus, an idea that it is worthwhile to highlight is that in America no large collections have been able to be found comparable to those on the peninsula, a circumstance that is explained by understanding that the interests of the New World society had a different basis.

A matter that should be raised is the existence of rooms dedicated to the exhibition of arms within the viceregal palace, starting from an obvious parallelism with the dispositions of court, where the Alcázar and later the Palacio Real de Madrid contained armouries in which munitions were preserved and pieces were exhibited. These rooms stood out for their evident structural simplicity, their diaphanous sense of space and their central location within the complex, constituting a particular model that was adapted to the American viceregal palaces, with evidence of the existence of armouries in Mexico City, Lima and Bogotá. This last example was specifically recognised by the engineer Juan Jiménez Donoso in the plan that he drew up in 1781 of the Santa Fe de Bogotá palace, whose structure was finalised at the end of the eighteenth century, but unfortunately disappeared after the fire at the beginning of the past century (Fig. 4)[16]. In the drawing, it is seen that the armoury is a rectangular room located near the stables and on one side of the main courtyard of the barracks, forming a

16 Gámez, 2018: 125–138. About this document, one can also consult Romero/Panduro, 2020: 259–276.

sector within the palace intended for the use of the militia and the preservation of the viceroy's own arsenal.

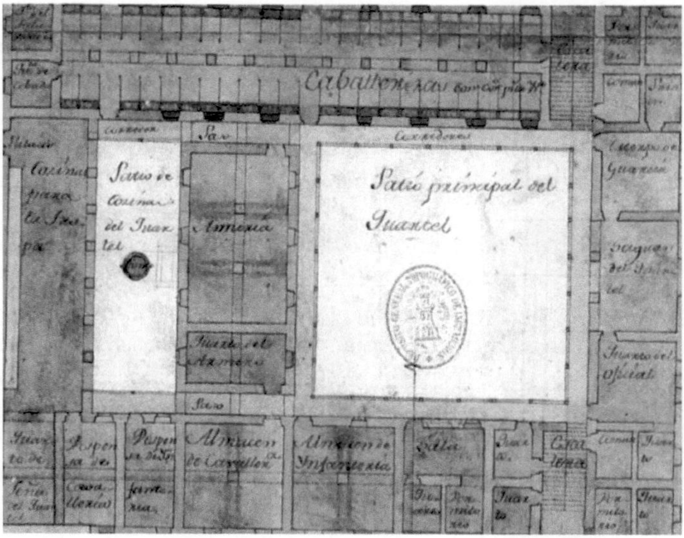

Fig. 4. *Plan of the viceregal palace of Santa Fe de Bogotá* (detail). 1781. Juan Jiménez Donoso. Madrid, Archivo General Militar (Wikimedia Commons).

On the other hand, different documents have been located relative to the renovation of the armoury of the viceregal palace in Mexico City. As early as the sixteenth century, we have documentary proof of the shipment of different consignments of arms, namely, the existence of a royal decree dated to 1597 and sent to the Casa de la Contratación so that Antonio de Armijo can carry 500 arquebuses to New Spain that the Count of Monterrey commissioned for Mexico's royal armoury[17]. This would be the earliest evidence of the specific transit of arms destined to the viceroy's private collection, which would be located in the palace armoury, a reflection of the empire's growing power on the other side of the Atlantic. Similarly, in 1578 there is evidence of a decree having been dispatched granting a licence to Pedro de Alvarado so that he could carry twelve swords with their daggers, six *Zweihänder* [two-handed swords], six

17 AGI, Gobierno, Indiferente General, Registros de partes para la Casa de la Contratación, Registro: Sevilla de partes, "Real Cédula", 1597, Sign. Indiferente, 1952, L.4, F. 167 V.

coats of mail, two double harnesses, two arquebuses and four rondaches[18]. These goods notability enriched the viceregal arms collection, intended both for the viceroy's private protection and for their particular exhibition as an element for the representation of power.

The royal collection of arms belonging to the viceroy of New Spain was of the greatest importance in all of Spanish America, both for its large number as well as the importance and the sophistication of the weaponry within it. Therefore, we know that, in 1636, when Lope Díez de Armendáriz, Marquis of Cadereyta, was viceroy, the private collection came to consist of almost 1,000 arquebuses and 500 muskets[19]. Also, it is worth highlighting the letter sent by the viceroy Antonio Sebastián de Toledo, Marquis of Mancera, about this expansion in 1670. In this letter, he recognises the ruin that the palace's armoury presented, requiring an important renovation and updating of the collection. For this, a year prior, it had been ordered that the senior master of the royal household, Francisco de Córdoba, make a report about the improvements that the palace would require, and a visit was organised to the Mexican armoury alongside the architect Luis Gómez de Trasmonte, who acted as senior master of the Cathedral of Mexico, the building engineer Rodrigo de Aguilera and the masters Miguel de Aguilera and Juan de Morales[20]. The architects recognised in their reports the need to repair the armoury given the poor state that its structure presented. Perhaps the most noteworthy recommendation was the adaptation of the flooring and the furnishings to the collection's demonstrative requirements that, although today is spread amongst different institutions, shows how a collection of arms displayed in this place truly existed. Thus, the architects designed new wooden beams to support the covering over the numerous imperfections that it presented and also created furniture intended for the weapon collection's placement. Nonetheless, it is striking that the plans preserved of the palace do not specify any room intended for an armoury, especially considering the importance of this building.

---

18 AGI, Gobierno, Audiencia de Charcas, Registros de oficio y partes, Registro de Oficio y partes para la Audiencia de Charcas, "Constancia de haberse despachado una cédula concediendo licencia a Pedro Alvarado", 1578, Sign. Charcas, 415, L.1, F. 13 R.

19 AGI, Gobierno, Audiencia de México, Cartas y expedientes del virrey de Nueva España, vistos en el Consejo, "Carta del virrey Lope Díez de Armendáriz, marqués de Cadereyta", 1636, Sign. México, 31, N.40.

20 AGI, Gobierno, Audiencia de México, Cartas y expedientes del virrey de Nueva España, vistos en el Consejo, "Carta del virrey Antonio Sebastián de Toledo, marqués de Mancera", 1670, Sign. México, 44, N.62.

In the case of Mexico, in addition to the army of its viceregal palace, it is also worthwhile to discuss that other edifices linked to the government had their own rooms intended for the accumulation of arms. One example was the Real Casa de la Moneda [Royal Treasury], an institution central for the viceregal system of governance and foundational for the consolidation of the new image of imperial power. Precisely its function's importance made it necessary to expand it on various occasions, highlighting here the expansion patronised by the Marquis of Croix in 1771. This work was assigned to the military engineer Nicolás de Lafora, who carried out a remodelling work of the old building in a detailed plan[21]. Lafora proposed that the armoury be situated on a corridor of the main courtyard, forming a right angle with the stables[22]. The model departs from the one used in the viceregal palaces, where the stables and armouries occupied nearby spaces with the intention of improving the internal organisation of the building. Lafora, who has estimated a total cost at 15,000 pesos, never saw his work finalised because the project was halted after the opinion of Pedro Núñez de Villavicencio, superintendent of the aforementioned Treasury, who suggested the execution of previous project credited to the engineer Ricardo Aylmer, which did not come to fruition. Ultimately, with the arrival of Viceroy Antonio María de Bucareli, this proposal was taken up, tasking to the engineer Miguel Constanzó the completion of the building, including its armoury. These novelties reveal the importance of the armouries within New Spanish governmental buildings, showing how these spaces also had their own identity as an image of power, going beyond the evident functionality of stockpiling the arsenal intended for the private defence of these royal institutions.

The example of the viceroy of Peru is similar to the previous case. The descriptions that we have about the viceregal palace in Lima allow us to claim that a collection of weapons exist in its interior, consisting of the accumulation of different pieces from as early as the sixteenth century. This was stated by, among others, Vázquez de Espinosa, who recognised that the palace was a massive construction that consisted of various rooms, such as those for hearings, tribunals and a weapon storehouse, currently missing[23]. In fact, it must not be forgotten that in the viceregal palaces, the position of *capitán de la sala de armas de palacio* [captain of the palace weapons room] was established. This office was

21 For more on this engineer, see Capel Sáez et al., 1983: 247.

22 AGI, Mapas, planos, documentos iconográficos y documentos especiales, Mapas, planos, dibujos, etc., México, Sign. MP-México, 580.

23 Vázquez de Espinosa, 1948: 130–145.

responsible for the management of the ammunition depot that was installed in the same viceregal enclosure and that could provide weapons to the city's residents during any emergency. A well-known case is that of Martín de Acedo, protege of the Peruvian viceroy, who was named *capitán de la sala de armas* of the Lima palace in 1619. Perhaps the most examined example is that of Tomás de Torrejón, musician of the highest level and composer of *La púrpura de la rosa*, considered the first opera composed in the Americas. This personage ascended from a young man in Peru's viceregal court, accompanying the Count of Lemos to the viceroyalty as a chamber musician until he was named to the office of *capitán de la sala de armas de palacio*, ultimately reaching the royal treasury and the *corregimiento* of Chachapoyas, one of the richest in the viceroyalty. These examples demonstrate the importance of armouries and collections of weapons within the viceroyalty itself, by influencing the court hierarchy and in the realm of social power in the West Indies[24].

The collections of weapons located in the office of the *alcalde* [municipal magistrate] of Natá, Panama, constitutes another important example within this context, because there were several families interested in the structure of armouries. Well known are the inventories of goods in which the pieces that were maintained within private collections linked to the powerful families of the aforementioned *alcalde*'s office are collected, comparable with other, more renowned, collections scattered throughout Spanish America, such as those of the Alburquerques in New Spain, the Amadors in Cartagena and the Almontes in Peru. Each of these families required a continuous transfer of weapons, as attested both by the inventories and by some pieces preserved in various museums and private collections. Thus, we know that in Natá in 1608 there were a total of thirty-nine swords, four arquebuses, thirty-three shotguns and two spears. These figures are striking for a small population with relative political and secondary importance in the viceregal political system. The rationale for this collection rests fundamentally in the defensive consciousness that these families had because they resided in a strategically important area for the geopolitical organisation of the Peruvian viceroyalty. Most of these weapons existed to be inherited by future generations, constituting a prestigious element for family genealogy.

24 Torres, 2006: 18–25.

CONCLUSION

The social structures consolidated in the Americas during the modern era had a fundamental resource in the use of weapons for the defensive organisation of the territory. The diverse use of these goods is summarised in two basic ideas. On one hand, they were employed as a resource of protection against any territorial threat that required the use of arms as a dissuasive element to allow the different militia bodies, fortifications and control points to be equipped. On the other hand, it was common to resort to the collection of these objects as a way of representing a certain image of the power achieved by certain New World elites, who had to have an armoury in imitation of the one inherited by viceroys and governors. Specifically, the political elite had armouries situated in the viceregal palaces as unequivocal proof of the importance of these pieces for heraldry. In short, what is presented here offers an initial approach to a topic of interest that requires a deeper analysis from a multidisciplinary perspective, integrating the use of weapons within the historical, social and political development of the Americas during the Early Modern Era.

BIBLIOGRAPHY

Appadurai, Arjun (ed.) (1988): *The social life of things: commodities in cultural perspective*. Cambridge, Cambridge University Press.

Capel Sáez, Horacio, *et al.* (1983): *Los ingenieros militares en España, siglo XVIII. Repertorio biográfico e inventario de su labor científica y espacial*. Barcelona, Universidad de Barcelona.

Chiva Beltrán, Juan (2012): *El triunfo del virrey. Glorias novohispanas: origen, apogeo y ocaso de la entrada virreinal*. Castellón, Editorial Universidad Jaume I.

Dacosta, Arsenio (2021): "Indios principales, nobles hijosdalgo: traducción cultural y jerarquización social en la América colonial hispana". In: Álvarez Cuartero, Izaskun/ Baena Zapatero, Alberto (eds.): *En compañía de salvajes: el sujeto indígena en la construcción del otro*. Madrid, Iberoamericana: pp. 31–49.

Gámez Casado, Manuel (2018): "Ingenieros militares y obras públicas. Algunos ejemplos de Nueva Granada en el siglo XVIII". In: *Ars Longa*, 27, pp. 125–138.

Gámez Casado, Manuel (2019): "Ingenieros militares en la Nueva Granada durante el siglo XVIII. Movilidad, proyectos y expediciones". In: *Revista de Indias*, 79, 277, pp. 765– 796.

Gruzinski, Sergei (2004): *Les quatre parties du monde: histoire d´une mondialisation*. Paris, Martinière.

Headrick, Daniel (2010): *Power over peoples: Technology, Environments, and Western Imperialism, 1400 to the Present*. Princeton, Princeton University Press.

Kamen, Henry (2004): *Empire: How Spain Became a World Power, 1492–1763*. New York, Perennial ed.

López de Gómara, Francisco (1552): *La conquista de México*. Zaragoza, Imprenta Audiencia.

Phillips, Carla Rahn (1991): *Seis galeones para el rey de España: la defensa imperial a principios del siglo XVII*. Madrid, Alianza.

Phillips, Carla Rhan (2003): "Mercado, modas y gustos: los cargamentos de ida y vuelta en el comercio atlántico de España". In: De Carlos Boutet, Guiomar (ed.). *España y América: un océano de negocios*. Madrid, Sociedad Estatal Quinto Centenario: pp. 187–202.

Romero Sánchez, Guadalupe/Panduro Sáez, Iván (2020): "El proyecto de Juan Jiménez Donoso para la construcción del palacio virreinal de Bogotá". In: *Archivo Español de Arte*, 93, 371, pp. 259–276.

Torres Arancivia, Eduardo (2006): *Corte de virreyes. El entorno del poder en el Perú en el siglo XVII*. Lima, Editorial Pontificia Universidad Católica del Perú.

Urquízar, Antonio (2011): "Imaginando América: objetos indígenas en las casas nobles de Renacimiento andaluz". In: *Historia y Genealogía*, 1, pp. 205–221.

Vázquez de Espinosa, Antonio (1948): *Compendio y descripción de las Indias occidentales*. Reed, Smithsonian Institution.

Werner, Michael/ Zimmermann, Bénedicte (2004): "Penser l'histoire croisée: entre empirie et reflexivité". In Werner, Michael/ Zimmermann, Bénedicte (eds.): *De la comparation à l'histoire croisée*. Paris, Seuil: pp. 5–36.

Jonathan Tavares

Art Institute, Chicago

SPANISH NOBLE ARMOURIES IN AMERICAN MUSEUMS AND THE CASE OF DOS AGUAS

At the turn of the twentieth century American collectors and museums sought the splendor of a Spanish golden age and the romantic remains of a Medieval past—a material past that Europe's aristocrats were striving to both maintain as well as sell, and restore as well as fabricate. The legendary avarice of newly minted American wealth is perhaps best summed up visually by a cartoon titled *The Magnet*, which lampoons the millionaire capitalist, J. P. Morgan, who attracts the world's treasures to America [Fig. 1]. Among the depicted treasures is a rare fifteenth-century Nasrid sword. Although Morgan did not personally own such an object, as a trustee he was involved in approving the Metropolitan Museum's purchase of the French Duc de Dino's collection in its entirety in 1904, which included such a Nasrid sword. Just how it got there and whether it is indeed fifteenth-century requires more research[1]. Nevertheless, its inclusion shows the heightened enthusiasm and zeal for Spanish antiquities among Americans during the early-twentieth century. With its romantic allure, arms and armour was placed high among the treasures coveted by America's wealthy captains of industry and the fledgling museums they supported, as it literally embodied the past. And where best to acquire authentic material but directly form the source, the ancestral armouries of the Grandees of Spain. This essay surveys some of the dispersed remains and sources of Spanish aristocratic armouries in American museums, and what can be garnered from the composition of these fragmented armouries regarding the patronage, taste and splendor of Spain's early-modern nobility. In the latter half of this study, focus is placed on the specific case of the ancestral armoury of the Marques de Dos Aguas, the only aristocratic Iberian armoury to be acquired and moved en bloc to America.

1 Metropolitan Museum of Art, 04.3.459ab.

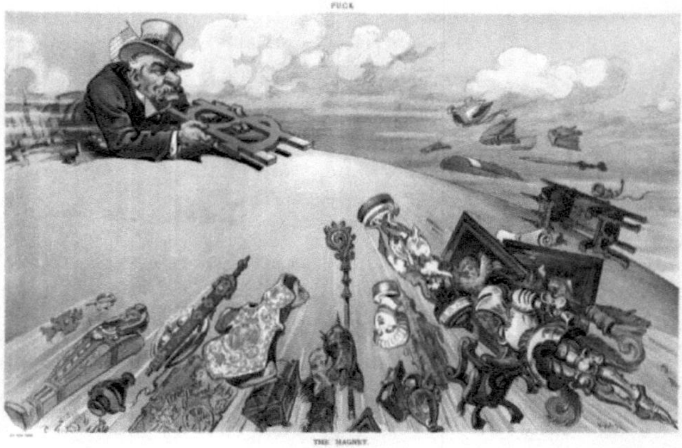

Fig. 1. *The Magnet* (cartoon from Puck magazine, 21 June 1911). 1911. New York, The Morgan Library and Museum.

In the field of collecting arms and armour at the turn of the twentieth century there was no American as voracious as Dr. Bashford Dean. A polymath scientist and historian, Dean's PhD from Columbia University was in the study of prehistoric fishes. Since boyhood, however, Dean had a passion for arms and armour and became a well-published authority on the subject. By 1912, he was made an official full-time curator of the Metropolitan's new department of arms and armour, at that time installing the department's galleries[2]. Aside from curating the Metropolitan's collection, Dean amassed a significant collection of his own at his home, Wave Hill, in the Bronx, New York. Additionally, he advised friends and patrons on their collecting, making many trips to Europe and establishing a substantial network of dealers, collectors and auction houses. It will be demonstrated throughout this survey that Dean was a lead catalyst for Spanish aristocratic pieces of arms and armour making their way to America. Sometime between 1916 and 1920, Dean gave an address to the New York Hobby

2 On the early history of the Metropolitan's arms and armour department and Dr. Dean see: Pyhrr, 2012.

Club on collecting arms and armour[3]. By chance, this speech was preserved through his family papers as an undated typed transcript. It gives unique insight into his methods; one particularly relevant excerpt reads:

> *I have already noted that armor is not easy to find, [...] Accordingly it may be of interest to the gentlemen of the Hobby Club if I should now speak of some of the ways in which armor may be collected. There is first of all ever the possibility of finding armor in its original "habitat" so to speak. Every ancient castle may yet have in or about it specimens which are not out of the collectors' reach.*
> *There are also castles which have been restored within the memory of man, which contain armor and which occasionally yield up their treasure to the amateur. I call to mind a visit I paid to one of these in Spain, [...] We went to the castle from Toledo, a long, hot, dusty ride and not a very interesting one, but when we came to our castle we had before us a picture which is photographed deeply in my mind. The arid plateau on which we had journeyed ended suddenly and below us lay our castle: It was built of richly colored brown sandstone, and was embowered in dark green foliage: all glowed in a purple and golden sun-set [...].*

The Castle was that of Guadamur, as captured in a photograph taken by Dean at that very moment about which he waxed poetically some years later. It was the recently renovated estate of Carlos de Morenes y de Tord, 4th Baron de las Cuatro Torres, and after 1890 Count consort of Asalto. A politician, antiquarian collector and lover of history, de Tord purchased the castle in 1887, which he then restored in a rather romantic fashion. Dean was treated as an honored guest when he visited, with the Count presenting his family to him for a picture that remains today in the archives of the Arms and Armor Department at the Metropolitan Museum. Dean further recounted in his speech to the Hobby Club: "The armory of this castle had many good things in it, and I am sorry I had not the means to purchase it en bloc instead of in fragments. I am glad though that some of the best suits came to this country and are now in Tuxedo in the cabinet of my friend Colonel Ambrose Monell". Four of these armours, which Monell purchased from the Count sometime before 1911, were lent for exhibition to the Metropolitan Museum[4].

In truth, the compositions of these armours are not as important as Dean suggested. While some rare fragments are mixed in, they are few. One example is a left-handed gauntlet for the joust (or manifer)—likely Spanish and from

---

3 This address given by Dr. Bashford Dean was re-discovered and transcribed by his great nephew, Dean K. Boorman, for the Society of American Arms Collectors: Boorman, 1994: 16–28.

4 Dean, 1911: 5, 8–11.

the second quarter of the sixteenth century—that is probably a relic from another ancestral noble armoury[5]. The composite tilting armour the gauntlet was associated with may have been purchased by de Tord from the armouries of the "Duke of Osuna," as suggested by Dean [Fig. 2][6]. The armour as a whole has been re-evaluated in recent years and found to be mostly made up with some reworked sixteenth-century parts[7]. De Tord apparently had a number of restorers working for him on the premises of the castle, including Nicomedes Cosentino (father) and Severiano (son), who together taught another generation of twentieth-century armourers in Guadamur, including Higinio Lorente and Segundo Martín[8]. De Tord's restorers may therefore have originated the trade in decorative armour produced in the Toledo area for tourists, as Lorente continued making armours well into the 1970s.

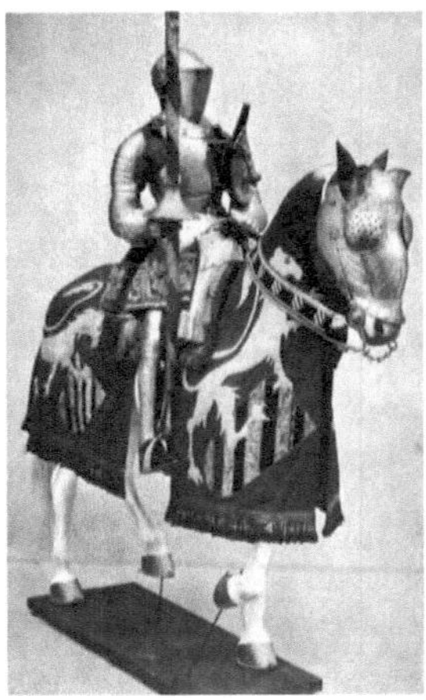

Fig. 2. *Composite Jousting Armour from the Castle of Guadamur when lent by Ambrose Monell to the Metropolitan Museum in 1911.* (Image taken from plate VII of the Catalogue of a Loan Exhibition of Arms and Armor).

5 The Art Institute of Chicago, 1982.2628.

6 Dean, 1911: 11.

7 The majority of this Spanish style tilting armour has been deaccessioned from the Art Institute of Chicago and now mostly in a private American collection, see catalogue entry for the sold portions of this armour in: *Thomas Del Mar Ltd Antique Arms, Armour & Militaria*, December, 2015: pp. 84, lot #175.

8 Information kindly offered by correspondence with Alejandro de la Fuente July, 2016.

By the mid-1920s, the Guadamur armours in America were sold by Monell through the London dealer, Hal Firmage, to the Chicago collector George F. Harding, Jr., whose collection eventually came to the Art Institute of Chicago. For Harding, these pieces would become the early centrepieces of his "castle" home in Hyde Park, Chicago, later torn down in 1965[9]. Little remains of the Guadamur armours in the Harding collection aside from some fragments like the aforementioned jousting gauntlet, as many pieces were sold by the Art Institute through auctions in 2007, 2015 and 2020 in recognition of their modern composition and heavily altered condition[10]. All this reveals that the state of connoisseurship was still fledgling among American collectors in the first decade of the twentieth century and has not stood the critical test of time. These Guadamur armors only represent the aspirational beginnings of what was to come as Dean honed his eye and expanded his network.

By contrast, however, a fragmentary armour from a large garniture[11] for field and tournament now in the Philadelphia Museum of Art is said to have come from the armoury of the "Count of Asalto," presumably Carlos de Morenes y de Tord. It is made up of etched and gilded pieces of the highest quality produced in Augsburg from about 1560[12]. By tradition it is associated with Don Sancho d'Avila[13]. It is not clear where this attribution originates, but the armour's modern surcoat, which stands in for the lost cuirass, is embroidered with heraldry related to d'Avila and has been associated with the armour since at least the late-nineteenth century. Before it was given to Philadelphia by the New York collector Carl Otto Kretzschmar von Kienbusch, this garniture was in the collection of the American media mogul and major armour collector William Randolph Hearst. It is likely that many of the rare period fragments from de Tord, like the abovementioned tilting gauntlet from Guadamur and this fragmented armour, may have come from other ancestral Spanish armouries collected by the Count, perhaps even from the armoury of the combined Dukes of Osuna and Infantado.

---

9 Karcheski, 1995: 8–15.

10 In addition to the jousting armour from Guadamur sold at Thomas Del Mar Ltd. 2 December 2015, lot 175, other pieces from Guadamur were in: Christies, London, 20 September, 2007, lot 52 and 73, and Thomas Del Mar Ltd. 2 December 2020, lots 97, 99 and 100.

11 A garniture is a group of like-decorated pieces for a patron to use in various forms for war as well as sport, on the fundamental purposes, composition and definition of sixteenth-century garnitures see: La Rocca, 2017:67–77.

12 The Philadelphia Museum of Art, 1977.1676.13.

13 Kienbusch, 1963: 28–30.

That illustrious armoury is likely the source for elements of a composite Spanish gothic armour owned by Bashford Dean, later purchased by Kienbusch and now broken up in pieces at the Philadelphia Museum of Art. Kienbusch has suggested the helmet (capacete)[14], bevor, and possibly the knee defenses (cuisses with polyenes) came from the armoury of the Ducal families of Osuna and Infantado, perhaps dispersed with the liquidation of the 12th Duke of Osuna's estate in 1882 or later[15]. This is likely the gothic armour Dean discussed acquiring in his speech to the New York Hobby Club:

> A collector no longer waits, as he once did, for a brother collector to die, in order that some particular object may find its way to his rival vitrine—he merely
> awaits the psychological moment when his friend wishes to get some other object and asks for the needful funds …
> I recall in this regard securing the suit of Spanish Gothic armor…, which a Madrid collector had been given by the Duke of Osuna; this object I want-ed keenly, but I had to wait seventeen years before the owner suggested letting me have it; he was then building a new home and needed ready money rather than armor![16].

It remains unclear who this Spanish collector intermediary was, but it could have been de Tord, who kept his collection in Madrid as well as Guadamur.

An embossed Milanese parade half-armour in the Metropolitan may be documented in the collection of the Dukes of the Infantado as originating from the Duke of Sessa, Don Gonzalo Fernández de Córdoba y Fernández de Córdoba [Fig. 3][17]. Armour of the Duke of Sessa is mentioned in an 1585 visit by a Dutchman, Henry Cock, to the Duke of the Infantado's armoury[18]. Significantly, the condition of the armour supports this identity as it appears to have sustained fire damage (losses and damage sustained by excess heat); the armoury of the Infantado at Guadalajara suffered a major fire in 1702[19]. At some stage in the late-nineteenth century, the Viennese-born French dealer and collector, Fredrick Spitzer, acquired, repaired and revamped this armour with new etched and gilt decoration. It was then sold to Maurice de Talleyrand-Périgord, Duc de Dino, who later sold it with the rest of his collection to the Metropolitan Museum of

14 The Philadelphia Museum of Art, 1977-167-3a.

15 Kienbusch, 1963: 15–16. On the dispersal of the Osuna and Infantado armory see: González Ramos, 2013: 11.

16 Boorman, 1994: 26.

17 The Metropolitan Museum of Art, 04.3.270a-o

18 González Ramos, 2013: 4.

19 González Ramos, 2013: 10.

Art in 1904. This armour remains today a curiosity of heavy restoration and is worthy of further study to untangle what is actually period and what is reworked.

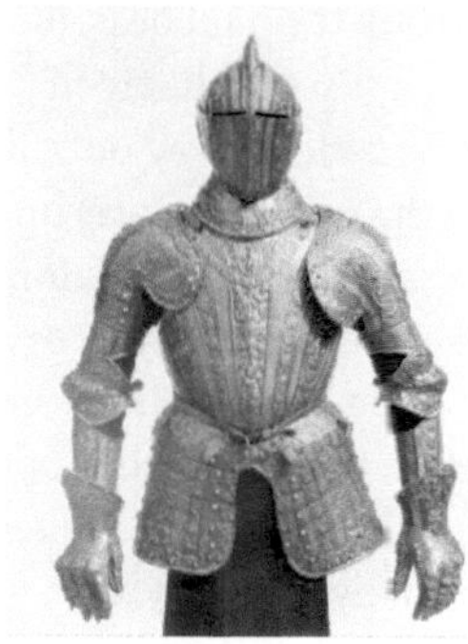

Fig. 3. *Half-Armour attributed to Don Gonzalo Fernández de Córdoba y Fernández de Córdoba, Duke of Sessa.* Partially mid- sixteenth century, comprehensively reworked in the nineteenth century Milan. Unknown artisans. New York, The Metropolitan Museum of Art.

A composite armour, thought to be made from perhaps two armours of the sons of the Duke of Alba, is yet another Milanese parade harness in the Metropolitan that has certainly sustained fire damage[20]. Unlike the Sessa armour, this one retains more of its decorative integrity despite its poor condition, with a horror vacui of combined embossed and damascened decoration. It is composed from armours of the highest class, following similar works that were produced for Alessandro Farnese, the Duke of Parma now in Vienna, a half-armour presumed to have been for the Duke of Ferrara in the Wallace Collection, and the child's armour of King Philip III in the Real Armería, Madrid, among others, all associated with the armourer Lucio Piccinino[21]. This composite armour reveals a level of patronage befitting the Alba family's stature in the court of Philip II. William H. Riggs, an American expatriate and armour collector in Paris, purchased this armour directly from the Alba family at their palace in Madrid in 1861. In 1914, Riggs donated his collection to the Metropolitan Museum.

Besides the armouries of the Dukes of Osuna-Infantado and Alba represented in American collections, there are also various objects associated with the Duke

20 The Metropolitan Museum of Art, 14.25.714a-h,j .

21 On this group of armours, see: Godoy/Leydi, 2003:17–19.

of Medina-Sidonia. One piece is a fine circa 1580 Milanese gauntlet in the Metropolitan for an armour partially preserved in the Real Armería, Madrid[22]. This gauntlet was apparently sold at the turn of the century to a New York collector, Amerigo Nicolosi, from the Duchess of Najera, Maria del Pilar de Zavala y Guzman or her daughter. It seems likely that the gauntlet may have been in the Real Armería but perhaps made its way out, like many other fragments in the period of 1834–1838, when the collection was under the care of Blas Zuloaga and his son, Eusebio[23]. More research should be done on the Duchess of Najera as a source for American collections, as there are several other fragments in the Philadelphia Museum of Art and the Metropolitan that also have this provenance.

Another group of objects related to the armoury of the Duke of Medina-Sidonia is a nearly complete Milanese armour in the Walters Art Museum, Baltimore, Maryland[24]. This armour is entirely covered with etched and partially gilded decoration in a continuous arabesque pattern framing letter M's surmounted by coronets. Dating to about 1590, it is an early example of all-over etched decoration with personalised devices that assimilated patterned silk, an ornamental style that became fashionable on Milanese armours for high-ranking nobles, especially after 1600. Henry Walters, an American railroad magnate and art collector, purchased this armour from the Marques de Mondejar in 1925, who supposedly purchased it from a sale of the Duke of Medina-Sidonia the year before[25]. The Philadelphia Museum of Art also has a Milanese close helmet for tournament on foot at the barriers that is etched with the intertwined letters forming the name "MEDINA" surmounted by a coronet, possibly from an armour of the Duke of Medina-Sidonia. This helmet, however, has uncertain origins before the mid-nineteenth century, when it can be traced to the William Meyrick collection, London. It has also been thought this cypher or monogram relates to a different Spanish Grandee, that of the Medinaceli family. Other fragments etched with this device are in the Armeria Reale in Turin[26].

These pieces represent just a short survey of some of the more significant tracible fragments of Spanish aristocratic armouries dispersed in American

---

22 The Metropolitan Museum of Art, 2002.507.

23 Pyhrr, 1999: 262.

24 The Walters Art Museum, 51.585.

25 Information from the Walters Art Museum Website: <https://art.thewalters.org/detail/17890/armor-for-the-duke-of-medina-sidonia/ > [23 June 2023]

26 Kienbusch, 1963: 70.

museums, excluding those of royal provenance[27]. Quite apart from these is the singular case of the armoury of the Marqueses de Dos Aguas, which was acquired and exported to America in its entirety, and is therefore more significant as a group for what it might tell of Spanish aristocratic patronage and the composition of ancestral armouries. Unlike most other case studies presented in this collective volume on Spanish aristocratic armouries, in the example of Dos Aguas there is an inverted situation where knowledge of the armoury relies very little on surviving documentation and mostly on the objects themselves[28].

Before its dispersal in 1927 to Chicago and New York, the armoury of the Marques de Dos Aguas was located in the Dos Aguas Palace, Valencia, on the second-floor mezzanine. Built over a late-fifteenth-century building that was remodelled in 1740–1744, the palace was the grandest in Valencia. Today it is home to the Gonzalez Martí National Ceramics Museum. Like many aristocratic armouries, that of the Dos Aguas was a hereditary accumulation, not a collection of pieces acquired in the antiques market, like Guadamur. It is an excellent case study of an aristocratic armoury framed as a trophy to the family's historic memory and patronage. This point is highly significant in relation to the armour's preservation and the rarity of its survival, seemingly untouched by the antiques trade and free of any severe restoration. Most especially, the armory uniquely preserved elements of several sixteenth-century armour garnitures for field and tournament, typical of Spanish court taste with South German as well as North Italian manufacture that will be discussed in better detail further below.

Looking broadly at the family history, the title Dos Aguas was first granted as a barony in 1496 to Giner Rabassa. During the seventeenth century, the Baron Giner Ramón Rabassa married Maria De Rocafull, establishing ties to another important noble Valencian family. In fact, in 1670, Doña Maria became the heiress to the Count of Albatera and Baron of Bétera and Lord of Peralada. When

27 On royal Spanish material, some of which is dispersed in American collections, see: Pyhrr, 1999.

28 I am grateful to Roberto González Ramos for finding and transcribing at least one inventory of assets of the Marques de Dos Aguas that includes pieces from the armoury in a letter dated 2 June 1843. The listings seem to be cursory and do not likely cover all that was in the armoury or at least what is known to be present later in 1927, but nearly all pieces described, including the armours now in The Art Institute of Chicago and The Metropolitan Museum, are identifiable. Unfortunately, no provenance for any pieces are suggested. From: Archivo HIstórico de la Nobleza, Osuna, Cartas, 538, D. 42. The lack of any further archival material from the Marques de Dos Aguas is further corroborated by Jose Luis Garcia Martinez. Garcia Martinez, 2015: 166–186.

a portion of the armoury was displayed for the public at the 1910 Exposición Nacional de Arte Retrospectivo, Valencia, a catalogue was published that serves now as the only record of what must have been centuries old family traditions[29]. For example, two of the armours now in the Art Institute of Chicago are said to be those of the Counts of Albatera[30] and Peralada[31], suggesting that it was indeed the Rocafull inheritance consolidated with the Dos Aguas after 1670 that formed the nucleus of the Dos Aguas armoury. Indeed, the catalogue notes about the Peralada armour that: "the Count of Peralada brought [it] to Valencia in the year 1689"[32]. This tradition, however, is slightly confusing, as the Lords and later Counts of Albatera and Lord of Peralada were one and the same. It seems reasonable then, that the armours could belong to members of the Rocafull family. At the time of the 1910 *Exposición* the collection was the property of the Marquesa de Dos Aguas, Rosalia Dasí Moreno, who would die only four years later, leaving the armoury and the palace Dos Aguas to her young daughter and husband, Guarmo Casanova Viace, the Marques de Galtero, who would sell the armoury in 1927.

The story of this dispersal begins, like many of the other objects discussed, with the travels of Dr. Bashford Dean. Sometime in 1925 or 1926, Dean was in Spain visiting the armoury and began negotiations with the Marques de Galtero through an intermediary agent, a vice counsel to Spain in Barcelona, Manuel Soler. Before Dean departed the palace, he not only sketched several of the armours that caught his eye, but made careful lists of each piece in several nineteenth-century vitrines built into the family armoury. Today, these notes and sketches, along with extensive correspondence, remain in the archives of the Arms and Armor Department of the Metropolitan Museum of Art in a correspondence file catalogued under the name of the agent, Manuel Soler.

Captivated by what he saw, but with very little funds at his disposal, Dean leveraged creative means to acquire the Dos Aquas armoury for the Metropolitan's collection upon his return. He managed to muster the requisite finances through a unique agreement to split the armoury between the Metropolitan Museum, backed by trustees George D. Pratt and Archer M. Huntington, and the Chicago

---

29 This catalogue of the 1910 *Exposición Nacional de Arte Retrospectivo*, Valencia is known to this author from a transcribed version in the correspondence files of the arms and armour department, The Metropolitan Museum of Art, as well as from excerpts cited in: García Martínez, 2015: 175–177.

30 The Art Institute of Chicago, 1982.2172a-o.

31 The Art Institute of Chicago, 1982.2102a-l.

32 As cited by: García Martínez, 2015: 176.

collector, George F. Harding, Jr., as well as the University of Illinois for their museum (the Spurlock). Harding, mentioned above in relation to the Guadamur armours acquired from Ambrose Monell, was a colourful Chicago real-estate magnate who also took up public office as a senator. And if these aspirations toward feudal power were not enough, in 1927 he built his own American "castle" as a stone addition to his otherwise ordinary home. Beginning in the 1920s and continuing until his death in 1939, Harding's collection numbered over 1,200 pieces, mostly of European origin. Of all the armour Harding owned, those from the Dos Aguas armoury are certainly among the most important. After his death, Harding left his collection as a museum, until his castle was pulled down in 1963. After much trial and tribulation, his collection, an American armoury of sorts, lives on today in the galleries of the Art Institute, where nearly all of it was deposited in 1982.

To negotiate the purchase with the Marques, Dean sent his very able assistant, Anita Reinhard, one of America's first women curators, to Valencia. A member of the Arms and Armor department since 1916, Reinhard was forty when she received her "marching orders" from Dean and embarked on a steamer on 19 February 1927 destined to Paris. From Paris, she boarded a train to Barcelona and travelled to Valencia, where she was instructed to have an initial interview with the Metropolitan's agent, Manuel Soler. Then she was to meet with the Marques de Galtero at the Palace to discuss the price and check on each piece as listed in Dean's notes from his prior trip. The correspondence record shows Dean opened the negotiations offering the Marques $50,000, with $5,000 commission to Soler, for the purchase of the entire armoury. After her meeting with the Marques on 4 March, Reinhard sent a telegram to Dean that the previous price was: "Refused flat…coldly received /claims offer here seventy net thinks we [are] able [to do] more…". And so Miss Reinhard's troubles began. By 9 March, she was cabling Dean to ask if $75,000 was possible. Dean then began to play a cat and mouse game with the Marques remotely through his telegrams to Miss Reinhard via their agent Soler, saying, "Now they ask seventy plus ten % commission/ nothing doing/Be prepared to cancel everything and return immediately". Dean then immediately sent a second telegram directly to Reinhard, stating, "Bluff pleasantly/Show Soler my cable sent today through him". In reality, Dean had no intention of cancelling and sending Reinhard home. Soler and the Marques seem to have called Dean's bluff though, as they continued to refuse. Reinhard was then instructed in another letter to raise the offer to $82,000. Soon after, Dean sent a telegram in clear terms to Reinhard: "Get everything purchased/Objects eternal/trouble temporary! -Dean". On 24 March, Reinhard wrote to Dean after the offer was finally accepted:

> *Do we, or rather I seem to have muddled thro' somehow after all, and the enemy [Marques] will accept 350,000 pesetas net for himself export taxes & commission extra.*
> *Yesterday the M. [Marques] put all the cards on the table to S. [Soler]. A dealer from Madrid had made an offer once or twice, but the housekeeper at V. [Valencia] did not like or trust him, & she has now advised the M. to accept your offer, which is sure! Hence this sudden about-face.*

Aside from this refusal to a Madrid dealer, there was an additional ploy that finally made the Marques capitulate and accept the Metropolitan's offer. Soler advised Reinhard to pretend to leave for Madrid, to make it appear that "a sure offer, with money on the table was seemingly about to be lost…". In reality, she only moved to a different hotel in Barcelona, but apparently the ruse worked, as the Marques rapidly instructed Soler to send word to Reinhard to return.

The next complication in the acquisition of the armoury was that the Marques did not want the public to know that he had sold it. Though it was entirely legal and he had received the requisite export licenses, the Marques was essentially selling his late wife and daughter's patrimony. He made the Metropolitan firmly agree to keep his name or that of Dos Aguas confidential for at least thirty years or until his death. To prevent public knowledge, the arms and armour were removed from the palace in several carloads under the cover of night through a back door. The Marques' chauffer drove the pieces to an apartment in Barcelona where they were documented in photographs prepared for the customs papers [Fig. 4].

After nine weeks of negotiation, on the 31 May the shipment was finally sent and made its way to New York City. The records show that on 9 April Harding cabled a payment to Barcelona of $58,512.40, more than half the funds for the armoury. In recognition for this, Harding was given first choice when the pieces arrived in June. Naturally his choice was for the gilded armours. In total, the purchase cost was $83,089.97. Below is a survey of the more significant pieces of the Dos Aguas armoury today in the Metropolitan Museum of Art and the Art Institute of Chicago. There is also a very small group of pieces in the Spurlock Museum of the University of Illinois, Urbana-Champaign. For a list of all the known pieces from the Dos Aguas armoury and their dispersal, see the appendix.

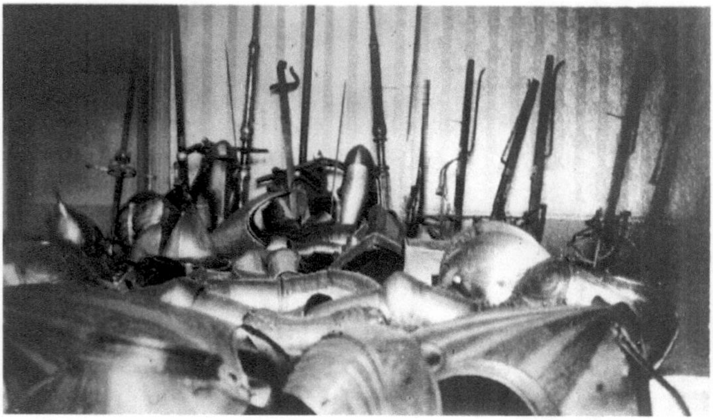

Fig. 4. *Photograph of parts of the Dos Aguas Armory taken 9 April 1927.* Most of what is in the picture is now in the Art Institute's collection.

If Harding took the flashy gilded pieces, the Metropolitan and Dean himself picked some of the fine rare early pieces. For example, a left-hand gauntlet (manifer) with a roundel over the lower cuff of either Spanish or North Italian production was purchased from the armoury privately by Dean and later purchased from his estate for the Metropolitan Museum[33]. Dating from about 1480 to 1500, this gauntlet was the single oldest fragment in the armoury. Chronologically, the next important piece from the Dos Aguas armoury is a closed burgonet that is attributed to the armourer most associated with the imperial court of Charles V (King Carlos I), Kolman Helmschmid of Augsburg. The burgonet is etched in the style of, if not by, the equally famous Daniel Hopfer of Augsburg, the master etcher who is generally credited as the first person to experiment with etching as a printmaking process[34]. This very fine fluted helmet is in excellent condition and dates to about 1525–30. A second falling buff—defense for the face—intended for a similar close helmet, is preserved with a piece of lining in the chin; it also came from the armoury and is now in the Metropolitan Museum[35]. Though no other related pieces of fluted armour seem to have remained in the armoury, both were clearly part of exceptional Augsburg field harnesses produced during the earlier reign of Emperor Charles V. They

33 The Metropolitan Museum of Art, 29.158.241.

34 The Metropolitan Museum of Art, 27.159.18.

35 The Metropolitan Museum of Art, 27.159.19.

demonstrate a Spanish court taste for German armourers in emulation of the King/Emperor.

This fashion for German armour continued for a few decades in the Dos Aguas armoury, as demonstrated by a group of elements from an armour garniture for field and tournament, circa 1545–1555, also produced in Augsburg[36]. These pieces, including a locking gauntlet, vamplate (or guard for the lance), pair of greaves, sabatons and cuisses, barded face defense, reinforcing buff, burgonet and left pauldron, are significant as they demonstrate a complex garniture for field and tourney that may represent a particular Spanish court fashion to dispense with heavy jousting elements in favour of jousting in tourney or reinforced field harnesses. This would reduce the pieces ordered, leaving out the heavy tilting helm, reinforces to the left side, and manifer; none of these elements are present in the armoury. It is likely that the core pieces—the cuirass, helmet, and vambraces (arms) —were separated generations before and perhaps remained in another family armoury, property or estate on display, and are therefore lost with time. Perhaps one day the prominent missing elements might emerge with their distinct etched decoration of fantastic trophies of arms[37]. The existence of an open face burgonet including a pink taffeta lining in the group of elements further suggests this garniture had a prolonged use, with the lining refreshed and the helmet worn in family pageantry perhaps into the seventeenth century. Here too, the etched bands on the helmet have been gilded, unlike the rest of the garniture, again conceivably for a later ceremonial purpose.

Indeed, the Dos Aguas armoury is most significant for the preservation of several nearly complete field and tournament garnitures all produced in Milan during the reign of Philip II from about 1570 to 1585. These armours do not represent the highest quality Milanese armourers had to offer in terms of form and decoration, but they do demonstrate a high-middle range befitting the rank of the presumed owners or single owner in the Rocafull family, the Lords of Albatera[38]. Though without any hard evidence, it is entirely possible these belong to the generation of the 7th Lord of Albatera, Ramón de Rocafull y Cascales and/or his two brothers Alfonso and Rodrigo, all presumably born between

---

36 The Metropolitan Museum of Art, 27.160.1–8. These pieces, their function, design and possible association with a distinct Spanish type of German manufactured garniture are the subject of an article: Tavares, 2014.

37 Tavares, 2014:124.

38 As noted above, the association with the Lords of Albatera and Paralada was published in the catalogue of the 1910 *Exposicíon Nacional in Valencia.* The quality of the garniture in the Metropolitan Museum is commented upon by LaRocca 2017: 37–41.

1550 and 1566, placing them at ages twenty to thirty at the time these garnitures were made in Milan. Further research into either personal inventories, military careers or participation in court tournaments of these individuals might lend some strength to this supposition.

One of these armours is the largest known Milanese garniture of its period, and is today in the Metropolitan Museum[39]. It is also among the most complete, which is likely why Dean selected this example. It includes elements, etched but ungilt, for use as a field armour for mounted use or on foot, as well as for two types of joust for sport. It may be the only Milanese garniture that survives with interchangeable reinforcing pieces over the left arm and shoulder (grandguard and pasguard), to serve either the Italian style joust with grandguard contoured over the shoulder or the German style with target-shaped granguard[40]. It suggests a patron who intended to participate in either type of joust depending on the court or event he was to attend. It also includes two pairs of saddle steels to fit out more than one horse, and even a pointed morion (helmet) for infantry use—the presence of which suggests that there are still some elements missing like a matching rondache for infantry. Its extensive composition as a garniture is explored in further detail in Donald J. LaRocca's 2017 work, *How to Read European Armor*. It forms an ideal exemplar to understand many other Milanese garnitures in a more fragmentary state.

The other two Milanese garnitures for field and tournament, both now in the Art Institute of Chicago, are quite different from the last. These pieces do not appear to have ever had components for the joust, rather, they are for field use and can be configured for foot tournament at the barrier with exchange helmets and additional right pauldrons cut away to allow freer lateral movement with blunted spears and swords[41]. These garnitures are etched as well as gilded, of the style found in innumerable portraits of Spanish and Italian nobility in the second half of the sixteenth and even into the seventeenth century as symbols of rank and courtly taste. In that period, tournaments on foot at the barriers were becoming an ever more popular noble sport at courts across Western Europe. It arguably reflects a shift towards infantry becoming the primary focus of military training as tournaments, though constructed for sport and spectacle, evolved to reference more serious battle[42].

39 The Metropolitan Museum of art, 27.159.1a–p, .11, .14a–e, .16–.17. This point is made by LaRocca, 2017: 37.

40 This point is made by LaRocca, 2017: 70.

41 The Art Institute of Chicago, 1982.2172a-o and 1982.2102a-l.

42 On this point and tournaments on foot at the barriers, see: Anglo, 2007: 91–106.

This sport required participants, either in teams or one-on-one, to wear half armours and stand behind a waste-high barrier, giving and receiving blows from blunted spears and swords. The spears were generally constructed of light fir, specially dried to easily splinter. One of the two garnitures has very clear evidence of use in this tournament, with various dents from blunted sword cuts on the left vambrace and helmet visor, as well as a large piece of delamination over the upper left brow of the helmet visor, where etched and gilt decoration have been sheared off, presumably from many strokes of a blunted spear[43]. This wear underscores that despite the lavish decoration, these garnitures are indeed working armors intended for physical use. They are as much sports equipment as they are for conspicuous display of family rank and wealth. The helmet of this armour also has the addition of several crude holes pierced through the raised centre comb to help affix the large panaches of feathers that usually embellished these courtly events. Since 2016, both armours are displayed in this foot combat configuration in the arms and armour gallery of the Art Institute with several field pieces, including exchange helmets and pauldrons, remaining separate [Fig. 5]. It is rare to be able to demonstrate this form of tournament with two complete armours from the same time and place. Further research into documented foot tournaments at the Spanish court may shed light into this grand context. Remarkably, a period lining of leather, wool and linen is still stitched into the foot-combat close helmet and a demi-shaffron (defense for a horse's head) that form part of one of the garnitures[44]. Surviving linings are very seldom encountered; they are perhaps more frequently found on objects maintained in family armouries such as this.

43 The Art Institute of Chicago, 1982.2172a-o.
44 The Art Institute of Chicago, 1982.2102a-l.

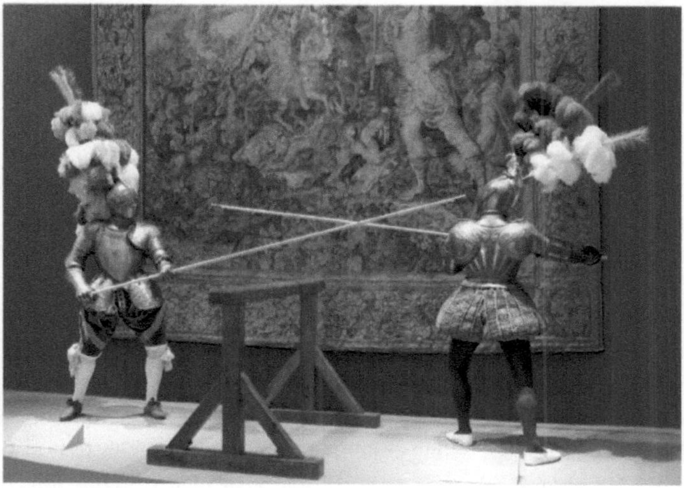

Fig. 5. *The two Milanese armours from the Dos Aguas Armory as displayed in foot combat at the barrier at The Art Institute of Chicago*

Other rare linings survive in several helmets and even a rondache from the Dos Aguas armoury in The Art Institute. The most perplexing object, however, is a late sixteenth-century Turkish helmet (shishak) made of gilded copper (tombac), with an intact blue linen lining that is stitched over both ear pieces, the suspended tail or neck defense and the interior conical bowl of the helmet[45]. The origin of this helmet—along with several pairs of Moorish/Berber prick-spurs[46]—is unclear. Could it have been a family trophy taken after a battle with the Ottomans or Berbers, like other important trophies such as galley lamps from the 1571 sea battle of Lepanto, or the banners and helmet of the Ottoman Grand Admiral, Ali Pasha, in the Real Armaría? Unfortunately, as noted above, documentation of the Dos Aguas armoury is limited, with the objects having to speak largely for themselves.

The mystery of this helmet remains, but perhaps it relates to other trophies that were kept in the family armoury. Returning to the few descriptions in the 1910 catalogue, a group of six crossbows and pellet crossbows with spanning

45 The Art Institute of Chicago, 1982.2238.

46 The Art Institute of Chicago, 1982.2210a-b, 1932.3127; at the University of Illinois Spurlock Museum 1927.05.0001AB; and The Metropolitan Museum of Art, 27.159.16,.17.

levers—two of which are now in the Metropolitan Museum and another two in the Art Institute of Chicago[47]—are described as belonging "[…] to the Moriscos archers who became strong in the Valley of Alhamar, protesting the expulsion of 1609". This likely refers to a revolt of the Moriscos in the Valencian municipality of Dos Aguas in 1609, who once defeated and expelled left the lands of Dos Aguas nearly abandoned. Recounting the earlier Morisco rebellions of 1568–1571, Diego Hurtado Mendoza noted that crossbows were used in great numbers in Granada by the Moriscos. These crossbows therefore imply members of the Dos Aguas family were connected with some of the engagements against the Moriscos, especially on the lands attached with their title. Likewise, the Ottoman shishak and Berber spurs may also have been obtained as family trophies through other conflicts against the Ottomans and Berbers.

There are a several other noteworthy objects from the Dos Aguas armoury. One is a high-quality Flemish zischägge helmet, about 1620–1630, that also retains a period silk lining and which is a typical piece of Flemish (possibly Brussels) armour used by Spanish commanders while fighting in the Netherlands[48]. And, representing holdings from the eighteenth century, the Metropolitan Museum retains a pair of luxury pistols made by one of the foremost Madrid gunsmiths, Gabriel de Algora (active 1733–1761), in the French Bourbon influenced *a la moda* style[49]. In a family with such wealth and distinction during the eighteenth century—as evidenced by the renovation of their Valencia Palace—one would expect vast quantities of luxury firearms in the family armoury; however, the examples of firearms from the Dos Aguas armoury number only about six objects. It can be surmised these pistols by de Algora were an anomaly for some reason, with the family firearms likely kept in a distinct gun cabinet in some other property or estate. This might allude to a distinction between the family armoury as a pure display of hereditary trophies and trappings of honour and the more functional *guardaroba*, or wardrobe, where functioning pieces like later eighteenth- and nineteenth-century personal arms, such as rapier, smallswords and firearms, would have been kept separately[50].

---

47 The Metropolitan Museum of Art, 27.160.18a, b; 27.160.19a, b; and The Art Institute of Chicago, 1982.3071; 1982.3073.

48 The Art Institute of Chicago, 1982.2235.

49 The Metropolitan Museum of Art, 27.160.12a-b.

50 The Dos Aguas armoury did have some examples of cup hilt rapiers, smallsword and daggers, but very few, numbering about seven small-edged weapons. See appendix to this essay.

To conclude, there is much to learn from the Dos Aguas armoury and more questions perhaps than answers in regard to provenance. However, one thing is certain: the agency of Dr. Basford Dean brought these pieces and many others to America. As a coda to this survey of dispersed Spanish arms in American museums, it is worth mentioning that some of Dean's letters to Reinhard at the time of acquisition of the Dos Aguas armoury were written on stationary from the Battle Creek Sanitarium, a health resort in the midwestern state of Michigan. Dean was there to help with his health issues. In 1928, soon after surgery there, he would die from what was likely pancreatic cancer. The Dos Aguas armoury was one the last major acquisitions he initiated. There can be no doubt from the review of pieces presented here from Dos Aguas, as well as several other great noble Spanish armouries, that Dean was the primary force that brought these Spanish treasures to American collections and institutions. This was noticed and lamented by Spanish collectors at the time, as recounted in one final amusing anecdote, worth presenting here at length, from Dean's own words to the New York Hobby club:

> *I recall purchasing a splendid Gothic headpiece, or salade, for fifty pesetas in a poor little shop in front of my hotel in Madrid. "Where in the world did you get that?" said Don George de Osma, coming into my room a few minutes later, picking up the object and examining it critically. Senor de Osma is a great connoisseur of ancient armor…, the son-in-law, by the way, of the late Count de Valencia de Don Juan, director of the Royal Armory of Madrid. When I told my friend where and how I found my prize, he shook his head gravely: he did not know that the little shop across the way had ever in it a piece really worth getting, and he had even seen the magenta-colored panel there bearing a trophy of sadly rusted and "bogus" helmets among which my salade had been hidden. Nor did he, I think, quite like the idea of a foreigner coming to town and picking up for a trifle an object that should have remained among Spanish amateurs—at least a mutual friend told me so, adding that Senor Osma "wished he had the power to lock that man Dean up 'til he would consent to leave Spain!".*

It is uncertain, but the Spanish sallet referred to by Dean here may be the same as one late fifteenth-century example in the Metropolitan Museum of Art from Dean's estate[51]. It resembles others known from Spanish armouries, particularly those associated with munitions now displayed in the Museo del Ejercito, Toledo, from the armoury of the Dukes of Medinaceli. The provenance of Dean's sallet is lost with time, but its form, with its low-swept cut brim and boxy shallow bowl, is distinctly Spanish. Dean's keen eye and quick action captured a relic from the golden age of Ferdinand and Isabella that early-twentieth-century

51 Metropolitan Museum of Art, 29.158.3a.

Americans so loved and romanticised. Hero or villain, his influence on American connoisseurship and collecting of arms and armour cannot be overestimated and continues to inspire attention and inquiry.

APPENDIX

Contents of The Dos Aguas Armoury and its Dispersal, partially taken from a 30 June 1927 list in the Archives of the Metropolitan Museum of Art compiled with current institutions and accession numbers where known, or auctions in which the object was deaccessioned. The Dos Aguas armoury also contained saddles, saddle cloths, holsters and other horse trappings that were not included in the purchase of the armoury, the whereabouts are unknown.

Purchased for the Metropolitan Museum of Art (MMA):

Armor Garniture for Field and Tournament, Milan, c. 1570, MMA 27.159.1–.15
Elements of an Armor Garniture for Field and Tourney, Augsburg, c. 1560, MMA 27.160.1–.8
Closed Burgonet attributed to Kolman Helmschmid, Augsburg, c. 1525, MMA 27.159.18
Falling buff attributed to Kolman Helmschmid, Augsburg, c. 1525, MMA 27.159.19
Round shield, Milan, late sixteenth century, MMA 27.160.9
Brigandine, sixteenth century, MMA 27.160.10
Fragments of a Brigandine, sixteenth century, MMA 27.160.11
Pair of Pistols by Gabriel de Algora, Madrid, c. 1735–40, MMA 27.160.12,.13
Carrousel Lance, MMA 27.160.14
Tortoiseshell Powder Flask, MMA 27.160.15
Pair of Prick Spurs, North African (Berber), sixteenth century, MMA 27.159.16,.17
Crossbow with Lever, Spanish, mid-sixteenth century, MMA 27.160.19a,b
Pellet Crossbow with Lever, Spanish, mid-sixteenth century, MMA 27.160.18a,b

Purchased by Dr. Bashford Dean:

Left-Handed Gauntlet (Manifer), Italian or Spanish, c. 1480–1510, MMA 29.158.241
Brigandine fragments in Buckskin, unknown
Three Breastplates and two Backplates (plain), unknown
Two Left-handed Daggers, unknown
Horse Bit with embellished bosses, unknown

Plain Horse Bit, unknown
Two Crossbows without Levers, Spanish, mid-sixteenth century, unknown

Purchased by University of Illinois (now University of Illinois, Spurlock Museum, SM):

Cup-hilt Rapier, SM 1927.06.0001
Cup-hilt Rapier, SM 1927.06.0002
Court Sword (small sword), SM 1927.06.002
Left-handed Dagger, SM 1927.06.0004
Pair of Pistols (plain), unknown
Pair of Prick Spurs, North African (Berber), sixteenth century, SM 1927.05.0001AB
Horse Bit with embellished bosses, SM 1927.05.0003

Purchased by George F. Harding Jr. (now mostly in The Art Institute of Chicago, AIC):

Armor Garniture for Filed and Tournament on Foot at the Barriers, Milan, c. 1570–1580, AIC 1982.2102
Armor Garniture for Tournament on Foot at the Barriers, Milan, c. 1570–1580, AIC 1982.2172
Round Shield, Pointed Morion (Cabaset) and Demishaffron, Milan, c. 1570–1580, AIC
1982.2229a-c.
Round Shield, Milan, c. 1570–1580, Deaccessioned: Christies, 20 September 2007, lot 26.
Round Shield, Milan, c. 1570–1580, AIC 1982.2260
Breastplate (etched), Milan, c. 1560, AIC 1982.2563
Shishak (Helmet), gilded copper, Ottoman, late sixteenth century, AIC 1982.2238
Zischagge, Flemish, c. 1620–1630, AIC 1982.2235
Two-handed sword, Spanish or Italian, late sixteenth century, AIC 1982.2837
Parade Sword with a gilded hilt inlayed with corral, sixteenth or seventeenth century, Deaccessioned:
Sotheby's, 1 June 1991, lot 440; Later restored and re-bladed by Peter Finer; sold to Berlin collector Karsten Klinbeil; sold Pierre Bergé/Herman Historica, 13 December 2011, lot 199.
Left-handed Dagger, seventeenth century, AIC 1982.3446
Matchlock Gun, seventeenth century, Deaccessioned: Sotheby's, 1 June 1991, lot 451.
Miquelet (*Patilla*) Carbine, Spanish, seventeenth century, AIC 1982.2894
Langue-de-Beouf (Partisan), Italian, early sixteenth century, AIC 1982.2345

Carrousel Lance, seventeenth century, Deaccessioned: Christies, 20 September 2007, lot 78.

Mace, sixteenth century, AIC 1982.3501

Horseman's Axe (gilded, formerly a poleaxe), early sixteenth century, AIC 1982.2109

Pair of Stirrups (brass), eighteenth–nineteenth century, Deaccessioned: Christies, 20 September 2007, lot 67.

Pair of Prick Spurs, gilded, North African (Berber), sixteenth century, AIC 1982.2210ab

Single Prick Spur, North African (Berber), sixteenth century, AIC 1982.3127

Crossbow with Lever, Spanish, mid-sixteenth century, AIC 1982.3073; 1982.3074

Pellet Crossbow with Lever, Spanish, mid-sixteenth century, AIC 1982.3071; 1982.3072

Horse bit with embellished bosses, late sixteenth century, AIC 1982.3122

Horse bit with embellished bosses, late sixteenth century, AIC 1982.2205

Horse bit with embellished bosses, late sixteenth century, AIC 1982.3118

BIBLIOGRAPHY

Anglo, Sydney (2007): "The Barriers: From Combat to Dance (Almost)". In: *Dance Research: The Journal of the Society for Dance Research*, 25, 2, (The Art that All Arts Do Approve: Manifestations of the Dance Impulse in High Renaissance Culture. Studies in Honour of Margaret M. McGowan), pp. 91–106.

Boorman, Dean K. (1994): "A Personal Reminiscence of Bashford Dean, and His Unpublished Talk, Circa 1920, 'The Hobby of Collecting Ancient Armor'". In: *American Society of Arms Collectors Bulletin*, 70, pp. 16–28.

Dean, Bashford (1911): *Catalogue of a Loan Exhibition of Arms and Armor*. New York, The Metropolitan Museum of Art.

*Thomas Del Mar Ltd Antique Arms, Armour & Milita*ria, Sale cat., London Wednesday 2nd December 2015, page 84, lot #175.

García Martínez, José Luis (2015): "El Legado de los Rabassa de Perellós, fundadores de la Casa de Dos Aguas". In: *erph, Revista electrónica de Patrimonio Histórico,* 17, pp. 166–186.

Godoy, José A. and Leydi, Silvio (2003): *Parate Trionfali Il manierismo nell'arte dell'armatura italiana.* Milan: 5 Continents Editions.

González Ramos, Roberto (2013): "The armoury of the Dukes of the Infantado. Collecting, prestige and meaning". In: Journal of the History of Collections, 25, 3, pp. 335–350.

Karcheski, Walter J. Jr. (1995): *Arms and Armor in The Art Institute of Chicago*. Chicago, The Art Institute of Chicago.

Kienbusch, C. Otto v. (1963): *The Kretzschmar von Kienbusch Collection of Armor and Arms*. Princeton, Princeton University Library.

LaRocca, Donald J. (2017): *How to Read European Armor*. New York, The Metropolitan Museum of Art.

Pyhrr, Stuart W. (2012): *"Of Arms And Men Arms and Armor at the Metropolitan 1912–2012". In: The Metropolitan Museum of Art Bulletin*, 70, 1.

Pyhrr, Stuart (1999): "'Ancient Armour and Arms Recently Recived From Spain': Eusebio Zuloaga, Henri LePage, and the Real Armeria in Madrid". In: *Gladius*, 19, pp. 261–295.

Tavares, Jonathan (2014): "Elements of an Augsburg Armor Garniture for a Spanish Client", In: La Rocca, Donald J. (ed.): *The Armorer's Art: Essays in Honor of Stuart Pyhrr*. Woonsocket, Mowbray Publishing: pp. 115–132.

Francisco Javier López Martín

THE KERNOOZER'S CLUB AND THE ESTABLISHMENT OF THE INSTITUTO DE VALENCIA DE DON JUAN[1]

THE KERNOOZER'S CLUB

During the eighteenth century, antique arms were representative of historical periods that were being idealised. When collecting sought new inspiration and markets to expand, arms and armour were a novelty, not only to buy and sell but to reproduce again with pencil and brush. In this period, the conduct of the studies was a fundamental feature and, unlike those developed previously, they began to focus on scientific or educational principles. Many of the pioneering publications were formed under the protection of study societies whose purpose was not only to investigate the origins of the antique pieces and their evolution but also to recover their ancient techniques of use and combat since it was only possible to understand them through practical knowledge. One of these societies that linked scientific study with collecting was the Kernoozer's Club. It was officially founded by one of the greatest experts in arms and armour of the late nineteenth century, Charles Alexander Baron de Cosson. His reputation was established in 1880 with the research he made for the catalogue of the exhibition *Ancient Helmets and Examples of Mail*, organised at the Royal Archaeological Institute of London[2]. His methodology set a standard that would be the norm of the following studies on arms and armour over the following years and placed him as the main authority in England. He maintained close contact with arms and armour experts, including the Spaniards Juan Crooke—Count of Valencia de Don Juan—and Guillermo de Osma. It is not surprising that de Cosson was the official founder of the Kernoozer's Club, the main group of lovers of arms and armour formed in Europe[3].

---

1 I must thank María Ángeles Santos Quer, Head of Collections and Librarian at the Instituto de Valencia de Don Juan,and David J. de M. Williams, Director, Department of Ancient Arms and Armour, Bonhams, London, for their invaluable help in writing this essay.

2 LaRocca, 2009: 197.

3 Spielmann, 1889: 361.

The British Library holds seven volumes relating to the Kernoozer's Club. Four are Minute Books bound in orange or crimson leather whose covers are engraved in black letters: "K.C. MINUTES Nº 1", or with gold letters: "KERNOOZER'S CLUB MINUTE BOOK Nº 2" (and so 3rd and 4th)[4]. Two volumes are Scrapbooks[5], and the last volume is a keyword index of the Scrapbooks compiled by Talbot Hughes in 1914[6]. The Minute Book Nº 1 includes sessions held between 17 December 1881 and 28 May 1889. It is the most unique from the point of view of the information since it has the charm of novelty. The freshness is also present in the first Scrapbook, which includes photographs of the members, their most notable pieces, views from their armouries, postcards and photographs from other European collections up to 447 items.

The idea of forming a club of expert collectors of arms and armour took place at an informal meeting held in the studio of the painter John Seymour Lucas in the late 1880s[7]. It was at this gathering that the young painter suggested the idea of forming a club. The proposal was received with enthusiasm and was accomplished immediately. The curious name of the Club arose from an anecdote lived by one of the partners in the Christie's auction room. There was a man who in his youth had served as model for artists. He was in the auction room before a major armour sale when one of the "hammers" of Christie's, who was cajoling a potential buyer, approached him and asked his opinion on a piece of armour. "Well, —replied the man— 'taint no use asking me anything about armour and sech-like. As the Frenchman says, I'm no kernoozer (*e.g.* connoisseur)". This fun play on words made the manager of Christie's smile. When during the informal meeting in the winter of 1880 the future partners were looking for a name for their club, Robert Alexander Hillingford recalled the scene, which he had witnessed, and the name was unanimously chosen by acclamation. The exact day on which the Club was established is unknown since no express note was made. However, the first surviving informal invitation was issued by de Cosson on 17 December 1881, to summon the members to a second meeting that would take place on 28 (Wednesday)[8]. The Club was already founded though the name of the Club was not yet fixed because de Cosson wrote it on the invitation in its original form: "Connoisseur's Club".

4 The British Library, London (hereafter BL), Add. 40.678, 40.679, 40.680 and 40.681; 19×23 cm.
5 Vol. I, BL, Add. 40.682; and vol. II BL, Add. 40683.
6 *Catalogue of Scrap Book. Kernoozer Club*. BL, Add. 40.684.
7 Spielmann, 1889: 363. Hardie, 1908: 196.
8 BL, Add. 40.678: 4r.

That meeting was set at Frederik K. Weekes' home, 119 Charlotte Street, London, being present the Baron de Cosson as Chairman and Secretary, Weekes as host and Vice-Chairman of the meeting, R. Hillingford, Thomas Bush Hardy, H. Wentworth Huyshe, Sir James D. Linton and Seymour Lucas. It was then when, in addition to the official name, a series of rules and proposals were approved in accordance with the resolutions approved on the 17th. Among other issues, it was approved to have a Minute Book for the memory of the Club. The number of members would be limited to twenty, and the meetings would be held monthly from October to June in members' homes in rotation. After checking ordinary affairs, any extraordinary matter could be discussed and interesting study excursions would be proposed[9].

The next meeting was held on 18 January 1882, at Cornwall Lodge, Hillingford's home, and on 22 February it was Hardy's turn, a meeting that discussed the technique of blueing and damascening on different armours. It was not only a matter of stimulating and increasing knowledge of arms and armour, but also of registering the greatest number of original pieces. Thus, on 30 March, at Linton's house, the members exposed their latest acquisitions to expert scrutiny. De Cosson reported that he already had the budget to print the Club's rules and circulars, which were finished drafting at that time. They were composed of twenty-five articles. The first article stated that the Club "is formed for the purpose of promoting friendly intercourse amongst Gentlemen who study or collect Ancient Armour & Arms". It would be composed only of collectors of these objects "or who have written some published work on the subject". The original founding members of the Club were Baron de Cosson; Thomas Bush Hardy; Robert Hillingford; Wentworth Huyshe; James D. Linton; Seymour Lucas and Frederik Weekes. Distinguished collectors living further than twenty miles from London would be eligible as Corresponding Members. An annual Club dinner would be held in the summer, around June, when, if possible, a property containing armour would be visited. The location would be decided at the Club meeting in May, and also a three-member committee would be appointed to make dinner arrangements. The Club would also have a scrapbook that would include sketches or photographs of arms and armour that members might own or have seen.

The Kernoozers billed themselves as the world's foremost arms and armour club with the stated goal of stimulating and increasing their knowledge and professionally displaying their use. This significant inclusion of researching the

9 BL, Add. 40.678:5r.

use of ancient arms distinguished the Club from other arms and armour societies. In accordance with the 2nd and 8th articles, no arms and armour merchant or dealer had a place in the Club, they could not even be invited to the meetings, since the information discussed by the members would serve to improve sales. To be a member it was a requirement to possess ancient arms or armour and to have written some work on the subject, giving importance to both collecting and the written compilation of knowledge. In fact, the essential requirements were to have "Good-fellowship and sound scholarship"[10]. Article 24th explains that the most chivalrous spirit should govern the Club's code, which, if violated, would lead to its immediate expulsion, deleting all records from the Books of the expelled member. It is noteworthy that the nucleus of the founding members was made up of historicist painters, with a profound taste for the neo-Gothic and scenes of medieval knights or cavalry with a profusion of arms and armour.

Early historicist painters had paid little attention to the historical accuracy of the objects they painted, sometimes filling romantic scenes with shocking anachronisms. In paintings depicting components of arms, armour or costumes, realism and accuracy were critical in recreating the past splendor of objects that had been used in the Middle Ages or the Renaissance but were no longer commonly used. Since the end of the eighteenth century, it was possible to find these objects in antique dealers' shops and auction rooms, making them more accessible to artists to be used as models. In order to represent faithfully arms and armour in their paintings, artists such as Théodore Gericault, Paul Delaroche, Edwin Henry Landseer, Richard Cook (illustrator of Walter Scott's novels) or Eugène Delacroix had already begun to visit Romantic interiors during the 1820s, whilst Eugène Viollet-le-Duc was studying the objects exhibited at the Louvre for his *Dictionnaire raisonné du mobilier français de l'époque Carolingienne à la Renaissance* (1858–1875). Artists were not alone in this search for truth, but scholars, antiquarians and collectors joined their goal. For these reasons, the inclusion of scrapbooks in the Club's statutes was especially important since each piece acquired or studied had to be included to appreciate certain technical or artistic aspects. Scrapbook I includes photographs of the collections of the members, drawings of singular pieces, souvenirs or engravings.

In 1874, during his apprenticeship, John Seymour Lucas studied in detail the collection of military-style Meissoniers at the Wallace Collection, and in 1877 he was elected a Fellow of the Royal Institute of Painters in Water Colours. In 1880 he made a three-week study trip to Madrid where he was trapped by Velázquez's

10 Hardie, 1908: 197.

La Rendición de Breda. Learning from masters such as Velázquez and Van Dyck his painting reached a high level, especially in the scenes he painted in relation to the 1588 Armada and the English Civil War (1642–1651). The following year he returned to Madrid for a much longer visit of which nothing is known[11]. Since Kernoozer was already founded, it is possible that he visited the Real Armería due to his interest in this type of collection. In 1886 he was elected member of the Royal Academy and soon after, according to a prestigious art magazine, "No one in the civilised world knows better than he how tall felt hats had grown to be in 1650, or the exact stature of a riding-boot of 1580"[12]. The same could be applied to the fidelity with which he reproduced a garment, a sword or outlined armour. In 1898 the Royal Academy made Lucas Associate (A.R.A.) and the Kernoozers honoured him with a dinner.[13] In 1899 he acted as a consultant to the Victoria and Albert Museum (V&A) for the acquisition of the Samuel Isham collection from Lamport Hall in Northamptonshire.

Lucas maintained a strong friendship with Guy Laking[14]. In addition to being the curator of the Royal Armouries at Windsor, Laking was in charge of the Wallace Collection and since 1902 was Keeper of the Museum of London, where he added his status as an expert in arms and armour to that of a specialist in furniture and porcelain. Lucas may have met Laking through their mutual interest through the Kernoozer's Club, though Lucas disposed of part of his own collection of arms and armour at an auction held at Christie's on 27 May 1903[15], where Laking was acting as a consultant, which ultimately prevented him from

11 Gosse, 1888: 6.

12 Gosse, 1888: 3.

13 Invitation sent by Joseph Grego from his home, 23 Granville Square, on 10 Feb. 1898, to celebrate dinner on Saturday, 5 March at 7 o'clock at *Pagani's*, a bohemian restaurant frequented by artists and musicians, which was in 40 Great Portland Street. David J. de M. Williams collection.

14 Both were members of the Junior Kernoozer's Club, which on 1 May 1890, split from the Kernoozer's Club to welcome young enthusiasts and which would eventually become The Meyrick Society, still in force today. The inaugural meeting of the Junior Kernoozer's Club was held at Lucas studio, although he was not present and was not a member of the Club until the following year. Extensive details about the Junior Club and the Meyrick Society are given in Blair 1991. In 2003 I had the honour of attending one of the Meyrick Society meetings at the Wallace Collection as the personal guest of Mr. William Reid (FSA, CBE), former Curator of the Tower of London and later Director of the National Army Museum between 1970 and 1987.

15 Another part of the collection was auctioned at Sotheby's on 7 December 1923. Cripps-Day, 1925: A.60, A.76.

joining the Kernoozers. As curator of the Museum of London, it is very likely that Laking was trying to convince his friend to keep his textile collection in England as Lucas had been approached by the Metropolitan Museum and was threatening to take the collection to New York through a substantial offer. Lucas declined the proposal and the V&A was able to buy it for a lower price in 1911[16]. Lucas must be understood not only as a painter of historical scenes, but also as an artist interested in multiple aspects of collecting in close contact with the museum world. He was not the only Kernoozer who was a painter interested in collecting arms and armour.

Robert Alexander Hillingford was a painter specialising in military history, as was Thomas Bush Hardy, a renowned watercolourist who had sold part of his armour collection in 1876[17]. Wentworth Huyshe was a painter as well as a historian, another lover of the medieval who had also started an armour collection. Sir James D. Linton was elected a Fellow of the Royal Institute of Watercolorists in 1870 and its President from 1884 to 1899. Frederik Weekes, the son of a renowned Victorian sculptor, was another important painter. His works reflected medieval aesthetics and contributed to the revival of medieval Gothic as he did in the castles of Cardiff, Castell Coch, and in the reconstruction of Saint Finbar's Cathedral, buildings in which he worked in collaboration with W. Burges. He was an arms collector and would have joined the Kernoozer's Club had he not passed away in March 1881[18]. All these artists maintained close social ties and Lucas played a fundamental role between them.

These historicist painters developed antiquarian ambitions to a greater or lesser extent. Their bond was born from an interest in studying the details of arms and historical clothing to illustrate their paintings or first photographs, acquiring objects to paint on canvas or use for photographic recreations of the past. Members of the Kernoozer's posed for the snapshots in harness, gimped garb, ruffs, and headpieces [Fig. 1].

---

16 Petrov, 2008: 2, 4.

17 *Catalogue of the collection of ancient armour of T.B. Hardy, Esq., comprising several fine suits of the time of Henry VIII. and Queen Elizabeth; helmets, shields, chanfrons, halberds, swords, rapiers, and daggers, some of which were purchased from the Bernal, Meyrick, and Fortung Collections; also, three splendid sets of beauvais and gobelins tapestry, the property of Sir Richard Wallace…* Londres, 1876.

18 Blair, 1991: 1.

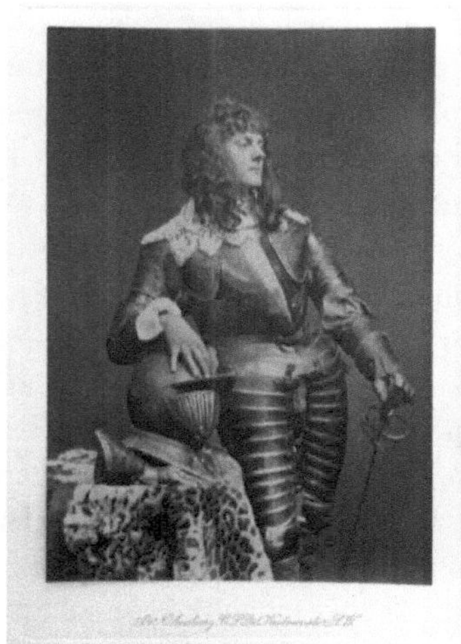

Fig. 1. *John Seymour Lucas with hairstyle and armour from the seventeenth century.* Late 1880s. David J. de M. Williams collection.

On 29 June 1882, the meeting was at Pyrcroft House, de Cosson's home in Chertsey, Surrey. The guests were received by the host and his wife Celia. Huyshe presented the wooden block that had been ordered made with the figure of a fourteenth-century basinet and the Club's motto "*We Seek for This*" contained in a phylactery according to de Cosson's design, who had drawn the whole set, perhaps with the assistance of his wife, an accomplished watercolourist. The figure was used in the ex libris to mark members' study books and correspondence. They also had their own anthem, "*The Kernoozer's Call*", a paean to the brotherhood of friends with a common interest.

Other entries in the first Book of Minutes offer further details of the excursions carried out. For example, at the meeting the following month, agreed to 12 July, the annual dinner and outing were planned. They went to Penshurst Place, Kent, to examine the armour arranged in the superb fourteenth-century Gothic hall. During the examination, the members made sketches of the most notable helmets and other pieces of special interest to Lucas, Hillingford, and de Cosson, which were inserted into the first Scrapbook for record-keeping[19]. In the evening they went to the hotel where an excellent dinner had been ordered, after which,

19 BL, Add. 40.682: 2.

as was the custom, came the liqueurs and toasts. The members returned to London satisfied with the success of the Club's first outing and its annual dinner. In other meetings, the possibility of giving entry to other collecting specialties was discussed, but after a short debate, the proposal was rejected outright. Sometimes the entry of new partners was discussed, as happened with Joseph Grego, proposed by Linton. They also discussed the printing of the Club's bases and that each member submitted his portrait for the scrapbook.

On 14 July 1883, it was approved to include the polymath Richard Burton, and Sir Joseph Noël Paton, the Scottish painter who was a member of the Royal Scottish Academy and a great connoisseur and passionate about arms and armour, whose collection was exhibited in his apartments at 33 George Square in Edinburgh. The annual visit to the castle of Warwick was also planned. Arrangements had begun in June but the visit took place seven days after Paton and Burton's admission, on 21 July, when the annual dinner was also to be held. In this case, the entourage met at Paddington rail station. The following day they went on an excursion to Kenilworth Castle and on the 23rd to Warwick Castle, where the agent of the Earl of Warwick received them. The Kernoozers studied the collection and made numerous sketches before returning to the hotel, where dinner and inevitable toasts awaited them[20] [Fig. 2]. On the 28th, a local newspaper reported about the visit and the members that made up the Club and their guests, for example, Paton, R. Hillingford, J. D. Linton, President of the Royal Institute of Watercolourists and his colleague from the same Institute George G. Kilburne. Among the guests was the renowned sculptor Charles Bell Birch.

20 A freehand drawing of the meeting in Scrapbook I with the notation "Kernoozers Kenilworth, July 22–83", signed G. G. Kilburne, along with some other armour sketches by Hillingford and Lucas. BL, Add. 40.682: XXII.

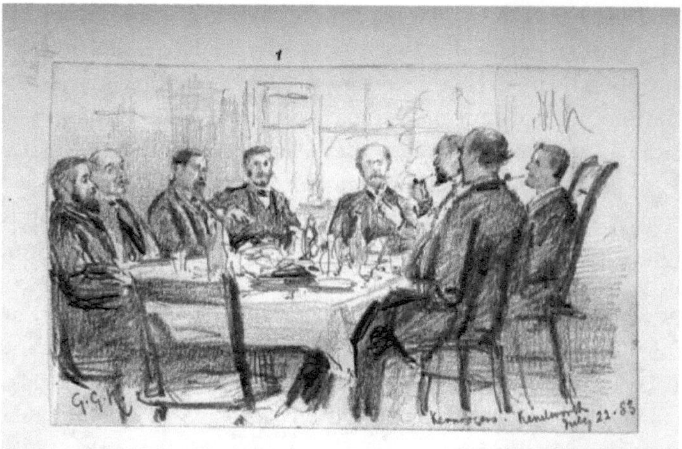

Fig. 2. *A George G. Kilburne's sketch of the Kernoozer's Club dinner after the visit to Kenilworth Castle, 22 July, 1883.* London, British Library, Ms., Add. 40.682, f. XXII.

On 8 July 1884, the meeting was again at the home of de Cosson, who presented his collection in a new arrangement. He wanted to show the members new acquisitions he had made in Paris. The summer excursion would be made to Wilton House, Lord Pembroke's castle near Salisbury, which housed a rich collection of armour. Among them were those of the Dukes of Montmorency and Bourbon captured by the Earl of Pembroke at the Battle of San Quentin along with the Earl's own armour worn in battle. The draft of the request for the visit was drawn up by de Cosson and sent to Wilton House[21]. However, they still had not received an answer at the time of the meeting and the subject of the summer excursion was postponed. In this meeting, Seymour Lucas was elected Vice-President of the Club[22].

On 1 July 1885, the Club was invited to participate in the Cutlers' Company's Industrial Exhibition of fine metalwork to be held in Sheffield. The best of Lord Londesborough's collection and those of various Kernoozers such as Linton, Lucas, Kilburne and the architect Sydney Lee were also sent[23]. The Club's Minute Book includes the invitation dated 30 June 1885, for the inauguration, which

21 BL, Add 40.678: 78v.

22 BL, Add. 40.678: 80v.

23 Some photographs in BL, Add. ms 40.682. *Sheffield Daily Telegraph*, Saturday, 18 July 1885.

would take place on 1 July[24]. Perhaps the Exhibition served as an excuse for another very particular visit that took place to the Wallace Collection in London. On 20 July 1885, Juan Crooke—Count of Valencia de Don Juan— the Count of Villagonzalo[25], the Count of Estrada[26], and Guillermo de Osma signed in the Wallace Guest Book. Three days after Osma and the Count of Villagonzalo returned to Hertford House accompanied by María Francisca Palafox Portocarrero y Kirkpatric, Duchess of Alba, wife of Jacobo Fitz-James Stuart Ventimiglia, 15th Duke of Alba (died in 1881)[27]. Visits to such an important collection would serve to increase Crooke and Osma's interest in collecting, especially arms and armour. Nevertheless, it shows that Crooke and Osma were together in London in 1885. Undoubtedly, contacts had to be made with the English collecting and, surely, with the Kernoozers.

Some of the distinguished visitors to the Wallace in 1885 met the Kernoozers three years later. On 31 January 1888, they were assembled to the home of Edward Bellamy. At the proposal of Lucas, Juan Crooke, Guillermo de Osma and José Argaiz were elected Corresponding Honorary Members of the Club. The proposal was approved by unanimity. According to the 3rd article of the Club, the appointment of Corresponding Members was the procedure to include collectors who lived twenty miles from London or abroad, as in the case of the three Spanish gentlemen, who therefore could not attend meetings regularly. They were exempt from paying membership and subscription fees but had the right to attend their meetings when they were in London.

Crooke lived at 38 Carrera de San Jerónimo, Madrid. It is particularly significant that Osma's address is listed a few numbers down the same street, No. 45. Crooke may have arranged for the accommodation of his then-young friend on his return from England. It is also revealing that the three Spaniards were proposed and accepted to be Corresponding Members at the same time. Thus, it can be assumed that they came from the same source. However, in light of the existing documentation, it is not possible to venture whether the link could be Lucas or de Cosson. Lucas stay in Madrid in 1880 probably included a visit to the Real Armería, and there is no reason to suppose that he did not repeat it in

24 BL, Add. 40.678: 97–98.

25 Mariano Maldonado y Dávalos, VII Count of Villagonzalo, collector, senator, and later Spanish ambassador to Czar Nicholas II.

26 Mariano Fernández de Henestrosa y Ortíz de Mioño, IV Count of Estrada, Vice-President of the Spanish Senate, gentleman of the chamber of King Alphonse XIII, Spanish ambassador to Japan, and Mayor of Madrid (1900–1901).

27 Visitors' Book, Wallace Collection, AR2/44, p. 96, July 20 and 23, 1885.

1881, when Crooke chaired the Spanish Art Commission at the V&A and when the Kernoozer's Club was founded. De Cosson was its President from 1881 and Lucas Vice-President from 1884, so both were in the best position to favor the Spanish collectors. The third Spaniard, José Argaiz, was also a collector from Madrid. He was a regular at Carrera de San Jerónimo and the Liria Palace. His collection was already famous in 1888, although its origins must go back to the beginning of the decade, since the IVDJ keeps a letter dated 18 January 1883, on the purchase at auction of the widow Duchess of Osuna and the Infantado for a value of 25,800 francs in the name of Argaiz[28]. Argaiz was *Chargé d'Affaires* of the Spanish Embassy in London, and its first secretary between 1870 and 1874, so in addition to his status as a collector he was able to add a link through the embassy. He passed away on 6 July 1888. After his death, part of his collection was transferred to José Estruch, which passed entirely in 1903 to M. Georges Pauilhac. At the same session, on the initiative of H. Spiller, it was also proposed that Frederick Pollock, Professor at Oxford and author of the article "*Sword*" in the Encyclopedia Britannica, would be admitted to the Club. In the session of 27 March of the same year (1888) his brother, Sir Walter Pollock, was admitted at the request of Egerton Castle seconded by de Cosson, who in turn proposed the admission of the Parisian collector Jean-Jacques Reubell, who was also in contact with de Cosson and it would be since then with Osma, to fill the vacancy left by Argaiz[29]. From then on, the following members of the Kernoozer's Club gave way to a type of collector closer to swords, probably due to their closeness to the Oxford Fencing Club.

Minute Books II, III and IV continue to reflect the Club's meetings, but the names of the members change over time. In November 1895 John G. Waller is listed as President[30], while de Cosson is listed as "Past President", as he had gone to live in French Brittany at the Château de la Foresterie[31]. The sessions

---

28 Instituto Valencia de Don Juan (hereafter IVDJ), Caja Verde, Armas, Hierros.

29 Phyrr, 2012: 188. Santos Quer, 2021: 30. Reubell was the mediator in the famous purchase of the Fortuny tile. After Fortuny's death in 1874, his collection was auctioned at the Hôtel Drouot in Paris on April 26, 1875. A few years later it passed into the collection of the Duke of Dino until his bankruptcy forced him to sell it. In May 1894 it was auctioned again at the Hôtel Drouot. Osma did not want to personally attend the auction so as not to raise the price of the tile, but he gave Reubell precise instructions to bid for it against other collectors who wanted to take it to Berlin and the V&A.

30 BL, Add. 40.679:49.

31 Probably from 1893 to 1898. A printed list of members insert into the Minute Book No. 2 quotes the Château as his residence. BL, Add. 40.679:35. A letter sent to Lucas

held between 1905 and 1914 reflect the new spirit of the Club with which it broadened its horizons to other types of collecting, especially clothing, furniture, footwear and household utensils[32]. This spirit fitted very well with the collecting of Guillermo de Osma.

JUAN BAUTISTA CROOKE Y NAVARROT

A figure of capital importance for the studies of arms and armour in Spain was Juan Bautista Crooke y Navarrot. On 27 September 1855, he married Adela María del Pilar de Guzmán y Caballero (1827–1901), daughter of the 14th Count of Oñate. Along with some works of art, she received the title of 23rd Countess of Valencia de Don Juan on 17 December 1850, while Crooke began to build his own collection according to his personal tastes[33], being arms and armour one of his favourite sections. From 1852 he received positions in the legations of London, Paris, The Hague and Lisbon. In 1865 he was in London, as evidenced by a daguerreo type made by William Edward Kilburn, a photographer whose studio was at 234 Regent Street[34]. In 1881 he chaired the committee of the Spanish and Portuguese Ornamental Art held at the V&A in South Kensington, which was continued at the *Exposição Retrospective de Arte Ornamental Portuguesa e Hespanhola* held in Lisbon, whose committee included Crooke. The exhibition presented 1,576 pieces of all kinds, but the loans of arms and armour from Spain came exclusively from the Real Armería and from the personal collections of Crooke and M. Rico y Sinobas. In 1881 he was in charge of the arrangement of the Real Armería (an equivalent position in England would be the later appointment of Sir Guy Laking as Keeper of the Royal Armouries at Windsor)[35]. Crooke remained in the post until his death, combining the work at the Real Armería with the formation of his own collection. In July 1884, after the fire at the Real Armería, Crooke made an inventory and began to organise the collection.

Crooke's friendship with de Cosson and probably also with Seymour Lucas was of paramount importance. The beginning of the relationship, when, how and why are still unknown, but first contact can be ventured through a Parisian

congratulating him on his appointment as an Associate of the Royal Academy (A.R.A.) is dated 8 February 1898, at La Foresterie. David J. de M. Williams collection.

32 BL, Add 40.680: 18.

33 Nebreda Martín, 2017: 90.

34 Sánchez Vigil *et al.*, 2015: 72–73.

35 López Martín, 2011: 198.

dealer. In 1880 and 1881 Lucas travelled to Madrid and a meeting with Crooke at the Real Armería would not have been a surprise. It was Lucas who proposed the entry of Crooke and Osma as Corresponding Members of the Kernoozer's Club. Over the years Crooke came to buy pieces from de Cosson's collection. An undated handwritten note from J. Mª. Florit, Crooke's successor at the Real Armería, lists the provenance of thirty-two pieces in the IVDJ collection, five of which were from Pyrcroft House[36]. On other occasions, it is not possible to determine the provenance of the objects. This is the case of the helmet of Sultan Mehmed II, The Conqueror (1430–1481), which might have been taken along with other items at the mid-nineteenth century from the arsenal of St. Irene, Istanbul, and then probably acquired by Crooke rather than Osma[37]. Crooke's and de Cosson's friendship and their common interest in collecting arms and armour promoted the organisation of meetings in Paris each autumn to exchange impressions and study notes. The meetings began around 1885, when Crooke continued his work in Madrid and began to use the title of "Director of the Real Armería"[38], Osma was transferred to the Spanish embassy in London, and when they went together to the Wallace Collection. Both Crooke and de Cosson had begun to collect hallmarks and gunsmith's marks independently, so at their annual meetings, they exchanged notes and provided each other with the information obtained. In addition to learning from each other, they managed to keep a backup copy of the records in case one of them lost them by accident. It was Crooke who proposed this method of study. In the summer of 1890 they went on a European tour to visit arms and armour collections in Central Europe, ending in Vienna with the company of Wendelin Böeheim, the director of the imperial collections[39]. During the trip they had the opportunity to examine a large number of pieces, such as Toledo, Italian and German swords; Gothic armour and equipment from the Kolman and Missaglia workshops, leaving

36 They are thirty-two objects in twenty-four entries. IVDJ, Caja Verde, Armas, Hierros. At least one fifteenth-century sword (No. 58 in Florit's catalogue, p.54), was among the pieces that Cosson took with him to the Château de la Foresterie and is photographed among the *Album of Drawings, Photographs, and Printed Ephemera relating to Baron Charles Alexander de Cosson*, purchased by the Metropolitan Museum from the Peter Parsons Collection. It does not appear in the auction catalogs of the De Cosson collection of 1890 and 1893. An early sixteenth-century headpiece (No. 16 in Florit's, p.20), is also probably from the Cosson collection.

37 Florit, 1927: 16. Girouard, 1981: 51. Tavares, 2013: 183–187.

38 Cabello, 2014: 266.

39 Cosson, 1891: 121.

hints of their scholarly as they described pieces from their rivals, the Negroli, for example, rondaches kept at the Real Armería "worthy of the chisel of a Greek sculptor". On their journey, they were received by many directors and museum curators, scheduled meetings with archaeologists, examined manuscripts, measured plaques and drew helmets. According to de Cosson, the "instructed and indefatigable" director of the Real Armería continually identified pieces or parts of armour scattered throughout Europe. In 1892, Crooke lent pieces for the *Exposición Histórico Europea* commemorating the 400 years of the discovery of America for which he had the assistance of his son-in-law, Osma[40].It was at 38 of Carrera de San Jerónimo where Crooke began to organise, probably in the mid-1890s, the famous "Tertulias domingueras" (Sunday gatherings) similar to the Kernoozers meetings, from where the idea must have arisen. Accompanying Crooke and Osma, his son-in-law since 1 May 1888, the most renowned Spanish intellectuals and scholars attended. The participants were immortalised in a sketchbook made by Florit covering the years between 1896 and 1904[41]. In these gatherings, the roast beef, beer and whiskey offered at Kernoozers meetings were exchanged for something more traditional such as *azucarillos* (a typical sweet from Madrid) and chocolate. After Adela's death in 1896, Crooke took the title of Conde Viudo de Valencia de Don Juan. In 1898 the Real Armería opened after the fire and Crooke published his life's work, the *Catálogo Histórico-Descriptivo de la Real Armería de Madrid*, a landmark on the subject[42]. At the 1900 Paris Universal Exhibition, he acted as curator of the Royal Collection, and his services were awarded the French *Légion d'honneur*. He was awarded other distinctions such as honorary member of the Royal Academy of Archeology of Belgium and, probably, honorary member of the Society of Antiquaries of London, since he was not a full member as is often cited[43]. He passed away on 2 May 1904.

---

40 Santos Quer, 2021: 23.

41 *Recuerdo de la tertulia dominguera del conde de Valencia de Don Juan: Arqueólogos, Anticuarios y Bibliófilos más o menos chiflados* (Madrid, 1904). IVDJ, Biblioteca, 7–3.

42 Until an in-depth study is carried out, the echo of his malpractice in the reorganisation of the Real Armería should not be attributed to Crooke, but perhaps to the gunsmith and goldsmith Eusebio Zuloaga, responsible for sending a series of shipments of arms and armour from various Spanish noble armouries—including the Real Armería—to merchants throughout Europe in the 1830s. Phyrr, 1999: 266, 276. Tavares, 2013: 69.

43 There is no record of him as a member of the Society of Antiquaries of London.

GUILLERMO JOAQUÍN DE OSMA AND THE INSTITUTO DE VALENCIA DE DON JUAN

On Crooke's death his daughter Adela Crooke—Adelita—inherited his art collection and the title from her mother, becoming 24th Countess of Valencia de Don Juan and her husband became Count consort, though he rarely used the title. Osma's biography has been discussed recently and will not be remarked here more than necessary[44].

Born in Havana on 24 January 1853, Guillermo de Osma y Scull studied High School at the Sorbonne in Paris and went on to Oxford University, being the first Spaniard to enrol after the opening of the University in 1871 to non-Anglican students. In *Alumni Oxonienses* of 1888, he is listed as enrolled at Pembroke College on 27 May 1871, at the age of eighteen and graduated in 1874 in what would be today Modern History, though the degree in 1879 was simply M.A. (*Magister Artium*)[45]. At the IVDJ there are several photographs of Osma's years at the University, of his classmates and the Isis boat races, their crews and views of the College. Photographs of his bedroom show his study table, with the Pembroke pennant at one end and two crossed foils at the other. The study table, the pennant, and other personal belongings he took with him to Madrid are today in his office at the IVDJ just as he left them. There are also photographs of the main hall of the exclusive Vincent's Club, founded in 1863, closely linked to Pembroke College Boat Club and the University rowing club, its oldest sports club. Rowing photographs are the most numerous. A couple of "Scratch" tankards that he kept in his office, reveal him occupying the cox position. These are commemorative memories, not trophies, for casual races made up of experienced and non-experienced rowers in a single-college race; that is, between members of the same college, in this case Pembroke. They were delivered to each participant, who received an identical one. One photograph is especially relevant because it shows the 1872 *Torpids* rowing team, a category initiated in 1841. Osma features again as cox along with his eight fellow rowers, all with their weights in stones and pounds. Osma's weight recorded in the photograph (8.5 stones) made him the perfect size to be a cox, a key place that governs the boat, motivating his crew and setting the pace of their strokes, as well as steering the boat. Coxes were required to be as light as possible to avoid

44 Pérez de Arcos, 2020. Santos Quer, 2021.

45 "*Osma, Guillermo Joaquin de, Is. John Ignatius, of Havannah, Cuba, arm. Pembroke Coll., matric. 27 May, 1871, aged 18; B.A. 1874, M.A. 1879, attached to the Spanish Embassy in Paris*". Foster, 1891: 1047.

adding extra weight to the boat and making it harder for the rowers. Osma perfectly fitted at Pembroke, otherwise, he would not have played such a pivotal role in boat racing. In other photographs, he appears with the members of his Common Room of 1873 and 1874. Before First World War, CR was organised as a London gentlemen's club. It was an elected society, so not all of its members were university students like Osma. For this reason, some of his companions, winners of other specialties, seem to be older [Fig. 3].

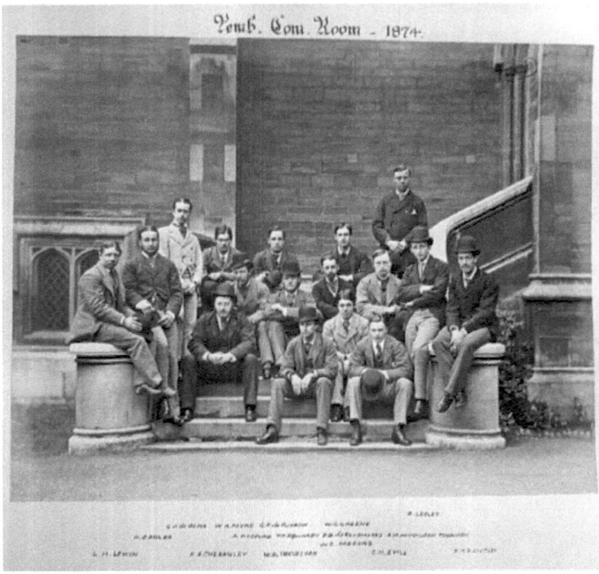

Fig. 3. *Guillermo Joaquín de Osma with his companions from the 1874 Common Room, Pembroke College, Oxford University*. Madrid, Instituto de Valencia de Don Juan, 02-23-0021-Pembroke College IMG_6053.

The philosophy of English imperialism was essentially elitist. The British people saw their elites of gentlemen as leaders who stood up for what they considered their genuine values. Within their ranks, British gentlemen tended to create small "brotherhoods" conscious of their traditions or blindly believing in their superiority. This perception had been fed since the end of the eighteenth century with a glorious past extolled by neo-Gothic and medievalism, and public schools, along with Oxford and Cambridge, were the training centres where values were transmitted to the young generations. The university colleges were in charge of outlining the character of British gentlemen spread throughout the world. By the end of the century, a gentleman had to be chivalrous in a wide variety of ways. A gentleman could lead because he was morally superior and it was his moral aura and glamour, which derived from the revival of chivalrous values and ideals during the Victorian era, that provided his chivalry as a goal to be pursued with the same ardor with which the Holy Grail would be sought[46]. The clubs that arose in this social context and that were dedicated to delving into a specific subject, such as the Kernoozers, were another way of transmitting values, and Osma had to learn, by simple absorption, everything that he experienced in one of the main centres of intellectual formation, such as Oxford, but also Cambridge, the Society of Antiquaries of London or the circles of collectors that moved around the Kernoozers. Keeping up with his studies, being part of the rowing team, or participating in conversations with other students who were preparing to occupy key positions in the administration of the Empire, had to deeply mark his character. Some of his closest friends reached those influential positions. Being a gentleman not only implied adopting his ethical and moral code but also looking and even speaking like one[47]. Osma's years at Oxford marked him deeply and shaped his character in all the aforementioned ways. What he learned from the College influenced him throughout his career as a man, as a gentleman and as a collector.

In 1877 Osma started working for the Spanish diplomatic service and lived in Paris attached to the Spanish embassy. At the Paris World's Fair of 1878, Osma received the First Class Cross for Naval Merit. Perhaps it was the occasion to meet Crooke for the first time. A receipt dated 23 May 1881, for a collection of 103 arms and armour that José María de Vera sold to Osma demonstrates his early inclinations for this type of collecting. At the beginning of 1883, Osma and Crooke already knew each other very well, as evidenced by four buffes and two shaffrons from the sixteenth century that Crooke bought at the Osuna auction as a gift to Osma. Crooke gave him the news from Madrid in a letter dated 13 January, telling him how, while Osma was absent from the capital, the entire Osuna armoury had been sold. Crooke was commissioned to buy some pieces for the Real Armería[48] and informed him that, given the possibility that he would not be able to spend the carnival with them in Madrid, he could send him the

---

46 Girouard, 1981: 233, 262.

47 Girouard, 1981: 260. "*Education in Victorian Britain*". En: <https://www.bl.uk/victorian-britain/articles/education-in-victorian-britain> [March, 29, 2023].

48 Florit, 1927: 33, 44 (nos. 36, 37 and 50), with receipt of the purchase of 9 January 1883, and giving notice of the purchase in a letter dated 13 January of the same year. IVDJ, Caja Verde, Armas, Hierros.

pieces he had purchased. Crooke uses distinguished forms of courtesy befitting two people who know and associate but have not yet established family ties. It has been proposed that after Crooke's encounter with Osma, Adelita met the man who would be her husband, possibly at a costume ball, horse race, a carnival dancing held in Madrid or, perhaps, in a *sarao* (a party with dance and music) organised close to 31 January, in which both were present. Perhaps the courtship began shortly after, since in 25 February 1884, they coincided at another carnival ball[49].

In 1885 Osma went on to serve as secretary of the Spanish embassy in London and it is known from the Wallace Collection Guest Book that Osma and Crooke paid a visit together to Hertford House on 20 July. They joined the Kernoozer's Club on 31 January 1888, Crooke being a mature man of fifty-nine years of age and Osma a youth of twenty-eight. It is possible to venture that Osma was admitted at Crooke's request and that it was thanks to his friendship with Crooke that Osma really began to court Adelita. It is also possible that it was at the Sunday gatherings held at 38 Carrera de San Jerónimo that the couple saw each other for the first time. The Kernoozer's entry of 31 January 1888, indicates Osma's address at No. 45 on the same street. It is clear that Osma had established his bachelor residence next to Crooke's prior to his wedding to Adelita, which occurred on 1 May 1888. After the wedding the couple lived at 16, Juan de Mena Street. Osma must have been an excellent match for young Adelita. Under Crooke's tutelage, Osma was also able to strengthen a desire for collecting supported by his studies in History. For the next few years, Osma served as Crooke's *alter ego* when he was in Paris, taking over the collection and showing it off to distinguished visitors.

Osma and Adelita acquired a piece of land at 43 Fortuny Street, Madrid, and commissioned the architect Enrique Fort to build a Neo-Mudejar-style building, which was completed in 1893. It is possible that the Spanish pavilion at the 1878 Paris Universal Exhibition, built in the same style, served as inspiration for the mansion that would house and display their art collection. It was a style that moved away from the French neoclassical canons, in vogue at the beginning of the twentieth century. When Crooke died in 1904, Adelita's parents' collection was moved to the new building, but faced with the need to prepare it to house the inherited pieces along with their owns, it was decided to enlarge the building. The remodelling took place between 1914 and 1916, but now the neo-Gothic style was chosen. Inside it incorporated a stone balustrade staircase in the Elizabethan Gothic style. It was the style that Osma had witnessed in London and Oxford

49 Nebreda Martín, 2019: 30. Salvador Benítez, 2020: 60.

since his youth. The neo-Gothic enlargement of the palace was far from the conventional style in force in Madrid and affirmed both Osma's personality and the couple's artistic tastes. On 15 March 1916, the IVDJ was formally established. However, its project was already outlined before 1905, as the deed of its foundation reveals. Together with the Counts, the Board of Trustees was made up of Antonio Maura y Montaner, President on several occasions of the Spanish Council of Ministers and President of the Royal Spanish Academy; Miguel Asín y Palacios, Professor of Arabic at Madrid Central University; Jacobo Fitz-James Stuart Falcó Portocarrero y Osorio, Duke of Alba; Archer Milton Huntington, founder and President of the Hispanic Society of America and Sir Charles Hercules Read, Curator of Medieval British Antiquities and Ethnography at the British Museum. Huntington restored Osma's honour by making him a member of his Society in 1918, the year of Adelita's death. Read was also President of the Society of Antiquaries to which many of Kernoozers belonged but not Osma, as he was not a Full Member, like Crooke, but perhaps just an Honorary Member. The IVDJ has a copy of a portrait of Read painted by Lucas in 1912, which Read dedicated to Osma[50].

Article 7 of the IVDJ's statutes, still in force, requires it to report its activities at least once a year to the Chancellor of the University of Oxford. If under exceptional circumstances Spain could not guarantee the conservation of the Institute, the University should take possession of the Institute's funds and transfer the entire collection to the University. This article, however, is no longer in force due to the new Spanish legislation. The founder's desire to maintain the Institute as an independent research centre linked to the University of Oxford was clear. The Osma Scholarship, created in July 1920 with the aim of promoting studies dedicated to Spanish culture and history is another link with Oxford. During the Spanish Civil War (1936–1939) the English flag flew in the IVDJ avoiding seizures by the Republican government[51].

Since its inception, the Institute's main objective has been to promote research into its collections and broaden knowledge of Spanish culture. The same desire to investigate, foster relationships with other centres, recover ancient combat

---

50 The British Museum, cat. 2002,0126.21. Portrait of Sir Charles Hercules Read. Red and black chalk on paper, 32.5 cm × 26.8 cm. It was exhibited at the Royal Academy in 1913. It was acquired in 2002. The IVDJ copy is dedicated in the lower right corner: *G. J. de Osma / from his friend / C. H. Read.*

51 Pérez de Arcos, 2020: 294.

techniques and learn from material objects embodied the spirit of the Kernoozers and Baron de Cosson [Fig. 4].

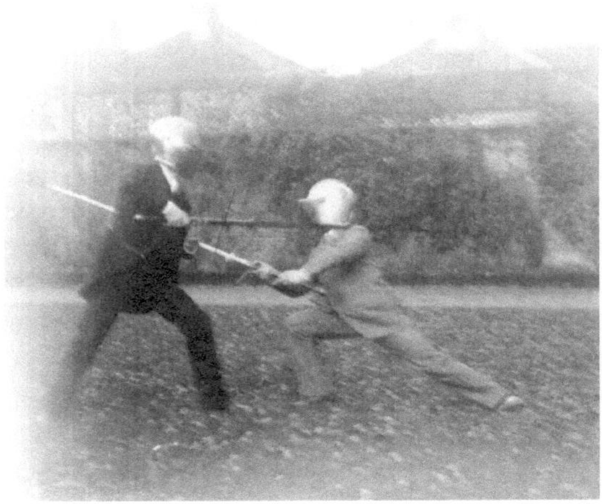

Fig. 4. *Two men engaged in a two-handed sword fight. Pyrcroft House.* Mid-1880s. Madrid, Instituto Valencia de Don Juan, 07-28-0097-Pyrcroft House-IMG_7364. Baron de Cosson in dark suit. The swords can be seen in the *Album of Drawings, Photographs and Ephemera Engravings Relating to Baron Charles Alexander de Cosson*, acquired by The Metropolitan Museum from the Peter Parsons Collection; also in some of the scenes at Pyrcroft House (fol. 202 and 203), at the Château de la Foresterie (fol. 77); and in a view from the library on Via Ghibelina (fol. 215).

Study trips carried out by Osma to foreign collections and museums, or the exchange of opinions and notes with other collectors and museum directors were nothing but the continuation of what was developed by Crooke, de Cosson, and other prestigious nineteenth century connoisseurs. Osma did not agree with giving free access to art collections. From his point of view, not all people were capable of appreciating them, only scholars. This criterion is still in use at the IVDJ and prevents free access to the public. Kernoozer's ethos regarding scholarly and researcher access to specialised collections is once again obvious. The meetings started by Crooke and followed by Osma were just a continuation of the spirited meetings of the Kernoozers. In short, the creation of a private museum was nothing more than the legal consolidation of a private collection as Osma knew it from his formative years in London. Osma must have taken all this as an example when thinking about the Institute. The patriotic feeling was the same driving force for others collectors, such as Richard Wallace or Frederick Stibbert who also legally formalised their collections. Osma knew how, on Wallace's death in 1897, his wife made his dream come true for the collection to open its doors as a museum in London in 1900. He also had to know news about the Florentine collection formed by Stibbert and it can be ventured that he would also have news of what happened to de Cosson in the last stage of his life. Osma had to know it since Crooke hadn't died yet. Due to his reputation, de Cosson worked on the study and cataloguing of the collection of the French aristocrat Charles Maurice Camille de Talleyrand-Périgord, Duke of Dino. In 1896 he worked at the Duke's castle in Montmorency, and in February 1897 the Duke proposed him to prepare a catalogue to exhibit the collection at the Paris 1900 World's Fair[52]. The following year de Cosson travelled to Florence. He intended to stay for a couple of months, but remained for the next twenty-seven years at 77 Via Ghibelina. He wrote the catalogue but it was a work more intended for commercial sale than for scientific exhibition. On 6 May 1904, he sold the collection for the astronomical sum of $250,000 to the Metropolitan Museum in New York[53]. In Florence, de Cosson learned of the extraordinary art collection formed by Frederick Stibbert, a wealthy Cambridge-educated Anglo-Italian merchant, whom he had probably already met in London in 1889. Upon Stibbert's death on 10 April, 1906, it was revealed that, in order to safeguard its integrity, he left his collection to the English government on the condition that it would remain intact in the place it was formed. If England rejected the legacy, the property would pass to the city of Florence. It was Guy Laking who pressured and convinced Stibbert to make this patriotic legacy, although England finally refused the inheritance. With other collections such as Seymour Lucas's textiles (bought by the V&A in 1911) Laking would be more successful. Both the Stibbert collection and the IVDJ were patriotic legacies and both were made to keep the collections intact, allowing their study by specialists. Osma's patriotic feeling was evident in a visit made by Bashford Dean, curator of Arms and Armour at the Metropolitan Museum of Art, to Madrid around 1920. When Dean bought a superb salade in an antique shop opposite the hotel where he was staying, Osma, who had gone to see him to his room (which shows that there was a professional friendship between them), made it clear that he did not like the

52 Phyrr, 2012: 202.
53 Cripps-Day, 1925: A.60.

idea of a foreigner coming to the city and acquiring for a trifle (fifty pesetas) an object that should have remained among Spanish collectors[54].

The inclusion of the Duke of Alba in the IVDJ Board of Trustees is important. Educated at Beaumont College, close to King Edward VII and a regular at Balmoral, polo matches and fox hunting, he had the opportunity to establish a close friendship with his cousin, Winston Churchill. Thanks to his links with the English aristocracy, he participated in a unique event, a medieval tournament that would take place among other great festivities that would pay tribute to Shakespeare's tercentenary: The Elizabethan Triumph Tourney, opened on 11 July 1912, in the gardens of Earl's Court in London. There were four judges in the parade, preceded by the knight Sir Guy Laking himself, armed in shining armour.

It is not surprising that Alba gave a touch of distinction to the IVDJ, not only because of his enormous social status but also because of his background, having participated in a tournament that recalled the ancient Middle Ages in the best neo-Gothic style, intimately close to the Neo-Mudejar style that Osma would use in the enlargement of his palace. The Sunday gatherings that had begun at Crooke's home continued at the IVDJ, changing sugars and chocolates for something more English such as tea with pastries and pickle sandwiches. J. Mª. Florit returned to compile a new album of portraits: *Memoria de la tertulia dominical del Excmo. Sr. D. Alejandro Pidal: años 1904–1913*. Finished in 1919, it shows the cast of personalities who passed by on Sundays by Fortuny Street. The same year de Cosson signed the foreword to Laking's *A record of European Armour and Arms through Seven Centuries*. Research work on the collections resulting from the first stage of the IVDJ was the catalogue of arms and armour. In 1921 Osma entrusted it to Florit. As previously seen, it is not surprising that at the beginning of the nineteenth century, a painter of historic themes was interested in a collection of arms and armour. Given that he had to know the trends in vogue in the presentation of European art collections, being Crooke's successor at the Real Armería, and that he attended the IVDJ, it could be assumed that at least part of that knowledge was obtained from certain collections of arms and armour that were consolidated on the continent, for example, the Wallace and shortly after the Stibbert. Florit's sudden death in 1924 forced the *Catálogo de las armas del Instituto de Valencia de Don Juan* to be published in 1927. Simple

54 Boorman, 1994: 27. I am grateful to Jonathan Tavares, curator at the Art Institute of Chicago, for calling this passage to my attention and providing the bibliographical reference.

in its preparation, the catalogue does not offer great details for the origin of the objects, although it does allow us to glimpse that a good part of them came from the inheritance of the Count's house of Oñate, in the auction of the ducal house of Osuna, and in the direct Crooke's or Osma's purchases from different private collectors, including Baron de Cosson. Osma could not see the catalogue printed. He died unexpectedly on 7 February 1922; Cosson on 8 February 1929.

BIBLIOGRAPHY

Blair, Claude (1991): "The Meyrick Society 1890–1990". In: *The Meyrick Society Bulletin*, pp. 1–35.

Boorman, Dean K. (1994): "A personal reminiscence of Bashford Dean, and his unpublished talk, circa 1920, The Hobby of Collecting Ancient Armor". In: *The American Society of Arms Collectors, Bulletin*, 70, pp. 16–23.

Cabello Carro, Paz (2014): "Del Patrimonio de la Corona hasta el actual Patrimonio Nacional (1819–1950)". In: *Patrimonio Cultural y Derecho*, 18, pp. 249–285.

Cosson, Charles Alexander de, Baron de (1891): "Arsenals and Armouries in Southern Germany and Austria". In: *The Archaeological Journal*, 48, pp. 117–136.

Cripps-Day, Francis Henry (1925): *A Record of Armour Sales 1881–1924*. London, G. Bell and Sons.

Florit y Arizcun, José María (1904): *Recuerdo de la Tertulia Dominguera del Conde de Valencia de Don Juan. Arqueólogos, Anticuarios y Bibliófilos más o menos chiflados*. Madrid, Fototipia Hauser y Menet.

Florit y Arizcun, José María (1919): *Recuerdo de la tertulia dominguera del Excmo. Sr. D. Alejandro Pidal: años 1904 - 1913*. Madrid, Fototipia Hauser y Menet.

Florit y Arizcun, José María (1927): *Catálogo de las armas del Instituto de Valencia de Don Juan*. Madrid, Editorial Reus.

Foster, Joseph (1891): *Alumni Oronienses. The members of the University of Oxford, 1715–1886, in 4 vols*. Vol. III. London, Parker and Co.

Girouard, Mark (1981): *The Return to Camelot. Chivalry and the English Gentleman*. London, Yale University Press.

Gosse, Edmund (1888): "Seymour Lucas, A.R.A.". In: *The Magazine of Art*, Jan. 1888, pp. 1–6.

Hardie, Martin (1908): *John Pettie*. London, Adam and Charles Black.

LaRocca, Donald, J. (2009): "A William Burges sketchbook of Helmets, ca. 1880, Bashford Dean and G. F. Lawrence, and Renaissance leatherwork from London". In: *The Journal of The Arms & Armour Society*, XIX-5, pp. 197–207.

López Martín, F. J. (2011): *Esculturas para la guerra. La creación y evolución de la artillería hasta el siglo XVII*. Madrid, CSIC-Ministerio de Defensa.

Nebreda Martín, Lara (2017): *Documentación sobre arte y arqueología en el Instituto de Valencia de Don Juan.Análisis de la colección andalusí a través de sus documentos.*PhD. Dissertation, Madrid, Universidad Complutense.

Pérez de Arcos, Marina (2020): "Individual, Institution, and Impact: The Untold History of the de Osma Studentship in Spanish Studies at Oxford". In: *Hispanic Research Journal*, 21, 3, pp. 276–320.

Petrov, Julia (2008): "The habit of their age. English genre painters, dress collecting, and museums, 1910–1914". In: *Journal of the History of Collections*, pp. 1–15.

Phyrr, Stuart W. (1999): "'Ancient armour and arms recently received from Spain' Eusebio Zuloaga, Henry Lepage, and the Real Armería in Madrid". In: *Gladius*, 19, pp. 261–290.

Phyrr, Stuart W. (2012): "Armour for America: The Duc de Dino Collection". In: *Metropolitan Museum Journal*, 47, pp. 183–230.

Sánchez Vigil, Juan Miguel/Olivera Zaldua, María/Salvador Benítez, Antonia (2015): *Los daguerrotipos del Instituto Valencia de Don Juan*. Madrid, Unniversidad Complutense.

Salvador Benítez, Antonia (2020): "*Mujeres a bordo*". In: Sánchez Vigil, Juan Miguel (ed.): *El viaje del Thistle Diarios de a bordo, 1897. Fotografías de Adela Crooke*. Madrid, Universidad Complutense: pp. 31–70.

Santos Quer, María Ángeles (2021): "Adela Crooke y Guillermo de Osma: Coleccionistas, mecenas e intelectuales". In: *Grupo de Investigación Fotodoc / Universidad Complutense de Madrid* (ed.): *La fotografía en el Instituto de Valencia de Don Juan. Guía-Inventario*. Madrid, Universidad Complutense: pp. 15–33.

Spielmann, M.H. (1889): "Glimpses of Artist-Life VII. The Kernoozer's Club". In: *The Magazine of Art*, 12, pp. 362–368.

Tavares, Jonathan James (2013): *Whence the splendour of our ancient Baronial Halls may be revived. Samuel Luke Pratt And the Arms & Armour Trade in Victorian Britain*. PhD. Dissertation., New York, Bard Graduate Center, Decorative Arts, Design History, Material Culture.

Wainwright, Clive (1989): *The Romantic Interior. The British Collector at Home, 1750–1850*. London, Yale University Press.

DOCUMENTARY APPENDIX

ALBA, DUQUES DE

Doc. 1.
1531, november 8.
Post-mortem inventory of Fadrique de Toledo, 2nd Duke of Alba.
Archivo Ducal de Alba, Alba, C. 23, nº. 29.

[unfoliated, the digitized folio number is entered]
(fol. r)(nº. 4)

En la villa de Alba de Tormes a ocho días del mes de nov^{e} año del nascimiº de nro Señor Jhuxº de mill e quiºs e treinta e un años ante el señor bachiller Diego Xi^{e}s corrºr en la dha villa e por ante my Pero Gonçález de Alba escnº e notº pu^{co} del número de la dha villa e tºs yuso escriptos pareció presente Juan de Portillo vº de esta villa de Alba en nombre e como pro^{or} q se mostró ser del muy Ill^{e} Señor don Hernando Álvarez de Toledo duq de Alba por virtud del poder q mostró e presentó de Su Señoría su tenor del qual es este q se sigue

Conoçida cosa sea a todos los que la presente vieren cómo yo don Hernando Álvarez de Toledo duque de Alba y marqués de Coria conde de Salvatierra señor de Valdecorneja en la ciudad de Huesca e Villa de Castilleja etc digo e conozco que doy e cargo todo my poder conplido libre e llenero e bastante a vos Juan de Portillo mi criado vº de la mi villa de Alba que soys ausente bien ansí como si fuese de presente para en todos mis pleitos e causas e negocios movidos e por mover podades pareçer ante sus Mag^{es} e ante los señores presidente e oidores de su Real audiencia de Valladolid e ante otros qualesquier juezes e justiçias e ante ellos e cada uno de ellos podades hazer e hagades ansí ende mandando como en defendiendo todos los autos e pedimientos que a mí convengan de hazer e responder a los en contrario hechos e que se hizieren e para que por mí y en mi ánima podáis jurar qualesquier juramentos ansí de calupnia (sic) como deçisorio e presentar qualesquier testigos e probanças e escrituras e pedir sentencia e sentencias ansí ynterlocativas como difinitivas e apelar e suplicar de ellas e las seguir e fenecer e acabar en los dhos grados e para que podáis por mi e en my nombre sostituyr un procurador dos e más con el mesmo o limytado poder e los revocar quando quisieredes e por bien [...] e quand cumplido e bastante poder yo tengo para de lo suso dho e para cada una cosa e parte de ello otro [...] cumplido (fol. v)(nº. 5) [...]

E ansí presentado el dho poder q de suso va incorporado por e dho Juan de Portillo en nombre de Su Sª dixo q por quanto el Illmo señor don Fadrique de

Toledo duque de Alba que haya Santa Gloria es fallecido e pasado de esta p^{e} vida e falleció a diez e nueve días del mes de ot^{e} p^{o}ximo pasado de este presente año e al tpo dexó çiertos bienes muebles mrs deudas dr^{o}s e açiones e semovientes e otras cosas de los quales dhos bi^{es} el dho Illmo Señor don Hernand Álvarez de T^{o} duq de Alba su parte como su heredero e sucesor e tenedor de los dhos bienes e él en su nombre quiere hazer ynventario por ser relevado (fol. r) de las penas e achaques que el dr^{o} pone por tanto que él en el dho nombre pedía al dho señor corr^{or} le dé lic^{a} para hazer el dho ynventario e lo mandase pregonar segund estilo se su Aud^{a} e reçibiese de él en nombre de Su S^{a} el jur^{o} que de dr^{o} en al caso se requiere el qual está presto de hazer para lo qual el ofiçio de Su md ynplora e lo pidió por testim^{o} testigos Pedro Vizcayno e Pero Fr^{ez} escri^{os} e Juan de Segovia v^{o}s de la dha v^{a}

E el dho señor corr^{o}r visto el dho pedimi^{o} a él fecho por el dho Juan de Portillo en nombre de Su S^{a} dixo q lo oía e que está presto de hazer en ello lo que fuere just^{a} e haciéndolo tomó e recibió juramento en forma debida e de dr^{o} de dho Juan de Portillo en nombre de Su Señoría por Dios nro señor e por Santa María e por las palabras de los Sanctos Evangelios e por una señal de cruz tal como esta + en que puso su mano dcha corporalmente que su illustrísima Señoría del dho señor duque de Alba a él en su nombre como buen xpiano temiente a Dios e a sus mandami^{o}s fazía el dho ynvenar^{o} de los bienes q del dho Illmo señor don Fadriq de Toledo duque de Alba quedaron ansí muebles como semovientes deudas dr^{o}s e açiones e que no dexaba de poner en él bienes ningunos de los que Su S^{a} e él en su nombre supiese e a su noticia vinyere e q si así lo hiziere que Dios nro Señor le ayudase en este mundo al cuerpo e en el otro al ánima donde más ha de durar e lo contrario haciendo que él se lo demandase mal e caramente e a las confusiones del dho juramento el dho Juan de Portillo dixo sí juro e amén e luego el dho señor corregidor hubo por fecho el dho juramento e dio lic^{a} al dho Juan de Portillo en nombre de Su Señoría para hazer el dho ynventario e lo començar a hazer e el dho Juan de Portillo lo pidió por testimonio testigos los sobredhos

(fol. v)(n°. 6)
Inventario

E luego el dho Juan de Portillo en nombre de Su Señoría por virtud de la dha liçençia que el dho señor corr^{o}r le dio començo a hazer el dho ynventario e puso en él un dosel de brocado con un çiervo de carmesí broslado (sic) en el dho dosel lo qual dixo q ponía e puso por ynventario con protestación que dixo q hazía e hizo de lo çerrar e acabar de hazer en el término de dr^{o} e el dho señor corregidor hubo por començado a hazer el dho ynventario e lo mandó pregonar según costumbre de su Audiª e el dho Juan de Portillo en nombre de Su Sª lo pidió por testimº testigos los dhos

Primero Pregón

E después de lo suso dho en la dha villa de Alba en audiª pública este dho día e mes e año suso dho estando el dho señor corrºr en audiencia pública oyendo e librando los pleitos que ante él pendían por ante mí el dho escribº e testigos ynfraescriptos de mandamiº de dhos señor corregidor e de pedimiº de dho Juan de Portillo en nombre de Su Señoría Alexo Mynz pregonero público de esta villa a altas vozes yntelegibles dixo e pregonó sepan todos los acreedores alegatarios fide comisarios y otras personas qualesquier que de dicho deban ser llamadas en cómo el muy Ill^{r} Señor don Hernad Álvarez de Toledo duque de Alba e Juan de Portillo en su nombre e como su procurador quiere hazer ynventario de los bienes e hazienda que del Illmo señor don Fadriq de Toledo duq de Alba quedaron por ende todas las personas que lo quisieren ver fazer vengan a esta audiª a verlo se hazer e dixo primero pregón testigos Pedro Vizcayno e Pº Fernández e Pº Mynez escribºs

Seg^{do} pregón [...]

(fol. r)

Trº pregón [...]

E después de lo suso dho en la dha villa de Alba a quatro días del mes de enero de mill e quiºs e treinta e dos años ante el Señor b^{e}r Diego Ximénez corrºr en la dha villa e por ante my Pero Gºz escribano e testigos yuso escritos pareció presente el dho Juan de Portillo e en nombre de Su Señoría presentó un memorial de Br^{e} de los bi^{es} e hazienda que del dho señor duq quedaron su tenor del qual de verbo a verbum es este que se sigue

Cosas de capilla y libros [...]
(fol. v)(nº 20)

Armas

-En un baúl viejo están las cosas siguientes

(fol. r)
-Un faldaje de armas dorado los bordes fecho a manera de faldaje de sayo con su funda

-dos quixotes con hebillas doradas y guarnición de terciopelo negro con sus fundas
-otros dos quijotes de la mesma manera con sus fundas
-dos grebas con sus escarpes y correas de carmesí raso con su funda
-un quixote con su guarnición de cuero blanco con su funda
-otro quijote con su guarnición de cuero blanco con su funda
Otro baúl
-otra barrata con su babera con su funda de acero
-otro barbote con su funda
-dos braçales con una guarnición de raso morado en sus fundas
-dos alças guarneçidas con una clavazón dorada en sus fundas
-una çelada blanca en su funda con su barbote
-dos manoplas con sus fundas
-otras dos alças con un correal de terciopelo negro
-una testera de caballo dorada toda vieja en su funda
-una babera con su funda de paño blanco
-un çastón (sic) aforrado en cuero
-unas platas de arnés guarneçidas con terciopelo morado con su clavazón dorada con su funda colorada
(fol. v)(nº 21)
-un yelmo para el dho arnés en su funda
-dos braçales y dos alças pa los mesmos braçales todo guarneçido del mesmo terciopelo en sus fundas
-dos quijotes guarneçidos de cuero leonado y en sus fundas
-dos grebas con sus escarpes en sus fundas
-dos manoplas guarneçidas con sus texillos morados y en sus fundas
-otros dos braçales guarneçidos de cuero
-dos guardas guarneçidas de oro a la redonda con unos texillos verdes con sus fundas de paño
-dos cañones guarneçidos de oro alrededor con sus texillos verdes con sus fundas de paño
-dos mosequies (sic) guarneçidos de oro alrededor e sus tejillos verdes con sus fundas de paño
-un alça con una giundadra (?) verde en su funda
-un codo dorado viejo
-un mosequíe aforrado en damasco negro viejo
-otro medio mosequíe aforrado en raso colorado
-un faldaje de coraças viejo guarneçido en terciopelo negro
-tres pieças de un cañón de escamas

-una punta de cobre dorada para un çapato
-un baúl de çaqueta (sic) aforrada en cuero negro
-una testera de caballo dorada vieja
-un almete con una correa de terciopelo negro y su clavazón dorada y en su funda
-medio quixote todo dorado con sus correas de cuero
-quatro musequíes de coraças guarneçidos en terciopelo y en raso negro
(fol. r)
-más otro musequí como los suso dhos guarneçido en terciopelo negro
-otros dos musequíes de coraças guarneçidas en damasco clanco con su clavazón dorada y unas puntas de malla
-otro almete con su barbote con su clavazón dorada en su funda
-dos quixotes guarneçidos todos en terciopelo negro y con su malla aforrado en terciopelo
-un peto de arnés aforrado todo en terciopelo negro
-dos alças con sus texillas coloradas
-una çebrida (sic) con una rosca alrededor
-un yelmo de justa
-otros dos yelmos de justa como el sobredho

Un serón
-En un serón un arnés entero viejo y mohoso
-una testera de caballo vieja
-una guarda vieja
-un barbote viejo
-dos escarçelas viejas
-un monsequí (sic) viejo
-una falda de malla gruesa muy vieja
-quatro frenos de la brida viejos
-tres asientos de penachos viejos
-en un serón viejo están unas platas de arnés viejas con unos quijotes
-unos braçales
-en un serón viejo un arnés entero sin manoplas ni barbote
(fol. v)(nº. 22)
-un arca vieja dos pares de platas muy viejas desguarneçidas con sus faldajes desguarneçidas (sic)
-otras platas viejas como las sobredhas
-dos capacetes viejos todos desguarneçidos
-dos çeladas viejas desguarneçidas
-un almete de torneo muy viejo
-unas manoplas desguarneçidas

Un serón
-en un serón un arnés viejo oriniento sin guarnición de braços
-otro serón viejo unas platas con su çelada y barbote con una greba y dos guardabraços todo viejo mohoso
-en otro serón viejo un arnés muy viejo entero sin manoplas
-ocho coseletes con sus espaldares
-unas platas de arnés sin espaldar con sus escarçelones
-dos piastrones de coseletes
-dos armaduras de cabeça vieja la una rota
-un espaldar de coselete oriniendo
-treynta e tres piastrones q están colgados en la recámara
-veynte e un coseletes que están colgados en la dha recámara
-tres çeladas que están colgadas cabe los dhos coseletes
-un peto con unos texillos negros que es colgado en una funda blanca

Un arca vieja hecha pedaços en q estaban las cosas sigui[es]
-unos çapatos de mula
-dos gozetes de jazerán
(fol. r)
-tres faldas de malla comidos
-un cuello de caballo guarneçido en brocado
-un jubón de malla nueva de faldas sin mangas
-dos gorjales el uno las charnelas de plata aforradas en damasco morado
-quatro pedaços de malla
-otros tres pedaços pa debaxo de las alças
-dos quartas de jubón de malla nueva
-dos pares de calças de malla
-otro cuello de malla pa caballo

Armas de malla
-unos gozetes de gozera nuevas con su guarnición de damasco colorado
-una falda y gozetes de jacerina guarneçidos en brocado raso que eran de R° de Bovadilla
-ocho arandelas en un çestón y onze gozetes y dos roquetes
-un peto con su espaldar de arnés de justa y un barbote del dho arnés una guarda y una pieça doble del dho arnés una manopla
-unas platillas e su espaldar de arnés de justa viejo pequeño y una llave para el dho arnés
-dos conteras de silla de caballo pa justa

Cubiertas de caballos
-Siete coplones pintados de las devisas del duque my señor
-dos pieças de cubiertas delanteras
-treze costelas de cubiertas pintadas de la dha devisa
-dos cubrelomos
(fol. v)(nº 23)
-dos fanqualetes todo dela dha devisa
-dos cabeçalejos para debaxo de las cubiertas

Un çestón grande que tiene las pieças y cosas siguientes
-çinco piastrones de arnés de justa viejos todos
-un espaldar de uno de ellos
-dos barbotes de los dhos arneses
-dos manoplas
Dos conteras de sillas
-una guarda con una pieça de quijote
-una arandela con un roquete
-seys hierros de lança de justa
-dos estribos de justa
-unas llaves con unos hierreçitos pa los arneses

Un caxón de madera
-unas cubiertas pintadas de colorado y oro sin devisa ni letras con unas forladuras (sic) verdes
-dos colchonçillos para debaxo de las cubiertas
-seys lanças con sus hierros con fundas
-dos astas de lança la una blanca buena y la otra verde y colorada sin hierros
-quatro lançones
-otro lançón a manera de dalle
-una alabarda
-otra alabarda dorada
-un borgaz (sic) con un asta negra quebrada
(fol. r)
-un arca de las ballestas ensayalada de paño verde con las cosas siguientes
-dos hierros de lança anchos de monte
-dos aljabas con una dozena de saetas con un cornezillo con yerbas y seys gafas
-otra gafa
-una caxita con çiertos tiros
-una horquilla de hierro
-dos quernos el uno vazío y el otro con yerbas

-nueve cabeças de caxquillos (sic)
-otros ocho casquillos quadrados
-diez fallones
-una caxa de cuchillos de monte con quatro pieças
-otra caxa con un puñal mellado
-otra caxa con un puñal mellado y un cuchillo
-una verga de ballesta
-un çepillo con un enguijuela
-una lima e una raspa
-quatro avanquerdas la una con sus anillos de hierro
-un çinto para hazer cuerdas
-dos pares de calçones de monte pardillos
-un paletoq (sic) de damasco verde
-un papahígo de damasco verde viejo
-dos caperuças de monte una verde y otra pardilla
-un paletoq (sic) pardillo de monte
(fol. v)(nº. 24)
-un caparaçón pardillo de monte con su capirote
-un (sic) aljaba vieja

Un arca en que están espadas y otras cosas
-una espada ropera dorada y negra
-una espada con una guarnición negra gallega
-un cuchillo con guarnición plateada y una correa negra
-otra espada dorada la vayna de cuero
-otra espada dorada la vayna de terciopelo
-un estoque dorado con una vayna de terciopelo morado
-otro estoq (sic) con una vayna negra sin guarnición
-un cuchillo alemán la guarniçión dorada y la vayna de terciopelo con tres cuchillos
-otro estoque ancho la guarnición barnizada con una vayna negra
-otros estoque ancho la guarniçión barnizada con una vayna negra
-otro estoq quadrado con una vayna redonda de gajos
-otro estoque con una vayna de cuero la guarniçión blanca
-otro estoq la guarniçión dorada y una vayna negra
-otro estoque los cabos de hierro y un cuchillo
-un estoque la guarniçión dorada y la vayna guarneçida de latón dorado
-una gomia con una vayna de cuero negra sin paño
-otra gomia con la vayna leonada

(fol. r)
-otra arma disforme vieja
-otro cuchillo alemán los cabos de un madero colorado e tres cuchillos en él
-una çimitarra con una guarniçión de cuero leonada
-otra çimitarra guarneçida la vayna de latón
-un estoq corto ancho dorado con una vayna negra la contera y brocal de plata
-una daga dorada con una vayna blanca e dos cuchillos en ella
-un estoq con una guarnición blanca con vayna con correas de cuero
-una espada de a dos manos el puño de hilo de latón
-una gomia con un cordón con dos borlas azules y el puño de cuerno
-una espada morisca sin puño ni pomo con un tejillo e borlas azules
-un terçiado turquesco guarneçido de plata con sus correas de cuero guarneçido de plata
-otra espada morisca con la guarnición dorada con un texillo y vayna colorada
-otra espada morisca con una guarnición blanca y dorada con su texillo y birlas
-una cruz y pomo de espada dorado
-otra cruz y dos pomos y una contera barnizados negros y otra cruz negra
-unas correas de cuero de espada con su clavaçón de plata dorada
(fol. v)(nº 25)
-dos hierros de açagaya largos
-quatro hierros para lança de monte el uno dorado y despuntado
-tres hierros de lança el uno grande y dos peqños
-unos açiones de carmesí viejos
-dos hierros de lança de armas dorados
-otras dos puntas de hierros de lanças de armas
-una vayna de terciopelo negro
-diez vaynas de espadas y estoques viejas
-un pavés pintado de oro y carmesí con un piñón en medio
-otra pavesina gallega qbrada de carmesí y oro con unos ramos verdes y unas letras en ella
-una rodela dorada y azul guarnezida de terciopelo negro y la clavazón dorada

Un arca ensayalada colorada e verde en que está lo sig[e]
-un coselete con su faldaje bueno del duque my señor
-un armadura de cabeça del mesmo coselete
-dos grebas con su malla del dho coselete
-dos manoplas
-un gorjal
-un barbote

-otro par de grebas con su malla
-un peto de arnés guarneçido de carmesí pelo en su funda

(fol. r)
En un arca blanca la tapadera quebrada está lo siguiente
-unas platas de arnés sin faldamento
-un almete dorado con su arnés de piernas entero dorado
-dos brocales dorados del dho arnés
-sus guardabraços con sus alças
-una pieça doble de yelmo
-está todo este arnés entero e dorado
-una manopla que no es de arnés
-una espada de esgremir y una calva de yelmo
-una espada de armas con una vayna de terciopelo negro y una guarniçión dorada con un paño de hilo de plata
-otra espada de armas con una guarniçión dorada con vayna y çinto de venado y en ella una daga dorada con dos cuchillos
-otra daga con una guarniçión dorada y una vayna de cuero e un cuchillo dorado que tiene los cabos de palo colorado
-dos hierros uno de azagaya con otro de [...]

Un caxón tumbado q tiene
-una testera de caballo guarneçida de plata con un escudo de los conpases
-unas platas de arnés con su malla y alpartazes
-dos grebas con sus escarpes
-dos quixotes
-dos guardabraços
-dos braçales
(fol. v)(nº. 26)
-un yelmo con su malla con su clavaçón dorada
-una pieça doble de dho arnés
-un arnés de piernas con sus escarpes
-están todas estas pieças en sus fundas

En un serón viejo
-está en este serón un arnés entero muy viejo
-una adarga de ante blanco labrado el envés
-diez y seys lanças de armas sanas y una qbrada
-dos pares de quixotes guarneçidos los unos de raso carmesí y los otros de cuero blanco

-una babera con una guarnición de latón dorada
-tres pares de falças guarneçidas de cuero blanco
-doze pares de guarnas (sic) guarneçidas de cuero blanco
-dos musequís con un caño guarneçidos de cuero negro
-un cuello de caballo de escamas con su malla guarneçido de damasco morado
-una tarjeta pintada guaneçida de raso carmesí con su funda
-una ballesta llana con su funda colorada guarneçida la qurueña (sic) de latón
-otra ballesta sin estribo quadrada con una funda blanca
-otra ballesta quadrada en una funda de frisa verde
-una ballesta ancha con su estribo grande con una curueña (sic) blanca
-otra ballesta sin estribo quadrada guarneçida la curueña de latón con una funda de paño pardillo
-Otra ballesta llana gruesa con una raqta por [...] guarneçida de hierro con un estribo grande en funda de cuero negro
-otra ballesta sin estribo quadrada con su qurueña (sic) guarneçida de latón con su funda de frisa verde
-otra ballesta quadrada sin estribo la qurueña peqña
-otra ballesta llana gruesa con su estribo grande guarneçida de hierro con una funda de cuero negra
-otra ballesta qbrada la verga guarneçida la qurueña de latón con un estribo peqño con su cuerda
-una qurueña de ballesta guarneçida de latón con su nuez
-dos vergas de azero quadradas
-otra verga de açero quadrada qbrada
-tres arcos turqueses sin querdas
-çinco cornequies de armar ballestas fuertes
-tres garruchas con sus aparejos para ballestas fuertes
-dos poleados los unos dorados
-quatro avancuerdas de ballestas e dos cuerdas viejas
-un casquete redondo sin funda
-tres bozinas de madera para andar a monte
-un portagrave (sic) de cuero para meter la lança con sus correas
(fol. v)(nº. 27)
-dos pares de guardafones (sic) los unos mayores q los otros en una bolsa de cuero
-un carcax de seda colorada para los arcos turqueses bordados de quero blanco
-un arco ynglés grande con quatro dozenas de pasadores largo guaneçido las enpulgueras de hueso y los pasadores con sus casquillos
-otra dozena de los mesmos pasadores con sus hierros
-otros veynte pasadores sin casquillos para el dho arco

-catorze pasadores un hierro (sic) para los arcos turquescos
-catorze pasadores de ballestas con unas plumas verdes
-una dozena de almazén con un carcax de cuero roto
-çinco dozenas de pasadores de xarilla sin hierros
-dos dozenas de virotes y rallones y paledoques y forquillos de ballestas

Armas y ballestas
-nueve pasadores gruesos de ballesta
-un carcax de hueso labrado el brocal con unas limas aforrado en cuero de puerco
-otro carcax de cuero de puerco grande con su çerradura
-otro carcax de madera aforrado en cuero de puerco más pequeño
(fol. r)
-otro carcax de cuero de puerco grande con su çerradura y llave e cadenilla
-un carcax de cuero redondo con hasta dozena e mª de tiros para los arcos turqueses
-otro carcax de cuero redondo pa tiros largos
-otro carcax de cuero leonado redondo
-otro carcax de cuero redondo
-otro carcax de cuero redondo negro con una dozena de virotes dentro y pasadores y paledoques y forquillas
-otro carcax de cuero con quatro senos y dos bolsas grandes
-veynte espelgas para virotes
-unas escales (sic) de seda negra
-una espada valençiana ropera dorada la guarnición y la vayna de cuero rota

Un arca blanca de pino chica q tiene lo sige
-uno como espurgada (sic) de quatro caños en una caxa de madero con quatro cebaderos
-una soga larga yngrinajada (sic) como sedas avinagrada
-uno como ydulo a manera de cuchillo
-una piedra verde hecha como cabeça de lagarto
-çinco pieças de palo a manera de ydulos
-un colmillo grande de elefante
-dos barquillas la una grande e la otra más pequeña la mayor negras e las otras blancas
-dos calabaças de barro
(fol. v)(nº. 28)
-quatro hierros de las Indias
-uno como dornilejo de madera negro

-noventa e tres quartas de madero
-un relox con su campanilla e todo su adereço
-unas escribanías de latón para de camiº
-un león de fuente con un mastelillo y un ramo de latón
-unas tijeras de espabilar
-unos hierros de almártiga
-un cayrel de caçador [...]
-unas coraças de raso carmesí y blanco con un clavazón dorada viejas y su alportaze de lo mesmo
-un piastrón con unas escarçelas dorado el borde en su funda
-dos aljamas (sic) de cuero de singano (sic) con sus cornezillos
-çinco hierros de lanças de armas
-dos hierros pequeños como garabatos e retorzidos
-diez e seys rexuelos de hasta mª libra de hierro cada uno
-dos armellas de çerrojos grandes

Cosas de paño y seda [...].

Doc. 2.
1540, september 7.
Document of Desiderius Helmschmid on arms made for the Duke of Alba.
Archivo Ducal de Alba, Casa de Alba, C. 111, nº. 8.

In nomine domini Amen Anno Naturitate eiusdem Millesimo quingentesimo quadragesimo, indictione tredecim a, die vero septima mensis septembris, Regnante Illustrissimo, Invictimoq, Principe et dno dno Carolo Romanorum Imperatore semper Augusto et dno nro Clementissimo, Maiestatis suis Romani Imperii, vigesimo primo, aliorum vero regnorum vicesimo quinto, In mei Notarii publici testisq infrascriptorum, ad hoc specialiter vocatorum et rogatorum pntia personaliter constitutus, honestus & providus vir Desiderius Helmenschmid, civis Augustensis, ac faber seu fabricator sive operarius armorum, in prae celebri urbe Augustensis Imperiali unide licorum, sponte et ex certa sua scientia confesus fuit ac in rei veritatis recognovit se a Nobili ac Magnifico viro domino Joanne Paungartner a Paungarten et Cesareae & Regiae Maiestatis Consiliario, habuisse & recepisse valorem summae ducentorum schutorum seu coronatorum aureorum in auro solis monetae regis Francorum, et hoc in defalcationem sive deductionem maioris summae precii, sive mercedis, quorundam armorum quae idem Desiderius Helmenschmid pro Illustrissimo Principe et Dno dno Dux de Alba et Regni Hispaniarum efficere sive fabricare habet atque suscepit, Sdens atq similiter confitensq praetacta summa ducentorum schutorum ipsi per prefatum domini Joannem Paungartner a Paumgarten et etiam in dicta civitate Augusten residens, computata e numerata fuit ad ordine Nobilis viri domini Leonardi Gentili mercatoris Genuensis, Antuerpiae agens, pro ac nomine prelibati domini Ducis de Alba vigore et in vmi quarunda literarum missoriarum et commissionis ipsi domino Leonardo Gentili ex Medina Campi dicti Regni Hispaniae, sub data diei Sexte Mensis Augusti, ultimae praeteriti, per nobiles viros Dominos Lucianum Spinulam et Leonardum Lomellinum, etiam mercatores Genuenses, illic[...]egociantes, ad praedictum oppidum Antuerpiensis, ducatis Brabantiae destinatarum et directarum, uti haec se intellexisse subiunxit, quitans propterea, liberans et absoluens de praedicta summa, tam praefactum dominum Joanem Paungarntner a Paungarten et ac domini Leonardum Gentilem, quae in primus praelibatum domini Ducem de Alba. Necnon quoscunq alios quos dictus dominus Leonardus Gentili inde quitatos, liberatos et absolutos ese voluerit, et censuerit, cum pacto [...] et expresso de rem habitam ulterius in perpetui non peteri, sub obligatione et Ypotheca omnium et singulorum bonorum suorum mobilium & inmobilium praesentium et futurorum, simil atque actionium et jurium in forma juris [...] ac omni juris & facti et auxilii cuinscutunq exceptione atque non habite et non receptae aut mimecatae tantae quantitatis pecuniarum, speique futurae mimerationis et receptionis, remuniration ad haec quomolt necessaria pariter & cautela, consentiens praefatus Desiderius Helmenschmid, pronta me Notario petita fuerunt super praemissis ómnibus fieri instrumenta publica unim vel plura in meliori forma, acta fuerint haec Augustar vindeliorum, et ibm in domo habitationis praefati dni Joanis Paungartner a Paungarten et sub anno, Indistion diae mense, Cesareoq Regimine quibus supra praesentibus ibm hosnestis viris Hieronimo Hueber, et Joanne Degenhart Augusten testibus as praemissa vocatis specialiter atq rogat

Et ego Gulielmus Schoffel in artibus magister civis Augustensis, Sacris Apostolica & Imperialis Auctoritatibus publicus Notarius, quia praemissis omnibus & singulis, sic du fuerent et agerentur, unacum paegminatis testibus praesens inter sui, eaq sit fieri vidi & auduri, ideo hoc praesens publicum instrumentum mani mea propia scripte exmide confeci, subscripsi, publicavi er in hanc publicam formam redegi, signoq & nomine meis solitis et consuetis, signavi, in fidem testimonium omnium & singolorum rogatus pariter & requisitus.

Doc. 3.
1558-1559.
Payments for cleaning and repair of arms property of the Duke of Alba.
Archivo Ducal de Alba, Casa de Alba, C. 164, nº. 23.

[unfoliated, the digitized folio number is entered]
(nº. 1)
Débensele a Maestre Enrique armero de Su Exª de lo q tiene hecho por estas dos q^{tas} trezientos y çinq^{ta} R^{es} de a 3 p^{s}- cada uno. Y más se le deben siete Ds y mº del partido de dos meses y mº que sirvió desde primº de agº 1558 hasta mediado oct^{e} siguiente

Tiene Rº para en q^{ta} de esto de Antº de Lada diez felipos de a 35 p^{s} cada uno- de cas^{r} 20 de mayo 1559
Digo yo Anriq Gamen armero q me doy por contento y pagado de esta sobredicha cuenta del S^{or} Juº Paulo tesorero del duq de Alba y porq es verdad y yo no sé escribir rogué con Antº de Madrigal lo escribiese y sumase por mí fecha hoy viernes 26 de mayo 1559

Por testigo/ Antonio de Madrigal
(signed)

(nº. 2)
Quenta con su Ezelençia duque de Alba mi señor
No se q^{ta}
-primeramente he pagado seys reales pa sacar la ropa que estaba en el palaçio pa llevar a Adrián el de la cámara__vi Rs
-más he limpiado un espada pa su Ezelençia barniçado- puesto vaina de quero y puño de seda quatro Rs__iiii Rs
-más por haber limpiado una daga y barnizado y vaina de cuero y puño de seda dos Rs__ii Rs
-más por haber limpiado otra espada negra y barnizado y puño de seda y vaina de cuero quatro Rs__iiijº Rs
-más por haber hecho una vaina de cuero baya de gamuza y limpiado la espada púsela sobre una espada plateada tres Reales__iii Rs
-más por haber dado dos contrafodios (sic) de vaqueta con sus botones por cada uno tres Rs__iii Rs
-más por haber hecho una vaina de cuero baya sobre la espada era la espada de armas y limpiado la espada dos Reales y medio__ii Rl mº
-más por haber limpiado la daga medio Real__Rl mº

-más por haber limpiado dos espadas un real por cada una__ii Rs
-más por haber limpiado un estoque plateado un Real__i Rl
(nº. 3)
-más por haber limpiado dos espadas doradas un Real por cada una__ii Rs
-más por haber limpiado un estoque tudesco un Real__i Rl
-más por haber limpiado un jado (sic) de malla con sus mangas seys Rs__vi Rs
-más por haber otros dos jados (sic) sin mangas diez Rs__x Rs
-más por haber limpiado dos pares de mangas seys Reales__vi Rs
-primeramente por haber limpiado las armas picadillas todas complidas con sus bardas de caballo tres escudos__xxxiiijº Rs
-más por haber limpiado otras dos armas complidas con todas sus pieças y con sus bardas de caballo que se hizo en Aranjuez por la costa quatro Escudos__ xl v Rs
-más por haber limpiado otras armas doradas que tenían en el pecho una ymajen de nra Señora eran con todas sus piezas de limpiallas dos escudos__xxiij Rs
-más por haber limpiado unas armas a prueba de pistolete blancas negras ocho Rs__viij Rs
(nº. 4)
-más por haber limpiado unas armas blancas a prueba de pistolete ocho Rs__ viijº Rs
-más por haber limpiado unas armas blancas de infantería vi Rs__vi Rs
-más por haber adrezado una rodela damasquina con sus bisajas y las haber limpiado y el morrión de lo mismo un escudo__xi Rs
-más por haber limpiado unas armas probadas que eran de don Hernando dos escudos__xxiij Rs

Hanse de pagar a Enrique de esta q^{ta} de tres planas 148 Rs_cxlviijº[...]

(nº. 5)

Quenta con su ezelençia
Primeramente
-por limpiar ocho sillas de armas a tres reales cada una que son veinte y quatro rrs__xxiiijº
-más por limpiadura y clavaçón adonde la había menester todas las armas cumplidas doradas y tres escudos treinta y tres rrs__xxxiii
-más por adereçar una manopla y puesto un guantelete de bufano blanco seis rrs__vi
-más por haber limpiado unas armas a prueba de arcabuz con su capaçete un escudo__xi

-más por haber limpiado unas armas blancas con su morrión de infante seis rrs__vi
-más por haber limpiado tres cofres de armas y una caja de bardas de caballo y tornallos a clavar y en las partes que era menester que estaban cerca del sanblón (sic) veinte y tres reales__xxiij
-más he pagado unos tornillos que se hizieron para las bardas un escudo__ci
-más he pagado tres quartas de raso diez rs para un peto de malla__x
Más por haber limpiado el peto de malla y unas mangas de malla y por haber puesto el raso un escudo__xi
(nº. 6)
-más he dado una funda de vaqueta para un murrión estofado en tela de oro ocho rs__viijº
-más he pagado un real por llevar los cofres__i
-más limpié dos cimitarras dos reales__ii
-más limpié una espada negra y barniçádola con su paño de seda y vaina quatro rs__iiijº
-más por limpiar una espada que tenía la guarnición parda y una vaina y puño de seda quatro rs__iiijº
-más he dado una hoja de daga y haberse hecho parda la guarnición y un puño de seda y una vaina de cuero seis rs__vi
-más por haber limpiado un estoque de armas y haber puesto un contrafodro (sic) de vaqueta tres rs__vi
-más por haber limpiado dos espadas de armas y dos vainas de cuero bayo de hechura y vainas seis rs__vi
-más por haber limpiado otra espada dorada y haber puesto una vaina baya tres rs__iij
-más di una daga dorada con su puño de oro y una vaina baya quatorze rs__xiiijº
-más he dado tres contra fodros (sic) de paño rojo y tres bolsas de cada una tres rs q son nueve__ix
-más de lo que se hizo en Mos de Henao tres contrafodros (sic) de hierros de ginetas por cada uno un real__iij
(nº. 7)
-más he pagado por aguxerar un peto a prueba de pistolete para meter el ristre quatro rs__iv
-más he pagado por dos palas para llevar las çeladas de su ezelençia un escudo__xi

Endorlan
-más hize dos cruzes para las armas de su ezelençia seis rs__vi

-más por haber aguxerado las armas a prueba de pistolete ocho rs__viijº
-más por haber hecho otros ocho agujeros a unas armas doradas que estaban en la tienda de su ezelençia y echar ocho rosetas__xi
-más por haber puesto unas correas y unos clavos seis rs__vi

Y unos clavos dorados
-más he puesto dos cubiertas a dos pomos de dos espadas doradas dos rs__ij
-más he hecho una vaina de gamuza para una espada de armas en el campo tres_ _iii
-más adereçé unas armas de un pariente de su ezelençia q se dize don ~~Garçía~~ [crossed out] Hr^do^ del laguila (sic)[del Águila] treinta y tres rs q el tesorero lo mandó__xxxiij

Hásele de pagar de las partidas de estas tres planas cci Rs ½
201 ½ [...].

Doc. 4.
1575, October 17.
[Copy, 1840].
Goods linked by the Duke of Alba to the entailed estate.
Archivo Ducal de Alba, Casa de Alba, C. 158, nº 15.

(fol. 15 r)
Efectos y alhajas agregadas por los Sres D. Fernando Álvarez de Toledo y Dª María de Toledo su mujer Duques de Alba a su mayorazgo por Escritura otorgada en Madrid a 17 de octubre de 1575 ante el escribano del númº Gaspar Testa.

-La artillería sus pertrechos y municiones
-Los vestidos de sus personas
-La tapicería que habían traído últimamente de Flandes compuesta de seis paños; los tres pintura de Gerónimo Boscho y otros tres de la Jornada de Alemania, todos con las armas de Toledo__600
-Dos espadas, una con guarnición llana dorada y la otra con guarnición negra de vueltas y la hoja con tres canales en los recazos q no pasan abajo
-Dos fuentes de plata que la ciudad de Nápoles había regalado al Sr duque siendo su virrey y capitán general en agradecim^{to} de la defensa que había hecho de aquel Reyno [...].

Doc. 5.
1637, April.
List of arms delivered to the king's service and artillery of the armoury of the town of Alba de Tormes.
Archivo Ducal de Alba, Ducado de Alba, C. 157, nº. 43.

[unfoliated, the digitized folio number is entered]
(nº. 1)
Fran^{co} de Aranguren ha venido aquí con comis^{on} de Su Mag^{d} en que me manda le entregue las armas q hubiere de provecho para llevallas a Fuenterrabía, haréis que les den todas las q fueren de provecho dexando las que no lo son como es las del duq^{e} de Saxonia las de la persona de mi abuelo y si hay algunos a prueba de arcabuz tomando de él su carta de pago nro Señor os guarde en la Abadía 30 de Abril 1637

El duq de Alba
(signed)

Fran^{co} Antonio

(nº. 2)
Armas para Fuenterrabía 1637
Carta del duq^{e} mi S^{r} de 30 de Abril 1637 para entregar las armas a orden del marqués de Castrofuerte
Llevaron treinta arneses con todas sus piezas
Nº 2º

(nº. 3)
En la villa de Alba a diez y ocho días del mes de mayo de mil y seiºs y treinta y siete años ante mí Manuel de Almeida escribº pú^{co} del número de la dha villa de Alba e su tierra e de la contaduría de Su Exª el duque de Alba mi S^{or} y tºs pareció pres^{te} Fran^{co} de Aranguren vºz de la villa de Madrid estante al pres^{te} en esta dha villa y en virtud de la comisión y orden a él dada por el señor marqués de Castrofuerte de los Consejos de Guerra e Yndias de Su Mag^{d} su veedor General de la caballería comiº Genl de la infantería y Capitán General de la infantería artillería que fue despachada en Madrid a dos de abril de mil y seiºs y treinta y siete años y tomada la razón por Ventura de Frías escri^{no} se Su Mag^{d} y contador de la razón General de la artillería que por ser notoria no va aquí inserta y de ella usando= dijo e confesó haber reçibido de Su Exª el Duq^{e} de Alba mi S^{or} y de su contaduría en su nombre que reside en esta dha villa de Alba por mano de Fran^{co} rrºs armero criado de Su Exª vºz de esta dha villa las armas e pieças siguientes=

Primeram^{te} recibió treinta arneses con las pieças sig^{es}= treinta petos= treinta espaldares= sesenta braçales= çinqu^{ta} y dos escarçelas= çinqu^{ta} y dos manoplas= treinta çeladas= unas de seguir e otras de encaje todas negras= y más seis petos y seis espaldares= blancos y negros= todas las quales armas son traídas= las quales dhas armas confesó haber rezibido e pasado a su parte e poder de la parte de Su Exª de que yo el pres^{te} escrib^{o} doy fee las recibió y pasó a su poder en presencia de los t^{o}s de esta carta y el suso dho las recibió para efeto de las conducir e llevar a la (n°. 4) villa de Fuenterrabía y a ello dijo obligaba la persona y bienes del dho marqués de que las dhas armas q ansí rezibe de Su Exª en esta dha vi^{a} las dará y entregará a la persona q en dha su comis^{ón} se le manda y en todo cumplirá con el tenor de ella pena q no lo cumpliendo quiere por ello ser castigado y para su cumplim^{to} dio poder a las justicias competentes de Su Mag^{d} de cualquier fuero e jurisdicción q sean e lo rr^{o} por ss^{a} definitiva de juez competente pasada en cosa juzgada e ansí lo otorgó ante mí el escrib^{o} dho día mes y año dhos t^{o}s Juan Juárez ensamblador Juan Flórez Juan Açedo vecinos de la dha villa e lo firmó el otorgante que yo el escri^{no} doy fee conozco Fran^{co} de Aranguren ante mí Manuel de Almeida= Yo Manuel de Almeida escribano público del número de esta villa de Alba e su jurisdic^{ón} con aprobado en el Consejo del rey nuestro señor y ansimismo escribano de la contaduría de Su Exª el duque de Alba mi Señor que fui presente y en fee de ello fice mi signo firma en testimonio de verdad

Manuel de Almeida
(signed)

Testim^{o} de armas sacadas de la Armería de S. E.

(n°. 5)
Relación de la artillería q está en la casa de ella del duque de Alba mi S^{r} en su villa de Alba y sus géneros de ella lo qual haze Fran^{co} García artillero de Su Mag^{d} q la ha venido a reconocer en 26 de Nov^{e} 1637
2-dos cañones de a treinta y ocho libras de calibo (sic) de boca y largas diez y siete diámetros y medio de la boca.
2- dos medias culebrinas de a diez libras y media de boca y largas treyn^{ta} y tres diámetros=
4- quatro piezas de a nueve libras de boca y veynte y ocho diámetros de largo=
4- otras quatro piezas e a cinco libras de boca y largo trey^{ta} y seis diámetros
3- Tres piezas de a seys libras de boca y largo 34 diámetros.
2- dos piezas de a seys libras de boca y treinta y ocho diámetros de largo.
2- dos piezas de a diez libras de boca y veynte y cinco diámetros de largo=
2- dos piezas de a dos libras y media de boca y largo treinta y nueve=

2- dos piezas de a tres libras y media de boca y largas quarenta y dos diámetros=
2- dos piezas de a dos libras y media de boca y largas treinta y seys diámetros.

—
25
(nº. 6)
4- quatro piezas de dos libras escasas de boca y de largo treinta y seys diámetros=
9- nueve piezas de a libra y media de boca y de largo treinta y seys diámetros=
3- tres piezas de a dos libras y media de boca y treinta y quatro diámetros de largo=
2- dos Piezas de a una libra de boca y quarenta diámetros de largo.
2- dos piezas de a dos libras y media de boca y de treinta y ocho diámetros de largo
2- dos piezas de a ocho libras de boca y de largo veynte y ocho diámetros=
2- dos piezas de a diez libras de boca y largas treynta y dos diámetros
1- Una pieza de ocho onzas de boca y larga.
10- diez piezas de hierro las quatro de a 12 onzas de boca las cinco de a libra y una de cinco onzas

—
60
-dos lombardas grandes de hierro pedreras
-tres servidores de hierro para lo mismo

(nº. 7)
1637
Mem^{a} de la artillería del duq^{e} mi S^{or}
Fecha en 26 de Novi^{e} 1637.

Doc. 6.
1839.
Inform on the armouries of Alba de Tormes, San Leonardo, Coca and Alaejos.
Archivo de la Casa de Alba, C. 195, nº. 7 (10).

(fol. r)
1839
Villas de Alba de Tormes=
de S^{n} Leonardo=
Coca=
y Alaexos=

Alba
Sobre las armerías que aparecen sitas en lo antiguo, en los Castillos o Fortalezas de dichas villas en el fin de fixarse para siempre lo que en su razón se describe.

Por lo que el Archivo pasa a manifestar no queda la menor duda en que en los Castillos o Fortalezas de las referidas villas hubo muchas armas y cañones y otros pertrechos de Guerra, ignorándose si habrá algunos (fol. v) en ellas, o si no existirán ya por las vicisitudes de los tiempos y ya por las órdenes superiores para que los Sres de vasallos no pudiesen ni reparar sus Castillos o Casas-fuertes, y menos tener en ellas piezas de artillería.
C. 1º. L. 2= Paquete de la villa de Alba nº. 3

Con respecto a 40 piezas de Artillería de bronce que hubo en el Castillo de la villa de Alba de Tormes, y que hubieron de ser traídas allí desde Flandes por el Gran Duque de Alba Dn Fernando Álbarez de Toledo resulta, que se las aplicó S. M. en junio de 1705, mandando hacer pago de su valor por (fol. r) dos R^{s} Cédulas expedidas en dicho año que existieron en el Archivo, y según minutas de ellas parece, que se mandó pagar en las Cajas del Perú y México unas cantidades que se debían a la casa de S. E. y entre ellas el valor de dicha Artillería

Al libro de salidas en general

En Junio de 1756 se hubieron de entregar al Agente de la Casa las indicadas R^{s} Cédulas, y sin duda hubo de ser para solicitar las cobranzas, más no aparece nada acerca de si se realizaron o no, lo q^{e} podría aparecer en las Cuentas del citado año o en las sucesivas.

(fol. v)
S^{n} Leonardo
Acerca de la Armería de S^{n} Leonardo resulta lo siguiente
de q^{a} de fundación de May^{o}

En la letra de fundación del Mayorazgo que el año de 1567 otorgaron los S^{res} D^{n} Juan Manrique de Lara Capitán de Artillería y Dª Ana Faxardo su mujer metieron en ella 12 piezas de artillería de bronce mayores con todos sus aparejos y otros pertrechos de Guerra que tenían en S^{n} Leonardo, ordenando q^{e} se colocasen en la Fortaleza q^{e} estaban haciendo entonces en dicha Villa, imponiendo la obligación a sus sucesores de conservarlas siempre, reponiéndolas en caso necesario
Leg^{o} de invent^{s} de Alhaxas &^{a} Pliego n° 4, Año de 1620

En una copia simple del ynventario de alhaxas &^{a} que se practicó en este año de las q^{e} existían en la fortaleza resultan especificadas varias armas y entre ellas unos arcabuces, pistoletes, cotas de malla y otros pertrechos de guerra, los más colocados en armarios y cofres, una porción de tiros de artillería en sus carretones

En el año 1656 existían las mismas armas

Consta en un testimonio dado por Juan Alonso de Herrera, escribano R^{l} en Sn Leonardo el año de 1641 que el Corregidor de Soria comisionado en virtud de una R^{l} Cédula (fol. r) pedía se le diese inventario de las armas que había en la fortaleza de S^{n} Leonardo y así lo dejó de disposición de S. M.

El año de 1693 el nuevo Alcalde mayor recibió del Sor Marqués p^{a} renovar el inventario de alhajas y pertrechos existentes en dicha fortaleza y de él resultó que entre pertrechos de guerra había 26 piezas de artillería cabalgadas en sus carretones grandes y pequeños; cerca de 500 arcabuces con sus cerrojos: tres culebrinas con sus llaves, una porción de venablos con unos pistoletes y caxas de rueda, mosqueteros y ballestas, muchas armas de punta con varios morriones y alguna cota de malla

Nota
En el año 1651 existe un recibo original dado por los Alcaides de S^{n} Leonardo de una Campana que había en la fortaleza p^{a} colocarla en la Yglesia Parroquial añadiendo q^{e} la debían tener siempre y cuando que se la pidieren por parte del S^{or} de la villa

Lo mismo por otro inventario (fol. v) del listado de Sn Leonardo desde su compra h^{a} en 1563 y fundación de Mayorazgo en 564 resulta que no obstante el ynventario hecho el año de 1641 por el Corregidor de Soria en virtud de la R^{l} cédula ya indicada, de todas las armas que había en la fortaleza de S^{n} Leonardo, y que las dejó a disposición de S. M. qe el año de 1658 tuvo el nuevo Alcalde mayor orden del Sor Marqués para renovar el ynventario de alhaxas y efectos existentes en dicha fortaleza, y de él resultó que entre los pertrechos de guerra había 26 piezas de Artillería cabalgadas en sus carretones grandes y pequeñas cerca de 300 arcabuces con sus cerraxas 3 culebrinas con sus llaves; una porción de venablos con unos pistoletes y caxas de rueda, mosqueteros y ballestas, muchas armas de punta con varios morriones y alguna cota de malla

La existencia en este espresado año demuestra que S. M. todavía no había dispuesto de las armas y pertrechos de guerra que ha(fol. r)bía en la fortaleza de S[n] Leonardo ni q[e] tampoco lo había hecho en los años de 1713 y 714 porque en las posesiones tomadas en estos años por los causanes de S. E. fueron extensivas a las armas qe a la sazón había en dicha fortaleza.

Es cierto que en otras posesiones hasta la tomada p[r] S. E. actual el año de 1836, nada aparece ya de esistir armas y sí desde la tomada en 1733 y sucesivas se ve relatado q[e] la fortaleza estaba bastante destruida y arruinada, y por lo tanto no se descubre el cómo y cuándo y por quienes desaparecieron las armas y pertrechos de guerra q[e] en lo antiguo hubo en ella.

Alaexos
Con respecto a la Armería q[e] hubo en la fortaleza de la villa de Alaexos en Castilla la Vieja lo que aparece es lo siguiente.

Hay pues una orden original del año de 1604 de la S[ra] entonces de Alaexos mandando entregar a D[n] Pedro de Luna a saber una media culebrina de bronce encabal(fol. v)gada= tres cañones de hierro colado encabalgados= cuatro sacres de bronce encabalgados dos medianos y dos pequeños= cuatro piezas de hierro pequeñas; y resulta la entrega y q[e] al Alcalde se le pasaría en cuenta con la carta de pago y q[e] el apoderado del Sor Luna dio la tal carta de pago.

Ya anteriormente se había entregado la artillería pues consta en forma fehaciente que los cañones, armas y pertrechos de Guerra que había en la fortaleza de dicha villa de Alaexos fueron trasladados a Salamanca en virtud de una R[l] [...]

ALBURQUERQUE, DUQUES DE

Doc. 1.
1643, February 5.
Testimony of the removal of arms from the Cuéllar castle to deliver them to the Ledesma castle on the occasion of the war in Portugal.
Archivo Ducal de Alburquerque, 160, Leg. 4, add. nº. 36.

Cuéllar 1643
Recibo de más alhajas
Corresponde a la armería
(fol. r)
5 de febrero de 1643
Yo García Álvarez de la Vega escribano público de esta v^{a} de Cuéllar y su tierra aprobado por el Rey nro Señor doy fee y verdadero testimonio a los que el presente vieren cómo hoy día de la fha de este Alonso Hernández Procurador General de la villa de Ledesma y su tierra estando en el Palaçio y fortaleza del Duque de Alburquerque mi Señor que tiene en esta villa y en virtud de la carta que presentó de la Exma Señora Duquesa de Alburquerque mi Señora recibió de mano de D^{n}. Rodrigo Herrera su Contador Mayor doze mosquetes y en ellos quatro baquetas las dos con sus porqueçuelas çiento y setenta y seis arcabuzes y en ellos noventa y dos baquetas con sus porqueçuelas y en ellas cinco rascadores, çiento y ochenta y ocho frascos grandes, y otros çiento y ochenta y ocho pequeños; setenta turquesas de arcabuzes; y otra turquesa para mosquetes y postas; y se tasaron los mosquetes con sus frascos a diez ducados, y los arcabuzes con sus frascos a cinco ducados; y en virtud del poder que presentó de la Justiçia y Regimiento de la dha villa de Ledesma que pasó ante Diego Hidalgo del Campo escribano del número y Ayuntamiento de la dha villa se obligó con su persona y bienes y de la del correxidor y regidores y demás vecinos y propios y rentas de la dha villa yn solidum a los volver a su costa a esta villa en pasándose la ocasión de las guerras de Portugal o cumplidos dos años y a voluntad de Su Exª y a costa del suso dho y sus partes y villa; y porque del que faltare pagar cada uno lo que va tasado y valuado; el qual dho poder se otorgó en la dha villa de Ledesma a treinta y un días del mes de enero de este presente año (fol. v) de quarenta y tres; y otorgó obligaçión y carta de pago y rez^{bo} en forma como esto y todo lo demás consta y pareze de la dha escript^{a} que en mi poder queda a que me remito que fue

azeptada por el dho D^{n} Rodrigo de Herrera y para que de ello conste de pedim^{to} del dho Alonso Hernández di el presente en Cuéllar a cinco días del mes de febrero de mill y seisz^{os} y quarenta y tres años, y en fee de ello lo signé y firmé, en testimonio de verdad, Garçía Álvarez de la Vega.

Doc. 2.
1671, April, 21.
Inventory of the armoury of Cuéllar.
Archivo Ducal de Alburquerque, 160, Leg. 4, add. nº. 42.

21 de abril de 1671
Armería de Cuéllar

Ynventario que se hiço de las armas y otras cosas que hay en ella el qual se hiço por orden del Exmo Sr Don Melchor de la Cueva y Enríquez mi Sr Gentilhombre de la cámara de su Mag^{d} General de la armada Real del mar océano hallándose su ex^{a} en la dicha villa en 21 de abril de 1671

-Primeramente treinta y cinco cuerpos enteros con guantes
-Setenta y ocho medios cuerpos sin ellos
-Ocho partesanas en la frontera de la pieza ençima de los espexos
-Diez y seis alabardas en el mesmo sitio con cuchillas dorado y blanco como las partesanas
-Quatro rodelas en la mesma parte
-Otra vieja
-Un alfanxe corvo con funda de terciopelo encarnado
-Una espada en triángulo de tres esquinas
-Una bengala de junco con remate de plata que entró ahora Su Exª el señor Don Melchor de la Cueva ni Señor
-Dos guantes blancos y dorados de mujer
-Un turbante
-Ocho morriones de diferentes géneros en la mesma parte
-Dos espejos de Armar
(fol. v)
-Diez cabezas de armas caballos de diferentes xéneros
-Una ballesta de saeta
-Dos turquesas de balas
-Una lanza de enristrar que está dorada con azul y encarnado
-Tres tiros de bronze con sus carritos pequeños
-Diez y seis tiros de bronze sin carros
-Dos trabucos
-Çiento y ochenta y ocho morriones de diferentes géneros
-Dos muy antiguos y estraordinarios
-Dos hachas de desarmar
-Dos alabardas una blanca y otra dorada

-Quatro partesanas= otras tres partidas están junto con tres estandartes pequeños a la primera ventana de junto a los espejos
-Veinte y seis chuzos
-Otra partesana
-Diez y siete picas
-Veinte y seis lanzas de arristre algunas con hierros
-Unas sillas de terciopelo verde guarneçidas de blanco y dorado
-Otra de terciopelo azul guarnezida de blanco
-Otra leonado guarnición blanca
(fol. r)
-Otra de la mesma color y guarnición
-Otra de terciopelo encarnado guarnición blanca y dorada
-Veinte y dos sillas de diferentes colores las cubiertas
-Un sillón de mujer
-Un peto aforrado en terciopelo encarnado guarnezido con tachuelas doradas
-Otro blanco guarnezido con tachuelas de bronze
-Diez y nueve petos viejos el uno tiene un tornillo sobre el pecho
-En un arquetón entre la primera y segunda ventana como se entra algunas piezas deshermanadas de cuerpos viejos
-Dos adargas y otra de madera
-Un arquetón con piezas deshermanadas de cuerpos viejos
-Un cofre con aderezos negros de terçiopelo y clavazón dorado para caballos
-En otro arquetón plumas viejas apolilladas con algunos guantes deshermanados
-En otra arca setenta y quatro hierros de picas grandes y pequeños y más en ella diez y ocho hierros de alabardas de diferentes hechuras
-Quatro hierros de donde descansa el brazo quando se enristra la lanza
-Más veinte recatones grandes y pequeños
(fol. v)
-Más una mesa de pino con su guardamazil
-En un caxón a la entrada de la armería a mano izquierda piezas dehermanadas de faldones y otras cosas
-A la mesma parte tres tiros con sus carros
-Otros tres sin ellos
-Quatro cañones de bronze
-Dos almirezes de bronze
-Quatro medidas de lo mesmo
-Una cuchar (sic) de echar pólvora con su tapador
-Un atacador
-Dos caxas de tocar con sus baquetas

-Dos atadillos de cargas
-Una funda de estandarte
-Un estandarte paxiço con cordones blancos y negros y paxiços con las armas de su exª y otras que dize Ysabel
-Uno de Damasco encarnado con armas Reales y de su exª el Duque mi señor
-Otro blanco con armas reales colocado por su exª quando tomó los bajeles que trayan socorro a Tortosa
-En un caxón otras piezas deshermanadas
-En otro caxón veinte pares de estribos de jineta de ponteçuelas y en él unas cubiertas viejas con flores de lis
-Otro caxón y en él docientos y catorze turquesas pequeñas
(fol. r)
-En otro caxón treinta y quatro faldones de terçiopelo negro con guarniçión de plata falsa y en él había onze faldones paxiços guarniçión negro y blanco y en tafetán viexo paxiço y blanco
-En otro caxón unas borlas viejas con cordones viejos y unas cubiertas de caballos
-Dos faldones de terciopelo encarnado
-En otro caxón unos frascos viejos sin ser de provecho
-Quatro caxones sin cosa ninguna
-Dos bolsas de pistolas sin ellas
-Noventa y siete mosquetes de cuerda en la pared que está al patio y con ellos veinte y dos frascos grandes con cordones de seda verdes azules y encarnados y en ellos siete pequeños pa zebar
-Ziento y noventa mosquetes en la pared de mano derecha como se entra en el quarto en diez y nueve hileras a diez cada una con sus frascos que solo les falta dos
-Noventa y quatro guantes diferentes unos de otros
-Seis arandelas con otras quatro piezas deshechas sin poder hermanarse
-Diez y nueve recados para frenos de coyundas blancas atado cada uno de por sí
-Medio cuerpo sin morrión ni guante que dio a su exª Don Manuel Velázquez
-Treçientas y setenta espadas de muniçión colgadas en las paredes y puestas donde habían de estar mosquetes
(fol. v)
-Una escalera de pino
-Una gradilla de tres pasos
-Dos mecheros de los tiros
-Un çedaçillo de cerner esmeril

Esto se halló en la armería en 21 de Abril de 1671 años=

Arcabuzes

En 11 de Mayo 1674 remitió el contador Matheo Flores a Pedro Ramos recau^{or} de Cuéllar los dos arcabuces que le dexó el ex^{mo} s^{or} D. Melchor de la Cueva mi s^{or} al tiempo de su partida a Cádiz a servir el puesto de Jeneral de la armada real ambos de chispa con sus llaves y cañones de boca ancha el uno grande con funda de vaqueta colorada de más de dos varas de largo= el otro de vara y media sin funda y le falta una argolleta por la parte de arriba= escribió el dho contador al dho Pedro Ramos los pusiese en la recámara donde están cosas del mesmo género=

Más cosas remitidas

-En 23 de n^{re} de 1678 una escopeta barcelonesa embutida toda la caxa en bronce y el cañón rayado por la parte de afuera con su llave la qual tiene una cabeza de perro por perrillo

-Una carabina de secreto con llave de fuiro (?)

-Un alfanxe corvo con su paño (sic) grab^{do} de madera

-Una espada ancha con guarnición antigua de tres q^{tas} de largo grabada la hoxa

Lo q^{l} le entregó el Duq^{e} mi s^{or} a P^{o} Ramos el dho día

(fol. r)

-Más tres pequeñitos de bronce con las armas de su ex^{a} el Duque mi S^{or} en sus cureñas coloradas

-Más cinco tiros pequeñitos con sus carritos blancos tres grandecitos y dos más pequeños

-Más diez pijotes de bronce sin culatas los ocho, y los dos con armas de los S^{res}

-Más dos trabucos de bronce con sus cajas quebradas

-Más tres tiros medianos en sus carros

-Más otros tres tiros sin ellos

-Más quatro cañones de bronce

-Más dos pretardos (sic) de bronce

-Más quatro medidas de lo mismo

-Más tres tiros grandes sin carros de bronce

-Más una espada y daga que envió el Duque mi S^{or} con dos fundas una encarnada y otra de ençerado verde

-Más una muleta (sic) de campaña

-Más un saco colorado con corchetes de plata con sus calçones grandes açules y su gorrete

-Más un cuerpo entero grabado de oro con las armas Cerdas- y dos arandelas- y cabeça de caballo con dhas armas

-Más medio cuerpo grabado de oro con una rodela de lo mesmo
-Más otra rodela lisa blanca
-Más una bandera blanca de lienço con flores de lis y encomienda de San Juan
-Más dos cuchillos corvos con sus conchas plateadas
-Más otros quatro alfanjes corvos con las guarniçiones plateadas
-Más otros quatro alfanxes por platear las guarniçiones
(fol. v)
-Más otra espada negra de a caballo morisma (sic) con guarnición antiguas (sic)
-Más un venablo de alférez
-Más una partesana lisa
-Más un cuerpo entero el peto de muelles que se juega todo
-Más medio cuerpo peto y espaldar con marxenita (sic) que envió el Duq^{e} mi S^{or} con Andrés Montero
-Más un cuerpo entero dado de negro con su rodela del propio género y quatro partes sanas las dos pavonadas y doradas, las otra dos doradas y blancas de enfrente
-Más una mergenita (sic) con su falda atrás en forma de sombrero
-Más otra morxineta (sic) aforrada en raso carmesí y cordobán negro por afuera guarnecida con galón de plata
-Más una caja con tres cuchillos de mesa un thenedor y un trinchete (sic) ancho todo con cabos de ámbar
-Más otra caja con dos cuchillos de mesa y un tinchante ancho con cabos de marfil embutidos con espejuelo
-Más quarenta y çinco pares de guantes
-Más diez y siete petos de prueba de mosquete
-Más seys petos çençillos (sic) y cinco espaldares de lo propio
-Más quatro espaldares de prueba de mosquete
-Más tres petos uno blanco grabado otro con su espaldar aforrado en terçiopelo carmesí
-Más un medio cuerpo barniçado de negro
-Más diez y ocho alabardas con sus astas las seis barniçadas de negro y las doçe blancas
-Más setenta y ocho picas con sus astas y hierros blancos que están en la techumbre y las alabardas
(fol. r)
-Más en dho sitio veynte y quatro partesanas con sus astas y recatones (sic) pavonadas de negro y doradas y blancas

-Más seys escopetas en la fhada (sic) de enfrente sobre los espejos de diferentes hechuras labores estas escopetas se ponen de por sí al fin de este ynventario para más claridad
-Más tresçientas y veinte espadas con sus guarniçiones de puentes de diferentes géneros y algunas sin ellas
-Más çiento y ochenta y nueve mosquetes con sus frascos que están en la fhada (sic) donde están los cuerpos enteros
-Más ochenta mosquetes sin frascos en la fhada (sic) de las ventanas
-Más diez y seis mosquetes con frascos en la mesma fhada y los dos sin caxa de diferentes géneros
-Más un mosquete de veynte y çinco libras largo con su frasco en dho sitio
-Más en la fhada (sic) de enfrente una bandera pequeña con armas
-Más un estandarte Real ençima de la ventana primera de damasco carmesí con las armas Reales
-Más otro estandarte Real blanco con armas Reales
-Más otro estandarte pequeño de tafetán pagizo con las armas de su Exª el Duque mi S^or^
-Más dos cajas de campaña con sus baquetas
-Más una paila grande de açofar
-Más en el arca primera como entramos a mano izquierda doze recatones pequeños, y çinco hierros de picas diferentes ocho muelles de cuerpos y una funda de vaqueta Moscovia
-Más en el arca que se sigue que es segunda diez pares de guantes viexos y veinte y quatro plumajes viejos y çinquenta y un capillos de raso de flores carmesí con tres flores de list doradas por armas guarneçidos con listones dorados en su caja (fol. v)
-Más en el cofre que se sigue al arca seg^da^ unas guarniciones de caballerías aforradas en terciopelo con hierros dorados
-En el arca que se sigue que es quarta catorçe pares de capillos de hierro para los pies y onçe pieças de diferentes géneros de armas tres pieças para una silla y más diez pares de faldillas de diferentes géneros más otras cinco piezas de diferentes (sic)
-Más en dha arca una visera de morrión y dos golas viejas, más un par de çapatos aforrados en terçiopelo negro, más dos piezas de los codos de los bracaltes (sic) más tres pares de çapatos grabados de diferentes xéneros
-Más un par de guantes más dos adargas de cuero blancas
-En el arca quinta seis pieças de morriones diferentes dos golas viejas onçe hombros de braçaletes de diferentes géneros una nariz de cabeça de caballo y otra pieça de diferente (sic)

-Más en el arca que se vuelve a mano derecha como entramos siete pares siete pares (sic) de guantes y un arco de flecha
-Más diez frascos y çinquenta y una cargilla (sic) de mosquete en la fachada donde está el cuerpo negro
-Más en el arca que se sigue segunda haçia dha mano derecha siete pares de çinchas de caballos y cinco tiros de vaqueta blanco de traer espada en çinta y más doçientos y ocho pares de sacatrapos y lavaderos de escopetas o mosquetes
-Más dos pieças de silla de armar de caballo y dos baquetas de caja de guerra
-Más el arca tercera vaçía
-Más en la quarta arca treçe banderas pequeñas de lienço teñido de açul con flores de lis
-Y más en dha arca diez y nueve pares de estribos de hierro de brida y doçe fundas de colas de caballos de vaqueta blanca
-En el arca quinta duçientas y nueve turquesas de haçer balas de mosquetes y tres llaves viexas de mosquetes.

Doc. 3.
1786, May 16.
Inventory of the Cuéllar castle and its two armouries.
Archivo Ducal de Alburquerque, 161, nº 18.

(fol. r)
Yo Antonio de Figueroa es^{no} del Número, y Ayuntam^{to} de esta villa de Cuéllar, y su Jurisdición (sic), y de rentas del Exmo. S^{or}. Duque de Alburquerque mi S^{or} doy fee q^{e} ante mí se hizo el Ynventario del thenor siguiente

En la villa de Cuéllar, a diez y seis de mayo del año de mill setecientos, ochenta y seis, el Sor Lizenciado D^{n}. Fran^{co} Sánchez Santiago abogado de los R^{s}. Consejos alcalde mayor de esta dha villa y su tierra ante mí el es^{no} dijo, q^{e} habiendo fallecido D^{n}. Antonio Martínez Pardo recaudador que fue de las rentas, y efectos pertenecientes, al ex^{mo} S^{or}. Duque de Alburquerque mi S^{or} en este su Marquesado de esta dha villa, y su tierra, recogió su mrd las llaves del Palacio que su ex^{a} goza en ella, y está en la Parroquia de S^{n}. Martín las que ha tenido en su poder hasta ahora, y en atención a haber mandado dho ex^{mo} S^{or} que para entregarlas al Lizenciado D^{n} Santiago Blanco Cavallero, abogado de los R^{s}. Consejos como recaudador nombrado para la percepción de las rentas y efectos se haga ynventario de todas las alhajas que se hallan en dho Palacio y de los peltrechos (sic) de guerra, y demás muebles, que están en las armerías, alta y baja, existentes dentro de él, cumpliendo con dho mandato, se proceda al expresado ynventario, poniéndolo con la formalidad correspondiente, el que se ejecute por ante mí el dho es^{no} que lo soy del número, y Ayuntamiento de la dha villa, y su tierra, y de rentas de dho ex^{mo} S^{or} arriba mencionadas, con asistencia del expresado Liz^{do} D^{n}. Santiago (fol. v) Blanco Cavallero, a cuio fin se le haga saber, y a él se principió en este día, y lo firmó su mrd de que doy fee = entre renglones= de esta dha villa y su tierra= valga= Liz^{do}. Sánchez= Ante mí Antonio de Figueroa

En la villa de Cuéllar dho día diez y seis de mayo, y año de mill setecientos ochenta y seis, el expresado S^{or}. Lizenciado D^{n} Fran^{co} Sánchez Santiago abogado de los R^{s} Consejos Alcalde mayor de esta mencionada villa, y su tierra en virtud al auto antez^{te}, con asistencia del referido D^{n}. Santiago Blanco Cavallero, y de mí el es^{no}, pasó a dho Palacio y en una pieza, que está en él, llamada la Armería grande, que es en la baja se inventariaron las alhajas sig^{tes}

Primeramente treinta y nueve partesanas= Ocho rodelas= Una espada en triángulo de tres esquinas= Un turbante viejo= Dos espejos de acero el uno quebrado con marcos negros= Una ballesta para disparar flechas= Dos hachas de desarmar= Una (sic) cuchillo, su hechura de alfanje puesta en un palo= Dos sillas para caballo, las que son de terciopelo negro bordadas de oro, y plata muy destrozadas= Una silla para caballo de terciopelo verde, con flecos de seda encarnada y plata descompuesta, y sin el casco de hierro de atrás= Otra de terciopelo azul con franja de seda= Otra pajiza guarnecida con hierro negro, y dorado vieja= Otra de terciopelo alimonado con flecos de seda= Diez y ocho de color pajizo de diferentes telas y alamar las falta la guarnición de hierro= Un casco de brida, de hierro en forma de silla= Un faldón de terciopelo carmesí= Una silla muy vieja para caballo= Tres tiros de bronce medianos en sus carros de madera el uno quebrado, y cada uno tiene (fol. v) su medida a la culata= Trece tiros de bronce con sus carros encarnados, y dos de ellos con dos ruedas quebradas= Otros catorce mayores que los antecedentes en sus carros los dos de estos quebrados= quatro cañones de bronce medianos sueltos los dos torneados, y los otros ochavados= Otros ocho cañones ochavados sin carros, y sin culatas= Otro redondo como de vara y media un poco más grande de los antecedentes con culata= Otro mayor de las misma hechura que la antec^{te}= Otro más pequeño reventado= Otro de cobre de bastante corpulencia aunque de menor longitud que los antec^{tes}= Dos petardos o morteros pequeños de bronce y otros tres mayores de hierro= Una bandera de lienzo blanco con las armas de Francia rota= Un venablo de alférez con una borla de seda encarnada con su asta quebrada= Quarenta y ocho petos, y espaldares en los quales entra uno forrado en terciopelo carmesí; y el otro con badana blanca= Doscientas y sesenta, y siete escopetas, y algunas de ellas sin llave= Seis cajas de escopeta= Un cañón de hierro largo sin caja= Una media bandera de damasco encarnada con una águila, y en el palo tiene una lanza, y otra pequeña sin lanza pendiente en el techo= Un estandarte Real de la misma tela de damasco encarnado viejo= Otro de tafetán pajizo y otro con las armas de dho ex^{mo}. S^{or}.= Dos cajas de campaña con los parches rotos= Una paila de cobre grande con quatro leones en el borde de lo mismo= Otro estandarte de lienzo blanco con las armas R^{s}.= Diez y seis capillos de damasco carmesí y once faldellines de raso pajizo muy viejos= Otros once faldellines de terciopelo muy viejos guarnecidos algunos de plata falsa= Una piedra jaspe de cinco quartas de largo, y tres y media de ancho= unos huesos que se dice ser de gigante= Un atacador de hierro grande para los cañones= Un barrón corto de hierro de hechura del exe de un carro= Un escudo con las armas de su ex^{a} pequeño fondo encarnado bordado= Çiento y treinta (fol. r) y tres baldosas de alabastro de pie en quadro y algunas quebradas= Dos gradillas de lo mismo de cinco quartas de largas cada una= Una tabla pintada con las armas de su ex^{a} y las ánimas= Una caldera grande de cobre hecha de diferentes piezas= Siete arandelas dos doradas con escudos de las armas R^{s} y el nombre de

D^{n}. Hernando de la Zerda, y las cinco blancas= Once guarniciones de cabezas de caballo las cinco enteras y las restantes sin guarnición del hocico= Unas correas de cuero blanco a modo de cabezadas= Y por ahora mandó su mrd cesar en este ynventario con protesta de porseguir en él, y lo firmó, y dho D^{n}. Santiago Blanco Cavallero= Ante mí Antonio de Figueroa

En diez y siete de dho mes de mayo del año de mill setecientos y ochenta y seis, el mencionado Sor Alcalde Mayor con asistencia del citado D^{n}. Santiago Blanco Cavallero y de mí el es^{no} pasó a dho Palacio en continuación de este ynventario de las alhajas que se hallan en la Armería grande, y se ejecutó de las siguientes

Doscientos y sesenta casquetes de hierro= Trescientos y once frasquillos para pólvora= Veinte y tres cuerpos enteros y de ellos el uno sin zapatos= Siete cuerpos con algunas piezas de muslos y piernas= Sesenta y nueve medios cuerpos y algunos les faltaban piezas= cinquen(fol. r)ta y un morriones y algunos de ellos sin visera y faltos de alguna otra pieza en los que (sic) y en los casquetes se incluyen los que están puestos en los cuerpos y medios cuerpos= Veinte y seis palos de estandarte= Quarenta, y seis lanzas= diez picas, y veinte y siete alabardas= Y por ser tarde mandó su mrd cesar en este ynventario, y lo firmó, y el mencionado D^{n}. Santiago Blanco doy fee= Liz^{do} Sánchez= Santiago Blanco Cavallero= Ante mí Antonio de Figueroa

En diez y ocho de dho mes de mayo y año de mill setecientos ochenta y seis por el mencionado Sor Alcalde mayor, se prosiguió con asistencia del referido D^{n}. Santiago Blanco Cavallero y de mí el es^{no} a hacer ynventario de las alhajas que existían en dha armería grande y se practicó de los siguientes

Trescientas y diez y siete espadas las más sin vaina= Ciento, y noventa guantes de hierro la mayor parte faltos de algunas piezas inclusos los que tienen dhos cuerpos, y diferentes pedazos de armar= Una muleta de campaña y la mazeta es de hierro= En una arca vieja cinquenta, y tres penachos apolillados de diferentes colores, algunas piezas de armar, y dos pares de espuelas= En otra arca vieja trescientos y veinte y dos sacatrapos y raederas de escopeta= Quince pares, y medio de estribos de hierro= En otra arca vieja, ciento y cinquenta y seis turquesas para hacer balas pequeñas, y otra grande (fol. v) para lo mismo y diferentes piezas de armar= en un cajón diferentes piezas de armar, y un cañón de trabuco, y varios pedazos de vainas de espadas, y un armador de ballestas= En una arca vieja, un pedazo de bronce hueco a modo de atacador con varias piezas de armar= Diferentes espadas quebradas= Dos fundas de escopeta viejas y quatro hierros de lanzas= En otra arca diferentes piezas de armar= En un baúl viejo varios pedazos de caparazón de sillas, y otros trapos que por su mal estado no se conoce lo que han sido= en otra Arca vieja diferentes pedazos de vaqueta blanca con corchetes de alambre gruesos= Varias piezas de armar, y carteras de badana apolilladas= En dos arcas diferentes piezas de armar= En otro cajón varios pedazos de armar y el uno de bronce a modo de atacador, y un pedazo de tafetán viejo que parece haber sido bandera= Dos cañones, o tiros pequeños de hierro como de dos quartas de largo cada uno y de calibre de a diez= Un paño encarnado plateado y dorado hechura de pañuelo= Y en otro cajón se hallaron diferentes piezas de armar, un pedazo de bronce a modo de atacador, y un badajo pequeño de hierro= con lo qual se finalizó el ynventario de las alhajas que se hallan en la citada armería grande, y lo firmó su mrd y dho D^{n} Santiago Blanco de que doy fee= Liz^{do} Sánchez Santiago Blanco Cavallero= Ante mí Antonio de Figueroa

En diez y nueve de dho mes de mayo, y año de mill setec^{tos} y ochenta, y seis dho S^{or}. Alcalde mayor con asistencia de dho D^{n} Santiago Blanco Cavallero y de mí el es^{no} pasó a dho Palacio a hacer ynventario de las alha(fol. r)jas q^{e} se hallan en la armería chica y se ejecutó de las sig^{tes}

Primeram^{te} una silla de mano de seda llamada ermosilla (sic) fondo verde, y flores Blancas antigua bastante marrocada (sic), y a los lados tiene cada uno su cortina de lo mismo= Unas guarniciones muy viejas para mulas de coche= Dos pares de morillos de bronce grandes son las barras de hierro para sostener la leña= Una silla de mula con su asiento de vaqueta de moscobia, y caparazón de lo mismo sin correaje de cinchas retranca, ni pretal= Otra silla vieja para el mismo fin forrada en [...] liso encarnado= Dos tablas forradas de felpa o terciopelo color carmesí acolchado= Una gualdrapa de lo mismo= Una gradilla de pino con cinco pasos= Un escudo grande de embajador sin pintura alguna de armas= Tres pares de fundas viejas antiguas para pistolas= Dos brujacas (sic) grandes de vaqueta= diferentes palos de flechas todos sin puntas a excepción (sic) de una que la tiene= Dos pares de bridas de vaqueta antiguas= Ocho barriles de haya= Quatro barriles de corcho, y dos de vaqueta de diferentes tamaños= Cinco morrales, o especie de alforjas de cuero apolillados= Un frasco para pólvora claveteado= Cinco frascos de asta también para pólvora= Tres botas muy viejas para vino= Un cofrecillo viejo claveteado, y en él dos bolas grandes de jaspe, y dos pequeñas= Dos cocos de Yndias sin guarnición= Una especie de caxuela de vaqueta de moscobia= Una turquesa para hacer balas pequeñas= En una arca vieja veinte libros impresos de mano escritos grandes, y pequeños= Una aljaba forrada en vaqueta con varias flechas= Dos petos, y dos respaldares (sic) de hierro muy negros= Once pilastras de jaspe pequeñas, y las más (fol. v) quebradas= Una bola de lo mismo grande= Seis escopetas y mosquetes con unas llaves antiguas de mecha, y rueda= Un trabuco= Una caja de escopeta sin cañón con llave antigua, y una de las escopetas antec[tes] no tienen llave= Una daga con una pistola unida sin vaina= Un cuchillo con mango de asta de ciervo con vaina verde, y vieja= Dos cornatillos de ciervo= Dos crucifijos de madera viejos= Una cruz de madera con su peana de negro= Dos tiros de bronce pequeños, y al uno le falta una pieza en el exe= Veinte y seis pistolas de diferentes tamaños, y a algunas las falta la baqueta, y de dhas pistolas tres están embutidas de hueso, y de estas la una no tiene llave= Ocho bolsillas de Holanda de mitan (sic) azul pequeñas= Nueve ballestas para disparar flechas la una quebrada y todas ellas faltan de algunas piezas= Una turquesa grande para hacer balas= Una hacheta pequeña con mango de madera= Una turquesa de hacer bodoques= Una cuchara para derretir plomo muy mala= Dos hostiarios de charol= Dos fuentes de peltre un aguamanil de lo mismo= Dos jarros de lo mismo= Un casco de lámpara y una gaveta, y un incensario de lo mismo= Dos lamparillas de peltre= Una pilita de alabastro= qua(fol. r)tro hebillas grandes de hierro doradas= Un brocal de plata sobredorado de una bota y le falta el agarradero para abrirle= Una araña de madera con quatro mecheros dada de barniz negro= Una espada quabrada con la guarnición de hierro dorada= Otra espada con guarnición de hierro también dorada quebrada la hoja= Una veleta con una cruz de hierro= Una caja con tres cuchillos, y un trinchante con mangos de ámbar, y hueso= Otra caja con dos cuchillos anchos y un trinchante con los mangos de hueso, o marfil= Un puño de plata para bastón, y en él grabada una batalla= Cinco palos arqueados dados de colores= Una jarra verde de barro con el asiento agujereado= Dos bericuer (sic) de terciopelo negro viejos= Dos cruzes de madera la una dorada toda, y la otra también dorada, y la peana dada de colores= Doce canes o repisas doradas pequeñas y pintadas el fondo blanco= Quatro bolillos encarnados y pajizos= Once candeleros que han sido plateados muy negros de pino pequeños, y dorados, quatro jarrillas de pino doradas dos pequeñas y las otras mayores= Un atril de madera de pino con las armas pintadas de dho ex[mo]. S[or].= Dos tiritos de bronce pequeños, y otro mayor sin cureña= Un relox de sobremesa descompuesto= Un cetro Real de hierro dorado= (fol. v) Ocho espuelas doradas= Diez armadores de ballestas= Ocho armadores de llaves de escopeta antiguas= Dos velones pequeños de latón de quatro mecheros cada uno= Un martillo de hierro de relox de torre= Y por ahora mandó dho Sor Alcalde mayor cesar en este ynventario con protesta de proseguir en él, y lo firmó su mrd y dho D[n]. Santiago Blanco de que doy fee= entre renglones= Un= dos jarros de lo mismo= Un casco de lámpara= valga= Liz[do]. Sánchez= Santiago Blanco Cavallero= Ante mí Antonio de Figueroa

En veinte de dho mes de mayo y año de mill setecientos, y ochenta y seis en prosecución de dho ynventario habiéndose concluido en el día de ayer de las alhajas que se encontraron en dha armería chica dho S^{or}. Alcalde mayor con la citada asistencia del nombrado D^{n}. Santiago Blanco Cavallero, y de mí el es^{no} pasó a las demás oficinas del referido Palacio, y en ellas se encontraron e inventariaron los bienes siguientes

En la sala que está pintada la cabeza del Gigante, siete quadros de diferentes pinturas sin marcos= [...] Y en el castillo ocho tiros de hierro de diferentes calibres= [...]= un tiro de hierro grande en una caja de madera que está debajo del corredor frente de la puerta para entrar al patio de dho Palacio; y no se halló otra cosa que inventariar en cuya vista el mencionado S^{or}. Alcalde mayor, hizo entrega de las llaves de dho Palacio, y sus oficinas al expresado D^{n} Santiago (fol. v) Blanco Cavallero quien se dio por entregado de ellas y de todos los bienes y peltrechos que quedan inventariados para tener a disposición de su ex^{a} siempre que le sean pedidos a lo q^{e} se obligó en toda forma y lo firmó junto con su mrd de que to el es^{no} doy fee= en m^{do}= of= ci= i= Alcalde m= valga= Liz^{do}. D^{n} Fran^{co}. Sánchez Santiago= Santiago Blanco Cavallero= Ante mí Antonio de Figueroa en m^{do}= seda encarnada y plata= enteros y de ellos= b= v^{a} [...].

ALCALÁ DE LOS GAZULES, DUQUES DE

Doc. 1.
1637, November 17.
Inventory of the armoury of the Duke of Alcalá in his Palace, Seville.
Biblioteca Francisco de Zabálburu, Altamira, 215, D. 1.

[unfoliated, the digitized folio number is entered]
(nº. 2)
Memª y entrego General que se hiço a Don Juan de Arroyo de todas las pinturas en lienzo tablas láminas yluminaziones y de las estatuas de bronce grandes y pequeñas y demás cosas curiosas que el Exmo Señor Duque de Alcalá mi señor dejó en sus casas Palaçio quando salió de esta ciudad para exerçer el cargo de Virrey de Nápoles y en esta memoria van inclusas todas las pinturas y demás cosas que en la villa de Madrid entregó su Exª al dho don Juan de Arroio en veinte y quatro caxas

En el oratorio alto pieça primª [...]

(nº 38)
En el aposento del Armería

1- Un lienzo copia de las armas q estaban en el comedor
2- Dos me^{os} cuerpos de armas grabados y dorados el uno sin morrión
3- Dos frentes de caballo de hierro grabadas y doradas
4- Un armadura de falta de lo mismo
5- Tres petos y espaldares con sus faldones dado de barniz nuevo a prueba de pistola y hechas en Barzelona que allí llaman armillas
6- Quinze adargas
7- Una adarga grande de jugar cañas
8- Tres mosquetes turquescos
9- Una escopeta de dos varas
10- Un arco de la ordena

(nº. 39)

11- Otra rodela de azero pavonada y dorada con fluecos verdes
12- Un tiro de bronze pequeño con su cañón
13- Dos pares de estribos de jineta de piezas grandes q fueron del S^{or} marq^{s} de Tarifa D. Fadriq
14- Tres guar^{es} de espada de una mano viejas

15- Una caxa de tres quartas de largo con un radio estronómico (sic)
16- Una gu^{on} viexa dorada de una mano- Dióla Marn de Valdés a un pobre criado de casa de su Exª Valdés
17- Quatro hojas de dagas sin g^{nes}
18- Dos hojas de daga con vaynas
19- Una espada ancha de dos canales con g^{ón} de dos conchas francés por m^{ca} m. o. n. mno
20- Un verduguillo ordin^{o} con gu^{on} antigua y dorado y sin çelada con un cuchillo en la bruna (sic)
21- Una espada ancha corta con esta marca [...] y g^{ón} negra con unas cabesillas
22- Una daga ancha q sirve de sierra con su g^{ón}
23- Una espada ancha de tres puentes a lo flamenco con tres canales por el lomo
24- Un alfanje como (sic) con g^{ón} aventanada sin m^{ca}
25- Otro alfanje corto con g^{ón} aventanada dorada con un letrero en la canal de la antigua gastada
26- Una espada maior de marca ancha por marca una R [with a crown on] y más abajo una R [with a crown on] y un letrero grande gastado
27- Un alfanje con un cabo de hueso con tres canales en el lomo y un letrero q diçe ampara
28- Dos almaradas en sus vainas
29- Una bolsa de terciopelo negro con una caxa con unos instrumentos de matemática y falta un hierro

(nº. 40)

30- Una caxa de quatro cuchillos con cabos de marfil
31- Otro alfanje con cabo de hueso g^{do} de plata con tres canales por el lomo la vaina con contera y canal y hebillas de lo mismo
32- Un alfanje con g^{ón} dorada y una canal por el lomo
33- Otro alfanje con g^{ón} de hueso y me^{os} hoçes por marca
34- Una espada corte de lomo con g^{ón} de hierro
35- Un me^{o} alfanje del perrillo con g^{ón} aventanada
36- Un alfanje pequeño de dos lomos sin g^{ón}
37- Una cuchilla aventanada con g^{ón} de melcocha (sic) embarnizada
38- Una espada ancha de a caballo con un letro (sic) en la canal mui antiguo que no se conocen las letras
39- Una espada con g^{ón} negra que alarga a montante
40- Un estoque con un letrero dorado de paulo 4º con estos versos

Hunc en sem demore dedit

Dum Labitur annus
Quartus ei Paulus Bentus
Tunc Papa Secundus

41- Un alfanje con una cruz + de S. Ju^{o} por m^{ca} con g^{on} de hierro
42- Una espada ancha antigua que dizen fue de Di^{o} Gómez de R^{a} [Ribera] 2° adelantado maior de la frontera con g^{ón} embutida de plata con su vaina y funda
43- Unas armas viejas pasadas
44- Sobrepeto de una de las armas
45- Dos alfanjes turcos con sus cabos y clavazón dorada

Concuerda con su original que queda en la contaduría del Exmo (n°. 41) Prin^{e} de Paternó Duque de Alcalá mi S^{or} salvo en los n^{c}s çiento y treinta y siete y çiento y sesenta y siete digo y cinco; y doçientas y treçe y doçientas y catorce del camarín grande y n^{o}. diez y seis del Armería que están anotados de mano y rúbrica del Sor Martín de Valdés fecho en Sevi^{a} en diez y siete de novi^{e} de mill seis^{o}s treinta y siete años=

Martín de Valdés
Gonzalo Caballero
Don Ju^{o} de Arroyo
(signed)

ARCOS, DUQUES DE

Doc. 1.
1630, March 13.
Inventory of the 3rd Duke of Arcos, Rodrigo Ponce de León.
Archivo Histórico de la Nobleza, Osuna, C. 125, D. 176.

Diligencias originales y autorizadas del inventario, tasación y almoneda de los bienes muebles que deja el duque agregados a su mayorazgo y estado.

(Fol. 4 r)
Inventario de los bienes de su exª
En la ciudad de Marchena en trece días del mes de março de mil seºs treinta aºs, en la [...]
Estando en las casas palaçio de su exª ante el lic^{do} Juan Muñiz de los Ríos asistente de esta dicha villa se entró en su recámara y en ella se comenzó a hacer el inventario de los bienes que en ella están en la manera siguiente

Inventario
esta y las que siguen
-Los quatorze montantes y dos rrodelas negras
-Tres espadas anchas de caballo con sus vaynas plateadas
-Quatro alfanjes turquescos con sus vainas
-Otro alfanje más de la misma forma
-Otro alfanje más con el puño de plata con su tahalí asul bordado que fue del rey Avenavo
-Un montante mediano de tres esquinas con su vaina
-Una espada ancha de a caballo con la vaina guarnecida con barreta de plata y la guarnición como de montante
-Otra espada de a caballo ancha grabada la hoja y grabada la guarnisión y el puño de hilo de oro
-Otra espada de a caballo con la guarnición dorada y la vaina baya
-Una cuchilla de alfanje sin vayna ni guarnisión
(fol. 4 v)
-Dos hojas de espada en una vaina sin guanisión
-Dos espadas sin guarnisión con sus vaynas negras
-otras dos cuchillas de espadas de a caballo sin guarnisión y con sus vainas bayas

hasta esta
-Tres cañones de arcabuzes sin caxas ni otra cosa

-Una mochilla (sic) de tersiopelo carmesí bordada de oro con sus fluecos de oro y seda carmesí
-Unas cubiertas de caballo de ante bordadas sobre faxas negras y sobre el mismo ante con su pretal de lo mismo
-Otra cubierta de caballo de tela blanca guarnesida con sus pasamanos de plata y guarnecida de seda negra sobre la tela de carmesí con sus fluecos y borlas pardas de plata y seda [...]
(fol. 5 r)
De estas dos cotas dio el duque mi s^{or} que Dios g^{de} al marqués de Villanueva la una cota [...] y la otra tiene en Guadalupe el armero pa limpialla
-Dos cotas con sus mangas y guantes de malla que son dos [...]

(fol. 16 r)
Caballerizas
-Lo prime^{o} una jaca ruzia vieja
-Un caballo rusio tordillo de edad de siete años
-Otro caballo castaño quatralbo de edad de siete años
-Otro caballo castaño sayno capón serrado
-Quatro machos viejos del coche
-Tres sillas biejas las tres de jineta y la otra de la brida de vaqueta viejas
-Ocho mantas de xerga viejas que sirven a los caballos y machos
-Dos coches, el uno negro de cordobán con sus enserados y cortinas de lanilla negra y el otro coloradillo de badana viejo en forma de carrozilla con sus enserados y cortinas de xergueta colorada
-Quatro guarnisiones viejas de los machos del coche
-Cinco cubos y seis almohadas
-Una arca donde se recoje la cebada
-Dos sillones de los machos de la litera con sus guarniciones viejas
Y en esta forma se acabó el ynventario delas dhas caballerizas para proseguir en los bienes que hubiere. Lic^{do} D. Juan de los Ríos Muñiz, Fran^{co} Ximénez ss^{no}

(fol. 18 r)
Armas
En Marchena en catorze de mayo del dho año de mil y S^{o}s treinta se continuó el dho inventario y se pusieron las armas que están a cargo de don Fran^{co} de Saavedra en la manera siguiente
-Una silla de armas asul bordada de tersiopelo
-Otra silla de tersiopelo llana
-Otra silla blanca de tersiopelo
-Otra silla de ante

-Otra silla de tersiopelo asul
-Dos sillas de tersiopelo amarillo
-Otras dos sillas de tersiopelo negro que no son de armas sino de la brida
Esta y todas las armas que se siguen
-Sinco rrodelas las tres negras y otra veneziana dorada y la otra con una encomienda de San Ju°
(fol. 18 v)
-Un arnés dorado con dos petos y falbesales (sic) y espaldarcete y espaldar y su gola y escarselas y quixotes rodilleras y sus grebas y manoplas y su selada borgoñona y un manoplón de justa
-Otro arnés dorado peto y espaldar y su gola y brasales y espaldarcete y escarselas y sus quixotes y rodilleras y grebas manoplas y selada borgoñona y su bragueta y soto gola de armas
-Otro arnés de pasta de Colmenar con todas sus pieças y aderesos
-Dos arandelas y otra gola borgoñona
-Otro arnés de pasta milanesa con todas sus pieças y aderezos
-Otro arnés pequeño de pasta de Agusta de un menino con sus escarselillas y selada
-Un volante blanco de las armas de Colmenar
-Un manoplón de pieça doble y otro manoplón de la mesma forma forma (sic)
-Otra pieça doble de las armas de su Exª que es contra selada y otra pieça doble que así mesmo es contra selada de las armas doradas
-Un seladón de campo abierto del prim° arnés
(fol. 19 r)
-Otro seladón grande de campo abierto
-Una tarjeta de las armas de su Exª
-Otra tarjeta de montería y otra contra selada
-Dos espaldarcetes borgoñones de seguir de medio brasso
-Una cabeça de caballo de las armas de su Exª
-Otras dos cabeças de caballo de las dhas armas
-Unas coraças de tela de oro con su morrión y falgas (sic) de malla
-Otra cabeça de caballo larga
-Una pieça doble de justar y otra pieça doble del braço
-Una selada de torneo
-Un brasal de gala para tornear
-Un manoplón de campo abierto
-Otra pieça doble de braço para justas
-Otra rodilla de asero
-Un peto y espaldar negro y gola y una selada con pieças dobles todo de prueba

-Una almilla negra de costilla y otra almilla negra entera
-Otro espaldarsete dorado
(fol. 19 v)
-Otra pieça doble de seladón
-Otra pieça doble de las armas de Colmenar y una tarjeta
-Otra pieça doble de justar del primº arnés
-Otra pieça doble de selada y un manoplón de las armas negras dobles
-Un morrión de seguir de las armas de Milán
-Dos cofres timbrados y quatro arcas en que están y quedan recogidas todas las armas y pieças sobredhas
-Y otra arca más que se dixo era del cargo de Juº del Olmo que todo queda en la dha armería
Armas que están en la silla
Esta – yten se pone por ynventario las armas que están en la silla del Pan de su Exª que son las siguientes adelante
Pan y m^{es}
Yten se pone por inventario unas creces de pan y más que resultaren por las quentas que se tomaren a el tesorero y mayordomos de esta villa y de los demás lugares de los estados de su Exª de que se pondrá rasón y testimonio con este inventario en que se dio por acabado el inventario en el dho día catorze de mayo de mill y Sºs y treinta con protestasion de que cada e quando que (fol. 20 r) parezcan otros qualesquier bienes que hayan quedado por muerte de su Exª, el dho don Christobal prometió de los manifestar para que se pongan con los de este inventario: y así lo juró en forma de dro. Licen^{do} don Juan de los Ríos Muñiz. Fran^{co} Ximénez scrno.

Doc. 2.
1658, December, 22.
Post-mortem inventory of the 4th Duke of Arcos, Rodrigo Ponce de León.
Archivo Histórico de la Nobleza, Osuna, C. 1591, D. 49

(fol. 1 r)
Inventario de los bienes del Sor Duque D. R^{o}. 4 y autos que en raçón de ellos se hiçieron.

(fol. 3 r)
Este es un traslado bien y fielmente sacado de un quaderno de autos de el ynventario apreçios y sumario del caudal que se hiço por fin y muerte del Exmo Señor Don Rodrigo Ponçe de León mi señor (que santa gloria haya) Duque que fue de la Ciu^{d} de Arcos= Y asimismo de las deudas y cargas de los dos bienes y balance y resumen que se hiço del valor de dos vienes y deudas y los autos que le siguieron todo como ello está escripto que es en la forma y manera siguiente

En la villa de Marchena en veinte y dos días del mes de diciembre de mil y seisçientos y çinquenta y ocho años de orden y mandado de Su Exª Don R^{o} Fran^{co} Ponçe de León Duque de la Ciu^{d} de Arcos mi señor y de su pedimiento a su mr^{d} el Liçençiado Don Juan Françisco de Rivera asistente de esta villa con asistencia de Don Juan de Villegas Tello Cavalleriço maior de la casa de Su Exª se pidió inventario de los bienes libres que quedaron por fin y muerte de Su Exª el Exmo Señor Don R^{o} Ponçe de León Duque que fue de la Ciu^{d} de Arcos su padre y mi señor que Santa gloria haya para que en todo tiempo conste y se haga de ellos división y partición con la Exª señora y su madre y con sus señorías los señores D. Manuel Ponçe de León y Don Josephe Ponçe de León sus hermanos y con quien más fuese parte para ello y se començó a haçer e hiço en la forma y man^{a} sig^{te}

Plata a cargo de Juan Blazquez Oliber [...]

(fol. 6 v)
Carabinas, espadas anchas y montantes
-Dos carabinas conformes de Bressa las llaves de ruedacon formes con sus cubetes labrados de tres cuartas de largo
-Otras dos pistolas de Bressa de a media vara de largo grabadas de açero con casquete de acero embutido
-Otras dos pistolas de tres quartas de largo llaves de rueda conformes
-Otras dos pistolas catalanas de vara menos ochava de largo de rueda grabadas
-Otras dos pistolas catalanas de dos terçias de largo con llave de chispa

-Otras dos pistolas las caxas enbutidas de marfil conformes con llaves de rueda de más de dos tercias de largo
-Otra pistola de dos tercias de largo= de dos cañones y dos llaves de rueda la caxa enbutida de marfil
-Otra pistola de largo la caxa llana llave de rueda
-Dos bolsas de pistolas negras con sus conteras de plata forradas de terciopelo carmesí las tapaderas del mismo terciopelo bordadas de oro
-Otras dos bolsas de pistolas de vaqueta negra llanas
-Una espada ancha de a caballo la guarnición de cruz dorada antigua
-Otra espada ancha de a caballo guarnición antigua de cruz el pomo amelonado
-Un montante ancho guarnición de cruz el puño de badana
(fol. 7 r)
-Otro montante más largo y más angosto guarnición de cruz
-Otro montantillo pequeño como estoque triangulado guarnición de cruz
-Una carabina de algo más de tres quartas el cañón ochavado con cos de hierro embebida en la madera llave de golpe
-Otra carabina de dos terçias algo más de largo el cañón ochavado llave de golpe
-Un arcabucillo de vara de largo el cañón reforçado y llave de rueda que llaman el toledano
-Otro arcabuz de tres quartas de largo el cañón reforçado con su mira y punto y llave de rueda que llaman de Raposillo
-Una carabina de tres quartas de largo algo más el cañón reforçado mira el punto de la Caro Caminas (sic) llave de golpe de Julián Core enbutida de la madera
-Otra carabina de vara de largo el cañón reforçado sin mira ni punto llave de golpe
-Una escopeta de cinco quartas de largo el cañón de Gaspar la llave de golpe de Simón [...]
(fol. 8 v)
-Otro coche grande de vaqueta por de fuera y alamares de oro con botones de lo mismo forradas de terciopelo verde con dos coxines de terciopelo verde cortinas de damasco con los alamares de oro y botones que se truxo de Nápoles
-Otra carroça que vino de Madrid de terciopelado almohadas de lo mismo y estribos de vaqueta y encorado las cortinas de damasco
-Otra carroça viexa colorada toda de vaqueta cortinas de paño colorado con su encerado verde
-Otro coche viexo verde de vaqueta colorado con sus cortinas de paño verde
-Un carrocín con sus vidrieras de vaqueta colorado por de dentro con unas cortinas de damasco hechas pedaços verdes [...]

(fol. 9 v)
Xaeçes
-Quatro mochilas açules y plata
-Quatro encaladas con sus cinco borlas de seda açul y plata con todas sus plaças con sus siete sinientales para las encaladas
-Quatro cabeçadas açules y plata cumplidas
(fol. 9 v)
-Quatro reatas de seda açul y plata
-Dos pares de estribos de plata
-Una espuela de plata
-Quatro pretales açules y plata con quinçe borlas açules y plata todo con sus platas cumplidas
-Otra mochila de terciopelo açul bordada viexa
-Un caparaçón de terçiiopelo açul con dos faxuelas alrededor de brocado amarillo viexo todo
-Otro caparaçón de terciopelo carmesí tres francas de bocado amarillo viexo todo
-Unas cabeçadas viexas con hierros de bronce dorado
-Un boçal viexo de metal dorado
-Un pretal bordado de seda con hierros de metal dorados [...]

(fol. 20 r)
Tiendas de Campaña
-Cinco tiendas de campaña de lienço metido en esteras de esparto
-Dos pabellones del mismo lienço
-Una vela que sirve en el patio principal de veinte y quatro varas de largo y nueve varas de ancho [...]

Espadas de librea
-Trese espadas las guarniciones redondas cerradas dorados los puños de hilo de latón con sus hojas de Alemania y vainas
-Diez dagas con las guarniçiones y puños de la misma calidad que las espadas de la partida de arriba
-Otra espada de librea compañera de las trece de arriba ygual en todo con ellas

(fol. 20 v)
-Un alfanxe con vaina pespuntada y su tahalí de vaqueta angosta la [...]da de Xénova la hoja el puño negro de cuerno [...]

(fol. 26 v)
Sota Caballeriço
Cargo de Josephe Sánchez sota caballeriço de las cosas tocantes a su oficio

-Primeramente seis guarniciones de terciopelo verde con galón de oro y cada uno con tres vorlas grandes de seda y oro con sus riendas
-Seis sotacolas de terciopelo verde
-Dos guarniciones de terciopelo con galón de oro y hierros dorados
-Un aldabón de lança de hierro de lança de hierro (sic)
-Un coxinillo de terciopelo verde de silla
-Ocho mantas viexas de caballos tocadores y conchas todo viexo
-Tres guarniciones de mulas de coche de vaqueta y seis frenos
(fol. 27 r)
-Más otras guaniciones de tiro de coche como las de arriba
-Quince bocados de freno cinco de gineta y los demás de brida con riendas y cabeçadas
-Unas cinchas viexas de brida que por serlo mucho no se contaron y quatro guruperas de brida con sus pendientes
-Cinco cabeçadas de freno y seis pretales
-Dos cordones viexos de seda
-Dos almártagas
-Tres reatas
-Dos pares de antoxos y otra gurupera de gineta
-Un envuelto de correas viexas y riendas
-Un envuelto de clavos viexos
-Diez y seis frenos de coche con sus cabeçadas bocados y riendas
-Ocho cabeçadas de freno de coche
-Dos pares de riendas viexas
-Y dos pares de estribos
-Unas cabeçadas con sus patenas de freno de acémila
-Dos sillas de brida con sus estribos= y otro par de estribos de p^{r} sí
-Más una guarnición de coche
-La silla del catalán con sus estribos y cubierta gurupera y pretal
-Más otras dos guarniciones de mulas de coche con sus tirantes
-Y dos frenos de coche con sus cabeçadas y riendas
(fol. 27 v)
-Más dos sillas de coche de terciopelo verde y galón de oro con sus cubiertas de jerga
-Más otras dos sillas de coche con sus estribos
-Quatro pares de estribos de gineta
-Cinco estribos de brida viexos
-Quatro almohadas las dos sin cabo y un bridón
-Quatro brocas y quatro peines

-Dos cojines de litera de Damasco
-Dos porteras y un estribo y tornillo de litera
-Dos guarniciones de caballo de coche de vaqueta con fluecos de seda y oro
-Tres frenos de coche con sus cabeçadas de seda y oro y sus borlas
-Quatro guarniciones de mulas de coche con sus tirantes
-Más otras quatro tirantes de por sí
-Una silla de vaqueta y coxinillo de terciopelo açul
-Quatro encerados un brilloso y guardapolvo
-Tres adereços de sillas de brida dos bordados y uno negro= con sus estribos riendas cabeçadas y almártagas con cordones de seda y borlas guruperas con sus pendientes
-Tres pares de antoxos
-Otros antoxos de otro adereço
-Los adereços de arriba con çinchas y encerados y otra cubierta más
-Quatro guarniciones de caballos de coche de terciopelo verde con galón de oro y herraje de bronce dorado
-Dos frenos con sus cabeçadas y borlas de hilo verde
-Más dos guarniciones de lo mismo sin borlas
-Otras dos guarniciones y otros dos frenos con sus cabeçadas y borlas de hilo verde como las de arriba
(fol. 28 r)
-Tres guarniciones de caballos de coche de vaqueta negra
-Seis frenos con cabeçadas y riendas de vaqueta negra con sus guías y cinco medias guarniciones
-Quatro bocados de caballo de brida
-Seis sotacolas de vaqueta
-Dos sillas de brida con sus estribos y guruperas la una redonda
-Otros estribos de brida llaman postillones con sus acciones
-Tres medias guarniciones de coche y dos enteras
-Una silla de brida con su gurupera estribos y cincha y un lengín de cochero quatro cabeçones
-Quatro pares de riendas y unas cabeçadas de gineta
-Unas tenaças de martillos uno de ellos de herrar dos tijeras
-Dos guías de vaqueta cinco tirantes dos amarros viexos
-Una cadena de coche
-Dos hierros para los balancines
-Una haquina de dos correas
-Otras dos guaniciones de vaqueta con herraje dorado de oro y seda
-Un freno con tres borlas de seda verde y oro

-Tres cabeçadas de riendas largas
-Una guarnición de vaqueta negra guarnición digo clavaçón de hierro con fluecos açules y una tirante de cáñamo viexas
-Otras dos guarniciones de coche y otras dos medias
-Dos sillas de brida con sus guruperas una cincha y una guarnición de coche
-Dos frenos cabeçadas y riendas con tres borlas verdes y oro y sus riendas con sus flueques (sic) de seda y oro
-Dos guarniciones de herraje dorado fluecos de seda y oro
-Quatro [...] con sus fluecos de lo mismo
(fol. 28 v)
-Dos sillas de brida con sus estribos y cinchas
-Tres guruperas trece fiadores y dos cinchas de gineta
-Otras tres guarniciones de coche de vaqueta negra= una media
-Dos frenos con sus cabeçadas= Unas cabeçadas blancas
-Dos borlas de seda de hiladillo con su cordón de lo mismo
-Otras dos sillas de gineta con sus caparaçones de vaqueta negra y un açote de coche
Lo que estaba en el guardarnés viexo
-Ocho bocados de freno de brida
-Once bocados de frenos de mulas de coche
-Dos frenos de mulas de coche con sus cabeçadas y bocados
-Nueve cabeçadas viexas de monte con un bocado a la gineta quabrado
-Cuatro correones de seda de coche
-Dos almártagas a la morisca una con cordón y otra sin él
-Siete sillas viexas que no pueden servir de la brida y gineta
-Cinco sillas de gineta la una con caparaçón y las demás sin él sin pretales ni guruperas ni acciones ni estribos
-Cinco adargas
-Una silla viexa de gineta hecha pedaços
-De guarda contones y hierros de coche= trece piezas
-Nueve boxes de ruedas de coche viexas y nuevas
-De clavos viexos de rueda quatro llantas y demás hierro viexo media arroba
(fol. 29 r)
-Un pretal de cascabeles de terciopelo verde que le falta la mitad
-Una ballesta de herrado con sus gafas y arlaba
-Quatro guarniciones de coche hechas pedaços que no pueden servir
-Unos açicates
-Dos cabeçadas de paño verde viexas que no pueden servir
-Tres amarros de coche y una tirante de cáñamo

-Ocho riendas berberiscas coloradas
-Una coraça colorada
-Un par de estribos de gineta que no son hermanos
Obra que se hiço después
-Un freno nuevo que hiço Diego Cabrera para una mula en que D. Pº Francisco a Bailén
-Más dos cadenas para las haquinas de dos mulas de coche
-Dos almohadas
-Unas tenaças
-Dos arriseres para una silla de los caballos que llevó Joseph
-Dos pares de espuelas de pico de gorrión
-Seis hebillas con sus hebillones para unas guarniciones de silla
-Una manta y tocados de la haça del Marqués mi Señor de jerga
-Una jaquina de vaqueta blanca
-Un bocado de gineta
-Unos antoxos negros
-Una silla xineta con su caparaçón de baquea colorada
-Una concha y unos estribos con sus acciones y el adereço blanco y negro pespuntado con seda encarnada que es caparaçón con reatas pretal gurupera y una cubierta encerada y las correas con que (fol. 29 v) se çiñe el caparaçón mandil y una almohaça [...] herrada
-Una silla de terciopelo açul bordada de plata con todas sus guar^{nes}
-Otra silla de terciopelo verde bordada de plata con su terliz de terciopelo verde
-Dos sillas bordades de plata y acero
-Seis sillas de gineta con sus frenos
-Ocho sillas de brida con sus frenos [...].

AUÑÓN, MARQUESES DE

Doc. 1.
1604, May 17.
Inventory of the Marquis of Auñón
Archivo Histórico de Protocolos de Madrid, Protocolo 1990.

(fol. 227 r)
El marqués de Auñón Escrip[a]
Gaspar de Zeballos carta de p[o]

En la villa de Madrid a diez y siete días del mes de mayo de mill y seisçientos y quatro años ante mí el escribano y testigos Martín González veçino de esta villa y guardarropa del señor marqués de Auñón nombrado por tal por çedula del dho marqués fecha en esta villa en seis de este presente mes y año por la qual permite y consiente le entregue por Gaspar de Ceballos su criado a cuyo cargo está la dha guardarropa todas las cosas que tiene oblig[on] de entregar y dar quenta las quales le ha entregado y entrega y reçibe y son de este tenor [...]

(fol. 230 v)
-dos báculos el uno tiene un estoque dentro [...]
-Un caracol de la mar
-Una guarnición de espada dorada antigua [...]
-Onze fundas de frisa en que estaban las armas [...]
-Una pica en que estaba una horquilla [...]
-Una tienda de campaña de armar de anjeo cortada [...]
(fol. 231 r)
-Un hierro de lança, otro hierro de silla de respaldar dorado [...]
-Una llave de escopeta que tiene un barreno y una guarnición de daga vieja [...]
(fol. 233 r)
-Un báculo de nácar con un remate de plata [...]
(fol. 234 v)
-Un pistolete como de tres quartas que se compró quando Su S[a] fue a Valladolid tiénele el liçençiado [...]
(fol. 236 r)
-Un aderezo de espada con su pretina y talabarte negro liso sin daga
-Dos espadas de esgrima [...]

(fol. 237 v)
-Una silla de terciopelo viejo con pasamanos de brida

-Otra silla de brida de paño vieja
-Dos cubiertas de cuero de estas dos sillas
-Una silla de jineta de cuero leonado con su cubierta
-Otra silla de cuero con fajas de terciopelo de brida q tiene la jaca
-Dos guarniciones de brida de paño faltan las açiones para los estribos
-Otro aderezo de brida de terciopelo guarneçido de pasamanos y entorchados por los cantos y belloticas en todos los remates es nuevo y tiene pretal cabeçadas riendas gurupera y açiones y la clavaçón de él dorada
-Unas cabeçadas de terciopelo viejas con la clavaçon vieja
-Unas cabeçadas y riendas de cuero leonado que sirven a la jaca viejo
-Un adereço de brida todo de cuero que sirve a la jaca
-Un pretal de cascabeles de terciopelo carmesí guarneçido de fluecos de oro y una bolsa en la punta de la roseta aforrado en cuero colorado
-Unos açicates dorados
-Dos pares de estribos de brida con unos q tiene la jaca
-Tres bocados de jineta el uno tiene la jaca
-Otros dos bocados de brida el uno con sus copas doradas
-Un candado grande que era de almoneda
-Una silla de terciopelo negro con pasamanos y fluecos anchos de seda nueva que la tiene el S^r^ don Lorenço para su caballo
-Una adarga tiénela don R^o^
-Unos estribos dorados, un manojo de llaves
(fol. 238 r)
-Veinte cuerpos de armas de ristre de seguir con braçales enteros manoplas ristres y çeladas
-Otro arnés de seguir con perfiles de oro con braçales enteros y manoplas y su gola y çelada y este y los de arriba tienen escarcelas y golas
-Quarenta pieças de armas para cubrir las piernas desde las escarzelas a la rodilla
-Treinta y nueve pieças de armas para guarnecer sillas
-Otro arnés que tiene peto y espaldar gola çelada braçales y manoplas y çelada con sus escarçelas y espineles hasta la garganta del pie todo grabado y trençado
-Otro arnés que tiene peto y espaldar gola çelada braçales y manoplas con escarçelas y sus piernas enteras con una espuela al talón y todo él labrado con unos ramos grandes
-Otro peto fuerte y un sobre brocal y otra pieça para sobre la gola que llega a la visera de la çelada y un plumajero
-Una arandela para la lança
-Una tarjeta y quatro pieças para la cabeça del caballo y otras dos pieças para los muslos que todo parece es de labor del arnés de arriba y otra pieça para la

bragueta de las calças y otra peçeçuela para la mano de la lança y dos tornillos de este arnés que todas estas pieças son diez y ocho del cargo de Çeballos en el arnés de arriba
-Otro peto fuerte trançado con su ristre que parece hermano del arnés grabado y trançado que está cargado arriba
-Un morrión sincelado grabado y dorado en dos pieças que es como el que se dio al S^r don Lorenzo
-Tres pieças de armar mohosas
-Una escopeta de rueda que tiene por señal una tenaça
-Quatro espadas de tornear con sus tiros y vainas la una de terciopelo carmesí y las tres blancas
-Un pistolete de rueda metido en una funda de vaqueta
-Onçe sillas de armas de cordobán blanco las dos con sus pieças de armas enteras
-Dos frascos para pólvora el uno colorado y el otro de tarazea con (fol. 238 v) sus cordones ambos
-Una silla de terciopelo amarilla con sus pieças de armas grabada
-Una guarnición para acémilas de litera que es de terciopelo negro con fluecos de seda negra tiene sus cabeçadas y riendas y quatro correas del mismo terciopelo y ocho roscas de lo mismo para las varas de la litera
-Dos almártegas para acémilas de cuero negro
-Otras dos almártegas de cuero negro para acémilas
-Otra guarnición para dos acémilas de litera que es de terciopelo negro con fluecos de oro y seda y clavaçón dorada y tienen sus cabeçadas para ellas y sus bocados con sus copas doradas y las copas del freno de la acémila de cantera son vaçiadas de labor y el cabo de las riendas sus borlas de seda negra
-Dos almártegas para las acémilas del mismo terciopelo con sus flocaduras de oro y seda y sus cordones con borlas al cabo y dos guruperas de esta guarnición
-Dos pretales de lo propio para las acémilas
-Siete sobrecinchas de cáñamo con sus correas de cuero bayo cada una que son de los caballos de los hombres de armas
-Nueve pares de estribos que eran de los hombres de armas con sus açiones bayas
-Dos pares de çinchas de brida las unas de seda de colores que cada par tiene tres çinchas
-Ocho pretales de cuero bayo de los caballos de armas
-Ocho guruperas de lo mismo que eran cuellos y cabeçadas
-Un pretal y riendas y gurupera francesa de cuero blanco
-Seis riendas para los caballos de las armas de cuero bayo
-Ocho pares de çinchas de los caballos de las armas que cada par tiene tres çinchas

-Siete cabeçadas de cuero bayo y pespuntado de caballos de caza
-Un pretal de cuero negro y gurupera de brida
-Dos pares de riendas de jineta de cuero blanco
-Un bocado de caballo de jineta viejo
(fol. 239 r)
-Otra çincha de jineta
-Otras çinchas de brida
-Çinco pares de espuelas de los hombres de armas
-Siete talabartes de espada que eran de los hombres de armas
-Seis pretinas de cuero bayo conforme a los tiros
-Dos talabartes y una pretina de cuero negro con clavaçón dorada
-Otras riendas de gineta de cuero blanco
-Dos guruperas y un pretal de cuero negro de brida
-Otras riendas de jineta
-Tres çinchas de brida de cáñamo
-Otra çincha de cáñamo
-Otra çincha de cáñamo de jineta sin correa
-Çinco açiones de cuero negro
-Otra gurupera de las sillas de armas
-Un pretal de cascabeles de terciopelo carmesí [...].

BÉJAR, DUQUES DE

Doc. 1.
1592, May 2.
Inventory of the armoury of the Duke of Béjar, Francisco de Zúñiga.
Archivo Histórico de la Nobleza, Osuna, C. 229, D. 43.

(fol. 22 r)
Cabeza de inventario de los bienes del Señor duque de Béjar don Francisco de Zúñiga difunto

En la villa de Béjar a dos días del mes de mayo de mil e quinientos y noventa e dos años por ante mí el dicho escribano y testigos pareció presente la Señora doña Brianda de la Cerda duquesa de Béjar y en cumplimiento de la dicha comisión e de lo que por ella se le manda cerca de hacer el inventario de los bienes del duque don Francisco difunto su marido= dijo que empezando hacer el dicho inventario declaraba e declaró que en las casas de palacio que están en la dicha villa de Béjar al tiempo y cuando el dicho su marido y su Señoría se fueron de ella a la villa de el Escorial a hablar a Su Majestad con su gente e familia quedó en la dicha casa en el aposento de los retratos en el aposento de Su Señoría junto al retrete una arca de hierro con dos cerraduras y el [...] y cerrada y con más de cuatro mil ducados en dineros los cuales quedaron en ella al tiempo y cuando se salieron de la dicha casa e como el dicho duque don Francisco murió en la ciudad de Ávila sin poder volver a la dicha villa ni esta que declara volvió [...]

(fol. 37 r)
En la villa de Béjar en seis días de mes de mayo de mil y quinientos y noventa y dos años, la dicha duquesa doña Brianda de la Cerda en respuesta de lo proveído por el dicho juez dijo que ya su Señoría tiene pedido y requerido al dicho Gabriel de Busto juez susodicho que en virtud de la dicha su comisión haga, prosiga el dicho inventario y en él ponga los bienes que hallare del duque difunto porque su Señoría de presente no sabe de bienes del dicho duque más de los que tiene declarados ante Juan González ejecutor y él tiene llevados y embargados y depositados en virtud de la comisión que le fue dada en la real Chancillería de Valladolid y los que su Señoría supiere y tuviere en su poder protesta de los manifestar y declarar y hacer juramento que de derecho se requiere [...]

(fol. 38 v)
E después de los susodicho en la villa de Béjar en el dicho día, mes y año dichos continuando las diligencias tocantes al dicho inventario yo el dicho escribano fui

a la casa e palacio del dicho señor duque de Béjar para me informar quien tenía los dichos bienes y a cuyo cargo estaban y de algunos criados y gente de la dicha casa fui informado que en poder de Juan de Herrera cerero vecino de la dicha villa estaban ciertos bienes y cosas que en él había dejado depositado Juan Gz° Ejecutor de la Real Chancillería de Valladolid de los que habían quedado por fin e muerte del dicho señor duque difunto que dichos los que estaban en la dicha casa e palacio del dicho señor duque de Béjar que dicho es con el cual dicho Juan de Herrera hize la diligencia siguiente

Busto (signed)

Este dicho día, mes e año dichos, continuando el dicho inventario e diligencias de él por la tarde a las horas señaladas por mí el comisionado en la dicha casa e palacio hice parecer ante mí al dicho Juan de Herrera depositario de los dichos bienes que de suso se hace mención y con juramento que primero hizo en forma de derecho declaró que era verdad que el dicho Juan González ejecutor había depositado en él ciertos bienes y cosas de los que habían quedado por fin y muerte del dicho señor duque de Béjar de los que se habían hallado en la dicha casa (fol. 39 r) y palacio del dicho señor duque que hoy es de Béjar como parecía por un inventario y depósito de ellos que en el fuese el cual originalmente exhibió ante mí el dicho escribano para los poner en este inventario del cual a la escritura fue compuesto en ese dicho inventario en la forma e manera siguiente

Primeramente se puso en este inventario por mí el dicho Gabriel de Bustos escribano y testigos yusoescritos [...]
(fol. 39 v)
-Ocho serones con armas viejas [...]
-Dos candiles de latón para poner en las paredes
-Nueve espadas de torneo con sus guarniciones las tres y las seis sin guarniciones e tres talabartes
-Dieciocho lanzones con sus fundas en los hierros
-Una lanzuela pequeña
(fol. 40 r)
-Cuatro tiros grandes de artillería que están en el huerto de Andrés de Mora
-Un tiro de cámara de doce piezas que encaja la una en la otra
-Otros diez tiros pequeños
-Dos arcas con pelotas de hierro
-Treinta arneses con sus celadas, grebas y quijotes
-Treinta cruceros de palos en que están puestos los arneses

-Cincuenta arneses como los de suso y otro arnés de seguir con veinte y una piezas grandes y pequeñas, digo que faltaron unas manoplas para estar cumplidos los dichos cincuenta arneses
-Noventa y ocho alabardas y venablos
-Cien testeras de caballos con sus rosetas y puntas
-Ciento e cincuenta e dos piezas grandes y pequeñas de arneses de justar
-Ciento y dos guarniciones de espadas doradas e un pomo e una guarnición en las cuales faltó una contera y las dichas guarniciones están con sus puños de hoja
-Noventa y ocho pares de estribos. Los tres sin acciones y los demás con ellas blancas de vaqueta
-Veinte e cinco piezas de arneses de justar
(fol. 40 v)
-Otras setenta y siete piezas grandes y pequeñas de los dichos arneses
-Otras ciento e doce piezas de los dichos arneses de justar grandes e pequeñas
-Otras ochenta e siete piezas grandes e pequeñas de los dichos arneses
-Otras cuarenta e cinco piezas grandes e pequeñas de los dichos arneses
-Noventa y ocho frenos los noventa y seis con riendas blancas de vaqueta e los dos restantes sin ellas
-Mil y setecientas y ochenta e cuatro piezas de talabartes entre ellas cuales hay ciertos tornillos de las sillas de armas
-Quince arandelas de armas
-Treinta piezas de armas dobles
-Sesenta pares de cinchas de cáñamo dobles para los caballos de las armas
-Sesenta blancas de enristrar con dos que llevó el juez de la cámara del duque mi señor que Dios tiene que había venido delante para muestra
-Una arquilla con ciertas piezas menudas, clavillos y otras cosas para apretar los arneses, sin cerradura
-Setenta hierros para las lanzas
-Una asta de fresno para estandarte
-Cuatro varillas de hierro para la dicha lanza de estandarte
(fol. 41 r)
-Setecientos e cincuenta clavos de hierro para las setenta lanzas
-Unas tijeras para cortar armas
-Una bigornia e un martillo e doce punzones para las dichas armas
-Un punzón con su cabo torneado para guarnecer las armas
-Un portallave y un arandel de hierro para la lanza del estandarte
-Veinte y ocho pares de espuelas para los hombres de armas
-Una bolsa para el estandarte

-Catorce varas de jerga en cuatro pedazos en que están revueltas las dichas lanzas de enristrar
-Quince sillas de armas con sus arzones delanteros y traseros acerados con sus fundas de frisa colorada
-Cien cabezadas de caballo blancas de vaqueta
-Cien petrales de caballos blancos de vaqueta
-Cien gruperas de caballos blancas de vaqueta
-Siete sillas de la brida con sus conteras de acero
-Diez y ocho sillas con sus conteras de acero con sus fundas de frisa colorada. Las tres de terciopelo azul y las demás de cuero blanco
-Cuarenta y tres piezas de sillas
-Tres sillas de terciopelo azul viejas de pechos

(fol. 53 v)
Inventario de bienes
Del cargo de Bernal Ramírez
E después de lo susodicho en la dicha villa de Béjar a cinco días del dicho mes de mayo del dicho año continuando el dicho inventario de bienes, yo el dicho Gabriel de Busto escribano en cumplimiento de lo pedido por su Señoría de la dicha Duquesa de Béjar como se contiene en el pedimiento de esta otra parte contenido, hice parecer ante mí a Bernal Ramírez, vecino de la dicha villa y de él recibí juramento en forma de derecho que dirá verdad el cual habiéndole fecho declaró que era verdad que de dicho Juan González ejecutor había dejado depositados en él ciertos bienes y cosas de las que habían quedado por fin y muerte del dicho duque don Francisco como constaba por el inventario y depósito que en él se había hecho el cual exhibió ante mí para que saque los dichos bienes y los ponga en este inventario y fecho lo suso dicho se le vuelva originalmente para (fol. 54 r) dar cuenta de los dichos bienes siendo testigos Gaspar Díez y Francisco García vecinos y estantes en la dicha villa

Ante mí
Gabriel de Busto escribano
(signed)

En la dicha villa de Béjar este dicho día, mes y año dichos yo el dicho Gabriel de Busto escribano habiendo visto lo susodicho digo que estoy presto de lo cumplir y en su cumplimiento hice poner el dicho depósito en inventario que ante mí exhibió el dicho Bernal Ramírez en este dicho inventario los bienes del siguiente

Primeramente [...]

(fol. 64 v)
-Iten veinte y dos aderezos de guarniciones de caballos de cuero blancos de vaqueta con sus estribos estañados e además clavazón, e veinte y dos pares de espuelas con unas negras de posta
-Unas cabezadas de mula viejas e una cincha a gineta negra [...]
(fol. 65 r)
-Iten un aderezo de gineta de plata e negro que tiene mochila de terciopelo negro con flocadura franjas e cordoncillos de plata con reata de seda e pretal e cabezadas e almártaga de cuero con clavazón plateada, rienda e estribos, espuelas plateadas e su cuerda
-Iten otro aderezo de caballo a la gineta cumplido como el de suso antes de este [...]
-Iten un aderezo de gineta de campo bayo e leonado de cuero que tiene las piezas siguientes, cabezadas, clavazón dorada, pretal de la misma manera, reata e riendas de la misma manera, espuelas e cinchas y estribos dorados
(fol. 65 v)
-Otro aderezo de terciopelo negro e franjoncillos de seda negra que todo es de brida que tiene cabezadas, grupera e pretal e riendas e cinchas negras e amarillas y estribos negros nuevos barnizados
-Otro aderezo entero de brida de cuero negro pespuntado de verde con clavazón dorada que tiene cabezadas, riendas, pretal grupera, cinchas y estribos con acciones y espuelas doradas
-Otro aderezo de terciopelo negro que tiene cabezadas, riendas, pretal e grupera e almártaga e cinchas de negro e amarillo y estribos dorados sin acciones
-Otro aderezo de brida de cuero negro que tiene cabezadas, pretal, grupera e riendas e acciones e almártaga con clavazón dorada
-Iten otro aderezo de gineta que tiene caparazón de paño negro guarnecido de cuero pespuntado e pretal, cincha e reata de cuero
-Seis ballestas e cinco fajas e una horquilla de hierro e una abancuerda con su cinto e seis fundas de ballestas sanas e rotas e unas fundas de gafas e aljaba de terciopelo verde con cuerdas de seda
(fol. 65 v) [...]
-Iten otra aljaba de terciopelo verde como la de suso con sus jarras dentro [...]
-Iten una bolsa de ballestero para la cinta
-Unos acicatillos pequeños
-Iten un carcaj de cuero bayo con seis virotes
-Otro carcaj enterrado contiene virotes de palomear

-Una caja de madera blanca nueva chica con cincuenta y cuatro virotes e jaras de diferentes hechuras
-Iten cuatro cofres viejos en que van metidos las cosas de suso
-Cinco cinchas ginetas de diferentes colores
-Un pretal de terciopelo azul con sus cascabeles
-Unas acciones negras nuevas de estribos
-Iten dos bocados de frenos de brida dorados
-Una funda de badana blanca para espada
-Cuatro fundas de estribos de la gineta e sus caperucetas [...]

(fol. 124 r)
Posesión y Cargo
De Martín de Terrades
Armero
En la villa de Béjar a doce días del mes de junio de mil e quinientos e noventa e dos años, yo el dicho Gabriel de Busto hice parecer ante mí a Martín de Terrades criado que al presente es del señor duque de Béjar y armero que fue del señor duque difunto del cual fue recibido juramento en forma de derecho que dirá verdad y habiendo jurado e siendo preguntado diga y declare que instrumentos y cosas y piezas tocantes a la armería y armas del dicho señor duque de Béjar difunto tiene este que declara con que aderezaba las dichas armas para lo poner por inventario y dar le posesión de ello a la parte de la dicha señora duquesa de Béjar como a heredera del dicho señor duque de Béjar difunto el cual después de haber jurado y estando en la pieza donde están las dichas armas que para este efecto se abrió en la casa e palacio del dicho señor duque de Béjar que hoy es declaró que lo que está a cargo de este que declara y con que aderezaba las dichas armas es lo siguiente

Inventario de herramientas de las armas
-Primeramente seis limas tres chicas y tres grandes con sus mangos de madera
-Un bruñidor de acero
-Tres punzoncillos de acero y hierro
-Un martillo de acero quebrado sin cabo
(fol. 124 v)
-Unas tijeras medianas de cortar correas
-Un cincelillo pequeño
-Un raspadorcico para los bordes
-Un almirez de hierro pequeño con una mano calzada de acero para moler el esmeril
-Un saco roblones de acero y hierro

-Un asentador de hierro y acero
-Un rebotador de lo mismo
-Un contrapunzón pequeño de lo mismo
-Una lancetilla de acero con su mango de madera
-Una macetilla de roblar los clavos de la celada
-Una sufridera de hierro para desguarnecer en las celadas
-Una llave de armar los tornillos y armar
-Una estaca de hierro calzada de acero con su pie de madera

Todo lo cual es lo que estaba a cargo de este que declara e no otra cosa alguna que se sepa y entienda so cargo del dicho juramento e protestó que todas las veces que otra cosa venga a su noticia o sepa de ella lo manifestará lo cual entregó para que se cumpla lo que se le manda y firmólo de su nombre va testado= de terraz/ Un saque de roblones/ no vala

Ante mí Gabriel de Busto escribano
Martín de Terrades
(signed)

Este dicho día, mes y año suso dichos el dicho Martín de Terrades so cargo del dicho juramento declaró que en poder de Sebastián Ramírez repostero, había unos pedazos (fol. 125 r) de esmeril del cual dicho Sebastián Ramírez que estaba presente se recibió otro tal juramento en forma que dirá verdad y habiendo jurado declaró estar en su poder y ser a su cargo trece piedras de esmeril grandecillas, medianas y pequeñas, las cuales entregó para el efecto que lo demás del cargo del dicho Martín de Terrades y lo firmó de su nombre y esto dijo ser verdad so cargo del dicho juramento

Ante mí Gabriel de Busto escribano
Sebastián Ramírez
(signed)

(fol. 137 r)
Más inventario hecho en Madrid de los bienes del señor duque de Béjar difunto

En la villa de Madrid a siete días del mes de julio de mil e quinientos e noventa e dos años por ante mí el dio Gabriel de Busto escribano en cumplimiento [...].

Doc. 2.
1632, January 29.
[Copy of 1637].
Inventory and appraisal of the Béjar armoury.
Archivo Histórico de la Nobleza, Osuna, C. 244, D. 43–44.

(fol. 2 r)
Inventario de los bienes, joyas y dinero que quedaron por bienes libres de la disposición del excelentísimo señor Don Francisco Diego López de Zúñiga y Sotomayor, Duque de Béjar y de Mandas difunto, Año de 1637

Copia simple, Juan González

(fol. 136 r)
El licenciado Antonio Amigo de Arenillas, en nombre de mi parte digo que a mi noticia es venido que demás de los bienes inventariados a mi pedimiento se resta por inventariar la armería y cosas anexas a ella como bienes libres y sin embargo de que se diçe ser las dichas armas anexas a laca (sic) casa y mayorazgo de Su excelencia el Duque de Béjar y como tales vinculadas porque en todo tiempo se ha visto no haber habido del oficio de mi parte omisión ninguna a V.M. pido y suplico mande se haga el dicho inventario sin perjuicio de cualquier derecho que pueda haber en razón del dicho vínculo y mayorazgo a las dichas armas, y atento están tasadas judicialmente antes de ahora se saque un tanto del dicho inventario y tasación con citación de los interesados y se me vuelva el original que presento con el juramento necesario pues es justicia que pido y para ello etc., El licenciado Antonio Amigo de Arenillas

Auto
Que se ha por presentada la dicha petición y el dicho inventario y tasación de la armería y con las calidades que refiere la petición, sea por inventariado y tasada la dicha armería y se saque un tanto en pública forma= del dicho inventario y tasación que se presenta y se ponga con el inventario que hoy se va haciendo proveyólo el licenciado Don Juan de la Vega en ejecución de esta villa de Béjar en ella en diez días del mes de julio de mil y seiscientos treinta y siete años el licenciado Don Juan de la Vega el licenciado Antonio Amigo de Arenillas ante mí Juan González

(fol. 136 v)
Cumplimiento

En la villa de Béjar, en el dicho día mes y año dichos yo el presente escribano en ejecución del dicho auto hice sacar y saque el inventario y tasación de los bienes de la armería es como se sigue

Inventario y tasación
de la armería
En la villa de Béjar en veinte y nueve días del mes de enero de mil y seiscientos y treinta y dos años para hacer la tasación de la arme (sic) de la casa de su excelencia el duque mi señor que Dios guarde su merced del licenciado Baltasar Blázquez de Frías alcalde ordinario en esta dicha villa y don Alonso García de Aponte y don Fernando de Torres y Medrano y don García de Herrera y Castilla en nombre de sus partes juntamente con Bartolomé de Lomana armero, fueron a la armería de su excelencia e hicieron el inventario y tasación siguiente:

-Primeramente se pone por inventario sesenta arneses de seguir con sus testeras y todas piezas tasado cada arnés a veinte y cuatro ducados
-Iten se pone por inventario diecinueve arneses de justa con piezas dobles tasados por el dicho Bartolomé de Lomana armero a seis mil maravedíes cada arnés.
-Iten se pone por inventario quince arandelas de lanzas de ristre tasadas por el dicho Bartolomé de Lomana tasada a siete reales cada una
-Iten se pone por inventario una celada y dos brazales con sus guarda(fol. 137 r) brazos y gola y mandiletes tasado todo por el dicho tasador en cuarenta y cuatro reales.
-Un arnés de la persona de su excelencia el duque mi señor de la cinta para arriba con sus mandiletes y celada tasado en once mil maravedíes por el dicho tasador.
-Iten otro arnés pequeño del conde mi señor, tasado en cuatro mil maravedíes por el dicho tasador.
-Iten cuarenta y tres espadas de armas anchas con sus petrinas y talabartes tasada cada una por el dicho Bartolomé de Lomana a dos ducados cada una.
-Un portaclave tasado en dos ducados por el dicho tasador
-Iten se pone por inventario quarenta y cuatro guarniciones sobredoradas de espadas con treinta y cinco pomos de lo mismo y guarniciones de hebillas y hierros para talabartes y algunas conteras tasado por el dicho Bartolomé de Lomana en cuatro reales cada guarnición.
-Más otra espada tasada en dos ducados por el dicho tasador.
-Iten se pone por inventario cincuenta y siete alabardas con una quebrada tasadas por el dicho Bartolomé de Lomana a cuatro ducados cada una.
-Iten doce venablos de guarnición tasados cada uno por el dicho tasador a tres ducados.

-Iten se pone por inventario unos pedazos de armas viejas tasadas en veinte reales por el dicho tasador.
(fol. 137 v)
-Iten se pone por inventario un morrión de escaramuza con guarnición de plata y el dicho Bartolomé de Lomana tasó el morrión sin la plata en seis reales la plata se tasó en otros cien reales.
-Iten se pone por inventario dieciocho venablos de montería tasados por el dicho Bartolomé de Lomana tasador a once reales.
-Iten se pone por inventario doscientas picas tasadas en once reales cada una.
-Más una bigornia y unas tenazas de hierro tasado por el dicho Bartolomé de Lomana en doce reales.
Y en la forma dicha se hizo el dicho inventario y tasación por su merced el dicho alcalde y el dicho Bartolomé de Lomana tasador presentes las partes interesadas y el dicho Bartolomé de Lomana debajo del dicho juramento declaró estar fecha bien y fielmente sin agravio alguno y su merced el dicho alcalde lo firmó y el dicho tasador y partes testigos presentes a lo que dicho es don Antonio Castilla de Zúñiga y Diego Gómez y el bachiller Blázquez de Frías= don García de Herrera y Castilla= don Alonso García de Aponte= don Fernando de Torres y Medrano= Bartolomé de Lomana= pasó ante mí Antonio Álvarez Noguerol enmendado= doce= y testado= de lome= no valga

Fecho y sacado corregido y concertado fue este traslado con su original que volví a la parte y para que a ello conste en ejecución (fol. 138 r) del auto que va por cabeza de este traslado lo signé yo Juan González escribano público del número de la villa de Béjar y su tierra, con aprobación en el dicho día mes y año dichos siendo testigos Diego Manrique de Castro y Diego García vecinos de esta dicha villa= En testimonio de verdad Juan González

Sillas de armería
En la dicha villa de Béjar en el dicho día diez de julio de mil y seiscientos y treinta y siete años el dicho licenciado Arenilla en nombre de su parte presentó ante el dicho corregidor un inventario y tasación fecha de las sillas y otras cosas de la armería y pidió sea el pie en el inventario y se le vuelva original en la forma y debajo de la presentación de los arneses y demás cosas inventariadas de dicha armería y el dicho corregidor mandó se acopiase y desde luego lo hubo por inventariado y tasado y yo el escribano lo hice sacar y saqué a la letra que su tenor es como se sigue

En la villa de Béjar en los dichos veinte y nueve del dicho mes de enero del dicho año ante su merced del licenciado Blázquez de Frías, alcalde ordinario en esta villa los dichos don Alonso de Aponte, don Fernando de Torres y don García de Herrera y Castilla dijeron que para tasar los coches, literas, sillas de manos y guarniciones y otras cosas tocantes a este género nombraban y nombraban por tasador a Antonio Pérez Guarnicionero vecino de esta villa su merced le hubo por nom(fol. 138 v)brado y le mandó lo acepte el cual lo aceptó y juró a Dios nuestro señor y a la señal de la cruz de que bien y fielmente hará la dicha tasación sin que en ella haya fraude ni engaño alguno y así juntos ante mí el dicho escribano hicieron la tasación de los dichos bienes en la forma siguiente

-Primeramente se pone por inventario cinco sillas que están en la una sala de la armería que son de borenes de terciopelo de armas traídas tasadas por el dicho Antonio Pérez a doce ducados cada una unas con otras con sus cinchas y estribos.
-Iten cincuenta y cinco digo sesenta y cinco sillas de borenes buenas tasadas por el dicho tasador sin estribos ni aderezos ni cinchas a cinco ducados cada una.
-Iten se pone por inventario una silla guarnecida digo con caparazón de terciopelo vieja negra tasada en dos ducados.
-Iten veintiséis aderezos y guarniciones de sillas de armas que tiene cada aderezo cabezadas freno pretal grupera y estribos y cinchas y acciones tasado cada aderezo por el dicho Antonio Pérez tasador en dos ducados digo a treinta reales.
-Iten diez y seis pares de cinchas a tres reales cada par tasadas por el dicho tasador
-Iten se pone por inventario sesenta y seis pares de estribos tasados por el dicho tasador a seis reales cada par.
-Iten cincuenta y un aderezos de caballos que son cabezadas, bocados, riendas, pretal (fol. 139 r) y grupera y acciones tasado cada aderezo por el dicho Antonio Pérez tasador a veinte reales.
-Iten otros veinte aderezos de caballos de vaqueta vaya que son dobles pespuntadas que se entiende cabezadas riendas pretal acciones y gruperas con sus caídas tasadas cada una a treinta reales.

Doc. 3.
1682.
Inventory of the Béjar armoury.
Archivo Histórico de la Nobleza Osuna, C. 255, D. 143.

(fol. 2 r)
1682
Inventario de las alhajas que hay en la armería del Duque mi señor entregadas a Juan de Alarcón armero

(fol. 3 r)
Inventario de las piezas y alhajas que están en la armería del Duque mi señor y de los arneses que están limpios y de los que no lo están, hecho por la contaduría de su excelencia

-Primeramente quince arneses limpios y otro dorado que vino de Madrid que son diez y seis los cuales tiene entregados Juan de Alarcón armero y están a su cuidado para entregarlos cuando se le pidan
-Iten sesenta y nueve aderezos de hombres armados con petos, espaldares, golas y morriones con sus celadas salvo cinco que les faltan faldones y seis que les faltan muslos y piernas y todos están armados en sus palos
-Iten un peto y una borgoñota, piernas y muslos que es parte de otro hombre armado
-Iten cincuenta alabardas, las dos los palos quebrados con sus borlas y flocadura en que entra un chuzo grande
-Iten noventa y cinco frentes de caballo algunas con sus puntas
-Iten sesenta y cinco sillas de caballo con su herraje por delante y por detrás, las cinco de ellas forradas de terciopelo y las demás de ante
-Un par de estribos de hierro con sus acciones
-Un manojo de gavilanes de guarniciones de espada dorados tiene veinte y ocho
-Nueve hombrilleras viejas
-Cuatro morriones viejos, que les faltan a tres piezas cada uno
-Doce brazaletes sin hombrilleras
-Cuatro espinilleras
-Un peto y un espaldar con faldones
-Una espinillera entera con su zapato
(fol. 3 v)
-Algunas piezas descabaladas viejas para aderezar las buenas
-Ocho piezas de caballos
-Nueve barbilleras de hombres

-Ciento y cuarenta y cinco cabezadas, gruperas y riendas, y otras correas de vaqueta
-Diecinueve tiros de espada de vaqueta viejos
-Veintidós cinchas de caballo
-Iten una arca cajón vacío
-Iten un cofre barreado vacío
-Iten una mesa de castaño

Herramientas
-Un tornillo de hierro
-Una bigornia de hierro mediana
-Dos martillos pequeños
-Seis limas, cuatro grandes y dos medianas
-Cuatro rascadores
-Tres punzones de hierro
-Dos tenazas la una de muelle y la otra de tornillo
-Y más dos candiles
-Y unas tijeras grandes
-Una rueda grande que es instrumento para limpiar las armas

Todo lo contenido en este inventario está en la armería del Duque mi señor y a cargo de Juan de Alarcón armero el cual se obligó a entregarlo en la conformidad que va dicho cada y cuando que por el Duque mi señor u otra persona en nombre de su excelencia se le fuere mandado y lo firmó en Béjar a 2 de enero de 1682

Juan de Alarcón
(signed)

(fol. r.)
Cargo que se le hace a Juan de Alarcón maestro armero de todas las herramientas que se le entregan para el servicio de limpiar las armas de la armería de Duque mi señor que remitió de Madrid el licenciado Don Juan del Carpio son las siguientes

-Primeramente una bigornia de acero
-Dos entenallas de lo mismo
-Cuatro rascadores
-Unas tijeras grandes de la Puerta Cerrada
-Dos limas grandes
-Otras dos medianas
-Otras dos más pequeñas
-Dos martillos medianos

-Tres punzones
-Más tres bordones. Estos bordones son cuerdas tripas que andan la rueda y se gastaron

Todas las cuales dichas herramientas recibió el dicho Juan Alonso de Alarcón y se obligó de entregarlas cuando el Duque mi señor fuese servido de mandárselo y lo firmó en Béjar a 1° de Noviembre de 1680

Juan de Alarcón
(signed)

Estas herramientas y las de este otro inventario son todas unas

Más se le entregó al dicho Juan de Alarcón en 4 de enero de 1681 un tornillo de hierro grande con su maza para limar herramientas de que ha de dar cuenta

Juan de Alarcón
(signed)

(fol. v)
66-arneses demás del uno que está colgado
36 alabardas

(fol. 4 r)
Y asimismo se entregó de trece frentes de caballos que están con los arneses limpios

Juan de Alarcón
(signed)

En 1° de Marzo de 1682 se volvió a registrar todas las armas, piezas y demás alhajas contenidas en el inventario que han estado a cargo de Juan de Alarcón armero y habiendo visto y registrado como va dicho, parece no ha faltado cosa alguna y estar todo como se contiene en dicho inventario según lo cual ha cumplido y salido el dicho Juan de Alarcón de toda obligación
Don Sebastián de Potestad
(signed)

Después de haberse registrado se volvió a ver y parece hay en dicha armería setenta hombres con petos, espaldares, morriones, brazaletes y hombreras demás del que está puesto en la pared.

Doc. 4.
1712, December 29.
Inventory of the Béjar armoury.
Archivo Histórico de la Nobleza, Osuna, C. 258, D. 30.

(fol. 3 r)
Ynventario de la Armadura de cuerpos de hombres y otras armas y peltrechos (sic) de guerra que se hallan en la Armería de (sic) Duque mi señor

-Primeramente un cuerpo entero con la armadura de charol negro de la China con su morrión, orejas, y diadema, dorado el morrión
-Cinquenta cuerpos de hombres con armadura entera, de peto, espaldar, muslos, y piernas, con sus morriones todos, y cubierto cada uno con fundas de anjeo blanco
-Un cuerpo entero dorado todo y el morrión con su peto, espaldar, braços, muslos, y piernas
-Otro cuerpo pequeño como de paje de jineta con su armadura con al menos las manos y este y el dorado también tienen sus fundas
-Diez morriones sueltos son cuerpos que están a los pies de los otros cuerpos
-Otros seis morriones que están sueltos sin cuerpos puestos arriba en la cornisa de la pieza tres en una frente, y tres en otra
-Veinte y ocho petos que están colgados en la pared de mano derecha como en tramos
-Diez y ocho espaldares q están interpolados con los petos en el mismo sitio
-Sesenta y dos frontadas de caballos que unas tienen estrellas y otras no
-Quarenta y tres muslos sueltos
-Catorze canillas
-Veinte y tres faldillas
-Seis brazaletes
En este lienzo de pared están colgadas las dhas piezas que suman ziento y noventa y quatro. Estas son las piezas sueltas q están colgadas

(fol. 3 v)
En el lienço de pared que está a mano izquierda enfrente del otro, tiene las pieças siguientes
-Primeramente veinte botas en dha pared
-Quarenta y tres broçoletes (sic)
-Veinte frontadas de caballo unas con estrella y otras no
-Más veinte y quatro faldillas
-Doze muslos

-Nueve esclavinas
-Ocho canillas
Ymportan las partidas dhas ciento y treinta y seis piezas
En la pared que haze cabezera a la pieza es lo siguiente
-Dos frontadas doradas embutidas
-Dos faldillas doradas embutidas
-Dos hombres dorados también embutidos
-Una esclavina dorada embutida
-Un peto y espaldar azerado que vino de Flandes pavonados a tiro de mosquete forrados en felpa encarnada
-Más zinco petos y un espaldar
-Onze frontadas de caballos
-Tres esclavinas
-Dos brozeletes (sic)
-Quatro muslos
-Tres faldillas

Ymportan las partidas dhas treinta y ocho piezas

En la pared de los pies de la pieza es lo siguiente
-Primeram[te], veinte brozeles
-Cinco botas
-Quatro esclavinas
-Seis frontadas de caballos
-Quatro muslos
-Seis canillas
-Quatro faldillas

(fol. 4 r)
Ymportan las partidas dhas Quarenta y nueve Piezas

Espadas
-Una espada ancha sin guarniz[ón], que tiene en la canal unas letras, que dizen: timebunt Gentes: y esta se tiene por del Maiorazgo de la Casa de Zúñiga
-Otra espada chiquita de niño de medio talle que fue del Duque mi Señor Don Manuel quando lo era con guarnición de la puerta Zerrada, y puño de hilo de plata que está desbaratado, y tiene vaina
-Otra de marca de medio talle con guarnición calada, y puño de hilo de plata con su vaina
-Otra espada ancha de conchas burilada la hoja de canal, por una parte y por otra
-Otra espada del mismo género

-Otra espada ancha con una guarniz^{ón}, de conchas burilada
-Una hoja ancha amolada sin azicalar de Gómez Carbonel sin guarniz^{ón}, ni vaina
-Otra espada ancha corba, y acanalada hasta la puta guarniz^{ón} de conchas lisa por un lado
-Un cuchillo ancho de dos cortes, el pomo dorado sin guarnizión
-Más otra espada ancha con guarniz^{ón}, de varillas
-Otra espada ancha con conchas buriladas
-Una espada ancha con tres canales con unas letras que dizen, audazes fortuna iubat, timidos que repellet. con guarnición de varillas
-Una hoja ancha también con tres canales cortas con rótulo de Fran^{co}, Corder (fol. 4 v) sin guarnizión
-Una espada ancha de un corte que junto la cruz tiene las armas de la Casa de el Ynfantado, y un mote que dize Marchio Rodrigo de Vibar y un pomo antiguo dorado
-Una catana con su vaina de charol, y un puño a lo morisco de tres quartas de largo
-Otra catana de dos quartas, y media de largo puño morisco con su vaina de charol
-Un cuchillo de monte de media vara de largo, y tres dedos de ancho con puño de caoba y vaina de felpa negra y le falta el cañabete
-Otro cuchillo de monte de una terçia de largo con cañabete y abuja (sic), con brocal, y contera de plata, y tres asillas en el brocal
-Yten una espada chiquita que era de Don Pedro mi Señor quando se criaba con guarnición de la puerta zerrada, y puño de hilo de plata
-Otra espada estoque de marca con guarniz^{ón} de la puerta zerrada, y puño de hilo de plata
-Otra espada de marca de medio talle con guarnición calada de Milán
-Un alfanje corbo morisco con una guarnizión dorada de hechura de un martillo
-Un arma como chuzo de tres quartas y media de largo metida en una vaina de badana
-Un cuerno de unicornio de tres quar(fol. 5 r)tas, y media de largo guarnecida la cabeza con plata, y una cadenilla de que pende con su aldaba
-Otro pedazo de poco más de una tercia que de la cabeza faltan algunas astillas (sic)

Armas de Fuego
-Un trabuco que llaman el Cachorrillo que hizo Juan Gutiérrez de una tercia de largo que haze bala mosquetera, y tiene secreto dentro del cañón con su llave y caja

-Un par de pistolas con los caños grabados y dorados por partes, y unas llaves, muy antiguas de rueda
-Una carabina de tres quartas y quatro dedos de largo con llave inglesa y su caja
-Otra carabina de dos tercias de largo y la llave hecha en Bruselas con su caja
-Otra carabina del mismo largo con llave de Ripol (sic), y caja de nogal
-Otra carabina con llave española del mismo largo
-Otra del mismo largo con llave francesa
-Otra carabina de dos tercias de largo con la llave de Ripol (sic) grabada la culata con latón

Todo esto quedó metido en el cofre, y en la pieza de la armería y también una caja con tres hierros de metal que parecen de Mathemáticas y un palo con tres tornillos que también parece de lo mismo

-Una aljaba con sesenta y seis flechas y algunas no tienen hierro que trujo de Buda (Plaza del Reino de Hungría) el (fol. 5 v) Exmo, Señor Marqués de Valero, quando vino acompañando el cuerpo del Exmo, Señor Duque de Béjar Don Manuel mi Señor su hermano (q Dios en Santa Gloria haya) que falleció sobre la misma Plaza el año de mill seiscientos y ochenta y seis.
-Una escopeta larga con letras góticas y la recámara grabada con figuras y dicen las letras Dôns; Dôns, noster quam admirabile est nomen tuum
-Un par de pistolas de dos cañones y dos llaves cada una, las conchas, y los casquetes de plata, y las llaves son españolas
-Otro par de pistolas largas de arzón y las llaves de rueda
-Un trabuco corto rayado que dizen dio Don Antonio Ortiz con su llave española
-Otro par de pistolas de rueda
-Otro par de pistolas pequeñitas como de faldriquera con sus llaves de rueda
-Otra pistola de rueda pequeña
-Otra pistola larga deshermanada llave de rueda

Estas onze bocas de fuego están colgadas en el lienzo de pared que haze cabezera a la pieza

-Zinco tiros de campaña de metal de dos baras de largo
-Tres culebrinas de dos varas y tres quartas de largo
-Çinquenta y dos balas gordas de calibre (fol. 6 r) de diez libras, todas de hierro
-Diez y seis narangeras también de hierro
-Ochenta y ocho pequeñas
-Cinco cureñas de hierro sin ruedas

Alabardas y Partesanas
-Dos partesanas
-Ocho espontones
-Quarenta alabardas

Armería Vieja
-Veinte morriones enteros pasados
-Ciento y quarenta y nueve manoplas también pasadas
-Ocho brazoletes (sic) también pasados
-Quatro frontadas de caballo viejas pasadas
-Un peto descabalado pasado
-Quatroçientas, y çinquenta, y seis piezas descabaladas pasadas que no sirven en que entran pequeñas, y pequeñitas
-Una espinilla
-Una rueda con su tornillo y su puerca
-Más la rueda grande con dos repasaderas
-Dos bancos con su tornillo y vigornia y martillo de guarnecer y una mesa
-Sesenta sillas de caballo antiguas sin pretales, ni acciones forrados los arzones y cojines con chapas de hierro

Toda la Pieza está colgada de baieta encarnada y las dos ventanas, y dos puertas principales tienen cortinas de lo mismo

De forma q las piezas que hay en la Armería son las Sig[tes]

(fol. 6 v)
Cuerpos Enteros
Cinquenta y tres cuerpos armados con sus fundas en que entran el dorado, el de charol y el del paje de jineta__053

Morriones
Diez y seis morriones__016

Petos
Treinta y quatro petos__034

Espaldares
Veinte espaldares__020

Brazoletes
Setenta y un brazoletes__071

Hombros
Dos hombros__002

Esclavinas
Diez y siete esclavinas__017

Faldillas
Cinquenta y seis faldillas__056

Muslos
Sesenta y tres muslos__063

Canillas
Veinte y ocho canillas__028

Botas
Veinte y çinco botas__025

Alabardas
Quarenta alabardas y dos partesanas y ocho espotones__050

Espadas
Quinze espadas anchas y de medio talle__015

Alfanxes
Dos alfanjes corbos y dos catanas__004

Cuchillos
Dos cuchillos de monte__002

Cuernos
Dos cuernos de unicornio__002

Aljaba
Una Aljaba con sesenta y seis flechas__066

Carabinas
Siete carabinas seis cortas y una larga__007

Trabucos
Dos trabucos__002

Pistolas
Cinco pares de pistolas__005

Cañones
Cinco tiros de artillería y tres culebrinas__008

Balas
Ziento y çinq^{ta} y seis balas de hierro gordas y medianas y pequeñas__156

Cureñas
Cinco cureñas de hierro sin ruedas__005

Armería Vieja
Morriones
Veinte morriones enteros pasados__020

Manoplas
Ziento y quarenta y nueve manoplas también pasadas__149

Brozoletes
Ocho brazoletes también pasados__008

Frontadas
Quatro frontadas de caballo viejas pasadas__004

(fol. 7 r)
Peto
Un peto decabalado pasado__001

Piezas
Quatrozientas çinq^{ta} y seis piezas descabaladas pasadas q no sirven en que entran pequeñas, y pequeñitas__456

Espinilla
Una espinilla__001

Ruedas
Dos ruedas una grande y otra pequeña con su tornillo y puerca, con dos repasaderas__002

Bancos
Dos bancos con su tornillo y vigornia y martillo de guarnecer y una mesa__003

Sillas
Sesenta sillas de caballo antiguas__060

Este ynventario se hizo de orden del Duque mi S^{or}, Dios le Gu^{e}, por Don Sebastián de Potestad contador m^{or}, de la Casa y Estados de Exª (sic),= Y todas las armas y alhajas en él contenidas quedan a cargo de Don Gaspar de Zúñiga Narbáes como alcaide de este Palacio; y también a cargo de Don Miguel de Béxar Zenteno contador de la Casa y Estados de Su Exª, como persona q de su orden tiene una llave de dha armería, y lo firmaron todos tres en Béxar a veinte y nueve de diziembre de mill setez^{os} y dos= Don Sebastián de Potestad= Don Gaspar de Zúñiga y Narbáes= Don Miguel de Béjar Zenteno=

Concuerda con el inventario original que queda en el archivo del Duque mi Señor y así lo certifico en Béxar dho día=

Seb^{án} de Potestad
(signed)

(fol. 7 v)
Diferentes Ynventarios de las Alhajas de este Palacio de Béjar= q los dejó el Duque mi S^{or}= en 29 de Diz^{re} de 1712=.

BENAVENTE, CONDES DE

Doc. 1
1530, July 16.
Inventory of Alonso Pimentel, 5th Count of Benavente. Arms.
Archivo Histórico de la Nobleza, Osuna, C. 423, D. 52.

(fol. 13 r)

-Un hábito de Santiago [...] en una cruz y con una venera al pie esmaltada [...]

(fol. 57 r)
-Una espada gineta con pomo y cruz y contera y cabo y hevilla y pasador de plata y tres jarcias de plata esmaltadas en la dicha vaina y dos borlas de trama con un tejillo y dos cabos de plata en el tejillo.
-Una espada para este bracamarte con una vaina de terciopelo con tres cuchillos con cabos de hueso blanco

(fol. 58 r)

-Una espada de la gineta con una vaina labrada de hilo de plata que es la de las perlas con pomo y puño y cruz y brocal y contera y otras piezas en la vaina y una hevilla y tres cabos y tres tachones en un tejillo de oro y de grana y son todas estas piezas de plata con esmaltes de oro y con dos borlas de grana guarnecidas de alfojar y los tachones y la hevilla no están esmaltadas y no se contaron las perlas porque eran muy pequeñas va esta espada metida en una caja de cuero colorado.

(fol. 69 r)

Arneses de justa Real

-Ocho yelmos de justa real de hierro
-Dos grebas con sus escarpes un poquito dorado y cincelados.
-Cuatro yelmos de justa real de hierro el uno de ellos con un poco dorado.
-Una calva suelta, cuatro escudos.
-Cuatro tacles el uno guarnecido de altibajo carmesí con clavazón dorada
-Dos rolletes de lienzo
-Un brazalete con correas de raso carmesí y hebillas doradas.
-Seis caracoles con sus llaves para otros yelmos.
-Una llave para apretar los tornillos de los tacles.
-Quattro flaones de lienzo viejos de otros arneses de Real.

Armas de la gineta

-Cinco escarcelas de hierro de ellos con cadenas de hierro y otros con correas.

(fol. 69 v)

-Un capacete aforrado con raso carmesí con dos guirnaldas a la redonda de plata y encima una rosa de plata metido en una funda de cuero.
-Unos quijotes blancos con charnelas y hebillas doradas guarnecidas en raso carmesí.
-Unos capazos de malla sobre terciopelo negro y con puntas de acero.
-Unas corazas de tela de oro y de tela de plata con clavazón dorada y el hábito de Santiago.
-Unas corazas de raso carmesí muy viejas.
-Una lanza gineta esta dorado el hierro con funda de cuero.
-Otra lanza gineta.
-Otras corazas de raso negro muy viejas con el hábito de Santiago.
-Otra lanza gineta.

(fol. 71 v)

-Una ballesta con su gafa que era del conde [...]

Doc. 2.
1541, 1550.
Book relating to the arms owned by Antonio Alonso Pimentel de Herrera, 6th Count of Benavente, at the Hospital de Nuestra Señora de la Piedad of Benavente.
Archivo Histórico de la Nobleza, Osuna, C. 424, D. 32.

(fol. 1 r)
Armería 1541
Libro de las armas q el conde nro Señor tiene en el Hospital de Nra señora de la Piedad de benavte de q se cargó a Pedro de Grijalba, criado de Su Señoría

Año de jUdxlj a^{os}
Año de mill y qui^{o}s y quarenta años
(fol. 2 r)
Libro de las armas que el conde nro señor tiene en el hospital de nra señora de la Piedad de Benavente de que se haze cargo a Pedro de Grijalba criado de su señoría año de jUdxlj años
(fol. 3 r)
En el hospital de nra señora de la Piedad de la villa de Benavente a veynte y un días del mes de mayo de myll e qui^{o}s e quarenta e un años los contadores Juan de Valdivieso e Alonso Pérez e Bernabé Gr^{a} hizieron cargo a Pedro de Grijalba criado del conde nro señor de las armas que su S^{a} tiene así de su psona como de la gente de su casa que están en la sala grande y quadra del dho hospital en esta manera
Arnés q el conde compró de Luys Méndez
-Una coraza peto y espaldar cinçelado y dorado con sus escarçelas enteras synçeladas y doradas.
-Un volante blanco que está hecho para justar de tarjeta para con este arnés con su sobre barriga dorada y sinçelada.
-Un escarçelón grande blanco que sirve para justas de este arnés.
-Unos brazales sinçelados y doradas las navajas y bordes tiene una rondeleta de brazo blanca para justar de tarjeta en el braçal yzquierdo. Pidió don Di^{o} de Toledo la […] rondeleta y no se cargó a Canseco.
-Unas grebas con sus escarpes cincelados y dorados todo el cincel.
-Unos quijotes del mysmo arnés dorados y cincelados con unas cuchilladas cinceladas y doradas.
-Un almete de este arnés encajado con su gola y vista dorado y cincelado.
-Un placarte cincelado y dorado del mismo arnés para enarponado.

-Una çelada de borgoña con su vista para justar de tarjeta.
-Un placarte que hizo Juan Pedro en Segovia cincelado y dorado.

Recibí yo Pedro de Grijalba las cosas contenidas en esta plana.
(fol. 3 v)
-Hay para la celada borgoñona una vista y sobrevista para torneo de a caballo çincelado y dorado.
-Más tiene esta misma celada y baberoncillos para torneo de pie sinçelado y dorado.
-Más tiene este arnés una çelada de infante con su baberón toda cincelada de unos penachos y dorada tiene esta celada su ventalle.
-Más tiene otra gola cincelada y dorada para con la celada borgoñona.
-Más dos guardabrazos cincelados de torneo y dorados para justa de tarjeta.
-Más un baberón de este arnés para torneo de pie y de a caballo y otro sobrebarbote para lo mismo con una vista de almete cincelado y dorado todo.
-Más dos guardabrazos de este arnés para seguir la guerra cincelados y dorados.
-Más una rondeleta de este arnés para algún guardabrazo derecho y para seguir la guerra y torneo de a caballo.
-Más una cofia con sus barras dorada y cincelada que es del mismo arnés que se hovo de Luis Méndez.
-Más otra cofia sin celada y dorada que la hizo después Juan Pedro armero en Segovia.
-Dos manoplas del mismo arnés doradas y cinceladas.
-Unos zapatos de malla con las puntas dorados y cincelados del mismo arnés.
-Una manopla de combatir del espada dorada y sin celada del mismo arnés.
Recibí yo Pedro de Grijalba las cosas contenidas en esta plana.

(fol. 4 r)
-Más unas medias grebas sinçeladas y doradas del mismo arnés
-Más un tonelete del mismo arnés cincelado blanco que está por dorar.
-Dos arneses uno de seguir a la guerra y otro de justa
-Una llave dorada para este arnés con un cordón de seda naranjada
-Más dos piezas que se hizieron en Valladolid de piezas menudas blancas para cubrir los quijotes para torneos de pie digo que es para cerrar los quijotes.
-Otras dos piezas para los brazales para las sangraderas para torneo de pie.
-Una tarjeta para este arnés. Pdiola don Di° de Toledo a quien se pestó este arnés/ no lo recibió […] Canseco
-Una sobremano y la de justa blanca.
-Más una bufeta para torneo cincelada y dorada más otra rondeleta cincelada y dorada para este dicho arnés porque son dos las que tiene […]

(fol. 5 r)
Arnés que se compró de don Yñigo de Guevara
-Una coraza con un espaldar y sus escarcelas y su ristre de guerra.
-Dos braçales de guerra.
-Dos medias grebas y sus escarpes de lo mismo que son cortados con sus espuelas estaban pegados a los escarpes y quitáronse.
-Dos quijotes.
-Una celada de seguir la guerra.
-Unos guardabrazos de guerra.
-Un tonelete.
-Dos manoplas.
-Un baberón con placarte para tornear
-Un mandilete pequeño.
-Una gola de la celada encajada.
-Una coraza para justar con su espaldar y su ristre de justa que sube de peto y volante.
-Dos guardabrazos de justa.
-Otro guardabrazo del brazo derecho.
-Un brazal de justa con su sobreguarda grande la bufa que es pieza bajuela para justa de guerra
-Una manopla de justa.
-Un almete en astas que tiene dos vistas una de guerra y otra de justa.
-Una cofia de combatir del espada.
-Un placarte [...]
-Unas grebas con sus escarpes enteras.
-Una testera del mismo arnés dorada y negra.
-Una cubierta de cuero a piezas doradas y moradas con fleco azul guarnecida de terciopelo morado que son coplón y pecho y cabezadas y riendas y cuello de malla en angeo puesto.
Recibí yo Pedro de Grijalba las cosas contenidas en esta plana.

(fol. 5 v)
-Unas puntas de escarpes doradas y moradas.
-Una rondeleta del brazal izquierdo de justar de tarjeta.
-Más una tarjeta de hierro.
Este arnés y las piezas que tiene son doradas a bandas y en medio de ellas está pintado de morado.

(fol. 6 r)
Arnés que se compró de Enrique alemán

-Una coraza con su espaldar cincelada con su peto.
-Unas escarcelas de justa cinceladas.
-Otras escarcelas de guerra.
-Otras escarcelas de infante.
-Un volante para justa para encambronarse y para tarjeta.
-Un baberón para encambronar y una tarjeta.
-La gran bufa entera con su tornillo.
-Brazales de justa con sus sobreguardas dobles.
-Un baberón para encambronar.
-Una tarjeta.
-Dos brazales de justa con sus sobreguardas dobles.
-Una gola.
-Una celada encajada con su vista de justa.
-Dos guardabrazos de justa.
-Otra celada con su vista de justa.
-Dos medias grebas con sus quijotes.
-Dos grebas enteras con sus quijotes cerrados todo para combatir a pie.
-Dos brazales para combate a pie cerrados.
-Dos guardabrazos para combatir a pie.
-Dos guardabrazos de guerra.
-Una pieza doble para la guerra sin celada.
-Una gola de guerra.
-Un baberón para torneo a caballo.
-Una celada de infante y de seguir que se pone el barbote por si
-Dos pares de manoplas las unas para justa y otras para torneo.
-Una sobremanopla de justa.
-Dos rondeletas de los guardabrazos de seguir la guerra. Perdiólos don Juan Pacheco cuando fue a la guerra de Tremecén y no se entregaron a [...] las rondeletas
-Un placarte de combatir a caballo y a pie.
-Una cofia de combatir a caballo.
-Un tonelete cincelado.

(fol. 6 v)
-Una testera de caballo dio don Diego otra testera chiquita la cual se carga a Canseco. Pidióla don Diego Sarmiento hijo del conde de Salinas y dio otra chiquita.
-Un ristre de guerra.
-Otro ristre de justa.

-Una llave para encambronar.
-Dos medias escarcelas de infante cinceladas de blanco.
-Más una tarjeta acerada.
-Más seis puntas de escarpes cincelados.

Recibí[...]
(fol. 7 r)
El arnés que se compró para la guerra de Francia
-Unas platas enteras con sus escarcelones dorados y cincelados los bordes.
-Dos espaldarcetes dorados los bordes.
-Dos brazales de guerra con una rondeleta media navaja.
-Dos guardabrazos para seguir la guerra.
-Dos quijotes
-Dos grebas con sus escarpes los bordes dorados.
-Una celada borgoñona con su vista de seguir y otra vista con su sobrevista para torneo de caballo los bordes dorados.
-Una gola.
-Un almete para encajar en ella los bordes dorados.
-Dos manoplas dorados los bordes.
-Unas medias grebas dorados los bordes.
-Un tonelete de este arnés con su borde dorado.
-Más dos piezas bajuelas de justa de guerra y si en otro arnés faltare alguna pieza de estas dos entiéndese que es una de estas suya.
-Más dos sobreguardas de justa de guerra del brazal izquierdo y entiéndese lo mismo que si en otro arnés faltare alguna pieza de estos dos que una de ellas es del dicho arnés.
-Más otros dos quijotes entiéndese lo mismo.
-Unas grebas enteras con sus escarpes blancas.
-Más otras medias grebas sin escarpes blancas.
-Más otras medias grebas sin escarpes blancas.
-Más un placarte de justa blanco para encambronar.

(fol. 7 v)
-Un barbote de orejas para torneo de a caballo y de pie.
-Más una tarjeta.
-Más una rondeleta para el brazal izquierdo para justa de tarjeta.
-Un escarcelón de justa grande con su sobrebarriga.
-Otro volante para justar.
-Otra sobrebarriga sin escarcelón.

(fol. 8 r)
Arnés que el camarero Juan de Carvajal dio al conde
-Una coraza con su espaldar.
-Dos grebas con sus quijotes.
-Dos brazales con sus sobreguarda para justa.
-Un almete de justa.
-Un volante de justa.
-Un volante de justa.
-Un placarte para encambronar.
-La gran bufa.
-Dos guardabrazos de guerra y otra pieza doble para el guardabrazo.
-Un espaldarcete de justa digo guardabrazo.
-Un escarcelón de justa.
-Dos manoplas de guerra.
-Otras dos manoplas de justa.
-Una sobremanopla de justa doble.
-Una vista de guerra y un placarte de combatir.
-Un volante de guerra.
-Dos medios quijotes de justa.
-Más tiene este arnés una tarjeta que hizo de las armas viejas que vinieron de Túnez.
-Más una pieza del brazal izquierdo del codo para justa de tarjeta.

(fol. 9 r)
Arnés que Su Majestad dio al conde
-Las platas con sus ristres blancos.
-Un escarcelón pequeño.
-Un volante.
-Un placarte para encambronar.
-Dos brazales.
-Dos quijotes, perdiose el uno cuando se prestó a don Alonso de Castillo.
-Dos grebas con zapatos de malla.
-Un almete con su vista.
-Una sobreguarda de brazo.
-Una manopla con su sobremanopla doble.
-Una bufa y guardabrazo doble todo junto para justar.
-Otro guardabrazo.
-Una tarjeta de hierro.
-Una pieza para el brazal izquierdo para justa de guerra.

(fol. 10 r)
Arnés que fue de don Alonso Manrique que el conde compró de él q es pavonado.
-Un peto y espaldar con su ristre para seguir la guerra es dorado y cincelado de una color pavonada.
-Una gola de la misma suerte.
-Una celada borgoñona de la misma color.
-Dos brazales de lo mismo.
-Dos guardabrazos para seguir la guerra.
-Dos mandiletes.
-Unas grebas de la misma suerte sin escarpes.
-Otras medias grebas de la misma suerte sin escarpes trenzadas.
-Unos quijotes de la misma suerte.
-Unos brazales enteros de infante y con tres piezas de cada brazal.
-Dos escarcelas de infante.
-Una rondeleta del brazal izquierdo de infante.

(fol. 11 r)
Arnés que compró el conde de don Pedro Sarmiento
-Una coraza que es peto y espaldar blanco y cincelados los bordes.
-Dos escarcelones de justa de tres piezas cada una.
-Más otro escarcelón de una pieza.
-Un volante
-Dos quijotes.
-Dos grebas enteras con escarpes.
-Dos brazales
-Dos manoplas
-Una sobremanopla
-Dos guardabrazos
-Otro guardabrazal es de justa.
-Otra celada borgoñona
-Una sobrevista
-Un barbote de orejas
-Una escofia
-Dos grebas trenzadas
-Dos piezas para engerir en la coraza para justar sin gola.
-Una rondeleta que cuelga de un guardabrazo
-Una sobreguarda de brazal para justa de tarjeta.
-Otra sobreguarda de brazal de justa de guerra.

-Más dos piezas dobles de justa de guerra.
-Más dos testeras de caballo la una dorada y plateada y la otra dorada y blanca [later anotation:] perdió don Alonso de Castillo la testera

(fol. 11 v)
-Más una tarjeta.
-Más una vista de guerra.

(fol. 12 r)
El arnés que el conde ferió al duque de Náxara a un caballo en Villa Franca de Nyça
-Una coraza que tiene peto y espaldar con su ristre es dorado y sinçelado a bandas.
-Más una gola de lo mismo.
-Más una çelada borgoñona de lo mismo.
-Más un volante de la misma suerte.
-Más dos escarcelas largas de la misma suerte.
-Más dos medios quijotes de la misma suerte.
-Más dos medias grebas de la misma suerte.
-Más dos guardabrazos de la misma suerte el derecho con una rondeleta.
-Más dos brazales de la misma suerte.
-Más dos manoplas de la misma suerte.
-Una escofia de la misma suerte dorada
-Dos puntas de escarpe de la misma suerte.

(fol. 13 r)
El arnés que hizo Juan Pedro armero del conde en Milán cuando Su Señoría fue con Su Majestad para entrar en Françia
-Unas platas que es peto y volante dorados y cincelados los bordes con sus escarcelas.
-Un ristre con dos tornillos dorado.
-Dos brazales dorados los bordes.
-Dos guardabrazos dorados los bordes.
-Una gola.
-Una celada con su barbote borgoñona: Perdió esta celada don Luis de Sotomayor cuando fue la jornada de Argel y compró otra blanca la cual se carga a Canseco. Mandó el conde hacer otra celada para este arnés blanca sinçelada con su barbote que se entregó a Francisco de Canseco.
-Un barbote de piezas
-Dos mandiletes

-Dos quijotes
-Dos grebas con sus escarpes enteros todo dorado y cincelado.

(fol. 14 r)
Arnés que don Alonso Pimentel Maldonado dio al conde
-Una coraza que es peto y espaldar.
-Dos volantes uno de justa y otro de guerra.
-Tres braçales.
-Dos guardabraços.
-Una manopla yzquierda.
-Una sobremanopla yzquierda.
-Dos escarcelas de pieças.
-Un escarcelón de justa entero.
-Dos quijotes
-Otras grebas enteras con escarpes.
-Otras grebas trenzadas.
-Un almete con su alpartaz de Milán
-Un placarte de justa
-Una vista de guerra
-Dos barbotes redondos
-Medio barbote
-Una gran pieza que sirve de guardabrazales
-Una bufa.
-Una sobreguarda del brazal izquierdo de justa de guerra.
-Otra sobreguarda del dicho brazal.
-Una tarjeta.
-Una piececita del brazal izquierdo para justa de torneo.

(fol. 15 r)
El arnés que dio don Juan Pimentel comendador al conde
-Una gola
-Una coraça que es peto y espaldar
-Dos escarcelas de piezas
-Un escarcelón de justa.
-Un volante de justa.
-Un placarte.
-Un almete encajado.
-Tres braçales.
-Cinco guardabrazos tres blancos y dorados y los otros dos mandiletes.
-Una sobremanopla.

-Perdióse otra manopla de combatir que tenía este arnés cuando le llevaron a Maqueda.
-Una rondeleta del brazal izquierdo de justa de tarjeta.
-Más otra rondeleta que cuelga de un guardabrazo.
-Más una pieza bajuela de justa de guerra.
-Más una sobreguarda del brazal izquierdo para justa de guerra.
-Más una vista de guerra.
-Una tarjeta.
-Dos quijotes.
-Medio quiote.
-Dos grebas enteras.

(fol. 15 v)
Recibí yo Juan de Valdivieso el mozo las piezas contenidas en este arnés de esta plana presente escritas de Pedro de Grijalba por mandado del conde mi señor en 18 de abril de 1550. [...]
(fol. 17 r)
Las piezas que tiene el coselete de don Alonso Pimentel Maldonado que está en la armería
-Una gola.
-Una celada de infante con su barbote.
-Una vista de celada borgoñona.
-Una coraza que es peto y espaldar con su ristre.
-Dos escarcelas largas que sirven de quijotes que se abren hasta las rodillas.
-Dos medias esquinelas
-Dos brazales
-Dos guardabrazos con rondeletas cada uno
-Dos manoplas
-Tiene también una silla acerada guarnecida de paño blanco con franja que está [...] de amarillo. [...]

(fol. 18 r)
Las piezas que hay sin arneses en la dicha armería que trajo Ruxilay de Milán son las siguientes
-Dos volantes de corazas para encambronar el uno de ellos se puso en el arnés que el conde ferio al duque de Nájera y el otro se cargó en el arnés que compró para la guerra de Francia.
-Dos placartes para encambronar.
-Dos bufas para dos arneses para encambronar, la una se metió en un arnés de Milán y la otra está perdida

-Dos sobreguardas se cargan con el arnés que compró para la guerra de Francia en Milán.
-Tres guardabrazos el uno se cargó con el arnés de don Juan Pimentel y los otros dos están por sí.
-Una vista con su sobrevista para combatir del espada.
-Un escarcelón se carga con el arnés de Milán que se compró para la guerra de Francia.
-Dos toneletes.
-Dos brazales
-Dos pares de grebas encargadas con el arnés que se compró en Milán para la guerra de Francia el un par el otro se carga con este arnés a Francisco de Canseco.
-Dos quijotes están en el arnés de don Jo Pimentel por que llevaron los suyos a la guerra de Perpiñán a do se perdieron.
-Otros quijotes cincelados que se cargan con el arnés que se compró en Milán para la guerra de Francia.
-Dos manoplas con sus dedos de malla, la una tiene el arnés del emperador i la otra esta por si sin malla
-Un baberón se haya con el arnés que se compró en Milán para la guerra de Francia.
-Un espaldarcete […]
-Dos escarcelas.
-Dos bufetas dobles.
-Una guarda de brazal doble.

(fol. 18 v)
-Dos pantorrillas de guerra que son grebas.
-Una celada que se compró de maestro Jorge
-Cuatro piezas que están forjadas para arnés.

(fol. 19 r)
Arnés que fue de don Alonso Pimentel nuestro señor que haya gloria
-Unas platas doradas y cinceladas con su volante y ristre dorado y cincelado y sus escarcelas con guarnición carmesí y hebillas doradas y su alpartaz y bati-herrera.
-Un almete dorado con clavazón dorada y con sus piezas que son vista y barbote y su alpartaz de malla.
-Otro almete con su barbote y vista y alpartaz de malla dorado y cincelado con su clavazón dorada y guarnecida de terciopelo carmesí.
-Una gran bufa dorada que es de este arnés con su tornillo.
-Dos quijotes dorados y cincelados los bordes.

-Unos brazales dorados y cincelados con correas de terciopelo carmesí con sus guardabrazos.
-Una guarnición de brazos que son brazales y guardabrazos de dorado y cincelado y con hebillas doradas y correas de cuero.
-Dos manoplas doradas y cinceladas aforradas en terciopelo carmesí.
-Dos grebas y dos quijotes de hombres de armas con sus escarpes dorados y correas de cuero.
-Dos escarceloncillos de los costados.
-Dos quijotes blancos.
-Unas grebas con sus escarpes doradas y cinceladas.
-Unos zapatos de terciopelo negro cubiertos de trechos de malla y con sus puntas de acero.

(fol. 20 r)
Los dos arneses que el conde compró de Milán que los entrego Rafael Archiole (sic) tiene las piezas siguientes los quales vinieron a Bte este año de jUdXLj aºs
El un arnés
-Una coraza que es peto y espaldar dorados y cincelados los bordes.
-Más un volante de justa blanco es dorado el borde.
-Dos golas dorados y cincelados los bordes.
-Un almete enjado dorados los bordes con su vista.
-Una celada borgoñona dorados los bordes.
-Una escofia para torneo
-Un placarte de justa para encambronar dorado parte del.
-Un barbote de guerra dorado parte del.
-Una sobrebarriga con un escarcelón blanco es dorado y cincelado los bordes.
-Dos escarcelas dorados los bordes.
-Dos medias escarcelas que se añaden con pernos hasta las rodillas.
-Dos brazales dorados los bordes y cincelados.
-Dos brazales de infante enteros dorados y cincelados los bordes.
-Dos guardabrazos para torneo y para guerra dorados y cincelados los bordes.
-Un morrión dorado y cincelados los bordes
-Una pieza bajuela para justa de guerra dorado y cincelado los bordes.
-Una bufa dorada y cincelados los bordes.
-Una sobreguarda del brazal izquierdo para justa de guerra dorados y cincelados los bordes.

(fol. 20 v)
-Dos manoplas dorados y cincelados.
-Una sobremanopla izquierda para justa dorada y cincelados los bordes.

-Una manopla de combatir del espada de la mano derecha dorada y cincelados los bordes.
-Unos quijotes dorado y cincelados los bordes.
-Dos grebas enteras con sus escarpes dorados y cincelados los bordes.
-Dos medias grebas trenzadas sin escarpes dorada y cincelada los bordes.
-Más unas puntas de escarpes doradas y cinceladas con su malla para cubrir zapatos.
-Más un tonelete para torneo de pie dorado y cincelado el borde debajo.
-Más una tarjeta de hierro pintada por encima.
-Más dos ristres dorados y cincelados por encima, perdiósele un ristre a un criado (?) del rey de Bohemia
-Dos guardabrazos de justa dorados y cincelados los bordes.
-Vienen con este arnés dos arandelas pintadas que se hayan con este arnés.
-Vienen dos sillas armadas con este arnés una para justa y otra para seguir la guerra que se hayan con las sillas.

(fol. 22 r)
El otro arnés que trajo Rafael Archioli tiene las pieças siguientes
-Una coraza que es peto y espaldar dorados y sinçelados los bordes.
-Un volante de justa blanco.
-Dos golas doradas y cincelados los bordes.
-Un almete encajado con su vista dorado y cincelado los bordes.
-Una celada borgoñona dorados los bordes.
-Una cofia blanca para torneo.
-Un placarte de justa para encambronar dorado parte del.
-Un barbote de guerra dorado parte de él.
-Una sobrebarriga con su escarcelón blanco dorado y cincelado los bordes.-Dos escarcelas dorados los bordes.
-Dos medias escarcelas que se añaden con unos hasta la rodilla.
-Dos brazales dorados y cincelados los bordes.
-Dos brazales de infante enteros dorados y cincelados los bordes.
-Dos guardabrazos para guerra e para torneo dorados y cincelados los bordes.
-Un morrión dorado y cincelado los bordes.
-Una pieza bajuela dorado y cincelado los bordes.
-Una bufa dorada y cincelados los bordes.
-Una sobreguarda del brazal izquierdo para justa de guerra dorada y cincelada los bordes.
-Dos manoplas doradas y cinceladas los bordes.
-Una sobremanopla izquierda para justa.

(fol. 22 v)
-Una manopla de combatir de la espada de la mano derecha dorados y cincelados los bordes.
-Dos grebas enteras dorados y cincelados los bordes.
-Unos quijotes dorados y cincelados los bordes.
-Dos medias grebas trenzadas sin escarpes dorados y cincelados los bordes.
-Más unas puntas de escarpes dorados y cincelados los bordes con su malla para cubrir zapatos.
-Más un tonelete para torneo de pie dorado y cincelado el borde de vaso.
-Una tarjeta de hierro pintada.
-Más dos ristres dorados y cincelados por cima.
-Dos guardabrazos de justa dorados e cincelados los bordes.
-Vienen con este arnés dos arandelas pintadas cárganse con las arandelas.
-Vienen dos sillas armadas con este arnés una para justa otra para seguir la guerra cárganse con las sillas.

(fol. 23 r)
Las piezas que tiene el arnés que su señoría compró de un armero vizcaíno al duque de Nájera son las siguientes
-Una coraza que es peto y espaldar con su ristre
-Un volante de justa
-Una gola
-Dos escarcelas
-Una sobrebarriga de justa y un escarcelón de justa
-Un almete encajado
-Un placarte
-Una celada borgoñona
-Un barbote
-Dos grebas enteras con escarpes
-Dos quijotes
-Dos guardabrazos
-Dos brazales
-Una pieza del brazal izquierdo de justa
-Dos manoplas
-Una sobremanopla
-Más una gran pieza de justa.
-Un volante de guerra.
-Dos brazales de guerra.
-Un guardabrazo de guerra.

(fol. 24 r)
Los seis arneses que se compraron de mre Herdo vizcayno arm° del duque de Nájera de las armerías de Albayna que tienen las piezas siguientes
-Un arnés que tie peto y espaldar
-Una gola
-Un almete encajado
-Un placarte de justa
-Un volante de justa
-Una sobrebarriga de justa
-Una pieza bajuela de justa
-Un escarcelón de justa
-Dos escarcelas de guerra
-Un ristre
-Dos guardabrazos con dos rondeletas
-Dos brazales el izquierdo con una sobreguarda de justa
-Dos manoplas
-Una sobremanopla
-Dos grebas enteras con sus escarpes
-Dos quijotes enteros
-Dos medias grebas
-Una celada borgoñona con mismo barbote y seis pares de puntas para zapatos de malla

Recibió las piezas de este arnés el dicho Pedro de Grijalba.
(Fol. 24 v)
Los otros arnés que son cinco
-Un arnés cumplido con todas las piezas de este primero de esta foja
-Otro arnés cumplido con todas las piezas del primero de esta foja.
-Otro arnés cumplido con todas las piezas del primero de esta foja.
-Otro arnés cumplido con todas las piezas del primero de esta foja.
-Otro arnés cumplido con todas las piezas del primero de esta foja.

(fol. 25 r)
-Un arnés que compró el conde de los Fúcares que costó seiscientos ducados
-Dos celadas una borgoñona y la otra de infante con su barbote y ventalla
-Dos golas
-Dos corazas que es peto y espaldar
-Dos ristres el uno de tornillo y el otro de muelas.
-Cuatro escarcelas de piezas
-Cuatro brazales

-Cuatro guardabrazos
-Cuatro mandiletes
-Dos quijotes enteros y tranzados.
-Dos grebas enteras con escarpes
-Dos escarcelas largas de infante con sus navajas y en las rodillas con sus grebas tranzadas
-Una almete para justa real
-Un almete encajado para justa
-Una celada de infante con su ventalla
-Un placarte de justa
-Un volante con su sobrebarriga de justa
-Dos escarcelones
-Cuatro guardabrazos nos [nuevos or nielados]
-Dos alcallas para justa real
-Un brazal de Sant Antón para justa real
-Una sobremanopla de justa
-Un barbote y su medio barbote
-Una escofia para torneo
-Una sobreguarda para el brazo izquierdo
-Otra sobreguarda para justa de tarjeta
-Tres rondeletas para guardabrazos
-Dos piezas para las rodillas para añadir con pernos para escarcelas de infante.
-Una bragueta para torneo de pie

(fol. 25 v)
-Dos esquinelas
-Dos puntas con su malla para zapatos.
-Un brazal derecho para justa real
-Una pieza bajuela para justa de guerra
-Dos trances para bridarse para justa real
-Una tarjeta barrada y grabada
-Una testera entera de caballo y media
-Cuatro aceros para dos sillas puestas en sus fustes
-Dos arandelas para las lanzas de justa-Es todo este arnés dorado y grabado a bandas.

(fol. 26 r)
El arnés que el conde nuestro señor ferió al conde de Luna y dos caballos
-Una coraza que es peto y espaldar con sus escarcelas de piezas y un ristre
-Una gola

-Dos brazales
-Dos guardabrazos con sus rondeletas
-Un volante de justa con dos escarcelones
-Una celada de justa con celada en gola con un placarte de justa.
-Una pieza bajuela de justa
-Una sobreguarda del brazal izquierdo para justa.
-Unas grebas enteras con sus escarpes.
-Unos quijotes enteros trenzados
-Unas manoplas y unas sobremanoplas de justa.
-Unas esquinelas de pies de fuera con sus quijotes de piezas.
-Unos zapatos de malla con sus puntas de acero
-Una celada seguir con ventalla y un barbote que baja para abajo la vista.
-Media testera de caballo.
-Una bragueta de acero
-Otro ristre
-Unos aceros de sillas que son cinco piezas
-Este arnés es blanco y grabado y guarnecido con clavazón dorada de Alemania y dos cofres de madera blancos en que vino.

(fol. 27 r)
El arnés q PerezCozq (sic), gentilhombre del rey de Bohemia, dio al conde
-Un peto y espaldar con sus escarcelas largas
-Una gola
-Unos brazales de justa
-Unos guardabrazos de justa
-Un volante para justa
-Otro volante para seguir la guerra
-Una celada calada.
-Un baberón de justa
-Una pieza bajuela para justa
-Una sobreguarda del brazal izquierdo
-Otra rondeleta para justa de tarjeta del brazo izquierdo
-Unas grebas enteras con sus escarpes
-Unos quijotes enteros
-Una sobremanopla de justa
-Unos escarcelones de justa
-Otra sobremanopla
-Un mandilete izquierdo son dos mandiletes
-Dos manoplas la una de combatir

-Otras grebas enteras con sus medios quijotes y escarpes
-Más otros brazales
-Más otro par de brazales
-Otro medio quijote
-Más otras dos navajas de quijote que alargan las escarcelas para torneo de a pie.

(fol. 27 v)
-Más medias grebas trancadas
-Más otros dos guardabrazos uno grande para torneo
-Otra celada encalada
-Otra celada encalada
-Otra celada para torneo y seguir la guerra
-Una escofia
-Una bragueta
-Una pieza a manera de gola
-Una testera con un escudo en medio
-Todo este arnés es blanco y grabado y clavazón dorada esta dentro de un cofre guarnecido de acero negro.

(fol. 28 r)
Coseletes que están en la armería del Hospital de Nuestra Señora de la Piedad de Benavente de que tiene cargo Pedro de Grijalba

Coselete que dio al conde el duque Alexandre de Florencia

-Una coraza que es con peto y espaldar negro labrado de unos bordes ataujía dorados
-Una celada de infante
-Unos brazales de infante con su rondeletas
-Unos mandiletes
-Un ristre dorado
-Dos escarcelas largas
-Una gola

(fol. 28v)
Coselete que Antonio de Leyva dio al conde
-Una coraza con sus escarcelas y espaldar
-Una gola
-Un morrión
-Unos brazales de infante

-Dos manoplas falta la una de ellas por que la mando el conde dar a Ventura Beltrán todo este coselete es cincelado y dorado.

(fol. 29 r)
Coselete que dio el duque de Nájera al conde en ferias de un caballo
-Una coraza que es peto y espaldar con sus escarcelas largas
-Una gola
-Un morrión con ventalla
-Un barbote- perdiose en casa del duque [...] dos mandiletes por el
-Unos brazales de la una que tienen una navaja para el codo y un espaldarcete pequeño para el otro. Es todo cincelado y dorado

(fol. 29 v)
[crossed out:] Coselete que dio al conde don Alonso Pimentel Maldonado

diose esta coraza el conde a don Gonzalo de Ulloa la que le pertenecía por cédula del conde fechada en Benavente a XXVIII de agosto de mil quinientos cuarenta y tres.
-Una coraza de peto y espaldar y medias escarcelas es alta de cuello
-Un morrión de ventalla
-Es todo este coselete negro [end of crossed out text]

Coselete que dio al conde el conde de Nieva

-Una coraza de peto y espaldar y escarcelas largas es negro.

(fol. 30 r)
Coseletes que se compraron en Milán para la entrada de Francia los cuales son de la ligera que tenía se tiene y entregaron al dicho Pedro de Grijalba
-Cinco coseletes con ristres tiene cada coselete peto y escarcelón largo y espaldar y gola brazales y guardabrazos y celadas borgoñonas todos cinco y cuatro pares de manoplas son blancas todos.
-Cuatro pares de puntas de escarpes de hierro
-Más unos quijotes y esquinelas
-Más otro par de esquinelas cinceladas
-Más dos coseletes que tiene cada uno peto y espaldar con su ristre y tienen escarcelas largas y sus medios quijotes y gola y brazales y guardabrazo manoplas y una celada borgoñona que se compró de Milán a maese Hernando Vizcaíno armero del duque de Nájera

(fol. 31 r)
Coselete que el duque de Nájera dio al conde

-Que se le carga más un coselete que el duque de Nájera dio al q[e] que es dorado y cincelado a bandas que tiene
-Una gola
-Una celada borgoñona
-Una coraza que es peto y espaldar
-Un volante con sus correas atrás guarnecidas de piezas
-Dos brazales enteros de infante
-Dos mandiletes de dedo
-Dos escarcelas de piezas
-Dos medios quijotes
-Dos grebas trenzadas
-Un barbote
-Dos puntas para zapatos de malla
-Dos piececillas son de los quijotes
-Un ristre dorado y cincelado con sus tornillos.
-Una coraza que dio maese Hernando Vizcaíno armero para el señor don Alonso.

(fol. 32r)
Morriones que se cargan al Pedro de Grijalba en veinticinco de mayo de mil cuarenta y un años
-Un morrión guarnecido de terciopelo negro y pasamanos de oro por cima que el conde hubo en ferias de Alfonso de Silva
-Más ocho morriones negros barnizados y labrados de cuchillo por cima los siete y el uno negro sin barnizar que el conde mando comprar en Génova cuando fue con Su Majestad a Villa franca de Niza
-Dos cascos redondos acerados que se compraron con el arnés que vendió el armero del duque de Nájera

Mas se le cargaron al dicho diez e seis cascos que se compraron de maese Hernando Vizcaíno armero del duque de Nájera. De estos veintiocho cascos se dieron los quince al señor don Juan Pimentel según parece por su firma y los trece que están se cargan
(fol. 33 r)
Cotas y cosas de malla que se entregó a Pedro de Grijalba en XXV de mayo de (1541) años
-Una cota de malla jazarina con mangas guarnecidas con una faja de terciopelo por dentro con un gorjal de malla de puntas que es de los que entrego Bozmediano
-Más otra cota de malla con sus escarcelas largas jazarinas con sus mangas y sus guantes de malla la compró el conde estando en el bosque al Bautista hijo de

Bautista armero de Su Majestad que estaba aforrada en tafetán y las mangas por dentro en raso por diéronselos
-Más otra cota de malla jazarina con unas mangas de malla que no es jazarina
-Un gorjal de malla que dio Su Majestad al conde
-Dos gocetes de malla que dio Su Majestad al conde
-Otros tres gocetes de malla
-Más un cuello de malla para caballo que son cinco fajas de malla y cinceladas en bocací negro.
-Otro cuello de malla para caballo guarnecido de anjeo que se hubo con el arnés de don Íñigo de Guevara cuando el conde se lo compró este cuello está cargado con el dicho arnés
-Otro cuello de malla para caballo de terciopelo leonado que tiene tres bandas de malla con franjas de seda leonada y blanca.

(fol. 34 r)
Armas de la gineta de que se hace cargo a Pedro de Grijalba en 21 de mayo de 1541 años
-Una coraza de tela de oro y de plata con clavazón dorada con el hábito de Santiago.
-Unas corazas de raso carmesí viejas con un hábito de Santiago
-Otras corazas de raso negro.
-Veinte y tres cuerpos de corazas de terciopelo aceituno y verdes de ellas y de ellas de morado y colorado
-Otras corazas de raso negro con el hábito de Santiago. Estas corazas no dice que son coracinas y que están aforradas en tafetán negro van cargadas a Francisco de Canseco por coracinas
-Unas corazas de las de Milán con sus mangas enteras y sus escarcelas guarnecidas en fustán negro por de fuera y clavazón embarnizadas. De estas tiene las unas Ramírez Bosquero por mandado del conde por una cédula hecha a 7 de diciembre de 48
-Otras coracinas de las de Milán sin mangas con sus escarcelas guarnecidas en fustán negro y clavazón blanca estañada.
-Una guarnición de brazos de la gineta con sus mandiletes y guardas doradas y cincelada guarnecidas en terciopelo negro.
-Un capacete aforrado en raso carmesí con dos guirnaldas a la redonda de plata y encima una rosa de plata.
-Una celada con su barbote y con sus mantillos de malla aforrados en terciopelo negro.

-Un capacete con su babera guarnecida de terciopelo dorado que fue del cargo de Bozmediano.
-Una gola con un barbote encajado para la gineta.

(fol. 34 v)
-Cuatro grebas de arneses viejos y a la una falta el escarpe la punta del es del cargo de Bozmediano.
-Otra pieza de armadura de cabeza vieja de la [...]
-Un brazal de la gineta dorado y cincelado que falta el mandilete del conde, de Bozmediano.
-Más dos alzas de la gineta que son dos piezas para sobre los hombros del cargo de Bozmediano.
-Más un mandilete del dicho cargo
-Más medio brazal para (?) de la gineta con una pieza del otro guarnecida de raso carmesí del dicho cargo
-Más una manopla de arnés de justa vieja del dicho cargo

(fol. 35 r)
Cubiertas de acero de caballo cargo que se hace a Pedro de Grijalba en 21 de mayo de 1541
-Unas cubiertas de acero de caballo que son nueve piezas todas grandes y pequeñas que son cinceladas y historiadas de cosas antiguas. Guarnecidas de unas franjas y borlas de lana amarillas las cuales compró el conde en Valladolid de Colman armero alemán.
-Una guarnición de a caballo acerada para la brida cumplida con su freno en las cabezadas que compró su Excelencia del dicho Colman en Valladolid.
-Otras cubiertas hay que compraron con el arnés de don Íñigo de Guevara están cargadas juntamente con el dicho arnés y por esto no se ponen aquí.

(fol. 35 v)
-Catorce piezas de cubiertas de caballo grandes y pequeñas que son
-Un coplón con dos costeras
-Un petral con tres piezas
-Una sotacola
-Dos piezas debajo de los estribos
-Un cuello de caballo
-Una gorja y un testera con unos cuernos dorados por testera
-Unas riendas de piezas
-Todas estas piezas son doradas y cinceladas con fajas sombradas de unas emes que compró el conde al señor don Alonso su hermano

-Más se le cargan como piezas de sillas de acero con sus fastes de madero con las mesmas emes doradas y cinceladas.

(fol. 36 r)
Testeras de caballos que se cargan a Pedro de Grijalba en primero de junio de mil e quinientos y cuarenta e un años
[…]
(fol. 37 r)
Arandelas que se cargan a Pedro de Grijalba en primero de junio de mil e quinientos e cuarenta y un años
-Siete arandelas blancas
-Once arandelas blancas excepto una que esta teñida de negro
-Más nueve arandelas viejas que entrego el alcaide Bozmediano algunas de ellas doradas
-Cuatro arandelas pintadas de justa que son las que vinieron con los dos arneses que envió Rafael Archioli
-Cuatro arandelas de justa que se compraron con el arnés que vendió el armero del duque de Nájera.
-De estas treinta e una arandelas se deshizo una para los caballicos y las viejas y dos que se perdieron en la justa cuando la condesa mi señora y otra se perdió en Zamora que el conde mandó prestar a Luis de Mella y las demás recibí yo Juan de Valdivieso en 22 de abril de 1550 años.

(fol. 38 r)
Sillas de armas que se entregaron al Pedro de Grijalba en primero de junio de mil quinientos cuarenta y un año
[…]
(fol. 44 r)
Alabardas y partesanas y hierros de picas y lanzas que se cargan al Pedro de Grijalba en primero de junio de mil quinientos cuarenta y un años
-Seis partesanas de Milán enastadas dorados por y con todos hierros guarnecidos de terciopelo y franjas que compró el camarero Carvajal en Valladolid.
-Veinte y cinco partesanas doradas parte de los hierros enastados guarnecidas de terciopelo y franjas de a colores que compraron cuando el conde fue a Villafranca de Niza.
-Seis alabardas de las de Milán que el camarero Carvajal compró de las de Milán enastadas en Valladolid doradas las cuchillas y guarnecidas de terciopelo con franjas.
-Treinta alabardas que se compraron en Villafranca de Niza doradas las cuchillas y guarnecidas como las de arriba.

-Doce hierros de picas de los de Milán que compró el armero Carvajal en Valladolid
-Una lanza de armas que vino de Cazales (?) sin hierro que es dorada y plateada y azul.
-Otras treinta alabardas de hierros blancos que entrego el Alcaide Bozmediano las tres de ellas obradas las astas
-Otra alabarda vieja obrada que no tiene sola la cuchilla del cargo del dicho Bozmediano
-Un martillo de hombre de armas para pie del cargo de Bozmediano.

(fol. 44 v)
-Un hierro a manera de arpón dorado un poco del cargo del dicho Bozmediano
-Cuatro lanzones enastados con sus hierros del cargo del dicho Boz mediano
-Dos roncas enastadas del cargo del dicho Bozmediano
-Diez y seis hierros de lanzones grandes del cargo del dicho Bozmediano
-Ocho hierros de lanzas jinetas del cargo de Lezcano que están en ocho picas de pino pintadas de las que entrego Bozmediano.
-Un lanzón enastado de un hierro ancho dorado parte de el con franja de oro y seda parda.
-Un lanzón de un hierro largo enastado que es del cargo del Alcaide Bozmediano

(fol. 45 r)
-Y Más se le cargan al dicho Pedro de Grijalba doce roncas doradas con las armas del conde que se compraron de Cristóbal milanés.
-Más se lo cargan al dicho doce alabardas doradas con el escudo de las armas del conde que se compraron de Cristóbal milanés
-Más se le cargan doce espedos boloñeses con las armas del conde.
Mas se le cargan doce armas en astadas y cinceladas parte de ellas que se compraron del dicho Cristóbal milanés con los hierros diferentes.
-Más se le cargan once armas en astadas que parece que hubo de más de las que estaban cargadas al dicho Pedro de Grijalba que son por todo ciento y ochenta y cuatro. Las cuales recibió Juan de Valdivieso de [...] en 1550.

(fol. 46 r)
Rodelas que se cargan al Pedro de Grijalba en primero de junio de mil quinientos cuarenta y un años
-Dos rodelas de las de Italia pintadas que trajo Cueto del conde de Valladolid la una de ellas es turca con franjas verdes y amarillas y azules.
-Cinco rodelas las dos de las de Módena de Italia pintadas y las otras dos de las de Milán aceradas la una guarnecida de terciopelo negro y franjas de oro y seda y la

otra esta dorada y pintada como de miel y fáltanle ciertas piezas que están en una taleguilla la otra es de las de Barcelona con las armas del conde que haya gloria.
-Otra rodela acerada de las de Milán dorados los hierros por de fuera guarnecida de terciopelo amarillo la cual hubo el conde en ferias de Alfonso de Silva portugués
-Otra rodela que el conde compró en Benavente, Mucio Escévola pintado en el medio.
-Otra rodela guarnecida de terciopelo carmesí con piezas.
-Otra rodela de las Indias pintada en azul y blanco y colorado con un bollón en medio de Laón [or León (?)] que es del conde del Alcaide Bozmediano
-Otra rodela pintada de las de Nápoles que es del cargo de Bozmediano
-Otra rodela de las de Italia pintada.

(fol. 46 v)
-Doce rodelas de las de Salamanca que el conde compró y decoraron y pintaron en Benavente por Vallejo que las cuatro de ellas son doradas de unos cercos a la redonda y otra dorada a la redonda y en el medio un hombre pintado que le lleva un buitre en las uñas y otra dorada acerada un león en el medio y otra rodada así acericos y un león es en el medio y otra dorada así aceros y una muerte en el medio y otra dorada de la misma manera y con un unicornio blanco en medio más otra dorada de la misma manera con dos obres pintados en el medio otra dorada así con una espera pintada en medio otra dorada así con un ciervo en medio otra dorada así con un castillo en medio con que se cumplen las dichas doce rodelas [...]
-Más dos rodelas negras la una con dos cerquillos dorados y unas letras en medio y la otra con un escudo dorado en medio que se compró de Cristóbal milanés de Medina del Campo.
-Una rodela de acero con quattro escudos de las armas del conde que dio a su señoría don Alonso Pimentel su hermano.
-Que se le cargan más veintiocho rodelas de las de Módena de Italia que son las que su señoría hizo traer a Cristóbal milanés son doradas y pintadas y diferentes unas de otras y todas con escudos de las armas de su señoría en el medio de la que de fuera y den(fol. 47 r)tro pintadas y guarnecidas de manijas y correas de terciopelo negro
-Más otra rodela de acero barnizada y pintada de raspado de cuchillo guarnecida a la redonda de franjas moradas de sirgo morado guarnecida dentro con manijas y correas de negro.
-Cárgansele más al dicho dos rodelas de acero de color pavonada pintadas de raspado con la historia de Mucio Escévola guarnecidas de azul por dentro con sus manijas y correas.

(fol. 48 r)
Espadas y estoques y dagas con talabartes que se cargan a Pedro de Grijalba en primero de junio de mil quinientos cuarenta y un años
-Una espada de monte que trujo Cueto de Valladolid del cargo de San Martín.
-Un estoque viejo con guarnición de latón.
-Un montante dorada la guarnición y la hoja es ondeada
-Otro montante dorado barnizada la guarnición
-Otro montante dorada la guarnición y barnizada
-Una espada de mano y media que tiene la guarnición parada el puño de hilo de oro hilado y la vaina de terciopelo negro con un talabarte de terciopelo negro y los hierros dorados.
-Un bracamarte con la guarnición y puño de plata la vaina de terciopelo morado con el brocal y contera de plata con un hilo de plata retorchada a la redonda con un talabarte de cuero labrado de hilo de oro atado con los hierros dorados y labrados de ataujía que marqués del Vasto dio al conde.
-Otro estoque dorada la guarnición y la vaina de terciopelo azul con un talabarte de terciopelo azul y los hierros dorados.
-Otro que la guarnición dorada con vaina de terciopelo azul.
-Otro estoque la guarnición dorada y plateada con un talabarte de terciopelo negro los hierros dorados y plateados.

(fol. 48v)
-Una cimitarra con la guarnición dorada y la cabeza de turco la vaina de terciopelo negro.
-Otra espada de tornear con la vaina de terciopelo azul y un talabarte de terciopelo azul con los hierros dorados es la guarnición estañada digo plateada.
-Una espada ancha vieja con las armas Zúñiga en los recazos y una guarnición barnizada de negro que dio al q[e] Diego de Reinoso Maçuecos.
-Otro estoque dorada y plateada la guarnición con una vaina de terciopelo blanco con talabarte de terciopelo blanco y los hierros dorados.
-Otro estoque dorado con vaina y talabarte de terciopelo blanco digo plateado.
-Otro estoque con la guarnición dorada con la vaina de terciopelo blanco recamada de oro con un talabarte de terciopelo blanco sin hierros recamado de oro.
-Otra espada de torneo la guarnición dorada con vainas de terciopelo blanco y un talabarte de terciopelo blanco y los hierros dorados.
-Otra espada de torneo la guarnición estañada con vaina de terciopelo blanco que tiene un talabarte de cuero de ante con los hierros dorados que la trujo cuando vino de Villafranca de Niza

-Otra espada de torneo ancha a manera del estoque con guarnición estañada.
-Otro estoque pequeño dorada y plateada la guarnición con vaina de terciopelo amarillo y los hierros dorados
-Otra espada de torneo dorada la guarnición con talabarte de terciopelo amarillo y los hierros blancos.

(fol. 49 r)
-Otra espada de combatir de torneo la guarnición estañada con vaina de terciopelo amarillo y un talabarte de cuero dante con los hierros dorados que el conde trajo de Villafranca de Niza
-Otra espada de torneo con la guarnición estañada que tiene la vaina de raso amarillo y un talabarte de cuero de ante con los hierros blancos.
-Una vaina de terciopelo morado recamada de oro con un talabarte de lo mismo sin hierros
-Otra vaina de terciopelo negro
-Otra vaina de terciopelo negro
-Una espada que la hoja dio Santiago Fernández al conde labrado de unas letras que tiene una guarnición dorada y plateada con unas medallas que hizo Padierna guarnicero con un puño de hilo blanco con vaina de cuero blanco.
-Una espada con una guarnición dorada toda y un puño de hilo de oro y con vaina de terciopelo negro que la hubo su señoría del marqués de Villena.
-Otra espada que su señoría compró a un soldado en Túnez que tiene un letrero de una parte y de otra y una guarnición dorada y plateada de la señal de Padierna y con vaina de terciopelo negro y puño de oro y sirgo blanco.
-Una espada de una hoja vieja que su señoría hubo de Antonio Coco con una guarnición dorada y barnizada y el pomo y la hechura de botón y el puño de hilo de oro hilado y sirgo negro entorchado y vaina de terciopelo negro.

(fol. 49 v)
-Otra espada a modo de cuchillo que su señoría hubo de Villa Santa tomo (?) con una guarnición dorada e plateada e su puño de oro hilado y sirgo blanco con vaina de terciopelo negro.
-Una damasquina que hubo su señoría en Túnez con una guarnición dorada e plateada y el pomo de figura de cabeza de ave y el puño de oro y sirgo blanco con vaina de terciopelo negro.
-Otra cimitarra que dio Moradilla dorador a su señoría de cosas que era a cargo su Señoría tiene una guarnición dorada y el pomo una cabeza de pájaro y el puño de oro de reja una uva ricas doradas y una vaina de terciopelo pardo con un cuchillo y punzón dorado.

-Un arma a manera de espalda redonda y al cabo un hierro como de azcona y con pomo barnizado con una vaina de cuero el arma para montería.
-Una damasquina turquesca que el rey de bohemia dio al conde que tiene una cruz de plata lisa y el puño guarnecido de cuero negro con dos piezas ricas de plata dorados del mismo puño y dos hojalicos de plata y un capirote de plata de dorado liso por pomo una vaina de cuero negro con una contera de plata es lisa sobredorada y un bocalico de plata sobre dorado y un talabarte de tejilla de plata de y sesenta cinco piezas de plata de doradas que son entre hebillas con botones y acicates dorados.
-Otra damasquina turquesa que dio el dicho rey de Bohemia al conde que tiene una cruz de hierro bruñida con pomo guarnecido de cuero negro que tiene en lo alto del pomo una pieza de plata de dorada y una cabezica de toro en ella y una vaina de cuero negro con una contera de plata de dorada lisa. Un bocal de plata dorado con un talabarte de cuero dorado y labrado de sirgo de colores con quince piececitas de pata doradas.

(fol. 50 r)
Las dagas que se entregaron del conde mi señor al Pedro de Grijalba en tres días del mes de junio de mil quinientos cuarenta y un años
-Una daga la guarnición dorada con un puño de oro hilado y la vaina de terciopelo negro
-Otra daga la guarnición dorada y labrada de ataujía con puño de oro y seda morada y vaina de terciopelo morado.
-Otra daga con guarnición dorada y plateada con vaina de terciopelo amarillo
-Otra daga la guarnición dorada con puño de oro hilado con vaina de terciopelo morado.
-Otra daga la guarnición dorada y vaina de terciopelo blanco.
-Otra daga dorada y plateada la guarnición con puño de oro y seda blanca con vaina de terciopelo negro
-Otra daga la guarnición dorada y el puño de oro y seda negra con vaina de terciopelo negro
-Otra daga embarnizada la guarnición con vaina de terciopelo negro
-Otra daga la guarnición barnizada con vaina de cuero tapetado
-Ocho dagas barnizadas las guarniciones las tres con vainas de terciopelo y la una de cuero tapetado las cuales compró el conde.
-Otra daga barnizada la guarnición y la vaina de cuero tapetado que le dio don Alonso de Tejeda.

(fol. 51 r)
Las mazas que entregaron al Pedro de Grijalba en tres de junio de mil quinientos cuarenta y un años
-Una maza de hierro que esta dorada y plateada guarnecida en un palo que dio el marqués de Villafranca don Fadrique al conde
-Tres mazas doradas y barnizadas de negro con sus portamazas de terciopelo negro y caireles de oro y negro
-Dos mazas doradas con su portamazas de terciopelo amarillo
-Ocho más doradas y embarnizadas que trujo Rafael Archioli con los arneses de Milán
-Seis masas de hierro estañadas con sus portamazas de cuero negro que solían ser embarnizadas de negro son las que se compraron para la guerra de Francia
-Otra maza de hierro blanco toda dorados unos trocillos della
-Otra maza de hierro bruñido con una banda dorada retocada.
-Una hachuela de armas dorada guarnecida el asta en terciopelo carmesí es del cargo del alcaide Bozmediano.
-Otra hachuela de armas de hierro toda que es del cargo del alcaide Bozmediano
-Otra hachuela de armas que tiene el asta de palo
-Más se cargan al dicho Pedro de Grijalba cuatro más barnizadas de negro que se compraron de Cristóbal milanés de Medina del Campo.

(fol. 51 v)
Siete guarniciones de caballo de caballo de la estadiota con clavazón dorado que son una de terciopelo carmesí toda entera y otra de terciopelo azul y otra de terciopelo negro con franjas de oro anchas y cuatro de terciopelo negro lo cual todo de ello entregó al dicho Pedro de Grijalba en la fortaleza por Luis de Ontiveros que eran del cargo de Lezcano en 18 de junio de 1544 años [...]

(fol. 56 r)
Arcabuces que se entregaron al dicho Pedro de Grijalba en uno de junio de mil e quinientos cuarenta e un años
-Un arcabuz con un frasco negro guarnecido de cuero negro
-Otros dos arcabuces con un frasco los cuales parece que dicho pero de Grijalba dio por cedula del conde fechada a XXIII de [...] de MDXXXVIII años al marqués de Villena

(fol. 62 r)
Los arneses que se cargan al dicho Pedro de Grijalba de guerra que están en el Hospital de Nuestra Señora de la Piedad de Benavente a diez de junio de mil quinientos cuarenta y un años

Un arnés
-Una coraza peto y espaldar con su ristre y volante para guerra dos escarcelas enteras su alpartaz de malla dos quijotes e dos grebas enteras con sus escarpes dos brazales dos manoplas dos guardabrazos un almete con su alpartaz de malla un barbote
-Otro arnés complido con las piezas que el de arriba
-Otro arnés complido como el primero fáltale a este el volante
-Otro arnés con las piezas que el primo fáltale el volante
-Otro arnés como el primero sin faltarle nada.
-Otro arnés complido sin faltarle nada como el primero
-Otro arnés complido
-Otro arnés complido
-Otro arnés complido como el primo no tiene escarpes sino unos zapatos de malla con sus puntas
-Otro arnés complido como el primero fáltanle los escarpes
-Otro arnés complido como el primero sin faltarle nada salvo unas puntas de escarpes
-Un arnés complido como el primero sin faltarle nada
-Otro arnés complido como el primero
-Más otro arnés complido como el primero fáltale el volante e una pieza del barbote
-Más otro arnés como el primero falta el volante y ristre
-Más otro arnés complido como el primero
-Más otro arnés como el primero fáltale medio barbote
-Más otro arnés como el primero

(fol. 62 v)
-Más otro arnés complido como el primero
-Más otro arnés complido como el primero fáltale el ristre y la manopla izquierda
-Más otro arnés como el primero falta el ristre
-Más otro arnés como el primero fáltanle los dedos de una manopla
-Más otro arnés como el primero
-Más otro arnés complido como el primo
-Más otro arnés como el primero
-Más otro arnés que le falta el volante y el ristre
-Más otro arnés que le falta la manopla dicha y ristre y medio barbote
-Más otro arnés complido como el primero
-Más otro arnés complido como el primero
-Más otro arnés que le falta medio barbote de la parte de arriba

-Más otro arnés que le falta el volante y dos manoplas y medio barbote
-Más otro arnés que le falta la manopla izquierda y ristre
-Más otro arnés que le falta la manopla y alpartaz de malle del almete
-Más otro arnés que le faltan las manoplas y el volante y medio barbote y alpartaz del almete
-Más otro arnés que le falta la manopla derecha y los escarpes
-Más otro arnés como el primero falta el ristre
-Otro arnés como el primero
-Más otro arnés que le falta el alpartaz de la coraza
-Más otro arnés que le falta el alpartaz de malla de la coraza y ristre y medio barbote
-Otro arnés que le falta el alpartaz de la coraza y del almete
-Otro arnés complido sin le faltar nada
-Otro arnés que le falta con la manopla y ristre
-Otro arnés complido
-Otro arnés que le falta el ristre
-Otro arnés que le falta medio barbote y el volante
-Otro arnés complido
-Otro arnés que le falta la manopla izquierda

(fol. 63 r)
-Otro arnés complido como el primero
-Otro arnés que le falta medio barbote
-Otro arnés como el primero
-Otro arnés complido
-Más otro arnés que le falta la manopla izquierda
-Más otro arnés que no tiene barbote ni manopla izquierda
-Otro arnés que le falta la manopla izquierda
-Otro arnés que le falta el ristre
-Otro arnés complido
-Otro arnés complido
-Otro arnés que le falta los dedos de la manopla izquierda
-Otro arnés que le faltan los escarpes
-Otro arnés como el primero
-Otro arnés como el primero
-Otro arnés complido como el primero
-Otro arnés como el primero
-Otro arnés como el primero
-Otro arnés como el primero

-Otro arnés como el primero
-Otro arnés que le falta la manopla derecha
-Otro arnés complido como el primero
-Otro arnés como el primero
-Otro arnés complido como el primero
-Otro arnés complido
-Cuarenta bufas
-Unas platas sin volante y sin ristre con su almete y alpartaz de malla e dos grebas redondas e dos quijotes e un barbote del almete e dos guardabrazos
-Otras platas con su volante una celada con su barbote unos quijotes y unas grebas trenzadas redondas dos guardabrazos
-Otras platas sin volante y un almete con su barbote e alpartaz de malla
-Otras platas con su volante sin ristre y su almete sin malla e barbote con brazal dicho
-Cinco piezas de guardabrazos que son principios cada una de su guardabrazo no son conformes
-Otro guardabrazo
-Otras dos piezas de guardabrazo del hombro
-Cuatro orejas de testera de caballo

(fol. 63 v)
-Un escarpe de greba que le falta tres piezas
-Tres sobreguardas de brazales izquierdos de justa enteros
-Una vista de guerra entera
-Tres sobremanoplas izquierdas de justa
-Otra vista de seguir
-Cuatro quijotes que le faltan cinco piezas de hacia las grebas
-Un barbote en dos piezas de seguir diez y seis piezas ricas y grandes obradas diferentes todas de caballos y barbotes e quijotes
-Ocho testeras de caballo
-Un cuello de acero de caballo
-Más se le cargan al dicho Pedro de Grijalba un arnés que se compró de Francisco Briceño tiene las piezas siguientes
-Una coraza que es peto y espaldar
-Un volante y un ristre
-Dos escarcelas
-Dos brazales
-Dos guardabrazos
-Un almete con su alpartaz de malla

-Un barbote y sobrebarbote
-Dos manoplas
-Dos quijotes
-Dos grebas trenzadas
-Dos puntas de hierro de escarpes con malla
-Una bufa
-Un gorjal de malla

(fol. 64 r)
Cárgansele más al dicho Pedro de Grijalba veinte arneses de la ligera que su señoría compró de maese Hernando armero vizcaíno tienen las piezas siguientes cada uno de los veinte
-Una celada borgoñona
-Una gola
-Una coraza que es peto y espaldar
-Dos escarcelas enteras.
-Dos guardabrazo
-Dos brazales
-Dos manoplas
-Dos medios quijotes
-Dos esquinelas un ristre
-Cárgansele más al dicho Pedro de Grijalba cien testeras de caballo que su señoría compró de maese Hernando armero vizcaíno.
-Cárgansele al dicho Pedro de Grijalba cincuenta sillas de la brida de badana blanca para la gente de guerra que Cueto envió a caballo para la jornada de Piamte son las dos armadas con los aceros que su señoría compró de maese Hernando armero con los dos coseletes
-Cárganse le Más al dicho cien aceros para sillas de caballo de la brida los cuarenta y nueve se armaron en cuarenta y nueve sillas de las cincuenta sillas arriba dichas.

(fol. 67 r)
Cargo que se hace al Pedro de Grijalba de lanzas de justa hombros de armas y ginetas y astas y picas en XVI de junio de MDLXI años
-Seis lanzas de armas la una de ellas esta pintada y la una está obrada. Quebróse una de ellas
-Veinte y ocho lanzas de justa que son del cargo de San Martín. Gastáronse en fiestas
-Veinte y siete lanzas de justa del cargo del alcaide Bozmediano. Gastáronse en la justa de la Cuaresma dos años y más otras 23 que son

-Diez y siete lanzas de justa que están en la armería están pintadas alguna de ellas
-Dos lanzas de armas con hierros de guerra
-Cuarenta y siete picas sin hierros están las cuatro de ellas con hierros e ocho de ellas están pintadas de blanco y negro porque se hicieron para lanzas ginetas para un regocijo. Las XX se gastaron en partesanas y alabardas y espedos boloñeses y birotes que hizo Matilla para el conde
-Tres astas para estandartes
-Ochenta y cuatro lanzas ginetas sin hierros. Gastaron seis lanzas.
-Más se le cargan cincuenta lanzas de remo de pino con unos hierros.

(fol. 67 v)
Por manera que [...] sacadas de todas ellas ciento y quince que se han gastado como parece en la margen que dan que yo recibo de Valdivieso de Pedro de Grijalba por mandado del conde en 22 de septiembre de 1550 años

Juan de Valdivieso
(signed).

Doc. 3.
1541, 1571, 1581.
Book relating to the arms owned by Antonio Alonso Pimentel de Herrera, 6th Count of Benavente, at the Hospital de Nuestra Señora de la Piedad of Benavente.
Archivo Histórico de la Nobleza, Osuna, C. 424, D. 33.

(Fol. 2r)
Juan Rodríguez de Mercado
Resulta
de la cuenta que en su nombre
va dando el prior Francisco Rodríguez de Mercado su hermano de las armas
Año 1581

Arneses del señor don Alonso

Ha de dar cuenta el dicho Francisco Rodríguez en nombre dicho Juan Rodríguez su hermano de las piezas que percibió que tratan de los dos arneses del señor don Alonso Pimentel que envió al dicho don Antonio que esté en el cielo que se halló con Pedro de Rivas su criado las cuales son las siguientes
-Una coraza con su peto y espaldar
-Una gola
-Sus escarcelas
-Su ristre
-Sus brazales con unos guardabrazos
-Unos mandiletes
-Una celada encajada
-Unas grebas con sus escarpes y sin quijotes trenzadas y unas espuelas doradas
-Una celada de seguir
-Un guardabrazo para justa
-Media testera

(fol. 2 v)

-Los fustes de una silla. Se obraron en este arnés dos baberones de justa

Segundo arnés
-Una coraza que espeto y espaldar con unas escarcelas y una gola
-Unos brazales para seguir
-Unos guardabrazos

-Unos mandiletes de dedos
-Una celada para seguir con su ventalla y la vista tiene tres piezas con una red dorada por dentro.
-Unas grebas trenzadas con unos quijotes trenzados
-Una llave de armar con sus tornillos
-Una silla con sus fustes
-Otra silla con cuatro piezas cinceladas y doradas con sus fustes

Del arnés que el conde compró de Luis Méndez faltan las piezas siguientes

-Dos guardabrazos de este arnés para seguir la guerra cincelados y dorados

(fol. 3 r)
-Una pieza en la punta della manopla de este arnés de la mano derecha
-Dos ristres uno de seguir la guerra y otro de justa que son de este arnés
-Una llave dorada de este arnés con un cordón de seda naranjada.

Del arnés que fue de don Íñigo de Guevara faltaron al dicho Juan Rodríguez las piezas siguientes de lo cual ha de dar cuenta el dicho Francisco Rodríguez
-Un ristre de justa
-Unas espuelas que están pegadas a los escarpes
-Un brazal
-Una bufa que es pieza pequeña para la justa
-Unas puntas de escarpes doradas y moradas
-Una tarjeta de hierro
-Una gola de la celada encajada
-Un ristre de la coraza de justa

Del arnés que fue del conde don Alonso que este en el cielo faltaron las piezas siguientes de las que ha de dar cuenta el dicho prior Francisco Rodríguez
-Un volante
-Un ristre dorado y cincelado
-Unos quijotes blancos
-Una punta de acero negro de un zapato de terciopelo negro.
-Un vate hierro.

Sobró en este arnés una celada que parece ser del dicho arnés y no se le hizo cargo de ella al dicho Juan Rodríguez de Mercado

(fol. 3 v)
Del arnés que se compró de Enrique Alemán faltan las piezas siguientes de las cuales ha de dar cuenta el dicho prior Francisco Rodríguez de Mercado en nombre del dicho Juan Rodríguez de Mercado su hermano

-Dos brazales para combatir a pie. Borráronse porque parecieron y se entregaron y cargaron a Salazar
-Una pieza doble para guerra cincelada
-Un placarte de combatir a pie y a caballo
-Una tarjeta

En este arnés se hallaron tres piezas que no le iban cargadas al dicho Juan Rodríguez de Mercado y sobraron del dicho Juan Rodríguez las cuales son un baberón pequeño y un medio barbote y media sobreguarda

Del arnés que se compró en Mayorga al licenciado Guzmán el año de MDLXXIII que es dorado y grabado del cual faltaron las piezas siguientes de las cuales ha de dar cuenta el dicho prior Francisco Rodríguez en nombre del dicho su hermano Juan Rodríguez de Mercado
-Un mandilete
-Dos coracinas negras con medias golas y sus escarcelas y dos sombreretes y alcetillas que dice se llevó a Sanabria y el mandilete dice que lo perdió don Pedro Enríquez vecino de Zamora o Don Mendo

(fol. 4 r)
Las dichas resultas del dicho
Prior Francisco Rodríguez de Mercado

El arnés que su señoría compró de don Gabriel Zapata en Madrid que es dorado a bandas no lo entregó el dicho prior Francisco Rodríguez de Mercado resultase para que dé cuenta del cual tiene las piezas siguientes
-Una coraza que es peto y espaldar con sus escarcelas
-Una gola
-Una celada para seguir la guerra
-Dos brazales y guardabrazos con sus arandelas en los guardabrazos
-Grebas y quijotes y escarpes
-Dos mandiletes
-Un ristre
-Una celada engolada
-Un almete para justa real
-Una tarjeta
-Un placarte
-Un volante con un escarcelón de justa
-Un baberón
-Otro medio escarcelón

-Una sobremanopla
-Otras dos manoplas
-Otra sobremanopla
-Otros dos guardabrazos
-Un brazal derecho de Santo Antón
-Una sobreguarda del brazal izquierdo
-Una arandela
-Una testera de caballo
-Otras dos arandelas blancas
-Una escofia

(fol. 4 v)
-Un barbote
-Una alceta una navaja para el brazal izquierdo es blanca
-Dos medias grebas con sus quijotes de piezas
-Otra pieza pequeña para el guardabrazo izquierdo
-Dos piezas pequeñas de embridar
-Dos zapatos de malla con sus puntas

(fol. 5 r)
Antonio de Salazar armero
Mayor del conde mi señor

Cargo
De las armas de la armería de su señoría así de a caballo como de infantes armas de caballos sillas y guarniciones

En Benavente a ocho de mayo de mil quinientos ochenta y un años se tomó cuenta de la dicha armería al señor prior Francisco Rodríguez de mercado el cual la dio por Juan Rodríguez de mercado su hermano quien se había e hizo cargo de ellas y por su fin y muerte dióla dicha cuenta las cuales dichas armas como las iba entregándose vaya siendo cargo de ellas que es el que de nuevo las recibe y son las siguientes en esta manera

Los dos arneses del señor don Alonso

De los dos arneses que el señor don Alonso Pimentel envió al conde don Antonio que este en el cielo con Pedro de Rivas su criado que eran las piezas cinceladas y labradas hechos unos golpes en la misma obra del arnés con unas rosas a manera de cruces por medio todas doradas de los cuales se entregaron al dicho Antonio de Salazar las piezas siguientes en esta manera

-Dos rondeletas para los guardabrazos
-Una tarjeta de justa
-Un placarte de justa
-Un volante con su sobrebarriga
-Un escarcelón que cae sobre la escarcela para justa
-Dos sobremanoplas
-Otra sobreguarda para el brazal izquierdo
-Otra pieza pequeña para el brazal izquierdo para justa de tarjeta

(fol. 5 v)
-Una pieza bajuela para justa de guerra que es a manera de bufa
-Un almete de justar

En este arnés sobraron dos baberones de este arnés y se entregaron al dicho Antonio de Salazar que lo firmo el dicho día mes y año

-Las cuales dichas piezas del dicho arnés se entregaron al dicho día ocho de mayo de mil quinientos ochenta y uno al dicho Antonio de Salazar armero mayor y lo firmó de su nombre no se le entregaron más piezas de este arnés porque dijo el dicho prior Francisco Rodríguez que le había mandado por mandado del conde don Juan las piezas dichas piezas se ponen el pliego de resulta hasta que den cuenta de ello porque dijo que las tenía Luis de Quintanilla vecino de Medina del Campo

Las piezas aquí contenidas de segundo arnés del señor don Alonso se entregaron al dicho Antonio de Salazar
-Una bragueta
-Un barbote que vino con este arnés que no es del y es para torneo
-Un brazal que vino con este arnés y no es del
-Unas puntas de escarpes
-Tres arandelas de justa
-Los aceros de dos sillas de los dos arneses sin fustes

Las cuales dichas piezas se entregaron al dicho Antonio de Salazar en dicho día y lo firmo de su nombre porque las piezas que faltaron se resultaron para que en cuenta de ellas el dicho Francisco Rodríguez de Mercado

(fol. 6 r)
Las dos cubiertas de caballo de acero que vinieron con los dos arneses del señor don Alonso Pimentel tienen las piezas siguientes
-Un coplón con una cabeza de sierpe en la cola con dos costeras y una sotacola y un paño de tres piezas con dos bollones levantados

-Una gorja que prende en el pecho que llega hasta el hocico del caballo
-Una testera que tiene por orejas unos cuernos enroscados
-Un cuello de caballo
-Unas riendas de piezas
-Unos francaletes que prenden desde el pecho del caballo y hasta el coplón.
-Otras cubiertas para el mismo arnés con las mismas piezas que va dicho arriba. Las cuales dichas piezas todas son doradas y cinceladas de la misma manera que los arneses

Las cuales dichas piezas aquí contenidas esta plana se entregaron al dicho Antonio de Salazar y lo firmo de su nombre en dicho día ocho de mayo de mil quinientos ochenta y un años todo lo cual se recibió en cuenta al dicho prior Francisco Rodríguez de Mercado en nombre del dicho Juan Rodríguez de Mercado

(fol. 7 r)
Arnés que el conde compró de Luis Méndez Quijada
-Una coraza peto y espaldar cincelado y dorado con sus escarcelas enteras cinceladas y labradas y doradas
-Un volante blanco que está hecho para justar de tarjeta para con este arnés con su sobre barriga dorado y cincelada
-Un escarcelón grande blanco que sirve para justa en este arnés
-Unos brazales cincelados y dorados
-Las navajas y bordes
-Unas grebas con sus escarpes cincelado y dorado
-Unos quijotes del mismo arnés dorados y cincelados con unas cuchilladas cinceladas y doradas.
-Un almete de este arnés encajado con su gola y vista dorado y cincelado
-Una celada de borgoña con su vista para justar de tarjeta

(fol. 7 v)
-Una vista y sobrevista para la dicha celada borgoña para torneo de a caballo dorado y cincelada
-Un baberoncillo para la dicha celada borgoña de torneo de a pie dorado y cincelado
-Una celada de infante con su baberón cincelado y dorado con unos penachos y su ventalla
-Otra gola cincelada y dorada para la dicha celada borgoñona
-Dos guardabrazos cincelados dorados para este arnés para justa de tarjeta
-Un baberón de este arnés para torneo de a caballo y otro sobrebarbote para lo mismo con media vista de almete para lo mismo cincelado y dorado todo

-Una rondeleta de este arnés para el guardabrazos para torneo de a caballo y seguir la guerra dorado y cincelado
-Una escofia con sus barras es dorada y cincelada del mismo arnés
-Más otra cofia cincelada y dorada que hizo después Juan Pedro en Segovia

(fol. 8 r)
-Dos manoplas del mismo arnés cinceladas y doradas
-Unos zapatos de malla con las puntas cinceladas y dorados
-Una manopla de combatir de la espada dorada y cincelada del mismo arnés
-Unas medias grebas cinceladas y doradas del mismo arnés
-Un tonelete del mismo arnés cincelado blanco que está por dorar
-Dos piezas que se hicieron en Valladolid con menudas blancas para los quijotes para torneo de a pie para cerrar.
-Otras dos piezas para las sangraderas de los brazales para torneo de a pie
-Una sobremanopla blanca para justa
-Una bufeta para justa cincelada y dorada
-Una rondeleta cincelada y dorada/ son dos rondeletas

(fol. 9 r)
El dicho Antonio de Salazar pliego tres
Cargo
de las armas que recibió

El arnés que fue de don Íñigo de Guevara que el arnés y las piezas son doradas a bandas y en medio de ellas está pintado de dorado y pavonado que tiene las piezas siguientes
-Una coraza con su espaldar y sus escarcelas
-Dos brazales de guerra
-Dos medias grebas y sus escarpes cortados
-Dos quijotes
-Una celada de seguir la guerra
-Unos guardabrazos de guerra
-Un tonelete
-Dos manoplas
-Un baberón y un placarte para tornear digo barbote para tornear
-Un mandilete pequeño que es manopla para guerra izquierda
-Una coraza para justar con su espaldar
-Dos guardabrazos de justa
-Otro guardabrazo del brazo dicho

-Una manopla de justa que es sobremanopla
-Un almete engolado que tiene dos vistas una de guerra y otra de justa

(fol. 9 v)
-Una cofia de combatir de espada
-Un placarte para encambronar
-Unas grebas enteras con sus escarpes
-Una testera del mismo arnés dorada y negra pavonada
-Unas cubiertas de acero a ruedas doradas y moradas con fleco azul y morado lo cual se debe de haber desteñido y parece el fleco blanco y encarnado guarnecidas de angeo, aunque dice el cargo de terciopelo con franjas blancas y encarnadas y en el cargo viejo dice franjas de morado en que hay coplón y pecho y cabezadas y riendas de lo mismo

(fol. 10 r)
El arnés que fue del conde don Alonso mi señor que esté en el cielo tiene las piezas siguientes
-Unas platas que es peto y espaldar doradas y cinceladas los bordes con sus escarcelas la una aforrada en carmesí y hebillas doradas
-Un almete dorado con clavazón dorado con su vista y barbote los cantos de este almete solo están dorados, aunque dice en el cargo viejo dorado
-Otro almete con su barbote y vista dorado y cincelado los bordes y clavazón dorada y guarnecido de carmesí
-Una gran bufa dorada los bordes con un tornillo que no le vino
-Dos quijotes dorados y cincelados los bordes
-Unos brazales dorados y cincelados los bordes con correas de terciopelo carmesí con sus guardabrazos
-Una guarnición de brazos que son brazales y guardabrazos dorada y cincelada los bordes con hebillas doradas y correas de cuero
-Dos manoplas doradas y cinceladas los bordeas aforradas en terciopelo carmesí

(fol. 10 v)
-Dos grebas con sus quijotes de hombres de armas con sus escarpes dorados correas de cuero blanco
-Dos escarceloncillos de los costados
-Otras dos grebas con sus escarpes dorados y cincelados los bordes
-Unos zapatos de terciopelo negro cubiertos a trozos de malla y con sus puntas de acero negro
-Una celada que pareció ser de este arnés aunque no estaba en el cargo viejo ni está cargada al dicho Juan Rodríguez la entregó el dicho prior Francisco

Rodríguez de Mercado al dicho Antonio de Salazar para que dé cuenta con las demás armas contenidas en esta plana por ser de la misma pasta y guarnición y estar con las dichas armas

(Signed). A 1581
(fol. 11 r)
El dicho cargo al dicho Antonio de Salazar
de las dichas armas pliego cuatro

El arnés que se compró de Enrique Alemán que tiene las piezas blancas y grabadas y son las siguientes las cuales se cargan al dicho Antonio de Salazar para que dé cuenta de ellas
-Una coraza peto y espaldar
-Unas escarcelas de justa
-Otras escarcelas de guerra
-Otras escarcelas de infante
-Un volante para justa de encambronar
-Un baberón para encambronar
-Una gran bufa entera con su tornillo que no le viene
-Dos brazales de justa con sus sobreguardas la una doblada la otra sencilla
-Una gola
-Una celada encajada con su vista de justa
-Dos guardabrazos de justa
-Otra celada con su vista de justa
-Dos medias grebas con sus quijotes dos grebas enteras con sus quijotes cerrados para combatir a pie
-Dos brazales para combatir el uno cerrado

(fol. 11v)
-Dos guardabrazos para combatir a pie
-Dos guardabrazos de guerra
-Una gola de guerra
-Un baberón para tornear a caballo
-Una celada de infante y de seguir tiene el barbote por si
-Dos pares de manoplas las unas de justa las otras para torneo
-Una sobremanopla de justa
-Una cofia de combatir de espada
-Un tonelete
-Una testera de caballo pequeña guarnecida de angeo que no es de este arnés porque perdió la otra don Diego Sarmiento hijo del conde de Salinas

-Un ristre de guerra
-Otro ristre de justa
-Una llave de encambronar
-Dos medias escarcelas de infante
-Una tarjeta cerrada
-Dos puntas de escarpes aceradas

Las piezas que faltaron de este arnés son estas- digo las piezas que sobraron son las siguientes
-Un baberón pequeño
-Un medio barbote
-Media sobreguarda

Las cuales se entregaron al dicho Antonio de Salazar para que dé con las demás armas cuenta y lo firmó el dicho día

Antonio de Salazar
(Signed)

(fol. 12 r)
El Arnés dorado y grabado que se compró en Mayorga del licenciado Guzmán el año de MDLXXIII que tiene las piezas siguientes
-Gola peto y espaldar
-Un sobrepeto con dos ristres
-Dos escarcelas de launas
-Un escarcelón y otro pequeño
-Cuatro brazales con cuatro guardabrazos
-Un mandilete y dos quijotes porque el otro mandilete que parece por el cargo viejo que hay que son dos mandiletes se puso por resulta para que dé cuenta del dicho prior en nombre del dicho su hermano
-Dos grebas con sus escarpes
-Un baberón y un barbote
-Una sobremanopla
-Una sobreguarda de brazal
-Una celada borgoñona con su escofia
-Otra celada encajada par ajusta
-Una tarjeta
-Dos arandelas, un morrión de infante
-Una media testera [...]

(fol. 13 r)
El dicho cargo de las armas al dicho Antonio
de Salazar pliego quinto

Otros dos arneses pequeños no se cargaron al dicho Salazar porque el uno que era el del príncipe mandó el conde se diese al hijo mayor del marqués de Viana y el otro a don Bernardo hijo del marqués de Távara
Un arnés pequeño de niño de los dos que compró Francisco Arias tiene las piezas siguientes el cual se entregó al dicho Antonio de Salazar
-Un peto y espaldar y gola con sus escarcelas largas
-Unos brazales con sus guardabrazos
-Una celada de seguir
-Unos mandiletes de dedos
-Un morrión

(fol. 14 r)
Memoria de las piezas que yo Catalina de Arguelles he entregado al muy ilustre señor don Pedro de Solís de los tres arneses que don Alonso de Tejeda mi señor mandó que se diesen al ilustrísimo conde de Benavente a lo cual se halló presente Robles armero vecino de esta ciudad de Salamanca y en mi poder queda una firmada del dicho señor don Pedro de como su mozo y que si entre ellas fuere alguna que no sea de ellos dichos arneses me la mandara volver las cuales piezas son las siguientes

Arnés dorado nuevo de seguir
-Primeramente un peto
-Un espaldar
-Una gola
-Una celada
-Un baberón
-Unos brazales
-Unos guardabrazos
-Una manopla izquierda
-Otra derecha
-Dos quijotes
-Dos medias grebas
-Dos arandelas
-Otras dos pequeñas
-Una manopla de tornear

-Una escofia
-Una silla del mismo jaez e con cuatro piezas y una testera

Otro arnés de justa dorado
-Un peto
-Un espaldar
-Un volante
-Un guardabarriga
-Una gola
-Un baberón
-Una celada
-Una calva
-Unos brazales
-Unos guardabrazos
-Dos manoplas
-Dos escarcelones
-Un morrión con ventalla digo con piezas y no ventalla
-Dos esquinelas
-Una manopla izquierda de justar
-Un ristre
-Dos quijotes
-Un placarte
-Una bufeta izquierda
-Otra derecha [...]
-Dos guardabrazales
-Una tarjeta
-Una testera entera
-Una bragueta
-Unas puntas de zapatos

Otro arnés blanco de seguir
-Un peto
-Un espaldar
-Una gola
-Una celada
-Un morrión
-Unos guardabrazos
-Unos brazales
-Dos quijotes
-Dos medias grebas

-Unas puntas de zapatos
-Dos manoplas
-Dos llaves
-Ciertos tornillos y tuercas
-Van algunas piezas sin fundas

Digo yo Cat^{a} de Arguelles que es verdad que las piezas arriba cargadas se entregaron a su mozo del señor don Pedro y no más y porque es verdad lo firme de mi nombre fechada en Salamanca a nueve días del mes de junio de mil quinientos setenta y siete años

(fol. 14 v)
Digo yo Juan Rodríguez de Mercado armero mayor del conde mi señor que recibí y me cargo de los arneses y piezas de esta otra parte contenidas que fueron de don Alonso de Tejeda lo cual se me entregó a tres días del mes de julio de mil quinientos setenta y siete días y por la verdad lo firmé de mi nombre

Juan Rodríguez de Mercado
(signed)

(fol. 15 r)
El arnés que Juan de Carvajal compró en Milán para la guerra de Francia [...]

(fol. 15 v)
Recibí lo contenido en estas dichas planas en 3 de marzo MDLXXIV

(fol. 16 r)
Resultas de la cuenta del cargo de Pedro de Rivas diputado y haciéndose cargo de la armería de su señoría a Juan Rodríguez de Mercado

Año de MDLXXI
-Un montante de las ondas que se llevó a Valencia dicen que está en la recámara.
-Un gorjal de malla con sus almenas de latón del pliego de mallas.
-Dos placartes del arnés que se compró al Luis Quijada
-Del arnés de don Íñigo falta un cuello de malla y una rondeleta
-Dos espaldarcetes del arnés que compró Juan de Carvajal para la guerra de Francia
-Una rondeleta del arnés que fue de don Alonso Pimentel
-Del arnés que el conde don Alonso tenía faltan los zapatos de malla

-Falta una testera del arnés de Pez Cozque (sic)
-Falta una manopla de dos que tiene el arnés que fue de don Do Sarmiento y una sobremanopla y una rondeleta y dos piezas dobles de guerra todo del dicho arnés
-Del arnés de la guerra de Francia falta una sobrebarriga sin escarcelón
-Un barbote del coselete dorado del duque de Nájera

(fol. 16 v)
-Hablar a su Señoría sobre el coselete y cota de malla con mangas que se prestó a Juan de Carvajal yendo por corregidor de Écija
-De otra cota que se prestó a Alonso González de Anaya guarda que fue de aldea del dicho
-Falta una daga dorada y plateada la guarnición con un puño de oro y seda blanca con vaina de terciopelo negro
-Una daga barnizada con vaina de terciopelo negro
-Una espada de torneo la guarnición dorada con vaina de terciopelo blanco.
-Más otra espada de torneo que tiene la vaina de cuero blanco con un talabarte de cuero de ante con hierro alemán
-Otra daga negra que compró el conde de Padierna
-Más una espada que dice la partida a manera de cuchillo que su Señoría hubo de Villasante escribano con una guarnición dorada y plateada y puño de hilo de oro hilado y sirgo blanco con vaina de terciopelo negro.
-De los seis arneses que se compraron de maese Hernando armero del duque de Nájera faltan tres rondeletas de los guardabrazos

(fol. 17 r)
-Una rondeleta del brazal izquierdo de justa de guerra del arnés de don Íñigo de Guevara
-Del arnés que don Alonso Pimentel Maldonado dio al conde faltan unas grebas trenzadas
-Más al dicho una gran pieza que sirve también de guardabracillo falta
-Más al dicho arnés falta una bufa
-Del arnés que dio su señoría a don Juan Pimentel comendador falta una rondeleta que dice la partida que cuelga de un guardabrazo
-Del dicho arnés una que llaman la pieza bajuela
-Fáltale un coselete de dos que se compró de Cristóbal milanés de Medina del Campo es negro que tiene peto y espaldar y medias escarcelas. Dice Teza que se dio que se dio a un guarda [...]
-Fáltale dos corazas que es peto y espaldar sin escarcelas ni otra cosa que se compraron de maese Hernando armero del duque de Nájera de unas armas viejas que vinieron de Túnez y que estaban a cargo del Eschave.

-Dos morriones de la partida de ocho morriones que su señoría compró en Génova el uno mostró cédula del señor don Juan por la cual se dio a Cristóbal de Porras vecino de Zamora

(fol. 17 v)
-Un casco le falta a Rivas de la plana de los morriones
-De una partida de veinte y tres cuerpos de corazas faltan dos y en las márgenes de la dicha partida dice quedar unas se dieron a Ramírez guarda del conde por carta fechada a 7 de diciembre de 1548/ y la otra se dio a Pedro Sánchez por carta del conde de 23 de noviembre de 1556
-Otra partida de unas coracinas de Milán con mangas enteras y escarcelas guarnecido por de fuera de fustán negro con clavazón barnizada dice la margen que las tiene el señor don Juan Pimentel hijo
-Otras coracinas de Milán sin mangas con escarcelas guarnecidas de fustán negro y clavazón estañada/ dos (¿) que se dio a Pedro Sánchez guarda de aldea del conde por carta fechada a 23 noviembre de 1556.
-El capacete que se pide del pliego de ginetas le falto la guarnición que dice que tenia de terciopelo morado
-Fáltanle dos arandelas de la partida de nueve arandelas
-Faltole un caballete de armar de tres pies guardo de terciopelo negro con tres hilitos dorados que se hizo para armarse el conde cuando los casamientos del rey en Toledo el año de 1560

(fol. 18 r)
-Fáltanle estribos. Cuatro pares de estribos estañados de LXXIIII pares que recibió
-Falta al dicho Rivas unas espuelas blancas plateadas que las dio Bozmediano.
-Unas espuelas digo tres pares de espuelas para justar blancas.
-Faltan más otras espuelas tenidas que dice la partida que
-Una rodela puntada de las que trajo Cueto de Valladolid del cargo de San Martín con franjas verdes amarillas y azules
-Una rodela de acero pavonada con un Mucio Escévola compañera de otra que queda de dos que compró de Cristóbal milanés dice Teza que la llevó Cristóbal de Porras por mandado del señor don Juan Pimentel

(fol. 18v)
-Fáltanle al dicho Rivas dos rodelas de la partida de 24 con los armas del conde que esas dos dicen que las tiene el conde de Mayorga.
-Faltan unas riendas de una guarnición de terciopelo pardo de la estradiota que tiene la clavazón dorada y es mitad pardo

-De la guarnición que el conde hizo para entrar en Roma que es de terciopelo negro con franjas y borlas de oro faltan al dicho Rivas ciento y con cuarenta y cuatro borlas y entre ellas la grande del coplón que no tiene más de la bellota en él
-Una correa de largo de un palmo guarnecida de tela de oro y ribetes de terciopelo carmesí con su cabo que es de una grupera de tela de oro y carmesí
-Una reata de lino de colores vieja
-Una guarnición de terciopelo negro con franjas de oro fáltale el pretal
-Otra negra con clavazón dorada faltan las riendas
-Otra negra como la de arriba falta pretal y riendas
-Otra negra de terciopelo como las de arriba faltan las riendas son de la partida de tres guarniciones que están en una partida
-Fáltanle al dicho Rivas una cincha de cáñamo
-Y ocho sobrecinchas de cáñamo
-Y seis talabartes de cuero colorado de los hombres de armas
-Y cuatro sombreros de los de tafetán amarillo de hombres de armas
Todos estos del cargo de guarniciones de caballo

(fol. 19 r)
-El pliego de los arneses de hombres de armas falta a Rivas XXX bufas dicen que podría ser cargarse por hierro diciendo que serán bufas siendo guardabrazos altos como se pueden ver
-El mismo cargo un pliego adelante del dicho le falta lo siguiente
-Una pieza de guardabrazo- otro guardabrazo- cuatro orejas de caballo- dos sobreguardas- cuatro quijotes- un barbote- la malla de un cuello de un caballo

(fol. 19 v)
-Faltan a Pedro de Rivas tres hierros de picas son de almacén
-Fáltanle tres borlas de oro y seda encarnada de caparazón cuello y petral de brocado encarnado.
-Una flecha de un carcaj que dice que tiene XIX flechas de un arco turquesco
-Seis espadas viejas para torneo que le estaban cargadas al dicho Rivas
-De las piezas que trajo Rogelio de Milán sin arnés están un placarte para encambronar y una bufa.

(fol. 20 r)
-De la partida de LXXI lanzas de hombres de armas faltan XXIII para cuantas se dieron XIII a Santander caballerizo por mandado del conde de Mayorga y no se cobraron más de cinco-faltaron tres y otra que se dio por orden del conde a la Cofradía de la Cruz- vendrán faltar 13 otra que quebró Aranda faltan x12
-En el pliego de los arneses de los hombres de armas faltaron treinta bufas de la partida que pide cuarenta bufas

-Asimesmo falta en el dicho pliego de setenta y siete arneses de munición cuarenta y un volantes y cuarenta y dos alpartaces de malla
-Tres arneses de una partida que dice que los dos tienen peto y espaldar y golas y escarcelas largas y unas celadas con rejas y el otro arnés que tiene una coraza y su celada y unos brazales y dos guardabrazos cada uno de su norma y una celada vieja
-Otros tres arneses de seguir que su señoría mando comprar de maese Hernando armero del duque de Nájera que cada uno tiene peto y espaldar y su ristre y escarcelas y sus brazales y guardabrazos y manoplas y gola y sus celadas de seguir y medias grebas trenzadas con sus quijotes
-Faltáronle algunas plumas y de ellas no parecieron y otras estaban con el tipo (¿) comido de traza y perdidas
-Un tambor que dice el cargo que estaba en Astorga y se había de traer el año de LXIII

(fol. 34 r)
El arnés que fue de don Alonso Manrique y se compró de él es pavonado
En ocho de marzo de setenta y cuatro recibí yo Juan Rodríguez de Mercado el arnés en esta plana contenido/ estaba cumplido y entregóse a Antonio de Salazar
-Un peto y espaldar con su ristre para seguir la guerra es obrado y cincelado de una color pavonado.
-Una gola de la misma suerte
-Una celada borgoñona de lo mismo
-Dos brazales de lo mismo
-Dos guardabrazos para seguir la guerra
-Dos mandiletes faltan los de dos pulgares
-Unas grebas de la misma suerte sin escarpes
-Otras medias grebas trenzadas sin escarpes
-Unos quijotes de la misma suerte
-Unos brazales de lo mismo enteros de infante con tres piezas con cada brazal
-Dos escarcelas de infante
-Una celada de infante
-Una rondeleta del brazal izquierdo

[...]

(fol. 37 r)
Un arnés pequeño que envió el príncipe don Carlos al señor don Alonso y otros dos arneses pequeños que compró Francisco Arias en Valladolid tienen piezas siguientes

El arnés que envió el príncipe
-Un peto y espaldar con unas escarcelas largas en piezas
-Una gola
-Unos brazales con sus guardabrazos
-Unos mandiletes
-Una celada para seguir
-Unas medias grebas trenzadas con sus quijóticos

Todas las piezas de este arnés son obradas y grabadas con una guarnición a la redonda de raso carmesí y en la guarnición unos pasamanos de oro y seda.

Uno de los dos arneses pequeños que compró Francisco Arias tiene las piezas siguientes

Entregose y cargose a Salazar//
-Un peto y espaldar y gola con sus escarcelas largas
-Unos brazales con sus guardabrazos
-Una celada de seguir
-Unos mandiletes de dedos
-Un morrión

Todas estas piezas son blancas y grabadas con una guarnición de terciopelo carmesí a la redonda
(fol. 37 v)
El otro arnés pequeño que compró Francisco Arias tiene las mesmas piezas que el que está de oro trae todas blancas y grabadas con la misma guarnición de terciopelo carmesí a la redonda
(fol. 41 r)
El arnés que el conde mi señor mando comprar a Francisco Arias de Medina del Campo que fue de don Gómez Cárdenas que costó LI M maravedís tiene las piezas siguientes
(fol. 60 r)
Las armas que el Alcaide Bozmediano entregó Cristóbal de Ribas de la munición de la fortaleza que son armas viejas son las siguientes
(fol. 79 r)
El contador de vuestra señoría me dio la memoria de vuestra señoría por la cual me enviaba a mandar que enviase las armas y doce guarniciones para caballos y así conforme al mandato de vuestra señoría di al dicho contador treinta y siete armas que son doce alabardas doce roncas y trece espedos boloñeses las cuales armas he puesto a dos reales cada una de más de las otras que vuestra señoría

lo mandó tomar en Benavente por la mucha ventaja que tienen a las otras y costas, y las cuales montan a quinientos y cuarenta y ocho maravedís cada una veinte mil y dos y setenta y seis maravedís, y las guarniciones a dos ducados y medio cada una. once mil y doscientos cincuenta, las cuales he asentado a cuenta de vuestra señoría con más setecientos cincuenta maravedís por los dos jaras que di a Logroño en Valladolid por mandado de vuestra señoría que son por todo treinta y dos mil doscientos y setenta y seis maravedís, las cuales el dicho contador me dijo que vuestra señoría los mandará pagar en fecha de agosto, primera que vendrá y así suplico a vuestra señoría lo mande proveer si en lo de que vuestra señoría mandara que las diese barato bien puede vuestra señoría creer que no es menester a mandármelo porque siempre he deseado y deseo servir a vuestra señoría
-Las plumas que V. S. manda que se guarden así se hará, aunque no son tales como quisiera para el servicio de V.S.
-Coracinas de raso cremosil no tengo. Las tengo de negro son de mediana suerte, y valen a diecioho ducados cada una a lo menos,
-Las rodelas aun no son venidas viniendo la avisaré a V. S. cuya ilustrísima persona nuestro señor guarde y prospere con todo acrecentando de vida y estado como yo lo deseo. De Medina XI de agosto 1562

Ilustrísimo señor

Beso las manos de v. s. // Cristóbal de Rivas
(signed)

(fol. 84 r)
Pedro de Ribas armero

Cargo
de las cosas que se le entregaron en seis de marzo de MDLXXI
-Cárgansele al dicho Pedro de Rivas un bracamarte con un paño y pomo de marfil que es todo una pieza y con un bocal y contera de plata esmaltado de negro tiene en la contera dos caras de leones y en el bocal unas veneras con una vaina de terciopelo y un cuchillo con un cavo de marfil e hizo el dicho cabo una Lucrecia
-Más se le carga un puñal antiguo sobredorado los cavos y bocal y contera que son de latón y el puño por de fuera es de hueso negro tiene unas veneras. El bocal y los cabos y una vaina raída de terciopelo negro
-Más se le carga un atabalejo pequeño turco que anda al arzón es de azófar

-Que se le carga más un alfanje con una guarnición que es pavonada y los cavos blancos con una vaina roja con su cuchillo y tenedor con cabos plateados y su funda de cuero con la vaina baya de cuero.

Digo yo Pedro de Arribas (sic) armero mayor del conde nuestro señor que recibí las cosas que señoría en estas cuatro partidas de mano de Antonio de Espínola guardajoyas de su señoría por mandado de su señoría en Benavente a seis de marzo de mil quinientos setenta y uno.
(signed)

(fol. 105 r)
Relación de las cosas que se han metido en la recámara de la armería después que se hizo testigo al Francisco de Canseco
-Doce frenos que envió el conde de la recámara de Espinosa que son pequeños para curtagos
-Quatro lanzas jinetas. Todas en pliego de lanzas
-Un lanzón de monte. Todo en el dicho pliego
-Una rodela de plumas de colores de Indias con las armas del conde en medio y con un buitre en lo alto de las armas y guarnecida a la redonda de pellejo de tigre. Toda en pliego de plumas
-Unas mangas de malla que compró maestre Luis por mandado del conde por ocho ducados. Cargas en pliego de vto (?) estas mangas
-Seis espadas viejas para torneo. Fáltanle a Rivas estas seis espadas resultáronsele en 19 de marzo de 1574
-Un arco turquesco que el comendador de Xerez Mos (sic) dio al conde con sus cuerdas con un carcaj con diez y nueve flechas y una caja de madera con treinta flechas y en la caja está un batidero de marfil a manera de sortija para abrirlo y un candadito con que se cierra esta caja.
-Más cien lanzas de justa que se compraron para cuando paso por esta villa el Rey de Inglaterra. Cargáronse en el pliego de lanzas LXXI las demás se desbastaron en los torneos que se hicieron al Rey y en el desposorio de la duquesa de Huesca

Digo yo Pedro de Rivas que recibí las cosas contenidas en esta plana en 28 de junio de 1563
(fol. 105 v)
Recibí yo Juan Rodríguez de Mercado las cosas de esta otra parte que excepto las que van resueltas a Pedro de Ribas fechado en Benavente a 29 de marzo de mil quinientos setenta y cuatro años y firmado de mi nombre

Juan Rodríguez de Mercado
(Signed)

(fol. 107 r)
Plumas de armas que se cargan a Francisco de Canseco que tiene cargo de la armería de las que sobraron del torneo que el conde hizo cuando se casó el rey nuestro señor con la reina de Francia el año de MDLX años estas plumas son las que su señoría sacó en el torneo de Toledo
-Unas plumas que se pone sobre la celada que tiene veinte plumas amarillas y veinte moradas están de dos en dos una amarilla y otra morada y son un plumaje de redondo amarillo y azul el medio con un poco de oro y argentería que cuelga de lo alto de la pluma.
-Otra pluma que se pone en la celada de su señoría que tiene veinte y seis plumas la mitad amarillas y la mitad negras todas dobladas con argentería pendiente de unos rapacejos de oro con tres flores en el medio de ellas plumas amarilla y morada con unas flores en lo alto por remate y salen de ellas unas garecutas (?) amarillas y azules.

(fol. 107 v)
Plumas que envió al conde el señor de Tezcuco (sic) con Ontiveros por el mes de diciembre de 1562
[…]

(fol. 109 r)
Plumas para hombres de caballo que se encargan a Pedro de Grijalba en tres de junio de mil quinientos cuarenta y un años otras cosas menudas. Todas estas plumas se averiguó que se habían perdido y acabado en 30 de junio de 1563
(fol. 109 v)
-Más tres espejos uno que se compró en Génova cuando su señoría fue a Villafranca de Niza y el otro que le dio Pedro Fernández y el otro pequeño redondo que se trajo de la fortaleza
-Más una bandera de gente de caballo de tafetán blanco con un Santiago y sembrado de veneras y con una franja de seda blanca y verde y colorada.
-Otra bandera de gente de caballo blanca de tafetán sembrada de venera con una franja de seda blanca y verde y colorada
-Más otra bandera de gente de caballo aforrada de tafetán con un buitre en medio y unas letras de oro en el medio con franja de seda blanca y negra.
-Más dos banderas de infantería de tafetán de la una con […] con su encomienda de Santiago sembrada de veneras de plata con flecaduras coloradas y blancas y la otra colorada con veneras de plata y con el hábito de Santiago y franjas coloradas y blancas

-Tres atambores de gente de infantería los dos pintados y el uno no
-Diez y ocho cordones de cubiertas [...]

(fol. 110 v)

-Más se cargan al dicho Pedro de Grijalba ocho penachos pequeños que se compraron de Cristóbal milanés de Medina del Campo [...]

(fol. 111 r)
Los arneses dobles y sencillos que parece que hay en la armería del conde mi señor en tres de abril de MDLIII años son los siguientes

Dentro de la cuadra estos que son dobles
-Tres coseletes uno que dio el señor duque de Florencia y otro Antonio de Leyva y otro el duque de Nájera
-El arnés dorado que se compró en Mayorga
-Otro que compró Juan de Carvajal para la guerra de Francia
-Otro que se compró a don Gabriel Zapata
-Otro blanco que le compró Francisco Arias
-Otro que fue de don Luis Méndez
-Otro del duque de Nájera
-Otro que son dos envió el señor don Alonso
-Otro del duque de Nájera
-Otro que son dos cuerpos de los dedos
-Otro que fue del conde de Luna
-Otro que son dos cuerpos dobles que se compraron de Rafael Archioli
-Otro que fue de don Antonio Sarmiento
-Otro que fue del conde don Alonso mi señor
-Otro que se compró de Pez Coz que
-Otro pavonado que fue de don Alonso Manrique
-Otro de Enrique Alemán
-Otro de don Íñigo de Guevara
-Otro de Juan Pedro armero
-Otros dos pequeños
-Otro de don Pedro Sarmiento

(fol. 111 v)
De la sala los arneses de seguir
-El arnés que dio don Alonso Pimentel Maldonado al conde
-Otro que dio don Juan Pimentel comendador
-El arnés que se compró de maese Hernando

-Otros seis arneses que se compraron del dicho
-Otro que fue de don Juan de Carvajal
-Otro que dio Su Majestad al conde
-Otros dos que compraron de su mujer de Petijuán
-Cinco coseletes que se compraron para la entrada de Francia
-Otros dos coseletes de maese Hernando
-Otro coselete de don Alonso Pimentel
-Otros tres de seguir del dicho maese Hernando
-Otro del contador Valdivieso
-Setenta y siete arneses de seguir
-Otros XX de Mercado
-Otros tres cargados a Rivas

CXXV
(fol. 113 r)
Los coseletes que hay [...]

El del duque de Florencia
-Una coraza y su espaldar y sus escarcelas largas dorado y negro
-Una gola
-Una celada de infante
-Unos brazales de infante con sus rodanchas
-Unos mandiletes
-Y su ristre dorado todo este coselete es dorado y negro todo

El coselete de Antonio de Leyva de infante
-Una coraza con sus escarcelas y su espaldar
-Una gola y un morrión
-Unos brazales de infante
-Dos manoplas- todo este coselete es cincelado y dorado todo.

(Signed)
(fol. 116 r)

Los arneses que el conde tiene en su armería los cuales manda entregar a Pedro de Grijalba son los siguientes

El arnés que se compró de Luis Méndez [...]

(fol. 118 r)
El arnés de guerra que era del conde que haya gloria
-Sus platas doradas los bordes
-Un volante
-Sus escarcelas
-Dos brazales dorados los bordes
-Dos guardabrazos blancos
-Dos quijotes dorados los bordes de las navajas
-Las grebas con sus escarpes dorados
-La gran bufa
-Un almete
-Dos mandiletes dorados
-Un ristre dorado
-Unos zapatos de terciopelo negro cubiertos de malla a trozos y con sus puntas de acero

(Signed)
-Más dos guardabrazos dorados los bordes
-Más otro almete
-Más otros dos brazales dorados los bordes e cincelados
-Más dos baberones blancos
-Más dos escarcelón [...]
-Más un barbote redondo
-Más unas grebas con sus escarpes doradas y cinceladas
-Más otros quijotes blancos
-Más otros dos [...]

(fol. 121 r)

El arnés que era del camarero
-Una coraza con su espaldar
-Dos grebas con dos quijotes
-Dos brazales con sus sobreguarda
-Un almete de justa
-Un volante de justa
-El placarte para [...]
-La gran bufa
-Dos guardabrazos de guerra y otra para doble para guardabrazo
-Un espaldarcete de justa
-Un escarcelón de justa

-Dos manoplas de guerra
-Otras dos manoplas de justa
-Una sobremanopla doble de justa
-Una bufa de guerra y un placarte de combatir
-Un volante de guerra
-Dos medios quijotes de justa

(Signed)
-Más tiene este arnés una tarjeta que se hizo de las demás viejas que vinieron de Túnez
-Más una pieza del brazal izquierdo del codo para justa de tarjeta

(fol. 125 r)
El arnés de Antonio de Leyva que dio al conde a la ligera hízole Juan Pedro armero de su señoría en Milán cuando su señoría fue con Su Majestad a Francia
-Unas platas con sus escarcelas y su ristre dorados los bordes cincelados
-Dos brazales dorados los bordes
-Dos guardabrazos dorados los bordes
-Una gola
-Una celada con su barbote
-Un barbote de orejas
-Dos mandiletes
-Dos quijotes
-Dos grebas con sus escarpes enteros todo dorado y cincelado

(Signed)
(fol. 131 r)
Cosas de armas. Relación de las armas que su señoría mando llevar a Valencia año de MDLXVII

-Los dos arneses que el señor don Alonso envió de Milán llevan las piezas siguientes que uno es de seguir y el otro de infante

Arnés de seguir
-Una gola
-Unos espaldares con sus escarcelas y quijotes
-Sus grebas enteras trenzadas con sus zapatos de terciopelo encarnado cubiertos de malla con sus puntas
-Dos brazales con dos mandiletes y dos guardabrazos con dos arandelicas digo rondelicas que se ponen en los guardabrazos
-Una celada borgoñona
-Otra celada borgoñona barreada

Todas estas piezas de arriba son del arnés de seguir y son todas doradas y grabadas

Arnés de infante
-Una gola y peto y espaldar con sus escarcelas
-Dos brazales con dos mandiletes
-Dos guardabrazos
-Una celada de infantes dorada y en medio de lo dorado una grabadura blanca que es de otro arnés aforrada en tafetán azul con unos pespuntes colorados

Todo este arnés es dorado y grabado como de seguir
(fol. 131 v)
-Unos aceros de silla dorados y grabados de la misma labor de las armas que son cinco piezas
-Todas estas piezas van metidas en sus fundas y en un cofre
-Un coselete con su peto y espaldar con sus escarcelas de launas todo negro con su gola y morrión a manera de sombrero
-Otro coselete negro con su peto y espaldar y una gola clavada que es peto y espaldar con su morrión a manera de sombrero con sus escarcelas
-Dos mazas de guerra negras con unas listas doradas para hombres de armas la una tiene un cordón de sirgo amarillo y negro con dos borlones (?) para prender del arzón y la otra una rueda de terciopelo negro que cae sobre la mano con una correa del mismo terciopelo para colgar al arzón.
-Dos morriones blancos doradas las crestas y los bordes que son de ellos dos arneses que compró el conde de Rafael Archioli.
-Dos rodelas de acero blancas y doradas que tiene cada una de ellas las armas del conde y la una tiene unas cifras dorados los escudos y las orlas estas dos rodelas le envió el señor don Alonso de Milán a su señoría la una tiene de forro delante y la otra de terciopelo carmesí pespuntado.
-Otra rodela de terciopelo amarillo con unas capas doradas con un botón cada una tiene alrededor un franjón de sirgo y oro esta forrada en terciopelo amarillo

(fol. 132 r)
-Otra rodelica pequeña de acero dorados los bordes con un árbol en el medio y un puerco al pie doradas las hojas del árbol esta forrada en terciopelo verde
-Una sotacola de tafetán amarillo colchada sobre que se pone la gola que es para arnés de seguir o de infantes.
-Tres cotas de malla las dos son jacerinas della y otra de media boca que no es acerina están guarnecidas de raso negro con unos botones para atar las escarcelas y sus cintas a los lados es de gola una.

-Una espada de mano y media con una guarnición dorada con sola una puente dorados los gavilanes es ancha con vaina de terciopelo.
-Otra espada ancha con una guarnición plateada con unos botones dorados y cuatro medallas en el pomo y en los recados dos imágenes de cada parte la suya de Nuestra Señora y grabada y un rótulo hasta la punta con su vaina de terciopelo pardo.
-Otra espada ancha una guarnición de joyas doradas todas las que hay y lo demás blanco tiene los gavilanes vueltos hacia dentro y un escudo dorados los recazos que tiene una banda y una cadena y otras armas y el otro recazo tiene cinco veneras grabadas.
-Seis espadas boloñeses y seis roncas dorados todos los hierros con las armas del conde estos vinieron de Milán// estas espadas y todo lo demás va metido en otro cofre con sus fundas
-Un montante la hoja de ondas y el recazo dorado y la cruz y pomo con un rótulo en los recazos que dice ciertas palabras del Ave María tiene su vaina de cuero

Todo lo contenido en este pliego llevó el conde mi señor a Valencia y se ha de descargar a Pedro de Rivas armero mayor de su señoría fecha a VI de mayo de MDLXVII años
(fol. 132 v)
Digo yo Francisco de Canseco que recibí todas las cosas cargadas en este pliego que se trajeron de Valencia donde el conde las había mandado llevar las cuales se me entregaron en XXIII de noviembre de MDLXXIII años.

Doc. 4.
1593, November 4.
Inventory of the armoury of Benavente.
Archivo Histórico de la Nobleza, Osuna, C. 427, D. 111.

(fol. 1 r)
Inventario de los bienes muebles y semovientes que al presente tiene el excelentísimo señor don Juan Alfonso Pimentel y de Herrera conde de Benavente [...] se inventarió a cuatro de noviembre de mil quinientos noventa y tres años

Armería
-Primeramente, seis rodelas doradas.
-Item una rodela carmesí guarnecida toda de piedras.
-Item otra rodela acerada carada labrada de relevado vieja.
-Item cuatro rodelas negras a prueba de arcabuz.
-Item cuarenta y cinco rodelas de Milán.
-Item otra rodela de pluma con las armas Pimentel.
-Item otra rodela de seda azul y blanca que está en poder del licenciado Ximénez corregidor de esta villa.
-Item treinta y una armas que se llaman roncas (fol. 1 v) digo que son veinte e una e obra de Milán.
-Item treinta y una partesanas obra de Milán.
-Item treinta y una bisarmas de Milán enastadas a manera de venablos.
-Item seis lanzones encolados de astas cortas en diferentes hierros.
-Un venablo de palo alemán nudoso.
-Item un martillo de armas enastado.
-Item cuarenta y una alabardas de Milán enastadas.
-Item cuatro hierros de lanzones sin astas grandes.
-Item veinte y un hierros de lanza de jineta sin astas.
-Item diez hierros de guerra de lanza sin astas.
-Item tres hierros de pieles dorados y uno blanco sin astas.
-Item veinte y una mazas de armas.
-Item tres hachas de armas enastadas.
-Item otra masa nueva (fol. 2 r) colgada una boca de hierro en ella.
-Item un martillo de armas de alas.
-Item cuarenta y nueve espadas con guarniciones plateadas y vainas varias todas con veinte y nueve dagas de la misma guarnición y vainas.
-Item treinta y dos pretinas con hierros plateados y cueros varios para hombres de armas.

-Item cinco espadas viejas sin vainas.
-Item cuatro estoques de armas viejos sin vaina.
-Item otros dos estoques de armas que no tienen corsé sino es de punta.
-Item cuatro cimitarras turcas las tres guarnecidas e la una sin guarnición.

(fol. 2 v)
-Item cinco espadas de torneo tres guarnecidas y con vainas de cuero bayo, dos sin vainas.
-Item dos arcos turquescos con sus flechas
-Dos atabalillos turquescos.
-Item ciento y once frenos de la brida.
-Item un bocal de caballo con sus correas.
-Item veinte y ocho pares de estribos de la brida.
-Item una alabarda guarnecida de terciopelo carmesí con asta y con pasamanos de plata tachonada.
-Item dos sillas de armas de terciopelo blanco digo que son quattro con sus aceros.
-Item dos sillas de armas de terciopelo verde con sus aceros grabados.
-Item otra silla de armas (fol. 3 r) de terciopelo negro con sus aceros grabados y labrados.
-Item una silla de justa de terciopelo carmesí [...] con sus aceros.
-Item dos sillas de terciopelo carmesí sin aceros.
-Item otra silla de terciopelo azul turca.
-Item otra silla de terciopelo pardo [...] sin aceros.
-Item otra silla de justa de terciopelo encarnado con sus aceros.
-Item otra silla de justa de terciopelo pardo muy vieja con sus aceros.
-Item dos sillas de justa de cuero blanco con sus aceros.
-Item una silla de armas (fol. 3 v) de terciopelo amarillo con aceros dorados.
-Otra silla de armas de terciopelo azul con sus aceros.
-Item cinco sillas de justar las tres con aceros y las dos sin ellos.
-Item una silla de terciopelo negro con sus aceros y es de armas.
-Item otras dos sillas de armas que no tienen más de dos fustes con barras la una y los aceros grabados y dorados.
-Item diez sillas de armas guarnecidas de acero y sus aceros.
-Item otra silla de terciopelo negro con sus aceros.
-Item cuatro sillas moriscas de la brida.
-Item dos sillas de armas amarillas de terciopelo con sus aceros.
-Item tres petos negros fuertes a prueba de arcabuz con tres morriones.
-Item un coselete negro color de hierro que tiene peto y espaldar y brazales gola celada mandiletes y escarcelas de infante.

(fol. 4 r)
-Item otro coselete blanco y dorado grabado [...] las piezas siguientes.
-Dos celadas y tres guardabrazos y su gola dos brazales dos quijotes con su rodillera con dos escarcelas dos mandiletes.
-Item otro coselete que tiene gola peto y espaldar con ristre los brazales enteros con sus mandiletes y escarcelas es blanco y dorado y labrado escaques tiene este coselete gola y dos guardabrazos para infante y otra celada.
-Item otro arnés pavonado que tiene gola peto y espaldar con su ristre dos guardabrazos dos brazales dos quijotes hasta la rodilla dos escarcelas dos mandiletes celada de seguir dos arandelas una manopla de justa y otro guardabrazo de justa.
-Item dos coseletes negros color de hierro que tienen gola peto y espaldar y escarcelas de infante con dos guardabrazos e morriones negros no tienen brazales ni manoplas.

(fol. 4 v)
-Item tres almillas negras coloradas de hierro.
-Item treintaiún arneses de munición que tienen gola peto y espaldar guardabrazos y brazales escarcelas grebas y quijotes y manoplas y celadas.
-Item cuarenta y un arneses de munición sin golas.
-Item sesenta y siete coseletes sin gola con morriones antiguos pintados y los coseletes no demás de peto y espaldar y morrión.
-Item veinte y seis arandelas de munición.

(fol. 5 r)
-Item cuarenta testeras viseradas de caballo.
-Item diez y séis corazas harto maltratadas.
-Item dos coracinas una de terciopelo colorado y con hábito de Santiago otra con guarnición de tela blanca que es la túnica encomienda de Santiago y la colorada no.
-Item diez y siete alabardas antiguas.
-Item cincuenta y cinco lanzas de guerra.
-Item cuarenta y siete sayeses de armas de paño amarillo y colorados guarnecidos con unas franjas algunas coloradas de mismo paño y otros con unas veneras que son los viejos.
-Item treinta y cinco toneletes los treinta y dos de raso falso blancos negros y amarillo y los tres de raso falso azul aforrados en bocací los demás en anjeo.
-Item una piececilla pequeña de bronce que está en su carretón con unas armas reales.

-Item un casco para justa
-Item cinco arneses para justa de planchón con cuatro petos y espaldares (fol. 5 v) para las justas los tres guarnecidos de cuero y el uno de rojo carmesí.
-Item un acerillo de caballo acerado
-Item dos quijotes hasta la rodilla en coselete guarnecido de terciopelo carmesí.
-Item cuatro tarjetas de madera guarnecidas de hierro por dentro que son de justa de planchón.
-Item cuatro escarcelas para los dichos arneses de la dicha justa.
-Item cuarenta guarniciones de caballos coloradas antiguas.
-Item veinte y cinco guarniciones de terciopelo y raso blanco y leonado.
-Item veinte e ocho aceros para sillas de guerra de munición.
-Item once mosquetes de los que [...] son de guerra.
-Item once frascos y frasquillos de arcabuz y mosquetes.
-Item once cañones de arcabuz sin cajas.
-Item nueve arcabuces de los muy antiguos.
-Item veintiuna ballestas antiguas de garrucha trece de madera de tejo y las siete de abeto las vergas.

(fol. 6 r)
-Item trece tableros de ballestas.
-Item trescientos y cuatro mangos de cuadrillos antiguos que cada manojo tiene una docena.
-Item veintitrés paveses antiguos.
-Item ocho paveses pequeños para espada.
-Item dos cofres trincados para armas viejos guarnecidos.
-Item uno sin llave y cerradura y otro con dos cerraduras.
-Item un arca vieja encorada sin cerradura.
-Item otro arca de madera.
-Item una bigornia de armería y unas turquesas
-Item un hierro de una forchina obra de Milán.
-Item un hierro de pica sin tirantes.
-Item cuarenta y cuatro digo veintidós pares de grebas que quedaron de los arneses de los que son de armas con que su señoría sirvió a Su Majestad.

Cubiertas de caballo de terciopelo bordadas.
-Una cubierta de caballo de terciopelo morado bordadas bordada de tela de oro blanca y amarilla con unas hojas y troncos que tiene cuello con nueve borlas y petral de la misma suerte con cinco borlas (fol. 6 v) y ancas de la misma suerte con cinco borlas guarnecido todo con una franja de oro y seda de la morada.

-Item una cubierta de caballo de terciopelo amarillo guarnecida con unos verdugos de raso amarillos mismo que todas son cinco piezas.
-Item un faldón de lo mismo que todas son cinco piezas.
-Item tres faldones de terciopelo blanco guarnecidos con unos largos de raso amarillo bordados, sino que se llevaron el oro aunque le dejaron alguno.
-Item otro faldón de terciopelo pardo con veneras amarillas de raso amarillo con unas trepas por abajo.
-Item otro faldón de raso amarillo con veneras de terciopelo pardo con trepas a la redonda del dicho terciopelo.
-Otro faldón de terciopelo verde bordado de unas sierpes de terciopelo carmesí con una guarnición a la redonda del mismo bordado con franjón de seda a la redonda.
-Item una cubierta de la mesa obra que el tonelete de arriba de la partida antes de este con las dichas sierpes como el dicho faldón tiene tres piezas una grande y dos pequeñas.

(fol. 7 r)
-Mas cinco piezas de terciopelo pardo con veneras de raso amarillo que son de una cubierta de caballo.
-Item tres piezas de otra cubierta de caballo digo cuatro de raso amarillo con veneras de terciopelo pardo.
-Item una cubierta de caballo de terciopelo azul y blanco guarnecida de seda azul y blanca tiene tres piezas de cuello y pecho y anca que es entera.
-Item una cubierta de caballo de terciopelo blanco vieja acuchillada sobre tafetán blanco que tiene espuelas.
-Item un sayo de tela de oro negra sin ninguna guarnición por que se le quitaron.
-Item dos sayos de armas de terciopelo amarillo guarnecidos con una franjilla de hilo.
-Item un sayo de diferentes colores que es de bordado de plata
-Item un sayete de terciopelo blanco acuchillado que es de bordado.
-Item otra cubierta de terciopelo morado con un franjón de oro y seda que es petral y anca.

(fol. 7 v)
-Item un tonelete de terciopelo amarillo con un franjón de seda negra y amarilla acuchillada.
-Item un sayete de armas amarillo de terciopelo con unos cordoncillos de seda encarnado con una manga de mismo terciopelo y de la misma guarnición.
-Item una cubierta de caballo de terciopelo amarillo sin aforro con tres piezas acuchilladas como tonelete de la partida antes de esta.

-Item una pieza de una cubierta de caballo de terciopelo amarillo con tres bandas de malla guarnecida con un franjón de seda blanca y leonada.
-Item una cubierta de caballo de paño amarillo que son cuello y petral y cubierta de ancas acuchillado sobre bocací negro con unas franjas negras a la redonda.
-Item otra cubierta de paño verde que es anca y cuello con una franja a la redonda de seda verde aforrada la cubierta en bocací verde y el pescuezo sin aforro que se lo llevaron.
-Item un petral de terciopelo amarillo aforrado en lienzo con guarnición de anca de seda y cuatro borlas de la dicha seda.

(fol. 8 r)
-Item diez y siete borlas de seda blanca.
-Item un petral de cubierta de caballo de raso azul guarnecido de terciopelo azul acuchillado sobre tafetán que el petral tiene cinco borlas de seda azul y blanca con seis borlas de la dicha seda azul.
-Item tres borlas de seda amarilla y negra con sus botones otra borla amarilla con su botón.
-Item tres borlas una parda y otra de seda azul y otra de seda amarilla y encarnada azul.
-Item cuatro varas de franja de seda parda.
-Item otra cubierta de terciopelo verde acuchillado que son tres piezas cuello ancas y petral.
-Otra cubierta de bayeta negra petral y anca de lino al pelo acuchillado sobre bocací amarillo el de las ancas se llevaron.
-Item un petral de raso blanco con unos largos del mismo raso acuchillado las tiras aforrado en lienzo con una franja de seda blanca.

(fol. 8 v)
-Item un tahalí de cuero pespuntado y labrado de plata el cabo con sus hierros labrados de ataujía.
-Otra tahalí de cuero con los hierros de plata que son cuatro.
-Un espejo grande de a media vara guarnecido de nogal con unas hojas de madera doradas y extremos dorados.
-Otro espejo de vara de largo la luna de alerzo con su guarnición de nogal con un frontispicies [...]
-Otro espejo de cristal redondo con otros dos espejos a los lados que son [...] con su guarnición en guado con una letra de Ave María.
-Otro espejo de acero [...] guarnecido de madera de nogal con otra laja en que se mete.
-Item una caja de atambor sola la madera [...]

-Item un arnés grabado de bisel y relevado las lavaduras doradas con listas en el peto y tres en las espaldas que tiene gola peto y espaldar (fol. 9 r) Guardabrazos y brazales celada de seguir escarcelas y grebas con su quijote y escarpe esto todo es para seguir tiene para justa un volante con su laderón con una fuerte silla hallado con una cofia tiene su navaja grande de justa con su manopla de justa · tiene una pieza que llaman gran bufa es para sobrebrazal izquierdo tiene quijote y tres piezas aceros de silla tiene otra gola con una celada engolada de justa y dos guardabrazos con sus puñetas y dos escarcelas dos arandelas y dos medias grebas y una testera entera otra testera entera otro volante con un espaldar dos guardabrazos con sus brazales.
-Item otro arnés grabado de bisel con tres listas en el peto las listas por el medio blancas biseladas y a los lados doradas y grabadas tiene gola, peto y espaldar guardabrazos brazales celada quijotes largos con medias grebas sin escarpias tiene otro brazal con dos medias cubierta con la parte del brazal alto tiene una pieza con seis correones que es de la espalda y una testera entera.
-Item otro arnés pequeño que tiene peto y espaldar y gola guardabrazos y brazales y celada grabado tiene tres listas en el peto otras tres en la espalda blanco todo clavazón dorada dos falsetes en el peto con escarcelas.

(fol. 9 v)
-Item otro arnés que tiene cinco listas grabadas doradas de despojos que tiene peto y espaldar y gola brazales enteros tiene dos escarcelas dos quijotes hasta la rrodillo (sic) y otras dos piezas esta guarnecido de terciopelo azul dos digo un volante entero con su celada de la misma labor.
-Item otro arnés pequeño que es del marqués de los Vélez con cinco listas grabadas y doradas y perfiles coloradas tiene gola peto y espaldar brazales enteros con una celada engolada.
-Item otro arnés dorados los remates y con florones en el guardabrazo que tiene gola peto y espaldar guardabrazos y brazales.

(fol. 10 r)
Los falta de las armas que fueron a León.
Para el cumplimiento de veinte y cinco arneses faltan las manoplas y peto y espaldar. Para el cumplimiento de seis espadas faltan dos y faltan tres guarniciones enteras que se llevaron por sí.
(fol. 10 v)
Yo Victorio Vázquez secretario de la Junta miembro de la ciudad de León recibí en nombre de la dicha ciudad por orden y mandato de su excelencia el conde de Benavente de su armería por mano del señor Luis de Vivero armero mayor de su

excelencia veinte y cuatro arneses con todas sus piezas que son peto y espaldar · brazales manoplas · gola y celada de estas calidades.
-Los dos arneses dorados y grabados de Milán.
-Otros dos arneses blancos de Milán.
-Los veinte arneses restantes para los veinte y cuatro son diferentes que son pintados de azul y plata y otros un blanco y si estos pintados se quisieren limpiar sin tocar lo grabado se permite que se haga.
-Seis espadas. Con sus cuchillas de torneo y con sus guarniciones = y otras seis guarniciones de por si son cuchillas todo lo cual con hierro que es preciado y me obligo de lo volver a la fortaleza de su excelencia de donde los saque y lo firmé en Benavente a veintitrés días del mes de julio de seiscientos veintisiete años.

Las manoplas no son más de veinte y tres pares los que recibo.

Doc. 5.
1603, September 4.
Inventory of the arms that are put in charge of Juan de Castro, "*armero*" of Juan Alfonso Pimentel de Herrera, 8th Count of Benavente.
Archivo Histórico de la Nobleza, Osuna, C. 427, D. 108.

(fol. 1 r)
Juan de castro Armero
Cargo que se hace a Juan de Castro armero, de todas las cosas de la armería de su Exª el cual se le hace en la forma siguiente
-Primeramente dos arneses Grabados dorados y el uno relevado para asegurar la guerra y para justar. Con dos sillas con sus aceros y con sus bardas de la misma labor y los dichos arneses están cumplidos
-Otros dos arneses grabados y dorados con las propias piezas que los de arriba
-Otros dos arneses grabados y dorados con las propias piezas que los de arriba.
-Otro arnés dorado acanelado y pavonado y blanco con todas sus piezas de seguir y de justar con sus aceros de silla y un tonelete de hierro y cubierta de caballo.
-Otros cincos arneses dorados y grabados con todas sus piezas de seguir.
-Otro arnés chiquito que llaman del marqués de los Vélez con su peto y espaldar y gola y celada con sus platas.
-Otro arnés pequeño del vizcaíno de los señores pequeños con las mismas piezas que el de arriba.
-Otros tres arneses del señor Don Alonso Pimentel el vero pavonado con sus piezas de justar y arandelas pavonadas y celada engolada con su ristre y manoplas con todas sus platas y los otros dos dorados y con dos celadas cada uno con todas sus piezas de seguir.

(fol. 1 v)
-Tres arneses grabados blancos con todas sus piezas de seguir y justar con sus aceros de sillas
-Un arnés pequeño blanco bandeado engolado
-Otro arnés acanelado con todas sus piezas de seguir.
-Más tres arneses del Vizcaíno Blancos con sus platas y morriones, guardabrazos, cañones.
-Otro arnés dorado grabado con todas sus piezas de seguir y justar y sus aceros de silla y las piezas dobles diferentes.
-Otro arnés grabado Blanco como el de arriba con todas sus piezas de seguir y justar y las piezas dobles diferentes.
-Dos coseletes con sus morriones

-Más un arnés negro del señor don Alonso Pimentel con todas sus piezas de seguir.
-Otro arnés pavonado y dorado con una imagen de Nuestra Señora de la Anunciación en los pechos con todas sus piezas de seguir
-Otros dos arneses dorados viejos con petos y espaldares y golas y morriones y brazales y el uno con sus cañones.
-Más otro peto y espaldar negro dorado con su gola y morrión y brazales y guardabrazos.

(fol. 2 r)
-Otro alfanje quebrado.
-Diez y ocho mazas de armas.
-Dos hachas de armas.
-Cuatro martillos de armas de diferentes maneras.
-Diez y ocho hierros de lanza para toros.
-Cuatro hierros y venablos mayores.
-Dos dagas y un puñal.
-Ocho hierros de encuentro para torneos de a caballo.
-Una gineta con sus balas.
-Una hacha de armas larga con su estoque dentro dorada.
-Dos arcos turquescos con sus carcajes uno de madera y otro de cuero.
-Dos atabalicos turquescos.
-Dos espejos de acero. Uno muy grande y otro pequeño de armar.
-Seis tarjetas grandes.
-Dos talabartes con sus pretinas de cuero blanco.
-Una gola de malla.
-Veintiún faldones de raso blanco, negro dorado y azul.

(fol. 2 v)
-Un vaquero viejo de tela de oro negro con sus mangas.
-Un cuchillo de monte cóncavo de una cabeza de carnero de marfil y su cuchillo que esta con el mismo de la misma manera guarnecido de plata
-Más un montante.
-Más un hierro de lanza africano.
-Doscientas y ochenta docenas de cuadrillos grandes y pequeños.
-Más cinco caballos de madera
-Una ballesta de marfil de Brasil contrahecho con la verga de acero y la nuez de marfil.
-Más otra ballesta de ataujía labrada de hueso de marfil con la verga de nervios sin nuez.

-Más seis ballestas de madera con las vergas de acero y las dos sin nuez.
-Más dos garruchas de ballestas, la una dorada y la otra negra con unas letras.
-Más otras cinco garruchas de hierro viejas.

(fol. 3 r)
-Más treinta y una archas todas doradas y gravadas.
-Más otra partesana.
-Más veinte y cuatro chuzos dorados.
-Más cuatro venablos Blancos enastados.
-Otro venablo blanco y dorado.
-Otro venablo enastado en asta alemana.
-Más otro venablo de asta alemana.
-Más dos hierros de bisarma en astas lisas.
-Más otros dos hierros de caza de venados en astas lisas.
-Más una lanza gineta de caza.
-Más ocho alabardas viejas con sus astas.
-Una coraza de tela blanca y pajiza con su hábito de Santiago.
-Otra coraza con la cubierta de carmesí.
-Doce corazas de terciopelo y raso de diferentes colores.
-Tres sillas de terciopelo colorado.
-Otra silla de terciopelo azul.
-Otra silla de terciopelo dorado vieja.
-Treinta y dos sillas de terciopelo viejas y las diecisiete con sus aceros.

(fol. 3 v)
-Ciento y quince bocados de frenos de todas suertes.
-Más un freno con once bocados y cuatro copas y tres pares de tiros.
-Más cincuenta y siete pares de espuelas de hombres de armas viejas.
-Más seis pares de estribos viejos.
-Más once guarniciones de caballos de terciopelo de diferentes colores viejos con clavazón dorada.
-Más sesenta y dos guarniciones de caballos algunas no enteras todas de cuero con su clavazón.
-Una pica inglesa.
-Una lanza para el estandarte de los hombres de armas.
-Una lanza que dicen ser de Bernardo del Carpio.
-Diez espadas de hombres de armas con sus vainas bayas.
-Tres espadas de torneo de los señores pequeños con dos dagas.

-Seis espadas de armas sin vainas digo once y los dos son esto que es dos alfanjes turquescos.
-Con las guarniciones y con tenas de plata y los tahalíes de plata tirada.

(fol. 4 r)
-Más dos petos fuertes con sus morriones y rodelas fuertes.
-Más tres almillas de hierro.
-Más veinte y cuatro morriones lisos vizcaínos.
-Más otro arnés blanco de seguir sin grebas, sin quijotes ni escarcelas.
-Más dos morriones dorados y grabados.
-Más un capacete guarnecido de terciopelo carmesí que es de Leon de más de treinta y cuatro arneses de seguir Blancos con todas sus piezas.
-Más veinte y cuatro petos y espaldares viejos.
-Más quince celadas, cinco viejas y diez nuevas. Son morriones.
-Más cinco celadas borgoñonas.
-Más cincuenta y una testeras de caballos enteras.
-Más una celada que llaman de gar de Herrera blanca.
-Más catorce sobrepetos viejos.
-Más una rodela y un morrión fuerte a prueba de arcabuz.
-Otra rodela dorada gravada con tres flores de Indias.
-Otra rodela blanca guarnecida con flecos de seda azules.
-Otra rodela blanca lisa.

(fol. 4 v)
-Otra rodela negra de terciopelo negro con siete florones grabados dorados.
-Otra rodela de relieve dorada de figuras.
-Otra rodela de terciopelo carmesí guarnecida de pedrería.
-Otra rodela negra de obra de Yndias dorada.
-Una rodela de pluma con las armas del conde que es de Indias.
-Otra rodela india guarnecida con flecos verdes a la redonda y en el medio.
-Otra rodela de la India del señor don Juan con un escudo en medio con cinco veneras.
-Otra rodela pequeña labrada de tauxía con sus flecos verdes de hierro.
-Otra rodela pequeña de mimbres de seda de colores.
-Más otra rodela vieja de hierro.
-Más cuarenta rodelas de madera doradas y negras algunas con armas del conde y guarnecidas por dentro con terciopelo negro.
-Más treinta y siente alabardas.

(fol. 5 r)
-Cuatro guarniciones de espadas de torneo de a pie.
-Más otra guarnición burilada de espada dorada.
-Más otras dos garruchas de madera
-Más un carcaj de cuero de una funda de un pistolete de vaqueta.
-Más once arcabuces viejos que están en el corredor sin llaves.
-Más dos mosquetes de posta.
-Más una alabarda guarnecida de terciopelo colorado y clavazón dorada.
-Más siete mosquetes con sus cajas.
-Más once cañones de munición.
-Más un banco con un tornillo para el arsenal de la armería y una bigornia.
-Setenta y cuatro balas de hierro colado.
-Más un esmeril de bronce con su carro.
-Más catorce ballestones de palo.
-Más cinco tableros de ballestas.

(fol. 5 v)
-Más diecinueve paveses.
-Seis arcas las dos llenas de despojos de armas, brazales y manoplas y otras cosas.
-Más treinta y tres sayos de media grana colorada y paño amarillo con sus veneras de los hombres de armas que fueron a Túnez.
-Más otros dos de terciopelo amarillo.
-Item en los tres cuartos de la armería, todas las paredes de ellos están cubiertas de despojos de armas que son brazales y manoplas y quijotes y guardabrazos y grebas y morriones, y arandelas y otras cosas.
-Item un arnés blanco liso y gola, peto, espaldar, brazales, guardabrazos y celada engolada.

Todo lo cual yo el dicho Juan de Castro daré como se declara en esta relación y lo firmo en Benavente a cuatro de septiembre de mil seiscientos tres años siendo testigo García de Rojas y de la Puerta oficiales de la contaduría de su excelencia.

Doc. 6.
1614, February 30.
Inventory of the Count of Benavente.
[Copy, 1633].
Archivo Histórico de la Nobleza, Osuna, C. 3916, D. 1.

(fol. 1 r)
Inventario de los bienes libres que tenían los señores condes D. Juan Alfonso Pimentel y Doña Mencía Fajardo de Requesens su segunda mujer en conformidad de los conciertos generales hecho con citación de los Señores Conde de Luna D. Antonio y del señor Marqués del Villar sus hijos en 25 de enero de 1614 ante Agustín de la Vega sacado con autoridad judicial el año de 1633 signado de Juan de Losada

Benavente
Diciembre…20
Año…1633

(fol. 2 r)
[…]
Inventario de todos los bienes del conde D. Juan y Dª Mencía en conformidad de los conciertos generales/
En la villa de Benavente a veinte y un días del mes de diciembre de mil seiscientos treinta y tres años ante el señor licenciado Francisco de Comontes corregidor de esta dicha villa y su tierra por ante mí el presente escribano público del número de ello pareció presente Luis de Olmos contador mayor que es de Su Excelencia el conde de Benavente mi señor y vecino de esta dicha villa y dijo que en el año pasado de mil seiscientos catorce ante Agustín de la Vega escribano que fue del número de esta dicha villa ya difunto (fol. 2 v) pasó un inventario de los bienes de Su Excelencia el conde don Juan que esté en el cielo del cual tiene su excelencia necesidad de un traslado para en guarda de su derecho por lo cual pidió a su merced y suplicó se sirva de entre[..] información que se ofrece a dar de la legalidad del dicho Agustín de la Vega y de los demás requisitos necesarios le mande dar un traslado signado y en manera que haga fe de todo el dicho inventario con pie y cabeza signado y en manera que haga fe pues es justicia que pide y cuanto necesario (fol. 3 r) Auto/ Su merced visto el dicho pedimiento dijo que mandaba y mandó que dé información de lo en él contenido y dada provea justicia la cual se comete a mí el escribano y así lo proveyó y mandó ante mí Juan de Losada […]

(fol. 17 v)
Cabeza de inventario
En la villa de Benavente a veinte y cinco días del mes de enero de mil seiscientos catorce años el licenciado Gerónimo de Bonifaz corregidor en esta dicha villa de Benavente y su tierra en presencia y por ante mí Agustín de la Vega escribano público (fol.18 r) del número de ella y de los testigos de uso escritos fue a las casas de Su Excelencia el señor don Juan Alfonso Pimentel y de Herrera conde de Benavente y en ellas comenzó açer (sic) el inventario de los bienes libres de Su Excelencia el dicho señor conde de Benavente y primeramente se pusieron e inventariaron las joyas y piezas de oro y plata dorada y blanca que adelante irán declaradas las cuales se pesaron por Domingo Rodríguez vecino de esta dicha villa […]
(fol. 218 r)
Otro inventario
En la dicha villa de Benavente a treinta días del mes de febrero del dicho año de mil seiscientos catorce por ante mí el dicho escribano público testigos y del licenciado Gerónimo de Bonifaz corregidor sobredicho prosiguiendo en el dicho inventario se hizo e inventarió lo siguiente
(fol. 218 v)
Arcabuces
-Primeramente, cinco mosquetes turquescos en sus cajas y con sus vaquetas doradas y gravadas
-Quince pistolas con sus llaves doradas y pavonadas y cajas gravadas y retocadas doradas
-Otras dos pistolas la una embutida la caja en marfil y la otra labrada y grabada a partes
-Otra pistola con la caja de Brasil
-Otras cuatro pistolas más pequeñas las dos todas de hierro y el uno con dos bocas y el otro con caja de mader
-Dos frascos de pólvora de búfalo
-Otros seis frascos de marfil cinco pequeños y uno grande

(fol. 219 r)
-Otro frasco de pólvora grande colorado
-Diez y nueve ballestas grandes y pequeñas con sus gafas
-Otras cinco ballestas sin gafas y la una sin arco
-Veintidós fundas de cuero de pistolas
-Seis arcos de bodoques con dos bolsas una de cuero y otra de terciopelo

-Dos cajas doradas cada una con una docena de cuchillos de Barcelona los cabos dorados
-Una espada con su daga de la guarnición dorada lisa con adrezo de cuero bajo

(fol. 219 v)
-Otra espada con daga dorada colorada con el hábito de Santiago por guarnición tiros y pretina de terciopelo carmesí
-Otra espada dorada y negra tiros y pretina de terciopelo negro
-Otra espada negra con puños dorados y tiros de zapa todo el adrezo
-Otra espada de ámbar negra con tiros de cuero
-Otra espada de guarnición negra amelonada con tiros y pretina de terciopelo
-Otra espada negra lisa llana tiros y pretina de cuero
-Otra espada negra acanelada con tiros y pretina
-Un estoque sago (?) guarnición negra

(fol. 220 r)
-Otra espada ancha de a caballo con guarnición dorada y su daga e tiros
-Otra espada blanca ancha y con guarnición dorada
-Una espada de dos manos negra la guarnición
-Otra espada ancha con guarnición dorada vieja

Todos los cuales dos bienes se inventariaron por el dicho corregidor en presencia de mí el dicho escribano según que de su uso va declarado en el dicho día mes y año dos de que doy fe estando presentes por testigos el capitán Francisco de Neira camarero mayor del señor marqués del Villar y (fol. 220 v) García de Rojas secretario de la contaduría mayor de Su Excelencia el conde de Benavente y Juan Ulloa de Gaona oficial de clavelinos de esta dicha villa y lo firmó el dicho corregidor el licenciado Bonifaz ante mí Juan de Losada digo Agustín de la Vega.

Otro inventario
Guarniciones
En la dicha villa de Benavente a dos días del mes de marzo de mil seiscientos catorce años el licenciado Gerónimo de Bonifaz corregidor en esta dicha villa por ante mí el dicho escribano público y testigos fue a las caballerizas de Su Excelencia el conde de Benavente y en ellas se inventarió lo siguiente
-Primeramente, se puso por inventario dos guarniciones (fol. 221 r) de caballo negras de terciopelo engandujadas con pasamanos y sus estribos
-Un caparazón de terciopelo azul con dos fajas de tela encarnada muy viejo con su cuerda de seda azul y un petral viejo

-Otro caparazón de terciopelo azul con dos pasamanos y el asiento de trencillas de oro engandujado con su petral y cabezadas
-Otro caparazón de terciopelo negro con dos fajas de tela de oro amarilla con un petral y cuerda con borlas negras
-Otro caparazón de terciopelo (fol. 221 v) azul muy viejo con unas fajas de tela de plata con su cuerda y petral
-Otro caparazón de terciopelo verde muy viejo con una faja de tela de oro con su petral y cuerda
-Otro caparazón de terciopelo azul con una faja de tela de oro y le falta el arzón postrero con una cuerda y balas azules y su petral
-Otro caparazón de terciopelo amarillo con dos fajas de brocado muy viejo y roto con su cuerda y borlas amarillas
-Otro caparazón de terciopelo verde con dos fajas (fol. 222 r) de brocado muy viejo y roto
-Otro de terciopelo carmesí con dos fajas de tela verde viejo con cuerda y borlas carmesíes
-Otro caparazón de terciopelo negro andazado de trencillas negras viejo con sus cabezadas y petral viejos
-Otro de terciopelo liso negro sin guarnición viejo con cabezada y petral
-Otros tres caparazones de cuero de colores pespuntados para el campo -Dos caparazones de cuero negro cordobán viejos con sus petrales de lo mismo

(fol. 222 v)
-Item más se pasó por inventario una mochila de terciopelo encarnado vieja con cuerda y borlas de seda encarnada
-Otra mochila de terciopelo amarillo vieja
-Otra mochila de terciopelo blanco con su cerda y borlas de terciopelo blanco
-Otra de terciopelo blanco sin guarnición con su cuerda y borlas de seda blanca
-Otra mochila verde rota y vieja engandujada de oro con cuerda y borlas de seda verde
-Otra de terciopelo azul engandujada de oro vieja con su cuerda petral y borlas

(fol. 223 r)
-Tres terlices que son cubiertas de caballos de terciopelo negro viejo aforrados en normesi negro con sus borlas y flecos de seda negra
-Una gualdrapa de terciopelo liso negro nueva con flecos de seda negra
-Otra gualdrapa de lo mismo casi tan nueva
-Otra gualdrapa de lo mismo como la de arriba
-Otra de terciopelo liso negro con cuatro fajas de terciopelo labrado y ribetes de otro terciopelo labrado y ahusado

-Otra de terciopelo liso negro nueva con tres fajas de terciopelo labrado y seis ribetes de otro (fol. 223 v) terciopelo labrado y flecos de seda
-Otra gualdrapa de terciopelo lisa guarnecida de otras tres fajas de terciopelo labrado y otras cuatro fajas angostas de otro terciopelo de otra labor sin flecos
-Otra gualdrapa de terciopelo liso negro nuevo von tres fajas de terciopelo labrado y flecos
-Otra de terciopelo liso nuevo con seis ribetes de terciopelo labrado y flecos
-Otra de terciopelo liso con cuatro fajas de terciopelo labrado y diez ribetes y flecos y ahusados
-Otras tres gualdrapas de paño negro la una llana y las dos con fajas (fol. 224 r) de lo mismo paño ya viejas
-nas cabezadas de cuero negro con su petral y clavazón barnizada
-Otras cabezadas y petral de la misma manera
-Seis cinchas nuevas de la jineta
-Unos estribos plateados de la brida
-Doce pares de estribos de la jineta
-Cuatro pares de estribos de la brida los tres plateados y unos dorados
-Otros dos pares de estribos negros de la brida
-Once pares de acicates con diferentes correas

(fol. 224 v)
-Un estoque de jineta con el cabo de marfil
-Otro estoque de jineta con guarnición dorada digo de plata y encajes de taria (sic)
-Otro de la misma manera
-Un estandarte de damasco azul con las armas de Su Excelencia y una nuestra señora de la Vega
-Diez y ocho adargas
-Una carroza de terciopelo negro por de fuera y lo de dentro de tela de plata y oro con diez coronas de la misma tela y todo el maderamiento guarnecido de plata con cinco manzanillas de plata sobre la misma carroza y en las cuatro sillas de dentro en cada una otras dos manzanillas de plata

(fol. 225 r)
-Seis guarniciones de terciopelo negro para sus caballos de la dicha carroza con sus frenos cabezadas y borlas de seda negra y oro y las guarniciones con franjas de oro y las conchas de los frenos doradas
-Otra carroza grande de vaqueta de Flandes leonada aforrada en damasco labrado negro y leonado por dentro y con diez cortinas de damasco con sus franjas de seda de la dicha color y cuatro sillas

-Seis guarniciones de la dicha carroza con sus cabezadas sin freno de vaqueta (fol. 225 v) de la misma color para seis caballos
-Otra carroza grande de vaqueta negra con cuatro sillas de vaqueta con diez cortinas de damasco morado y negro
-Otras seis guarniciones viejas de la misma carroza para otros seis caballos
-Un carrocín de vaqueta leonada para el campo con pilares de hierro
-Otro coche pequeño viejo con su cubierta de encerado sobre la madera
-Otro coche viejo de pesebrón con cuatro al(fol. 226 r)mohadas de terciopelo carmesí con trencillas de oro viejo y nueve cortinas de damasco carmesí con pasamanos de oro y aforrada de dentro con el mismo damasco con el abazón dorada
-Un carricoche de dos ruedas con su cubierta de encerado sobre la madera
-Una litera de terciopelo negro liso por de fuera y tachonada con clavos dorados sobre pasamanos de oro y por de dentro bordada de oro y plata sobre tercio(fol. 226 v)pelo leonado con franjas y cenefas de oro y plata de dentro y fuera con dos sillones y cuatro almohadas las dos bordadas y sus cortinas de tela de oro y plata leonada con alamares de lo mismo
-Dos caparazones de terciopelo negro y dos guarniciones de la dicha litera para los machos y sus arlanartixos (sic)
-Otra litera de camino de vaqueta negra con sus varas y sillones de vaqueta
-Una silla de manos de (fol. 227 r) terciopelo negro liso por de fuera y tachonada con clavos dorados sobre pasamanos de oro y por de dentro bordada de oro y plata sobre terciopelo leonado con franjas y cenefas de oro y plata con sus almohadas
-Otra silla de manos de terciopelo negro liso guarnecida por de fuera como la de arriba aforrada en tela de oro por de dentro y sus cortinas
-Otras dos sillas de manos de vaqueta de Flandes leonada con cortinas de da(fol. 227 v)masco azul
-Cuatro vaqueros de los cocheros de terciopelo liso guarnecidos de pasamanos negros y ahusados […]

(fol. 228 r)
-Un jaez bordado de oro y plata en terciopelo leonado liso con su petral con cuatro bolsas de seda leonada y oro e plata y en él doce medallas de plata y el petral es de cuero bordado de oro y plata con seis piezas de plata y dos hevillas y dos conteras de plata dorada y sus acicates plateados con diez y ocho piezas de plata y cuatro hevillas y sus cabezadas de cuero con piezas de plata enteras y una gola de piezas (fol.228 v) y medallas de plata adornadas los extremos con siete

borlas leonadas y oro y plata de deda y estribos de bronce plateados de jineta con medallas de plata con su caja de madera en que está todo dicho jaez

-Dos cabezadas de terciopelo carmesí con sus borlas y cordones guarnecidos de oro y tres partes más en cada uno de plata los cuales de acémilas [...]

CARPIO, MARQUESES DEL

Doc. 1.
1661, November 24.
Post-mortem inventory of Luis Méndez de Haro, Marquis of El Carpio and Count-Duke of Olivares.
Archivo Ducal de Medinaceli, Medinaceli, Leg. 104, nº 2.

(fol. 1 v)
En la villa de Madrid el dho día veinte y quatro de noviembre y año de mill Seis^{os} y sesenta y uno se comenzó el ynventario de los bienes que quedaron por fin y muerte del Exmo Señor Don Luis Méndez de Haro y Guzmán Marqués que fue del Carpio Duque de Montoro Conde Duque de Olivares y caballeriço Maior de Su Mag^{d} en la forma siguiente [...]

(fol. 28 v)
Bandas bordadas
Del cargo de Goetens
-Una banda carmesí toda quajada de oro pasada y puntas de oro a los cabos
(fol. 29 r)
-Más otra de tafetán açul de oro y plata pasado toda quajada
-Otra de tafetán carmesí de medio ancho de tafetán bordada al canto de plata pasado
-Una de tafetán carmesí de plata pasado toda quajada y un flueco de rapacejo a los cabos

(fol. 29 v)
Cosas de recámara del cargo de Goetens
-Seis bengalas de concha y caña con remates de oro y a una le falta un casquillo de oro [...]
-Más dos alfanjes turcos con las guarniciones de oro
+Más una espada con guarnición de cristal y caja de terciopelo carmesí
(fol. 30 r)
+Más dos espadas con sus guarniciones de Su Exª [...].

Doc. 2.
1669, December 12.
Inventory of Marquise of El Carpio.
Archivo Ducal de Medinaceli, Medinaceli, Leg. 104, nº 1.

(fol. 51 v)
En la Recámara del Sor Marqués [...]

(fol. 53 r)
Alcabuces (sic) y pistolas
-Diez y siete pares y medio de pistolas con diez y seis fundas para ellas-868
-Catorce alcabuzes (sic) y una caravina de diferentes maestros-869 [...]
-Seis espadines y dos espadas-871
-Un espadín de ágata guarnecido de rubíes y turquesas con su vayna de zapa en su caja-872
-Otro espadín esmaltado verde y blanco guarnecido de diamantes con sus hierros de tala (sic) del mismo género en su caja-873 y 874
-Once bastones con guarnición de marfil-875
-Dos de caña guarnecido de plata-876
-Otros tres con guarnición pavonada-877
-Otro con guarnición dorada-878
-Un cuchillo con su espadín y sus hierros -879 (fol. 53 v) de plata cubierto de cachunbo (sic)
-Cinco bolsas con dardos y flechas y siete ballestas y quatro tijeras-880
-Más dos cañas para bastones-881
-Más tres bajones (sic)-882
-Una cartera con ocho plumajes-883
-Una caja con su geringa-884
-Ocho bolsas de arcabuzes-885
-Una caja con cinco campanillas y otros instrumentos-886
-Seis collares para perros con sus cordones-887 [...]

(fol. 127 v)
En el Palacio de Loeches [...]

(fol. 129 v)
En la recámara del Sr Marq^{s} [...]

(fol. 130 r)
-Una espada con la guarnición de Xpistal con su caxa de terciopelo carmesí-1913
-Quatro espadines dos de açero y dos de Françia-1914
-Seis bengalas de concha y caña con remates de oro-1915 […].

Doc. 3.
1695, March 29.
Inventory and sale of arms of Gaspar de Haro y Guzmán, Marquis of El Carpio,
Archivo Histórico de Protocolos de Madrid, Protocolo 9893.

(fol. 191 r)
Aprobazión de la q^{ta} dada p^{r} Dª Hipólita Correa
viuda de D^{n} Joseph Pelerin de la depositaría
De los b^{es} del Sor Marq^{s} del Carpio D^{n} Gaspar

29 Marzo
En la villa de Madrid a v^{te} y nueve días del mes de Março de mill seiscientos y noventa y cinco ante mí el escribano y testigos el señor Don Ysidro de Camargo caballero del Orden de Santiago del Consexo de Su Magestad en el de Castilla en n^{re} de la Exma Señora Doña Cathalina de Haro y Guzmán Marquesa del Carpio Duquesa de Montoro con benefizio de ynventario del Exmo Señor Don Gaspar de Haro y Guzmán Marqués del Carpio Duque de Montoro del Consexo de Estado de Su Magestad Virrey y Capitán General del Reyno de Nápoles y en virtud del poder que de dha Señora Marquesa tiene otorgado en presencia con asistencia y lizenzia del Exmo Señor Marqués Don Franzisco de Haro y Guzmán y Toledo su marido ante mí el escribano en cinco de Marzo del año pasado de mill seiscientos y ochenta y ocho que me pide le yncorpore en esta escriptura yo lo hize así que su tenor es como se sigue

Aquí el poder

El qual dho poder el dho Sor Don Ysidro de Camargo dixo tiene aceptado y siendo necesario de nuevo le aceptar y declara (fol. 191 v) no le está revocado ni limitado en cosa alguna y del usando dixo que a veinte y seis de Mayo del dho año de mill seiscientos y ochenta y ocho ante mí el escribano nombró Su Señoría por depositario y tesorero de los bienes y efetos (sic) que quedaron por muerte del dho Señor Marqués Don Gaspar a Don Joseph Pelegrín vezino que fue de esta villa el qual exerzió el cargo de tal depositario y tesorero y en su poder entraron todos los que se inventariaron en esta corte por muerte de dho Señor Marqués Don Gaspar y otros efetos y de ellos con órdenes del Señor otorgante y en virtud del dho poder vendió y remató el dho Don Joseph Pelegrín algunos bienes y pagó las libranzas que sobre él dio el señor otorgante hasta veinte y ocho de enero del año de mill seiscientos y ochenta y nueve que falleció, con cuya causa Dª Hipólita Correa Romano como madre y curadora de D^{a}, Mariana de

Luna y abuela y tutora de Don Pablo Nicolás Pelegrín hixo heredero del dho D[n] Joseph Pelegrín dio relación jurada y firmada de las cantidades de maravedís que entraron en poder del dho Don Joseph Pelegrín como tal depositario y de las que por quenta de ellas pagó cuyo cargo importa diez y ocho quentos novezientos y treinta y quatro mill quatrozientos y sesenta y seis maravedís y la data la misma cantidad con que no resulta alcanze contra ninguna de las partes: y vista y reconozida la dha relación por Don Juan de Chavarría contador de Su Magestad en su contaduría mayor de quentas y de la Casa y estados del dho Señor Marqués del Carpio no puso reparo alguno a las partidas del cargo y data de dha relación como en ella pareze que es del tenor siguiente

Relazión jurada y firmada que yo D[a] Hipólita Correa (fol. 192 r) Romano doy como madre tutora y curadora de Doña Mariana de Luna y abuela de Don Pablo Nicolás Pelegrín de Villarreal hijo y heredero de Don Joseph Pelegrín de Villarreal depositario que fue de los bienes y efetos pertenecientes a la testamentaría del Exmo Señor Marqués del Carpio Don Gaspar de Haro y Guzmán que Santa Gloria haya de las cantidades de maravedís que entraron en su poder prozedidos de dhos bienes y efetos y lo que de ellos pagó en virtud de libranzas del Señor Don Ysidro de Camargo caballero del Orden de Santiago del Consexo de Su Magestad en el Real de Castilla en virtud del poder general que tiene de los Ex[mos] Señores Don Fran[co] de Haro, Guzmán y Toledo y Doña Cathalina de Haro y Guzmán, Marqueses del Carpio Condes Duques de Olivares su fha en esta villa de Madrid en cinco de Marzo de mill seiscientos y ochenta y ocho ante Andrés de Caltañazor (sic) escribano del núm[o] de ella, así para poner cobro en los dos efectos bienes y rentas de la dha testamentaría como para hazer pago a los acreedores a ella desde primero de julio del dho año de ochenta y ocho que entró a servir la referida ocupación hasta veinte y ocho de enero de este presente a[o] de seiscientos y ochenta y nueve q falleció que con cargo y data es en la forma y manera siguiente

Cargo […]

(fol. 193 r)
Alcabuzes vendidos al S[or]
Duque de Alba
-Más me hago cargo de seis mill setezientos y cinquenta reales de vellón que valen duzientas y veinte y nueve mill y quinientos maravedís por los mismos en que se vendieron quatro arcabuzes al señor Duque de Alba (fol. 193 v) los tres con cañón de Mirueña [Juan Sánchez de] y llave de Gaspar a cien pesos cada uno

y el otro rayado en ciento y cinquenta pesos de orden de dho señor D^{n} Ysidro Camargo__229U500 [...]

Alcabuz rayado vendido
Al S^{or} marq^{s} de Zara (sic)
-Más me hago cargo de dos mill duzientos y cinquenta reales de vellón que valen setenta y seis mill y quinientos maravedís por los mismos que rezivió del Exmo señor Marqués de Zara de un arcabuz que se le vendió cañón de Gaspar llave de Mirueña rayado y sobre escrito de orden de dho Sor Don Ysidro__76U500 [...]

(fol. 199 r)
Más cargo de las alaxas que dexó de entregar y estaban en su poder [...]

(fol. 199 v)
Arcabuz con rótulo
De Lázaro Cominazo
-Cárgasele más duzientos reales de vellón que valen seis mil y ochozientos maravedís por los mismos en que estaba tasado un arcabuz con rótulo de Lázaro Cominazo y llave guarnezido en el Reino de Murzia que dejó de entregar__6U800

Otro arcabuz catalán
-Cárgasele más cien reales de vellón que valen tres mil y quatrozientos maravedís por los mismos en que estaba tasado un arcabuz catalán de Coma el cañón que dexó de entregar__3U400

Un par de pistolas
Cañones y llave de Simón
-Cárgansele más mil y quinientos Rs de v^{on} (fol. 200 r) que valen cinquenta y un mil maravedís por los mismos en que estaban tasadas un par de pistolas cañón y llaves de Simón que se dexaron de entregar al nuevo depositario__51U

Otro par de pistolas
Catalanas
-Cárgansele más seiscientos reales de vellón que valen veinte mil y quatrozientos mrs por los mismos en que estaban tasadas otro par de pistolas cathalanas las culatas guarnezidas con unos mascarones de plata que se dexaron de entregar al dho depositario__20U400

Una media carabina
Valenziana
-Cárgansele más trezientos y sesenta reales de vellón que valen doze mil duzientos y quarenta maravedís por los mismos en que estaba tasada una media carabina valenziana con un rótulo que dize Sepúlveda y se dejó de entregar por el dho depositario Don Joseph Pelegrín al nuevo Don Martín de Narbarte__12U240

Un par de pistolas
Valenzianas
-Cárgansele más trezientos y sesenta Rs de vellón que valen doze mil duzientos y quarenta maravedís por los mismos en que estaban tasadas un par de pistolas valenzianas con casquillos de concha y los cañones con letreros de Pascual que dejó de entregar__12U240

Una pistola suelta
-Cárgansele más ciento y veinte reales de vellón que valen quatro mil y ochenta mrs (fol. 200 v) por los mismos en que estaba tasada una pistola suelta guarnezida en latón con felpa debaxo que se dexó de entregar al dho depositario__4U080

Frasco pª póvora
-Cárgansele más sesenta reales de vellón que valen dos mil y quarenta mrs, por los mismos en que estaba tasado un frasco pª pólvora guarnezido de Lorenzo que se dexó de entregar [...]

(fol. 201 v)
Espadín de plata
-Cárgansele más setenta y cinco reales de vellón que valen dos mil quinientos y cinqᵗᵃ maravedís por los mismos en que se tasó un espadín de plata que pesaría cinco onzas q dexó de entregar__2U550

Alfanxe con el puño
De cuerno de venado
-Cárgansele más noventa reales vellón que valen tres mil y sesenta maravedís por los mismos en que estaba tasado un alfanje con el puño de cuerno de venado guarnezido de plata que dexó de entregar__3U060 [...]

(fol. 203 v)
Más cargo de diferentes alhajas vendidas de que da en data su prozedido [...]

(fol. 204 v)
Espadín de azero dorado
-Cárgansele más ciento y veinte reales de vellón que valen quatro mil y ochenta mrs por los mismos en que estaba tasado y se vendió al dho D^{n} Bartholomé de Vega un espadín de azero dorado de molido__4U080

Arcabuz rayado
-Cárgansele más dos mil duzientos y cinquenta reales de vellón que valen setenta y seis mil y quinientos maravedís por los mismos en que estaba tasado y se vendió al dho un arcabuz cañón de Gaspar y llave de Simón rayado__76U500 [...]

(fol. 227 r)
D^{n} Martín de Nabarte
Su quenta
De la depositaría de los bienes y efectos de la testam^{ría} del Marq^{s} D^{n} Gaspar de Haro y Guzmán mi S^{or} que estuvo a su cargo desde pr^{o} de julio de 1689 hasta fin de Diz^{re} de 1691 que çesó [...]

Cargos [...]

(fol. 228 v)
A P^{o} de Ibarguren [...]

Un par de pistolas con
sus fundas de Arçón v^{das} al dho= 8
-Más hágome cargo de cien Rs que valen 3U200 ms por los mismos en que se vendió al dho D^{n} Pedro de Ibarguren unas pistolas de arzón con sus fundas tassadas en 120 Rs [...]

(fol. 229 r)
Arcabuz sin marcas ray^{do} con llave
De Lucas de Ros v^{do} al dho=12
-Más hágome cargo de setecientos y cinq^{ta} Rs de ve^{on} que valen 25U500 ms p^{r} los mismos en que se vendió al dho Gabriel Requero una arcabuz sin marcas rayado con la llave de Lucas de Ros hecha en Ma^{d} tasada en la misma cantidad [...]

(fol. 236 v)
Un par de pistolas cañones de Molas
V^{das} a Thomás de Pazos mro tapizero= 65

-A Thomás de Pazos maestro de tapizería un par de pistolas catalanas cañones de Molas guarnezidas las cajas de hierro embutido tasadas la misma can^{d}__ 10U200 [...]

(fol. 239 v)
Al lizen^{do} D^{n} Ju^{o} Domínguez [...]
Pistolas cathalanas un par guarnezidas
Las cajas n° 151 v^{das} al dho
-Más me hago cargo de doszientos y setenta Rs que valen 9U180 mrs por los mismos en que se vendió bajada la quarta parte un par (fol. 240 r) de pistolas cathalanas guarnezidas las cajas de plata tasadas en 360 Rs

Carabina de Lázaro Cominaz n° 159
ven^{da} al dho, 96
-Más me hago cargo de ciento y ochenta Rs que valen 6U120 mrs por los mismos en que se vendió al dho una carabina maltratada de Lázaro Cominaz con un rótulo que dize Pasqual bajada la q^{ta} p^{te} tasada en 240 Rs [...]

Un florete con tres pájaros en medio
Ven^{do} al dho= 99
-Más me hago cargo de trezientos y treinta Rs de valen que valen 11U220 mrs por los mismos en que estaba tasado y se vendió al dho D^{n} Matheo de Roa un florete con tres pájaros [...]

(fol. 247 r)
6 banderolas de Dam^{co} ver^{e} bordada
A 2 haces con las armas de Haro
Y Guzmán V^{das} al dho convento= 168
-Más me hago cargo de mil y quinientos Rs que valen 51U mrs por los mismos en que estaban tasadas y se vendieron al dho convento seis banderolas de Damasco ver^{e} bordadas a dos hazes con las armas de Haro y Guzmán tasada cada una a 250 Rs [...]

(fol. 250 r)
Dos espadines d^{os} al dho= 195
-Más me hago [...] que valen 23U800 mrs por los mismos en [...] al dho dos espadines de azero labrados en Milán maltratados los puños y guarniziones tasado en [...]

(fol. 251 r)
Arcabuz v^{do} a Mª Sánchez
viu^{da} de José Rodríguez casero que fue de Loeches- 204
-Más me hago cargo de seiscien^{tos} Rs valen 20U400 mrs por los m en que estaba tas^{do} y se vendió a la dha un arcabuz el cañón sin marca ray^{do} con la llave de Simón hec^{a} en M^{d}

7 par^{s} de pistolas de
rueda v^{da} a la dha 205
-Más me hago cargo de quatrozientos y veinte Rs que valen 14U250 mrs p^{r} los mismos que importan siete pares de pistolas de rueda que se vendieron a la dha tas^{das} a 60 Rs cada par

Un par de pistolas de dos cañones v^{do} a la dha= 206
-Más me hago cargo de setecientos y veinte Rs que valen 24U480 mrs por los mismos en que estaba tas^{do} y se vendió a la dha un par de pistolas de dos cañones que dan vuelta para disparar

Un par de pistolas romanas
V^{do} a la dha= 207
-Más me hago cargo de setecientos y veinte Rs que valen 24U480 mrs por los mismos en que estaba tas^{da} y se vendió a la dha otro par de pistolas romanas labradas la caja y cañón

Un p^{r} de pistolas los cañones
de Molas v^{da} a la dha= 209
-Más me hago cargo de trzientos Rs que valen 10U200 mrs p^{r} [...] en que estaba tas^{do} y se vendió a la dha otro par de pistolas los cañones de Molas guarniz^{das} las caxas

(fol. 251 v)
Arcabuz v^{do} a Mª Sánchez
viu^{da} de José Rodríguez casero que fue de Loeches- 204
-Más me hago cargo de seiscien^{tos} Rs valen 20U400 mrs por los m en que estaba tas^{do} y se vendió a la dha un arcabuz el cañón sin marca ray^{do} con la llave de Simón hec^{a} en M^{d}

7 par^{s} de pistolas de
Prueda v^{da} a la dha 205
-Más me hago cargo de quatrozientos y veinte Rs que valen 14U250 mrs p^{r} los mismos que importan siete pares de pistolas de rueda que se vendieron a la dha tas^{das} a 60 Rs cada par

Un par de pistolas de dos cañones v^{do} a la dha= 206
-Más me hago cargo de setecientos y veinte Rs que valen 24U480 mrs por los mismos en que estaba tas^{do} y se vendió a la dha un par de pistolas de dos cañones que dan vuelta para disparar

Un par de pistolas romanas
V^{do} a la dha= 207
-Más me hago cargo de setecientos y veinte Rs que valen 24U480 mrs por los mismos en que estaba tas^{da} y se vendió a la dha otro par de pistolas romanas labradas la caja y cañón

Un p^{r} de pistolas los cañones
de Molas v^{da} a la dha= 209
-Más me hago cargo de trzientos Rs que valen 10U200 mrs p^{r} […] en que estaba tas^{do} y se vendió a la dha otro par de pistolas los cañones de Molas guarniz^{das} las caxas

(fol. 251 v)
Un par de pistolas largas
V^{da}. A la dha= 210
-Más me hago cargo de çinto y ochenta R^{s} que valen 6U120 mrs por los mismos en que estaban tasadas y se vendió a la dha un par de pistolas largas de cañones redondos y llaves italianas […]

(fol. 272 r)
Pistola pequeña de faldriquera de rueda
V^{da}. al dho= 405
-Mas me hago cargo de quarenta y cinco R^{s} que valen 1530 mrs por los mismos en que se vendió al dho Mateo de Roa una pistola con llave de rueda pequeña de faldiquera tas^{da} en 60 R^{s} […]
Trompetilla de caza
V^{da}. al dho=409

-Más me hago cargo de doze R^{s} que valen 408 mrs por los ms en que se vendió al dho una trompetilla de caza tazada en 30 R^{s} [...]

(fol. 274 r)
Media carabina valençiana con rótulo
Que dize Sepúlveda V^{da} a Ph^{e} de Zayas=428
-Más me hago cargo de treçientos y sesenta y cinco R^{s} y m^{o}, que valen 12U427 mrs p^{r}. los ms en que se vendió a D^{n} Phe Garçés de Zayas una media carabina valençiana con rótulo que dize Sepúlveda tasada en 360 R^{s} [...]

(fol. 276 v)
Dos alfanjes turcos
V^{dos}, a Man^{l} Serrano=
-Más me hago cargo de mill trezientos y veinte R^{s} que valen 44U880 mrs por los mismos en que se vendió a Man^{l} Serrano dos alfanxes turcos damasquinos tasados a diez y seis doblon^{s} cada uno [...]

(fol. 279 r)
Dos pistolas y una carabina
De dos bocas
V^{das}.a D. Jacobo de Alza y Arana
-Más me hago cargo de quinientos y setenta R^{s} que valen 19U380 mrs por los ms en que se vendió un juego de pistolas y carabina con dos bocas tasadas en 750 R^{s} a D^{n} Jacobo de Alza [...]

(fol. 281 r)
Dho depositario D^{n} Mrn de Narbarte
Dho su cargo= [...]

(fol. 281 v)
Arcabuz cañón y llave de Gaspar
V^{dos}, a D^{n}, Juan Sendín= 49
-Más me hago cargo de mill y quinientos R^{s} que valen 51U mrs por los mismos en que se vendió a D^{n} Juan Sandín un arcabuz cañón y llave de Gaspar Fernández tas^{do} en çien R^{s} de a ocho [...]

(fol. 283 r)
Dho depositario D^{n} Mrn de Narbarte
Dho su cargo [...]

Carabina catalana con cañón de Coma
V^{da} al conde de Buenavista= 62
-Más me hago cargo de dozientos R^{s} que valen 6U800 mrs por los mismos en que se vendió al conde de Buenavista una carabina catalana con cañón de Coma tasada en 240 R^{s}

Escopeta corta con un rótulo en el cañón
que dize Domini Bonomi
V^{da} al dho= 63
-Más me hago cargo de duzientos R^{s} que valen 6U800 mrs por los ms en que se vendió al dho una escopeta corta con un rótulo que dize Domini Bonomi tasada en lo mismo

Un par de pistolas romanas labradas
de relieve V^{das} al dho= 64
-Más me hago cargo de quatrozientos y quarenta R^{s} de vellón que valen 14U960 mrs por los mismos en que se vendieron al dho conde un par de pistolas romanas tas^{das} en 720 R^{s} [...]

(fol. 285 r)
Dho depositario D^{n}, Mrn de Narbarte
Dho su cargo [...]

Dos frascos de pólvora uno grande y
otro de Garçía y dos collares de perros
V^{dos} a D^{n} Pº. Calderón
-Más me hago cargo de ochenta y dos R^{s} y mº, que valen 2805 mrs por los ms en que se vendieron a D^{n}, Pº, Calder^{n}, dos frascos de pólvora el uno grande guarnezido de Garçía y dos collares para perros [...]
(fol. 286 v)

Un par de pistolas de Simón
V^{das} al Marq^{s} de Canales= 97
-Más me hago cargo de 900 R^{s} que valen 30U600 mrs p^{r} los ms en que se vendier^{n} al Marq^{s} de Canales un par de pistolas de Simón cañón y llaves tasadas en 1500 R^{s} [...]

Arcabuz de Mirueña
V^{do}, a D^{n}, Pº, Calderón= 100

-Más me hago cargo de 1500 R^{s} q valen 51U mrs p^{r} los ms en q se vendió a D^{n}, P^{o}, Calder^{n}, un arcabuz de Mirueña con letrero que dize añ^{o}, de 1626 tas^{do} en 1500 R^{s} […]

(fol. 287 r)
Dho depositario D^{n} Martín de Narbarte
Dho su cargo […]
Arcabuz murçiano con llave de
caracolillo
V^{do} a Esteban Garçía
-Más me hago cargo de 200 R^{s} q^{e} valen 6U800 mrs p^{r} los ms en que se vendió a Esteban Garçía una escopeta murçiana con llave de caracolillo tas^{da} en lo mismo

(fol. 287 v)
Espadín liso de azero labra^{do} en
Milán V^{do} al dho y una papelera
de nogal pequeña= 110
-Mas me hago cargo de 120 R^{s} q^{e} valen 4080 mrs por los mismos en que se vendió a Esteban Garçía un espadín liso de azero labrado en Milán tas^{do} en lo mismo y una papelera de nogal de var^{a} de largo y m^{a} de alto tas^{da} en 44 R^{s}

6 fronteler^{s} de máscara p^{a} caballo
V^{dos} a Juan Serrete
-Más me hago cargo de 138 R^{s} que valen 4692 mrs p^{r} los mismos en que se vendier^{n} a Juan Serrete 6 fronteroles de máscara para cab^{s} tasados los 2 en 6 R^{s} de a 8 y los a uno […]

(fol. 288 r)
Frasco de pólvora
V^{do} a Matías de Olm^{do}= 118
-Más me hago (sic) de 15 R^{s} que valen 510 R^{s} por los ms en que se vendió a Matías de Olm^{do}, un frasco de venado para pólvora tas^{do} en 30 R^{s} […].

CASTILLA, ALMIRANTES DE; MEDINA DE RIOSECO, DUQUES DE

Doc. 1.
1540, November 8.
Inventories of arms, fortress of Medina de Rioseco.
Archivo Ducal de Alba, Casa de Alba, C. 79, nº 19.

[unfoliated, the digitized folio number is entered]
(nº. 6)
1540
-Primeramente un tiro traçido en su carretón e adereços de fruslera
-otro cañón de flor de lis y cerca de la culata una corona de broz con su carretón y aparejos
-una media colebrina (sic) de brozno (sic) con sus flor de lises encabalgada en su carretón con sus aparejos como las otras piezas
-dos sacres de metal encabalgados con sus carretones y aparejos
-un falconete de metal con sus aparejos
-Seys servidores de hierro dos grandes y quatro peqños/ el un servidor está reventado
-una lombarda vieja
-un morteruelo de hierro
-quatro sacres de metal encabalgados y a punto
-tres falconetes pequeños de brozno encabalgados y a punto
-otros dos sacres pequeños desencabalgados en sus curueñas (sic) de metal
-Syete lombardas de hierro la una pequeña en sus curueñas tendidas en el suelo
-diez falconetes de hyerro desencabalgados
-catorze falconetes de hierro menores desencabalgados
-dos moldes para hazer pelotas para falconetes
-tres yugos para los carretones de la artellería (sic)
-quatro lombardas grandes de hierro viejas con sus curueñas syn servidores/
-en la sala primera de las armas se contaron que había colgados doçientos y veynte y un cosoletes (sic) con sus çeladas y braçales y sus escarçelas limpios con petos y respaldares (sic) y están limpios
-en la segunda sala se contaron setenta y nueve cosoletes de la mesma manera que lo de la primera sala eçebto (sic)(nº. 7) q los diez de ellos estaban sin braçaletes y todos los otros con braçales doblados y entre todos ellos hubo quinze gorjales
-diez y nueve petos de la infantería sin respaldares con diez y seys çeladas y sus braçales

-otros cinco petos de lo mismo con sus braçales sin celada
-quarenta y cinco coçoletes (sic) con un braçal cada uno y sus morriones y quarenta barbotes syn gorjales algunos de ellos syn escarçelas
-más otros dos cosoletes q estaban
-veynte y siete ballestas de garrucha y armatostes fuertes
-otras dos de palo y una quebrada
-çinquenta cajas de cubiertas en q hay en cada caja dos cubiertas q son por todas çien cubiertas de caballo pintadas y doradas de bufano/
-en la una arca no hay más de un coplón viejo y un fanalete
-çiento y onze escopetas de las de Flandes las mismas pintadas y con sus bocas serpentinas y dos caños desencabalgados de lo mismo
-treçientas y treinta y ocho escopetas de muniçión vizcaínas
-diez y seys arneses y cosoletes enteros de caballo en sus serones y los cosoletes son sin celadas/ hallóse falta un [...] dizen q se dio por orden de su S^a a un vizcayno
Otrosí se halló en otra cámara sesenta y ocho seras de arneses los quales son llanos y algunos de ellos hay falta de algunas piezas y son pocas las faltas segund lo juró el armero
-treynta alabardas viejas pesadas
-hasta seteçientas pieças poco más o menos q están en la sala de las armas de fresno y de haya
-diez y ocho cañones quebrados de escopetas reventadas desencabalgadas
(fol. r)
-quinientas pelotas de munyçión de hierro colado grandes y pequeñas
-un sacre de fruslera q está en la torre de homenaje
-una lombarda vieja desencabalgada

-otrosí declaró Ju^o de Lizama mayordomo del pan de Su S^a e está en la cámara baja de la dha fortaleza en dos hojas la harina q salió de setenta cargas de trigo q se hizo harina en la villa de Melgar la qual dha harina no se medió al presente por la dilaçión q en ello podía haber maquilose de cada carga tres celemines e más las harchaduras q del dho trigo salieron

-Asymesmo se halló en la dha fortaleza una tienda real q era del rey Católico y un pabellón labrado y otras dos tiendas de anjeo con su pasadiço de una pa otra
-otro pabellón viejo q era coçina de campo
-dos tendejones rotos viejos
-çinquenta y dos costales de salitre e piedra açufre
-veynte y quatro pipotes de pólvora e açufre y salitre

-çiertos paveses viejos y ballestas viejas de madera e pasadores del tpo viejo q están en la cámara de la pólvora

(nº. 8)
-esto q se [...] queda en el ynventº puesto q di synado e cierto las dos escaleras q quedan con los tiros.

(nº. 9)
Alcaydía de la fortaleza de Ríoseco
En la noble villa de Medina de Ríoseco ocho días del mes de noviembre año del señor de mill e quiºs e quarenta años en presencia de mí RodrigoVelázquez escribano de Su Mag[t] de la dha villa e de los testigos de yuso escritos el Illmo don Hernando Enríquez almirante e duque de Medina et[a] mi señor dixo que porque al presente la fortaleça de esta su villa de Medina de Ríoseco está sin alcaide por fin e muerte de Vítores de Villasante alcayde que hasta aquí ha sido de la dha fortaleça daba e dio a Alonso de Heredia su criado la dha fortaleça para que sea alcayde de ella por el tpo que la voluntad de su Señoría fuere e goze de todas las libertades e sanciones que por razón del dho cargo se le deban guardar según que de ello su Señoría le tiene dada mçd firmada de su nombre e sellada con el sello de sus armas e refrendada de Diego de Espinosa su secretario e que para le tomar e recibir el pleito homenaje que en tal caso se requiere daba e dio su poder e facultad bastante al comendador Gaspar de Villarroel su alguacil mayor en la dha villa e para que así tomado la dé y entregue las llaves de la dha fortaleça e que le sea dado y entregado y el dho Aº de Heredia reçiba todas las armas e otras qualesquier cosas e tyros e munyçión y escopetas ballestas e todo lo demás que en la dha fortaleça está por ynventario que ante mí el dho escribano se haga aquí e dho el dho pleito homenaje e entregada la dha fortaleça con lo que en ella está mandaba e mandó que los vºs de esta dha villa le hayan e tengan por tal alcayde de como su señoría dio la dicha facultad al dho Gaspar de Villarroel fueron tºs el licº Juº de Villarroel alcayde mayor de su señoría e Diº de Espinosa su secretario

E después de lo suso dho en la dha villa día e mes e año suso dho en presencia de mí el dho escribano e tºs de yuso escritos el dho comendador Gaspar de Villarroel alguacil mayor en la dha villa fue a la fortaleça de la dha villa y llevó consigo al dho Alonso de Heredia y estando cerca de la torre del homenaje el dho corregidor por virtud de la dha facultad a él dada por su señoría tomó al dho pleito (nº. 10) homenaje al dho Alonso de Heredia el qual él hiço en pública forma con juramº e palabra que dio una e dos e tres vezes como hijodalgo que no rendirá con la dha fortaleza a otra persona alguna salvo si su señoría del dho señor don Hernando Enríquez almirante e duque mi señor o a quien su señoría mandase e que como le fue demandado por su señoría dexase la dha fortaleza la dexará libremente e acudirá con llaves en mano a su señoría o a quien su señoría mandase o fuese servido so pena de caer en caso de a leve e de las otras penas en que caen los que retienen semejantes fuerças contra voluntad de sus señores e así fho el dho pleito homenaje luego el dho Gaspar de Villarroel le entregó al dho Alonso de Heredia çiertas llaves que dixo ser de la dha fortaleça e el dho Alonso de Heredia las recibió e de cómo lo suso dho pasó el dho Gaspar de Villarroel lo pidió por testimº de lo qual fueron tºs Sebastián de Vijanda e Diº de Espinosa secretarios e Pº Requexo e Juº Conzellón vºs de la dha villa

E después de lo susodho en la dha villa día e mes e año suso dho en presencia de mí el dho escribano e tºs de yuso escritos el noble señor lic^{do} Juan de Villarroel alcalde mayor de la dha villa por su señoría y el secretario Diego de Espinosa fueron a la dha fortaleça y por mandado de su señoría entregaron al dho Alonso de Heredia alcayde toda la artillería de munyçión arneses corsoletes (sic) picas ballestas escopetas e todo lo que más se halló en dha fortaleça e aquello que se halló e dio y entregó al dho Alonso de Heredia es lo siguiente

-Primeramente un tiro traçido en su carretón e adrezos el qual es de fruslera
-Otro cañón de flor de lis y çerca de la culata una corona de broz (sic) con su carretón y aparejos
(fol. r)
-una media culebrina de brozno (sic) con sus flor de lises encabalgada en su carretón con sus aparejos como las otras pieças
-dos sacres de metal encabalgados con sus carretones y aparejos
-un falconete de metal con sus aparejos
-seys servidores de hierro el uno está reventado dos grandes y quatro pequeños
-una lombarda vieja
-un morteruelo de hierro
-quatro sacres de metal encabalgados y a punto
-tres falconetes grandes encabalgados y a punto
-dos falconetes pequeños de brozne encabalgados y a punto
-otros dos sacres pequeños desencabalgados en sus cureñas de metal
-syete lombardas de hierro la una pequeña en sus cureñas tendidas en el suelo
-diez falconetes de hierro desencabalgados
-catorze falconetes de hierro menores desencabalgados
-dos moldes para hazer pelotas para falconetes

(fol. v) (nº. 11)
-tres yugos para los carretones de la artelliría (sic)
-quatro lombardas grandes de hierro con sus cureñas sin servidores
-En la sala primera de las armas se contaron que había colgados duçientos e veynte y un cosoletes (sic) con sus çeladas e braçales e sus escarçelas limpios con sus petos e respaldares (sic) están limpios
-En la segunda sala se contaron setenta y nueve corsoletes (sic) de la mesma manera que los de la primera sala eçebto (sic) que los diez de ellos estaban syn braçaletes y todos los otros con braçales doblados y entre todos ellos hubo quinze gorjales
-diez y nueve petos de infantería sin respaldares con diez y seys çeladas e sus braçales
-otros cinco petos de los mismo con sus braçales syn çeladas
-quarenta e cinco corsoletes (sic) con un braçal cada uno e sus morriones e quarenta barbotes syn gorxales algunos de ellos syn escarçelas
-Más otros dos corsoletes (sic) que estaban en el suelo que tienen dos barbotes
-más veynte y syete ballestas de garrucha y armatostes fuertes
-otras dos de palo y la una quebrada
-dos bastos de cubas vaçías

Va testaºs en esta plana las dos cubas vala

(fol. r)
-çinquenta cajas de cubierta de caballo en que hay en cada caxa dos cubiertas que son por todas çien cubiertas pintadas y doradas de un faño (sic) en la una arca no hay más de un coplón viejo e un flancalete
-çiento e onze escopetas de las de flan (sic) las cureñas pintadas y con sus bocas serpentinas y dos caños desencabalgados lo mismo
-treçientas e treinta y ocho escopetas de monyçión Vizcaynas
-diez y seys arneses y corsoletes enteros de caballo en sus serones y los corsoletes son çinçelados hallóse de esto faltó en un serón un corsolete me dize que se dio por mandado de su señoría a un vizcayno
-otrosí se halló en otra cámara sesenta y ocho seras de arneses los quales son llanos y algunos de ellos son faltos de algunas pieças que son pocas las faltas segund lo juró Çeballos armero que los puso en los dhos serones
-veynte y quatro alabardas viejas pegadas
-fasta seteçientas picas poco más o menos- que están en la sala de las armas de fresno o de haya

-diez y ocho cañones quebrados de escopetas reventadas e desencabalgadas que están en la sala p^{ma} de las armas
(fol. v)(nº. 12)
-quinientas pelotas de monición de hierro colado grandes y pequeñas- son de plomo
-un sacre de fruslera que está en la torre del homenaje
-una lombarda vieja desencabalgada que está en la cámara de la pólvora
-una tienda real que fue del rey Católico y un pabellón labrado y estas dos tiendas de angeo con su pasadiço de una para otra
-otro pabellón viejo que era coçina del campo
-dos tendejones rotos viejos
-quarenta y ocho costales de salitre e piedra açufre
-veynte y quatro pipotes de pólvora y açufre y salitre
-çiertos paveses viejos y ballestas viejas de madera y pasadores del tpo viejo que están en la cámara de la pólvora

Están sí reçibidas y entregadas las dhas armas tiros y munyçión y cosas en el dho ynventario contenidas al dho Alonso de Heredia en la manera que dha es e pues el dho Alonso de Heredia vistas luego el dho Alonso de Heredia dixo que se daba e dio e otorgaba e otorgó por entrego de todo ello a su voluntad por quanto lo recibió realmente y con efecto en presencia de my el dho escribº e tºs de mano del dho señor alcayde mayor e del dho secretario Diego de Espinosa en nombre de su señoría de la qual dha entrega yo el dho escribº doy fee e ansy entregado el dho Alonso de Heredia dixo que debaxo (fol. r) del pleito homenaxe.

Doc. 2.
1647, August 4.
Inventory and appraisal of Juan Alfonso Enríquez de Cabrera, Admiral of Castile.
Archivo Histórico de Protocolos de Madrid, Protocolo 6233.

(fol. 323 r)
En la villa de Madrid a quatro días del mes de agosto de mill y seisçientos y quarenta y siete ante mí el s^{no}: Gerónimo Perón sastre, Marco Antonio Catilino y Manuel de la Rúa bordadores, nombrados pa tasar los vestidos libreas cojines vorlas y espuelas y espadas de pajes y lacayos, en virtud de los nombramientos en ellos fechos y delante del señor teniente hicieron la tasación de lo que conforme a ella les toca en esta manera.

Vestidos, libreas, cojines, votas, y espuelas, y espadas de pajes, y lacayos,
Pajes [...]

(fol. 324 r)
217
-Veinte y siete tahalíes bordados de oro y el uno que falta a cumplimiento de los v^{te} y ocho tiene la mujer del caballeriço
218
-Veinte y siete pretinas bordadas de oro y la una tiene la mug^{r} del caballeriço

(fol. 324 v)
Por manera que todo lo referido son veinte y ocho vestidos enteros a saber cada uno calçón, ropilla, ferreruelo, jubón, mangas de terciopelo, medias de seda y ligas, tahalíes y rejos todo guarnecido con alamares de oro bordados= Tasaron cada vestido destos a quar^{ta} ducados montan en todos doçe mill treçientos y veinte reales-12U320 [...]

222
-Yten vieron diez pretinas y diez tahalíes negros y pardos inventariados a nº 222= tasáronlos cada uno a dos ducados montan ducientos y v^{te} reales-220 [...]

(fol. 325 v)
Espuelas para pages
1598

-Yten vieron veinte y tres pares de espuelas sin correas para paxes inventariadas a nº. 1598= tasaron cada par a tres reales montan sesenta y nueve reales-069

1599

-Yten vieron otros ocho pares de espuelas con correas pa pajes inventariadas a nº 1599= tasaron cada par a tres reales montan v^{te} y quatro reales-024

Espadas y dagas pº lacayos

1600

-Yten vieron veinte y dos espadas con sus dagas (fol. 326 r) cada una las diez y nueve mexores y las tres de peor calidad inventariadas a nº 1600= Tasaron las diez y nueve a veinte y dos rs cada una = y las tres a diez y seis reales montan todas quatroisº y se^{ta} y seis rs- 466

(fol. 363 r)

En la villa de Madrid a ocho días del mes de agosto de mill y seisçientos y quarenta y siete años ante mí el presente escribano se juntaron Marco Antonio Catilino y Manuel de la Rúa bordadores tasadores nombrados para tasar las cosas pertenecientes a sus ofiçios del guardarnés que quedó por fin y muerte del Exmo Sor D. Juan Alfonso Enrríquez de Cabrera Almirante de Castilla= y con ellos Andrés de la Fuente guarnicionero y Juan Ruiz sillero también nombrados para tasar el dho guardarnés en todo lo tocante a sus ofiçios los quales ansi juntos hiçieron la tasación del según el orden del ynbentario en la manera siguiente

Guardarnés [...]

708

-Yten vieron una silla bordada de oro sobre terciopelo negro liso con sus pendientes y gurupera cabeçadas y todo adereço de lo mismo ynventariado a nº 708= tasaronlo los dos en quatro mil reales-4U000

709

-Yten vieron quatro sillas blancas de picar usadas con sus adereços del mismo color inventariadas a nº. 709= tasáronlas (fol. 363 v) los dos Andrés de la Fuente y Juº Sillero en quarenta rs cada una montan ciento y sesenta rs-160

710

-Yten vieron quatro pares de estribos de hierro ordinario usados inventariados a nº 710= tasáronlos los dos a cinco reales cada par montan veinte reales_020

711

-Yten vieron diez cabeçadas de cuero para atar los caballos inventariados a nº 711= tasáronlas los dos a quatro reales cada una montan quarenta reales-040

712
-Yten vieron onçe bridones por usar de cuero blanco inventariados a nº. 712= tasáronlos los dos a seis reales cada uno son sesenta y seis reales-066 [...]
715
-Yten vieron quatro frenos de picar inventariados a nº. 715= tasáronlos los dos a ocho reales cada uno montan treinta i dos reales-032
716
-Yten vieron quatro sillas de picar blancas con todos sus adereços de lo mismo inventariados a nº. 716 tasáronlas a siete ducados cada una montan treçisº y ocho reales-308 [...]
718
-Yten vieron ocho frenos inventariados a nº 718 tasáronlos los dos a ocho reales cada uno montan sesenta y quatro reales-064
719
-Yten vieron dos pares de estribos inventariados a nº 719 tasáronlos los dos a cinco reales cada par son diez reales-010
720
-Yten vieron seis bridones inventariados a nº. 720= tasáronlos los dos a seis reales cada uno son trª y seis reales-036
721
-Yten vieron diez cabeçones de picar inventariados a nº 721 tasáronlos los dos a ocho reales cada uno son ochenta reales-080
(fol. 364 r)
726
-Yten los dos Marco Antonio y Manuel de la Rúa bordadores vieron junto con los demás una silla de lama bordada de plata con su adereço de lo mismo inventariada a nº. 726= y aviendo cada qual por lo tocante a su ofiçio hecho tanteo de su valor la tasaron en duçientos y veinte reales-220
727
-Yten todos los tasadores vieron otra silla de terciopelo negro bordado de plata y oro con todos sus adereços inventariada a nº 727 y habiendo hecho cada qual tanteo de lo tocante a su ofiº la tasaron de conformidad en mill y quinientos reales-1U500
728
-Yten vieron otra silla de terciopelo negro bordada con hylo de oro con todo su adereço inventariada a nº 728 y habiendo hecho cada qual tanteó por su ofiçio la tasaron de conformidad en quatroçientos y quar^{ta} rs- 440

729
-Yten vieron otra silla de terçiopelo negro liso con una frangilla de oro con todos sus adereços inventariada a nº 729 y hecho el tanteo referido lo tasaron en çiento y diez reales- 110
(fol. 364 v)
733
-Yten vieron quatro sillas de terciopelo liso negro guarneçidas con galón de seda negra y franjas de lo mismo inventariadas a nº 733 tasáronlas los dos a çien reales cada una son quatroçientos reales- 400
734
-Yten vieron dos pares de estrivos negros= inventariados a nº 734= tasáronlos los dos a cinco reales son diez reales-010
735
-Yten vieron quatro sillas de caballo; la una de terciopelo carmesí guarneçida con puntas de plata con todos sus aderezos= inventariadas todas a nº 735= todos los dhs tasadores las tasaron
-Esta primera en veinte ducados-220
-La otra de terciopelo açul guarneçida con puntas de plata con todos sus aderezos en treinta ducados-330
-La otra de terciopelo verde guarneçida toda con puntas de oro con todos sus aderezos de lo mismo en otros treinta ducados- 330
-La otra de terciopelo negro liso con todos sus adereços de cuero usados en çient reales- 100
737
-Yten los dhos Andrés de la Fuente y Juº Ruiz vieron los seis frenos inventariados a nº 737= tasaron cada uno a ocho reales= son q^{ta} y ocho reales-048
738
-Yten vieron dos estrivos de color de açofar con sus correas inventariado a nº 738= tasásonlos los dos en ocho reales- 008
739
-Yten vieron quatro sillas de cuero de coche con sus estrivos y correas= inventariadas a nº 739= tasáronlas los dichos a ocho reales cada una son treinta y dos reales- 032 [...]
740
-Yten vieron seis guarniciones de cuero con sus hevillas pa tiro de coche con sus tirantes de [...]= inventariadas a nº. 740= tasáronlas los dos en ciento y diez reales- 110

741
-Los dos estrivos de la carroça del Prínçipe de Ásculi, inventariados a nº 741, no se tasan porque van con la carroza
742
-Yten vieron quatro sillas blancas de vaqueta con sus çinchas de picar pespuntadas, con seda, cada una con su aderezo de lo mismo= inventariadas a nº 742= tasáronlas los dos a ochenta reales cada una son treçientos y veinte reales-320
743
-Yten vieron seis cabezadas de cuero blanco para atar los caballos= inventariadas a nº 743= tasáronlas los dichos a quatro rs= son v^te^ y quatro rs- 024
(fol. 365 v)
745
-Yten vieron seis guarniciones de caballos de coche de terciopelo negra (sic) guarneçido con franja de oro tachonado con hevillas de bronze con seis cabezadas de los mismo con borlas de seda negra y oro y todo surcado de lo mismo= inventariado a nº 745= y habiendo hecho tanteo de lo q a cada cosa por su oficio toca los tasaron en dos mil y docientos reales- 2U200 [...]
Carroza
-Los recados de la carroça del Príncipe de Ásculi que están desde el nº del inventº 747= hasta el nº 758 no se tasan aquí, porque se an de tasar con toda la carroza; y así hirán en blanco los millares de cada partida por menor pª mayor claridad= en esta manera,
747
-Quatro pilares guarneçidos
748
-Ocho almoadas de los asientos
749
-Quatro cojines de lana y lienzo
750
-Los lados de vaqueta del rodapiés
751
-La vaqta de la cenefa del çielo con su franja
(fol. 366 r)
752
-La franja de la p^te^ de dentro de seda verde
753
-Otra cenefa del çielo de la p^te^ de arriba de los mismo

754
-Seis respaldares de terciopelo verde
755
-Más todos los pedaços de terciopelo de la dha carroza
756
-Las quatro cenefas de damasco con su flueco de la p[te] de adentro
757
-Ocho cortinas de damasco
758
-Los antepechos de vaqueta aforrada en damasco
759
-Yten vieron quatro sillas de vaqueta blanca de picar inventariadas a nº 759= tasáronlas los dos Andrés de la Fuente y Juº Ruiz a quarenta reales cada una son ciento y sesenta reales- 160
760
-Yten vieron quatro frenos de picar= inventariados a nº 760= tasáronlos los dos a ocho reales son treinta y dos reales
761
-Yten vieron quatro pares de estivos= inventariados a nº 761= tasáronlos los dos a cinco rs son veinte reales- 020
762
-El aderezo de las sillas de vaqueta blanca inventariado a nº 762= va tasado con ella no se puede tasar aquí= y asi se nota y pone el millar de esta partida en blanco
763
-Yten vieron quatro sillas de terciopelo negro guarneçido con galón franja negro con guarniciones de cuero liso= inventariados a nº 763= tasáronlas los dos a cien reales cada una son quatrocientos reales- 400
764
-Yten vieron ocho guarniciones ordinarias con tirantes de cuerda con sus cabezadas, y riendas inventariadas a nº 764= tasáronlas en ciento y treinta reales- 130
(fol. 366 r)
765
-Yten vieron dos sillas de estas guarniciones de coche inventariadas a nº 765= tasáronlas los dos en doçe reales cada una= son veinte y quatro reales- 024
766
-Yten todos los dos quatro tasadores vieron quatro sillas de terciopelo carmesí: las tres con galones de ojuela de plata: y la una con galón de oro con guarniciones de lo mismo= inventariadas a nº. 766 y hecho el tanteo de conformidad las tasaron

a treinta docados cada una= montan todas quatro mil y trecientos y veinte reales- 4U320

767

-Yten vieron tres pretales de lama los dos con todos sus cascabeles guarneçidos de galón de plata y el otro de terciopelo negro con galón de oro con cascabeles= inventariados a nº 767= tasáronlos los dos a diez ducados cada uno= montan trescientos y treinta reales- 330

768

-Los dos estrivos de la carroza del Príncipe de Ásculi y otros dos de la carroza de España de vaqueta= inventariados a nº 768= van con las carrozas tásanse con ellas y no aquí y así el millar va en blanco-

769

-Yten vieron veinte pares de estrivos de brida ordin^{os} inventariados a nº 769= tasáronlos los dos Andrés de la Fuente y Juº Ruiz a cinco reales cada par: son çient reales- 100

(fol. 367 r)

770

-Yten vieron siete frenos de picar, inventariados a nº 770= tasáronlos los dos a ocho reales: son cinquenta y seis reales-056

771

-Yten vieron un par de estrivos de gineta= inventariados a nº 771= tasáronlos los dos en seis rs- 006

772

-Yten vieron un caparazón de gineta: inventariado a nº 772 tasáronle los dos en seis reales-006

773

-Yten vieron ocho cabezones de picar inventariados a nº 778= tasáronlos los dos a seis reales son quar^{ta} y ocho rs- 048 [...]

775

-Yten vieron cinco sillas las dos de teriopelo carmesí con su franja al canto; una de vaqueta blanca hecha a la francesa; y las otras dos de terciopelo carmesí llanas: inventariadas a nº 775= tasaron las dos primeras carmesíes a cient reales cada una= la de vaqueta blanca en ocho duºs= las otras dos cada una a tres ducados= suman todas trecientos y cinquenta y quatro reales- 354

776

-Yten vieron seis guarniciones de la carroza de Milán con sus frenos y cabezadas= inventariados a nº 776= tasáronlas todas ellas en cient rs- 100

(fol. 367 v)

777
-Yten vieron seis cabezadas con sus borlas y frenos de la carroça del conde de Siruela, pero no se tasan con las guarniciones= y por estos se pone el millar de esta partida en blanco-
778
-Las seis cabezadas con sus frenos y borlas de las guarniciones de la carroza grande açul= inventariadas a nº 778 no se tasan a este nº porque se tasan con las guarniciones-
779
-Yten vieron dos sillas de coche inventariadas a nº 779= tasáronlas los dos a diez reales cada una= son veinte reales- 020
780
-Yten vieron quatro sillas de picar blancas con todas sus guarniciones= inventariadas a nº 780= tasaron los dos= las dos a ochenta reales cada una= las otras dos a quarenta reales= montan todas docientos y quarenta reales- 240
781
-Yten vieron diez y ocho cabezadas para atar los caballos inventariadas a nº 781: los dos tasaron cada una a quatro reales son set[a] y dos reales- 074
782
-Yten vieron tres pares de estrivos ordinarios: inventariados a nº 782: tasáronlos los dos a cinco reales cada par= son quinze reales- 015
783
-Ytem vieron tres guarniciones con otras tres cabezadas borlas y frenos de la carroza azul bordada inventariadas a nº 783= estas y las otras tres conpañeras inventariadas a nº 746 y no se tasaron allí reservando p[a] este lugar= y con los demás adereços a ellas tocantes= las tasaron los dichos en dos mill y docientos reales- 2U200
(fol. 368 r)
784
-Yten vieron dos sillas de la dha carroça= inventariadas a nº 784= tasáronlas los dos a quatro ducados cada una: son ochenta y ocho rs anbas- 088 [...]
786
-Yten vieron doçe bridones inventariados a nº 786= tasáronlos los dos a seis reales cada uno son setenta y dos reales- 072
787
-La silla de manos açul inventariada a nº 787 no se tasa aquí en este nº porque se tasa con las carrozas y coches y asi en este guardarnés se pone el millar en blanco_

788
-La madera de la carroça del Príncipe de Ásculi inventariada a nº 788: no se tasa aquí porque se pone en las carrozas su precio-
789
-Dos antepechos de la carroza negra bordada no se tasan aquí porque van con la carroza donde llevan su estim^{on}-
790
-Otros dos de la carroza açul bordada inventariada a nº 790 no se tasan aquí porque van con la carroza-
791
-Otros dos de la carroza açul ordinª y banquillos de dha carroza= inventariados a nº 791= no se tasan aquí porque van con la carroça
792
-Dos banquillos y dos almohadas y ocho cortinas de la carroza negra bordada= inventariadas a nº 792 no se tasan aquí porque van con la carroza-
793
-Otros dos banquillos, ocho cortinas y dos almoadas de la carroça açul bordada= inventariadas a nº 793= no se tasan aqui porq van con la carroza-
(fol. 368 v)
794
-Dos almoadas y ocho cortinas de la carroza del conde de siruela: cubierta de vaqueta aforrada en damasco verde de la carroça del Príncipe de Ásculi= inventariado a nº 794: no se tasan aquí que van con las carrozas-
795
-Yten vieron de quatro sillas de vaqueta blanca con coxines de terciopelo verde de mulas de regalo [...]
804
-La silla bordada de terciopelo= inventariada a nº 804= no se tasa aquí porque no está en ser= y dice la partida del inventario se dio al Marqués de Morata-

Más Guardarnés en el Jardín
A los Recoletos Agustinos
1140
-Yten vieron quatro cajas de guardarnés viejas en que hay dos petrales de cascabeles= borlas de seda cabezadas y otras correas y restos viejos= inventariadas a nº 1140= tasaron (fol. 369 r) los petrales de cascabeles cada uno a ocho ducados q montan ducientos y stª y seis reales_ Lo demás que hay en estas cuatro cajas son trastos viejos de poco probecho= tasáronlos todos en ciento y sesenta reales= q junto todo monta trecientos y treinta y seis reales- 336

1141
-Yten vieron tres sillas de caballo viejas inventariadas a nº 1141= tasáronlas a dos ducados cada una montan quar^{ta} y quatro reales- 044
1142
-Yten vieron otras dos de galóncillo de oro y p^{ta} y seda a la brida= inventariadas a nº 1142 son de quatro borrenas la una de lama azul con pasamanillo de plata y la otra de lama blanca y guarnecida con pasamanillo plata y negro= tasaron cada una a seis ducados valen anbas ciento y tr^{a} y dos reales- 132
1143
-Yten vieron un montón de guarniciones de tiros de coches cabezadas frenos tirantes y otras muchas correas y todo está muy viejo y maltratado= inventariado a nº. 1143= tasáronlas todas en quatro^{os} rs- 400
1145
-Yten vieron dos adargas= inventariadas a nº 1145= tasaron cada una en quatro ducados i ambas montaron ochenta y ocho rs- 088
1146
-Yten vieron dos lanzas de correr sortija inventariadas a nº 1146= tasaron cada una a doçe rs- 024

(fol. 369 v)
Más de guardarnés en poder
De Se^{án} de Araujo sota caballerizo
1627
-Yten vieron una silla bordada de conchuelas de p^{ta} sobre tercionela noguerada obscura toda quajada: con petral, grupera y cabezada de lo mismo: con botones= inventariada a nº 1627= tasáronla en tres mill reales vellón- 3U000

Más guardarnés en poder de
Sebastián de Araujo sotacavall^{zo}
1628
-Yten seis pretales de cascabeles de diferentes colores: inventariados a nº 1628= tasáronlos cada uno a diez ducados= montan sesenta ducados valen reales seiscientos y sesenta- 660
1629
-Yten vieron quatro caparazones a la gineta viejos con sus cabezadas= el uno de terciopelo azul con franja de plata= el otro de tirela blanca= el otro de terciopelo negro= el otro de cuero bordado de hilo= inventariados a nº 1629= tasaron cada uno a cient reales montaron todos quatrocientos reales- 400

(fol. 370 r)
1630
-Yten tres pares de estrivos de gineta= inventariadas a nº 1630= tasáronlos en dos docados cada par valen sesenta y seis reales- 066
1631
-Yten vieron otros estrivos de brida= inventariados a nº 1631= tasáronlos en diez y seis reales- 016
1632
-Yten vieron diez y seis barbas de pita turcas para caballos= inventariadas a nº 1632= tasolas en doce reales cada una montan ciento y cinquenta y dos reales- 192 [...]

(fol. 441 r)
Domingo de
Murueta
En la villa de Madrid a vte y dos días del mes de agosto año de mill y seiscientos y quarenta y siete ante mí el scrº Domingo de Murueta vecino de esta dha viª persona nombrada por parte de los sses testamentarios del Excelentíssimo señor Don Juº Alonso Enrríquez de Cabrera Almirte de Castilla difunto= para efecto de tasar las pistolas= votas y espuelas, armaduras de hombre, cortes de zapatos votas de cordobán, jeringa de latón, chinelas, frasquera de agua de la piedra= espuelas doradas y aceradas= caravinas= fundas de pistolas= fundas de espadas= otra jeringa de latón= hierro de vigotes=que quedaron entre los bienes muebles inventariados de Su Excelencia= hizo la tasación de ellas en la manera sigte

Pistolas
1005
-Lo primero vio dos pistolas con sus fundas de bayeta inventariadas a nº 1005= tasó cada una en veinte ducados valen ambas reales quatrocientos y quarenta- 440

Botas y espuelas
1039
-Yten vio unas votas y espuelas de vaqueta de Flandes inventariadas a nº 1039= tasolas en setenta y siete reales= y porque están con ellas otras votas menores tasó estas en cinco ducados= ambos pares montan çiento y trª y dos- 132

(fol. 441 v)
Dos armaduras de hombre
1136

-Yten vio dos armaduras de onbre de hierro colado inventariadas a nº 1136= tasolas en [blank]
Cortes de çapatos de cordobán
1369
-Yten vio tres cortes de çapatos de cordobán ynventariados a nº 1385= tasó las unas en quar^{ta} rs las demás a veinte montan ochenta reales_ 080
Jeringa de latón
1386
-Yten vio una jeringa de latón inventariada a nº 1386 tasola en treinta reales- 030
Chinelas
1395
-Yten vio unas chinelas de terciopelo negro y inventariadas a nº 1395 tasolas en seis reales- 006
Frasquera de agua de la piedra
1396
-Yten vio una frasquera de agua de la piedra en que hay un poco inventariada a nº 1396= tasola en treinta y dos reales- 032
Espuelas doradas y aceradas
1440
-Yten vio unas espuelas doradas y otras açeradas inventariadas a nº 1440= tasolas en veinte reales- 020
1442
-Yten vio otras espuelas doradas inventariadas a nº 1442= tasolas en doçe reales- 012
(fol. 442 r)
Dos carabinas
1683
-Yten vio dos caravinas largas inventariadas a nº 1683= tasó cada una en veinte d^{o}s valen reales ducientas y v^{te}- 220
Carabina con frasco
1684
-Yten vio una caravina corta con su frasco ynventarida a nº 1684= tasola en ciento y cinq^{ta} rs- 150
Fundas de pistolas
1685
-Yten vio quat^{o} pares de fundas de pistolas de todas colores inventariadas a nº 1685 = tasolas en ocho ducados- 088
Fundas de espadas

1686
-Yten vio dos fundas de espada inventariadas a nº 1686= tasolas en ocho rs- 008
Otra jeringa de latón
1687
-Yten vio otra jeringa de latón inventariada a nº 1687= tasola en treinta rs- 030
Hierro para los bigotes
1688
-Yten vio un hierro para los bigotes inventariado a nº 1688= tasole en dos rs- 002

En la forma y precios q van referidos y en cada partida declarados el dho Domingo de Murueta tasador nombrado hiço la dha tasación (fol. 442 v) y juró a Dios y una cruz en forma de drº como esta + haverla hecho bien y fielmente a todo su saber y entender sin haçer agravio a ninguna de las partes interesadas y lo firmó

Domingo de Murueta
Ante mí Frº Núñez de Guen[ta]
(signed)

(fol. 443 r)
Domingo de Murueta
Guarnicionº de espadas
En la villa de Madrid a v[te] y dos días del mes de agosto año de mill y seiscientos y quar[ta] y siete ante mí el Sno. Domingo de Murueta guarnicionero de espadas vezino de esta villa nombrado por parte de los ss[es] testamentaºs del Excelentísimo señor Almirante de Castilla Don Juº Alfonso Enrríquez de Cabrera q haya sancta gloria para tasar las espadas y espadines de la persona de su Excelenzia= hiço conforme a su […] delante del señor teni[e] de correg[or] la tasación q se sigue

Espadas y espadines de la pers[a]
del Almirante q haya gloria
1033
-Primeramente vio cinco adereços de espada y un espadín y entre ellos una guarnición de plata inventariado todo a número 1033 y hiço la tasa sig[te]
-Una espada guarnición açulada bruñida el pomo es chato y el puño de pilares acanalado tasole con la hoja en quatrociºs rs- 400
-Otra espada con guarnición de p[ta] y la hoja de Tomás de Ayala contrahecha tasola en tresciºs rs- 300

-Otra espada con ocho veneras bruñidas en blanco (fol. 443 v) el pomo abierto con hoja de Toledo de lomo tasola en duci°s rs- 200
-Otra espada con dos beneras doradas puño de plata dorado fino la hoja ancha del perrillo tasola en çiento y cinq^{ta} rs- 150
-Un espadín de oro y negro con su puño y hoja ancha tasolo en duci°s rs- 200
-Una guarnición con dos veneras grandes açulado y grabado con su pomo y puño tasolo en çien rs- 100
-Una guarnición con su pomo puño y contera con las armas del Almir^{te} que haya gloria bruñido en blanco esta tomado tasolo en sess^{ta} y seis reales- 066

1U416

Todas las dhas partidas suman mil quatroci°s y diez y seis reales en cuyo preçio se tasan-1U416

1373
-Yten vio una espada blanca con su puño de plata y su hoja inventariada a n° 1373 el puño es cerrado y tiene una memoria la hoja es de Sahagún el viejo= tasola en duci°s y v^{te} rs- 220
(fol. 444 r)
1374
-Yten vio una espada negra con su hoja inventariada a n° 1374 la guarnición es pavonada de ochavas encontradas las veneras en cadenadas con su hoja estranxera tasola en sesenta y seis reales- 066
1375
-Yten vio otra espada de acero con su hoja y espuelas inventariado a n° 1375 tiene junto a la caxa dos ruedas y dos botones= tasola en mill y duci°s rs- 1U200
1376
-Yten vio un adereço de espada con su hoja con puño dorado inventariado a n° 1376 tiene veneras jaspeadas hoja extranjera buena acanalada hasta la punta tasola en çineto y sesenta y cinco reales- 165
1377
-Yten vio un espadín de acero inventariado a n° 1377 tiene dos mascarones a la caxa y dos abuxeros (sic) hallado tasole en quatroci°s rs- 400
1378
-Yten vio una hoja de espada de Sebastián Hernández inventariada a n° 1378= tasóla en çien reales- 100
1381
-Yten vio tres adereços de espada hechos en Nápoles inventariados a n° 1381 el uno es de memorias con su pomo y contera taso este en ochenta y ocho rs.

El otro es de veneras açules enlacado de ochavas encontradas tasole en setenta y siete reales. El otro es de chapa cerrada tasole en ciento y diez rs. Todos tres ontan duciºs y setenta y cinco rs- 275
1382
-Yten vio un espadín de oro francés inventariado a nº 1382 es de ataugía de oro sicelado tasole en quatrociºs rs- 400

(fol. 444 v)
En los precios y forma que van decarados el dicho Domingo de Murueta tasador nombrado hiço la dha tasación y juró a Dios y a una cruz en forma de drº haverla hecho bien y fielmente sin agravio de ninguna de las partes y lo firmó de su nombre

Domingo de Murueta
Ante mí Juº Núñez de Guen^{ta}
(signed).

Doc. 3.
1691, December 24.
Inventory and appraisal of firearms of Admiral of Castile.
Archivo Histórico de la Nobleza, Osuna, C. 498, D. 2.

(fol. 31 r)

-Otra pintura en lienzo que tiene dos varas y media de (fol. 31 v) de alto, y quatro varas y seis de dedos de largo en que se ve diferentes armas petos y espaldares y murriones alabardas una pieza de artillería cabalgada un caballo que solo se le ve la cabeza y pescuezo Marte junto a él, Venus le tiene de la mano y la otra puesta sobre el hombro y está desnuda y solo tiene un manto colorado un cupidillo junto a ella y detrás un sátiro con un hacha enzendida en la mano y otro cupidillo entre la armas en siete mill R^{s} [...]

(fol. 380 v)

Tas^{ón} del arcabucero

En la villa de Madrid dho día veinte y quatro del mes de diciembre año de mill y seis^{os} y noventa y uno, ante mí el escribano y testigos pareció Julián Verendas maestro arcabucero de esta villa tasador nombrado para tasar las cosas tocantes a su oficio que quedaron por muerte del Exmo Señor Don Juan Gaspar Enríquez de Cabrera Gran Almirante (fol. 381 r) que fue de Castilla (que Santa Gloria haya) y habiendo jurado por Dios nuestro Señor y una señal de cruz en forma y prometido decir verdad hizo la tasación siguiente

-Primeramente tasó cinco arcabuzes cortos de fábrica antigua con llaves de rueda, las quatro, y la otra ordinaria tasada cada una en veinte reales__U100

-Çinco carabinas descompuestas todas en sesenta reales__U060

-Un frasquillo pequeño de terzia de largo en treinta reales__U030

-Diez llaves e carabinas y arcabuzes muy maltratados todas en quarenta reales_ U040

-Veinte y tres bolsas de vaqueta (fol. 381 v) antiguas en treinta reales__U030

Y en la forma referida hizo la dha tasación y declara haberla hecho bien y fielmente a su leal saber y entender sin hazer agravio a parte alguna so cargo del juramento que lleva fho en que se afirmó y ratificó y dixo ser de edad de treinta y ocho años poco más o menos y lo firmó a quien doy fee conozco siendo testigos Don Agustín de los Reyes, Don Diego de Miranda y Miguel de Eguzquiza residentes en esta corte= Julián Verendás: ante mí Juan de Medina

CASTILLA, CONDESTABLES DE; FRÍAS, DUQUES DE

Doc. 1.
1628, August 17.
Inventory of the armoury of the Constables of Castile.
Archivo Histórico Provincial de Burgos, Protocolo 6191.

(fol. 467 r)
Entrega de la armería de su Ex[a] el Sr Condestable de Castilla
a Diego Pasqual su armero en 17 de agosto de 1628

1628
En la ciudad de Burgos a diez y siete días del mes de agosto de mil y seisçientos y veinte y ocho años en presencia y por ante mí Toribio Díez de Real escrib[o] del rey nuestro señor y del núm[o] y [...] de esta dha ciu[d] y t[o]s paresció pres[te] Juan de Maeda Salaçar mayordomo de su ex[a] del señor don Bernardino de Velasco condestable de Cast[a] y de León en el partido de esta çiu[d] y dixo que él tiene orden de su ex[a] del dho s[or] condestable para entregar la armería y pertrechos y demás bienes y cosas a ella tocantes que el dho señor condestable tiene en sus casas principales de esta çiu[d] que es lo que tuvo a su cargo Pedro de Sahavedra armero que fue de su ex[a] difunto por cuya muerte vacó la dha plaça y el dicho señor Condestable la ha proveydo y nombrado por su armero a Diego Pasqual vz[o] de la dha çiu[d] y para efeto de tomar la quenta de las dhas armas y demás bienes a Ysabel de Bárcena viuda muxer que fue del dho Pedro de Sahavedra y entregarla al dho Diego Pasqual y que se pongan por ynventario las armas y demás bienes de la dha armería que estaban por quenta del dho Pedro de Sahavedra y lo han de estar de hoy en adel[te] del dicho Diego Pasqual que la ha de dar de ellos cada y quando que se le mande y para que conste de los bienes y armas que la dha Ysabel de Bárcena entrega y el dho Diego Pasqual reçibe en presencia de mí el presente escribano se hiço (fol. 467 v) el ynventario y entrego y reçibo en la manera siguiente

De la persona de su ex[a]
-Primeramente unas armas doradas con llamas de fuego y cruces de borgoña coloradas y en medio del pecho la hechura de la Madre de Dios con seis estrellas por corona y al contorno un cordón de San Francisco que tienen las pieças siguientes

-Un peto y espaldar con escarçelas y su ristre y un poco de cota de malla entre las dos escarçelas
-Un guadarrenes grande
-Braçales y espaldarçetes enteros
-Dos manoplas
-Una gola y çelada borgoñona
-Otra çelada a la ferreruela
-Otro pecto de infante con la misma ymaxen y cruçes de Borgoña y llamas de fuego
-Una testera de caballo con las mismas armas y llamas
-Dos grebas y quixotes con rodilleras

De la pers[a] de su ex[a]
-Más se le haçe cargo de otras armas doradas y grabadas de seguir que tiene por cifra en medio del pecho María y más abaxo el nombre de Pompei que tiene las pieças siguientes
-Peto espaldar y escarçelas çelada borgoñona braçales y espaldarçetes manoplas quixotes con rrodilleras

De la pers[a] de su ex[a]
-Más se le haçe cargo de otras armas doradas y pavonadas con una ymagen de Nuestra Señora de la Concepción con siete estrellas por corona y llenas las armas de bastones y ençima llamas de (fol. 468 r)
fuego y çifras de María
-Peto y espaldar y gola
-Çelada borgoñona aventanada
-Braçales espaldarçetes y manoplas e
-Quixotes largos que sirven de escarçelas con rodilleras
-Guardarrenes grande
-Testera de caballo

Alfanxe pavonado
-Un alfanxe pavonado que diçen le entregó el contador Cuéllar en Valladolid que tiene la vayna pavonada y en ella una paletilla que es de cobre y está en lugar de cuchillo

De la per[a] de su ex[a]
-Más otras armas grabadas y doradas guarneçidas de terçiopelo carmesí con pasamanos de oro con una figura en medio del peto y listas blancas de trofeos de guerra en medio que tienen las pieças siguientes

-Peto de ristre espaldar y çelada borgoñona gola dos espaldarçetes uno diferente del otro y una Nuestra Señora dorada
-Manoplas y guardarrenes
-Una arandela con una bufeta tarxeta
-Un morrión que es de de la misma doradura forrado con las armas
-Otro peto de ynfante y espaldar sirve ambos petos con escarçelas dos manoplas çelada dos guardas una mayor que otra un guardabraço del lado derecho rodela
-Una rodela forrada de terciopelo carmesí con trençillas de oro con una rosa y punta en medio de la misma labor y trofeos y arandela para la lança que estas armas dicen vinieron de Milán

(fol. 468 v)
-Otras armas doradas y grabadas de seguir guarneçidas de terciopelo açul con las listas blancas y dos medallas en el pecho junto a la gola que tiene las pieças siguientes
-Peto espaldar gola escarçelas
-Una çelada engolada
-Manoplas. Un morrión con orejeras forrado. Braçales y espaldarçetes
-Testera entera con hoçico
-Una arandela

-Yten otras armas doradas y grabadas de seguir que tienen las pieças siguientes
-Pecto y espaldar y gola
-Escarçela con malla
-Çelada de justar
-Dos manoplas
-Braçales y espaldarçetes y el izquierdo está forrado en terciopelo colorado con pasamanos de oro
-Una tarxetilla para la nabaxa del lado yzquierdo
-Quixotes y queriecaes (?)
-Unas puntillas con escarçelas de malla otra manopla de justar
-Unos aceros de silla

-Más se le hace cargo de otras armas antiguas retortixadas con bordes grabados y dorados que tienen las pieças siguientes
-Peto espaldar y gola
-Braçales enteros con espaldarçetes manoplas con sus dos (sic) pegados
(fol. 469 r)

-Morrión con orejeras
-Escarçelas y quixotes todo de una pieça
-Una bragueta
-Una trabaxilla para el codo grabada

-Item se hace cargo de otras armas doradas y grabadas relebadas a listas que parecen medias lunas que tienen las pieças siguientes
-Peto espaldar de seguir con gola
-Un volante con su alilla para justar
-Una çelada engolada con su barbote para justar
-Unos braçales con sangraderas
-Dos espaldarçetes pequeños
-Otra çelada engolada para seguir
-Grebas enteras y quixotes grandes
-Escarçelas y una sobremanopla grande
-Una arandela y un placarte y espaldarçetes
-Dos manoplas pequeñas
-Una sobreguarda de la nabaxa de braçal y su testera de caballo entera con hoçico todo de la misma labor que las demás pieças de arriba

-Más otras armas llanas y negras de seguir que tienen los bordes un poco dorados con las pieças siguientes
-Un peto espaldar y gola con otro sobrepeto que tiene ristre braçales y espaldarçetes
-Manoplas y escarçelas
-Çelada borgoñona
-Quixotes y esquineras
-Un açero para silla de armar
(fol. 469 v)
-Una bufeta
-Una testera de caballo
rodela
-Una rodela forrada de terciopelo negro y franxas negras de seda y oro

-Yten otras armas chicas llanas y negras sin çelada que paresçe haber sido de tornear con las pieças siguientes
-Peto y espaldar y gola y escarçelas braçales y espaldarçetes
-Otro peto y espaldar chico
-dos manoplas

-Más otras armas de torneo con listas grabadas color de agua que tienen las pieças siguientes
y las listas doradas de pintura
-Peto espaldar y gola
-Una çelada de encaxe
-Braçales y espaldarçetes enteros
-Dos manoplas

-Otras armas hechas en Flandes que son de color de agua llenas de trofeos labradas con buril guarneçidas de ante y galón de oro que tiene las pieças siguientes
-Peto y espaldar con gola
-Escarçelas y quixotes con rodilleras braçales y espaldarçetes
-Manoplas
-Otra gola y morrión tudesco
-Otro espaldarcete de ynfante que son nuevos
-Yten más un peto blanco a prueba hecho en Alemania con gola y escarçela (fol. 470 r) y un morrión y un barbote guarneçido de terciopelo colorado eçeto (sic) el morrión
-Más otro peto y espaldar fuerte a prueba con gola escarçelones y morrión con orexeras negras y lisas y blancas guarnecido el morrión de raso colorado

-Más se le hace cargo de otro peto y espaldar fuerte y negro con una imaxen de Nuestra Señora dorada en el mismo peto al lado izquierdo y tiene dos balaços en ambos lados y otro en la espalda con listas blancas

-Más otras armas grabadas y blancas antiguas de medias cañas que contienen lo siguiente
-Peto espaldar y gola
-Braçales y espaldarçetes y manoplas
-Çelada y faldón a manera de verdugado
-Grebas enteras y quixotes
-Una navaxa pequeña para el lado izquierdo
-Una arandelilla que es guardabrazo
-Un peto volante blanco con una alilla que se ve al lado derecho con dos escudos uno mayor que otro de justa grabados con su çelada con un braçal y con espaldarçete pequeño

-Una çelada que llaman almete antiguo blanco y grabado en la cabeça entero para justa y unos botones dorados con agujeros para ir dos arandelas grandes blancas y grabadas que tienen letrero don Juan de Cardona año de 1552

-Otra arandela más chica grabada y blanca
(fol. 470 v)
-Un barbotillo pequeño muchos aguxeros grabado
-Otro barbotillo blanco grabado
-Un morrión negro con orexeras doradas antiguo que dicen fue del gran capitán y en la cresta tiene una sierpe con el rostro negro de mena forrado en raso carmesí pespuntado con oro y tiene su funda de badana con cordón verde
-Un peto a prueba con unas listas blancas con una ymaxen de plata de la Madre de Dios que tiene siete estrellas por corona de quando subió a los cielos y al pie de la ymaxen un querubín

-Más se le hace cargo de unas armas blancas grabadas que dicen se hicieron en Peñacerrada y tienen las pieças sig[es]
-Petos y dos espaldares con sus golas tres çeladas de seguir y una de justar con una escofia
-Dos murriones (sic) grabados
-Quatro braçales y sus espaldarçetes
-Dos manoplas
-Un placarte que es a manera de tarxeta suelto que sirve para justar
-Otra tarxeta
-Una sobre guarda que es navaja del braçal
-Unas sobremanoplas para justar
-Otra para justar
-Otra manopla de combatir
-Dos escarçelas chicas para de a caballo
(fol. 471 r)
-Unas grebas enteras con sus quixotes
-Dos pares de açeros de sillas unos de seguir y otros de justar
-Tres arandelas y un ristre
-Una bragueta
-Unas medias grebas
-Un barbote
-Una cresta de çelada aventanada

-Más otras armas chicas de seguir de la çinta arriba con todas pieças que se diçe son hechas en Segovia y que eran blancas y para un torneo se pintaron de blanco y amarillo y la una manopla de las armas es de la misma pintura de ellas y la otra de por pintar

arneses
-Más se le hace cargo de diez y seis arneses de Milán de caballos lixeros todos con sus pieças de seguir y en cada uno tiene las pieças siguientes
-Peto espaldar y çelada borgoñona
-Escarçelas y medios quixotes e medias grebas y testeras de caballos e estoques pretinas y talabartes e lancas con sus hierros banderolas de tafetán açul y amarillo y sus maças de hierro

-Más otros cinco arneses blancos que son de justar y de seguir que diçen los quatro de ellos se hicieron en Peñacerrada y que el otro era del capitán don Juan de Velasco que cada uno tiene peto y espaldar braçales y guardabraços çelada con faldamento largo que sirve de gola grebas y quixotes estoque (fol. 471 v) pretina y talabartes maça de hierro y testera de caballo y el uno de estos arneses tiene más una bufeta que sirve de espaldarcete y la çelada es almete
-Más otros treinta y nueve arneses de seguir blancos que tiene cada uno de ellos las pieças siguientes
-Peto espaldar braçales y guardabrazos y escarçelas y quixotes enteros çeladas de pabellón y los quatro son almetes testeras de caballo y maça de hierro estoque y pretina y talabartes lança con banderola de tafetán amarillo y açul y en todos estos arneses faltan quatro grebas y los mandiletes y todo lo demás tienen

-Yten mas se le entregaron otros tres arneses antiguos blancos que cada uno tiene las pieças siguientes
-Peto y espaldarcete
-Grebas y quixotes
-Escarçelas
-Almetes puntiagudos que están algo maltratados

*armas del S^or^*
conde de Haro
-Unas armas grabadas de color cielo que diçen fueron del S^or^ conde de Haro con listas blancas y son de torneo que tienen las pieças siguientes
-Peto espaldar y gola
-Braçales y guardabrazos
-Dos manoplas y la derecha con dos (fol. 472 r) dedos de malla y una çelada engolada

-Otras armas negras de color de hierro para torneo que tienen las pieças siguientes

-Peto espaldar y gola
-Çelada braçales y guardabraços
-Dos manoplas la çelada de estas armas es de pabellón

-Otras armas negras color de çielo con las pieças siguientes
-Peto espaldar braçales
-Y espaldarçetes gola y çelada
-Dos manoplas y la una tienen dos dedos de malla

-Más otras armas fuertes de color de hierro que tienen las pieças siguientes
-Peto espaldar a prueba
-Y el peto tienen en los pechos la Anunçiaçión de la Madre de Dios
-Golas y murrión (sic) con sus orexeras
-Espaldarçetes a prueba
-Manoplas y cañones

-Más se le hace cargo de otras armas blancas de seguir grabadas con el clavaçón dorada en los guardabraços
-Y unos ramos en el peto que en el cargo viexo diçe se compraron en Madrid y tienen las pieças siguientes
-Peto y espaldar escarçelas y ristre çeladas y gola braçales y espaldarçetes una tarxeta redonda (fol. 472 v) al lado derecho greba entera y quixotes y una manopla sola

-Más otras armas blancas con listas doradas y las armas de su esçelençia (sic) en el peto que tienen las pieças siguientes
-Peto y espaldar
-Gola y çelada
-Braçales y espaldarçetes e manoplas diferentes de las armas

-Unas armas chicas de ynfante blancas de la çinta arriba con clavaçón dorada que tienen las pieças siguientes
-Peto y espaldar gola y escarçelas murrión braçales espaldarçetes y manoplas

-Otras armas blancas chicas grabadas con clavaçón dorada que tienen las pieças siguientes
-Peto espaldar y gola
-Çelada borgoñona un morrión blanco braçales y espaldarçetes

-Otras armas doradas y pavonadas pequeñas de seguir con clavaçón dorada guarnecidas de terciopelo carmesí que tienen las pieças siguientes
-Peto y espaldar escarçelas y rristre gola y çelada braçales y espaldarçetes y manoplas e medios quixotes con rodilleras falta el ristre

(fol. 473 r)
-Más otras armas chicas doradas y grabadas con listas blancas peto y espaldar con escarçelas gola y morrión braçales y espaldarçetes y manoplas espada y daga dorada con tiros y pretina de cuero y trençillas de plata

-Unas armas negras fuertes guarneçidas de terçiopelo carmesí con pasamano de oro que tienen las pieças sigu[es]
-Un peto y espaldar con listón blanco en el peto hecho con el mismo y al pie de él una ymaxen grabada de la Madre de Dios del Rosario tiene las pieças siguientes
-Un escarcelón con rodilleras gola y murrión espaldarçetes para el codo un guanton y izquierdo para el codo, un guardarrenes muy grande hecho de pieças una çelada borgoñona aventanada la vista lixera, una manopla chica de la mano derecha.

=Armas que dicen estaban en Pedraça=
Pectos (sic)
-Catorce petos antiguos viexos
-Cinco peto volantes viexos
-Un espaldar
-Quatro capaçetes monteros viexos
-Tres çeladas almetes
-Veynte testeras de caballos
-Tres morriones grabados biexos e dos esquineras
-Tres caxas con diversas pieças viejas de armas
-Yten mas un peto negro y espaldar con su gola muy chico con unas listas de oro en medio y en los bordes
(fol. 473 v)
-Un peto y espaldar negro muy chico con una venera relebada y una cruz hecha de clavos en los pechos
-Otro peto y espaldar chico de un xeme blanco
-Otro pecto y espaldar fuerte sin otra señal
-Otro peto y espaldar negro de laonas para traer de secreto
-Otro peto y espaldar forrado en tefetan negro
-Una coraça barnizadas negra con figuras doradas

Rodelas

-Una rodela dorada que tiene en medio el nudo de Salamón (sic) y un letrero alrededor que diçe para el Condestable de Castilla mi S^or^ forrada en terciopelo negro y unas sirenas doradas en ella

-Una rodela fuerte con un balaço bollado forrada de quero roxo roto

-Otra rodela de higuera dorada dentro y fuera que tiene un arco que le tiran dos sierpes y otro trofeos

-Una rodela blanca dorada al friso y en medio un escudo con esta cifra RA y está forrada en terciopelo rroxo con pasamanillo de oro y plata

-Una rodela negra fuerte con un Xpo de plata en medio y ocho clavos con que están guarnecidas las correas que haçen muestra en la misma rodela forrada de camuça con pasamanos negros

-Otra rodela de madera pintada y dorada con figuras

-Otra rodela con un escudo dorado y [...] con una cifra de María y una corona (fol. 474 r) encima forrada de terciopelo carmesí con trencillas de oro y su fleco de oro y seda

-Otra rodela grabada con su punta en medio y unos florones y quatro clavos rayados forrados de terciopelo verde y flecos del mismo color

-Otra rodela pequeña grabada y dorada con su florón en medio y una espiga a manera de piña y otras figuras y unas cabeças de clavos de latón dorados forradas en terciopelo carmesí con trencillas de oro y flocadura de oro y seda

-Otra rodela de madera pintada y dorada con quatro veneras de latón y una fuente y ninfas del Parnaso y otras labores letras y figuras

-Otra rodela fuerte con un Xpo en medio plantado en un calvario forrada en terciopelo colorado

-Un broquel antiguo de hierro aperdigonado con su empuñadura de madera

-Otro broquel pavonado con un mascarón con oro en medio y clavaçón dorada alrededor forrado en terciopelo carmesí con su manera

Pistolas

-Una pistola de dos cañones con rueda y sus dos baquetas y su funda de cuero forrada de terciopelo carmesí y en el pomo de la caxa tres medallas de madreperla sin llave de armar

-Otra pistola pavonada con trofeos de encaxe de plata en la caxa ynserto en el pomo molde para balas con su funda de terçiopelo morado y bordado

-Otras dos pistolas chicas que dio el señor duque de Saboya con sus caxas de atauxía de hueso blanco muy fina de ruedas de rastrillo con sus baquetas y funda dorada de quero (fol. 474 v) y en ellas las armas de Saboya y cordón de seda verde forrada de terciopelo verde con su frasquillo para balas

-Otras dos pistolas con sus caxas de hueso de color verde colorado y blanco y sus fundas de cuero negro forradas en terciopelo sin llaves de armar y con sus llaves de ruedas
-Otras dos pistolas medianas con las caxas de hueso labradas de atauxía de Alemania y los cañones pavonados y los remates dorados en una funda ambas de cuero negro forrada de terciopelo verde sin llaves de armar
-Otras dos pistolas chicas las caxas de ellas de atauxía de hueso de Alemania los cañones dorados y pavonados y en la coz de ellas una figura de latón y su funda para ambas de cuero forrado con cuero negro forrada en terciopelo negro
-Otra pistola con caxa de nogal llana quebrada la coz con funda de quero negro forrada de terciopelo negro y su cordón de oro y seda negra hay llave de armar
-Otra pistola el cañón muy mal dorado y la caxa negra con hueso blanco y en ella unos conexos y su funda forrada en terçiopelo negro con su molde para balas y llave para armar
-Otra pistola de hierro blanco con caxa de nogal labrada de hueso blanco quebrada la coz y su funda con bolsa y molde para balas y cargas de hoxalata sin llave de armar
-Otras dos pistolas los cañones dorados muy pequeños y en el remate unos (fol. 475 r) pies de cuervo dorados con sus baquetas y dos fundas de terciopelo verde con pasamanos de plata sin llaves de armar
-Otras dos pistolas pequeñas los cañones dorados y caxas labradas de hueso en Alemania con dos rostros a manera de micos en los cabos y su llave para armar y funda de cuero negro forrada en su funda verde
-Otras tres pistolas los cañones dorados y las caxas labradas atauxía de hueso los pomos redondos con sus fundas de vaqueta negra y solo una tiene la llave de armar
-Otra pistola con caxa de nogal y en el remate de la coz un botón de hierro con su funda de cuero negro forrada en terciopelo roxo y una bolsa colorada para balas y no tiene llave de armar
-Dos pistoletes de hierro el uno de un palmo y el otro de medio que el uno fue dorado no tienen llaves de armar
-Otra pistola chica con caxa de nogal labrada de atauxía de hueso blanco en una funda de badana colorada y su llave de armar
-Otras dos pistolas los cañones pavonados las caxas labradas atauxía de hueso en Alemania ambas en una funda forrada en terciopelo morado con sus baquetas y unas llaves de armar la una tiene en la coz cara de león y la otra una ardilla en hueso blanco

-Otras dos pistolas raytas las caxas negras en su funda de vaqueta ambas y en el remate de la coz dos rosas blancas de hueso con una llave de armar
(fol. 475 v)
-Otra pistola el cañón blanco la caxa llana de nogal con funda de cuero negro forrada en terciopelo morado esta pistola no tiene más de cañón y caja y sin llave de armar
-Otras dos pistolas las cajas de nogal labradas de verde y hueso blanco con cañones medio dorados sin fundas ni llaves de armar
-Otra pistola con caja de ataujía de hueso y en el pomo un mascarón y cuatro mascaroncillos de latón en el mismo remate de la coz y su funda de vaqueta y su llave de armar
-Otra pistola como la de arriba y el lugar de los cuatro mascaroncillos tiene cinco de hueso blanco sin caxa de hueso y sin llave de armar ni baqueta
-Otra pistola con cañón pavonado con oro en los remates la caxa de nogal con mascarón cara de León de azófar sin funda y tiene llave de armar
-Una pistola con su caja de nogal labrada de hueso blanco y una dama flamenca bajo la caxa y una medalla con las fuerças de Hércules y en el pomo de la coz de hueso blanco con sus fundas sin tapador con llave de armar
-Una pistola con el cañón blanco para la cinta con su caxa de nogal labrada de atauxía hueso blanco tiene dos hombres del medio cuerpo o arriba junto a la baqueta con sombreros grandes entallados en hueso blanco no tiene llave de armar
(fol. 476 r)
-Otra pistola con caja de nogal con listas de hueso blanco y unas planchas de metal de bronce en la caxa y el pomo esta descolado no tiene funda ni llave de armar
-Otras dos pistolas estriadas la cámara con unos estríos en la boca y caxa de nogal embutidas de nácar con frascos de nácar y sus llaves de armar no tienen funda
-Una pistola de rueda con los remates del cañón y en el medio dorado y un número 1594 sin llave de armar
-Otras dos pistolas de rueda labradas las cajas de nácar y en la coz de ellas un chiflo sin llaves de armar
-Otras dos pistolas de rueda labradas las cajas de marfil con dos bolas en la coz estriadas las cámaras de los cañones sin llaves de armar
-Dos pistoletillos pequeños el uno todo de hierro dorado y el otro blanco con caxa de nogal y dos flores embutidas de hierro sin llaves de armar
-Otra pistola con dos cañones con sus dos llaves guarnecida la caxa de nácar sin llaves de armar ni funda

-Una pistola larga de rueda con caxa de nogal con un letrero en el cañón que dice Láçari Cominas sin llave de armar
-Otra pistola que envió don Luis de Velasco la caxa de nogal labrada de colores y el cañón pavonado (fol. 476 v) y dorado con una culebra junto a la mira y una funda negra vieja sin llave de armar
-Otras dos pistolas con los cañones que parecen pedreros y las cajas labradas de marfil con sus llaves de armar
-Una pistola larga con el cañón pavonado y dorado a listas
-Otra pistola de rueda dorada la culata y en el medio y la boca

Arcabuçes
-Un arcabuz grande de rueda muy largo con su funda de lienzo y bajo de la rueda tiene un número iiii de hueso blanco tiene en la coz secreto de poner balas y pólvora y sus cargas de hojalata este arcabuz que está reventado por medio y le falta un pedazo del cañón de cuarta de largo
-Otro arcabuz largo con su rueda y caxa de nogal y bajo la rueda un título que dice furioso con su funda de lienzo blanco y no tiene llave de armar
-Otros dos arcabuces de rueda con sus caxas de nogal y funda de vaqueta sin llaves de armar
-Otro arcabuz de rueda con dos canes la caxa labrada de hueso blanco
-Otro arcabuz de rueda con la caxa de nogal labrada de hueso blanco con sus medallas de lo mismo y en la culata le falta algunas que debía te(fol. 477 r)ner y no tiene llave de armar
-Otro arcabuz con caxa de nogal labrada de labor de hueso tiene el brocal del cañón grande con borde tiene llave de armar
-Otro arcabuz la caxa antigua con secreto en ella para pólvora y balas no tiene llave de armar
-Otro arcabuz largo de caça con dos canes y un pellejo que los cubre con su escudo de armas y su funda no tiene llave de armar
-Otro arcabuz de rueda con caxa llana de nogal y en el punto un bolsillo blanco sin llave de armar
-Otro arcabuz largo de seis palmos de rueda con unas tachuelas clavadas debajo la llave con llave de armar
-Otro arcabuz de chispa y guarda de más de cuatro palmos con caxa de nogal
-Otro arcabuz de rueda de tres palmos y un xeme rayado por dentro y tiene la llave con muchos follajes y no tiene llave de armar
-Otro arcabuz de rueda largo acanalado el cañón por encima hasta la mitad con encajes en la caxa de marfil no tiene baqueta ni llave de armar

-Otro arcabuz de rueda de tres (fol. 477 v) palmos y de un xeme de largo el cañón con unos hierros en la caxa para ponerle un cordón y colgarle a la espalda tiene su llave de armar
-Otro arcabuz de rueda con una corona y unas palmas en la cámara con su llave de armar
-Otro arcabuz de rueda de quatro palmos y quatro dedos el cañón con una tachuela detrás en la coz algo hendida y tiene su llave de armar
-Otro arcabuz de rueda de cinco palmos y en el cañón un rótulo que diçe soy de don Pedro Xirón y en la caxa quebrada por detrás de la llave
-Otro arcabuz de rueda de cuatro palmos y un xeme el cañón y junto a la llave está la caxa hendida
-Otro arcabuz de rueda corto que se da la vuelta levantando el gatillo
-Otro arcabuz de chispa largo y delgado el cañón con caxa de nogal sin baqueta
-Otro arcabuz de rueda de cuatro palmos con un rótulo en el cañón que dice Juan Salado
-Otro arcabuz corto con su llave de rueda y la caxa de nogal que tiene un agujero en la mira de tornillo y el cañón es ochavado
-Otro arcabuz de rueda con un mas(fol. 478 r)carón a la culata y hoxeado en medio y a la boca
-Otro arcabuz de rueda con caxa de nogal que está hendida junto a la llave de hasta cinco palmos de largo el cañón

Llaves
-Yten çinco llaves de armar las ruedas de los arcabuces
-Yten otras treçe llaves de armar los arcabuces y pistolas que se acomodaron con los mismos arcabuces y pistolas y así no se han cargo aparte y se dan por entregadas y recibidas
-Más otras llaves de rueda para arcabuz
-Más otras cinco çinco llaves de tornillos

frascos
-Un frasco labrado atauxía de hueso blanco con los remates dorados de nogal con sus cordones de seda verde y plata con su llave
-Otro frasco de hueso liso que ha sido dorado y está algo saltado por arriba con cordón de seda verde
-Otro frasco de cuerno blanco grande que los hierros han sido dorados con cordón de seda azul y cabellada

-Otro frasco negro liso que sale en punta por abaxo donde tiene su muelleçillo, con cordón de seda açul y cabellado con su llave
-Otro chiquito blanco y otro también pequeño redondo con cordón de seda verde y cabellado
-Otro frasco negro con un escudo de armas con unas águilas en él con cordones y borlas de seda negra
-Otro frasquillo del cuerno blanco redondo con sus llaves cordón de seda cabellada
(fol. 479 r)
-Más seis cargas de hoxa de lata
-Doçe digo çinco tirillos pequeños de cuatro a cinco dedos de largo y otros hierreçillos de arcabuçes y pistolas de poco valor
-Diez y ocho pieças chicas que contienen saca balas rascadores para limpiar arcabuçes en una bolsilla de cuero vieja blanca y doçe moldes de hierro para balas grandes y pequeñas
-Çinco ruedecillas de pistolas sueltas sin otras herramientas viexas
-Otros cuatro moldes de hierro para perdigones grandes y pequeños

Arcabuçes 19
-Más diez y nueve cañones de arcabuçes sin caxas ni llaves y el uno parece se puso en un caño en la fuente de casa
-Un hierro para sacar lumbre de piedras en campaña que al tomar de la mano hynche el puño
-Otras tres bolsillas pequeñas la una de lienço dos de fustán colorado y negro para balas y perdigones

espadas y estoques
-Un báculo con caxa de ébano y en el remate de él una cabeza de turco de latón dorada con su espada y en ella una pistola con su llaveçita para armar la pistola y al remate del báculo su brocal dorado con punta de hierro
-Un alfanxe turquesco con su vayna colorada y estremos de plata
-Otro alfanxe turquesco que en la guarnición ha sido dorada y la vaina es de çapa
(fol. 479 v)
-Otro alfanxe turquesco con vayna de çapa dorado el puño y guarniçión y contera y en la vayna hay estremos de plata dorada y en la cruz de la guarnición un xaçinto y otro encima de la empuñadura con talabarte roxo
-Otro alfanxe con guarniçión dorada y el puño de plata y el pomo pico de gorrión y la vaina baya

-Una espada y daga corta poco ancha guarnición dorada antigua y un puño de hueso blanco vayna de terciopelo colorado y tiros de cuero bayo prespuntados y forrados en terciopelo verde y su tahalí
-Otra espada y daga de a caballo con guarnición atauxía dorada de un corte con su vayna baya y tiros de terciopelo açul contera de oro amarilla y dos punzones y un cuchillo
-Otra espada y daga aventanada dorada con su tahalí el cuero bayo prespuntado con oro
-Otra espada corta ancha guarniçión antigua dorada con vayna de terciopelo negro y tiros açules
-Otra espada de estoque con una media luna por guarniçión y en el puño unas armas, y el puño de hiladillo naranxado y la vaina baya
-Otro estoque de hierro de cuatro esquí(fol. 480 r)nas para enristrar y la guarniçión solo cruz dorada y el puño esquinado vayna y talabarte y puño de hiladillo naranxado
-Ocho estoques con guarniciones negras y vaynas bayas a lo antiguo
-Una espada de guarnición dorada y el puño de seda blanca y vayna negra
-Otra espada ancha la guarnición plateada y el puño de seda y plata vayna negra
-Otra espada muy ancha y corta acanalada la hoxa un palmo la guarnición aventanada quebrado el gavilán vayna baya
-Otra espada con guarnición alemana de un corte y vayna baya
-Otra hoxa ancha de tres dedos con esta señal en ella [...] y vayna negra
-Más diez y siete espadas de a caballo para hombres de armas estañadas las guarniciones y vaynas bayas
-Más quarenta y dos dagas de las sobredichas espadas
-Un estoque con guarnición de atauxía dorado y en el pomo cuatro hombres armados con vaina de quero baya y tahalí de ante blanco bordado de oro y plata y cinto de lo mismo
-Más otro estoque que es más largo que el de arriba cuadrado con vayna de cuero negra y guarnición dorada y el pomo ahovado con unos sexitarios (sic) y hombres armados en él y al remate en media luna
-Más una espada ancha dorada la guarnición y el pomo alcachofado (fol. 481 r) vaynas y tiros y las dos dagas son diferentes las guarniciones
-Un arma de hechura de daga que en la cuchilla tiene una pistola y una punta que sale de la misma daga y por esta parte se carga la pistola con su vayna negra
-Otra daga que en lugar de guarnición tiene a manera de cabos de cuchillo dorados y labrado de marfil y un encaxe para el dado a manera de horquilla
-Otra que tira a la misma hechura de hierro dorado

=Estandartes y guiones=

-Un estandarte de hombres de armas de damasco açul con cruz roxa y franxas açules y coloradas y en medio una ymaxen de San Andrés y de la otra parte armas de Velasco

-Otro estandarte açul con aspas de San Andrés y armas de Velasco

-Otro estandarte de armas açul con cruz roxa y una ymaxen de San Andrés y armas de Velasco con su borla

-Otro estandarte de damasco roxo con la cifra de María es el guion de Milán y Borgoña

-Otro estandarte de damasco roxo con franxillas de oro

-Otro de amarillo con dos aspas de San Andrés y franxas açules en él

-Una banderola de damasco açul de trompeta con armas de Velasco y franxas açules y amarillas con cordones y borlas

-Un quadrete de damasco roxo con su escudo de armas para trompeta

(fol. 481 v)

-Un guion de damasco leonado oscuro con aspas de San Andrés de la una y otra parte con cifra que diçe Domingo de Velasco y alrededor armas de Velasco en tela de plata y açul y en las esquinas leones y castillos con dos borlas de seda leonadas y oro

-Tres toneletes el uno de raso blanco con pasamanos de seda y los dos de tefetán doble

-Otro de raso pardo con galones de plata y oro

-Otro de grana con trencillas de plata

-Otro de raso carmesí con pasamanos

-Otro de tela de oro açul con pasamanos de plata

-Un pedaço de terciopelo negro colchado para silla

=Lancas venablos y armas de asta=

-Çinquenta y cinco lanças xinetas con sus hierros y astas blancas y al pie su punçón

-Catorce lanças de ristre las doçe con banderolas y las dos sin ellas

-Tres medias lanças de ristre quebradas y sin hierros

-Mas otras veinte y dos lanças de ristre algunas de ellas sin hierros y banderolas y todas estas lanças de ristre son de los mismos arneses

-Ocho astas de lanças que están pintadas de amarillo y plata

-Mas diez y ocho venablos con hierros blancos y astas blancas

(fol. 482 r)

-Otros dos venablos con las astas retortixadas y grabados y dorados los hierros con correas negras

-Otro venablo blanco grabado con dos borlas verdes con algo de plata en ella y doçe tachuelas
-Dos astas de lança de pino sin hierros
-Mas ocho hierros de a palmo y a palmo y medio de largo para medias picas sin astas
-Otros dos hierros grandes para venablos
-Otros dos hierros más chicos para cayados
-Una xineta de capitán con borla y cordón de seda negra y plata y asta de pino teñida de negro
-Una alabarda dorada y aventanada con asta guarnecida de terçiopelo negro tachonada con tres borlas de seda y oro las dos con oro y la otra sin ello
-Una asta de la estatura de un hombre con un martillo y al otro lado su punta de pico quadrado con una punta larga que se mete con tornillo
-Un venablo guarnecido en terciopelo açul y oro y la cuchilla labrada de atauxía con dos pistolas que se desparan por medio del asta del venablo
-Una hacheta con el asta de ébano y marfil con tres borlas y debaxo de la mayor su llave y cañón de pistola
-Un bastón de hierro que tiene dentro un estoque labrado el bastón atauxía con funda de bayeta
(fol. 482 v)
-Un montante para jugar la esgrima con el puño verde de hiladillo
-Otro montante largo con vaina negra
-Otro montante dorado en la cuchilla con letrero de cada parte y la guarnición labrada a modo de atauxía y el pomo roxo sin vayna
-Dos báculos de caña de la India que parecen bordones el uno está guarneçido los cabos de latón
-Un bastón de xeneral de hierro y llamas de fuego con cordón de seda y oro
-Otro bastón de xeneral de hierro con muchos números dorados con su funda de vaqueta
-Otros cuatro bastones de xeneral de madera de fresno los dos con cordones roxos y pardo de oro y seda y el otro verde y amarillo con hierro pequeño
-Un hierro para arma de asta blanca y dorado y la mitad de él hacia el bastón una sierpe con alas
-Un asta de palo
-Un xinetón largo con borlilla açul
-Una alabarda antigua con una estrella medio del hierro
-Dos astas con sus hierros blancos
-Una çagaya (sic) con una borla de oro y açul junto al hierro y en el asta revuelto de arriba a abaxo un hilo de cobre

-Otra çagaya (sic) larga y parte de ella dorada y grabada
-Dos astas de palo muy chicas con dos hie(fol. 483 r)rros y un listón verde que tiene poco más de media vara

Fundas
-Yten siete fundas de frisa morada para armas diferentes
-Otras çiento y diez fundas chicas y grandes de diferentes colores para cosas de la armería y algunas de ellas viejas y rotas
-Otra funda para rodela de vaqueta forrada en bayeta amarilla
-Otra funda de terçiopelo verde con pasamanos de oro y cordón de seda y oro
-Otra funda de vaqueta forrada en terciopelo negro con cordones de seda y plata
-Otra pequeña de pistolete forrada de terciopelo verde con cordones de seda verde
-Otra funda de arcabuz de vaqueta forrada en terciopelo negro y cordones de seda negra
-Otra funda de vaqueta forrada en terciopelo negro con cordones de seda negra y borla de lo mismo
-Otra funda para dos pistoletes de vaqueta forrada en frisa colorada
-Mas veinte fundas de arcabuces y pistolas de cuero diferentes unas forradas y otras por forrar y algunas de ellas sin tapas y otras con ellas
-Otras nueve fundas de cuero viexas y algunas rotas de diferentes cosas de la armería
-Veinte y nueve fundas de cordellate y frisa de diferentes colores y armas
-Otras quatro fundas de vaqueta para arcabuces de rueda sin cubierta

Estribos
-Un par de estribos de xineta dorados
-Otros tres pares negros de la xineta
(fol. 483 v)
-Otros tres pares de estribos negros y pavonados a la brida
-Otros cinco pares de estribos a la brida los unos dorados otros estañados dos pares blancos otros dorados antiguos
-Otro par de estribos dorados a lo antiguo
-Otros veynte y quatro pares de estribos de la brida diferentes unos de otros y algunos de ellos viexos que en tres o quatro faltan algunas pieças
-Otro par de estriberas de latón de la xineta buenas

=Diversidad de armas y otras menudençias y adornos de la armería=
-Una maça de hierro dorada

-Un morrión blanco antiguo con falda larga guarnecido de terciopelo roxo sin orexeras
-Una tarxeta de justar blanca
-Una çelada blanca grabada de seguir
-Un almete blanco llano antiguo
-Cinco escudos de xusta de hierro blancos llanos grandes para de a caballo con clavazón blanca alrededor
-Una bufeta de hierro blanco
-Un medio placarte de hierro blanco con su barbote en una pieza para justa y tiene su ventanilla el placarte para sacar la ristre
-Un morrión negro fuerte forrado en terciopelo colorado con pasamanos de oro y plata sin orexas
-Otro morrión con tres crestas doradas y blancas cubierto entre la una y la otra de terciopelo negro con orexas fo(fol. 484 r)rrado de lo mismo y por dentro colorado
-Un morrión que diçen es de Calatayud blanco y dorado las cresta y del medio hasta la falda con clavaçón dorada con raso carmesí con estrellas, el forro de seda amarilla y la falda de terçiopelo colorado sin orexas y dos tafetanes para atarle debaxo la barba
-Una coraça de costillas para hombre sobre cuero blanco en tres pieças con malla
-Quatro planchas de hierro blanco que sirven para sobre armas de caballo
-Quarenta y dos penachos desgastados y maltratados
-Un rretrato del Gran Capitán hasta la çinta con bastón de xeneral y corona de laurel
-Dos arcos tudescos con quarenta y una flechas y la mitad de ellas tienen su funda
-Un costalillo con unas pocas de piedras
-Dos mangas de malla con dos medias limetas y bragueta todo de malla
-Una cota de malla grande guarnecida con raso pardo y botones de seda
-Tres cuchillos con los cabos de hueso blanco en una vayna de cuero verde
-Tres martillos para traer a caballo que son para desarmar los dos con cabos blancos y cordones de seda colorada y amarilla y el otro con bastón cubierto de cuero roxo con estrellas doradas sin cordón
(fol. 484 v)
-Una hacheta con martillo y mango de nogal
-Una ballesta con sus gafas y funda verde con bastón que sirve de horquilla
-Una ballesta pequeña reforçada con el tablero labrado de marfil y armas y con un torno que va con la dicha ballesta
-Otra ballesta maás pequeña embutida una faxa de marfil debaxo la nuez

-Otra ballesta muy pequeña de palmo y medio sin gafas
-Otras seis ballestas con sus gafas
-Siete fundas las quatro de ballestas de cordobán y las tres de gafas
-Unas gafas y horquillas de ballesta en su funda de cordobán
-Cinco carcaxes moriscos llenos de saetas y con algunos virotes
-Más seis arcos tudescos
-Una sotagola de terciopelo carmesí forrada en raso
-Otra sotagola de terciopelo paxiço aforrada en tafetán del mismo colorada
-Otra sotagola de anxeo blanco un bonete de tela de oro blanco amarillo y colorado
-Otro bonete de terciopelo colorado con canutillo de vidrio
-Otro del mismo terciopelo a lo turquesco
-Un collar de malla
(fol. 485 r)
-Un cuerno de marfil para la brama blanco con armas de Portugal en su funda de terciopelo verde y blanco
-Ocho ristres sueltos para petos e quatro banderolas de hoxa de lata con las armas de Velasco
-Treynta y siete paveses que eran al contorno de la armería para adorno los cinco de ellos con armas de Velasco y los demás con diferentes armas y señales
-Un compás de latón grande de largo de un palmo y quatro dedos y las puntas de acero
-Un espexo grande de madera de perla bueno para la armería
-Un martillo de hierro sin asta para desarmar
-Más veynte çintas de cordón de seda colorada largas y pequeñas
-Otras cinco cintas açules
-Otras siete verdes y pardas
-Un aderezo de silla de terçiopelo carmesí que son las pieças siguientes cabeçadas y pretal riendas almártiga acciones grupera con estribos y espuelas dorado
-Una ropilla de terçiopelo açul con mangas largas y pasamanos de oro y veynte alamares de lo mismo para el paxe del guión
-Otras tres ropillas de paño açul con mangas con pasamanos anchos açules y amarillos para caballos lixeros
(fol. 485 v)
-Un martillo de hierro grande con cañón y llave de arcabuz
-Una hacheta con el cabo de seda açul y plata y dentro de ella una sierra y un cuchillo lima y barreno y escoplo pónense los hierros en el cabo para usar de ellos
-Una testera de caballo negro con listas doradas
-Dos escarçelas y unos açeros de sillas que parecen de la misma labor

-Más siete banderolas de tafetán açul y amarillo que son de las mismas lanças
-Dos guantes viexos y rotos de malla
-Una guarnición de espada antigua sin pomo
-Un jaco de malla jaçerina con manga de los mismo guarnecido de malla de latón que parece dorado y raso pardo
-Una almilla de malla con sus magas forrada de raso colorado con sus botones de seda en las mangas cintas de oro y seda con sus corchetes
-Una bragueta de malla forrada en raso leonado
-Una malla en dos pieças cada quarto por sí guarnecido en raso carmesí con sus oxales
-Un tafetán verde con una arandela de oro y plata quadrado viexo y roto.
-Çinco hierros de lanças de ristre los quatro en unos pedaços de asta de lança quebrados
-Tres baquetas de arcabuçes las dos medianas y la otra mayor
-Cinco palillos que pareçen baquetas de pistolas
(fol. 486 r)
-Un sacatrapos de hierro para arcabuçes
-Dos hachetas para de arçón con unas listas de latón embutido en la cuchilla y en el martillo y los martillos son picados las bocas a punta de diamante con sus puntas arriba
-Otra hacheta con su martillo que el martillo tiene picado como a punta de diamante
-Una vayna de terciopelo verde con su tahalí de lo mismo bordado en oro
-Unos tiros de terciopelo amarillo con trencillas de oro
-Tres pares de tiros de cuero bayo viejos
-Tres pretinas la una de terciopelo açul con pasamanos de oro paxiço y la otra de lo mismo con trencilla de plata y la otra de terciopelo verde con cayrel de seda verde y blanca
-Dos acciones de estribos blancas con sus hebillas
-Otra pretina de cuero bayo pespuntada
-Otra pretina de cuero verde pespuntada con seda de la misma color con hierros a lo antiguo
-Una ación de terciopelo carmesí con un hierro redondo dorado en medio de ella
-Unos açeros de sillas con una medias lunas grabadas y doradas con listas blancas con una escarçela de lo propio
-Una manopla blanca para el codo de justa
-Una gola dorada y grabada con listas blancas
-Otra gola negra con almenillas de terçiopelo negro con trencilla de seda y oro
(fol. 486 v)

-Un faldón de coraça de tela de oro con tachuelas de lo mismo y sus hebillas de latón doradas con ribete de terçiopelo carmesí
-Una arandela blanca y grabada alrededor
-Un pabellón de campaña para de a caballo a manera de guardasol con su ençerado forrado en fustán verde con flocadura de amarillo y verde con su polea y cordón verde y en el remate un hierro hueco para meter la asta y estandarte
-Una horquilla para descolgar cosas de la armería
-Quatro marcos de lienzo de las ventanas de la armería y puerta del corredor
-Unos cordones de seda colorada con sus borlas pendientes de seda oro y plata
-Los arneses con todas sus golas eçeto catorçe que no las tienen
-Unos estantes grandes con quatro puertas que sirven para cerrar dentro algunos arneses y armas
-Una escalera para descolgar cosas de la armería es de banços de madera
-Las esteras que están colgadas para adorno de la armería
-Un cofre viejo con çerradura y llave
-Un arca grande y redonda donde están las fundas
-Otro cofre viexo largo quebrada la tapa
-Otras dos arcas de madera blancas viejas y rotas de poco valor
-Quatro cortinas de bocaçí colorado que toman desde lo alto a lo baxo con (fol. 487 r) que se cubren las armas doradas de su esçelençia
-Otra cortina que parece de campaña de lienço muy larga bordadas de figuras y forradas en fustán y forrada que sirve de guardapolvo en la armería para las armas de su exçelençia que parece ser tienda de campaña
-Un navichuelo de madera que está colgado en la armería
-Quatro cascos de galeras rotas
-Otro casco de nave sin xarcias
-Una horquilla de mosquete con su horquilla de ciervo de cuerno
-Dos adargas la una con las armas de su esçelençia
-Quatro pabellones grandes de campaña
-Una tienda de campaña
-Dos cortinas grandes con figuras
-Más otras quatro cortinas de campaña
-Veynte maderos para armar dichas tiendas y pabellones que los ocho no tienen hierros
-Dos cofres forrados el uno en bayeta colorada y el otro anxeo blanco sin llaves
-Una cubierta de caballo antigua de cuero colorado y pintado de colores todo en nueve pieças
-Otra cubierta de caballo antigua en seis pieças que por encima de ellas es una tela de colores azul y blanco y colorado con franxas de hyladillo muy viexo

-Una barda y media barda para caballo amarilla con sus escarchas de hierro blanco en tres pieças
(fol. 487 v)
-Unas guarniçiones de terçiopelo negro con pasamanos de oro que son cabeçadas y grupera con diez pendientes pretal con franxas y unas riendas con borla y botón y una almártiga
-Una manta de raso blanco encarnado y amarillo que sirve de cubierta de caballo en quatro pieças
-Quatro frenos con sus caxas doradas muy buenas y en uno de ellos falta una copa
-Otro freno grande sin bronchas que se llama cañón para potro
-Otro fierro a la xineta para caballo y su bocado muy antiguos
-Tres frenos y xinetas de los muy viexos
-Un freno xinetón con su cabeçada sin riendas
-Otro freno xinetón sin riendas ni cabeçadas
-Unas guarniciones para camuça digo de camuça para caballo con pasamanos de badana negra prespuntada buenas que son cabeçadas riendas gurupera pretal y lo demás necesario
-Un señuelo para llamar el açor
-Las redes con sus cuerdas y las cascabeles para caçar
-Unos morillos que son de hierro con su pedestral (sic) y sus molduras y quatro bolillas de latón y su peramida (sic) que sube arriba y por remate bolas de latón con su cerco paleta tenaças escarbador y un tenedor con (fol. 488 r) dos puntas grandes y todos las quatro pieças con sus remates de latón

=Armas viexas que eran aparte fuera de la armería en otro aposento que las ha limpiado el armero y sacado de lo viexo que estaba puesto por perdido=
-Un arnés negro de laonas con sus botones y listas doradas con una cruz de San Juan en el pecho debaxo la çelada que tiene las pieças siguientes
-Peto y espaldar
-Gola y çelada
-Braçales y guardabrazos pequeños
-Y escarçelas y quixotes con su estoque
-Otros seis arneses blancos que son peto y espaldar y algunos de ellos tienen braços y guardabraços y con algunos quixotes y les faltan otras pieças en las vistas de los almetes que como se iban juntando y componiendo de lo viejo no se acabaron de juntar todas las pieças y solo tienen las que aquí van declaradas y están en sus cruces de pino como las demás
-Otras doce cruces de pino nuevas que sean hecho para poner armas

-Yten armas viejas deshechas que son las que no están por inventario sino por mayor que contienen mandiletes testeras quixotes esquineras petos y de (fol. 488 v) todo xénero de arneses que alguna parte de ello paresçe se ha començado a limpiar y que se pone por mayor por ser todo viejo y deshecho de la armería y no se sabe lo que pesa
-Una rrexa de hierro çerrada en un marco de palo de tres quartas en quadro poco más o menos

=Pertrechos para trabaxar el armero=
-Una rueda de limpiar armas que está en tres ruedas con çigüençillas de hierro para andarla y la una rueda es nueva y hay más otra rueda vieja de salçe
-Un martillo de moler esmerilados
-Dos martillos de piconar
-Una estaca de hierro
-Una bigornia de hierro
-Unas tixeras para cortar hierro
-Un tornillo de limar con su banco
-Dos bancos largos de madera para sentarse a trabaxar
-Siete bastones de madera para limpiar las armas
-Un velador de madera
-Dos cuchares (sic) de hierro para aceytes y esmeril para limpiar armas
-Dos pares de tenaças de la armería viexas para arrancar clavos
-Una lima de hierro viexa
-Un arca viexa con cantidad de menudencias dentro de ella de poco valor
-Un baúl barreado viexo y la tapa es diferente
(fol. 489 r)
-Una arquilla de madera larga viexa
-Una alimilla (sic) chica sin cabos vieja
-Un alargador de hierro
-Un çincel de hierro
-Tres punçones de hierro
-Dos desclavaderas de hierro
-Un abotonador de hierro
-Doce hebilletas
-Dos caxuelas para echar clavos y cosas tocantes a la armería
-Unas tixeras a modo de sastre
-Un alicatillo y una lesna
-Unos tornillos hierros y otras menudençias tocantes a la armería
-Dos çedaçillos para cerner esmeril

-Un marco de lienço grande para la ventana de la pieça donde se limpian las armas
-Seis ristres de arneses para la lança
-Çinco cuerdas de ballesta viejas
-Un barreno viexo de hierro
-Tres o quatro pedaços de malla viejos en un arca viexa

=Sillas bordadas y de hombres de armas y de caballos lixeros y otras cosas=
-Primeramente una silla de a caballo de terciopelo açul con franxas açules con cubierta de cordobán negro forrada de bayeta açul y un rrebitillo de la misma bayeta con pasamanos de oro anchos y trencillas de lo mismo y flecos de seda açul y plata con sus açeros grabados y dorados con llamas de fuego y su caballete de pino
-Dos sillas de terçiopelo colorado con sus açeros de hierro grabados que el uno fue dorado (fol. 489 v) y el otro no con sus cubiertas de badana negra viexas con su caballete
-Otra silla de camuça blanca con su pasamanos de oro y plata con cubierta negra forrada en bayeta amarilla
-Otra silla de terçiopelo colorado con pasamanos de plata para la brida con cubierta de badana y el forro amarillo
-Nueve sillas de caballos de terciopelo de tripa amarilla las siete con sus aceros viexas no tienen estribos ni otro recado
-Veynte sillas de cordobán blanco con guarnición negra y sus cubiertas de badana colorada con los arçones negros los quatro con clavaçón dorada y diez y seis clavaçón y blanca de hierro
-Otras seis sillas como las de arriba que tienen las sobrecubiertas de badana naranxada
-Otras tres sillas como las de arriba con clavaçón dorada con las cubiertas negras las dos de badana y las otras dos de paño blanco
-Tres sillas viexas con sus aceros los dos de terciopelo colorado y la otra de paño colorado escarlata
-Diez y siete sillas de badana blanca armadas con sus cueros las quatro de ellas muy viexas y maltratadas y la una sin fustes y las dos sin açeros no tienen çinchas ni otra cosa
(fol. 490 r)
-Una silla de quartago negra viexa sin arçón ni otra cosa
-Quatro sillas xinetas pequeñas las tres con caparaçones viexas y maltratadas
-Un sillón grande negro cubierto de badana para fiambreras sin otros recados
-Otra silla pequeña para jaca que no es de provecho por ser muy viexa
-Otra silla de terciopelo negro viexa muy maltratada

-Otra silla grande de las amarillas que no es de provecho
-Otra de terciopelo açul que solo los fustes son de provecho
-Un armaçón de un sillón grande antiguo y más otras cuatro pieças de sillas antiguas viexas que no son de provecho
-Todas estas sillas están sobre sus caballetes y maderos de pino que cargan sobre quatro bancos altos y otro banco largo que todo toca a la armería

Frenos
-Veinte y ocho frenos bridones con sus cabeçadas y riendas de cuero blanco buenos y algunos les faltan en los bocados algunas pieças
-Y entre todos estos hay tres o quatro diferentes
-Otro freno bridón sin cabeçadas ni riendas
-Otro freno de mula sin cabeçadas ni riendas
-Otro freno son sus cabezadas y riendas negras quebrada una punta donde se pone la rienda
-Otro bocado xinete con cabeçadas y riendas negras
-Quatro cabeçadas de frenos negra con tres pares de riendas y una grupera
-Yten treynta y tres gruperas blancas con sus pendientes blancos
(fol. 490 v)
-Veynte y nueve pretales de cuero blanco
-Çinco cabeçadas de frenos de cuero blanco
-Veynte y ocho pares de açiones para estribos algunas blancas y buenas
-Veynte y tres tahalís blancos y entre ellos cinco negros
-Çinco pares de riendas que en algunas de ellas les faltan algunas hebillas y las medias puntas
-Otros çinco pretales de cuero negro viexos
-Nueve conchas para llevar las lanças en el arçon de cuero blanco y negro
-Veynte y cuatro pretinas de cuero bayo con sus hebillas
-Çiento y ocho pares de cinchas de cáñamo viexas con sus hebillas
-Unos pedaços de correas quebradas de açiones y algunos tiros de espadas todo ello viexo que pro no ser de valor no se ponen de cargo y se quedan para cosas necesarias al armero

Casacas bordadas y guarniciones
que eran en poder del mayordomo
-Yten se declara que las casacas bordadas y otras guarniciones que estaban en el armería además de las conthenidas en este ynventario se entregaron a Ju° de Maeda Salaçar mayordomo como pareçerá del ynventario y reçibo de bienes que se hyço quando entró en el ofiçio de tal mayordomo y para que no se haga cargo

de ello al armero presente ni a su antecesor se pone aquí esta declaración y los bienes que son los siguientes
-Una casaca de terciopelo açul con mangas bordadas con la cruz de Borgoña con diez alamares y seis en las mangas y lo bordado y escarchado y llamas de oro y plata forrada de terciopelo amarillo
-Otra casaca para el paxe del guión de terçiopelo açul con pasamanos grandes de oro y plata y quinçe alamares con los de las mangas
-Un xirel açul de terciopelo en cinco pieças con veynte y dos borlas pendientes de oro y seda (fol. 491 r) bordado de oro y seda escarchada con llamas de fuego y cruçes de Borgoña
-Las guarniciones de la silla de caballo de terciopelo açul bordado y tiene las pieças siguientes
-La gurupera con diez pendientes y veynte y dos borlas de oro y seda
-Y la cabeçada con otras nueve borlas
-el pretal con sus franxas de lo mismo y sus riendas con una borla grande
-Un botón de oro y seda
-Una almártica (sic) con su cordón de oro y seda y borla grande
-Dos açiones de lo mismo para los estribos
-Una casaca de terciopelo carmesí bordada de oro en tela de oro amarillo con mangas y sus alamares de oro y plata
-Un xirel de terciopelo carmesí en tres pieças bordadas como lo de arriba con veynte borlas pendientes
-Una casaca de terciopelo negro con diez y ocho alamares de seda negra guarneçida con un pasamano ancho forrada en tafetán negro

Todo lo qual está a cargo del dho Ju° de Maeda Salaçar y por su quenta con los demás bienes que tiene a su cargo de que es obligado a dar quenta y así no corren por esta más de que van aquí declarados porque conste de la verdad y todos los demás conthenidos y espresados en este ynventario los entregó la dha Ysabel de Bárçena viuda del dho Pedro de Sahavedra al dho Diego Pasqual armero que al presente es con yntervençión del dho Ju° de Maheda y por su orden y mandado como tal mayordomo y asimismo (fol. 491 v)entregó las llaves de la dha armería y las pieças donde están las sillas y frenos y donde se limpian las armas todo lo qual recibió el dho Diego Pasqual armero en presençia de mí el pres^te^ escrib° y de ello doy fee y a mayor abundamiento el susodho renun° las leyes de la entrega y demás de este caso como en ellas se contiene y se obligó con su persona y bienes habidos y por haber de dar de todo ello buena quenta con pago leal y verdadera cada y quando que por su ex^a^ del S^or^ condestable de Castilla o su mayordomo u otra persona en su nombre que para ello sea p^te^ se le mande y lo tendrá bien puesto limpio y reparado a uso de armería en todo tiempo para lo que debe y es obligado al ofiçio de tal armero y además de esta escritura y sin la innovar ni alterar antes aprobándola y ratificándola hará otra con fiança a satisfaçión de su exª y del dho Juº de Maeda Salaçar en su nombre y a ello se obligó en forma renunº las leyes de su favor con la que proybe la x^{al} renun^{ón} y todas las partes lo otorgaron y firmaron los que supieron y por la dha Ysabel de Bárçena y a su ruego lo firmó un tº a los quales dhos otorgantes yo el escribº doy fee conozco y la dha entrega se acabó de haçer a seis de setiembre de mill y seiºs y veynte y ocho años siendo tºs Juº de Silanos y Juº de Arçe y Juº Martínez de Bárçena estantes y residentes en Burgos va entre rs de ristre guarneçido de terciopelo roxo sin orexeras// y ss^{do} de riendas [...] navaja y en la contera unas armas tiros de terciopelo// de oro em^{do} ss valga=

Diº Pasqual
Juº de Silanos
Juº de Maeda Salazar
Ante mí Toribio Díez del Real
(signed).

CHINCHÓN, CONDES DE

Doc. 1.
1617, February 3.
Inventory of arms and artillery of the fortress of Chinchón.
Archivo Histórico de la Nobleza, Frías, C. 1436, D. 25.

(fol. r)[loose document]

Los arneses que se llevan por mandado del conde mi S^{or}

-Un arnés negro fuerte a prueba de mosquete peto espaldar y escarzelas brazales enteros con manoplas gola y zelada quijotes enteros

-Otro arnés negro y oro a prueba de mosquete peto y espaldar brazales enteros gola y zelada medios quijotes en tres piezas

-Otro negro a prueba como los pasados peto espaldar y medios brazales y la gola junto con ellos morrión grande borgoñota

(fol. r)[document]

Artillería

-Primeramente una pieza de artillería cañón de batir con un letrero que diçe don Pedro de Bobadilla conde de Chinchón fundióse año de 1559 y el n^{o} del peso es dos mill y ochoçientos y treinta y dos libras el escudo de armas tiene dos águilas y dos castillos y por orla castillos y leones__28 q^{es}-32 ls

-Otro cañón de batir con el mesmo letrero y armas fundiçión del mesmo año y el número del peso es dos mill y quarenta y quatro libras_20 q^{es}-44ls

-Otro cañón de batir con el mesmo letrero y armas fundación (sic) del mesmo año que el número del peso es dos mill y noveçientas y quarenta y ocho libras_ _29 q^{es}-48 ls

-Otro cañón de batir con el mesmo letrero armas y fundación (sic) que el número del peso es dos mill y veinte y ocho libras__20 q^{es}- 28 ls

-Una pieza de campaña con un letrero que diçe don P^{o} de Bobadilla conde de Chinchón y un escudo de armas con dos castillos y dos águilas y por orla castillos y leones fundóse año de mill y quinientos y çinquena y nueve y el número de peso es treçientas y sesenta y quatro libras__3 q^{es}-64 ls

-Otra pieza de campaña con el mesmo letrero y armas fundiçión año mill y quinientos y çinquenta y ocho y el número del peso es treçientas y sesenta y siete libras__3 q^{es}-67 ls

(fol. v)

-Otra pieza de campaña con el mesmo letrero armas y fundación (sic) el núm^{o} del peso es treçientas y çinquenta y seis libras__3 q^{es}-56 ls

-Otra pieza de campaña con el mesmo letrero y armas y fundación año de mill y quinientos y çinquenta y nueve y el número del peso es treçientas y çinquenta y quatro libras__3 q^{es}-54 ls

-Otra pieza de campaña con el mesmo letrero y armas fundióse año de mill y quinientos y çinquenta y nueve y el número del peso es treçientas y quarenta y una libras__3 q^{es}-41 ls

-Otra pieza de campaña con el mesmo letrero y armas fundióse año de mill y quinientos y çinquenta y ocho y el número del peso es treçientas y çinquenta y una libras__3 q^{es}-51

-Otra pieza de campaña con el mesmo letrero y armas fundióse año de mill y quinientos y zinquenta y nueve y el número del peso es treçientas y çinquenta y quatro libras__3 q^{es}-54 ls

-Otra pieza de campaña con el mesmo letrero y armas fundación y peso__3 q^{es}-54 ls

-Otra pieza de campaña con el mesmo letrero y armas fundióse año de mill y quinientos y çinquenta y siete y el número del peso es treçientas y çinquenta y quatro libras__3 q^{es}-54 ls

(fol. r)

-Otra pieza de campaña con el mesmo letrero y armas fundióse año de mill y quinientos y çinquenta y ocho y el número de peso es treçientas y quarenta y ocho libras__3 q^{es}-48 ls

-Otra pieza de campaña con un escudo que diçe don Diego Fernández de Bobadilla y Cabrera conde de Chinchón con un escudo de armas de Cabreras y Bobadilla y por orla castillos y leones fundióse año de mill y quinientos y çinquenta y nueve pesa quatroçientas y çinquenta y dos libras__4 q^{es}-52 ls

-Otra pieza de campaña con el mesmo letrero y armas y fundaçión (sic) pesa quatroçientas y treinta libras__4 q^{es}-30 ls

-Otra pieza de campaña con el mesmo letrero y armas y fundaçión (sic) pesa quatroçientas y sesenta y seis libras__4 q^{es}-66 ls

-Otra pieza de campaña con el mesmo letrero armas y fundaçión (sic) pesa quatroçientas y treinta y nueve libras__4 q^{es}-39 ls

-Otra pieza de campaña con el mesmo letrero armas y fundaçión (sic) pesa quatroçientas y quarenta y seis libras__4 q^{es}-45 ls

-Otra pieza de campaña con el mesmo letrero armas y fundaçión (sic) pesa quatroçientas y veinte y una libras__4 q^{es} 21 ls

(fol. v)

-Otra pieza de campaña sin letrero con las armas de Cabreras y Bobadilla y ençima del fogón unas flores de lis relevadas y el número de la otra de la mesma manera y por asa dos leones pesa a seiscientas libras__6 q^{es}

-Otra pieza de campaña sin letrero y con las mesmas armas pesará seisçientas libras__6 q^{es}

-Otra Pieza de campaña corta sin letrero con un escudo de armas de Cabreras y Bobadilla y orla de castillos y leones toda la cámara de ella de unas labores relevadas con su caja y ruedas de nogal labradas de relieves pesará doçientas libras__2 q^{es}

-Otra pieza de campaña gruesa con un escudo de Cabreras y Bobadilla orla castillos y leones y alrededor del letrero un escudo que diçe Don Diego Fernández de Cabrera y Bobadilla conde de Chinchón fundióse año mill y quinientos y noventa pesará ochoçientas libras__8 q^{es}

-Otra pieza de campaña con el escudo de armas de la mesma forma y fundiçión y pesará lo mesmo__8 q^{es}

(fol. r)

-Otra pieza de campaña con un escudo de armas de Cabreras y Bobadilla orlas castillos y leones y ençima del fogón un letrero que diçe Neapuli y no tiene el año que se fundió con su caxa y ruedas de nogal con algunas labores pesará seteçientas libras__7 q^{es}

-Otra pieza de campaña con el mesmo escudo y armas y de la forma mesma pesará lo mesmo__7 q^{es}

-Otra pieza de campaña gruesa con las armas de Cabreras y Bobadilla y orlas castillos y leones toda la cámara de ella hasta la mitad de labores de relieves por asas unas ninfas caxa ruedas de nogal con muchas labores sin letrero ni año que se fundió pesa seteçientas libras__7 q^{es}

-Otra pieza gruesa de campaña con un escudo de armas de Cabreras y Bobadillas por orla castillos y leones y un letrero ençima del escudo que diçe Don Diego Fernández de Cabrera y Bobadilla conde de Chinchón fundióse año (fol. v) de mill y quinientos y noventa toda ella muy labrada de labores de relieve pesará noveçientas libras__9 q^{es}

-Otra Pieza de campaña de la mesma forma y peso__9 q^{es}

-Otras tres piezas esmeriles pequeñas con un escudo cada una con un ramo y tres alas con un letrero que diçe opus Nicolay de Bolo sin cajas pesará çinquenta libras cada una__1 q^{l}- 50 ls

-Otras dos piezas pequeñitas encabalgadas

Otras tres pieçiçuelas por encabalgar de diferente hechura

-Otras dos peziçuelas labradas a lo tosco que todas estas siete peçiçuelas pesarán cinquenta libras poco más o menos__q^{es}- 50 ls

-Una bala enramada de bronce con unas balas de plomo por la parte de fuera

E yo fran^{co} Calderón escrib^{o} del rey nro S^{or} pu^{co} del núm^{o} en la villa de Chinchón vz^{o} de ella presente fuy a lo que dho es y lo signé

En testim^{o} de Verdad

Fran^{co} Calderón escrib^{o}

(signed)

(fol. r)

En la villa de Chinchón a tres días del mes de febrero de mill y seysçientos y diez y siete años Garçi Díez de Liana alcayde de la fortaleza de Su Señoría del conde de Chinchón mi Señor confessó haber resçibido las armas y tiros y municiones en estos treinta y quatro capítulos para los dar de manyfiesto cada que se le pida a ley de tal alcayde y otorgó depósito en forma y firmó

Garci Díez de Liana

(signed)

E yo el sobredho escribº [...] my signo que es a tal en testimº de verdad

Fran^{co} Calderón

(signed).

Doc. 2.
1665, October 27.
Inventory of Count of Chinchón. Armoury.
Archivo Histórico de la Nobleza, Frías, C. 1435, D. 4.

(fol. r)
Mi Señora Dios la guarde
Resulta
Contra su Exª del ynventario de los bienes y alhaxas que había en Chinchón de que se le ha de haçer cargo en su quenta con la testamentaría

Ynventario y tasaz^{on} de los bi^{es} de Odón
Cargo [...]

(fol. r)
En la villa de Odón a veinte y siete días del mes de otubre de mil y seisçientos y sesenta y cinco años ante el señor Juan de Zuaço theniente de alcalde mayor en ella y por antemí el escribano pareció Don Manuel de la Fuente maestre sala que fue del conde de Chinchón mi señor que santa gloria haya y dijo que entre las juntas que han hecho los señores testamentarios de SE es una para que se haga ynventario de los bienes muebles y raíçes que parezieren ser de SE que quedaron en esta villa así en el castillo casa del monte huerta y otras partes como consta en la dha orden y junta de que hiço presentación por la qual pidió se haga el dho ynventario y vista la dha orden por el dho señor theniente de alcalde mayor m^{do} se haga el dho ynventario según y como se refiere en la dha orden para cuyo efeto se fue al dho castillo y se pidieron las llaves de él de las salas y demás pieças a Juan Sánchez trompeta casero y conserje (fol. v) del dho castillo y el suso dho las entregó y por ante mí y con asistencia de su mrd y del dho Don Manuel de la Fuente se hiço el dho ynventario que es del tenor siguiente

Pinturas [...]

(fol. v)
Armería
-En la pieça del armería y en otra del quarto alto se halló quatro pistolas de un cañón y llaves de rueda
-En dha pieça otra de dos cañones y dos llaves
-En dha pieça otra pistola de tres cañones que todos se disparan con una llave
-Una llave de rueda de escopeta

-Una adarga blanca
-Una horquilla de hierro
-Unas armas grandes grabadas que tienen las pieças siguientes, dos petos espaldar dos golas dos brazaletes dos manoplas un faldón dos musleros dos çeladas un hombro dos cascos de hierro negros tres morriones con faldones
-Quinçe ballestas de virotes y flechas muy maltratadas
-Onze arcabuzes cortos y largos algunos con cajas y llaves y otros sin ellas
-Dos arcas de pino sobre dos banquillos cada una y en ellas seis llaves de rueda de arcabuz
(fol. r)
-Un carro pequeño con dos ruedas y un alcabuz grande para matar gurullas (sic)
-Diez gafas de armar ballestas
-dos tornos con sus aspas y dos asnillas p[a] ellos para subir madera a lo alto [...]

CIFUENTES, CONDES DE

Doc. 1.
1593, February 15. 1590, May 21.
1590, May 21.
[Copy of 1670].
Inventory of the Count of Cifuentes.
Archivo Histórico de Protocolos de Madrid, Protocolo 10801.

(fol. 1 r)
El liz^{do} Don Diego de Riberto Báñez de Segovia del q^{o} [...] en el Real de Castilla y juez particular para con[...] de los negocios tocantes a el estado y condado de Cifuentes y hacer pago a sus acreedores= Diego Miguel Ibañez ss^{no} del número de esta villa sucesor en el oficio y papeles de Gabriel de Rojas, su anterior, luego que con este mandamiento compulsorio se ha requerido por parte de los Marqueses de Malagón le dé un traslado de las particiones que se hicieron entre Dª Blanca de la Cerda segunda mujer del conde D. Fernando de Silva y sus hijos con todos los demás recados de ellas del testamento del dho Conde y de las capitulaciones matrimoniales y escritura de dote otorgada a [...] de la dha condesa Dª Blanca de la Cerda por [...] Don Fernando de Silva signado y en forma [...] q haga fee que le p^{e} para presentar [...] q^{e} los dos Marqueses de Malagón tratan con [...] Conde de Cifuentes y su estado q para ante [...] cumpla que por mi auto así está mandado [...] ello ha sido citada la parte fho en M^{d} a seis de sep^{e} de mill y ss^{os} y setenta años= [...]

(fol. 2 v)
En la villa de Madrid a quinze días del mes de febrero de mil e quinientos y noventa e tres yo el presente sno de pedimiento de Doña Blanca de la Zerda condesa de Cifuentes ante el S^{r} liz^{do} Tamayo teniente de correxidor en la dha villa de Madrid y su tierra por Su Mag^{d} leí la petición de suso contenida= y visto por el dho teniente mandó que se haga la dha partición de los bienes que quedaron por fallecimiento del dicho Conde de Cifuentes y q se den las requisitorias que se piden= e nombró por curador ad liten para hazer la dicha partición por los dichos Conde de Zifuentes Don Joan menor y los demás sus hermanos contenidos en el dicho pedimiento a Balthasar Ortiz residente en esta corte del qual mando que parezca ante su mrd a hazer juram^{to} e oblig^{on} e dar la fianza de dere^{o} neces^{a} y ansí lo proveyó e m^{dó}= ante mí Roxas

(fol. 3 r)
En la villa de Madrid a nueve (diez)[between lines] días del mes de febrero de mil y quinientos e noventa y tres años ante el dicho señor teniente lizenciado Tamayo th^{e} de correxidor en la dha villa de Madrid y su tierra por Su Magestad pareció el dicho Balthasar Ortiz y dixo que azepteba y aceptó el dicho cargo de curador ad liten del dicho Conde D. Ju^{o} Balthasar de Silva y de los dos Don Francisco (fol. 3 v) y Dª Inés y Doña Ana de Silva sus hermanos todos ellos menores hijos de los dichos Conde Don Fernando y Condesa Doña Blanca para el dicho efecto de la dicha partición pidió al dicho señor teniente reciba del juramento que está presto de hazer y la obligación y dar la fiança en tal caso necesaria y fho le diszierna el dicho cargo y dé poder (fol. 4 r) en forma= y el dho señor teniente tomó y recibió juramento del dicho Balthasar Ortiz el qual juró por Dios nro Señor e Santa María su madre y a la señal de la cruz a tal como esta + en que puso su mano derecha y por las palabras de los Santos Evangelios según en forma de derecho so cargo del qual prometió que como buen curador ad liten de los dichos menores [...]

(fol. 159 v)
Los bienes que el Conde Don Fernando tenía de su capital al tiempo que se casó con la señora Condesa Doña Blanca [...]

(fol. 173 r)
-Yten todas las armas que están en la fortaleza de Cifuentes contenidas en la tasación que se hiço ante Gaspar Díaz escribano que está firmado de su nombre y de los tasadores y del Dotor Texas que por ser partidas muchas y muy menudas se ponen todas en esta (fol. 173 v) partida juntas, la qual dicha tasación queda con esta partición, y sumadas todas las partidas de la dha tasación, montaron tresçientas y sesenta y un mil ochoçientos y setenta y cinco mrs___361U875 [...]
(fol. 176 r)
-Yten dos cañones de bronce que llaman sacres por otro nombre quarto de cañón que pesaron çien arrobas bien tratados (fol. 176 v) y de buena fundiçión hechos en Molinas (sic) y otras dos petacas de la misma fundación y del mismo Maestro que pesaron duçientas y quarenta y cinco arrobas, y dos falconetes que pesaron nueve arrobas y seis libras con doze carretones y sus cureñas y herraxe, y las piezas pertenecientes a ellos tasado todo en dos quentos de mrs___2.000U
-Yten dos mil y cien (fol. 177 r) balas de hierro colado de más de a libra cada una y otras treçientas y veinte y cinco balas del mismo hierro que pesaron seisçientas y cinquenta libras tasadas en noventa y tres mil y seteçientos mrs__93U700

-Más ocho lombardas de hierro con sus cureñas y culatas tasadas en diez y siete mill mrs como todo consta por la tasación hecha en Barçiençe con requisitoria de la Just[a] de esta villa de Madrid (fol. 177 v) que queda con esta partición__17U
-Yten catorce paños de tapicería que diçen de galerías que tienen [...]
(fol. 178 r)
En todo lo qual suma y monta nueve quentos ochocientos y noventa mil treçientos y quinçe maravedís= Baltasar Ortiz= Fran[co] de Quintana escribano__
_Monta todo el capital 9 q[s]. 890U315 mrs

Ynventario de los bienes que
quedaron del Conde en Cifuentes
En la villa de Cifuentes a veinte y un días del mes de mayo de mil y quinientos y noventa años Su Señoría (fol. 178 v) de D[a] Blanca de la Zerda Condesa de Zifuentes gobernadora del estado de D[n] Juan de Silva Conde de Zifuentes alférez mayor de Castilla, señor de las villas de Escamilla y Barçiençe su hijo & dijo al D[or] Fortea su correxidor que por muerte de Don Fernando de Silva Conde de Zifuentes su señor que haya gloria quedaron mucha cantidad de bienes muebles y raíces derechos y acciones, de todo lo qual quiere haçer inventario para que haya quanta y razón y luz de los dos bienes y para (fol. 179 r) con más rectitud se haga juró por Dios nro Señor que bien y fielmente Su S[ría] hará el dho inventario sin incubierta (sic) de ningunos bienes y protestó fenecerle y acabado dentro del término de la ley, e pidió justicia y testimonio testigos Antonio de Salinas y Murçe, Francisco, criados de Su Señoría

-El dicho correx[or] dijo que lo oye y que Su Señoría comience y fenezca el inventario que diçe y feneçido se haga el pedimiento necesario que su merced (fol. 179 v) Proverá (sic) justicia testigos los dichos= Gaspar Díez
-E luego Su Señoría comenzando haçer (sic) el dicho inventario puso en él los bienes sigui[tes] [...]

(fol. 211 r)
Bienes de Barçiençe
Fortaleza
-Onze tiros de bronze con las armas del Conde mi S[r] marcados y encabalgados en sus carros fuertes
-Otro tiro del mismo xaez desencabalgado y el carro quebrado
-Un barril de madera para (fol. 211 v) pólvora sin pólvora
-Otro barril sin pólvora
-Seis arcabuzes con sus cajas cascadas viexas

-Dos cañones de arcabuzes
-Nueve cántaras de hierro
-Otra cántara de hierro quebrada
-Tres cántaras de bronce
-Dos atacadores de pino
-Una balança de peso grande de madera con cuatro cadenas
-Veinte y un trebalos de los carros de los tiros
-Una cuchilla y una canasta grande llenas de pelotas
-Una escalera de litera
-Doze sillones antiguos
-Doze cobreras con sus guarniciones viexas
-Dos banquillos que están en (fol. 212 r) las xarcias
-Una maza de palo de moler pólvora
-Una asta con una bandera viexa
-Dos puertas de la pieza más adentro de la sala grande de abajo
-En la cozina dos piedras labradas redondas
-Otra piedra redonda a la puerta de la cozina
-Un mortero de piedra para moler pólvora
-Nueve tinaxas de vino tener y la una caída
-Otro tiro de bronze a la (fol. 212 v) cámara de ello
-Setenta petos y espaldares y guardabrazos y piezas de costillas
-Otras quinçe armaduras de brazos de ellas sanas y de ellas quebradas
-Cinco almetes
-Diez capacetes
-Un morrión
-Una cuchara de atacar los tiros
-Una cántara de hierro quebrada
-Otra cantarilla de hierro quebrada
-Diez y seis arcabuzes con sus caxas
(fol. 213 r)
-Otro arcabuz sin caxa
-Una caxa grande de arcabuz
-Una ballesta con su tablero
-Un ballestón encabalgado
-Dos ballestones desencabalgados
-Dos faxas de los ballestones
-Un potro de madera
-Una maza de moler pólvora
-Dos puertas pequeñas

-Una puerta del potro
-Otro tiro pequeño con su caxa enzima de la mazmorra
-Más otros siete petos respaldares (sic) y algunos con sus faldones
-Un peto y una espalda de costillas
-Quatro yelmos
-Tres manoplas
(fol. 213 v)
-Cinco armaduras de muslos
-Tres cántaras de hierro la una quebrada y la otra reventada que no hay más de la media
-Otro tiro de hierro en la torre del león
-En la torre mocha un tiro en la tronera
-Otro tiro en otra tronera
-En el alxibe dos tinaxicas de tener azeite
-Un molinillo con sus piedras
-Una tolva del pastor
-El ziguñal (sic) del molino de madera
-Quatro ballestones de palo
-Tres sogas antiguas
-Quatro tablas enzima de la torre
(fol. 214 r)
-Un madero atravesado con nueve escarpias
-Otro de la misma manera del alxibe
-Otro madero de la misma manera
-Otra puerta
-Unos pedazos doze de ellos de armas decompasadas
-Un carrillo de palo
-Una litera viexa
-Dos tablas anchas de a nueve pies de largo
-Unos pies de mesa grande
-Una zerradura
-Tres tinajas en la mazmorra
-Cinco tiros grandes de hierro con sus caxas debajo de la escalera (fol. 214 v) del corredor
-Una puerta enzerada del alxibe de la escalera de él
-Una bacía

Bienes de la fortaleza
de Cifuentes
-Un arnés grabado y dorado que tiene peto con medias escarzelas espaldar gola braçales enteros guardabrazos, y otro guardabrazo de por sí, dos manoplas y una zelada con sus fundas de frisa
-Otro arnés negro cumplido con peto y espaldar brazales y guardabrazos gola zelada manoplas
-Otro arnés de prueba grabado que tiene peto y espaldar gola e dos guardabrazos, e morrión e dos escarzelas (fol. 215 r) sus fundas de frisa verdes
-Otro arnés negro que tiene peto y espaldar gola y çelada manoplas brazales y un guardabrazo
-Otro arnés blanco grabado con su peto doble celada gola brazales manoplas y escarpes
-Un peto y espaldar y gola negro de infante
-Otro peto y espaldar viexo a lo antiguo con zelada doblada dos brazales y otro brazal con guardabrazos quatro manoplas viexas y dos grebas viejos y dos escarpes viexos y ocho piezas pequeñas y el mismo arnés tiene dos canilleras viexas en medio guardabrazos viexos
(fol. 215 v)
-Una almilla pavonada y dorada de la insinia del conde
-Diez corazinas con sus mangas en fustán negro
-Una rodela e morrión grabado y dorado con su cerradura de oro y seda verde y aforrada en terçiopelo verde con cordonzillos de oro y en ella hecha una historia de relieve
-Otra rodela grabada y dorada aforrada en terçiopelo carmesí con cordoncillos de oro
-Tres morriones grabados y dorados
-Otro morrión grabado y dorado y guarnezida con una flocadura de seda verde
-Dos morriones negros antiguos de prueba
(fol. 216 r)
-Una rodela de hierro negro
-Ocho piezas de una silla de armas con unos hierros grandes de estandarte con su cadena e llave y cinco tornillos
-Dos pedazos pequeños de malla
-Un escopeta (sic) de perdernal con su llave y funda cumplida
-Otra de la misma suerte
-Un pistolete de dos llaves con sus caxas grabadas de marfil e funda con frasco e frasquillo de la misma suerte y guarnición
-Otro pistolete de Alemania con su caxa grabada de marfil

-Otro de la misma manera
-Otro pistolete pequeño con (fol. 216 v) su llave
-Su venablo con tres pistoletes es grabado y dorado
-Dos venablos grabados y dorados los hierros y guarnezidas las astas de cuero negro
-Otros dos venablos dorados e grabados
-Otros dos de la misma suerte con fundas
-Seis venablos de montería con las astas de palo blanco
-Un tiro de artillería pequeño encabalgado con un carretón y ruedas
-Dos cofres viexos
-Un cajón de pino
-Tres alfanxes turquescos guarneçidos de plata (fol. 217 r) esmaltada con borlas de seda carmesí y las vainas guarnezidas de la misma plata
-Otro alfanxe negro turquesco con una vaina de zapa
-Un cuchillo con la guaniçión de atauxía dorada y con una vaina de terciopelo carmesí
-Un alfanxe grande antiguo muy ancho con el pomo partido por medio
-Otro alfanxe con la guarnición de marfil
-Tres espadas antiguas con las guarniciones barnizadas y puños
-Otra hoja de espada ancha sin puño metida en una vaina baya
(fol. 217 v)
-Otra espada con un pomo redondo con dos leones en ella
-Un alfanxe pequeño con otra vaina baya
-Dos espadas una ancha y otra angosta con guarniciones negras e llanas barnizadas
-Dos hojas de espadas anchas con guarnición metidas en unas vainas bayas
-Seis espadas viexas y una daga con las guarniciones plateadas
-Un puñal buido con una guarnición de palo pintado
-Otro puñal buido con una guarnición de cuernos y unos botoncillos
-Dos cuchillos de monte con vainas verdes el uno con (fol. 218 r) un cordón de seda y otro cuchillo en él y en el otro cuchillo e punzón
-Quatro espadas y un montante de esgrima con sus guarniçiones
-Dos raquetas (sic) la una sin cuerda
-Dos palas para jugar a la pelota
-Dos puños de hierro para lo mismo
-Una corneta de marfil grande
-y otra negra guarnecida de plata
-Un arco turquesco sin cuerda y un manojo de saetas grandes
-Un zamarrillo lleno de corzueles de yerba

-Dos pares de estriberas doradas de atauxía
-Otro par de estriberas chico negras (sic)
(fol. 218 v)
-Dos cargas de arcabuz de marfil
-Y de los dos arcabuzes que llevan cargados se le entregó un frasco con cordones de seda parda guarneçida de marfil y dos llaves de los dos arcabuzes y una bolsa viexa, un molde de pelotas
-Un carcax de cuero bayo, digo tres axabas de cuero bayo con alguno otro tres (sic)
-Un carcax de cuero con doze tiros
-Dos hojas de espada juntas en una vaina con guarniçiones en una pavonadas
-Otra espada la hoja volteada con una guarnición plateada abierta e tiros e pretina (fol. 219 r) baya con cordoncillos de plata
-Una guarnición de caballos a la francesa de cuero bayo
-Una cabeza de pretal y su ropera de cuero blanco
-Un tablero de ballesta con sus gafas y de vergas de ballesta quebradas
-Sesenta quadrillos de jaras de montería

Bienes q hay
en Toledo […]

DOS AGUAS, MARQUESES DE

Doc. 1.
1843, Juny 2.
Letter from Victoriano Morera to Manuel González de Espada on the testamentary of the Marquis of Dos Aguas, with an inventory of the armoury.
Archivo Histórico de la Nobleza, Osuna, Cartas, 538, D. 42.

Puebla y Junio 2, de 1843
Sr. Dn. Manuel González de Espada
Muy S^or^ mío y de mi mayor atención: tan luego como recibí su apreciable de 23 de Mayo, me valí de mi hijo político Dn. Juan Ant° Pérez por constarme que mantiene estrecha amistad con Dn. Miguel Gombau dependiente antiguo de la casa del difunto Sor Marqués de Dos Aguas para adquirir las noticias, que S. E. desea, y en efecto supe por de pronto, que la descripción de bienes era judicial, que no estaba concluida, y los efectos se hallaban por justipreciar, y finalmente, que no se procedería a la almoneda a menos que no se remitiese de Roma la disposición testamentaria del difunto, y poderes del señorito, de quien no se tenía noticia, ni de persona alguna de aquella casa, por cuyo motivo había ya partido para Roma (fol. v) el hijo de Dn. Miguel.
Este llegó de su expedición el primero del que rige con el testamento, poder general del Señorito en favor del citado Gombau y demás instrucciones necesarias, y en su vista nos ha facilitado la descripción de lo más notable de la que acompaño copia, manifestando que entre otras de las órdenes recibidas una de ellas es, que no se venda pintura alguna= que se va a continuar la descripción de los demás bienes, el justiprecio de todos para principiar luego la almoneda; todo lo qual espero pondrá V en la consideración de S. E mi señor
Con este motivo tengo la satisfacción de ponerme a su disposición, deseando ocasión de poder manifestar el afecto que le profesa este su obligado e invariable servidor

Y S. M. B.
Victoriano Morera
(signed)

(fol. r)
Salones

Una porción de sillas de brazos de nogal grandes con el respaldo y asiento de terciopelo labrados de varios colores, pero estropeado.
Otra íd mucho mayores con respaldo y asiento de otra tela antigua, pero ordinaria: un canapé de la misma hechura.
Tres mesas grandes con mármol armazón de talla dorada.
Una porción de cuadros antiguos.
Nueve arañas de cristal antiquísimas.
Cuatro escritorios de dos cuerpos con muchísimos cajoncitos, las pantallas con lunas de espejos.
Cuatro armarios pequeños librerías de dos cuerpos de nogal con embutiros de mucho mérito.
Las pinturas y demás adornos dorados de las piezas están fijos sobre las paredes, guarniciones de puertas y atajados de cristales.

Armería
Tres armaduras, una de ellas entera de acero con su rodela: Otra de media gala de acero y oro; y (fol. v) otra de gala más lujosa, estas dos de medio cuerpo con sus rodelas, machetes, y esta última con varias piezas para el caballo.
Una silla de hierro forrada de terciopelo verde, está muy deteriorado.
Una porción de dagas de derecha e izquierda.
Una espada de dos manos de diez palmos, que llamaban mandoble.
Otra con guarnición de coral y bronce dorado hoja ancha y de mucho mérito.
Otra de hechura árabe, pero estropeada.
Muchas piezas sueltas de otras armaduras doradas: varios cascos de diferentes hechuras: botines: manoplas de acero: Un vestido de malla y cuero: Otro ídem de terciopelo con brocado de oro, que está muy estropeado.
Cinco arcabuces con llaves para mechas.
Cuatro ballestas para tirar flechas.

Desvanes
Una gran porción de cristal de arañas antiguas: muebles de todas clases, que no se pueden enumerar por (fol. v) estar hacinados que se almacenaban al paso que se hacían antiguos o inservibles.
Cortinajes de damasco y terciopelo con bordados, franjas y galones de oro: Varias alfombras.
Una porción de guitarras y laúdes trabajados con el mayor primor de ébano y marfil, hechuras las más raras, pues muchas tienen clavijas para veinte y seis cuerdas.

Cocheras
Una carroza antigua trabajada de talla la más perfecta del famoso Ignacio Vergara, con un grupo al presente y otro a la trasera, varias figuras y adornos, todo y la escala puesto el oro sobre la misma madera; la caja con siete portillas que se cierran por cristales de mucho mérito en particular el de la testera que es grandísimo y de media pulgada de grueso; el centro de la caja y almohadones guarnecido de terciopelo con bordados de plata.

Guarnés (sic)
Unas guarniciones para seis mulas por el mismo (fol. r) estilo y antigüedad de la carroza con el hebillaje dorado. Todo debe considerarse está estropeado atendidos los años.

FERIA, DUQUES DE

Doc. 1.
1552, November 4.
[Copy of 19th century].
Inventory of the Duke of Feria. Arms in the town of Zafra.
Archivo Ducal de Medinaceli, Archivo Histórico, Leg. 82, D. 65.

S. XIX
Relación de armas y otros objetos q existían en el ducado de Feria en 1552 y 1625.

(fol. r)
Zafra 4 de Nov^{e} de 1552
En el inventario practicado ante el escrib^{o} Luis Maldonado de los bienes que quedaron al fallecim^{o} del Sr Pedro de Córdoba y de Figueroa conde de Feria resultan las armas y armaduras sig^{tes}

-Dos adargas
-Dos lanzas con sus hierros
-Una barjuleta y aljaba
-Un arma como partesana
-Dos venablos con sus astas
-Un arcabuz pequeño barnizado con su funda
-Seis armaduras de cabeza como celadillas de infante forradas en seda de colores con sus fundas
-Siete pares de estribos de estradiota dorados y traídos
-Veinte pares de riendas de caballo de cadena para la guerra. Son de hierro estañadas y barnizadas
-Un freno de brida con las copas plateadas
-Seis pares de copas, otras de freno de brida doradas
-Diez hierros de lanzas
-Nueve partesanas en sus fundas sin astas dorados los encajes y parte de ellas
-Cinco martillos de armas con sus fundas
-Tres hachas las dos enfundadas
-Otra maza de armas pequeña

(fol. r)
Ducado de Feria Cap. 8.

En el libro titulado de la Joya a fol°. 477 se halla un inventario de muebles que se hallaron en el alcázar de la villa de Zafra, el cual se empezó en 10 de Nov^e^ de 1624 y concluyó en 11 de enero de 1625, practicado por el contador D. García Ramírez Durán, para que se hiciera cargo el guardarropa de S. E. Fernando de Carvajal, ante el escrib° Jerónimo Sánchez.
Este inventario contiene varias ropas, muebles, armaduras, cuadros y libros.
Los libros no se reseñan más que por su número y forro.
Las armaduras se hallaban guardadas en cajas de hierro.
Entre los cuadros se reseñan dos retratos de D. Gómez Suárez y uno de D. Lorenzo.

Doc. 2.
1571.
Inventory and appraisal of arms and saddles of the Duke of Feria.
Archivo Histórico de Protocolos de Madrid, Protocolo 742.

(fol. 1 r)
Sillas de caballos
Primeramente una silla vieja de tr°pelo de quatro borrenas tasó e apreció en doze reales__Uj[...]
-Yten otra silla de quatro borrenas de tr°pelo en veynte reales__Udclxx
-Yten otra silla de tr°pelo vieja de quatro borrenas en ocho reales__Ucclxx
-Yten una silla nueva de tr°p° de quatro borrenas se tasó e apreció en diez y seis ducados__vjU
-Yten una silla guarneçida de tr°p° de cuero se tasó y apreció en tres ducados e medio con sus estribos y adreço__iUccc
-Yten otra silla como la de arriba en otro tanto con sus estribos y adreço__iUccc
-Yten otra silla de cuero con estribos e guarnición y çinchas lamas nuevas en quatro ducados__iUd
-Otra silla de la misma manera q las [...]
(fol. 2 r)
-Yten dos guarniciones de cuero pespuntadas se tasaron ambas en un ducado__[...]
-Yten dos corazas de sillones de literas de tr°p° morado con las rapazeras [...] de oro se tasaron ambos en veynte ducados__[..]
-Yten dos coraças de tr°p° carmesí de unos sillonçillos a ducado cada uno__ Udccl
-Yten una guarnición de paño nuevo de caballo se tasó en dos ducados__Udccl
-Yten otra guarnición como el de arriba en otro tanto__Udccl
-Yten una guarnición de caballo doblado de los de Mantua en un ducado__ Uccclxx
-Yten una cubierta de paño guarnecida de cuero y otra jerguilla se tasaron ambos en veynte reales__Udclxxx°
-Yten unos pedaços de guarniciones y otros pedaços de cueros tasaron en quatro reales__Ucxxx
-Yten una lança pequeña con un hierro dorado se tasó en quatro reales__Ucxxx

E después de lo suso dho en la dha villa a diez e seis días del mes de noviembre del dho año los dos Sebastián Marcos y [...] Valdevieso prosiguiendo la dha tasación tasaron y apreciaron los bienes siguientes [...]

(fol. 19 r)
E después de lo suso dho en la dha villa el mismo día mes e año [...] ante mí el escribano pareçieron Diego de Rueda espadero maestro en esta arte para tasar y apreciar las cosas de espadas [...] e otras cosas que quedaron del dho duque del qual se tomó e recibió juram° en forma de dr° so cargo del dho juramento tasó e apreció en la forma e manera sigui[te]

Espadas
-Yten una espada larga pa de noche con una guarniçión lisa se tasó e apreció en diez e ocho reales
-Yten una espada e daga con su guarniçión lacada dorada y puños de plata dorada con sus talabartes metida en una funda verde se tasó doze du°s
-Yten otra espada con su guarniçión de hierro quemado con unos botones labrados en la guarniçión con sus talabartes de tr°pelo metida en una guarniçión se tasó en catorce du°s
-Yten otra espada y daga con sus guarniçiones doradas y talabartes con botones labrados y talabartes de tr°pelo pespuntados e puños de plata metida en una funda verde se tasó en diez y ocho ducados
-Yten otra espada con guarniçión plateada e puño de plata y talabarte de cuero metida en una funda verde de bayeta se tasó en quatro du°s
-Yten otra espada con guarnición [...] e talabarte de tr°pelo metida en una funda se tasó en cinco ducados
-Yten otra espada con su daga y guarniçión torchada con talabarte de cuero y puños de hierro se tasaron en seis ducados
(fol. 19 v)
-Yten una espada con su guarniçión barnizada y talabartes de cuero con su funda en dos ducados
-Yten otra espada con guaniçión barnizada syn talabartes se tasó en catorce reales
-Yten otra espada e daga con guarniciones doradas con puños de plata dorados y talabartes de cuero y cayrel de plata y seda verde se tasó en doze ducados
-Yten un cuchillo de monte con unos hierros dorados con guarnición de cuero verde pespuntada se tasó en tres ducados
-Yten una daga ancha con una guarniçión de atauxía con cuchillo e punzón y puño de seda y petrina de cuero en tres ducados
-Yten una bolsa grande de holandilla con un hierro grande de atauxía se tasó en dos ducados e medio

-La qual dha tasación el dho Diego de Rueda la hizo en la forma e manera que va declarado e dixo que so cargo del dho juram° va bien y fielmente hecha a todo su saber y entender e a lo qual [...] y lo firmó de su n^{e} t^{o}s Diego Rodríguez e Agustín [...] estantes en esta corte [...]

(fol. 45 r)

Espuelas doradas

-Primeramente se tasó quatro pares de espuelas doradas de la brida a siete reales cada par unas con otras

-Yten tres pares de espuelas de color de hierro se tasaron a cinco reales cada par las unas con otras

-Yten un par de espuelas pavonadas nuevas en seis reales

-Yten un par de espuelas de media posta barnizadas en seis reales

-Yten otras quatro pares de espuelas unas con otras a quatro reales el par

-Yten tres sobrevaynas de espada a tres reales una con otra

E después de lo suso dho en la villa de Madrid a diez días del mes de ene^{o} de myll qui^{o}s e setenta y dos años Ordóñez librero de esta arte tasó los libros que quedaron del dho S^{r} duque en la forma sigui^{e} [...].

FICALLO, CONDES DE

Doc. 1.
1600, June 27.
Inventory of Juan de Borja, Count of Ficallo.
Archivo Histórico de Protocolos de Madrid, Protocolo 933.

(fol. 379 r)
27 de junio de 1600 aºs
Su Mag^{t} ynbentario de los bienes de don Juº de Borxa

En la villa de Madrid a v^{te} e siete de días del mes de junio de mil e seiscientos años estando en las casas que se nombran de Buenavista que fueron de Don Juan de Borxa conde de Ficallo comendador de Çuaga que es de la orden de Sanctiago q las dhas casas son en esta dha villa en las parroquias de San M^{n} e San Xinés a los caños de Alcalá por ante mí el scrivº pu^{co} el señor Sebastián Hurtado behedor y contador de las obras Reales de Su Mag^{d} del Alcáçar de esta dha villa y casa Real del Pardo= dixo q las dhas casas y lo que en ellas está de menaxe de casa, pintura [...]treros y otras cosas de diferentes suertes es de Su Mag^{d} del rei nro señor e por haçienda suya e para que de ello se haga ynventario en presencia del dho Sebastián Hurtado en nombre de Su mag^{d} conforme a lo acordado por los señores de la junta de obras e bosques ques por decreto de v^{te} y çinco días de este dho mes de junio y año de mil y seisçientos que esta rubricado de los dhos señores de la junta y el dho decreto quedó en poder de los dhos Esteban Hurtado Arias de Abrea en n^{e} del dho don Juan de Borxa como persona q los a tenido a cargo declaró e manifestó las cosas sig^{tes}

En una pieça de la dhas casas que nombran el estudio había
(fol. 379 v)
-Un lienço grande con su marco dorado e su corrediça de tafetán carmesí que es un paso de la Pasión quando alçaban a nro señor del suelo [...]

(fol. 381 r)
En el camarín de las armas hay lo siguiente
-Quatro rodelas de la China
-Dos alabardas doradas aforradas en tzºpelo verde
-Dos hachetas Húngaras guarneçidas de plata
-Otra cubierta de çapa
-Quatro hachetas húngaras con palos blancos

-Dos maças de armas húngaras
-Una espada ancha con su canal dorada y la guarnición de hierro nyelada
-Dos coldres (sic) turquescos guarnesçidos de plata dorada
-Una espada de montear
-Dos hachetas de Alemania con sus espadas dentro
(fol. 381 v)
-Una espada con dos hoxas nyeladas de Alemania
-Dos alfanxes
-Otro alfanxe damasqueño a manera de cuchilla
-Otro alfanxe pequeño con un cabo de marfil
-Una daga con su puño dorado
-Otro alfanxe con la guarniçión plateada
-Otra espada a lo antiguo dorada
-Una daga que sirve de estrolago (sic)
-Una espada que sirve de estrolago (sic)
-Una ballesta de virotes
-Otra de bodoques labrada de marfil
-Una caxa de herramienta de huerta
-Una alxaba de cuero con sus xaras
-Tres arcabuces de Alemania con sus fundas de vaqueta
-Dos pistoletes el uno de dos llaves y el otro de una con sus fundas de tr^{o}pelo
-Un murrión dorado
-Dos estribos húngaros con sus espuelas

En la galería había las cosas sigui^{tes}
-Un retrato del rey don F^{do} el Católico
-El retrato de la reina doña Ysabel
-Otro del Rey don Felipe primero
-Otro de la rey^{a} Dña Juana
-Otro del Emperador don Carlos
-Otro de la Emperatriz su mujer
-Otro del rey don Felipe segundo
-Otro de la reyna doña Ana su mujer
-Otro del rey don Ju^{o} tercero de Portugal
(fol. 382 r)
-Otro de la rey^{a} doña Catalina su mujer
-Otro del rey don Al^{o} de Nápoles
-El rey don Alonso el Sabio
-El rey don Enrique de Portugal que fue cardenal

-Otro del rey don Alonso de Nápoles
-Otro de Roberto rey de Nápoles
-Renato rey de Nápoles
-Matías rey de Hungría
-Carlos duque de Borgoña
-Otro de Carlos duque de Borgoña
-Otro de Jacob rey de Escoçia
-Godofe Gullón (sic)
-Otro de Tutila
-Otro de Tristán de Acuña portugués
-Munseur del Lautracha
-Otro del marqués de Pescara
-Próspero Colona
-Otro de Antonio de Leyba
-Otro de Colón
-Otro de Andrea Doria
-Otro de Her^{do} Cortés
-Otro de don Fer^{do} de T^{do} duque de Alba
-Otro del marqués del Gasto (sic)
-Otro de Fr^{co} Gonzaga
-Jacobo marqués de Mariñán
-Pº Estolfa
-Otro de Vorgan
-Otro del Gran Capitán Gº Frez de Córdoba
-Lorenço de Médiçis
-Otro de Agustino Barbárico cap^{an}
-Otro de Ugo conde
(fol. 382 v)
-Otro de Pandulfo tirano
-Otro de Gatamelata
-Otro de Basilios prinçipe de Mascolva (sic)
-Otro de Pº Meloques
-Otro de Ermolios Barbolo
-Otro de Sagadinos sultán
-Otro de Lorenço de Médicis p^{e} de León déçimo
-Otro de Julián de Médiçis padre de Clem^{te} octavo
-Otro retrato de Oraçio Doria
-Otro de Br^{me} Cordero
-Otro del duque Balentino

-Lucreçia de Borxa duquesa de Ferrara
-Otro del cardenal de Guissa
-Otro de Enrique duque de Guissa
-Don Juº de Borxa duque de Camarin
-Doña Luisa de Borxa duquesa de Villahermcsa
-Otro del duque de Gandía don Fran^{co} de Borxa
-Don Juan de Borxa conde de Ficallo
-Otro de doña Fran^{ca} de Aragón su mujer
-Otro del cardenal don Enrique de Borxa
-Otro de don Juº de Borxa marqués de Nabange
-Otro de don Alonso de Borxa
-Otro del maestre de Montessa
-Otro de Solimán
-Otro de Xurife rey de Marruecos
-Otro del gran Taborlan (sic) rey de Tartaria
-Otro de Xácome de Treço
-Otro de León Alberto
-Otro de Magalean
-Jacobo Salazar
-Otro de Troca Totas
-Otro de Ludribico Canoso Danti
-Otro de Juan Bocaçi
(fol. 383 r)
-Otro retablo Cinos Pistoriensis
-Otro de Picos Mirandulanus
-Otro de Alexandro Ybolenesis
-Otro de Manos Arcuçis
-Otro del dotor Navarro
-Otro de Paulo Castrensis
-Otro de Valamos
-Otro de Bartulus
-Otro de Paulo Jobio
-Otro de Niculao de Saxi
-Otro de Laurencios Patriarca de Venençia
-Otro de Tomás Morus
-Otro de Fran^{co} Sebier
-Otro de Ynaçio de Layola
-Otro del cardenal de Médiçis
-Otro del cardenal Ganbela (sic)

-Otro de Systo Quinto
-Otro de Greg° déçimo terçio
-Otro de Pío quinto
-Otro de Pío quarto
-Otro de Paulo quarto
-Otro de Marçelo segundo
-Otro de Julio terçio
-Otro de Paulo terçio
-Otro de Clemente sétimo
-Otro de Adriano sesto
-León dézimo
-Jullio segundo
-Otro de Pío terçero
-Alexandro sesto
-Otro de ynoçençios otavo
-Systo quarto
-Otro de Paulo segundo
(fol. 383 v)
-Pío segundo
-Otro de Calisto terçero
-Otro retrato de la princesa doña Juana entero
-Otro de la Sª ynfanta doña Ysabel también entero
-Otro de la infanta doña Catalina de la misma manera
-Otro del príncipe don Carlos
-Otro de don Juan de Austria
-Otro del Emperador Maximiliano en un óvalo redondo
-La emperatriz su mujer de la misma mª
-Otra de doña Antª hija de don Grª de Alvarado entero
-Un retrato pequeño de Alberto Duredo (sic)
-Quatro tablillas de Flandes de los quatro sentidos
-Doçe tablillas de villanas alemanas
-Doçe ynpresas de las de don Ju° de Borxa
-Los quatro tiempos del año en quadros pequeños
-Çinco quadros de baçan las tres con marcos dorados y las dos syn marcos
-Un quadro con un cardenal y un viejo con él
-Tres tablas de vendedoras de fruteras y pescaderas de Milán
-Otro lienço de la coçina de Flandes
-Tres cabeças de la gran bestia y los cuernos naturales
-Otras tres cabeças de venados

-Tres bolas de candiles de açofar
-Unos morillos de açofar en una chimenea

(fol. 384 r)
En la pieça de más afuera que cae sobre la huerta nueva había lo sig[e]
-Un retrato del archiduque Alberto con un tafetán carmesí
-Otro retrato del archiduque Alberto
-Un retrato de una hija del duque de Florençia
-Otro retrato de Catalina Rey[a] de Chipre

En la saleta junto a la escalera principal había

-Diez lienços de Flandes con sus marquillos de madera

En una sala grande más adentro había
-Veynte y seis cabeças de emperadores
-y dos de Eradico y de Modico
-Un muchacho pintado en un lienzo que sopla un tiçón
-Seys quadros de Flandes con sus marcos dorados- Los tres de hombres de la fama e los tres de mujeres de la fama
-Quatro quadros de paysaxes de Flandes con sus marcos dorados
-Diez cabeças de venados con sus cuernos naturales
-Dos bolas de candiles de açofar
-Seys sillas de cuero de respaldar
-Dos morillos de açofar con su paleta y tenaça
-Un palanquín de la China
-Una mesa ochavada de xaspe con sus pies de madera
-Un lienço de Flandes sobre la chimenea [...]

FUENSALIDA, CONDES DE

Doc. 1.
1593, March 8.
Inventory of the armoury of Count of Fuensalida in Guadamur Castle.
Archivo Histórico de la Nobleza, Frías, C. 24, D. 113.

(fol. 4 v)
E después de lo susodho en la dha villa de Guadamur a ocho días del dho mes de março del dho año de mill e quiºs y noventa y tres años el dho Tomás de Gándara por vrd de la dha comisión en presencia del dho Juº de Terán alcayde de la dha fortaleza hizo inventario de los tiros armas y ballestas paveses y otras cosas q están en la dha fortaleza que es hacienda de Su Señoría e de todas las cosas q se hallaron se hizo el inventario sigte

-Primeramente una ymajen de Nra Señora grande q está al olio en una tabla
-ytem dos culebrinas de bronce con su carretoncillo
-ytem quatro tiros de hierro colado todos ellos con una culata y caballos de madera de toco de naranjo poco más o menos
-ytem dos falconetes
-ytem quinze escopetas y quinze cañones de escopetas
-ytem catorce ballestas grandes y dos cureñas y más quatro cureñas q son seis
(fol. 5 r)
-ytem dos gafas de tornillo de palo y siete otras de hierro todas de garracha
-ytem cinco paveses y una cadena gorda y un yunque
-ytem diez y ocho pelotas de piedra para tiros
-ytem un relox qbrado un hierro de alabarda
-las pieças para armar dos personas
-unas coraças viejas
-veinte saetillas de palo sin casquillos
-ytem sesenta y tres grebas y braçaletes
-ytem treinta golas y veinte y tres pieças grandes de armar
-ytem otras muchas pieças de armas muy pequeñas y una manopla y una testera de caballo
-ytem un çepo de palo y unos grillos
-ytem dos atabales de cobre
-ytem nueve balas grandes de piedra para tiros en los cubos del terrado
-ytem un retrato de Su Señoría en la sala a mano derecha encima de la puerta
(fol. 5 v)

-ytem dos atahonas con sus tolvas e piedras a palos para medallas (sic) salvo un palo q falta
-ytem dos falconetes q hay más en la pieça de abaxo y un pavés
-ytem un cajón grande sin suelo para tener trigo o harina
-ytem un candado de la sala de las armas con una llave hueca
-ytem más tres llaves del castillo grandes

De los quales dhs bienes el dho Juº de Terán se dio por entregado en ellos y se obligó de los dar e volver a Su Sª [...] e quando y en qualquier tiempo q Su Sª se los pidiere debaxo de dho pleito o [...] e juramento q tiene hecho e obligación q hizo de su persona e bienes habidos y por haber dio poder a las just^{as} renunº las leyes q sean en su favor [...]

Doc. 2.
1637, March 31.
Order issued by the king to the Count of Fuensalida for the delivery of arms.
Archivo Histórico de la Nobleza, Frías, C. 26, D. 23.

El Rey
Conde de Fuensalida pariente los avisos de las prevenciones que se hacen en Francia para recobrar los puestos que en la provincia de Labort han ocupado mis armas y inquietar las provincias confinantes a ella obligan a disponer lo necesario para su defensa con suma brevedad, y porque una de las cosas que más lo dificulta es la falta que hay de armas en particular para la caballería que se está levantando por las muchas que de las fábricas y magaçenes de estos reynos se han sacado y son menester para las ocasiones presentes aunque se ha prevenido mandándose se traygan de fuera de España y porque en tanto que llegan conviene socorrer esta necesidad y he entendido tenéys algunas, os encargo hagáis entregar a la persona que nombrare el Marqués de Castrofuerte de los mis consejos de Guerra y Indias Capitán General de la artillería en ínterin y veedor General de la caballería en España todas las pistolas carabinas arneses coraças y otras armas de a caballo que tuviéredes tomando recibo de él a persona que el Marq^{s} enviare a conducirlas con declaración de la calidad y cantidad de cada género para que pasada la ocasión se os restituyan en la misma forma con toda promptitud lo qual executaréis entregándolas sin dilaçión alguna por lo que conviene que no la haya en la dispusiçión de la defensa de estos reynos y estoy cierto de vra atención y çelo de mi servicio que con vra asistencia se ha de repar (sic) el daño que podría resultar a estos reynos si faltasen estas armas para su defensa y avisaréis de lo que dispusiéredes para tenerlo entendido, de M^{d} a 31 de Març o de mil y seysçientos y treinta y siete años.

Yo el Rey
(signed)

Por manto (sic) del Rey nro Señor [...]

(fol. v)
El Marq^{s} de Castrofuerte de los Consejos de la guerra y Indias de Su Mag^{d} su veedor g^{l} de la cab^{ría} comis^{o} g^{l} de la Inf^{a} y Capitán gen^{l} de la Artillería
En conformidad de lo que Su Mag^{d} manda por su R^{l} cédula retroescrita nombro a Juan Pérez para que reciba las armas que por orden del señor Conde de Fuensalida se le entregaren en virtud de ella para el efecto que en la dha cédula

y para que las haga conducir y llevar de la parte que las recibiere a la villa de Fuenterrabía según y en la forma que le he ordenado en este día q habiendo tomado la raçón de ella y de la presente Ventura de Frías S^{rio} de Su Magd y su contador de la raçón gen^{l} de la Artillería y carta de pago del dho comis^{o} Ju^{o} Pérez serán bien dadas sin otro recado alguno M^{d} dos de Abril de mil seisçientos treinta y siete años

El marqués de Castrofuerte
(signed)

Tomo la razón de la c^{a} de su md y de esta orden

Ventura de Frías
(signed).

Doc. 3.
1637, May 19.
Receipt for delivery of arms from the Count of Fuensalida.
Archivo Histórico de la Nobleza, Frías, C. 26, D. 31.

(fol. 1 r)
En la villa de M^{d} a diez y nueve días del mes de mayo de mil y seisçientos y treinta y siete años ante mí el escribº y testigos paresçió presente Juan Pérez esº de esta villa de Madrid y residente en ella= Comisario nombrado por el S^{r} marqués de Castrofuerte del Consejo de Guerra de Su Mg^{d} y Capitán General de la artillería de España que le nombró en virtud de la çédula real que el dho S^{r} marqués tiene del rey nro Señor= para resçibir las armas espresadas en la dha çédula firmada de su real mano refrendada de don Fernando Ruiz de Contreras su secretario su fecha en Madrid a treinta y uno de março de mil y seisçientos y treinta y siete= y el nombramiento fecho por el S^{r} marqués en el dho Juan Pérez parece fue en esta villa de M^{d} a dos días del mes de abril de este dho año= de que tomó la raçón Ventura de Frías conforme a lo que por la dha real çédula se manda= y otorgó que usando de la dha comisión a él dada= que resçibió del Sr Conde de Fuensalida Gentilhombre de la cámara de Su Mag^{d} por mano del Capitán Juan Ortiz su caballeriço residente en esta corte treinta querpos de cosoletes (sic) de infantes blancos y listados de negro que cada coselete tiene las pieças cabales que son las siguientes=

Primeramente= murrión= gola= peto (fol. v) y espaldar= escarçelas= braçales= cañonçetes= manoplas= todo bien acondiçionado sin faltar un clavo= ni una hebilleta todo a satisfaçión del dho Juan Pérez= todo lo qual recibió en virtud de la dha çédula real y comisión a él dada en presençia de mí el escribº y testigos de esta entrega doy fe= como el dho cap^{án} Juan Ortiz caballeriço del dho conde de Fuensalida dio y entregó al dho Juan Pérez como tal comisario todos los dhos treinta querpos de armas en la forma y manera que va declarado y los pasó a su poder realmente y con efeto= y como contento y entregado otorgó esta carta de pago y recibo en bastante forma en favor del dho S^{r} conde de Fuensalida por la dha mano del dho capitán Juan Ortiz de lo que fueron testigos don Juan de Rivera y Juº de Casares y don Alonso Hernández vzºnos y estantes en la corte y el otorgante que doy fe conozco lo firmó de su nombre= Juan Pérez= ante mí Ýñigo López de Salaçar= e yo el dho Ýñigo López de Salaçar escribº del rey nro Señor residente en su corte y provincia y en fe de ello sine de mi sino (sic)

En testimonio de Verdad
Ýñigo de Salaçar
(signed)

Reçibo de las armas que se llevaron a Fuenterrabía.

Doc. 4.
1641, January 11.
Order issued by the king to the Count of Fuensalida for the delivery of arms.
Archivo Histórico de la Nobleza, Frías, C. 26, D. 24.

El Rey
Conde de Fuensalida primo la falta de armas que hay en estos Reinos para las ocasiones que se ofreçen es grande y porque se ha entendido que tenéis cantidad de armas ha parecido encargaros que luego que reçiban este despacho den la orden neçesaria pª que la persona que las tuviere a su cargo las entregue a la que nombrare el Marqués de Castrofuerte de mi Consejo de Estado y Capitán general de la Artillería en el ínterin por inventario en que se declare el número y calidad de ellas para que acabada la ocasión se os puedan restituir que en ello reçibiré de Vos servicio particular, de Madrid a onze de enº de *1641*

Yo el Rey
(signed)

Por mand^{do} del Rey nro Señor
Don Juº Ruiz de León
(signed)

Al C^{de} de Fuensalida Ser q entregue las Armas q tuv^{e} pª la ocass^{ón} pres^{te} de Portug^{l}.

Doc. 5.
1662, January 21.
Inventory of Count of Fuensalida made in Madrid.
Archivo Histórico de Protocolos de Madrid, Protocolo 9801.

Inventario de los bienes del Sr Conde de Fuensalida Don Bernardino López de Ayala Velasco y Cárdenas. 1662
Juez el s^{or} liç^{do} Don Fr^{co} Ramírez

(fol. 16 r)
Doña Ana María de Velasco Condesa de Siruela y Fuensalida viuda de D. Ber^{no} de Velasco y Cárdenas Conde de Fuensalida su marido y como una de sus testamentarios y en n^{e} de los demás= digo que el dho Conde mi marido murió como es notorio y conviene hacer inventario de todos los bienes que quedaron por su fin y muerte en esta villa y otros lugares para que en todo tiempo conste a los herederos y acreedores de los que aquesto hubiere montado= Pido y sup^{co} a Vm mande se haga el dho inventario y que se cometa al pres^{te} escribano o a otro qualquiera de Su Mag^{d} para que lo ex^{te} con citación de los dos testamentarios y herederos dando principio a él dentro del término de los nueve días inmediatos a la muerte del dho Conde mi marido pido just^{a} &^{a}=

La Condesa de Siruela
(signed) [...]

(fol. 16 v)
Inventario
En la villa de Madrid a veinte y uno de en^{o} año de mill y seis^{o} y sesenta y dos estando en las casas donde vivió el Exmo Sor Conde de Fuensalida y Siruela estando en el cuarto del Sr Conde de Fuensalida su hijo se empeçó el inventario de los bienes del dho Sr Conde difunto en la m^{a} sig^{te} [...]

(fol. 25 r)
Prosigue el invent^{o}
En la villa de Madrid a catorze días del dho mes de febrero del dho año estando en las casas donde murió el dho Sor Conde de Fuensalida por ante my el escribano se fue continuando el dho inventario en la forma siguiente [...]

(fol. 27 v)
Recámara
Un cofre de vaqueta colorada forrado de bocací açul y en él lo sig^{te} [...]
-Dos pistolas de Flandes iguales doradas
-Un adereço de espada y daga de almenillas pavonado de (fol. 28 r) blanco hecha en Madrid con su funda colorada colorada (sic) hoxa de Toledo [...]
-Una guarniçión de espadín de memorias con su puño y pomo
-Una guarniçión de espada labrada con puño pomo y contera
-Una llave de arcabuz de Simón suelta
-Unos tiros y pretina de felpa negra
-Otros tiros y pretina de tela bordado de tercal negro [...]
-Dos cordones açules y uno pardo p^{a} bolsas de arcabuz [...]
-Otro cofre con dos cerraduras cubierto de encerado y aforrado en holandilla amarilla y en él lo siguiente
-Un adereço de espada y daga liso blanco de la puerta cerrada con hoxa de tres cortes con pretina y tiros
-Otra espada con guarnición vizcaína y la hoxa de Juanes de la Orta [...]
(fol. 27 v)
-dos espadines con guarniciones de plata con hoxa el uno más pequeño que el otro [...].

GANDÍA, DUQUES DE

Doc. 1.
1588, January.
Inventory of Carlos de Borja, 5th Duke of Gandía.
Archivo Histórico de la Nobleza, Osuna, C. 568, D. 127.

(fol. 1 r)
Gandía, 2602, 5to Duque D^n^. Carlos. Inventario de sus bienes. Año 1588, desde enero.
Copia simple del inventario de bienes que quedaron por muerte del S^or^ D^n^. Carlos de Borxa, 5to Duque de Gandía, tocantes a la Guardaropa, Caballeriza, Cámara, Plata y Ropa blanca que estaban en Gandía, y Castellón, en poder de algunos criados. Fecho ut supra.

(fol. 2 r)
Gan^a^ n. 2602
Ynventario de cosas de guardaropa y de la cámara y de la caballeriza y plata y ropa blanca y otras cosas que están aquí en Gandía y Castellón en poder de algunos criados que lo tienen a cargo. Hecho en el mes de henero 1588

A 28 de Abril 1588 me dio este ynventario el canónigo Ambros (sic) Martínez a mí el Cont^or^ Melchor Clua

(fol. 3 r)
Armería
C^tas^ de Ortega
Gandía
-Yten tres testeras de caballo viejas
-yten seis mazas de armas a manera de martillos que se trageron con los coseletes de Mylán
-yten otra maza de armas con unos bacíos en medio
-yten un estoque con un puño de plata no tiene contera
-yten seis morriones grabados que ynbió el capitán Bernia
-yten seis sillas de armas muy antiguas, están las quatro en Castellón y no pareçen
-yten en el amería hay las piezas que son menester por veynte y quatro sillas de armas las quales trageron con los coseletes de Mylán en el año 1566
-yten tres adargas viejas

-yten dos atanbores
-yten tres pares de atabales
-yten quatro sobre petos viejas falta de una el broçel y una pieza
-yten una corneta con un cordón de seda
-E más dos tiendas y dos pabellones de campo de cotonia blanca y madera por todo
-yten un estoque pequeño con una vayna de terciopelo blanco muy viejo
-yten dos espadas antiguas la una que está quebrada
-yten un montante de esgrima
-yten una ballesta vieja
Castellón
-E más en Castellón un venablo con dos arcabuçillos diçen se llevó a Castilla con los pilares
Castellón
-yten una partesana con tres puntas está en la guardarropa de Castellón
Castellón
-yten en la recámara en Castellón un alfanje muy pequeño con vayna colorada y cabos de latón

(fol. 3 v)
Armería
Ortega
Castellón
-yten en la guardarropa de Castellón seis partesanas más guarneçidas de verde
Gandía
-Yten en la guardarropa de Gandía ocho partesanas las quales dio el capitán de Bernia a su Exª
Gandía
-yten tres venablos pequeños blancos el uno tiene unas berrugas
Castellón
-yten en Castellón doçe lanzones medianos
Gandía
-yten un lanzón pequeño que dio el Sor Don Al^{o}
Gandía
-yten otros dos lanzones medianos bastardos el uno está quebrado
Gandía
-yten en los torrejones tienen las guardas otro lanzón

Castellón
-yten tiene Viçente de Marques una alabarda dorada con terciopelo azul
Picas
-E más en la guardarropa de aquí de Gandía y con veynte y quatro que ha de haber en la sala de Castellón, quatroçientas picas hanse contado y faltan diez y ocho que se han quebrado desde el año 66 hasta hoy y han aprovechado pa cargadores del artillería y pa barrear la garropa (sic) en jaraco (sic) y pa otras cosas
Lanzas ginetas
-E más aquí en Gandía veynte lanzas ginetas algunas sin hierros y en Castellón ha de haber otras cinco y otra qbrada
Lanzas de justa
-yten en la guardarropa diez y ocho lanzas de justa
Lanzas de torneo
-yten en la guardarropa doçe lanzas de torneo de a pie

(fol. 4 r)
Guardarropa/Ortega
Estandarte
-yten un estandarte de tafetán leonado y azul con sus cordones de la mesma color
Cuera de ante
-yten una cuera de ante doble con su jubón de lienzo de armar y un sonbrero fuerte aforrado de tafetán negro
Extrumentos (sic) de música [...]
Gandía/Ortega
-E más tiene Ortega en Gandía dos tiendas de campo grandes y un tendejón pequeño con su madera vieja [...]

(fol. 14 r)
Arcas y cofres
Gandía
-yten una arca en q están los flascos (sic) de arcabuçes que se abre la cubierta a dos partes eran dos y la una se perdió en el campo de Génoba
Gandía y Castellón
-yten cinco cofres blancos en que vinieron las armas de Mylán eran seis y el uno se quedó en las galeras por no poder venir en carga las dos están en Castellón
Gandía y Castellón
-yten siete arcas encoradas viejas las dos están en Gandía y quatro en Castellón y una se dio a don S° por m^{do} de su Exª
Castellón en la guardarropa

-E más tienen las mujeres en la guardarropa de Castellón dos cofres con aderezos de camas

(fol. 16 r)
Aderezos de la caballeriza
Gandía
Ortega en Gandía
-Primeram[te] tiene Ortega dos pares de estribos dorados de la gineta el uno es diferente del otro
-yten dos guarniciones de paño de lucto pa caballos faltan riendas
-yten tres pretales y tres guruperas de terciopelo negro pa caballos viejas
-yten quatro almártegas de terciopelo negro muy viejas con sus cordones de seda las tres de ellas
-yten quatro sotacolas de caballos
-yten dos petrales (sic) el uno de tela de oro viejo y el otro de cuero blanco
-yten otro de cuero blanco
-yten un pretal y unas cabezadas de tela de oro colorado viejas
-yten una almartega de cuero negro y otro pretal viejo
Gandía/ojo
-E más en Gandía tres sillas de armar de cuero blanco faltan otras tres q han de estar en Castellón
-yten (sic)

(fol. 16 v)
Aderezos de caballeriza en Castellón
Sillas de caballos/ Castellón
-Primeram[te] una silla pa la haca en q va su Ex[a] con un caparazón y su guarnyçión de cuero de tigre
-yten otra silla pa el macho q era de Méndez con su guarnición
-yten dos sillas de picar pespuntadas con guarniciones blancas
-yten otras dos sillas viejas pa las hacas con sus guarniciones
-yten dos sillas blancas pa caballos de caza con guarniciones blancas
-yten una silla de vaqueta pa el macho/ no tiene guarnición
-yten un sillón pa las burras sin guarnición
-yten una silla blanca pa caballo de caza
-yten una silla pa el machuelo sardo con su guarnición
-yten tiene Méndez una silla gineta con su guarnición diçe Megía que no la ha vuelto

En la guardarropa de Castellón
-yten una silla nueva con fajas de cuero bayo pespuntada con seda morada es blanca
Castellón
-yten otra silla nueba blanca con fajas de cuero negro pespuntada de amarillo y clavazón dorada
-E más en la guardarropa en Castellón dos sillas nuevas de cordobán amarillo
-yten en dha guardarropa otra silla de cordobán gamuzado con fajas de cuero negro pespuntada de seda amarilla nueva
-yten en dha guardarropa otra silla de cuero negro con sus fajas de terciopelo negro y franjas de seda ya traída
-yten dos sillones pa mujeres
-yten dos pares de anganillas (sic)

(fol. 17 r)
Adrezos de caballeriza en Castellón
Guardarropa
-yten en la guardarropa unos estribos de posta
-yten una guarniçión de la brida y unos estribos dorados la guarniçión de terçiopelo negro
-yten quatro gualdrapas de paño con fajas del mismo paño
-yten otra gualdrapa de paño con fajas de terçiopelo
-yten tres gualdrapas e terçiopelo negro buenas
-yten siete cadenas pa atar caballos
-yten ocho çinchas pa mantas de caballos

En la Recámara de Su Ex[a] adrezos
En poder de Ysabel Ana/ Castellón
Castellón
-Primeram[te] en un cofre se halló un caparazón de la gineta de terciopelo carmesí bordado y aforrado en raso azul
-yten otros dos caparazones de la gineta de terciopelo negro
-yten unos estribos dorados con esmaltes muy buenos son de los antiguos
-yten otros estribos de la gineta dorados con esmaltes muy buenos
-yten otros estribos de la gineta viejos y con esmaltes
-yten otro par de estribos dorados de la gineta
-yten otro par de estribos de la gineta plateados
-yten unas espuelas negras de la gineta

-yten otro par de espuelas plateadas de la gineta
-yten unas cabezadas de la gineta con hierros y borlas coloradas y un pretal de la gineta de terçiopelo negro con una reata y tafetán pa el caballo
-yten otras cabezadas de la gineta con clavazón muy buena
-yten tres çinchas de la brida por guarneçer
-yten más en la recámara unos estribos de la brida dorados con açiones de terçiopelo negro

(fol. 17 v)
Caballeriza
En cofres de la cámara
Castellón
-yten en un cofre se hallaron nueve guarniciones de caballos de la brida muy buenas con todo lo que es menester
-yten dos pares de cabezadas con sus riendas pa la gineta
-yten dos almártegas muy buenas pespuntadas de diferentes colores con sus cordones
-yten dos pares de espuelas de la brida
-yten una bolsa de caza de terciopelo carmesí digo de telilla amarilla y colorado y el señuelo es de dho terçiopelo carmesí
-yten otra brujaca de caza de telilla colorada no hay señuelo
-yten unas cabezadas de cuero amarillo con sus riendas
-yten unas çinchas de la brida de telilla colorada y amarilla y blanca con sus hierros
-yten cinco o seis dpas (sic) de xaras por labrar
-E más en otro cofre en dha recámara una guarnición de cuero bayo aforrada de cuero de lobos marinos
-yten dos pares de riendas bayas con un petral de cuero de […]
-yten otro adrezo de cuero negro pa mula o macho no hay riendas
-yten una guarniçión de cuero de tigre negra
-yten dos guarniçiones de terçiopelo negro con una almartega son ya viejas
-yten dos pares de açiones y riendas de vaqueta blanca
Recámara
Frenos en la Recámara
-Hallóse más en la recámara sesenta y un frenos de la brida grandes y pequeños el uno de ellos dorado
-E más de hallaron otros siete frenos de la gineta

(fol. 18 r)
Cueros de venados y vaquetas y cordobanes en Castellón
En la guardarropa
En Castellón [...]
(fol. 19 r)
Cofre nº 2 en Castellón [...]
(fol. 19 v)
-yten diez pares de espuelas de la brida [...]
Cofre nº 7
-E más dos talabartes de terciopelo negro [...]

(fol. 20 r)
Cofre nº. 8
Castellón [...]
(fol. 20 v)
-yten un cuchillo de monte con guarniçión dorada y un puño de jaspe pardo y blanco
-yten otro cuchillo de monte con una vayna verde con dos cuchillos los cabos de cuernos de benado
-yten dos cuchillos turquescos con una vayna negra [...]

(fol. 24 r)
Castellón
Recámara fuera de los cofres
-Primeram[te] una espada pavonada con vayna de terçiopelo negro sin talabarte
-yten otra espada con guarniçión dorada con vayna de terçiopelo negro con su talabarte
-yten otra espada pavonada la guaniçión ochabada con una vayna de zapa con talabarte de cuero negro
-yten otra espada plateada con vayna de vaqueta baya con su talabarte
-E más dos muletas y otros muchos juncos y báculos
-yten una rodela muy buena dorada señalada con un escudo de las barras de Aragón con terçiopelo y de por de dentro
-yten un flasco (sic) turquesco pa agua pintado de muchas colores
-yten un saquillo dos maneras de hierros
-yten una brujaca verde con su señuelo de halcones
-yten un carcax lleno de flechas turquescas
-yten una bolsa grande de cuero bayo con correas de venado
-yten otra bolsa grande de cuero colorado

-yten dos cajas redondas de hoja de Milán y dentro unos retratos de los papas
-yten un papel con muchas líneas
-yten tres bolsillas turquesas pa beber agua
-yten diez y seis çirios grandes dorados y otros pequeños todos blancos y dorados
-yten un flasco (sic) de barro blanco de Talabera
-yten unas alforjas verdes
-yten un chapín pa agua
-yten una bolsa de caza con dos señales la bolsa es de telilla blanca y colorada
-yten otra brujaca de caza de estameña
-yten una bolsa de terçiopelo negro pa libros
-yten dos calzadores de latón
-yten dos caretas pa cortar colmenas

(fol. 24 v)
Castellón
-yten un axedrez con sus piezas
-yten un cuchillo grande de monte con su çinta de seda verde [...]

(fol. 25 r)
Cámara/ Castellón
-yten un espejo grande de açero
-yten tres ballestas con sus gafas
-yten un arco de bodoques con su funda de lienzo [...]
-yten seys cajuelas con capirotes de halcones
-yten un agedrez (sic) de marfil con sus piezas [...]

(fol. 39 r)
Guardarropa
Cargo de Ortega
Artillería
-Primeram[te] dos piezas de artillería que se nombran sacres el uno se nonbra Sanctiago y el otro el de los Cascabeles-2
Ortega
-yten otras siete piezas largas que se nonbran medios sacres-7
Ortega
-yten quatro pieçeçillas largas que se nonbran falconetes-4
Ortega
-yten otras quinçe piezas cortas que se nonbran cañones pedreros que se hiçieron en Val[a]- 15

Castellón
-E más en Castellón hay quatro pieçeçillas largas que se nonbran esmerejones-4
Castellón
-yten en Castellón otra pieçeçilla de hasta tres o quatro palmos de largo como pedrero-1
Gandía
-yten en la guardarropa diez y nueve mosquetes de hierro y de metal con siete flascos (sic) grandes-19
Gandía
Hierro/Ortega
-yten de hierro viejo de artillería y pa ruedas de carro y pa trapichos hasta siete u ocho quintales de hierro
Ortega
-yten una naveçilla de bronze pa fuente esta en casa de mrs Viçente Calderero

(fol. 39 v)
Arcabuçería
A c[ta] de Ortega en Gandía
Barreros Verdeguer
-Primeram[te] a c[ta] de Myn de Ortega çiento y çinc[ta] y seis arcabuçes de munición, están en Castellón en la guardarropa a c[ta] de Viçente Marques seis de ellos -156
Ortega
-E más pa los arcabuçes tiene çiento y quarenta flascos (sic) grandes de ellos de terçiopelo y de ellos de cuero y otros çiento y quarenta flascos (sic) pequeños pa dos arcabuçes
-E más ochenta bolsas de terciopelo negro pa dos arcabuçes son viejas.
Guardarropa
Castellón
-E más hay en la guardarropa que tienen las mugeres en Castellón seis arcabuçes largos de pedernal y mecha
Castellón
-yten otras dos caxas de arcabuçes sin cañones
Castellón
-yten dos llaves de arcabuçes y un flasco (sic) grande biejo y un flasquillo de hierro
Recámara
Cofre nº 20
-E más se hallaron en la recámara de Su Ex[a] çinco arcabuçes de pedernal de marca menor con dos flascos (sic) de macho montés y un flasquillo de hierro y una llave

-yten un cuchillo de monte con otros dos cuchillos y un tenedor
-yten otro cuchillo de monte con una vayna verde con un cuchillo y punzón
-yten dos dagas la una pavonada y otra dorada
-yten dos candadicos pequeños
-yten tres piedras pa amolar
-yten dos maletas de cordobán negro viejas
-yten dos flascos (sic) de arcabuz de terçiopelo amarillo
Recámara
En Castellón
-yten un flasquillo de hierro y otro de madera [...]

(fol. 42 r)
Pólvora y plomo
A c^{ta} de Ortega
-Primeram^{te} en la torre de la pólvora se halló hasta ochenta arrobas de pólvora de arcabuçería y artillería granujada (sic)
Gandía
-E más en dha torre hasta quarenta y ocho çinq^{ta} arrobas de pólvora de artillería sin granujar (sic)
Gandía
-yten en dha torre habrá hasta quatro o cinco quintales de açufre
Gandía
-yten en dha torre hasta tres doçenas de linternicas pa el artillería de los cañones pedreros
Gandía
-yten en dha torre hasta catorce arrobas de mecha pa arcabucería
Gandía
-yten en dha torre hasta 650 pelotas de hierro y plomo de diversos moldes
Gandía
-yten en dha torre otras 650 pelotas de piedra grandes y pequeñas
Gandía
-yten dha torre quatro moldes de pelotas pa el artillería
Gandía
-yten en palaçio hay doçe cajas de artillería grandes y pequeñas
Gandía
-yten algunas ruedas de dha artillería
-yten otras doçe cajas pa los cañones pedreros
Plomo
Gandía

-E más en la guardarropa de Gandía se halló hasta setenta arrobas de plomo poco más o menos en 17 tortugas algunas comenzadas
Hierro/
En Castellón
-yten se halla en Castellón una bigornia de hierro y más dos martillos y unas tenazas y unos fuelles pa la fragua y un tornillo de hierro
Castellón
-yten arriba en la guardarropa de mugeres una caja con herraduras y una bigornia pequeña y un tornillo y tenazas y otros hierros pa çerraduras y una sierreçilla pequeña
Castellón
-yten se halló hasta dos quintales de hierro viejo pa carros

(fol. 42 v)
Armería
A c^{ta} de Ortega
Gandía
-Primeram^{te} hay en el armería veynte y quatro coseletes menos de los que se trageron de Milán los dos de ellos con piezas dobles/ faltan dos/ porq el uno se dio al marqués mi S^{or} y otro se dio a Montoya por m^{do} de su ex^{a} quedan 22
Castellón
-yten otro coselete q tiene el duque mi S^{or} en Castellón
Gandía
-E más se halló aquí en Gandía veinte y quatro coseletes de los biejos y más otro pequeño que son 25/ estos les faltan manoplas de los doçe de ellos
Gandía
-yten tres espaldares y dos petos de los coseletes biejos faltan los escarçelones y las demás piezas q son escarçelones y grebas y cujotes (sic) y manoplas
Gandía
-yten un arnés grabado q dio el capitán de Bernia a Su ex^{a} con todas sus piezas
Gandía
-yten una rodela de la prueba y quatro petos fuertes
Gandía
-yten cinco rodelas negras pa muralla con las armas del duque mi señor
Gandía
-yten una rodela grande negra y otra pintada
Gandía
-yten seis morriones de los viejos y una medio çeladica negra
Gandía

-yten otros treinta y nueve morriones de los 40 que se trageron de Milán eran quarenta y se dio uno al marqués de Alcañiçes quando pasó a Roma por m^{do} de su ex^{a}
Castellón
-yten hay en Castellón en la guardarropa dos petos a la ligera estan a c^{ta} de Viçen Marqués
Gandía
-yten quatro corazas las tres blancas y la una colorada eran çinco y se dio una a Ju^{o} Bort de Oliba por m^{do} del duqe mi señor
Gandía
-yten otra coraza pequeña de un [...]rano
Recámara en Castellón
-E más en la recámara un peto negro hecho de papel de estraza y de [...].

HÍJAR, DUQUES DE

Doc. 1.
1631, January 4.
Inventory of the Duke of Híjar, Count of Salinas
Archivo Histórico de la Nobleza, Osuna, C. 3910, D. 149.

(fol. 1 r)
Inventario de las colgaduras, tapicerías= Alfombras= estrados= silla, bufetes y pinturas de la Casa del conde de Salinas Duque de Híxar mi S^{or} hecho por mandado de mi S^{a} la Duquesa después de la q^{ta} de Gaspar Escudero, escribióle Lorenço Copero contador= Y lo q se compró del marqués de Castañeda. Falta lo del Conde de Montalvo.

(fol. 3 r)
Copia (en forma distinta) de un ynventario que en quatro de enero del año pasado de seiscientos y treinta y uno se hiço con asistiencia de Don Christóbal de Mallea Cam^{ro} del Conde de Salinas Duque de Híxar mi S^{or} de toda la Guardarropa de Su ex^{a}, el qual por haverse hecho con alguna prisa tiene poca distinción, y para que cada cossa del quando se buscare se halle con más presteça yo Lorenço Copero Contador de Su ex^{a} lo puse en la manera Sig^{te} [...]

(fol. 25 r)
Cosas tocantes a Caballeriça y Guardarnés y armería
Adereço de Carroça
-Todo el adereço de la carroça del Prínçipe de Guastala (sic) q su ex^{a} le compró que son respaldos, almohadones, estribos grandes y chicos y ocho cortinas con todos sus alamares y fluecos sin faltarle nada. Los respaldos y estribos están fueran de las caxas, las cortinas en la n° 17 y lo demás en el n° 11.
Adereço de Palafrén
-Un adereço de Palafrén de tela rica Primavera labrada en ondas con tres guarniciones la Gualdrapa de quatro dedos de ancho cada una bordada de canutillo de oro y plata sobre raso carmesí con flueco de oro y plata a la redonda forrada en lana de colores primavera con flores todo nuevo= Tablillas pa subir a caballo q son de terciop^{o} carmesí y ençima del chapa de plata cortada en laços y en los cabos sus remates de plata maçiços, respaldo del sillón bordado de la mesma suerte sobre el mesmo raso= Gurupera con sus pendientes de terçiop^{o} carmesí [...]

(fol. 27 r)
Silla de caballo
-Una silla de caballo que es (fol. 27 v) caparaçón de ella es de terçiop° amarillo y negro guarneçido de pasamanos amarillos y negros y franja a la redonda con una funda de badana
Silla negra
-Otra silla de terciopelo negro bordada de seda negra, con su funda de badana negra. Está en el Guardarnés hase de volver
Silla verde
-Una silla de cuero verde de Borrenas cuajada toda de pasamanos de oro y plata falsos q tiene cabeçada, pretal, gurupera, riendas y açiones
Silla de Tornear
-Una silla de Borrenes de tornear con los arçones de Açi[...] y terciopelo naranjado y clavaçón dorada, q tiene cabeçada de terciopelo naranjado con pasamanos a los cantos de plata y oro, pretal y gurupera, sin cinchas, una almártega de guadamecí dorado, bordada con seda blanca con un cordón de seda
Silla de Armas
-Otra silla de Borrenes de Armar con los açiones de azero dorados y el caparaçón de tela de oro y plata (fol. 28 r) fina rica con un franjón a la redonda, con cabeçadas, pretal y gurupera, y una ación sola guarneçido todo de oro, unas riendas de cuero bayas, çinchas de hiladillo de color, bordadas las puntas sobre terciopelo negro con su funda.
Arnés
-Un Arnés de hierro liso sin dorar con veinte y quatro pieças grandes y pequeñas
Otro
-Otro Arnés de justar grabado y dorado todo q tiene, peto y espaldar, braçales, dos golas, una de justar, dos çeladas, una de tornear, y otra de justar, y una testera de caballo, dos manoplas, dos faldones sueltos, dos grebas con sus quixotes

Jaez açul
-Una jaez de terciopelo açul bordado de ojuela de plata y canutillo realçado todo cuajado con sus fluecos de oro y seda a la redonda aforrado en tafetán açul con las pieças siguientes
Cuerda
-La cuerda con veinte y tres pieças de plata grandes y pequeñas y quatro borlas (fol. 28 v)
Barbada
-La barbada con veinte y dos pieças de plata grandes y pequeñas y cinco borlas. Vídose después este jaez y le faltan dos flores de lis de la Bar^{da}

Cabeçadas
-Unas cabeçadas con treinta y quatro pieças grandes y pequeñas doradas las flores
Pretal
-Pretal de cuero açul bordado de seda y oro con veinte y dos pieças de plata
Reata
-La reata con una borla
Açicates
-Unos açicates con quarenta pieças
Estribos
-Estribos con sus chapas de plata
Pretal de Cascab^{es}
-Pretal de cascabeles de cuero açul, todo metido en su caxa sin llave

Otro Jaez negro
-Otro jaez de terçiopelo negro bordado de canutillo y ojuela de plata realçado cuajado con su flueco alrededor aforrado en tafetán negro
Cuerda
-La cuerda con veinte y tres pieças de plata y quatro borlas las dos pequeñas
Cabeçadas
-Las cabeçadas con treinta y dos pieças de plata grandes y pequeñas
(fol. 29 r)
Bartada
-La bartada con veinte y dos pieças de plata y cinco borlas
Pretal
-Pretal bordado de plata con veinte y dos pieças y quatro borlas
Açicates
-Açicates con treinta y ocho pieças de plata grandes y pequeñas
Estribos y pretal
-Estribos y pretal de cascabeles todo metido en su caxa

Otro jaez nogue^{do} o naranj^{do}
-Un jaez de terciopelo naranjado bordado de canutillo y ojuela de plata fina realçado con las pieças sig^{tes}
-La mochila como queda referido
Cuerda
-La cuerda con v^{te} y tres pieças de plata y quatro borlas de seda y plata
Barbada
-La barbada con veinte y dos pieças de plata y cinco borlas, las quatro pequeñas
Cabeçada

-La cabeçada de cuero con treinta y quatro pieças de plata chicas y grandes
Pretal
-Pretal de cuero naranj^{do} bordº de plata con veinte y dos pieças de plata y quatro borlas
(fol. 29 v)
Açicates
-Açicates con treinta y ocho pieças de plata chicas y grandes
Estribos
-Unos estribos con sus chapas de plata
Reata
-Una reata con su borla
Pretal de cascabeles
-Pretal de cascabeles de terciopelo naranjado todo metido en su caxa

Otro jaez blanco
-Otro jaez de rasso blanco la mochila bordada de oro y ojuela de plata realçado con su flueco de oro y seda aforrado en naranjado
Pretal
-Pretal de cuero blanco bordº con diez y ocho pieças de plata doradas, unos mascaronçillos en medio y tres borlas
Cuerda
-Cuerda con veinte y tres pieças de plata y quatro borlas, las dos chicas
Barb^{da},
-Barbada con veinte y dos pieças de plata y quatro borlas las tres chicas
(fol. 30 r)
Cabeçada
-Cabeçada de cuero blanco con treinta y quatro pieças de plata
Estribos
-Estribos con sus chapas de plata doradas las flores en sus fundas de cuero
Açicates
-Açicates con quarenta pieças de plata
Reata
-La reata con su borla
Pretal de cascabe^{s}
-Pretal de Cascabeles de rasso blanco y todo metido en su caxa sin cerradura

Otro jaez leonado
-Un jaez de terciopelo leonado la mochila bordada de ojuela de oro y plata con su flueco alrededor aforrado en leonado

Cabeça[as]
-Unas cabeçadas de cuero bordadas con treinta y quatro pieças de plata, doradas las flores
Cuerda
-Una cuerda con veinte y tres pieças de plata con quatro borlas guarneçidas de plata y seda
Barbada
-Barbada con veinte y una pieças de plata y cinco borlas
(fol. 30 v)
Pretal
-Pretal de cuero leonado bordado de oro y plata con veinte y dos pieças de plata y quatro borlas
Reata
-La reata
Açicates
-Unos açicates con veinte pieças cada uno chicas y grandes
Pretal de cascabeles
-Pretal de cascabeles
Estribos
-Estribos con sus chapas de plata todo metido en su caxa

En la caxa nº 18 están los cinco
Nueve pretales de cascabeles
-Nueve pretales de cascabeles los dos de raso amarillo, otro de rasso blanco, otros de terciopelo açul, quatro de terciopelo leonado y otros de terciopelo negro
En la caxa nº 18
Dos vestidos de moros &ª la silla
-Dos jaquetas y dos calçones de paño colorado con sus pasamanos por las costuras de plata falsa y oro, con medias mangas de los mismo y otras mangas de raso nácar con sus trençillas y dos bonetes del mismo paño y ligas de tafetán colorado
Dos Guar[s] de coche carmesíes
-Dos guarniçiones de caballos de coche de terciopelo carmesí (fol. 31 r) con galón o franxa de oro fino al canto y calavaçón dorada, y nueve correas de la misma suerte, y quatro cabeçadas de la mesma manera y las dos no tienen clavaçón dorada. Las quales dixo Don Christóbal de Mallea son las q se hiçieron pª la carroça de mi Sª doña Luisa de Cárdenas
Diçe Escudero se entregaron a Domº de Barrio

Dos Guarn^{s} de cuero negro
-Otras dos guarniçiones de caballos de coche de cuero negro guarneçidas de claso o bollones dorados muy espesos y cabeçadas de la mesma manera
Dos Adargas
-Dos Adargas, la una pespuntada por la parte de la manilla, y por la haz una banda negra y dos muertes pintadas y la otra forrada en cuero con labores doradas y ribete de terciopelo, y un cairel amarillo y colorado
Otras dos Adargas
-Otras dos adargas blancas forradas en cuero, y metidas en otra funda
Otra blanca
-Otra adarga blanca con una banda negra y un unicornio pintado con un letrero
(fol. 31 v)
Otra Adarga plateada
-Otra adarga plateada con forro de cuero labrado de oro, y manja de terciopelo y banda açul en la haz
Otras dos Adargas blancas
-Otras dos adargas blancas nuevas que se compraron de Françisco Galán p^{a} las cañas que se jugaron en el naçim^{to} del príncipe nro S^{or}
159º Dos Rodelas
-Dos rodelas de Charaum de la India

Sirvieron en la librea q su ex^{a} dio p^{a} la jura del príncipe
Diez espadas
-Diez espadas doradas de lacayos con seis dagas y nueve tahelíes y nueve pretinas
Silla de manos
-Una silla de manos de terciopelo de nácar fondo de oropalla (sic) con toda la calvaçón dorada con tres cortinas de tabi nácar y plata y tres marcos de vidrieras, dos de los lados con doze vidrieras y la de delante más grande, correones del mismo terciopelo
Sillón
-Un sillón de palafrén del qual se quitó la plata p^{a} el bufete de estrado que hiço G^{mo} del Vado
Tres cojines
-Tres cojines y tres maletas de cojines cuero bajo
(fol. 32 r)
Onçe faldones de Armar
-Once faldones de armar sobre paño açul, bordados de tela amarilla con torcales de oro fino
Seis cortinas de litera

-Seis cortinas de litera de Dam^{co} colorado con sus alamares
Otras seis de Dam^{co}
-Otras cinco cortinillas de damasco verde sin guarn^{on} ninguna
Un çielo de carroça
-Un çielo de carroça de tela blanca rica bordada toda con sobrepuestos, q diçen fue la con que se casó el Marq^{s} mi S^{or} con mi S^{a} Doña Luisa de Cárdenas con flueco de oro, y debajo de él otro de seda, y le falta el flueco por un lado
Una cortina de lo mesmo
-Una cortina de lo mesmo con dos bordaduras en medio como escudos
Otras quatro cortinas
-Otras quatro cortinas de la misma tela sin bordar con v^{te} y dos alamares, forradas en tafetán carmesí q entre todas tienen diez y ocho anchos y franjón corto a la redonda
(fol. 32 v)
Diez y ocho escudos
-Diez y ocho escudos de armas diferentes bordados sobre tela y terçiopelo de matiçes
Otros dos
-Otros dos escudos de armas de raso amarillo y blanco
Otros treçe
-Otros treçe escudos de armas de los Silvas tejidos al telar con oro y seda leonada
Otro escudo
-Un escudo de armas de la casa de dos haçes pintado sobre damasco açul
Otro escudo
-Otro escudo grande con las armas de la casa de paño leonado y amarillo
Cuatro cortinas
-Quatro cortinas de entrapadas que son de litera con sus fluecos alrededor y sortijas p^{a} colgarlas
Un escudo
-Un escudo de armas de los Silvas de tela blanca bordado el león con torçales de oro
Quatro cortinas dos alm^{das} y dos asientos
-Quatro cortinas de coche de paño colorado y dos asientos de terçiop^{o} y dos almohadas de lo mesmo grandes también de coche
(fol. 33 r)
Tres Arcabuçes
-Tres arcabuçes de rueda los dos con fundas
Una pistola

-Una pistola casi vara de largo
Carcax
-Un carcax con virotes
Pistolete
-Un pistolete en caxa de marfil con su funda
Horq^{lla}
-Una horquilla de hierro
Espinille^{as}
-Unas espinilleras de hierro
Caxa pª cargas
-Una caxa de cargas de arcabuz de Marfil
Tres frascos
-Tres frascos los dos de marfil y el otro de hierro
Venablo
-Un hierro de venablo dorado
Hacha
-Una hacha de armas dorada
Mont^{e}
-Una hoja de montante
Faldones
-Los faldones de unas armas
Manopla
-Una manopla de hierro
Espada
-Una espada en vayna de terçiopº
Escopeta
-Una escopeta pequeña con llave de rueda con su funda
(fol. 33 v)
Escopeta
-Una escopeta muy larga en caxa de marfil
Espada
-Una espada corta de a caballo con una cruz por guarn^{on}
Espada
-Otra de la mesma suerte en vayna de terciopelo carmesí
Espada
-Otra espada más larga q tiene escudo de los Mendoças
Espada
-Otra espada antigua corba
Escopeta

-Una escopeta, o mosquete de mecha
Otra
-Otra escopeta larga con su llave
Otra
-Otra escopeta de llave de piedra
Dos pistoletes
-Dos pistoletes en caxas de marfil en una bolsa
Un maço
-Un maço pª jugar al mallo
Pistola
-Otra pistola con caxa de marfil con su funda
Carcax
-Un carcax con virotes
Dos frascos
-Otros dos frascos pª pólvora, uno de marfil y otro de cuerno
Quatro ballestas
-Quatro ballestas con sus gafas
(fol. 34 r)
Ballesta
-Otra ballesta pequeña quebrada
Mont^{e}
-Un medio montante con g^{on}
Espada
-Otra espada de torneo vieja con guarnición
Tres malet^{es}
-Tres maletones grandes de vaqueta aforrados en lienço blanco pª llevar camas de camino
Otro
-Otro más pequeño aforrado en bocací negro
Otro
-Otro de vaqueta con unas listas aforrado en lienço
Dos Cascos
-Dos cascos de hierro
Otro
-Una montera con un casco
Armas
-Unas armas de Charaum que son de cuerpo entero, en doze pieças
Guar^{on}
-Una Guarn^{on} de espada estañada pª torneo

Dos tiendas de campaña
-Dos tiendas de campaña de Catelufa de la India viejas con sus cordones
-Dos respaldos, dos estribos y dos banquillos del coche açul
-Un cojín y manguiolla de terciopelo negro viejo
-Una muçeta de fieltro verde aforrada la espalda en bocací colorado
-Çinco lanças de stafermo y tornear
-Dos almoadones de coche de vaq^{ta} vaçíos viejos
(fol. 44 r)
Razón de las cosas que el Conde Duque mi señor compró de la almoneda del Marqués de Castañeda i los precios de cada una i la forma en que se pagaron 20U310 Rs que montó todo ello [...]
-Dos frascos de bronce dorados en seiscientos y sesenta Rs___U660 [...].

INFANTADO, DUQUES DEL

Doc. 1.
1624, November 29.
Post-mortem Inventory of Juan Hurtado de Mendoza, Duke of the Infantado.
Armoury.
Archivo Histórico de la Nobleza, Osuna, C. 653, D. 3.

(fol. 151 r)
Primeramente un arnés de torneo de a pie del tiempo antiguo que tiene las pieças siguientes= çelada gola engolada peto y espaldar braçaletes y guardabraços y manoplas y la una antigua sus faldones antiguos de hierro con sus puntas de malla sus grebas y quixotes juntas y una arandelilla en el braço derecho y el guardabraço blanco tiene todas las sangrías quaxadas de pieças y es grabado de (fol. 151 v) Fuentidueña tasado todo en quarenta ducados
-ytem otro arnés de justa grabado con peto espaldar gola entera con su çelada y baberón braçales y guardabraços dos navajones manopla y manoplón tarjeta y ristre escarçela y escarçelón quixotes y grebas salvo que el derecho es medio quixote para justa su placarte entero y una çelada de seguir una arandela dos testeras de caballo unas tarjetillas de Mendoça y unos açeros de silla salvo que no son de la propia grabadura por señas tiene este arnés en los codos la corona de la cruz de Christo tasado todo en mil reales
-ytem otro arnés de justa grabado que tiene peto espaldar gola entera de justa con su çelada y baberón braçales y guardabraços manopla y manoplón con un navajón pequeño dos escarçelones pequeños grebas y quixotes un medio placarte con dos arandelas sus açeros de silla una testera (fol. 152 r) de caballo con un escudo con un mascarón que tiene dos agujeros por ojos este arnés tiene su tarjeta y ristre y alrededor de lo grabado unas quentas grabadas tasado todo en ochoçientos reales
-ytem otro arnés de justa grabado que diçen que era del Marqués de Poça que tiene la grabadura a escalones con su peto y espaldar gola entera y celada con su baberón tarjeta y ristre braçales y guardabraços y su navajón manoplón escarcelas y escarcelón con medio volante todo junto medias grabas (sic) y quixotes con dos sillas con sus açeros la una de terciopelo de tripa amarilla y la otra de cuero de Córdoba y dos testeras y dos arandelas y otra çelada de justa tasado todo en çien ducados
-ytem otro arnés de justa que tiene peto espaldar volante entero con sus escarçelas y trancado con su gola çelada baberón braçales y guardabraços dos placartes

chicos y es grabado todo el arnés y tres navajones dos grandes (fol. 152 v) y uno chico dos manoplas y un manoplón ristre y tarjeta y un placarte grande de otro manoplón de justa una calva dos escarçelas de seguir una testera de caballo dos guardabraços dos pares de quixotes con unas grebas un morrión de campo abierto con sus pieças una çelada de torneo y otra de seguir dos pieças de peto y espaldar para prender las escarçelas de seguir tasado todo en mil reales
-ytem otro arnés de justa que es grabado y tiene peto espaldar gola entera pegada al peto y espaldar y su çelada y baberón tarjeta y ristre braçales y guardabraços un navajón grande y un (sic) pequeño manopla y manoplón escarçela y escarçelón con medio volante un placarte entero una testera de caballo unos quixotes dos pares de grebas las unas con sus pies enteros una silla de terciopelo verde con sus açeros y testera tasado todo en mil reales
-ytem otro arnés de justa graba(fol. 153 r)do que tiene peto espaldar volante entero çelada y gola braçales y guardabraços tres navajones un placarte dos manoplones y dos manoplas con sus escarçelas y escarçelón con una pieça trançada tarjeta y dos ristres y dos arandelas grandes y su baberón y una pieça de la calva que está en la propia çelada un braçal trinfante (sic) dos placartes un morrión unas escarçelas de seguir una cofia un baberonçillo dos pieças de espaldar dos pares de quixotes con dos pares de grebas las unas medias las otras enteras con sus çapatos de hierro más unos çapatos de malla con puntas de açero tasado todo en mil reales
-ytem otro arnés grabado de justa que tiene peto espaldar gola entera çelada con su baberón tarjeta y ristre braçales y guardabraços con su navajón manopla y manoplón con sus escarçelones y un placarte grebas y quixotes las grebas con pies enteros dos arandelas una (fol. 153 v) testera de caballo unos açeros de silla y tiene más un navaxón chico tasado todo en ochoçientos reales
-ytem otro arnés de justa blanco que tiene peto y espaldar çelada con su baberón gola volante entero con sus escarcelones braçales y guardabraços manopla y manoplón con un navajón grande medio volante un placarte grande un (sic) arandela grabas (sic) y quixotes una testera de caballo unas medias grebas un navajón pequeño un manoplón de justa una cofia vieja para la celada una pieça de un quixote un baberonçillo una pieça de un guardabraço tasado todo en seisçientos reales
-ytem otro arnés de justa blanco que tiene peto y espaldar çelada con su baberón gola volante entero con sus escarçelones braçales y guardabraços manopla y manoplón con un navajón grande medio volante un placarte grande un (sic) arandela grabas (sic) y quixotes una testera de caballo unas medias grebas un navajón pequeño un manoplón de justa una cofia vieja para la celada una pieça

de un quixote un baberonçillo una pieça de un guardabraço tasado todo en seisçientos reales
-ytem otro arnés de justa blanco que tiene peto y espaldar gola, çelada baberón volante entero trançado con escarçelones chico y grande braçales y guardabraços una manopla y un manoplón y su navajón tarjeta y ristre grabas (sic) y quixotes con sus pies enteros y es(fol. 154 r)puelas una pieçeçuela para la gula (sic) un baberonçillo que es medio una arendela (sic) una testera de caballo unas escarçelas con su pieça de seguir una cofia otro manoplón un morrión de a prueba con su cresta y faldilla de montera y todos estos arneses tienen sus tarjetas y ristres tasado todo en seis° reales
-ytem otro arnés grabado de seguir que tiene peto y espaldar gola braçales y guardabraços ristre con sus escarçelas quixotes y medias grebas una çelada y un morrión de campo abierto con su cresta y faldeta de montera este arnés tiene por señal un Christo en el pecho con dos manoplas la una blanca sin grebas tasado todo en tresçientos reales
-ytem otro arnés blanco grabado y a prueba que tiene çelada de tronco con su gola un baberonçillo para torneo de a caballo braçales y guardabraços con una cufa (sic) con sus escarcelas unas grebas una manopla (fol. 154 v) Peto y espaldar con un sobrepeto que es el que sirve de prueba con su ristre tasado todo en tresçientos reales
-ytem otro arnés de seguir grabado que tiene peto y espaldar gola çelada medio baberonçillo sin faldillas braçales y guardabraços escarçelas grebas con pies de açero una manopla el guardabraço yzquierdo es liso con una testera grabada que es del arnés primero de justa tasado todo en duçientos y çinquenta reales
-ytem quatro arneses de seguir de Milán lisos con sus pieças q son petos espaldares golas braçales y guardabraços y manoplas bufas grandes y bufas chicas escarçelas grebas y quixotes y çeladas y sus ristres y testeras de caballos con sus puntas de açero tasados todos en çiento y veinte ducados
-ytem otro arnés de seguir blanco que tiene peto espaldar gola braçales y guardabraços manoplas, çelada escarcelas ristre quixo(fol 155 r)tes y medias grebas un tarjeta unos aceros de sillas unas orejas de un morrión de campo abierto con una falda de montera dos baberçillos un navajón un escarçelón una pieça de vista de morrión de campo abierto tasado todo en tresçientos reales

Sala grande
-ytem doçe arneses grabados de Milán de seguir los ocho guarneçidos de terçiopelo verde y açul y pasamanos de oro y los quatro de terçiopelo carmesí y pasamanos de oro que tienen petos espaldares, çeladas golas braçales y guardabraços mandiletes ristres escarçelas grebas quixotes nueve pares de çapatos

de malla y puntas de açero y doçe bufas chicas y doçe grandes y onçe testeras de caballo y onçe sillas de açero de caballo grabadas conforme los arneses tasados doçe arneses uno con otro a seisçientos y çinquenta reales cada uno
-ytem otro arnés de seguir blanco con unas listas anchas blancas y unas angostas a los largo que tiene peto espaldar gola braçales guardabraços manoplas (fol. 155 v) çelada escarçelas y grebas y quixotes grabados que son de otro arnés tasado todo en duçientos y çinquenta reales
-ytem otro arnés pequeño para un niño que tiene peto y espaldar, gola çelada de torneo abierto por la barbota con braçales y guardabraços escarçelas grebas y quixotes y unas manoplas antiguas tasado todo en diez y seis ducados
-ytem un arnés grabado y dorado de medias cañas tudescas de seguir que tiene peto y espaldar gola çelada de campo abierto y braçales y guardabraços manoplas escarçelas grebas con pies enteros y una bragueta con sus guardarenes (sic) un volante con su ristre dos sobreguardabraços y un placarte blanco liso y dos arandelillas de este arnés tasado todo en quatrocientos reales
-ytem otro arnés de seguir blanco dorados los perfiles del peto y de la çelada y de las escarçelas que tiene peto y espaldar gola braçales y guardabraços çelada manoplas y escarçelas con su brague(fol. 156 r)ta grebas y quixotes con sus pies enteros y un escarçelón redondo para por detrás con espuelas largas doradas y tiene el peto a modo de una venera de Santiago tasado todo en tresçientos y çinquenta reales
-ytem siete arneses blancos de Millán (sic) con unas ynsignias grabados en los petos y el uno tiene un Christo en un lado derecho y son de seguir que tiene las guarniçión de cuero de ante con sus pasamanos de Milán que tienen petos espaldares golas braçales guardabraços çeladas manoplas escarçelas grebas y quixotes con sus escarpias y puntas de açero y sus ristres y testeras de caballos con sus sillas con sus açeros y siete bufas chicas y siete bufas grandes tasado todo en dos mill y quinientos digo quatroçientos y cinquenta rs
-ytem quinçe arneses blancos de Milán de seguir guarneçidos de cuero de ante con pasamanos de oro de Millán (sic) que tienen petos espaldares golas çeladas braçales y guardabraços manoplas (fol. 156 v) escarçelas ristres bufas grandes y chicas grebas y quixotes y sus escarpias y testeras de caballos tasado cada uno a treinta reales
-ytem veinte y çinco arneses de Milán de seguir en el tablado alto blancos que tienen, petos, espaldares, golas, çeladas, braçales, guardabraços, manoplas, escarçelas y quinçe ristres, bufas grandes, y chicas, grebas, y quixotes seis escarpies con puntas de açero guarneçidos con cuero de ante las escalerones (sic) y pasamanos de oro tasado cada uno a dosçientos reales

-ytem veinte y quatro arneses de Calatayud los diez y ocho que tienen petos espaldares braçales y guardabraços, celadas, escarçelas, grebas y quixotes y los seis los tres (sic) tienen escarçelas y los otros tres no las tienen ya, todos seis les faltan quixotes y tienen todas las demás pieças que los diez y ocho primeros los diez y ocho primeros (sic) que tienen todas sus pieças tasado cada uno a çiento y çinquenta reales y los seis tasado cada uno a siete ducados
(fol. 157 r)
-ytem un morrión antiguo con unos mascarones y seis testeras de caballos tasado todo en çiento y setenta reales

Sala de lo dorado
-ytem un arnés grabado blanco del señor Duque don Joan digo del Duque don Yñigo que tiene peto espaldar gola braçales guardabraços manoplas escarçelas quixotes grabas (sic) unos açeros de silla una testera de caballo con su ristre dos çeladas una de seguir y otra de torneo un almilla (sic) de la unas (sic) con su gola y un morrión de campo abierto çapatos de açero tasado todo en mill reales
-ytem un arnés de seguir pavonado con unas listas de oro que diçen era del conde don Jorxe que tiene peto y espaldar, gola braçales y guardabraços manoplas y escarçelas, quixotes, celada y ristre y unas espinillas tasado todo en treinta ducados
-ytem otro arnés de la persona del señor Almirante de Aragón negro de color de hierro que tiene peto y espaldar gola braçales y guardabraços enteros escarçelas y manoplas y un escarçelón para las (fol. 157 v) Guarda renes (sic) con sus quixotes y espinillas un sobrepeto y espaldar para de mosquete con su gola forrada en raso pardo una testera de caballo entera con punta de diamante y una roseta con su çelada a modo de morrión de campo abierto y es el arnés a prueba y tiene un perro de traylla tasado todo en mill y quinientos reales
-ytem otro arnés con unas listas doradas y relevadas grabadas de atauxía guarneçido de terçiopelo morado con pasamanos de oro que diçen era del Duque de Sesa y tiene el collar del tusón en el peto que tiene peto espaldar gola braçales y guardabraços, manoplas escarçelas grebas y quixotes çelada con un baberonçillo y una testera de caballo tasado todo en dos mill reales
-ytem otro arnés dorado de seguir con unas labores anchas que salen unas medias ondas a las listas blancas que tiene peto y espaldar gola braçales y guardabraços manoplas y escarçelas con unos quixotes y una testera (fol. 158 r) de caballo con su çelada y sus escarpias y dos arandelas tasado todo en ochoçientos reales
-ytem otro arnés dorado de torneo de a pie y caballo con la labor de lo dorado unos laços a modo de nudo de çereço que tiene peto tendido el borde haçia la garganta espaldar, gola, çelada con medio baberón con braçales y guardabraços

una manopla y un manoplón una bufa, grebas y quixotes con su ristre y su testera de caballo y açeros de silla tasado todo en ochoçientos reales
-ytem otro arnés de justa dorado que tiene peto y espaldar gola entera, çelada con su baberón tarjeta con dos ristres braçales y guardabraços manopla y manoplón con dos escarçelones, dos navajones dos placartes tres arandelas otra, celada de justa con su gola dos testeras de caballo unos quixotes el uno trançado dos pares de grebas unas con pies enteros y otros con çapatos de malla con tres sillas con sus açeros dorados las dos de terçiopelo negro y la otra de terçiopelo naranjado (fol. 158 v) con unos pasamanillos açules y plata la una de estas sillas y una arandela son hermanas y no son del arnés tiene por señas unas olicas en lo blanco tasado en dos mill y quinientos reales
-ytem otro arnés de seguir dorado de siete listas con terçiopelo carmesí que tiene peto espaldar, gola braçales y guardabraços bufa grande y bufa chica manoplas y escarçelas grebas y quixotes çelada y ristre unos açeros de silla y una testera de caballo y este arnés es de Pompeo tasado en ochoçientos reales
-ytem otro arnés de torneo de a pie dorado que tiene peto y espaldar gola çelada, braçales y guardabraços y manoplas tasado en quarenta ducados
-ytem siete arneses dorados de Milán con terçiopelo morado dorados solo los excaques (sic) que tienen, petos, espaldares, golas, çeladas, braçales y guardabraços bufas grandes y chi(fol. 159 r)cas manoplas escarçelas ristres grebas y quixotes con sus escapes de malla y puntas de açero seis sillas con sus açeros y testeras de caballos la una testera es solo grabada tasado cada uno en seteçientos reales

Sala de la linter[a]
-ytem çinco arneses de Calatayud antiguos, que tienen petos y espaldares braçales y guardabraços, celadas todas antiguas con unos baberoncillos viejos tasado cada, arnés a cinquenta reales
-ytem dos arneses de torneo con las pieças siguientes, petos y espaldares, golas, braçales y guardabraços, celadas y manoplas adviértese que el uno era del Duque mi señor pavonado y dorado a modo de el (sic) conde don Xorje que le lleva el señor conde de Saldaña a pintarle y por no hacerse el torneo se quedaron en casa del pintor y se echaron a perder tasados en quatrocientos reales
-ytem otro arnés negro de seguir que tiene peto, y espaldar braçales gola y guardabraços escar(fol. 159 v)las (sic) y celada medios quixotes y medias grebas tasado en duçientos reales
-ytem otros tres arneses de prueba e arcabuz que tienen petos y espaldares golas dos çeladas y un morrión braçales y guardabraços con sus escarçelas un

par de quixotes dos pares de medias grebas unas manoplas dos ristres unos escarçeloncillos= otro par de quixotes y otro par de manoplas unos açeros de silla una arandela para una lança y unas espinillas tasado cada uno a dosçientos reales
-ytem otros dos arneses negros que tienen, petos, espaldares, golas, braçales, guardabraços unas escarçelas un morrión fuerte con faldeta de montera y su cresta y un morrión de campo abierto con tres rejillas delante de la vista y el uno de ellos tiene ristre tasados en tresçientos reales
-ytem tres arneses de las Yndias los dos con morriones negros de un barniz y el otro (fol. 160 r) con un casco dorado el uno tiene el peto de cuero colorado y en el espaldar pegada una aljaba para quatro saetas y el otro peto es negro como el morrión y el otro de seda de colores estos dos arneses tienen dos catanas de los propios arneses que pareçen chillas (sic) tasados todos tres con sus catanas en quatroçientos reales

Sala última
-ytem çinco petos de prueba de arcabuz con sus morriones el un peto tiene unas pieças en las hijadas otro tiene un baberoncillo y su gola y el morrión es de cresta tasado cada uno a çien rreales
-ytem un peto fuerte blanco tasado en cinquenta reales

Sala quarta
-ytem un peto espaldar y gola negro y tres pares de espinillas las unas pequeñas tasado en çien reales

-La qual dicha tassaçion el dicho Diego de Sanctos juró en forma de derecho haberla fecho bien y fielmente a su saber y entender sin haber hecho agravio a ninguna de las partes ni tener en ello más ynterés que (fol. 160 v) deçir verdad y lo firmó de su nombre =

Diego de Sanctos
ante mí Gaspar de Torres
(signed)

En la çiudad de Guadalaxara a treinta dias del mes de noviemvre de mill y seisçientos y v[te] y quatro años el dicho Diego de Santos armero veçino de esta dicha çiudad fue tasando y tasó las cosas del Armería que quedaron por fin y muerte del dicho señor Duque del Ynfantado en la manera sig[te]

Pieças sueltas
-Primeramente dos pares de manoplas las unas doradas y grabadas que no tienen arneses en toda la armería y el otro par blancas del arnés de verdugado de la primera sala tasado en çinquenta reales
-ytem quatro testeras de caballos las dos doradas con unas listas angostas la una con las armas de Mendoça y la otra con unos grifos y las dos de Milán la una grabada tasado todo en çiento y quarenta reales
-ytem veinte pares de grebas sueltas viejas en el mismo aposento tasadas todas en diez (fol. 161 r) y seis reales
-Más quatro pares de escarçelones antiguos tasados en doçe reales
-ytem diez y ocho bufas grandes y diez y seis bufas pequeñas tasado cada par uno con otro a quatro reales
-ytem ocho guardabraços antiguos de Calatayud tasados todos en doçe reales
-ytem çinco braçales viejos tasados todos en ocho reales
-ytem tres manoplones viejos antiguos en seis reales
-ytem seis baberones viejos los quatro de Calatayud y los dos grabados hermanos tasados en doçe reales
-ytem çinco pieças de vista de Calatayud viejas tasadas en seis reales
-ytem ocho pares de puntas las seis con escorpes (sic) de malla y las dos sin ellas tasado cada par a tres reales
-ytem tres pieças la una es una cofia y la otra una pieça de vista grabada y una bufa grande tasado todo en quatro reales
-ytem un tornillo de hierro tasado (fol. 161 v) en seis reales
-ytem quinçe ristres de seguir y quatro de justa doçe goçetes de lanças de justa tasado toda esta partida en seis ducados
-ytem seis mosquetes cada uno con su rascador moldes frascos y frasquillos forrado de terçiopelo de tripa amarillo con cordones y borlas de hyllo (sic) amarillo açul y blanco tasado cada uno en ocho ducados
-ytem un mosquete largo a manera de culebrina con sus llaves de pedernal y de mecha por si aparte con sus moldes de pelotas y perdigones en el mismo con su frasco y frasquillo rascadores y sacapelotas tasado todo en quinientos reales
-ytem treinta y ocho mosquetes de color de hierro con sus caxas coloradas con sus frascos y frasquillos y treinta y çinco tascadores y treinta y çinco sacapelotas tasado cada mosquete con su recado a seis ducados cada uno
-ytem quarenta y ocho mosquetes blancos con quarenta y ocho fras(fol. 162 r) cos y frasquillos y quarenta y tres turquesas y sacatrapos y quarenta y quatro baquetas que se compraron de Barnachea tasado cada uno con su recado a diez ducados

-ytem un mosquete grande con frasquillo rascador y baqueta tasado en sesenta reales
-ytem otro mosquete grande con frasco y frasquillo que tiene otro adereço tasado en çinquenta reales
-ytem nueve mosquetes turquescos con tres frascos de cuerno de buey tasado uno con otro a ocho ducados cada uno
-ytem un mosquete viejo la caja quebrada con chapas de hierro tasado en çinco ducados
-ytem çinco tirillos pequeños de bronçe con sus carillos (sic) de madera que están en la sala grande tasados en mill reales

Sala de linterna (sic)
-ytem una balesta (sic) de dos jaras con el tablero de serval y en el tablero un letrero que diçe Duquessa del Ynfantado y dentro Puebla en Madrid tasado en quatro ducados, digo en çinco
-ytem otra ballesta de una jara con el mismo tablero que diçe Pue(fol. 162 v)bla en Madrid y atrás mill y quinientos setenta y seis, treinta y quatro, onças tasado en çinco ducados
-ytem otra ballesta de dos jaras con un rótulo que diçe Jesús María y dentro Puebla en Madrid y detrás 1586 tasado en çinco ducados
-Más otra ballesta que tiene en la verga por marca un cordero y por detrás una .m. por marca tasada en quatro ducados
-ytem otra ballesta de dos jaras de vaga (sic) quadrada con una .M. por marca tasada en quatro ducados
-ytem otra ballesta de dos jaras que la verga diçe Jesús María y por de dentro Puebla en Madrid tasada en çinco ducados
-ytem otra ballesta de un virote con quatro marcas en la verga por de dentro unas flores de lises con coronas tasada en quatro ducados
-ytem otra ballesta de un virote que tiene por marca t° y dentro un .L. y una .M. tasada en çinquenta reales
-ytem otra ballesta que tiene dos marcas en la verga que diçe (fol. 163 r) Rodrigo tasada en quatro ducados
-ytem otra ballesta de un virote sin nuez que diçe en la verga Jesús María y por de dentro Luis me hiço en la Puebla tasada en çinquenta reales
-ytem otra ballesta de un virote que diçe un letrero Ecçe Agnus Dei y por de dentro diçe Joan Blanco tasada en çinco ducados
-ytem otra ballesta de una jara con un letrero que diçe don Yñigo de Córdova y por de dentro Luis me hiço en la Puebla tasada en cinquenta reales

-ytem otra ballesta con otro letrero que diçe, Ecçe Agnus Dei y por de dentro Joan Blanco tasada en quatro ducados
-ytem otra que diçe en la verga don Luis de Roxas tasada en çinquenta reales
-ytem otra que diçe, Agnus Dei y dentro Joan Blanco y es de dos jaras tasada en çinquenta rs
-ytem otra ballesta que diçe en la verga In Domino Confido tasada en seis ducados
-ytem otra ballesta de un viro(fol. 163 v)te que diçe un letrero Jessusxptus tasada en çinquenta rs
-ytem otra ballesta de un virote que diçe Jesús María y por de dentro Luis Moreno en la Puebla tasada en sesenta reales
-ytem otra ballesta de dos jaras que diçe en la verga Jesús María y por de dentro Puebla en M^{d} tasada en çinco ducados
-ytem otra ballesta de un virote que diçe en la verga Duçis Medina y dentro Puebla en M^{d} tasada en seis ducados
-ytem otra ballesta de un virote que diçe en la verga Duque del Ynfantado tasada en çien reales
-ytem otra ballesta de bodoqs que se arma con una ynvençión con un número que diçe veinte y çinco tasada en quatro ducados
-ytem otra con el tablero de ébano con un letrero que diçe Antonio del Castillo tasada en çinco ducados
-ytem otra ballesta de dos jaras que diçe don Joan de Acuña tasada en quatro ducados
(fol. 164 r)
-ytem otra ballesta de un virote que tiene por marcas dos .PP. con dos coronas tasada en quatro ducados
-ytem otra ballesta de un virote que diçe Jesús María y por de dentro Puebla en Madrid tasada en çinco ducados
-ytem otra ballesta que tiene por marca dos .tt. y dos o/o ençima tasada en çinco ducados
-ytem otra ballesta que diçe Jesús María y dentro una L. y una M. tasada en quatro ducados
-ytem otra ballesta el tablero de huesso que se arma con un torni (sic) la verga grabada tasada en çien reales
-ytem çien vergas sin tableros que están sobre la puerta de la sala de medio día tasadas en v^{te} ducados todas
-y tienen todas las ballestas arriba dichas treçe gafas

Arcos y carcaxes
-ytem çinco arcos turquescos uno grande y quatro pequeños y el uno tiene barniz dorado tasado cada uno a diez y seis reales (fol. 164 v) con dos manojos de saetas y tres cañutos (sic) de plata tasado en treinta reales
-ytem otro carcax de cuero verde con unos virotes y tres casquillos de plata con una horquilla de madera tasado en quarenta reales
-ytem otro carcax de cuero negro con unos cordones açules q tiene quetro casquillos barniçados tasados en ocho reales
-ytem otro carcax de pinavete (sic) blanco con dos flores labradas en medio lleno de saetas tasado en veinte y quatro reales
-ytem otro carcax de cuero turco leonado lleno de saetas tasado en veinte reales
-ytem otro carcax de cuero negro lleno de flechas turquescas con otro manojo de flechas tasado en ocho reales
-ytem otro carcax de cuero negro digo leonado con labores leonadas tasado en veinte reales
-ytem otro carcax de cuero negro con una çerradura sin llave tasado en doçe reales
-ytem otro carcax de cuero labrado (fol. 165 r) de colores con veinte y una flechas dentro tasado en doçe rs
-ytem un carcax de madera blanco sin tapador que tiene dentro ochenta y seis saetas tasado en doçe reales
-ytem un carcax de palma negra y blanca con çinquenta y seis flechas tasado en diez reales
-Otro carcax de cuero leonado con veinte flechas dentro tasado en diez y seis reales
-Dos horquillas de ballesta de hierro tasadas en seis reales
-Un arco con una ynvençión para torçer las querdas (sic) de las ballestas tasado en diez y seis rs
-ytem un potro de dar vueltas a las ballestas tasado en doçe reales
-dos caxas llenas de saetas labradas y por labrar tasado en seis reales
-Otra caxa con un cuerno de cabra monte (sic) y unos huessos blancos para labrar algunas nueçes tasado en ocho reales
-Tres çepillos y dos moldes para adereçar las dichas saetas tasado en ocho reales (fol. 165 v)
-Una caxa de punçones para las dichas ballestas tasada en seis reales
-Un talego de casquillos de saetas tasado en treinta reales
-Dos talegos de cuernos de tener yerva de ballesteros tasado en quatro reales
-La qual dicha tassaçion el dicho Diego de Santos armero juró en forma de derecho haber hecho bien y fielmente […]

En la dicha Çiudad de Guadalax[a] a dos días del mes de diçiembre de mill y seisçientos y veinte y quatro años el dicho Diego de Santos armero fue tassando y tasó las cossas de la armería tocante a su ofiçio en la manera siguiente

Pliego de picas y lanças
-Primeramente nueve lanças jinetas que hay en la sala primera (fol. 166 r) de la esquina tasada cada una en diez y seis reales
-ytem noventa y nueve lanças jinetas que están en la sala grande y hay dos o tres de dos hierros tasadas las noventa y seis a diez y seis reales cada una y las tres con dos hierros a veinte reales
-ytem treinta y quatro picas que están en la dicha sala grande tasada cada una a veinte y quatro reales
-ytem treinta y una lanças de ristre que hay en la dicha sala grande con dos que hay en los estandartes tasada cada una a catorçe reales
-ytem una lança jineta que dicen que era del rey Chico de Granada que es toda el asta de ébano tasada en seis ducados
-ytem una alabarda guarneçida de terçiopelo carmesí claveteada de clavaçón dorado tasada en ocho ducados
-ytem un venablo con el hierro grabado y dorado con dos pistoletes en el tasado en doçe ducados
-ytem veinte y çinco venablos de hierros vizcaynas (sic) con vueltas (fol. 166 v) de serpeçillas astas de fresno y regatones tasado cada uno a diez y ocho reales
-ytem catorçe chuços con sus astas las ocho de pino y las demás de fresno tasadas las ocho de pino a doçe reales y las seis de fresno a diez y seis reales
-ytem veinte y tres hierros de chuços sin astas ni regatones tasado cada uno a ocho reales
-ytem un hierro de chuço con un pedaço de asta que diçen es el del enquentro del conde don Jorje tasado en quatro ducados
-ytem çinco juncos de las Yndias sin hierros ningunos tasado cada uno a diez reales
-ytem mas un venablo blanco con el asta guarneçida con unas correguelas tasado en veinte y seis reales
-Más dos venablos los hierros dorados y grabados con unas orejas como çerpeçuelas (sic) y las astas negras tasado cada uno a çien reales
-ytem otros dos venablos dorados y grabados los hierros el uno tiene orejas y el otro no y el (fol. 167 r) con funda de terçiopelo carmesí tasado el de orejas en ocho ducados y el otro tasado en çinco ducados
-ytem una jineta dorada y grabada el hierro con una borla de seda negra tasado en quatro ducados

-Más un hierro de partesana sin asta tasado en doçe reales
-ytem en la sala de la fuente noventa y çinco arcabuçes y noventa y quatro frascos y frasquillos y ochenta y siete morriones tasado cada arcabuz con su frasco y frasquillo a seis ducados cada uno y cada morrión a diez y seis reales
-ytem diez y ocho arcabuçes doradas las llaves con diez y ocho frascos y frasquillos de terçiopelo açul con sus cordones de seda con sus turquesas tasado cada arcabuz con su recado a siete ducados
-ytem ochenta y çinco arcabuces y ochenta y dos frascos y frasquillos de terçiopelo de tripa amarillo y cordones de seda como los de la primera sala tasado cada (fol. 167 v) uno a seis ducados
-ytem treçe morriones blancos grabados tasado cada uno a dos ducados
-ytem quatro morriones grabados con unas listas doradas tasado cada uno a veinte y çinco reales
-ytem doçe morriones grabados y dorados con una floreçilla en la punta de cada uno tasado cada uno a veinte y seis reales
-ytem tres morriones dorados labrados de atauxía tasado cada uno a quarenta reales
-Más dos morriones a modo de çeladas de campo abierto dorado con unas figuras relavadas (sic) el uno con una sierpe en la cresta con cara de muxer y el otro en la frente un marte con un escudo en la mano tasado cada uno a siete ducados
-ytem un morrión con tres crestas doradas y relevadas con unos campos de terçiopelo morado y pasamanos de oro tasado en quarenta reales
-ytem quatro morriones dorados y grabados con quatro listas (fol. 168 r) cada uno y los tres entre lista y lista unas figuras a modo de medallas doradas tasado cada uno a quarenta reales
-Más dos morriones negros de prueba mosquete (sic) tasado cada uno a quarenta reales
-Más sesenta y morriones (sic) blancos como los de la primera sala tasado cada uno a diez y seis reales
-ytem nueve morriones dorados y uno blanco a modo de sombrero tasado cada uno a ocho duc[os]
-ytem otro de color de hierro de a prueba con una lista dorada tasado en quarenta reales
-ytem en la sala penúltima de medio día çiento y tres arcabuçes y noventa y çinco frascos y frasquillos tasado cada uno a çinquenta reales
-ytem en la sala última çiento y ochenta y ocho arcabuçes de los quales hay tres reventados y veinte y çinco caxas quebradas y çiento y noventa y tres frascos y çiento y quarenta y siete frasquillos tasados los çiento y ochenta y çinco a çinquenta reales (fol. 168 v) digo los ciento y ochenta que están sanos a los

dichos çinquenta reales y los veinte y çinco que hay sin caxas a quarenta reales cada uno y no se tasan los tres reventados porque no valen nada
-ytem diez y siete morriones blancos tasado cada uno a diez y seis reales
-Dos morriones quebrados muy viejos de Calatayud tasados en quatro reales
-y la dicha tasaçión el dicho Diego de Santos [...]

En la çiudad de Guadalax[a] a tres días del mes de diçiembre de mill y seisçientos y v[te] y quatro años el dicho Diego de Sanctos armero y Blas Garçía sastre veçinos de esta çiudad fueron tasando y tasaron (fol. 169 r) las cosas que hay en el armería que pareçe haber quedado del dicho señor don Joan Hurtado de Mendoça Duque del Ynfantado difunto en la manera siguiente

Sala prim[a]
-Primeramente çinco banderas de damasco verde y columbino con las armas de Mendoça y sus cordones tasada cada una a diez y seis reales

Sala grande
-ytem un estandarte blanco de tafetán y por ambas partes las armas del emperador Carlos quinto con un letrero alrededor y un flueco de seda colorada y blanca tasado en treinta y seis reales
-ytem otro estandarte grande açul con un Santiago de una parte y a la otra no tiene nada con un flueco alrededor de oro y seda açul tasado en ciento y veinte reales
-ytem çinco banderas de trompetas como las de la primera sala tasada cada una a diez y seis reales
-ytem quatro trompetas como las de arriba que están en la sala primera tasada cada (fol. 169 v) una a diez y seis Reales
-ytem otra bandera de trompeta de damasco açul que esta sobre la dicha sala sola y espexo que hay allí con armas de Tusson y cordones tasada en veinte reales
-ytem otra bandera de damasco açul compañera de las de abajo tasada en doçe reales
-ytem una bandera açul de tafetán con un Santiago vieja y rota tasada en dos reales
-ytem otra bandera blanca y colorada de damasco con la estrella y letras de Buena Guya mui vieja y manchada y pequeña en dos reales
-ytem otra banderilla morisca rota con unos quadrillos alrededor tasada en doçe reales
-ytem quarenta y siete banderetas de lanças açules blancas y amarillas que están en las lanças y en un arca de la sala de medio día tasada cada una a real

-ytem nueve banderillas de lanças pajiças verdes y pardas tasada cada una a real y medio
(fol. 170 r)
-ytem dos a modo de banderetas de blanco açul y naranxado de tafetán tasada cada una a seis reales
-ytem quarenta y nueve toneletes que son faldones de terçiopelo amarillo la guarniçión açul y blanca forrados en bocaçí las çinturas de gamuça tasado cada uno a setenta reales
-ytem quatro toneletes de torneo pequeños de raso carmesí y pestañas de raso tasado cada uno a ocho reales
-ytem otro tonelete de tafetán amarillo bordado de terçiopelo açul tasado en diez y ocho reales
-ytem un tonelete de terçiopelo blanco guarneçido de terçiopelo verde y colorado tasado en seis reales
-Otro tonelete de terçiopelo amarillo guarneçido de lo mismo tasado en diez reales
-ytem otro tonelete blanco guarneçido de terçiopelo blanco con dos pieças de cubiertas de caballos que es pretal y sobre ancas (fol. 170 v) tasado en çiento y çinquenta reales
-ytem otro tonelete de terçiopelo guarneçido de raso blanco falso con otras dos pieças de caballo como las de arriba tasado en çien reales
-ytem una casaca de terçiopelo negro con unas armas de San Quintín que son dos bastones en cruz con unas çenefas bordadas de seda amarilla y carmesí y más tres pieças para el caballo de lo mismo tasado todo en duçientos reales
-ytem dos sota golas una de raso carmesí y otra de lienço blanco tasadas entrambas en quatro reales
-ytem seis borlas grandes de seda con sus cordones las tres açules y amarillas y las otras tres amarillas y carmesíes con sus clavetes tasadas en ochenta y ocho reales
-ytem una cubierta de paño amarillo guarneçida de cuero para sobre un caballo tasada en treinta reales
-ytem un caparaçón de terçiope(fol. 171 r)lo carmesí digo negro de tripa con unos fluecos de hylo blancos tasado en ocho reales
-ytem dos gualdrapas de terçiopelo negro con fluecos de seda tasadas en ocho ducados
-Mas diez pieças de adereços de caballos para sortijas de bocaçí negro y amarillo tasada cada una a quatro reales
-ytem treinta y quatro pares de espuelas latonadas de los hombres de armas tasado cada par a dos reales

-ytem dos pares de espuelas doradas romas tasadas en seis reales
-ytem noventa pares de estribos latonados los ochenta y seis y los demás viejos tasado cada par a doçe rales
-ytem dos pares de estribos dorados y otro par de estribos de frailes tasados los dorados a diez reales y los de frayle a quatro
-ytem tres pares de estribos viejos tasado cada par a medio ducado
-ytem dos pares de estribos blancos tasado cada par a quatro rs
(fol. 171 v)
-ytem otro par de espuelas como las de los hombres de armas a tres reales el par
-ytem quatro martillos de guerra con sus picos de hierro tasados a treinta y dos reales
-ytem una hacheta guarneçida toda el astil de hierro con una media luna y un pico tasada en diez y seis reales
-ytem dos hachetas los astiles de madera el uno negro y el otro de fresno la una con una claraboya tasadas en veinte y quatro reales
-ytem una hacheta guarneçido el palo en terçiopelo dorado y morado y la cuchilla dorada tasada en veinte y dos reales
-ytem otro martillo con un pico largo dorado y el astil forrado en terçiopelo blanco y encarnado tasado en veinte reales
-ytem una hachuela vieja con unas puntas en el astil tasada en seis reales
-ytem un çetro todo de hierro dorado y pavonado tasado en veinte y quatro reales
-ytem una maça plateada y dorada (fol. 172 r) el puño de plata con unas borlas de seda y oro tasada en veinte y seis reales
-ytem una maça el astil de madera con una bola de hierro colgada de una cadenilla tasada en quatro reales
-ytem çiento y treinta y una maças de hierro unas redondas y otras de puntas tasada cada una a ocho reales
-ytem un almilla de launas (sic) con quatro cañones por guardabraços tasada en seis ducados
-ytem onçe cotas forradas en bocaçi por de dentro y por de fuera de fustán y doçe pares de mangas y diez cascos y diez caperuças negras y onçe guantes derechos de malla las nueve cotas con sus cascos mangas y guante una con otra a siete ducados cada una y las otras dos por ser viejas a quatro ducados
-ytem una cota con sus dos guantes y dos mangas jaçerinas forrada en tafetán negro tasada en tresçientos reales
-ytem una cota con sus claçones dos pares de mangas y dos guantes (fol. 172 v) todo de malla menuda clavada y dos goçetes tasada en seisçientos reales
-ytem unos goçetes de malla dorados para de torneo tasado en çinquenta reales

-ytem unas coraças de hoja de hierro forradas en terçiopelo negro con pasamanos de hylo de oro y un gorjal de malla tasado en veinte ducados
-ytem otras coraças que son peto espaldar, dos braçales dos faldas y dos escarçelas dos hombros pegados en el peto cubierto todo de tela de plata tachonado de clavos dorados quaxado de pasamamanillos (sic) de oro tasado en quatroçientos reales
-Más unas pieças de coraças de las Yndias negras forradas en tafetán açul çeleste tasadas en seis reales
-ytem un collar de lebrel de terçiopelo verde con unas veneras doradas tasado en doçe reales
-ytem tres coraças de perros la una de vaqueta la una colchada (sic) blanca y la otra de dos colores tasado todo en veinte y quatro reales
(fol. 173 r)
-Un casco forrado en tafetán a modo de sombrero viejo tasado en seis reales
-ytem otro casco en un sombrero de tafetán tasado en diez y seis reales
-ytem en la sala de mediodía hay quarenta y una casacas de paño amarillo de Toledo tostado guarneçidas de carissca (sic) blanca y raja açul con sus alamares por delante en el cuerpo tasada cada una en tres ducados
-ytem una rodela blanca grabada con ocho listas grabadas tasada en quatro ducados
-ytem una rodela labrada con tres figuras relevadas y el campo dorado forrado en terçiopelo negro bordada de cordonçillos de oro y plata que haçen labores de una rosa y con sus correas de terçiopelo labrado de oro de canutillo con su funda de vaqueta tasado en duçientos Rs
-ytem una rodela labrada de figuras de media talla de oro y taujía (sic) y de medio relieve y hay morrión y arcabuz de esta rodela cargado en sus pliegos está (fol. 173 v) forrada en terçiopelo açul digo de tela tasado en duçientos reales
-ytem una rodela de oreja de elefante con un çerco a la redonda de plata sobredorada y en medio una plancha de plata con las armas de los Guzmanes tasado en çiento y çinquenta reales
-ytem una rodela de hierro que se compró de masse (sic) Joan armero tasado en çinquenta reales
-ytem una rodela de concha de tortuga forrada por de dentro en damasco blanco tasada en çien reales
-ytem una rodela de cuero negro pintada de oro con unas armas que tienen seis raposos y ençima un timbre y una muxer ençima del tasada en tres duc^{os}
-ytem otra rodela negra dorada de oro molido labrada de yndios que tiene en medio una tarxeta y una figura de yndio y seis rosetas tasada en tres ducados
-ytem dos rodelas grabadas y doradas que son conforme a los arneses de su exelençia (sic) la una (fol. 174 r) forrada en terçiopelo negro y la otra en

terçiopelo colorado y entrambas con pasamanos de oro y seda tasada cada una a doçe ducados
-ytem una rodela labrada y dorada de atauxía forrada en terçiopelo colorado con unas sierpes en medio dos rosetas con un diamante tasada en diez ducados
-ytem una rodela negra con una lista dorada y el flueco de seda colorado y una piña en medio con dos rosetas aforrada en terçiopelo colorado con una çinta dorada alrededor tasada en sesenta reales
-ytem otra rodela el campo de color de hierro con una çiudad y un Hércules desquixarando un león relevada y forrada en terçiopelo carmesí y trençillas y flueco de oro y seda tasada en çiento y çinquenta reales
-ytem otra rodela de acero grabada y dorada con una estrella en medio con unas letras que diçe Buena Guía tasada en doce ducados
(fol. 174 v)
-ytem otra rodela negra con oro molido que tiene en medio las Ninphas del Parnasso en una fuente tasada en tres ducados
-ytem tres rodelas de madera pintadas con unos ramos de avellano tasada cada una en ocho reales
-ytem otra rodela de madera pintada con un hábito de Santiago en medio tasada en ocho reales
-ytem otra rodela de madera con una cruz de Montessa tasada en ocho reales
-ytem otra rodela de madera con seis rodeles y un tussón por orla tasada en ocho Reales
-ytem otra rodela con un campo amarillo y otro colorado con una raya en medio tasada en ocho reales
-ytem otra rodela con otro hábito de Montessa tasada en ocho reales
-ytem dos rodelas negras viejas tasadas entrambas en seis reales

-La qual dicha tassaçion el dicho Diego de Sanctos armero y Blas Garçía sastre juraron [...]

(fol. 175 r)
En la dicha çiudad de Guadalax[a] a quatro días del dicho mes de diçiembre de mill y seisçientos y veinte y quatro años el dicho Diego de Santos Armero de su majestad digo su ecçelençia la dicha señora Duquessa del Ynfantado fue prosiguiendo en haçer tasaçion de las cosas del armería en la manera siguen[te]
-Primeramente diez escopetas guarneçidas las caxas de marfil unas al largo y otras al través con algunas figuras las siete doradas y pavonadas los cañones y unos grabados la cámara y otros hasta el cabo a trechos y los tres por dorar

y todos son de rueda con dos llaves solas las seis tasada cada una a duçientos reales y las tres que son pequeñas una con otra a çiento (fol. 175 v) y çinquenta reales cada una y una larga en tresçientos Rs
-ytem una escopeta con una llave con su rueda y una serpentín para querda y otro para perdenal (sic) con una ynvençión en el cañón que se tira con carga de hierro con unas peçeçuelas de hueso tasado en doçe ducados
-ytem otra escopeta turquesca con un cañón labrado de atauxía que tiene treinta y una roseta (sic) de plata tasada en diez y seis ducados
-ytem dos escopetas de cuerda labrados los cañones de atauxía y la una caxa con unas pieças de hierro labradas de la misma labor de atauxía tasados entrambos en veinte ducados
-ytem seis alcabuçillos (sic) pequeños mayores que pistolas con sus llaves de ruedas y el menor de ellos es el que llaman del Jitano los çinco a çien reales y al que llaman del Jitano en quatro ducados
-ytem una escopeta turquesca labrada de atauxía hasta el cabo con su guardamonte tasada en duçientos Reales
(fol. 176 r)
-ytem otra pavonada labrada la cámara y la punta de atauxía y dorada y está sin llave tasada en çiento y çinquenta Rs
-ytem otra escopeta turquesca y tiene llave de rueda y una rosetilla en la cámara tasada en diez y seis ducados
-ytem dos escopetas turquescas la una lisa y la otra pavonada y labrada la punta de atauxía con unas rosetillas de plata tasadas entrambas en veinte ducados
-ytem quatro escopetas turquescas con llaves de rueda y una tiene la boca grande y rañada (sic) tasada cada una a diez ducados
-ytem otra escopeta turquesca con una listilla de plata hasta la punta y una llave de rueda con dos gatillos tasada en diez ducados
-ytem seis escopetas negras lisas de rueda tasadas las çinco a seis ducados y la otra que es vieja en treinta reales
-ytem tres horquillas de estas escopetas y un báculo labrado de hueso tasado todo en doçe reales
-ytem dos pistoletes chiquitos (fol. 176 v) de a palmo labrados y dorados las propias caxas de hierro tasados en diez y seis ducados
-Otros dos pistoletes pavonados y dorados y en medio de los cañones las armas de Mendoça y las caxas de box labradas con una funda de telilla de Nápoles tasados en duçientos Rs
-ytem un pistolete grabado y dorado con la caxa toda dorada y grabada con su llave y su caxa y bolsa de terçiopelo carmesí tasado en çiento y çinquenta reales

-ytem dos pistoletes pavonados y dorados las cámaras y entrambos tienen un letrero que diçe del Duque don Diego de Silva y Mendoça con sus fundas de vaqueta forradas en terçiopelo negro tasado cada uno en diez y seis ducados
-Dos pistoletes dorados y grabados con las caxas labradas de hueso y una bolsa de cuero en que están ambos tasados en veinte ducados
-Dos pistoletes de a terçia dorados y grabados las caxas de (fol. 177 r) Milán con dos fundas de vaqueta tasados en diez y seis ducados
-ytem dos pistoletes de chispa las cámaras estriadas y las caxas labradas de hueso con su llave en una funda de vaqueta tasado cada uno en siete ducados
-dos pistoletes los cañones de a tres quartas y las caxas labradas de hueso con su llave de armar y sus caxas de vaqueta tasado cada uno en seis ducados
-Un Pistolete de a tres quartas el cañón dorado y grabado toda la caxa dorada y grabada con su funda de vaqueta tasado en diez ducados
-Dos pistoletes de a tres quartas q tienen las caxas de madera labrada a manera de çapa en una bolsa de vaqueta tasados entrambos en diez ducados
-Un Pistolete de media vara pavonado con la caxa labrada de hueso bolsa de vaqueta tasado en seis ducados
-un frasco de terçiopelo negro guarneçido de plata con su frasquillo de lo mismo tasado en tresçientos y çinquenta Reales
(fol. 177 v)
-ytem otro frasco y frasquillo de Milán grabado y dorado con seis borlas y botones de oro y seda çeleste con unas figuras tasado en çiento y setenta y seis reales
-ytem otro frasco y frasquillo de Milán forrado en terçiopelo negro grabado y dorado de atauxía con sus borlas de seda negra y oro tasado en çinco ducados
-ytem otro frasco y frasquillo con su llave dorado y pavonado de blanco guarneçido guarneçido (sic) de marfil con armas de Mendoça y Silva tasado en tres ducados
-Otro frasco y frasquillo de hueso el herraje dorado con cordones de seda y oro açul tasado en veinte y seis reales
-Otro frasco y frasquillo de hueso dorado de atauxía con sus cordones de seda verde y oro tasado en dos ducados
-Otros frasquillos yguales de bronçe dorados con sus cordones redondos tasados en sesenta reales
-Otro frasco de bronçe dorado (fol. 178 r) a modo de cuerno con un relox en una caxa tasado en treinta reales
-ytem un frasco y frasquillo de Milán de terçiopelo negro guarneçido el canto de hierros dorados y pavonados tasado en diez y seis reales

-Otros dos frascos uno de hueso y otro de palo guarneçidos de hueso con hierros dorados de atauxía tasado en tres ducados
-ytem quatro frascos grandes y dos chicos de hueso guarneçidos de latón y el uno de los chicos es guarnecido de plata con las armas de Mendoça tasados todos en setenta y seis reales
-ytem quatro frascos de cuerno de cabra monte (sic) con cordones de seda diferentes y una llave de arcabuz tasados en çinco ducados
-ytem dos frascos de cuerno colorados con hierros dorados y cordones de oro y seda que tienen sus dos llaves tasados en treinta reales
(fol. 178 v)
-ytem tres frasquillos el uno de hierro el uno dorado y grabado con sus cordones de seda tasados en treinta Reales
-ytem tres frascos de cuerno turquescos esta partida está cargada en otra parte donde están los arcabuçes turquescos y así no se tasan aquí
-ytem un frasco y frasquillo forrado en piel de gamo con cordones de seda verde y plata tasado en tres ducados
-ytem un frasquillo de cuerno de çiervo que sirve de polvorín y llave tasado en diez y ocho reales
-Otro frasco de cuerno de venado con tres puntas tasado en diez y ocho reales
-ytem otro frasco de cuerno açul que sirve para pólvora y polvorín con hierros dorados tasado en veinte y seis reales
-ytem otro frasco de cuerno negro raiado (sic) con la herramienta dorada y una estrella tasado en diez y ocho reales
-ytem otro frasco de hueso (fol. 179 r) blanco que sirve de polvorín y llave con cordones de seda verde y parda tasado en veinte y seis reales
-ytem otro frasco de cuerno punta de toro tasado en quatro reales, digo en ocho
-Treçe llaves de arcabuz que son de las escopetas de este pliego tasadas en tres ducados
-ytem un frasco de cuer[...] tresco (sic) pespuntado con seis cadenillas de plata con tapadores de tornillos y una asa y dos cercos todos de plata con sus mascarones y su bolsa de cuero tasado en çien reales
-ytem otro frasquillo de ébano sin guarniçión tasado en quatro reales
-ytem un frasco pequeño con un pico de gorrión de hierro con un escudo dorado de las armas de Mendoça tasado en quatro ducados
-ytem otro frasquito viejo de latón dorado y relevado tasado en tres reales
-ytem quatro llaves de ruedas grandes de arcabuz la una (fol. 179 v) sin gatillo tasadas en sesenta reales
-ytem dos llaves de cuerda de chapa ancha la una grande para el mosquetón esta se tasó con el mosquete para tirar a grullas y el pequeño se tasa en quatro reales

-ytem seis llaves de los arneses de justa envueltas en unos cordones de seda tasada cada una a siete reales
-Dos llaves de armar la una de mosquete y la otra de arcabuz tasadas en çinco reales
-ytem ocho turquesas senҫillas (sic) y dos dobladas para pelotas y algunos sacatrapos y rascadores en una talegilla (sic) açul las ocho cençillas (sic) a tres reales cada una y las dos a seis reales
-ytem un frasco de cabra montés como otros que hay arriba tasado en diez y seis reales
-Una llave de armas de los arcabuçes de rueda tasada en seis reales
-Más siete turquessas que hay más de estas escopetas tasadas (fol. 180 r) cada una en tres reales
-Más un frasquillo de hierro para polvorín que sirve de llave tasado en diez reales
-ytem noventa y siete horquillas de arcabuçes tasada cada una a tres reales una con otra y están algunas pasadas de polilla
-ytem una horquilla de pallo (sic) de la Yndia con el hierro plateado tasado en ocho reales

-y en este estado se quedó la dicha tasaçión y el dicho Diego de Santos juró [...]

En la dicha çiudad de Guadalaxara a çinco días del dicho mes de diçiembre del dicho año de mill y seisçientos y veinte y quatro años el dho Diego de Santos armero y Jerónimo Ortiz platero y Lucas (fol. 180 v) Vázquez espadero veçinos de esta dicha çiudad fueron tasando y tasaron las cosas del armería que quedaron por fin y muerte del dicho señor don Joan Hurtado de Mendoça Duque del Ynfantado en la manera siguiente

-Primeramente veinte y ocho adereços de guarniçiones de caballo de terçiopelo negro tasado cada adereço a tres ducados
-ytem unas guarniçiones de caballo de terçiopelo açul tasadas en diez y seis reales
-Más un adereço de palafrén de terçiopelo negro con fluecos de oro y seda tasado en tresçientos reales
-Más dos guarniçiones bordadas la una de oro y la otra de plata vieja tasadas en ocho ducados
-Un adereço de caballo de una tela amarilla muy vieja tasada en diez y seis reales
-ytem sesenta y quatro guarniçiones de caballo de cuero blancas tasada cada una a veinte y seis reales
-ytem una guarniçión amari(fol. 181 r)lla de terçiopelo guaneçida de franjillas de seda açul tasada en çien reales

-ytem una guarnición de cuero baio tasada en diez y seis reales
-ytem ocho guarniçiones de caballos de cuero diferentes viejas y de colores tasada cada una a doçe reales
-Dos adereços de machos de literas leonados tasados en seis ducados
-Más siete cabeçadas de lebreles de cuero tasado cada uno a quatro reales
-ytem un manojo de correas viejas y una almártaga y çincha tasado todo en doçe rs
-ytem quarenta y tres çinchas y quarenta y tres sobreçinchas que están en el arca postrera tasada cada una a tres reales
-ytem cinquenta y un codones (sic) de cuero bayo con cordones tasado cada uno a quatro reales
-ytem quarenta y seis pares y medio de cadenillas para riendas tasada cada una a diez reales
-ytem diez y ocho guarniçiones (fol. 181 v) de adereços de caballos de corcho polacas labradas de flores de palma con seis sillas polacas con sus estribos de madera pintados con diez y ocho frenos tasada cada guarnición con su recado y silla y freno a veinte ducados cada adereço
-ytem un cordón de terçiopelo negro con cordones de seda negra y plata tasado en doçe reales
-Unas riendas de vaqueta nuevas tasadas en quatro reales
-ytem treinta y seis bolsas de las çeladas que son las q se hiçieron para los hombres de armas tasada cada una a diez y seis reales
-ytem una bolsa de caça de cuero de gamo guarneçida de grana y cuero negro pespuntada tasada en dos ducados
-ytem dos bolsas bayas (sic) de cuero turco la una con una çinta açul çeleste y otra correa de cuero tasada cada una en quatro ducados
-Una bolsa de cuero turco verde bordada de hylo de plata tira(fol. 182)da tasada en çinquenta reales
-ytem una bolsa con muchos bolsicos y una venera en medio tasada en tres ducados
-ytem una bolsa de cuero colorado cubierta de terçiopelo verde tasada en veinte reales
-ytem una bolsa de cuero vieja con unos rascadores y sacatrapos tasada en ocho reales
-ytem tres bolsas viejas una de terçiopelo negro y otra de raxa y otra parda tasadas en quatro reales
-ytem una bolsa de seda verde ya vieja con una arandilla (sic) de oro a la redonda tasada en çinquenta reales

-ytem dos escarçelones de caça de damasquillo açul y verde viejos tasados en veinte reales
-Más seis fundas de rodelas de cuero viejas tasada cada una a tres reales
-ytem nueve fundas de morriones de çelada viejas a dos reales cada una
-ytem treinta y una fundas de escopetas de cuero tasadas cada una a ocho Reales (fol. 182 v)
-ytem doçe fundas coloradas de cordellate (sic) para meter armas tasadas en tres ducados
-ytem çiento y çinquenta roscas de cuerdas tasada cada una a real y medio

Espadas dagas alfanjes
cuchillos guarniçiones
tiros y pretinas
-ytem una espada dorada con su daga guarnición antigua vainas y tiros de terçiopelo negro y espada y daga tienen dos pistoletes tasado en tresçientos reales
-ytem una espada y daga dorada labrada de atauxía con tiros y pretina de ante pasamanos de oro muy viejo tasado en çinquenta reales
-Más una espada ancha con quatro canales hasta en medio con una guarnición antigua negra y su vaina baya (sic) tasada en çien Reales
-ytem una espada ancha dorada la guarnición y aventanada del perrillo tasada en çien reales
-Una guarnición ancha digo espada aventanada y en la hoja un letrero que diçe recaderus (sic) gotorum res (sic) tasada en çin(fol. 183 r)quenta ducados
-ytem una espada ancha antigua que diçe senatus populosque romanos (sic) tasada en doçe escudos
-Otra espada ancha del perrillo con la guarnición negra dorada tasada en seis ducados
-ytem una espada negra que se alarga y sirve de montante tasada en çiento y cinquenta reales
-ytem una espada del perrillo con una guarnición antiquissima dorada tasada en çien reales
-Mas una espada y daga las hojas de Miguel Sánchez y las guarniçiones huecas negras tasado en ocho ducados
-ytem una espada ancha de un corte con unas letras arábigas guarnición plateada y hueca tasada en çien reales
-Una espada ancha con sola una cruz y una chapa grande tasada en seis ducados
-Una espada la hoja de maese Domingo la guarnición dorada antigua tasada en çien reales

-ytem una espada corta de Sahagún (fol. 183 v) con guarniçión dorada antigua tasada en dos ducados
-ytem otra espada ancha corta con unas letras doradas en el recaço y guarniçión dorada antigua tasada en seis ducados
-ytem otra espada ancha antigua que tiene por marca una rueda en la hoja tasada en quatro ducados
-Mas un estoque de a caballo del perrillo la empuñadura de montante tasada en quatro ducados
-Otra espada ancha y corta con dos, oo, en la hoja y la guarnic[on] antigua plateada y puño de plata tasada en seis ducados
-Una espada ancha y corta con unas letras que diçe vizcote guarniçión estañada antigua tasada en veinte y seis reales
-ytem otro estoque de tres filos guarniçión dorada antigua tasada en seis ducados
-yten otra espada ancha y corta con dos, oo, por marca y unas letras dentro de las oes guarniçión negra antigua vayna y tiros viejos tasada en quatro ducados
(fol. 184 r)
-ytem quatro espadas cortas y anchas sin guarniçiones con una .g. por marca tasadas todas en doçe ducados
-ytem una espada de çinta con una, v, en el recaço y una vaina de terçiopelo negro tasada en doçe reales
-Otra espada de çinta con una, v, en el recaço las puntas negras sin vaina tasada en doçe reales
-ytem una espada de çinta que tiene Toledo por marca tasada en ocho reales
-ytem una espada de çinta que la marca diçe Eliçalde con una guarniçión antigua negra tasada en dos ducados
-ytem una espada la hoja vizcayna con una guarniçión antigua estañada y la daga el puño de plata tasada en veinte reales
-Un estoque largo metido en un palo que sirve de báculo tasado en dos ducados
-ytem tres espadas de torneo la una la guarniçión dorada tasadas todas en treinta reales
(fol. 184 v)
-Dos estoques de ristre el uno la vaina de terçiopelo amarillo sin guarniçión el uno tasado en quatro ducados
-Más tres alfanjes damasquinos los puños de palo o cuerno con las guarniçionçillas de rabo de gallo tasados en çien reales
-ytem dos cuchillos grandes de monte con vainas verdes y sus cuchillos y martillos tasados en duçientos reales

-Más un cuchillo de monte de cuerno de venado con un cuchillejo tasado en veinte reales
-Un cuchillo de monte con la guarniçión de boca de serpiente con vaina y puño negro tasado en veinte y seis reales
-Un cuchillo damasquino con unas letras doradas y grabadas y el puño y guarniçión de hierro dorado tasado en tres ducados
-Otro cuchillo de monte con el pomo de cuerno y una chapa de latón sobre el tasado en doçe reales
-ytem un cuchillo pequeño (fol. 185 r) con un pistolete y la vaina de terçiopelo morado tasado en çinquenta reales
-ytem un cuchillo corto con las cachas de box guarneçido de latón y vaina de terçiopelo tasado en doçe reales
-ytem un cuchillo la vaina y guarniçión de plata tasado en diez ducados
-Un cuchillo damasquino con la vaina de las Yndias guarniçión de plata en çien reales
-Un alfanje damasquino de çapa negra con contera y brocal y otra pieça de plata labrado a lo mosaico y el puño y guarniçión de lo mismo con una çinta de seda morisca tasado en quinientos reales
-Un alfanje morisco con la guarniçión de plata y otras tres pieças de plata en la vaina sin contera y la vaina bordada de hylos de plata tasado en tresçientos reales
-Un alfanje corvo damasquino con vaina de terçiopelo carmesí con dos pieças de plata para el tahalí con su contera (fol. 185 v) y guarniçión de plata de cruz con una serpentilla y un puño de çapa con nueve peçeçuelas de plata en el tahalí tasado en tresçientos reales
-ytem un alfanje turquesco corvo con vaina de çapa y tres pieças doradas en la vaina tasado en quatro ducados
-Un alfanje turquesco sin vaina ni guarniçión tasado en ocho ducados
-Un alfanje turquesco corvo con puño y vayna de çapa guarniçión negra tasado en ocho ducados
-ytem un cuchillo rico que diçen que fue del rey Chico con el puño de hueso labrado pomo y guarniçión brocal y contera de plata dorado y otras quatro pieças en la vaina de plata doradas y tahalí con unas rosetillas esmaltadas y dos borlas de seda blanca y una çinta tejida de seda y oro carmesí tasado en çien ducados
-ytem quatro dagas y un cuchillo viejos tasado todo en doçe reales
(fol. 186 r)
-ytem treçe guarniçiones de espadas viejas tasadas en veinte y seis reales
-ytem seis pomos diferentes viejos en dos reales
-Ocho vainas de terçiopelo tres blancas y çinco negras tasadas en ocho reales

-Más çinco vainas de cuero viejas en dos reales
-Dos pares y tiros de pretina de terçiopelo negro y clavaçón dorada tasado en tres ducados
-ytem unos tiros y pretina de terçiopelo pardo y clavaçón de plata y tiene tahalí tasado en çien reales
-ytem dos pares de tiros y pretinas bayas (sic) con clavaçón dorada tasado en veinte reales
-y dos talabartes y dos pretinas de cuero baio y el uno sin hierros tasado en dos reales
-quatro fundas de espadas y un atado de correas viejas tasado en dos reales

Jaeçes y adargas y caparaçones
-ytem dos caparaçones de terçiopelo açul bordados el uno de oro y el otro de plata con dos agaderos (sic) que el uno tiene quaren(fol. 186 v)ta y quatro pieças de plata y el otro diez y nueve pieças de plata con borlas de oro y seda açul tasados en ochoçientos reales
-ytem un caparaçón de terçiopelo carmesí bordado todo de oro con tela amarilla y blanca con unas cabeçadas y dos borlas de seda que las cabeçadas tienen veinte y ocho pieças de plata grandes y chicas y las borlas ocho pieças de plata y otras dos borlas que las manijas están bordadas de aljófar y otras nueve borlas de seda encarnada grandes y chicas tasados en noveçientos reales
-ytem un caparaçón naranjado de terçiopelo bordado de oro y plata y un ahogadero con unas borlas y diez y nueve pieças de plata entre grandes y chicas las tres con remates de plata tasado en seisçientos reales
-yten unas cabeçadas de cuero naranjado con veinte pieças de cover (sic) dorado y esmaltado (fol. 187 r) tasado en çiento y çinquenta reales
-ytem un caparaçón de terçiopelo bordado de tela amarilla blanca y encarnada con sus borlas verdes tasado en tresçientos reales
-ytem unas borlas nacaradas que son solas tasadas en dos ducados
-ytem un pretal de cuero bayo (sic) bordado de oro con seis pieças de plata tasado en veinte ducados
-ytem un pretal de cuero verde bordado de oro con seis pieças de cobre esmaltadas tasado en seis ducados
-Un boçal de cobre dorado con cadenillas y campanillas cascabeles de plata y un mascarón con una punta y una campanilla al cabo tasado en duçientos reales
-ytem una barba de cola de caballo con una quenta grande de plata dorada tasado en seis du[os]

-Unas estriberas de bronçe doradas con quatro chapas de plata a los lados tasado en çien Rs
-ytem un caparaçón de terçiopelo negro bordado de abalorio con pretal y cabeçadas y la herramienta dorada tasado en (fol. 187 v) tresçientos reales
-ytem unas correas para açicates con treinta y siete peçeçuelas cada una de plata sin los remates tasado en duçientos y çinquenta reales
-Una cajuela con veinte y çinco y pieças de cobre y quadradas de un pretal deshecho tasado en ocho ducados
-ytem unos açicatillos de encaxe para el çapato y unas espuelas negras tasado en quatro reales
-Una gualdrapa de capichola (sic) negra guarneçida de terçiopelo labrado que todo lo dicho está en una caxa negra tasado en tresçientos reales
-ytem quatro adargas tres grandes y una pequeña aforrada en damasco verde tasada cada una a çien reales
-Un contadorçico forrado en terçiopelo negro tasado en seis ducados
-Una caxa negra de llevar jaeçes tasada en dos ducados

-La qual dicha tassaçion el dicho Diego de Santos armero y Jeró(fol. 188 r)nimo Ortiz platero y Lucas Vázquez espadero y Blas Garçía sastre Juraron [...]

-En la dicha çiudad de Guadalaxara a seis días del mes de diçiembre del dicho año de mill y seisçientos y veinte y quatro el dicho Diego de Santos armero fue tasando y tasó las cosas del armería que quedaron por fin y muerte del dicho señor don Joan Hurtado de Mendoça Duque del Ynfantado de la manera Sig^{te}

frenos y sillas
-Primeramente diez y ocho frenos de los adereços de coche del señor Almirante de Aragón no se tasan porque están tasados en otra partida
-ytem treçe bocados de frenos (fol. 188 v) de caballos viejos algunos dorados tasado cada uno en tres Rs por ser muy viejos
-ytem tres sillas de terçiopelo negro viejas con sus cubiertas sin estribos tasada cada una en tres ducados
-Una silla vieja de diamvrera (sic) digo de sillón tasado en treinta reales
-ytem una silla de terçiopelo amarillo esta se tasó con su arnés que es de justa
-ytem una silla de justa de terçiopelo negro vieja con sus açeros blancos con fluecos blancos y negros tasada en seis ducados
-ytem una silla de seguir con el arçón delantero grande dorada y grabada de terçiopelo negro con fluecos de plata tasada en siete ducados

-ytem tres sillas antiguas con sus açeros la una de raso amarillo rota y las dos de cuero blancas y la una tiene un encaxe de hierro para el estandarte tasadas en çien reales
-yten otra silla de justa dora(fol. 189 r)da y grabada de terçiopelo negro de tripa fluecos blancos y negros de hylo tasada en siete ducados
-ytem veinte y una silla (sic) de seguir de cuero blanco con sus açeros tasada una con otra a çien reales
-Una silla açul vieja con sus açeros tasada en veinte reales
-Otra silla de terçiopelo de tripa negro dorados y grabados los açeros tasada en quarenta reales
-ytem una silla muy vieja sin açeros ningunos tasada en doçe reales

Cosas diferentes
-Una arquilla con un barrilete de pólvora dentro tasado en seis reales
-ytem dos caxas de traer armas y cantidad de ençerados tasado en ocho reales
-Dos bolas de christal que sirven de espejos la una en diez y seis reales y la otra en quatro reales
-ytem tres clarines turquescos y unos atabales de cobre tasado todo en seis ducados
(fol. 189 v)
-ytem tres raquetas sin açerdas (sic) de jugar a la pelota tasado en tres reales
-Una tablilla de nogal con las armas de Mendoça tasado en quatro reales
-ytem una linterna chica de madera dorada tasada en dos reales
-ytem çinco palos para ballestas por labrar tasado en çinco reales
-Dos sartas de capirotes de halcones tasado en veinte y dos reales
-Unos hierros de una silla de campo tasado en doçe reales
-ytem tres estacas de hierro grandes con sus sortijas tasadas en tres reales
-ytem siete lienços de Flandes viejos arrollados tasados en diez y seis reales
-ytem una sierpe de lienço grande pintada hecha pedaços que es para echar en el ayre tasada en seis reales
-un hierro de vedriçias (sic) quadrado tasado en dos reales
-ytem dos pretales de cascabe(fol. 190 r)les viejos tasados en ocho Rs
-ytem una caxa de serviçio negro de las que se cojen tasada en seis reales
-ytem un asiento de terçiopelo verde de silla de campo tasada en tres reales
-ytem dos medias fullebas (sic) de ventanas tasado en dos reales
-Dos hierros de un bufetillo dorados en seis reales
-Una tabla de ajedrez vieja sin pieças no se tasa porq no vale nada
-ytem una piedra blanca de ara tasada en doçe reales
-ytem una campanilla vieja quebrada tasada en un real

-Dos pies de amigo uno con virote y otro sin el tasado en tres reales
-Una escala con sus hierros y otro hierro más de otra en veinte y seis reales
-Tres cerrajas la una buena en seis reales
-Una cuchar (sic) de hierro de haçer perdigones tasada en dos reales
(fol. 190 v)
-Una caxa a modo de achero con dos esparveles viejos tasado en dos Reales
-ytem un cuero de gamo con su pelo y unas pintas tasado en seis reales
-Una caxa de atambor tasada en tres ducados
-ytem veinte y un cofres y arcas en que está mucha de la haçienda de estos cargos los seis son encorados tasado cada uno a tres ducados y dos arcas grandes a çinquenta reales cada una y las demás arcas y caxas viexas que haçen cumplimiento a esta partida tasada cada una a diez reales
-ytem dos escaleras de madera la una de dos troços y la otra que se coje junta tasadas en dos ducados
-ytem doçe cabeças de venados con sus cuernos tasada cada cabeça a medio ducado
-ytem veinte y tres pieças de cubierta de caballos de ante tasada cada una a quatro reales
-ytem una caxa verde sin ga(fol. 191 r)vetas con los pies quebrados tasada en doçe reales
-Una regadera de cobre vieja en veinte reales
-Una rueda grande con sus pequeñas para repasar tasado en quatro ducados
-Una cigüeña y un çigoñón de hierro tasado en seis reales
-Dos pares de tenaças viejas que no sirven tasado en un real
-ytem otro par de tenaças que sirven tasado en veinte reales
-Dos enfriaderas y una maçeta de hierro tasado en dos reales
-Una estaca de hierro para clavar clavos tasada en siete reales
-ytem dos limas la una quebrada y un asienta clavos y dos çinçeles y dos puntones (sic) de hierro tasado en çinco reales
-Unas tijeras de cortar que no sirven tasado en un real
-Unas tijeras grandes de cortar que sirven de presente tasadas en diez reales
-ytem quatro bastones tasados en quatro reales
-ytem una bigornia grande (fol. 191 v) y dos pequeñas tasado en tres ducados
-Un macho y dos martillos uno grande y un (sic) pequeño tasado en doçe reales
-Una barra de hierro tasada en quatro reales
-Un banco de quatro pies con un tornillo en el tasado en veinte y seis Reales
-Un plomo para haçer agujeros ençima de seis libras tasado en tres reales
-Una llave para volver tornillos tasado en dos reales

-Una muela pequeña con su cigüeña tasado en seis reales
-Tres arrobas de esmeril no se tasan porque se gastan en el Armería
-ytem un banco grande de en q se trabaja en el armería tasado en ocho reales
-Un banquillo de pino en dos reales
-Un tonel para limpiar cotas en quarenta Reales
-ytem un arca grande vieja y un rascador tasado en dos Reales
(fol. 192 r)
-Un volvedor de culatas de alcabuçes (sic) hechos pedaços tasado en dos reales
-Tres candados de hierro tasados en veinte y seis reales
-y la dicha tassaçion el dicho Diego de Santos armero juró [...]

En la dicha ciudad de Guadalaxara a siete días del dicho mes de diçiembre del dicho año de mill y seisçientos y veinte y quatro el dicho Diego de Santos Armero y Blas Garçía sastre fueron tasando y tasaron las cosas que hay en el armería que quedaron por fin y muerte del dicho señor Duque del Ynfantado en la manera siguiente

-Primeramente ocho pistolas tres grandes con sus cajas (fol. 192 v) de atauxía y llaves de rueda
-Otras tres pequeñas todas de hierro con llaves de rueda y las dos pequeñas ansimismo digo con llaves de rueda tasada cada una en seis ducados se entiende una con otra
-ytem un frasco torçido de pólvora con cordones açules tasado en doçe reales
-ytem una espada y daga del Gran Capitán plateada con un cordón açul con que peleaba y un tahalí y pretina de terçiopelo açul bordada de plata escarchada tasada en duçientos reales
-Otra espada enrejada la guarniçión con vayna de terçiopelo que era del rey de Françia y se la dio al señor baylío de Lora por ser con la que se coronó tasada en çien reales
-ytem otra espada de a caballo con daga dorada tahalí y pretina de terçiopelo negro tachonada que era con que peleaba el señor Almirante tasada en duçientos reales
-ytem una hoja de espada ancha (fol. 193 r) de a caballo tasada en quatro ducados
-ytem una sabla con su vaina y tres escudos y una contera dorados tasado en veinte ducados
-ytem una espada de a caballo con la guarniçión de plata [...] y vaina de terçiopelo tasado en çien reales
-ytem una espada y daga con la guarniçión varniçada y la hoja de Thomás de Ayala y la vayna de çapa tasada en çien reales

-ytem otra espada y daga varniçada tiros y pretina de riço la hoja de Joan Martínez tasada en çien reales
-ytem otra espada y daga tiros y pretina dorados la hoja de Sebastián Hernández con vaina de terçiopelo tasada en çien reales
-ytem el bastón de general con que gobernó los exérçitos en Flandes el señor Almirante de Aragón tasado en nada porque no es más de la estima
(fol. 193 v)
-ytem una banda de tafetán carmesí vordada con las armas de Mendoça y Buena Guya y otras empresas con que se perdió en la batalla de las Dunas tasada en duçientos reales
-ytem unas chinelas de vaqueta coloradas que le dio el conde Mauriçio al señor Almirante para quitarse las botas q^{do} le prendió tasadas en dos reales
-ytem treinta y tres cordones gruesos de armas de diferentes colores con los cabos de plata de martillo tasados en ocho ducados
-ytem quatro bocados de la brida cañones uno dorado y otro plateado con sus copas tasados en veinte y quatro reales
-ytem una cuerda de la jineta de hylo de plata tasado en çinquenta reales
-ytem unos estribos a la jineta barniçados tasados en seis reales
-ytem dos pares de estribos a la brida tasados en v^{te} reales
(fol. 194 r)
-ytem siete banderetas que ganó el señor Almirante a los herejes de Holanda en Frissa (sic) tasado en tres ducados que estan hechas pedaços
-ytem dos banderolas de trompetas con sus borlas y cordones la una carmesí y la otra verde con las armas de Mendoça y Buena Guya que heran de las trompetas tasadas en seis ducados
-ytem un estandarte de damasco carmesí con una cruz blanca que hera del señor Bailío quando fue general de las galeras de Malta tasado en veinte ducados
-ytem siete pares de çapatos polacos de diferentes colores tasados en veinte y ocho reales
-ytem unas botas polacas amarillas tasadas en ocho reales
-ytem dos rodelas de hierro la una a prueba de mosquete tasada la de prueba de mosquete doçe ducados y la otra en tres ducados
-ytem un baúl en que va todo (fol. 194 v) lo referido tasado en dos duc^{os}
-ytem una jineta y un venablo tasado en diez ducados todo

La qual dicha tassaçion el dicho Diego de Santos armero y Blas Garçía sastre juraron […].

LA ELISEDA, MARQUESES DE

Doc. 1.
1616, September 25.
Inventory and appraisal of Marquis of La Eliseda.
Archivo Histórico de Protocolos de Madrid, Protocolo 4317.

Tas^{on} de los bienes muebles que quedaron por fin y muerte del Señor Rui Gómez de Silva y del Águila marqués que fue de la Lisera (sic)
Ante mí Ju^{o} Bap^{ta} de Ulpite

(fol. 637 r)
Nombramiento de tasadores
En la villa de Madrid a veyntiçinco días del mes de setiembre de mill y seiscientos y diez y seis años ante mí el escribano y testigos pareció presente Baltasar Escolano residente en esta q^{te} al qual doy fee que conozco en voz y en nombre y por virtud del poder que tiene de la señora doña Antonia Manrique de la Çerda marquesa de la lisera viuda mujer que fue del Señor Ruy Gómez de Silva y del Águila difunto marqués que fue de la Lisera que fue otorgado en la ciudad de Valladolid en tres días de este presente mes y año ante Miguel Becerra escrib^{o} de Su Majestad que originalmente entregó a mí el presente escribano para que ynserte en este escriptura que sacado a la letra es del tenor siguiente
-aquí el poder-
El qual dho traslado del dho poder va cierto [...] = Dijo que en el dho nombre nombraba y nombró por tasadores de la haçienda y bienes muebles que quedaron por fin y muerte del dho señor marqués de la Lisera a Alonso (fol. 637 v) del Campo tapicero [...] = a Gabriel Montero armero de Su Majestad [...] = a Simón de Valladolid Guardicionero = a Diego Gabilán dorador= [...]

(fol. 672 v)
Armas
En la villa de Madrid a seis días del mes de otubre de mill y seysçientos y diez (fol. 673 r) y seis años por ante mí el dho escrib^{o} y testigos pareció presente Gabriel Montero armero de Su Mag^{d} y dijo que en conformidad del nombramiento que en está (sic) fecho por el dho Baltasar Escolano en nombre de la dha señora marquesa y en virtud de su poder fue tasando las armas y demás cosas cosas (sic) tocantes a su oficio en la forma y manera siguiente

Nº 1
-Un arnés de justa grabado y dorado que sirve también para torneo de a pie y para seguir la guerra que tiene las pieças siguientes dos petos y dos espaldares= y dos golas y dos çeladas un morrión dos braçales con su guardabraços dos pares de grebas con sus quijotes un manoplón= una navaja= una testera de caballo un par de manoplas= una silla una arandela una sobrecalva dos arandelillas pequeñas un par de çapatos de malla con sus puntas de açero= dos braçales con sus guardabraços= otro par de manoplas dos guardabraçitos a lo tu(fol. 673 v)desco= una navaja chica dos navajas del codo= un baberón entero= una barbota= otra arandela grande= una tarjeta= un placarte= unas escarçelas= tres pieças para la çelada= una bufa chica= y una tarjeta= una bragueta= una taleguita donde hay dos ristres y tornillos de las dichas armas tasóse todo lo suso dho en mill Reales__1U000
2
-Una silla dorada y blanca se tasó en dos ducados__U022
3
-Otra silla blanca lisa en otros dos ducados__U022
4
-Una cota con sus mangas tasóse en quatroçientos reales__U400
5
-Una alabarda con unas borlas verdes se tasó en doçe reales__U012
6
-Otra alabarda con tres borlas verdes y el hierro dorado y grabado en doçe reales__U012
7
-Una escopeta larga con la llave de rueda y la caja quebrada y sin baqueta con una señal (fol. 674 r) de marfil adonde poner la baqueta se tasó en diez y seis reales__U016
8
-Otra escopeta larga como la dha sin baqueta y la caja roja en otros diez y seis reales__U016
9
-Otra escopeta larga como la dha con la caja quebrada por la culata con su baqueta se tasó en diez y seis reales__U016
10
-Un alcabuz (sic) con su llave de rueda y su caja y baqueta se tasó en veynte reales__U020
11
-Otro alcabuz como el dho en otros veynte reales__U020

12
-Otro alcabuz como el dho sin bqueta y a la llave le falta el medio gatillo se tasó en diez y seis reales__U016
13
-Un alcabuz de munición sin baqueta con su flasco (sic) se tasó todo en doçe reales__U012
14
-Una pistola el cañón digo la cama labrada de taujía y en la caja y unas chapas de mar (fol. 674 v)fil metida en su funda de vaqueta se tasó en dos ducados__U022
15
-Otra pistola más corta se tasó en doçe reales__U022
16
-Otra pistola más corta y la caja hecha pedaços se tasó en ocho reales__U008
17
-Una ballesta verga de Luis Moreno y tablero de alçudota (sic) y sus gafas con una çapatilla de cuero de ante se tasó en doçe ducados__U132
18
-Otra ballesta vizcaína con sus gafas se tasó en cinco ducados__U055
19
-Una caja de espejo con su pie labrado de ataujía y su escribanía en el dho pie se tasó en quinientos reales__U500
20
-Un brasero grande de hierro y açero los bajos pavonados y unas chapas sobrepuestas todo en él doradas con sus aldabones dorados y seis mançanillas arriba por remate= y otras seys (fol. 675 r) bolas abajo todas de metal se tasó con la bacía de cobre en ochoçientos reales__U800
21
-Otro brasero pequeño de hierro y açero con los bajos pavonados y sus chapas sobrepuestas doradas con seis bolas arriba por remate y otras seis abajo todos de metal sin baçía se tasó en trecientos reales_U300
22
-Un relox grande de campana falto de pesas le tasó en ducientos reales con la caja de nogal que tiene__U200
La qual dha tasa el dho Gabriel Montero declaró haber fecho a su saber y entender sin haçer agravio a ninguna de las partes y ansí lo juró a Dios y a una cruz en forma de dr^o^ y lo firmó de su nombre siendo t^o^s Alonso Moscoso el dotor don Jer^mo^ de Mendoça y don Ju^o^ de Salierna residentes en esta q^te^

Gabriel Montero
Ante mí Ju° Bap^{ta} de Ulpite
(signed)

(fol. 675 v)
23
-E después de lo qual el dho Gabriel Montero tasó una espada corta y ancha de a caballo con la guarnición dorada antigua y aventanada y puño de plata tiros y vayna de terciopelo liso negro se tasó en seis ducados__U066
24
-Otra hoja suelta angosta y el recaço dorado y una canal hasta la punta en una vayna vieja se tasó en dos ducados__U022
-Yten dos espadas para cocheros con sus tahalíes de terciopelo liso negro las guarniciones doradas viejas la mejor en diez y seis reales y la otra que está más gastada en doçe que montan veynte y ocho Rs__U028
Y lo firmó de su nombre testigos los dhos

Gabriel Montero
Ante mí Ju° Bap^{ta} de Ulpite
(signed)

(fol. 678 r)
Caballeriça
En la villa de Madrid a onçe días del mes de octubre de mill y seiscientos y diez y seis años ante mí el dho escrib° y testigos pareció presente Simón de Valladolid guarnicionero vecino de esta dha villa al qual doy fee que conozco y dijo que en conformidad del nombramiento que en él está fecho por el dho Baltasar Escolano para la tasación que de suso se hará mención tasaba y tasó las cosas tocantes a su oficio en la manera siguiente
N° 1
-Primeramente una silla de terciopelo negro de media [...] (fol. 678 v) guarneçida con diez pasamanos y su clavazón dorada amelonada con su cubierta de badana se tasó todo en doçe ducados__U132
2
-Yten otra silla de terciopelo guarneçida con nueve pasamanos de oro y plata anchos y angostos vieja con su clavaçón dorada con un adreço pretral y cabeçadas grupera y riendas y açiones y armatiga (sic) todo guarneçido de pasamanos de plata y oro como la silla y engandujado de cordonçillo de oro viejo se tasó todo ello en çiento y çinquenta reales__U150

3
-Yten otra silla de terciopelo vieja con unos pasamanos que es muy vieja se tasó en doçe reales_U012
4
-Yten otra silla de cordobán guarneçida con una faja de terciopelo y ribete de lo mismo y al canto de la faja tiene un pasamano y una cu(fol. 679 r)vierta de badana se tasó todo en çient reales_U100
5
-Una gualdrapa de paño negro guarneçida con doçe ribetes que está rayda se tasó en tres ducados__U033
6
-Un caparaçón de la jineta de vuelta de terciopelo negro guarneçido con dos fajas anchas bordadas de oro y plata= con un petral de cuero borcado con su clavaçón dorada y labrada con su cabeçada de banda= de pieça y otra que es una cuerda vieja con una borla vieja las cuerdas son dos con sus láminas bordadas todo lo tasó en duçientos Rs__U200
7
-Dos coraças de sillones de litera que son de terciopelo negro guarneçidas con un flueco de oro y seda a la redonda se tasaron en catorce ducados__U148
8
-Yten una guarnición de caballo de cuero con sus hebillas doradas que son quatro pieças riendas petral cabeçadas y gurupera se tasó todo en veyntiquatro reales__U024
(fol. 679 v)
9
-Yten otra gurupera y cabeçada de cuero negro pespuntado se tasó en diez y seis reales__U016
10
-Dos pares de estribos de la brida de verduguillo que en el uno de ellos le falta una floreta de delante se tasaron en doçe reales__U012
11
-Yten tres pares de estribos los dos pares dorados de metal y el otro par son de hierro dorados viejos se tasaron todos en veyntiséys reales__U026
12
-Yten una cabeçada y petral viejo de la jineta de terciopelo negro con su clavaçón se tasó en doçe reales__U012
13
-Unos açicates de la jineta con unas pieças de metal en las correas se tasaron en diez y seys reales__U026

14
-Un petral de cascabeles de terciopelo negro con su flueco de oro y plata viejo se tasó en diezyocho reales__U018
15
-Otro petral viejo de cuero con (fol. 680 r) dos cascabeles se tasó en dos reales_ _U002
16
-Yten dos petrales para acémilas de cuero negro con sus cascabeles gordos y sus garabatos en que se ponen se tasaron en diez y seys reales___U016
17
-Quatro bodones viejos de cuero negro con sus corchetes dorados con veyntiquatro borlas de seda negra con sus mançanillas de bronce doradas se tasó todo en seys ducados__U066
18
-Dos borlas viejas de seda carmesí para coche se tasaron en ocho reales__U008
19
-Quatro guarniciones de coche dos de silla y dos de la mano todas enteras de vaqueta negra guarnecidas con sus franjas de seda negra y sus cabeçadas con sus frenos y sus copas dorados y todas ellas llenas de bollones amelonados y dorados todo se tasó en seysçientos reales__U600
(fol. 680 v)
20
-Doçe clavijas de litera unas doradas y otras de hierro viejo se tasaron en dieciocho reales__U028
21
-Una guarnición de terciopelo rayda que son cabeçadas petral gurupera riendas açiones y estribos dorados de cordonçillo y su almártiga con sus cordones de seda todo ello rreydo se tasó en çiento y çinquenta reales__U150
22
-Çinco adreços de acémilas con sus borlas de lana y frenillos se tasaron con sus penachos se tasaron en setenta ydos reales__U072
23
-Unos estribos de hierro para la jineta de laço pequeño se tasaron en veynte reales_U020
24
-Yten unas riendas digo açiones de la jineta de Tánger se tasaron en seys reales_ _U006

25
-Un látigo de Tánger en dos reales se asó__U002
26
-Una çincha de cáñamo guarneçida con una faja ancha de (fol 681 r) cuero se tasó en doçe reales__U012
27
-Otra cincha de la jineta de tela verde y las floretas labradas de colores a la morisca con su látigo fino de Tánger y los hierros dorados se tasó en veynte reales__U020
28
-Un par de açicates plateados con chapas y junquillos en las correas se tasaron en diez y seys reales__U016
29
-Otro par de açicates negros pequeños se tasaron en ocho reales__U008
30
-Un cojín de terciopelo carmesí maletilla guardapolvo atrás con sus estribos dorados y plateados y todo ello arrençillado escamado todo quajado de trencillas de oro y plata la maletilla y guardapolvo toda quajada y en medio de las conchuelas unos manojicos de priscado (sic) de plata se tasó todo en seysçientos reales__U600
31
-Una hamaca de red berberisca de color amarillo y negra con sus cordones se tasó en ocho ducados__U088
(fol. 681 v)
32
-Dos amohadonçillos de cuero verde para el coche de labor de lana se tasaron en veyntiquatro reales__U024
La qual dha tasa el dho Simón de Valladolid declaró haberla fecho a su saber y entender sin hacer agravio a ninguna de las partes y ansí lo juró a Dios y a una cruz en forma de dr^{o} siendo testigos Alonso Nieto Andrés López Alonso de Moscoso estantes en esta q^{te} y lo firmó un t^{o} por el dho otorgante

Alonso Moscoso
Ante mí Ju^{o} Bap^{ta} de Ulpite
(signed)

Espadas
En la dha villa de Madrid en el dho día y año dhos por ante mí el dho escrib^{o} y testigos pareció presente Diego Gabilán guarnicionero digo dorador veçino de esta dha villa al qual yo el dho escribano doy fee que conozco y dijo que en

conformidad del nombramiento que en él está fecho por el dho Baltasar (fol. 682 r) Escolano para la tasación de las cosas tocantes a su ofiçio la qual dha tasación hiço en la forma y manera siguiente

Nº 1

-Un adreço entero que es una espada corta de a caballo de Bilbao con la guarnición dorada y aventanada con su daga tiros y pretina se tasó en onçe ducados__U121

2

-Yten se tasó una espada de a caballo con su guarnición dorada y aventanada con talabartes sin daga y sin petrina se tasó en cient reales__U100

3

-Yten se tasó otra hoja de espada sin guarniçión que es del perrillo de Çaragoça se tasó en seis ducados__U066

La qual dha tasación declaró haber fecho a su saber y enterder sin haçer agravio a ninguna de las partes y así lo juró a Dios y a una cruz en forma y lo firmó testigos los dhos

Diego Gabilán
Ante mí Juº Bapta de Ulpite
(signed).

LEGANÉS, MARQUESES DE

Doc. 1.
1637, June 20.
Inventory of Marquis of Leganés.
Archivo Histórico de Protocolos de Madrid, Protocolo 5993.

(fol. 218 r)
Bentura de Frías secretario de Su Mag^{d} y su contador de la Razón gen^{l} del artt^{a} en nombre del señor marq^{s} de Leganés Don Di^{o} Phelipez de Guzmán del Consejo de estado de Su Mag^{d} su gobernador y capitán gen^{l} en el de Milán y como testamentario de la señora marquesa de Leganés Doña Polizena Spinola difunta mujer del dho señor Marqs que del matrim^{o} de entrambos los dos señores marqueses quedaron por sus hijos ligitimos Doña Inés, don Gaspar don Anbrosio y don Di^{o} designan que todos están dentro de la edad pupilaria porque la dha Doña Inés que es la mayor tiene ocho años y a todas las partes conviene se haga inventario tasación partición y división de los bienes que han quedado de los dos señores marqueses al tpo que la dicha señora marquesa falleció que fue a catorze de junio de este año de mil y seisçientos y treinta y siete y pa el dho efecto se nombrará en su tpo los tasad^{s} y contadores necesarios y curador ad liten a los dos menores= Supp^{ca} a Vm^{d} mande que se haga el dho inventario y fho protesto pedir lo demás que convenga y pido justicia &^{a} [...]

(fol. 218 v)
Inventario
En la villa de Madrid a veynte días del mes de junio año de mill seysçienos y treinta y siete ante mí el es^{o} el señor secretario Don Ventura de Frías comenzó a hacer ynventario de los bienes de los dos señores Marqueses de Leganés en la forma sig^{te} [...]

(fol. 265 v)
Plata labrada dorada y blanca q parte de ella tiene en Milán el s^{or} Marqués y parte en sus casas de Madrid [...]

(fol. 328 r)
-Una planta de madera de un castillo de cinco baluartes [...]
-Una rodela con una flor de lis en medio [...]
-Dos pedaços de hoja de lata a modo de trompeta [...]

-Una argolla con virote, unas esposas y (fol. 328 v) Unos grillos todo de hierro
-Ocho balas pequeñas de hierro [...]
(fol. 329 r)
-Una carabina con su funda de bayeta carmesí
-Una cabeza pequeña del sabio Merlín [...]
-Quatro mosquetes y una horquilla y a los dos mosquetes faltan las baquetas
-Otros dos de llaves como carabinas de a una vara
-Çinco frascos grandes y uno pequeño con sus cordones de lana
(fol. 329 v)
-Dos pistolas con sus llaves
-Dos frasqueras con sus cordones de lana
-Una alabarda
-Dos chuços a modo de alabardas
-Dos medias picas la una con hierro [...]
-Una pieça de artillería de bronçe pequeña en su carretón de madera
-Otra de madera en su carretón [...]
-Una piedra para moler colores en su mesilla de pino [...]
-Dos hojas de espada sin vainas
-Dos guarniciones de por sí antiguas con sus pomos
-Tres hierros de arriba para picas
-Una pieceçuela pequeña de bronçe en su carretoncillo
-Seis moldes para balas
-Dos moldes menores para balas
-Dos hierros de abajo para picas
-Dos hierros de rejones
-Unos sacatrapos [...]
(fol. 347 r)
-Una pistolilla guarnecida de marfil [...]
(fol. 348 v)
Joyas vestidos y otras cosas que están en Milán
(fol. 349 v)
-Diez y seis pares de espuelas plateadas, doradas y negras y de otras colores
(fol. 350 r)
-Un par de estribos dorados y dos frenos de la misma manera [...]
+-Una gineta de maestre de campo con un casquillo guarnecido de oro
-Unos anteojos de larga vista
-Un adereço de rayadillo dorado con una hoja de lomo de Domingo Rodríguez que tiene color de hierro

-Otro adereço açul labrado de labor atornillado la guarnición manquilla y hoja de Joanes de la Orta
-Otro adereço de labor de dados y su color de hierro espada y daga la hoja de Fran^{co} Ruiz
+-Un par de pistolas de chispa y otro par de rueda [...]
(fol. 353 r)
-Unos anteojos de larga vista [...]
-Una caña con que tomó la posesión del cargo de general de la mar el Infante don Carlos [...]
-Un adereço de espada leonado la guarnición manquilla y hoja de Fran^{co} Ruiz de un corte
-Otro adereço pavonado de labor de tirela hoja de tres mesas de Ju^{o} Martínez
-Otro adereço noguerado con hoja del mismo de labor de huesos de muerto
-Onçe pares de espuelas doradas plateadas y negras
-Una espada de a caballo que dio el Duque de Ebuli [...]
(fol. 354 r)
-Un adereço de espada color de hierro labrado de grano de trigo y rayadillo
-Dos adereços de espadas doradas
-Otro adereço de espada plateada
-Un adereço de grano de trigo [...]
(fol. 354 v)
+-Quatro cañas de la India que sirven de gineta
-Dos adereços de brida que vinieron en una caja de Córdoba el uno de cordobán negro y verde cortado y pespuntado de seda blanca y el otro de terciopelo carmesí y cordobán açul pespuntado de seda blanca tienen todas sus pieças sino çinchas y estribos
-Un arcabuz que dio a Su Exª don Alonso de la Carrera
-Dos pares de estribos que se hicieron en Brusselas los unos açules y los otros negros
-Otro par de estribos dorados y traídos
-Un adereço de espada dorado de espada con unos nudos y puño de bronce
-Un forro [...]
(fol. 355 r)
-Una hoja de espada de Sahagún
-Tres hojas una de Fran^{co} Ruiz otra de Simón Rodríguez y otra de Silvestre Nieto
-Dos frascos de cuerno guarneçidos
-Un adereço liso barniçado con hoja de lomo de Joanes con tiros y pretina de cordobán
(fol. 355 v)

-Un adereço de sarmientillo con puño de plata y hoja de Ortuño con tahelí y rojo bordado de seda negra
-Un adereço de quatro puentes pavonado espada y daga y hoja de Lomo que fue del Duque de Medina
Lo q havía en la recámara de Su Ex[a]
(fol. 358 v)
-Un guión de compañía de caballos de damasco carmesí bordado de oro [...]
-Una banderilla de estandarte naranjada y bordada [...]
-Una bengala con cabos de plata [...]
-Dos lanças la una de sortija y la otra de juego de cañas
-Un modelo de madera con quatro torres y su chapitel en medio
(fol. 359 r)
-Quatro pistolas con sus fundas de terciopelo carmesí bordadas
-Un cuchillo de monte con su cabo verde y dorado
-Quatro frascos de cuerno de arcabuz el uno con hierro y los tres sin ellos
-Dos llaves de arcabuz
-Un ingenio como llave de arcabuz
-Una silla de a caballo de terçiop[o] anaranjado bordada de plata de borrenes con su funda
-Otra silla de gamuça con pasamanos de oro de borrenes con cubierta
-Otra de terciopelo negro guarnecida con pasamanos de seda negra con cubierta
-Otra de terçiopelo negro guarneçida con pasamanos y su cubierta
-Otra de vaqueta blanca lisa con borrenes y cubierta
-Otra silla de terçiopelo açul bord[a]
-Otra de terçiopelo negro [...] para la posta
-Otra silla de gineta con su caparaçón cabeçadas riendas gurupera (sic) y almártiga
-Dos caparaçones de gineta el uno de cuero verde y el otro de lama (sic) blanca bordado con todos sus adereços de cabeçadas riendas pretales çinchas y almártigas
-Una gualdrapa de terçiop[o] negro [...]
-Una adarga
(fol. 359 v)
-Tres guarniciones de las dichas sillas bordadas cabeçadas guruperas riendas y cordones
-Dos fundas de pistolas bordadas açules
-Un adereço açul de cabeçadas riendas y guruperas cordón y almártiga
-Un freno de la brida con sus riendas y cabeçada
-Otro freno de la brida

-Unas cabeçadas riendas y gurupera de la silla bordada
-Un açote de posta
-Dos llaves de arcabuz viejas
-Otras cabezadas y riendas de la silla de vaqueta
-Dos frenos grandes de la brida
-Unos estribos de palo
-Dos frascos de cuerno
-Otro frasco de terciop^{o} carmesí guarne^{do}
-Otro frasco de arcabuz con su cordón
-Una bolsa de cuero de munición
-Una funda de arcabuz
-Tres tahelíes para arcabuçes de campaña
-Una funda de espada
-Una funda de flechas
-Dos fundas de espadas
-Quatro fundas de arcabuz dos de vaqueta y dos de bayeta
-Dos cofres dos baules y una caja en que está esta ropa de lo arriba escripto
-Veynte y siete fundas de pistolas (fol. 360 r) las quatro de cañamaço y las otras de cuero negro
-Quatro pares de tiros los tres negros y el otro plateado
-Un mapa pequeño con un letrero que dice don Fran^{co} de Tejada con cubierta de tafetán
-Unos cañones de botas [...]
-Un tahelí y petrina noguerado bordado de oro con conchillas en medio
-Otro tahelí y petrina de terciopelo negro bordado de oro
-Otro tahelí y petrina bordado de seda negro con eses al contorno
-Otro con su petrina bordado de negro en cordobán
-Otro tahelí andado bordado de negro
-Otro y petrina bordado de oro con sus espejuelos y estrellas
-Otro y petrina de cordobán guarnecido con lantejuelas (sic) negras
-Otro y petrina de cordobán noguerado bordado con eses
-Otro y petrina bordado de negro sobre tafetán y clavaçón noguerada
-Otro y petrina de cordobán bordado de noguerado [...]

(fol. 361 r)
Armería
-Mas se ponen por inventario unas armas fuertes de a caballo enteras con todas pieças color de hierro con clabaçon dorada que tienen su sobrepeto y espaldar del mismo color

-Otras armas de ynfante enteras color de hierro y listas doradas con su rodela de lo mismo y murrión y braçales enteros con manoplas
-Unas armas fuertes de a caballo con todas sus pieças color de hierro con clavaçón dorada
-Un casquete de arcabuz^{ro} de a caballo
-Otro casquete de arcabucero
-Un martillo de armas con un cordón de seda ya traído
-Una maça de hierro de armas con el cabo de plata dorado
-Quince arcabuçes con sus llaves
-Otro arcabuz a modo de carabina con su caja guarneçida
-Más ocho arcabuçes y alguno de ellos labrado con su mira y culata dorada y con su frasco a modo de cuerno la guarnición de plata dorada
-Más onçe arcabuçes el uno grande a modo de mosquete y otro labrado al uso antiguo
-Más otros tres arcabuçes qbrados dos tienen los jardineros de la huerta de Leganés y el otro la guarda del bosque de Morata
-Más treinta pistolas
-Otras diez y nueve pistolas a modo de carabinas
(fol. 361 v)
-Más diez y seis pistolas grandes y chicas
-Quatro llaves de pistolas sueltas
-Más nueve carabinas
-Una bolsa con cinco hierros para perdigones cargas de arcabuz
-Tres pieças pequeñas de hierro blanco de martillos y abridero
-Tres frascos de cuerno
-Una caña grande con su horquilla de plata
-Un báculo abierto por tres partes con cabos de hierro dorados
-Un báculo pequeño con cabos de plata
-Siete cuchillos de monte
-Dos horquillas de caña con sus hierros
-Una ballesta
-Un montante con su puño de terciopelo negro
-Dos alfanjes grandes el uno con vaina de terçiop^{o} carmesí bordado y el otro con vaina de terciopelo negro con guarniçión dorada
-Un alfanje turco guerneçido con testa puño y cabos
-Diez y seis espadas con sus dagas de diferentes maestros todas con sus guarniciones negras y pavonadas
-Una espada con su daga y la guarnición dorada

(fol. 362 r)
-Más quatro dagas con sus guarniciones y vainas
-Quatro hojas de espada con sus vainas y sin guarniciones
-Más otra hoja grande y ancha también sin guarnición
-Catorçe tiros de bronce pequeños y encabalgados en sus carritos y otro chico de la mis[a] man[a]
-Una testera de caballo con unas listas doradas […].

Doc. 2.
1642, April 30.
Inventory and appraisal of Marquis of Leganés armoury.
Archivo Histórico de Protocolos de Madrid, Protocolo 5993.

(fol. 408 r)
Tasación de Armería
En la villa de Madrid a treinta días del mes de abril de mil y ss^{os} y quarenta y dos años ante mí el escribano del número, Manuel Luis armero y dorador residente en esta corte como persona nombrada por las partes para haçer tassaçión de la armería y otras cosas y debaxo de juramento que hizo por Dios nro Señor y una señal de cruz en forma de derecho tasó la dha Armería y más cosas que se le mostraron en la forma y manera siguiente
-Primeram^{te} unas Armas fuertes de a caballo enteras con todas pieças color de hierro con clavazón dorada que tienen su sobrepeto y espaldar del mismo color y tasó en treçientos ducados que valen çiento y doze mil y duçientos maravedís_ _112U200
-Más otras armas de infante enteras color de hierro y listas doradas con su rodela de lo mismo y murrión y braçales enteros con manoplas y tasó en dos mil reales que valen setenta y ocho mil maravedís__68U
-Mas unas armas fuertes de a caballo con todas sus pieças color de hierro con clavazón dorada y tasó en la misma cantidad__68U
-Más un casquete de arcabuzero de a caballo y tasó en quinientos Reales que valen diez i siete mil maravedís__17U
-Más otro casquete de arcabuzero y tasó en quatroçientos reales que valen trezemil y seiscientos mrs__13U600
-Más un martillo de armas con un cordón de seda ya traído y tasó en çien reales que valen tres mil y quatroçientos mrs__3U400
(fol. 508 v)
-Más una maza de hierro de Armas con el cabo de plata dorado y tasó en treçientos reales que valen diez mil y duçientos mrs__10U200
-Más quince arcabuzes con sus llaves y tasó cada una de ellas en ochocientos reales que montan doze mil reales que valen quatrocientos y ocho mil maravedís_ _408U
-Más otro arcabuz a modo de carabina con su caxa guarneçido y tasó en mil Reales q valen treinta y quatro mil maravedís__34U

-Más ocho arcabuzes y el uno de ellos labrado con su mira y culata dorada y con su frasco a modo de cuerno la guarnición de plata dorado y tasó este en cien ducados y los demás a ochoçientos reales que montan todos seis mil y seteçientos reales y valen ducientas y veinte y siete mil y ochocientos maravedís__227U800
-Más onze arcabuçes el uno grande a modo de mosquete y otro labrado al usso antiguo y tasó unos con otros todos ellos a ducientos y çinquenta reales que montan dos mil setecientos y cinquenta reales que valen noventa y tres mil y quinientos mrs__93U500
-Más otros tres arcabuçes que los dos de ellos tienen los jardineros de la huerta de Leganés y el otro la guarda del bosque de Morata y tasó cada uno de ellos a seiscientos reales que montan mil y ochocientos reales q valen sesenta y un mil y ducientos mrs__61U200
-Más treinta pistolas que hacen quinçe pares y tasó cada uno de ellos a quatrocientos reales q montan seis mil reales y valen ducientos y quatro mil mrs_ _204U
(fol. 509 r)
-Más otras diez y nueve pistolas a modo de carabinas que hacen nueve pares y una pistola y tasó cada uno de ellos a quatrocientos reales que montan tres mil y ochocientos reales que montan tres mil y ochocientos reales y valen ciento y veinte y nueve mil y ducientos mrs__129U200
-Más diez y seis pistolas grandes y chicas que hacen ocho pares y tasó cada uno de ellos a quatrocientos reales que montan tres mil y duçientos maravedís digo reales y valen ciento y ocho mil y seiscientos maravedís__108U600
-Más quatro llaves de pistolas sueltas y tasó todas en ducientos reales que valen seis mil y ochocientos maravedís__6U800
-Más nueve carabinas y tasó cada una de ellas a treçientos reales que montan dos mil y seteçientos reales y valen noventa y un mil y ochocientos mrs__91U800
-Más una bolsa con cinco hierros cargas de arcabuz para perdigones y tasó en cien reales q valen tres mil y quatrocientos mrs__3U400
-Más tres piezas pequeñas de hierro blanco de martillos y abridero y tasó en tres ducados todas q valen mil ciento y veinte y dos maravedís__1U122
-Más tres frascos de cuerno y tasó todos en sesenta reales que valen dos mil y quarenta mrs__2U040
-Más una horquilla grande con, digo una caña grande con su horquilla de plata y tasó en cinquenta reales que valen mil y setecientos mrs__1U700
-Más un báculo abierto por tres partes con cabos de hierro dorados y tasó en ochocientos reales q valen veinte y siete mil y ducientos mrs__27U200
(fol. 509 v)

-Más un báculo pequeño con cabos de plata y tasó en quatrocientos reales que valen treze mil y seiscientos maravedís__13U600
-Más siete cuchillos de monte y tasó en mil reales que valen treinta y quatro mil mrs__34U
-Más dos horquillas de caña con sus hierros y tasó en ocho ducados que valen dos mil novecientos y noventa y dos mrs__2U992
-Más una ballesta y tasó en seis ducados que valen dos mil ducientos y quarenta y quatro mrs__2U240
-Más un montante con su puño de terciopelo negro y tasó en quatrocientos reales que valen treze mil y seiscientos mrs__13U600
-Más dos alfanjes grandes el uno con vaina de terciopelo cremesí (sic) bordada y el otro con vaina de terciopelo negro y guarnición dorada y tasó el cremesí (sic) en quatrocientos reales y el negro en ochocientos que montan mil y ducientos reales y valen quarenta mil y ochocientos mrs__40U800
-Más un alfanje turco guarnecido con testa puño y cabos y tasó en mil reales que valen treinta y quatro mil maravedís
-Más diez y seis espadas con sus dagas de diferentes maestros y todas con sus guarniciones negras y pavonadas y tasó cada una de ellas en trecientos reales que montan quatro mil (fol. 510 r) y ochocientos reales y valen çiento y çinquenta y tres mil y ducientos mrs__153U200
-Más una espada con su daga y la guarnición dorada y tasó en mil reales que valen treinta y quatro mil maravedís__34U
-Más quatro dagas con sus guarniciones y vainas van tasadas con los aderezos__
-Más quatro hojas de espada con sus vainas y sin guarnición y tasó a çien reales cada una que montan quatrocientos reales y valen treze mil y seiscientos mrs_ _13U600
-Más otra hoja grande y ancha también sin guarnición y tasó en los dos quatrocientos Rs__13U600
-Más catorce piececillas de artillería de bronce pequeñas y encabalgadas en sus carritos y otra chica de la misma manera y tasó a seiscientos reales cada una que montan nueve mil reales y valen treçientos y seis mil mrs__306U
-Más una testera de caballo con unas listas doradas y tasó en quatrocientos reales que valen treze mil y seiscientos mrs__13U600
[al marge] en la recámara de la Marqs[a]
-Más una zimitarra chica con su vaina de terciopelo azul y pretina q estaba en la recámara de la S[a] Marquesa y tasó en çien reales que valen tres mil y quatroçientos mrs__3U400

-Más un adrezo de rayadillo dorado y tasó en trecientos reales que valen diez mil y ducientos mrs__10U200
(fol. 510 v)
-Más otro adrezo dorado y tasó en la misma cant[d] de los trescientos reales__ 10U200
-Más otro de labor de dados y tasó en la misma antidad__10U200
-Más otro adrezo leonado la guarnición manquilla y hoja de Fran[co] Ruiz de un corte y tasó en la misma cantidad__10U200
-Más otro adrezo pavonado de labor de tinela y hoja de tres mesas de Ju[o] Martínez y tasó en la misma cantidad__10U200
-Más otro noguerado de labor de huesos de muerto con hoja del mismo y tasó en lo propio__10U200
-Más una espada de a caballo que dio el Duq de Ebuli y tasó en ducientos reales que valen seis mil y ochocientos mrs__6U800
-Más otro adrezo de espada de color de hierro labrado de grano de trigo y rayadillo y tasó en docientos y cinquenta reales que valen ocho mil y quinientos mrs__8U500
-Más dos adrezos de espadas doradas y tasó ambos en seiscientos reales que valen veinte mil y quatrocientos mrs__20U400
-Más otro adrezo plateado y tasó en trecientos reales que valen diez mil y ducientos mrs__10U200
-Más otro adrezo de grano de trigo y tasó en ducientos reales que valen seis mil y ochocientos maravedís__6U800
-Más otro adrezo dorado con nudos y puño de bronce y tasó en trecientos reales que valen diez mil y ducientos maravedís__10U200
(fol. 511 r)
-Más un adrezo liso barnizado con hoja de lomo de Juanes con tiros y pretina de cordobán y tasó en trecientos reales que valen diez mil y ducientos maravedís_ _10U200
-Más otro adrezo de sarmientillo con puño de plata y hoja de Ortuño con tali rojo bordado de seda negra y tasó en la misma cantidad__10U200
-Más un adrezo de quatro puentes pavonado espada y daga y hoja de lomo que fue del duq de Medina y tasó en la misma cantidad__10U200
-Más una hoja de espada de Sahagún y tasó en ciento y cinquenta reales que valen cinco mil y cien mrs__5U100
-Más tres hojas de espada una de Fran[co] Ruiz otra de Simón Rodríguez y otra de Silvestre Nieto y tasó todas en quatrocientos reales que valen treze mil y seiscientos mrs__13U600

-Más una jineta de Mtro de campo con un casquillo guarneçido de oro y tasó en ducientos reales que valen seis mil y ochocientos mrs__6U800
-Más unos anteojos de larga vista y tasó en ciento y cinquenta reales que valen cinco mil y cien mrs__5U100
-Más un par de pistolas y tasó en veinte ducados que valen siete mil quatrocientos y ochenta mrs__7U480
-Más otro par de pistolas y tasó en la misma cantidad de los veinte ducados__ 7U480
Más unos antojos de larga vista y tasó en cinq[ta] treales que valen mil y setecientos mrs__1U700
(fol. 511 v)
-Más una caña con que tomó la posesión del cargo de General de la mar el infante Don Carlos q Dios haya, y tasó en trecientos reales que valen diez mil y ducientos mrs__10U200
-Más onze pares de espuelas doradas plateadas y negras y tasó a seis reales el par que montan sesenta y seis reales y valen dos mil ducientos y quarenta y quatro mrs__2U240
-Más quatro cañas de jineta digo de la India que sirven de jineta y tasó todas en ochenta reales que valen dos mil setecientos y veinte mrs__2U720
-Más un arcabuz que dio a Su Ex[a] Don Alonso de la Carrera y tasó en ochocientos reales que valen veinte y siete mil y ducientos mrs__27U200
-Más dos frascos de cuerno guarnecidos y tasó ambos en quarenta reales que valen mil y quatrocientos sesenta mrs__1U460
-Más dos pares de pistolas con sus fundas de terciopelo carmesí bordadas y tasó a seisçientos reales el par con fundas que montan mil y ducientos reales y valen quarenta mil y ochocientos maravedís__40U800
-Más un cuchillo de monte con su cabo verde y dorado fino y tasó en mil y ochocientos rea[s] que valen sesenta y un mil y ducientos mrs__61U200
-Más quatro frascos de arcabuz el uno con hierro (fol. 512 r) y los tres sin ellos y tasó todos en quarenta reales que valen mil trecientos y sesenta mrs__1U360
Más dos llaves de arcabuz de chispa y tasó ambas en ducientos reales que valen seis mil y ochocientos mrs__6U800
-Más un ingenio como llave de arcabuz y tasó en cinquenta reales que valen mil y setecientos mrs__1U700
-Más dos llaves de arcabuz viexas y tasó en cien reales ambas que valen tres mil y quatrocientos mrs__3U400
-Más una adarga y tasó en veinte reales q valen seiscientos y ochenta mrs__U680
-Más dos frascos de cuerno y tasó en la misma cantidad ambos__U680

-Más otro de terciopelo cremesi (sic) guarnecido y tasó en cinquenta reales que valen mil y setecientos mrs__1U700
-Más otro de arcabuz con su cordón y tasó en veinte reales que valen seiscientos y ochenta mrs__U680
-Más una bolsa de cuero de munición y tasó en quince reales que valen quinientos y diez mrs__U510
-Más una funda de arcabuz y tasó en veinte reales que valen seiscientos y ochenta mrs__U680
(fol. 512 v)
-Más tres bandoleras y tasó todas en treinta y seis reales que valen dos mil y di, digo (sic) mil ducientos y veinte y quatro mrs__1U224
-Más una funda de espada tasó en ciento y treinta y seis mrs__U136
-Más una funda de flechas y tasó en sesenta y ocho maravedís__U068
-Más dos fundas de espada y tasó en ocho reales que valen ducientos y setenta y dos mrs__U272
-Más quatro fundas de arcabuz dos de baq^{ta} y dos de vayeta y tasó en cinquenta reales todas que valen mil y setecientos mrs__1U700
-Más quatro pares de tiros y tasó en ciento y treinta y seis mrs todos p ser viexos__U136
La qual tasación declaró estar bien y fielm^{te} fecha a todo su saber y entender sin fraude ni engaño y ansí lo juro por Dios nuestro señor y a una cruz en forma de derecho y no lo firmó por no saber

Ante mí
Di^{o} de Ledesma
(signed).

Doc. 3.
1655, April 1.
Post-mortem inventory of Marquis of Leganés.
Archivo Histórico de Protocolos de Madrid, Protocolo 6267.

(fol. 472 v)
En la villa de Madrid a primero día del mes de abril de mil y seisçientos y çinquenta y cinco años los dos señores testamentarios del dho señor marqués de Leganés prosiguieron en el dho inventario en la manera siguiente

(fol. 473 r)
Arcabuzes
-Primeramente catorçe alcabuzes (sic) de chispa
Espadas
-Más diez espadas de diferentes maestros q se hallaron con los alcabuzes de arriba en la recámara del dho señor Marqués de Leganés que esté en el çielo
-Más seis alfanjes de diferentes maestros anchos que asimismo se hallaron en dha recámara
-Más diez dagas de las diez espadas de arriba que se hallaron en dha recámara [...]

(fol. 757 r)
Armería q está en la casa prin[1]

=Un estante de nogal con sus puertas de arambre (sic) emberjado (sic) con su cortina de tafetán pajizo y gayado (sic) y dentro del lo sigte
-Un cuerpo de armas entero con todas sus piezas
-Un alfanje corvo
(fol. 757 v)
-Dos rodelas de acero
-Una hacheta de desarmar
-Ocho pares de pistolas de rueda
-Dos turquesas pª hacer balas
-Seis martillos de pistolas

=Otro estante de nogal con sus puertas de arambre y una cortina de tafetán pajizo y gayado (sic) y en él lo sigte
-Un peto de armas con sus hombros braçaletes y morrión
-Dos medios braços de armas y una falda

-Un alfanje
-Un martillo de desarmar
-Una pistola de dos bocas larga
-Ocho pares de pistola de rueda

=Otro estante de nogal grande de la misma manera con su cortina, y en él lo sig^{te}
-Tres cuerpos de armas enteros
-Dos alfanjes corvos
-Una hacheta de desarmar
-Una maza de hierro
-Catore (sic) pares de pistolas de rueda
-Una pistola pequeña dorada

=Otro estante de nogal con su cortina y en él lo sig^{te}
(fol. 758 r)
-Dos medios cuerpos de armas
-Un peto
-Unas armas p^{a} la cabeça de un caballo
-Dos mosquetes vizcaínos
-Una maceta de bronce con un estoque dentro

=Otro estante como los dichos con su cortina y en él lo sig^{te}
-Un cuerpo de armas entero grabado
-Un peto y morrión
-Unas armas p^{a} la cabeza de un caballo
-Una rodela de acero grabada con flueco de seda y oro
-Otro medio cuerpo de armas
-Un montante dorado
-Otro montante negro
-Un alfanje sin guarnición
-Siete mosquetes grandes turquescos
-Quatro ballestas las tres con gafas
-Dos arcos y quatro bolsas con flechas y virotes

=Otro estante como los dhos y en él lo sig^{te}
-Un cuerpo de armas entero grabado con cifras de coronas y eses
-Un peto de armas grabado con las mismas cifras
-Un alfanje

-Una hacheta de desarmar
(fol. 758 v)
-Un mosquete con dos llaves
-Seis mosquetes
-Un par de pistolas con sus fundas bordadas de plata sobre campo verde
-Otro par de pistolas con fundas de terciopelo carmesí con galón de oro

=Otro estante con su cortina como los demás y en él lo sig^{te}
-Un cuerpo entero de armas negro y dorado
-Otro medio cuerpo de armas negro y dorado
-Un peto y un espaldar
-Dos armas p^{a} cabeças de caballos
-Dos alfanjes corvos
-Dos maças de hierro de desarmar con puntas de acero
-Una carabina de rueda la caja de ébano con embutidos de plata
-Dos carabinas de rueda con un martillo
-Quatro carabinas
-Una carabina la caja labrada de marfil y nácar
-Otra caravina la caja con alg^{as} labores
-Un arcabuz el cañón de dos piezas
-Dos pistolas con sus fundas de terciopelo carmesí bordadas de oro con las armas del (fol. 759 r) Marqués mi S^{r}
-Otras dos pistolas con sus fundas las vueltas de terciopelo bordado de oro

=Más se puso por imbent^{o} veinte y seis tiros de arti^{a} de bronce pequeños con sus encavalgamentos
-Una bala de bronce grande hueca

Y en la forma que dicha es el dho Ju^{o} Morales de Gamarra dijo haber hecho el dho imventario y no pareció haber más bienes q poner en él y cada y quando q a su noticia venga protesta declararlos y pidió q el dho alcalde le apruebe y dé por bueno y de él se le entregue un traslado escrito en limpio, y en pública forma p^{a} ponerle con el principal q de los bienes q dejó S E, en la villa de M^{d} se ha hecho y lo firmó Ju^{o} Morales de Gamarra ante mí Al^{o} Sánchez

Auto
En la villa de Morata en nueve de Julio de mil y seis^{os} y cinquenta y cinco años su mrd de Greg^{o} Ruiz de Vivero alcalde ordin^{o} hab^{do} visto el dho invent^{o} y a pedim^{to} hecho (fol. 759 v) por el dho Juan Morales de Gamarra dijo q le aprobaba, y

aprobó por bueno y bien hecho y mandó q de él se le dé un traslado signado y en pública forma al q^{e} p^{a} su validación interpone su authoridad y decreto judicial, y lo firmó, Gregorio Ruiz de Vivero, ante mí Alonso Sánchez
Yo el dho Alonso Sánchez escriv^{o} del número y ayuntamiento de esta villa de Morata presente fui a lo que dho es y en fee de ello lo signé

En testim^{o} de verdad
Al^{o} Sánchez
(signed).

Doc. 4.
1753, September 24.
Inventory of Morata de Tajuña palace.
Archivo Histórico de Protocolos de Madrid, Protocolo 30083.

(fol, 220 r)
Inventario de las pinturas y demás alhajas de Mayorazgo, que se hallan en la casa Palazio de la Villa de Moratta, hecho en 24 de septiembre de 1753 [...]

(fol. 221 r)
dé traslado para el Archivo de mi S^{or} en Sello seg^{do}

En la villa de Morata en veinte y quatro de Sep.^{re} de mill sett^{os} cinquenta y tres el Sor D^{n} Balthasar de Rivera corregidor de esta villa por ante mí el ss^{no} dijo que D^{n} Diego de Almazan Adm^{or} de las Alcab^{s} y demás rentas q en esta villa tiene el Exmo Sr. Conde de Altamira mi S^{or} ha rezivido una carta orden de la Exma S^{ra}. Condesa de Oñate mi S^{ra} para efecto de hacer inventario de los bienes y alhajas q existen en su Cassa Palazio y obedeziendo dha Carta orden y [...] en ex^{on} de dho mandato de la referida ex^{ma} S^{ra} como madre curadora del expresado Exmo Sor Conde de Altamira mi S^{or} se hizo dho Inventario en la forma y manera Sig^{te}.

(fol. 248 r)
En todo el casco del dho palazio se hallaron seis mesas con la armadura de la madera de pino y una piedra q las cinco tienen quebrada y una está entera y toda bien maltratada
Un quadro q estaba en el oratorio de la huerta de la pintura de Sto fran^{co} de Paula de cuerpo entero marco negro maltratado q^{e} se halla colocado en la casa de la plaza y puesto en el inventario de los bienes libres q^{e} están a cargo de dho D^{n} Diego de Almazán.
Una campana q^{e} está en tres palos en el tejado de la casilla de la huerta sin q^{e} haya otras Alhajas de las q^{e} expresa el inventario antig^{o} excepto un poco de Armería q^{e} ha quedado q^{e} solo hay quatro o cinco vestidos para armar hombres, una hoja de una espada, diferentes tiros de Artillería y un mortero, todo peque(fol. 248 v)ño algunos mosquetes con caños, muchas cajas de escopetas y pistolas sin llaves, ni cañones y otros muchos pedazos de Armaduras y morriones de q^{e} tiene la llave del q^{to} donde esta todo, el referido D^{n} Diego con seis armarios de nogal con redecillas de alambre dorado muy maltratados.

De todos los quales dos vienes â exzemcion de los de Armería ha nombrado su Exª p^{r} guardarropa de todos ellos a Pedro de Corpa y Antonia de Bustos su mujer, quienes se han hallado presentes a la formación de este inventario y después de repetir las más expresivas gracias a su ex^{a} p^{r} la nueva honra q se ha servido hazerles se dieron p^{r} entregados de los referidos bienes y se obligan bajo de la lizenzia mancomunidad y demás requisitos q^{e} de marido a mujer son nezesarios, a cuidarlos limpiarlos con las piezas patios y demás ofizinas del dho palazio sin dar lugar a q^{e} se espere menos (fol. 249 r) menoscabo en ellos [...].

LERMA, DUQUES DE

Doc. 1.
1604, May 12.
Testimony of the delivery of spears and pikes to the castle of Ampudia, for the Duke of Lerma.
Archivo Ducal de Medinaceli, Lerma, Leg. 85, Doc. 17.

Hempudia y Mayo 12 de 1604
Testimonio orig^{l}. de haber entregado en el Castillo y fortaleza de Hempudia Juan de Gamarra dos mil y una picas de Guerra y entre ellas una pica negra de 8 varas y terzia de largo para la propia mano del Sor Duque de Lerma

(fol. r)
Henpudia y Mayo 12 de 1604
En la villa de Hempudia estando en la fortaleza de ella que es del Exmo Señor duque de Lerma conde de esta dicha villa &^{a} miércoles a doze días del mes de mayo de mil y seisçientos y quatro años en presencia de mí Pedro de la Vega escribano estando presente don Joan de Azevedo gobernador y alcalde mayor de la dha villa y alcaide del dho castillo y fortaleza pareció Joan de Gamarra Señor de Yvarguen e pidió testimonio de cómo él ha entregado y puesto en el dho castillo en poder del dho don Joan de Azevedo=
Dos mil y una picas de guerra con todos sus hierros azerados a punto de pelear y entre ellas una pica negra de ocho varas y terçia de largo con todos sus hierros y clavazón dorado que ha de servir para la propia mano de Su Exª y más seis lanzas ginetas extraordinarias con todos sus hierros azerados a punto de pelear que ansí mismo han de servir a la propia mano de Su Exª=

Para que de ello conste a Su Exª y a sus contadores= E yo el dho escribano doy fee que vi en la dha fortaleza las dhas dos mil y una picas y seis lanzas ginetas y las conté una a una y el dho don Joan de Azavedo alcaide de la dha fortaleça que presente estaba se dio por entregado de las dhas dos mil y una picas y seis lanças ginetas en la forma arriba dha y de ello otorgó carta de reçibo en forma de derecho y lo firmó de su nombre y aquí se comprenden las mil picas que el año pasado entregó por el dho Joan de Gamarra Blas de Aguirre testigos Manuel de Tovar y Toribio de la Maza estantes en esta villa
Ju^{o} de Azevedo E yo el suso dho Pedro de la Vega esbano p^{co} [...]

(fol. v)
Joan de Gamarra
Armería de Ampudia
Don Juº de Acevedo
(fol. r)
Las picas quedan sueltas y bien y Juº de Gamarra desea servir a V Exª con otras armas lleva recibo suyo de dos mil picas que por su orden he recebido todas son buenas y parezen muy bien en la forma que las he puesto las lanzas ginetas pª la mano de V Exª son admirables todo lo de aquí va bien gracias a nro S^{r} y rogando a Dios por la salud y vida de V Exª a quien me guarde Dios como sus criados habemos menester en 12 días de mayo 604

Juº de Azevedo
(signed)

(fol. r)
Empudia
Juan de Azevedo a Su Exª
-Del Rescibo de 2U picas q ha entregado Juan de Gamarra, de que envía testimº
-Carg^{da}s al armería de Empudia.

Doc. 2.
1607
General Inventory of Francisco Gómez de Sandoval y Rojas, 1st Duke of Lerma. Armoury of Ampudia, Arms in Madrid and other places.
Archivo Ducal de Medinaceli, Archivo Histórico, Leg. 262, Doc. 15 (former, Lerma, Leg. 52,Doc.14).

(fol. 184 v)
Ventosilla
-Doze alabardas con las astas negras [...]
(fol. 185 r)
-Otras doze alabardas doradas con las astas aforradas en terciopelo carmesí tachonadas todas con tachuelas doradas. Tienen sus fundas de frisa colorada [...]
(fol. 186 r)
-Dos ballestas con sus fundas la una de cuero y la otra de bayeta colorada
-Tres gafas y una funda de cuero
-Veinte y nueve virotes y dos viras y una arca con dos cerraduras en que están
-Un frasco de cuerno de buey
-Otro frasquillo de hierro barniçado
-Dos moldes de pelotas
-Un arcabuz con su funda de sayal
-Tres arcos berberiscos los dos sin cuerdas

(fol. 189 r)
Las cosas que hay en Lerma [...]

(fol. 190 r)
Las cosas que hay en la casa de junto al prado de San Gerónimo [...]

(fol. 265 r)
Armas diversas
=*Lo que está en la armería de Empudia* (sic)
-Ocho cotas de malla finas
-Çiento y doçe frenos de caballo, bridones, sin riendas
-Doçe alabardas doradas aforradas en terciopelo carmesí con tachuelas doradas con sus fundas
-Otras doçe alabardas blancas con sus fundas
-Çiento y doçe pares de estribos blancos de hierro
-Veinte y dos rodelas blancas las veinte ligeras y las dos fuertes

(fol. 265 v)
-Çiento y veinte y un pares de çapatos de malla sola y de cuero que se les hiço después
-Çiento y diez y nueve maças de hierro de desarmar
-Çien murriones blancos de infante
-Çiento y setenta testeras de caballo blancas
-Çien pistolas con sus fundas de cuero de vaqueta con su adereço de molde y llave
-Dos petos negros fuertes de prueba açerados
-Treinta y nueve açeros de silla blancos
(fol. 266 r)
-Doçe cascos negros con sus aforros
-Doçe montantes con sus vainas negras y falta a uno el pomo
-Çinquenta arneses de infante
-Çien arneses de caballo ligero y dos de ellos son grabados y dorados con todas sus pieças
-Çien espadas con guarniçiones blancas con adreço de tiros y petrina
-Dos mil y çiento y quarenta y ocho picas con sus hierros y argollas clavados
-Dos picas negras la una sin hierro
-Seis lanças ginetas con hierros açerados
(fol. 266 v)
-Dos cubiertas de hierro de caballos grabadas y blancas
-Dos caballos de bulto hechos a propósito en que están puestas las armas de los caballos
-Dos sillas de cuero blancas con cabeçadas riendas y açiones q[e] se hiçieron a propósito
-Çinquenta y quatro máscaras diferentes una de atrás para adorno de los arneses de infante
-Çiento y veinte pares de espuelas de muniçión
-Diez y ocho espadas de tornear con vainas, tiros y petrina blancas
(fol. 267 r)
-Veinte vaqueros de terciopelo negro con pasamanos de plata, y no son finos, con mangas de raso corta, con trençillas de lo mismo, y manga larga de terciopelo aforrados en tafetán blanco
-Veinte sombreros de terciopelo nº con trençillas de plata no finas los ocho con plumas y los doçe sin ellas, las quales se gastaron en una fiesta que los pajes de Su E. hiçieron a los Reyes en la dha Vª
-Una pieça pequeña de artillería de hasta diez libras de metal con su caxa de madera y guarnición de hierro
-Dos murriones viejos

-Un çestón de plumajes de Indias
(fol. 267 v)
-Seis arcos de madera de indios con sus flechas de cañas
-Dos rodelas de Yndias de madera
-Treçe arneses de infante de a caballo, todos ellos viejos sino uno que era del Rey de Françia, de justa con todas pieças, grabado y blanco
-Dos arneses que eran del S^{r} marqués de Villamiçar, el uno grabado y dorado con solas las pieças de tornear de a pie, y el otro blanco y dorado de color al olio de amarillo y plata con las mismas pieças, de tornear de a pie
-Un arnés a prueba dado de color, al fuego grabado de dibujo, con peto y espaldar con braçaletes gola y murrión y rodela todo a prueba
(fol. 268 r)
-Dos arneses grabados acanalados y dorados con todas pieças y una rodela de la misma labor y hierro, el uno de a caballo y el otro de a pie, y el de a caballo tiene unas armas de caballo de la misma obra y labor con una silla de caballo de terciopelo amarillo, y pasamanos de plata con sus açeros
-Un xirel de terciopelo negro con la lavor de plata escarchada con sus borlas de seda y plata aforrado en damasco blanco y son tres pieças, pescuezo, pecho y ancas
-Ocho sillas de terciopelo negro de respeto, las dos de ellas las dos de ellas (sic) labradas de plata escalchada (sic) de labor de xerer y las seis con pasamanos de plata todas ellas sin açeros
(fol. 268 v)
-Doçe sillas de cuero blanco con guarnición de cuero negro pespunt^{do} y faltan tres pares de çinchas
-Un arcabuz de rueda con el cañón de plata, y las abraçaderas de oro y en la llave cuya chapa es de plata, tiene dos pieças de oro la sobrerueda, y pieça del gatillo que sustenta el tornillo y ansimismo la mira, y el cabo de la vaqueta de oro y está sin punto
-Otro arcabuz de rueda hecho en Alemania con la caxa labrada de hueso
-Otro arcabuz de chispa largo con el cañón turquesco
-Otro arcabuz de guerra el cañón labrado de tauxía con su frasco y frasquillo
(fol. 269 r)
-Otro arcabuz de guerra liso
-Tres arcabuçes turquescos
-Ocho pistolas hechas en Alemania la una de dos cañones con la caxa de madera, con perfiles de marfil y otra con un cañón y la caxa, toda de hierro y los estremos (sic) dorados y las seis labradas las caxas de marfil
-Dos montantes dorados

-Diez espadas de hombre de a caballo, y entre ellas la del Rey de Françia y la de Jullio Çéssar
-Siete alfanjes turquescos ricos el uno con guarnición de oro
(fol. 269 v)
-Quatro ballestas con sus gafas y una alfamilla
-Un arco y aljaba turquesco
-Dos rodelas estraordinarias (sic) la una negra y la otra de colores de la India
-Dos petos fuertes entrambos antiguos, el uno tiene un Cristo pintado
-Nueve báculos de Nra Señora de Monserrate
-Quatro bolsas de muniçión
-Un martillo de armas
-Un mosquete de rueda de tirar a grullas
-Dos arcabuces turquescos
(fol. 270 r)
-Una ballesta sin cuerda y sin gafas
-Una horquilla de campo
-Quatro frascos grandes
-Dos frascos pequeños
-Quatro frasquillos uno de hierro y los otros de madera
-Un montante buido de tres esquinas
-Un cuchillo damasquino con las guarnición dorada, y la vaina de caba (sic) [caoba] y el cabo de marfil que (sic)
-Quatro fundas tres de pistola y la una de arcabuz
-Ocho coseletes de ynfante con golas tudescas y murriones
(fol. 270 v)
-Un arco turquesco con sus flechas y dos cuerdas
-Una llave de armar de uno de los arcabuçes de rueda

=*Lo que hay en Lerma*
-Diez y seis pieças de artillería pequeñas de bronce con sus carretones las quinçe tienen a dos ruedas y la una a quatro
-Tres morteretes de bronce
-Diez y seis cuchares (sic) de metal para cargar las dhas pieças
-Catorçe tacos para el dho efecto
-Tres sacapelotas y trapos
-Una alabarda

(fol. 271 r)
Armas que tiene Su Ex[a] en Madrid

-Un arnés de torneo de acero dorado tiene gola, peto, espaldar, guardabrazos y brazales enteros dos manoplas y su çelada engolada quaxado con sobrepuestos de plata çinçelado (sic) de medio relieve con faldón bordado de canutillo sobre terciopelo negro y franjón de plata tasóse en [...]. Véndase

-Otro arnés de a caballo pavonado de diversas labores que tiene gola peto y espaldar dos carçelas (sic) con sus quixotes y guardabraços brazales de por sí y manoplas y la zelada abierta tiene otra çelada para torneo de a pie y un guardabrazo derecho tasóse en Mill y duçientos reales. Véndase

-Otro arnés negro con sus listas blancas que sirve de seguir la guerra de torneo de a pie y sus piezas de infantería que tiene peto gola y espaldar y çelada y restre (sic) guardabrazos y brazales y manoplas y esarçeles (sic) quijotes y grebas todo junto guadarren (sic) y otro peto y espaldar con su gola y dos escarçelas (fol. 271 v) y sus brazales y guardabrazos otro guardabraço para tornear y una çelada de encaxe y un murrión de infantería y un par de manoplas tasóse en dos mil reales todo. Véndase

-Otro arnés negro y dorado trençado que sirve de infantería que tiene las piezas siguientes, peto espaldar gola pegado en ellos y su murrión dos brazales enteros a lo tudesco con sus manoplas y escarçelas largas y medias quijotes tasóse en çinquenta ducados. Véndase

-Otro arnés negro fuerte que tiene peto y espaldar gola escarçelas y quixotes tranzados y braçales enteros y manoplas y murrión de terciopelo tiene su rodela negra ligera con franjas de seda negra a la redonda en quinientos reales

-Otro arnés de justa, q sirve de seguir blanco con su ristro (sic) escarçelas y placarte para justar, espaldar, gola, celada, dos braçales el yzquierdo con su navaja pequeña y dos guardabraços dos escarcelones de justar una çelada con su taberón (sic), dos tarjetas y una çelada de seguir con su escofia (fol. 272 r) un baberón de justa, dos manoplones dos grebas con sus quixotes, una navaja grande, un placarte, un par de manoplas, una testera de caballo entera, una escofia, una arandela, una bragueta, dos medias grebas, un morrión, unos azeros de silla, un par de escarçelas lijeras, dos puntas de azero, una falseta pequeña, una bufa pequeña, una taleguilla con un ristre de justa y tornillos de muchas suertes tasóse en duçientos y çinq^{ta} du^{o}s. Véndase

-Unas corazas viejas de terciopelo carmesí en çinquenta R^{s}

-Una rodela parda de higuera con su roseta y punta pavonada en medio en sesenta R^{es}

-Diez alfanjes de la Yndia con sus vaynas y cabos negros y unos engastes de cobre dorados a çient R^{s} cada uno

-Una lança de gineta de la Yndia con hierro largo y vayna de vaqueta. Véndase

-Un arcabuz de cinco palmos poco más o menos con caxa de çereço de maestre Simón con llave de rueda con frasco de cuerno y llave tiene funda de frisa colorada en quatroçientos R^{s}

(fol. 272 v)
-Dos pistolas françesas de tres palmos de llaves de rueda con sus fundas de vaqueta con unos bolsillos aforrados de terçiopelo carmesí con pasamanos de oro en doçientos R^{s} cada una
-Otras dos pequeñas con caxas de hierro grabado y lo mismo los cañones hechos en Aragón de quatro palmos con llaves de rueda y fundas de frisa a doçientos R^{s} cada una
-Otras dos de la misma suerte y hechura de dos palmos y medio a çiento y çinquenta reales cada una
-Otras dos pistolas de la misma suerte y hechura de palmo y medio a çiento y çinquenta R^{s} cada una
-Un cuchillo damasquino con el cabo de marfil y el remate de taraçea y la vayna de çapa con contera y brocal y medio todo de plata y un rubinillo y turquesas en treçientos R^{s}
-Una caxa de tres cuchillos con los cabos de hueso y la vayna de çapa y el brocal y contera de plata dorada en çiento y çinquenta R^{s}

(fol. 273 r)
-Un cuchillo de monte el cabo de madera barniçada y el resguardo de cobre y en el puño unas correas negras en çinquenta R^{s}
-Un puñal con dos cortes calados de taugía de oro y la vayna y puño de oro nielado lleno de turquesas y rubíes cabujones malos en seisçientos R^{s}
-Un cuchillo de mesa damasquino guarnecida la vayna de oro y turquesas y rubíes y el cabo de la propia labor los rubíes son cabujones en çient escudos
-Un puñal de cobre con la vayna de madera amarilla y el cabo de cobre liso en çinquenta R^{s}
-Una caxa con dos cuchillos y la caxa guarnecida de cobre barniçada en veinte y quatro R^{s}
-Un puñal de acero en forma de hierro de alabarda de taraçea de oro y el cabo de cobre y el […] de en medio lleno de rubíes guarneçidos de oro con la vayna de madera amarilla en çiento y sesenta $d^{o}s$
-Un cuchillo despuntado con vayna y cabo de marfil grabado todo en doçientos R^{s}

(fol. 273 v)
-Un alfanje damasquino guarnecido de plata contera y brocal con un cuchillo y punçón, la vayna de çapa morada en ocho $d^{o}s$

-Un machete corvo con vayna y cabo de plata dorado en treçientos R^{s}
-Un cuchillo damasquino con vayna de terciopelo carmesí guarnecido de plata y marfil y el cabo de jaspe en seis d^{o}s
-Una caxa de cuchillo con tres cuchillos y un tenedor la caxa grabada de S. Fran^{co} y S. Ju^{o} de marfil en quarenta R^{s}
-Quatro arneses de seguir, de Milán blancos, y grabados y dorados los cantones, cada uno su çelada, una gola, un peto con su ristre y espaldar, dos carçelas (sic), dos quixotes, dos medias grebas, dos puntas de çapatos con su malla, dos guardabraços con sus braçales y sus dos mandiletes, una bufa, una testera de caballo, cinco pieças de silla de caballo
-Otro arnés dorado y grabado de siete listas, una çelada y gola, peto con su ristre, un espaldar, dos carçelas (sic), dos quixotes, dos espinillas, una testera de (fol. 274 r) caballo, dos guardabraços con sus braçales, dos mandiletes, cinco pieças de silla de caballo
-Otro arnés pavonado, de seguir, labrado de relieve, y dorado adamasquino (sic), una çelada, una gola, un peto con su ristre y escarçelas, un espaldar, dos quixotes, dos medias grebas, dos guardabraços con sus braçales, dos pieças de bufa, dos mandiletes, cinco pieças de silla de caballo, una testera de caballo, más tiene otro peto de infantería con dos carcelas (sic) y una rodela grande y un murrión [...] (fol. 298 r)
Cosas que están por inventariar [...]

-Lo que está en el castillo de Denia [...].

Doc. 3.
1608, March.
Memorial of the arms that were to be taken to Ampudia, which had been presented to the Duke of Lerma in March 1608.
Archivo Ducal de Medinaceli, Archivo Histórico, Leg. 262, Doc. 2 (former Lerma, Leg. 53, Doc. 8).

(fol. r)
Memoria de las armas q se han de llevar a Ampudia
Son las que presentaron al duque mi Sr en março de *1608*

Invento ffo 387
Asentadas en la copia del inventario general
y se han de cargar a Pierres en su qta

+Primeramente quatro arneses de seguir de Milán blancas y grabadas (sic) y dorados los cantones con cada uno su selada (sic) una gola un peto con su ristro (sic) y espaldar dos carselas (sic) de quijotes (sic) dos medias grebas dos puntas de sapatos (sic) con su malla dos guardabrazos con sus brasales y sus dos mandiletes una bufa una testera de caballo zinco piesas de silla del caballo. Vendiéronse los dos 2
+Otro arnés dorado y grabado de siete listas primero una selada y gola peto con su ristro (sic) un espaldar dos carseles (sic) dos quijotes dos espinillas una testera de caballo dos guardabrazos con sus brasales dos mandiletes- Zinco piesas de silla del caballo
-Otro arnés pavonado de seguir labrado de relibes (sic) y dorado a damascino una selada. Vendiose
+Una gola un peto con su ristro (sic) y escarselas un espaldar dos quijotes dos medias grebas dos guardabrazos con sus brasales dos piesas de bufa dos mandiletes sinco piesas de silla de caballo una testera de caballo- más tiene otro peto de enfatería (sic) con dos carseles y una rodela grande un morion (sic)

(fol. v)
Andrés de Laredo
Ojo
-Las armas que presentaron al duque mi Sr en março de *1608*
-Están inventariadas a ffo 387/388

Están en caxas en la g^{da} Ropa
P^{a} llevarlas a Empudia

Un arnés blanco grabado y dorado de cinco listas-.

Doc. 4
1614, December 11.
Inventory of the armoury of the Duke of Lerma in Ampudia.
Archivo Ducal de Medinaceli, Secc. Lerma, Leg. 55, Doc. 16.

Hempudia 11 de Diciembre de 1614
Inventario authentico hecho de orden del Sor Duq^{e} de Lerma Marqués de Denia de la Armería que S.E. tenía en dha villa de Empudia
Acompañan otros papeles sobre lo mismo
Inventarios de Hacienda-Hempudia Armería

(fol. 1 r)
En la villa de Hempudia a once días del mes de diciembre de mill y sei^{o}s catorce años ante mí Pedro de la Vega escribano público del número de esta villa e rentas de su Exª el duque de Lerma e testigos pareció press^{te} El señor Juan García Gil Ramírez mayordomo de su Exª estando presente el señor Toribio de la Maça secretario del Santo Oficio de la inquisición de Valladolid como persona que tiene poder del ylustrísimo señor don Fernando de Azebedo arçobispo de Burgos su tío como testamentario que quedó del señor Don Juan de Acebedo que sea en gloria su hermano del hábito de Santiago y alcayde que fue de la casa y castillo y fortaleza de esta villa y el dho señor Juan García Xill hiço demostración de una orden por escrito que tiene de su Exª firmada de su nombre y refrendada de Juan López de Oçaeta su contador y secretario de sus estados su fecha en Torresandino en nueve de noviembre de mill y seiscientos y catorce su tenor de la qual es como se sigue

Por quanto por fin y muerte de Don Juan de Azebedo del hábito de Santiago alcayde de la fortaleça de mi villa de Hempudia está vaca la dha alcaldía y porque conviene poner cobro en las armas y demás cosas que estaban en el dho castillo a cargo del dho Don Juan de Açebedo mando a vos Juan García Xil mayordomo de mis rentas del estado de la dha Hempudia su tierra e partido que habiendo citado a los herederos del dho Don Juan a que se hallen presentes reçibais por ynventario ante escribano que de ello dé fee el artillería peltrechos (sic) de ella coseletes de a caballo y de infantería mosquetes alcabuces y pistolas con frascos y frasquillos moldes sacapelotas y rascadores picas casacas de hombres de armas lanças dardos alabardas espadas rodelas adargas sillas sombreros y todas las demás cosas mías que en qualquier manera se hallaren en el dho castillo y las llaves de él y habiéndolas recibido las ternéis en guardia y custodia como más convenga a su conservación viendo por vista de ojos cada pieza de por sí y haciendo relación por menor si converná limpiar las dhas armas o parte de ellas y la necesidad que tuviere de ello y me la ynviaréis a mi contaduría para que lo tenga entendido y provea lo que conviene y porque Juan López de Oçaeta mi contador que por mi mandado visitó las dhas armas e castillo el qual dize tiene necesidad (fol. 1 v) de reparalo y señaladamente los tejados que se están hundiendo antes que el daño sea mayor os mando que ynvíes la carta que se os ynvía con esta a Pedro Gutiérrez Ramírez veedor de las obras reales de Vall^{id} en que le aviso vaya a essa villa a ver la dha casa y castillo para que lo vea y haga relación de los reparos que sean menester se hagan en él y lo que costarán y me ynvíe la raçón= Tenerlo eis así entendido y quando fuere ahí le abriréis el dho castillo para que vea los dos tejados y las demás partes donde sea nezerario haçer reparos y el conduto de las aguas que se recojen en el patio para que puedan correr sin detenerse y porque de ninguna cosa tiene necesidad la casa como de habitarse y de encender lumbre en ella y para esto conviene que haya casero en ella nombraréisle que sea persona tal qual conviene que viva en la dha casa que la tenga limpia y aseada al qual le señalaréis el aprovechamiento de las tierras que llaman del castillo que solía tener y goçar el dho Don Juan de Açebedo que yo desde luego se las señalo para el dho efecto por esta mi cédula que os mandé dar firmada de mi mano sellada con el sello de mis armas refrendada del infrascripto mi secretario de que se ha de tomar la raçón en los libros de mi contaduría dada en Torresandino a nueve de noviembre de mill y seiscientos y catorce años= El duque y marqués de Denia por mandado de su Exª= Juan López Oçaeta

Conforme a la qual dha ynstruzión el dho señor Juan García requirió al dho señor Toribio de la Maça para que como tal persona que tiene poder del dho señor arçobispo de Burgos como testamentario del dho señor Don Juan de Açebedo difunto alcayde que fue de la dha fortaleça y castillo le dé y entregue por ynventario verdadero las armas artillería peltrechos coseletes de a caballo e infantería e todas las demás armas y cosas y llaves de la armería y demás puertas de la fortaleza que fue a cargo del dho señor don Juan difunto que está presto de lo recibir como su Exª del S^{or} duque mi señor manda haciéndolo ansí hará lo debido en otra manera protesta todo lo que protestar puede y debe y lo pidió por testimonio= y luego incontinente el dho (fol. 2 r) Señor Toribio de la Maça dijo e respondió que él no es heredero del dho señor Don Juan de Açebedo más de que tiene poder del Illmo señor don Fernando de Açebedo arçobispo de Burgos testamentario del dho don Juan su hermano y como tal se hallaba en cumplimiento de la dha orden e institución de su Exª de lo entregar por ynventario siendo de ello testigos Esteban Sánchez y Miguel de Ortega v^{o}s de esta villa= ante mí Pedro de la Vega.

Como se entrega
En la dha villa de Henpudia estando en la casa y fortaleza de ella a los dos onçe días del dho mes de diciembre del dho año de mill y seiscientos y catorce se hiço el entrego de la dha armería y cosas que en ella se hallaron y recibió en sí el dho Sr Juan García Xill que son las siguientes

Cotas
Primeramente se entregaron ocho cotas de malla finas__Viii[o]
Frenos
Yten anssí mesmo recibió y está en la dha armería ciento y doze frenos de caballo bridones sin riendas solo el hierro__cxy
Estribos
Yten recibió y está en la dha armería ciento y doze pares de estribos blancos de hierro sin otra cossa__cxy
Rodelas
Yten veinte y dos rodelas blancas las veinte lixeras y las dos fuertes__xxy
Zapatos
Yten ciento e veinte y un pares de zapatos de mallo (sic) con su cuero__Cxxj
Maças
Yten ciento y diez y nueve maças de hierro de desarmar (sic)__cxix
Murriones (sic)
Yten ciento y treinta y seis murriones blancos de ynfante con las armas reales y aunque en el ynventario antiguo no estaban más de ziento se hallaron treinta y seis más__cxxxvj
Testeras
Yten reçivió y está en la dha armería ciento e quarenta y nueve testeras de caballo blancas__cxlix
(fol. 2 v)
Pistolas
Yten noventa y nueve pistolas con sus fundas de cuero y de vaqueta con su adreço de molde y llave de las quales faltaron dos llaves quando por mandado de su Ex[a] se dieron a don Bernardino de Velasco para la xornada de Valencia__xcix
Petos
Yten recibió y está en la dha armería dos petos fuertes negros de prueba açerados__y
Aceros
Yten treinta y nueve aceros de silla blancos__xxxix
Cascos
Yten once cascos negros con sus aforros__xi

Montantes
Yten doze montantes con sus vainas negras__xy
Arneses
Yten çiento y veinte y seis arneses blancos con el Rey de Françia y los dos de los caballos todos ellos con celadas gola peto espaldar gebras (sic) y bracetes (sic) y manoplas y faldones__cxxvi
Arneses
Yten cinco arneses grabados y dorados y los dos acancelados (sic) con todas sus piezas exceto el uno que le faltan las grebas__v
Lo dho
Yten un arnés açul fuerte con su rodela y murrión peto y espaldar gola y faldones de lo mismo__i
Lo dho
Yten cinquenta arneses de infantería__L
Espadas
Yten cien espadas guarniciones blancas con su adreço de tiros y petrina__c
Picas
Yten dos mill y çiento y quarenta y ocho picas con sus hierros y argollas clavados_ _yUcxlviijº
Picas
Yten otras dos picas negras la una sin hierro__y
Lanças
Yten seis lanças xinetas con sus hierros açerados__vj
Armas de caballo
Yten dos cubiertas de hierro de caballo grabados (sic) y blancas__y
Caballos de bulto
Yten dos caballos de bulto hechos a propósito en que están puestas las armas de los caballos__y
(fol. 3 r)
Sillas
Yten dos sillas de cuero blancas con sus cabeçadas y riendas y açiones que se hicieron a propósito__y
Máscaras
Yten treinta y quatro máscaras diferentes unas de otras para adorno de los arneses de ynfante__xxxiiijº
Espuelas
Yten ciento y veinte pares de espuelas de mulición (sic)__cxL
Espadas de tornear

Yten diez y ocho espadas de tornear con vainas tiros y petrina blancas y faltaron unos tiros y petrina__XViij°
Baqueros
Yten ocho baqueros de terciopelo negro con pasamanos de plata no finos con manga de raso corta con trencillas de plata no finos y manga larga de terciopelo aforrado en tafetán blanco al uno le faltan las mangas__viij°
Sombreros
Yten ocho sombreros de terciopelo negro con trencillas de plata no finas de lo mismo de los vaqueros__viij°
Pieça de artillería
Yten una pieça pequeña de artillería de hasta diez libras de metal con su caxa de madera y guarnición de hierro__j
Murriones viejos
Yten dos murriones viejos__y
Arcos
Yten seis arcos de madera de yndias con sus flechas de cañas que están en la escalera__vj
Rodelas
Yten dos rodelas de yndias de madera__y
Xerel
Yten un xerel de terciopelo negro con clavos de plata escarchada con sus borlas de ~~plata~~ seda y plata aforrado en damasco blanco y son tres pieças pesqueço (sic) pecho y ancas-T[do] plata__j
(fol. 3 v)
Sillas de respeto
Yten ocho sillas de terciopelo negro que llaman de respecto las dos de ellas labradas de plata escarchado de la labor del jerel (sic) y las seis con pasamanos de plata todas de ellas sin aceros son las siete negras y la una naranxada con sus aceros__viij°
Sillas
Yten doze sillas de cuero blanco con guarnición de cuero negro pespuntado y faltan tres pares de zinchas__xy
Alcabuz (sic)
Yten un arcabuz de rueda que vino de las yndias con el cañón de plata y las abraçaderas del cañón de oro y en la llave que ansimesmo tiene la chapa de plata tiene dos pieças de oro la sobrerueda y pieça del gatillo que sustenta al tornillo y ansimismo la mira y el cavo de la baqueta de oro y está sin punto y la guardamano es de oro esmaltada__i

Otro
Yten otro arcabuz de rueda hecho en Alemania con la caxa labra (sic) de hueso_
_i
Otro
Yten otro arcabuz labrado de chispa largo con el cañón tu~~desco~~quesco__i
Otro
Yten otro arcabuz de guerra e cañón labrado de taujía__i
Otro
Yten otro alcabuz (sic) de guerra liso__i
Otros
Yten otros tres arcabuzes turquescos__iij
Pistolas
Yten ocho pistolas hechas en Alemania (fol. 4 r) la una de dos cañones con la caxa de madera con perfil de marfil y otra de un cañón con la caxa toda de cuero y los estremos dorados y las demás que son seis labradas de marfil las caxas__
viijº
Montantes
Yten dos montantes dorados__y
Espadas
Yten diez espadas de hombre de a caballo y entre ellas la del Rey de Françia y la de Julio César__x
Alfanxes
Yten siete alfanxes turquescos ricos el uno con guarnición de oro__vij
Ballestas
Yten quatro ballestas con sus gafas una alfamilla__iiijº
Arco
Yten un arco y aljaba turquesco__i
Rodelas
Yten dos rodelas extraordinaria (sic) la una negra y la otra de colores de la yndia_
_y
Petos
Yten dos petos fuertes entrambos antiguos el uno tiene un Cristo pintado__y
Báculos
Yten nueve báculos de nra S^{ra} de Monserrate está el uno quebrado__ix
Bolsas
Yten tres bolsas de munición las dos de vaqueta de moscabra (sic) leonado y la otra blanca__iij
Martillo
Yten un martillo de armas__j

Mosquete
Yten un mosquete de tirar aguillas (sic) de rueda__j
Arcabuzes
Yten dos arcabuzes turquescos__y
(fol. 4 v)
Ballesta
Yten una ballesta sin cuerda y sin gafas__j
Arquilla
Yten una arquilla de campo__j
Cuerno
Yten un cuerno turquesco pª tener pólvora__j
Frascos
Yten dos frascos uno de cuerno de ziervo y otro de marfil con sus hierros__y
Frasquito
Yten otro frasquito negro con su guarnición dorada que parece ser de ébano__j
Frasquito
Yten otro frasquito ~~negro~~ [crossed out] dorado con sus borlas de seda colorado y de oro__j
Polvorines
Yten dos polvorines pequeños el uno está dorado y el otro llano__y
Montantes
Yten otro montante buido de tres esquinas__j
Cuchillo
Yten un cuchillo damasquino con la guarnición dorada y la vaina de zapa y el clavo de marfil__j
Fundas
Yten quatro fundas las tres de pistola y una de arcabuz__iiijº
Arco
Yten otro arco turquesco con sus flechas__j
Maças
Yten ciento y diez y nueve maças de hierro de armas con sus cordones de hiladillo colorado__cxix
Tocados
Yten se halló en la dha armería en la escalera vᵗᵉ y tres tocados de Yndias de plumas amarillos y colorados__xxiij
Casco
Yten se halló en la dha armería un casco de conchas de la mar__j
Petral
Yten se halló en la dha armería un petral de caracoles de la mar__j
Tabla

Yten una tabla a manera de rabel pintada de diferentes colores__j
Paletas
Yten quatro paletas a manera de peines de Yndias__iiijº
Rodelas
Yten dos Rodelas de madera de Yndias__y
Cofres
Yten se halló cinco cofres viejos encorados__v
(fol. 5 r)
Arca
Yten una arca de pino con su llave e zerradura__j
Cabeçadas
Yten veinte y tres cabezadas de cuero blanco con sus riendas para caballos y guruperas__xxiij
Caldero
Yten un caldero de cobre para servicio del poço__j
Llaves
Yten dos llaves doradas que se dieron a su Exª la primera vez que entró en esta villa__y
Bigas
Yten diez bigas de pino gruesas que están en el portal junto a la caballeriça__x
Tiro pedrero
Yten un medio tiro pedrero__j
Estera
Yten se le entregaron y se hallaron en la dha fortaleça las esteras de zinco salas que están puestas en ellas y más otras dos viexas que están abajo yten más la pieça de la armería que está esterada.

Todo lo qual que dho es el dho Sor Juan García Xil recibió e se dio por entregado de todo ello e cada una cosa en virtud de la dha cédula y orden de su Exª y con esto y de los dos bienes quedan libres los herederos del dho Don Juan de Acebedo de lo que fue a su cargo conforme al ynventº para que no se le será pedido no demandado otra vez todo lo qual recibió de mano del dho Sr Toribio de la Maça es se obligó por su persona y bienes en forma de derecho que agora ni en ningún tiempo no será pedido ni demandado cosa ninguna de las contenidas en este ynventario al dho señor Toribio de la Maça ni a los herederos del dho Sor Don Juan de Acebedo y si lo fuere les sacará a paz y a salvo a su costa indemne y assí ambas partes lo dieron (fol. 5 v)por bueno y lo firmaron de sus nombres en Henpudia a trece de Diziembre de mill y seiscientos y catorce años tºs Pedro de Briçuela y Esteban Sánchez e Miguel de Ortega vºs de esta villa= Jhoan García

Gil Ramírez= Toribio de la Maça de la Cruz y Meneses= Passó ante mí Pedro de la Vega

Va en m^{do} texados y t^{do} tex

Concuerda con su orixinal que queda en mi poder y lo signé

En testimonio de verdad
Pedro de la Vega
(signed).

Doc. 5.
1641, March 8.
Documentation on the arms removed by order of the King from the Ampudia armoury.
Archivo Histórico de la Nobleza, Osuna, C. 1965, D. 25.

Amp^{a} 2
Ynventario de las armas que se sacaron con orden de Su Mg^{d} de la armería de la villa de Hempudia en el año de 1641

(fol. r)
Testimonio dado por Pedro de la Vega es^{no} de Ampudia en 8 de marzo de 1641 con preescrito de la carta orden escrita por el Duque de Lerma Adelantado mayor de la Frontera al concejo de aquella su villa mandándole que a la persona que diputare el Marq^{s} de Castro-fuerte capitán Gral de Artillería le entregase todas las armas que hubiere en su fortaleza y castillo como igualm^{te} del ante dho Marqués nombrando para recibirlas al capitán Christóbal Cabrerizo y Escamilla, y la diligencia de inventario de las que se le entregaron

(fol. 1 r)
Carta
Concejo Justicia y Reximiento de mi villa de Hempudia por carta de 15 de este se sirve Su Majestad Dios le guarde de mandarme que las armas de todos géneros que hubiere en la fortaleça que tengo en esa villa se entreguen a la persona que para rezibirlas tuviere orden del Sr Marqués de Castrofuerte Cap^{án} General de la Artillería de España y aunque en virtud de otra que tuve de Su Majestad de 31 de março de 637 se llevaron las que había a Fuenterrabía por si quedaron algunas os mando que las que hubiere de qualquier género que sean se den a la orden del señor Marqués que os la doy a vos por estar vaca la tenencia de esa fortaleça por muerte del que la servía executado con puntualidad y como Su Majestad quede mexor servido y de las que se sacaren tomaréis rezibo que me emitiréis Dios guarde Luçena 26 de di^{e} 1640 el duque Adelantado Mayor= Villa de Hempudia

Al Conçexo Justiçia y Reximiento de mi villa de Hempudia

El Marqués de Castrofuerte de los Consejos de Estado Guerra y Indias de Su Majestad y Capitán General de la artillería de España= Por quanto Su Majestad que Dios guarde se ha servido de mandar al Sr Duque de Lerma Adelantado Mayor de Castilla entregue las armas que tuviere de todos géneros a la persona que yo nombrare para efectos de su Real servicio por tanto en virtud de la presente ordeno al capitán Xpobal Cabreriço de Escamilla entretenido çerca de mi persona que luego que lo reziba vaya a la villa de Ampudia y de la carta que el dicho señor Duque escribe al Conçejo Justicia y Reximiento de ella que se le entregará con esta y reciba las armas que de todos (fol. 1 v) géneros hubiere en la armería y fortaleza de su Exª que sean de muniçión reservando las grabadas y que fueren de las personas de los señores de esta Casa y para más que servir de muniçión y las picas y las conduçirá a la ciudad de Mérida donde los ha de entregar a Juan Maçías en cuyo poder entran las que se llevan a ella o de la persona que estuviere nombrada por la Justicia y Reximiento de la dicha Mérida con sabiduría del gobernador de aquella ciudad y con yntervençión suya tomará carta de pago del dicho Juan Maçías o persona a quien las entregare y para los portes de las dichas armas y demás gastos que hiçiere con ellas se valdrá de los mill reales de vellón que por orden mía de este día le he librado en el pagador del artª Don Diego Ximénez de Vargas de que dar (sic) quenta a Ventura de Frías secretº de Su Mag^{d} y su contador de la raçón general de ella el qual la tomará de la presente Datª en M^{d} a nueve de enero de mill y seisçientos y quarenta y un años= El Marqués de Castrofuerte= Tomó la raçón Ventura de Frías

Requerim^{to}
En la villa de Hempudia en veinte y un días del mes de enero de mill y seisçientos y quarenta y un años el cap^{án} Xpobal Cabreriço de Escamilla en virtud de la carta del señor Duque de Lerma Adelantado Mayor y de la orden del Sr Marqués de Castrofuerte que son las atrás contenidas requirió a los señores Miguel Rodríguez alcalde y Antonio Camaçón rexidor por ante mí el escribano para que en virtud de ellas le entreguen las armas que hubiere en el armería de esta villa= y vista por sus mds dixeron las obedecían y estaban prestos de acudir al servicio de Su Majestad pero que las armas no corrían por quenta de la villa sino del deán de Palençia y que los demás oficiales de Justiçia y Reximiento no estaban en esta villa no vendrían hasta pasado mañana que se les hiçiese notorio y que en el entretanto no les pase perxuiçio y esto respondieron de que doy fee y lo firmé= Miguel Rodríguez= Antonio Camaçón= ante mí Pedro de la Vega

(fol. 2 r)
En la villa de Hempudia en veinte y tres días del mes de enero de mill y seisçientos y quarenta y un años estando en la sala del ayuntam^{to} de esta villa el dho capitán Escamilla por ante mí el escribano en virtud de la carta mesiva del Exmo Señor Duque de Lerma Adelantado Mayor y orden del señor Marqués de Castrofuerte que son las atrás contenidas requirió a los señores Alonso Martín teniente de

Corregidor de esta villa y don Juan de Morales y Miguel Rodríguez alcaldes ordinarios y Françisco de Puelles y Antonio Camaçón y Françisco Cantarín y Bernabé de Bertavillo Rexidores y Sebastián de Villaverde y Blas de Hempudia procuradores generales del conçexo de esta villa= y visto por sus mds dixeron que están prestos de haçer y cumplir lo que por su parte les toca al servicio de Su Majestad= y las órdenes de su Ex[a] y señor Marqués de Castrofuerte pero que las llaves de la fortaleça y armería de esta villa no corren ni están por quenta de la villa ni las tiene ni nunca las ha tenido la just[a] y reximiento de ella que corre por quenta de don Françisco Garçía Xil Ramírez deán de Palençia por haber muerto Don Baltasar Garçía Xil Ramírez su sobrino alcayde que fue de la dha fortaleça hasta dar quenta de ellas que la villa está presta de dar aviso a dho deán para que en todo se cumpla el servicio de Su Mag[d] y lo que por las dhas órdenes se manda y esto dieron por su respuesta y lo firmaron= Alonso Martín= don Juan de Morales= Miguel Rodríguez= Françisco de Puelles= Antonio Camaçón= Françisco Cantarín= Bernabé de Bertavillo= Sebastián de Villaverde= Blas de Hempudia= ante mí Pedro de la Vega

(fol. 2 v)

Rezibo de Armas

En la villa de Hempudia en primero día del mes de febrero de mill y seisçientos y quarenta y un años ante mí el escribano y testigos el capitán Xpobal Cabreriço de Escamilla en virtud de la carta del Exmo Señor Duque de Lerma adelantado mayor y del Orden del Señor Marqués de Castrofuerte general del artillería de España que son las atrás contenidas confesó haber recibido del l[do] don Françisco Garçía Xil Ramírez deán de la ciudad de Palençia por cuya quenta están las armas de la armería de esta villa por mano de Santos de Uña su criado las armas siguientes

-Primeramente çiento y treinta morriones grabados de Milán blancos labrados
-Yten quarenta y seis petos con sus respaldares (sic) y cada uno con su gola y braçales hasta sobre el codo= de ellos blancos y algunos negros coraças de hombres de armas con sus correas y hebilletas
-Yten veinte zeladas de caballería negras y blancas
-Yten seis rodelas con su punta de diamante dorada tachonadas con fluecos de seda y clavo dorado limpias y nuevas
-Yten dos petos fuertes de a prueba de mosquete con sus correones

Todas las quales dhas armas confesó haber rezibido en virtud de dhas órdenes para las conducir a la ciudad de Mérida según se le manda y confesó haberlas rezibido del dho deán por mano del dho Santos de Uña su criado y en razón de su entrega aunque es bien zierta a mayor abundamiento renunçió las leyes de la entrega prueba y paga y demás de este caso como en ellas se contiene y de ellas otorgó carta de pago y rezibo en forma ante mí el escribano y testigos siéndolo Martín de Tobar y Rodrigo Sánchez y Hernando de Valverde vecinos y estantes en esta villa y el otorgante que yo el escribano (fol. 3 r) doy fee conozco lo firmó= Xpobal Cabreriço de Escamilla= ante mí Pedro de la Vega= e yo el dho Pedro de la Vega escribº de Su Mag^{d} público del número de esta villa de Empuª presente fui a lo que dho es e se sacó del original que queda lo que ante mí pasó en papel e [...] para despachos de oficios en Hempudia en ocho de março de mill y seisçientos e quarenta e un años y va en tres hoxas con esta y lo signé

En testimº de verdad
Pedro de la Vega
(signed).

Doc. 6.
1798.
Report made by Rafael de Floranes y Encinas on Ampudia, including its Castle and armoury, made by order of the Duke of Berwick.
Archivo de la Casa de Alba, Caja 190, nº. 16.

(fol. r)
Reservado
Informe, Noticia, y descripción del Estado de Ampudia
Por el apoderado que ha tomado las posesiones

Motivo del viaje

La villa de Ampudia es la Capital del Estado de este nombre, de que fundó Mayorazgo Regio el Sr. Rey D. Juan II de Castilla para su Mariscal Pedro García de Herrera y sus descendientes legítimos, en consideración a sus grandes servicios, por Real Privilegio en Segovia a 14 de setiembre de 1419 y es el mismo de que se acaba de declarar la tenuta a su XIº nieto el Exc^mo^ Sor D. Carlos Miguel Fitz, James, Stuard, Silva, Stolberg, Toledo, Colón de Portugal, Ayala, Manrique, Fajardo y Dávalos, Ulloa, Fonseca, Castro, Henríquez, Osorio, Andrade, Valcárcel, Roxas, y Herrera, Duque de Berwick y de Liria, Conde de Ayala, Gelbes, Lemos, Villalba y Andrade, Marqués de la Mota, S. Leonardo, y Sarria, Grande de España de Primera clase, &ª; la tenuta, digo, por sentencia de los señores del Consejo de 18 de diciembre de 1797 y R^l^ carta Executoria librada con su inserción en 31 de este siguiente año 1798 refrendada de D. Juan Manuel de Reboles su Es^no^ de cámara, en cuya virtud se han tomado por mí a su nombre y de su Exma madre y tutora la S^ra^ Duquesa viuda de los mismos títulos Dª María Teresa de Silva y Palafox, y con el poder especial que S. E. se sirvió otorgar a mi favor en Madrid a 29 de dicho mes de Enero, por testimonio de D. Gervasio Fernández Yzquierdo Esc^no^ de S. M. y su Notario público en todos sus Reynos y Señoríos, las posesiones reales y efectivas, generales e individuas, quietas, pacíficas, y sin contradición q^e^ van originales en las diligencias a continuación de la misma R^l^ Carta Executoria, y así en dicha Villa de Ampudia, como en los demás Lugares y pertenencias del Estado, con general aprobación de los mismos Pueblos y de todas las gentes.
(fol. v)

Camino a Ampudia

Dista esta Villa, de Valladolid 6 leguas al Nordeste, caminando hacia lo interior de Campos. Hállase entre las Ciudades de Rioseco y Palencia, casi a la mitad del camino de una a la otra, que es de 8 leguas, Palencia al Norte y Rioseco al Poniente y forman con Valladolid como una especie de trángulo. A las dos leguas de esta Ciud se pasa por la Villa de Cigales, y no hay en el medio más población hasta Ampudia.
(fol. r)

25. *Castillo y Fortaleza*

Paso a otra cosa: a decir algo del Castillo y Armería, porque se deseará, aunque esto toca ya más a la curiosidad que a la necesidad de lo urgente en el día; y seguramte yo no puedo hacer ahora una descripción completa porque no le ví todo en su interior, si solo (sic) las piezas más principales. El Alcázar pues, Castillo y Fortaleza de Ampudia (qe así se llama la Fundación) habitación qe en otro tiempo fue a los señores, y qe no ha mucho lo era del antepenúltimo de ellos D. Esteban de Herrera canónigo de Toledo qe vivió aquí a temporadas, esta Fortaleza digo, es famosa, y se hace de ella mucha memoria en las guerras civiles de las Comunidades de Castilla, por los años 1521. Véase al Sr. Sandoval en la Historia de Carlos V lib. VIII, ʃ. 16. Se halla situada al extremo de la Villa hacia Poniente arrimada a una cuesta por donde se va a Valoria. Cerca tiene algunas heredades propias, la Huerta de hortaliza con anoria (sic) y una Iglesia de Santiago, antiguamente Parroquia del Castillo y Barrio de él, y hoy ermita, donde se les dice Misa los días festivos, que antes era señaladamte a las 11: hoy lo hacen los Señores del Cabildo qe corren con todo, a la hora qe se les antoja, aunque sea temprana y muy incómoda; lo qe en los principios de la novación no debiera haberse tolerado, porque vendría sin duda de asiento con los Señores o Fundación particular, estando como está en su Territorio castelar.

El Castillo es un edificio vasto y soberbio, con sus dos torreones almenados, foso, contrafoso, petriles, antepechos, cubos, fortines, muros alares hasta alcanzar a dos de las Puertas de la Villa y sus Castilletes pa guarnecer la entrada: Y en fin, todas las demás circunstancias y seguridades de este género de fortificaciones y baluartes, al uso antiguo: en qe se conoce bien el poder de los señores pasados y su pericia militar, pues hoy solo un Príncipe fuera capaz de aquella obra; porque aunque a mano el material, solo de la mano supera la imaginación. Padece ya varias y lastimosas ruinas por partes, unas antiguas, otras modernas; aquellas no sé por qué; estas por la damnable condescendencia de los Administradores en permitir a los particulares excavar piedra de la mejor labrada y la más pronta en los cimientos pa sus ruines edificios, que todos juntos no valían o que este; aunque no fuera más q^{e} por conservar su hermosura y un monumento de antigüedad tan grandioso y honorífico a los Señores. Hombres bárbaros sin gusto de nada, que no sea su tacaña y devoradora lucralidad torpe e infame. Yo confieso los hubiera colgado de las almenas para perpetuo escarmiento. En fin, con estas encetaduras él con el tiempo se irá viniendo a tierra, y no tardará si no se impide con rigor la continuación ¿Qué sabemos lo que les habrá valido el material? Es regular lo hayan vendido a buen precio.

26. *Armería y su Sala*

Lo más precioso y memorable en él es la Sala de Armería, digna de haberse mirado como un relicario de la antigua paciencia, fortaleza, coraje, valor, denuedo, genio, y pericia militar de nuestra invicta Nación, infatigable siempre a las faenas de la guerra, siempre en el campo lidiando por nosotros y la Religión Santa q^{e} puede hacernos felices; siempre comprándonos a azumbre de sangre el palmo de tierra q^{e} pisamos, y nos dexaron por herencia sempiterna y pacífica. Dios les habrá dado el premio, y le dará a los que no le hayan recibido, por su ardiente y católico zelo. Los instrumentos con q^{e} se obraron tantas y tan insignes hazañas, en petos, spalderas (sic), canillares, espadas, lanzas, mosquetes, morteros, cañones, yelmos, morriones, escudos, cotas, corazas, casquetes, alfanjes, cuchillos, hachas, caballos encubertados, y otras armas y armaduras, así eqüestres, como pedestres, de hierro y azero, son en esta Armería, o por mejor decir Fragua de Marte, en tan grande número, q^{e} p^{a} recontarlas, y dar a cada una su propio nombre, definición, descripción, oficio y mérito, especialmente el fabril o de la hechura (¡manos preciosas!), sería bien necesario que una persona inteligente se estuviese allí encerrada algunos meses. La lástima es que falte casi la mitad, porque todo el q^{e} ha ido a pretexto de verlo, ha llevado algo, por la demasiada franqueza de los Administradores en entregar la llave y no zelarlo. Una carretería entera me dicen (fol. r) haber visto trasladar los nacidos de esta Armería a la famosa de Cuéllar, propia del Sor Duque de Alburquerque, sin saberse con qué título o motivo, ni en virtud de qué donación: Y esto sin lo q^{e} pasó a Zamora para las guerras de 1762 con Portugal, y otras extracciones diferentes.

Alhaja pues, tan preciosa, y que tanto autoriza la fama y nombre de los Señores del Estado, de q^{e} hizo tanta vanidad y gala el Cardenal Duque de Lerma en su testamento, en medio de ser solo un poseedor transeúnte, y de q^{e} debe hacerla todo aquel q^{e} se precie de estimar semejantes Antigüedades Militares, y unos monumentos tan brillantes de sus pasados, requería y parecía estar exigiendo de justicia del zelo del actual y de la cuidadosa Ilustre Madre que le rige, se hiciere un Inventario formal con la posible especificación; de que una copia estuviese en su Archivo, y otra en la misma Oficina, asegurando las puertas y candados, y duplicando o triplicando las llaves p[a] q[e] no se pudiese abrir con tanta facilidad y al arbitrio de un hombre solo como hasta aquí, en q[e] ha consistido su perdición. El Sr. Rey D. Felipe III es regular hubiese tenido el gusto de ver y admirar este copioso Almazén de Armas, quando estuvo en Ampudia en Enero del año 1602; y situada la Corte en Valladolid, el Duque de Lerma su Primer Ministro, le sacó de ella p[a] divertirle en sus Estados. [...].

LOS VÉLEZ, MARQUESES DE

Doc. 1.
1579, April 1.
Inventory and appraisal of Marquis of Los Vélez assets.
Archivo Histórico de Protocolos de Madrid, Protocolo 992.

(fol. 1 r)
1 abril 1579
Tasación de los bienes del Illmo Sr Marq^{s} de los Vélez

En la noble villa de Madrid a primero día del mes de abril de mill y quiºs y setenta y nueve aºs en presençia de mí el presente escribano público yusoescrito los señores don Gaspar de Mendoça canónigo en la Santa iglesia de la ciudad de Toledo y licen^{do} Gonçalo del Castillo fiscal de la cárçel real de la corte de Su mg^{d} como albaçeas testamentarios del Illmo S^{or} marqués de los Vélez don Pedro Fajardo mayordomo mayor de la reina nra Sª ya difunto q sea en gloria, dixeron como tal testamentarios que conviene q se tasen los bienes muebles ansí joyas de oro y plata y todos los bienes que quedaron y fincaron por fin y fallesçimiento del dho S^{or} marqués en sus bienes y hazienda, y se averigüen y los [...] y valor de ello y de cada [...] de ello [...] necesº y por agora y en todo tiempo [...]

(fol. 18 r)
-Yten una caxita de madera dentro de ella muchas piedras de mynas de arcabuz en un ducado

Ballestero
-Yten más una ballesta con una verga de dos empulgueras y el tablero de serbal la gafa redonda y pavonada en cien R^{es}
-yten otra ballesta de çerbal con una verga hecha en Madrid y una gafa pavonada redonda en ciento y treinta y dos R^{es}
-Yten otra ballesta con una verga de la Puebla y el tablero de madera de Indias sin gafa tasada en treinta y tres R^{es}
(fol. 18 v)
-Yten otra ballesta de ébano con sus gafas tasada en quatro duºs
ballestº Pº de la Fu^{te}
-Yten otra ballesta pequeña con sus gafas tasada en quatro duºs

-Yten otra ballesta de çerbal con una verga de la puebla sin gafas tasada en otros quatro du°s
-Yten más un tablero y una verga por encabalgar tasada en tres du°s
-Yten más una horquilla de ballesta tasada en quatro R^{es}
-Yten dos vergas de la Puebla tasadas en quarenta R^{es}
-Yten otra verga de Çamora tasada en doze R^{es}
-Yten una aljaba de cuero blanco con una doçena de jaras y tres virotes herrados tasados en diez y ocho R^{es}
-Yten un carcax verde con diez jugaderas tasado en v^{te} y dos R^{es}
-Yten una ballesta alemana de garrucha e ynvençión con su garrucha y caja de madera y jugaderas y çiertas herramientas dentro tasada en cien R^{es}
(fol. 19 r)
-Yten un carcax de terçiopelo azul bordado de oro con cinco virotes sostrados y quatro tragazetes tasado todo en çinquenta y cinco R^{es}
-Yten otro carcax vede bordado de oro con cinco virotes y tres tragazetes y dos jaras tasado todo en cinquenta y cinco R^{es}
-Yten un cuchillo de monte mediano metido en una caja con su cuchillo y punzón y martillo tasado en catorze Reales
-Yten otro cuchillo grande de monte metido en su caxa verde con su cuchillo y punçón tasado en diez y ocho R^{es}
-Yten una caja de cuchillos de Alemania con tres cuchillos pequeños y un punzón tasado en diez y seis R^{es}
-Yten otra caxa de cuchillos de Alemania el uno grande e quatro pequeños y un punzón tasado todo en v^{te} y dos R^{es}
-Yten sesenta e quatro virotes sostrados tasados en v^{te} y dos R^{es}
-Yten dos palos de ballesta tasados en ocho R^{es}
(fol. 19 v)
-Yten un pistolete de cámara con un cañón de dos palmos y medio con su funda y bolsa con quatro cargas de cámara tasado en çiento y diez R^{es}
-Yten otro arcabuz de tres palmos y medio de cañón labrada de hueso blanco con su funda tasado en ocho du°s
-Yten otro arcabuz de tres palmos de cañón tasado en seis du°s
-Yten otro pistolete con una llave para dos tiros cada uno de por sí de dos palmos de cañón con su funda tasado en ochenta y ocho R^{es}
-Yten otro pistolete con dos cañones en una llave con dos ruedas con su funda tasado en ocho du°s
-Yten otros dos pistoletes de un palmo de largos metidos entrambos en una funda tassados en ocho du°s
-Yten otros dos pistoletes metidos en otra funda pequeños tasados en ocho du°s

-Otro arcabuz de tres palmos t^{e} su funda tasado en ocho du^{o}s
-Otros dos pistoletes sin fundas (fol. 20 r) de menos de un palmo de larga tasados entrambos en seis du^{o}s
-Otro pistolete de ocho dedos de largo tasado en dos ducados
-Yten una llave de arcabuz de Alemania yzquierda tasada en tres du^{o}s
-Yten otra lave de Alemania rueda por de dentro tasada en quarenta R^{es}
-Más otra llave de Alemania tasada en quarenta y quatro R^{es}
-Yten otra llave pequeña de Alemania tasada en v^{te} y dos R^{es}
-Yten otra llave de mecha y de pedernal tasada en v^{te} y dos R^{es}
-Yten un frasquillo de hierro para arcabuz tasado en onze R^{es}
-Una ballestica p^{a} un niño con su gafa de palo tasada en ocho R^{es}
-Yten un frasquillo de hierro dorado tasado en diez y seis R^{es}
-Yten un venablo con su cuchilla y dos cañones de arcabuz a los lados tasado en sesenta y seis R^{es}. Tiénelo don Gaspar
-Yten otro venablo con hierro llano tasado en v^{te} y dos R^{es}
-Yten quatro hachas polacas en quatro du^{o}s
-Yten seis barricas de açero en ocho R^{es}
-Un alfanje turquesco en ducientos R^{es}
-Un çinto de monte en tres R^{es} [...]

(fol. 24 r)
Hallóse en la recámara de su S^{a} lo sig^{te}
-Yten en un cofre tumbado viejo se halló lo sig^{te} [...]
-en otra arca
(fol. 30 r)
Morrión
-Yten otro morrión de castor forado en felpa viejo tasado en quatro R^{es} [...]

(fol. 32 r)
Espadas
-Yten una espada q dizen del marqués don Luys tasada en doze du^{o}s (fol. 32 v) la guarniçión dorada e su talabarte y petrina de terciopelo negro y la clavaçon dorada con su daga de diferente labor tasada en doze du^{o}s
-Yten otra espada con su guarniçión labrada de color de hierro y con su daga y los puños de hilo de oro y con vaynas de terciopelo negro y talabartes de lo mismo con la clavaçón de la mesma labor tasado en tresçientos R^{es}
-Yten otra espada de color de hierro lisa con su vayna de çapa y su daga tasada en dos du^{o}s
-Yten otra espada barniçada negra con su daga tasada en quatro du^{o}s

-Yten otra espada ochavada negra con su vayna de çapa vieja tasada en dos du°s
-Yten otra espada ancha con una guarnición barniçada negra con vayna de çapa tasada en tres du°s
(fol. 33 r)
-Yten otra espada ancha su guarniçión negra tasada en dos du°s
-Yten una hoja de espada de Toledo sin guarniçión con su vayna de çapa que es ancha y corta tasada en dos du°s
-Yten siete hojas de espadas de las muy viejas anchas sin vaynas envueltas en un ovillo tasadas en siete du°s
-Yten otra espada con una guarniçión muy vieja ancha y corta de las muy viejas tasadas en dos du°s
-Yten un alfanxe labrado a la morisca la guarniçión de plata con su tahalí de seda carmesí y con sus letras de oro tasado en seis du°s
-Yten otro alfanxe a la turquesca tasado en dos ducados
-Yten un cuchillo grande con la guarniçión de plata y con su vayna de terçio(fol. 33 v)pelo verde pespuntada tasado en dos du°s
-Yten otro cuchillo de monte grande la guarniçión de hierro y puño de cuero tasado en dos ducados
-Un talabarte dorado viejo en quatro R^es
-Yten un puñalejo a la turquesca de atauxía tasado en un ducado
-Yten unas mangas de malla tasadas doze du°s
Frenos
-Yten diez y seis frenos de caballos hechos en Mylán tasados a tres R^es cada uno
Copas de hierro doradas de frenos
-Yten veynte y tres pares de copas de hierro doradas de frenos tasadas a dos R^es cada una
Herradura
-Yten una herradura de caballo con su tornillo tasada en un ducado
Espuelas
-Yten cinco pares de espuelas hechas en Mylán tasadas en dos Rs cada par
Estribos
-Yten un par de estribos hechos en Mylán tasados en cinco Rs
(fol. 34 r)
Espuelas
-Yten cinco pares de espuelas sin pico para ençima de hielos tasadas a real cada una
Barriles de pólvora
-Yten dos barriles pequeños de pólvora de Alemanya a dos Rs y medio la libra
Piedra blanca

-Yten una piedra blanca quadrada para ara tasada en dos Rs
Frasco
-Yten un frasco turquesco pa agua tasado en seis Rs
Caxa de peines […]
(fol. 55 r)
-Ocho florines de Alemania y sesenta y qtro pieças de monedas de Alemania […]
-Quarenta y tres medallas antiguas pequeñas de plata
-Doze medallas pequeñas de plata
(fol. 55 v)
-Yten un estrum^{to} (sic) matemático de latón a manera de estrolabio (sic) tasado en sesenta y seis Rs
-dos caxuelas de retrato de plomo la una del rey y la otra de la reina se tasan en ocho Rs […]
(fol. 56 r)
-más una piedra ymán rica en una caxica pequeña en quatrocientos reales
-yten otra piedra ymán pequeña buena en diez ducados
-yten otra piedra ymán q es muy ruin en ocho reales
-otra piedra ymán que es muy ruin en otros ocho Rs
-Una caxa de lata con un instrum^{to} de latón en diez reales
-Una caxa de lata con un abecedario de letras de açero en cinquenta reales
-Un ynstrumento de papeles de trazas en quatro Rs […]
-yten una caxita chiquita con çera de las Yndias en un real
(fol. 57 v)
-Un papel con ocho compases de hierro y latón en treinta Rs […]
(fol. 58 r)
-una caxa de madera blanca con un estrolabio en diez du^{o}s
-Un quadrante de estrolabio en çien reales
-Una caxa negra de cartón con tablas de cartón de astrolabio tasadas en quarenta Rs […]
-Unos espolones de hierro para andar por el hielo en tres Rs […]
(fol. 59 v)
-Un pedaço de cera de las Yndias […]
(fol. 60 r)
En otros dos cofres había lo sig^{te}
-Un instrum^{o} de astrolabio largo y angosto tasado en çinquenta Rs
Arcabuz
-Un arcabuz a manera de mosquete de pedernal en siete du^{o}s
Ballestas
-dos ballestas con tres gafas en diez y seis du^{o}s

Bolsas
-dos hierros de bolsas para caça tres Rs
Sombrerillo
-Un sombrerillo de hierro guarneçido de terciopelo es de las armas con lo qual ha de ir tasado
Lo mismo
-Otro sombrero de lo mismo lo mismo
Frenos
-once frenos de la brida hechos en Milán a dos Rs y medio cada uno […]
-ocho compases de hierro grandes tasados en quarenta Rs […]
Compás
-Una bolsica con un compás y otros herrecillos en dos Rs
Diente de jabalí
-Un diente de jabalí en dos Rs
(fol. 60 v)
Arcabuz
-Otro arcabuz pequeño en cinco du^{o}s […]
-Un torno de hierro en tres du^{o}s
-una caxa con perdigones de arcabuz en medio real […]
Cofres de armas
-dos cofres de armas pavonadas y grabadas […]
-Un arca mediana forrada en cordobán y por de dentro en bayeta verde y dentro un escritorio en que había tres hábitos de Santiago de tr^{o}p^{o} otros tres de rana otros tres de razo y una tabla […]
-y más un ynstrumento de astrolabio en ocho Rs
-y en un caxón del dho escritorio había compases reglas y quadrantes de latón y hierro y un velero pequeño […]
Yten en otro caxón tiene lo mismo […]
(fol. 61 r)
Carcaxes de virotes
-dos carcaxes de ballestas llenos de virotes […]
papeles de traças
-Algunos papeles de traças de marca mayor y otro papel de estampas
Agujas de marear
-dos agujas de marear […]
Estampas
-Un libro pequeño de estampas- tres Rs […]
Guar^{on} de daga
-Una guarnición de una daga […]

(fol. 61 v)
-Un instrumº a manera de astrolabio dorado de latón en seis duºs
-Otro instrumº menor que es de plata sobredorada en quinze duºs
-Otro instrumº grabado de latón dorado para tomar la altura del sol en diez duºs
-Un quadrante de latón de oronçio en diez Rs
-dos compases de latón sobredorados en una caxa metidos en tres duºs
-Un compás de plata se tasó en el peso y en diez Rs de hechura
-unos hierros de traçador q son onçe hierros en v^{te} Rs
-Otros hierros q son compases q son diez pieças en diez Rs […].

MEDINACELI, DUQUES DE

Doc. 1.
1594, July 6.
Inventory of Juan de la Cerda, 5th Duke of Medinaceli. Inventory of the Palace of Cogolludo.
Archivo Histórico de Protocolos de Madrid. Protocolo 628.

(fol. 647 r)
En la villa de Cogolludo a seis días del mes de julio de mil q^{o}s e noventa e quatro años ante el doctor Molina alcalde mayor en la dha villa e su marquesado e por ante mí el presente escriva^{o} pu^{co} e testigos de yuso escritos pares^{o} presente Juan Apariçio alguacil mayor de la dha villa e pres^{to} esta carta requisitoria atrás cont^{da} del tini^{te} de correg^{or} de la villa de Madrid y pidió cumplimiento de ella e guer^{a} e testimonio siendo testigos Miguel López y Antón Fernández v^{o}s de esta villa El dho doctor Molina alcalde mayor habiendo visto la dicha requisitoria dixo que mandaba y mandó se cumpla como en ella se contiene y en su cumplim^{to} se haga el dho ynventario e que parezcan a lo hazer Lucas de Çavala mayor^{mo} de hazienda de Su S^{a} del dho don Ju^{o} de la Çerda duque de M^{a}zeli difunto y Pablo de Estrigana su guardarropa y las demás personas que convengan e tengan noticia de los bienes e hazienda del dho duque con la solenidad y Ju^{to} necesa^{o} e fecho se proveerá just^{a}= otrosí m^{do} que asistan al dho ynventa^{o} Ju^{o} del Castillo y el dho Pablo de Estrigana v^{o}s de esta villa- Los quales por el pres^{te} nombra por tales depositarios de los dhs bienes a quien se entreguen y se obliguen conforme a la dha req^{a} y lo cumplan so p^{a} de cada cinco mil mrs p^{a} la cam^{a} de Su S^{a} ansí lo proveyó de m^{do} e firmó t^{o}s dho

El D^{or} Molina
ante mí Morales scr^{o}
(signed)

Este dos día yo el dho escri^{o} lo notifiqué a los dos Ju^{o} del Castillo y Pablo de Estrigana en sus pers^{as} t^{o}s de

Morales scr^{o}
(signed)

En Cog^{do} en el dho día mes e año yo el dho escriv^{o} notifiqué al dho Lucas de Çavala en su pers^{a} y de ello doy fee

Morales scrº
(signed)

Después de lo suso dho en el dho día seis de julio del dho año de nov^{ta} e quatro ante el dho D^{or} Molina alcalde mayor estando p^{te} el dho Juº del Castillo depositº por ante mí el pres^{e} scrivº e tºs parec^{ó} el dho Pablo de Estigana a hazer el dho ynbenº y se le tomó el pº jur^{to} por Dios nro S^{or} e sobre una señal de cruz en forma de drº so cargo de [...] se le m^{dó} y él prometió de manifestar todos los bi^{es} muebles e raíces que quedaron por fin y muerte de Su Sª el dho duq don Juº Luis de la Çerda [...](fol. 647 v) sigura y amen tºs dhs los bienes que mani^{tó} son los siguientes

-Primeram^{te} una media tienda de campo liada en honçe líos de angeo
-Yten seis banquillos con dos tableros sobre que está la dha media tienda
-Otra tabla con otros dos banquillos de pino
-Un ajedrez de nácar en una caja sin pieças
-Una arquilla negra con clavazón de armas y hebillas e otras cosillas de hierro
-Dos caparazones de terciopelo negro viejos
-Una almohada de trºpelo verde vieja con damasco por abajo
-Otra almohada de damasco verde viejo
-Tres gualdrapas de paño negro viejo
-Dos pedazos de bayeta negra vieja de hasta vara y media
-Una escarcela de cuero bayo pª munición
-Quatro arpilleras viejas de angeo
-Quatro piezas de cubiertas de caballo de armas de cota enlaçada y sobredorada
-Dos cotas de malla con medias mangas con otros dos pedazos de malla de por sí a modo de faldones
-Quatro servillas blancas de cuero con puntas açeradas
-Unas coraçinas aforradas en tela de brocado con franjas de trºpelo negro
-Otras corazinas aforradas en raso carmesí con tachuelas doradas
-Otras corazinas ricas enteras aforradas en trºpelo carmesí con la misma clavazón con dos medios hombros
-Otros dos hombrillos aforrados en trºpelo carmesí
-Tres trompetillas quebradas
-Un cofre viejo sin llave donde están las dhas corazinas
(fol. 648 r)
-Yten un cajón de madera de pino dentro del qual se halló lo sig^{e}
-Una guarniçioncilla de espada dorada con otras dos guarniciones de espada viejas
-Unas cubiertas viejas de caballo de librea de frisa blanca e bocacín amarillo

-Yten una arca vieja dentro de la qual había lo sig^e
-Un atril quebrado de azofar
-Veinte e tres garabatos de hierro de [colgar] paños
-Unos papeles viejos e rotos que parezen ser traças
-Dos palas de jugar a la pelota
-Unos cartapacios desquadernados y un libro también desquadernado de los mandamientos de la ley de Dios
-Yten la Elíada de Homero en ytaliano
-Yten otra arca vieja en la qual había lo sig^e
-Una caxa redonda y en ella un puño de espada de hilo de oro, quatro alamares negros de seda, un cubilete de marfil, una gorra de tafetán, otra gorra de raja aforrada en raso viejas, una gorra sombrero de tafetán negro viejo
-Un libro desquadernado que llaman ynstruçión e dotrina de cómo todo xpiano debe oyr misa de fr Br^me de Miranda arçobispo de Toledo
-Quatro mançanillas de caña
-Una bolsa en que estaba un caracol de perla quebrado en que hay unos pocos de dientes
(fol. 648 v)
-Dos pares de medios borceguíes blancos
-Quatro cordones de seda negra para dosel
-Un mosqueador de palma con un palillo de palo vinoso
-Dos tinteros de cuerno
-Dos retratos de Jesuxpo nro S^or e de la Virgen nuestra señora de Brones (sic) en unas cajitas de nogal
-Un par de zuecos italianos
-Unas servillas a la morisca coloradas
-Una cazuela de barro con tachuelas y rosillas de hierro para litera
-Tres pieças de dosel de raso verde bordadas de tela de oro
-Un bonete de lienzo colchado
-Un sayete de raso blanco para armas guarnezido de tr^opelo blanco
-Un cordón de fraile de seda blanca e negra
-Un listón de tela de oro e seda carmesí
-Una colcha de tafetán carmesí con franjas de oro
-Un bonete de turco apolillado labrado de seda verde
-Dos aziones de cuero nuevas
-Una cerraja de picaporte nueva y buena
-Una espuela de palo granadino
-Dos almohadas de tr^opelo morado viejo
-Un faldón pequeño de tr^opelo morado vejo

(fol. 649 r)
-Yten una zuela (sic) de madera con unos papeles de traças viejos
-Ocho tornillos con quatro codaduras de hierro
-Una bigornia pequeña
-Una puente de hierro de juego de trucos
-Un hierro de lança viejo
-Un retrato de cera negra y blanca puesto en una piçarra redonda con un letrero que diçe Juanes de la Çerda
-Tres cazuelas de madera con retazos de cuero e palillos e cosas de hierro viejos

Por hoy se quedó començado el dho ynven^{o} siendo t^{o}s Miguel López e Antón Fern^{dez} v^{o}s de Cog^{do}

Ante mí Morales scriv^{o}
(signed)

E después de lo suso dho en la dha v^{a} de Cog^{do} a siete días del dho mes de julio del dho año de mil q^{o}s y nov^{a} y quatro a^{o}s en pres^{a} del dho alcalde mayor y por ante mí el dho escriva^{o} e t^{o}s pres^{te} el dho Ju^{o} del Castillo depositario e dho Pablo de Estrigana prosiguiendo el dho ynventa^{o} puso de manifiesto haber los demás bienes sig^{tes}

-Una arquilla vieja de coche encorada
-Dos candeleros de azofar viejos
-Dos pedazos de unos morillos de alatón
-Un zerrador de puerta de hierro
-Otro cerrador de la misma forma quebrado
-Diez e seis piezas de hierro [...] de pescar
-Unas tijeras de despabilar
-Una [...]
(fol. 649 v)
-En una cazuela unos hierros viejos
-Dos fuelles chicos de mano
-Una redezuela de pájaros
-Una montera de paja y dos cantillos viejos
-Una arquilla de nogal vieja con cerradura
-Una caja vieja con su tapa de pino
-Un modelo de monesterio
-Una aguja de marear vieja quebrada

-Quatro atabales rotos
-Un estoque ancho dorado
-Una espada y daga con cuchillo e punzón dorada
-Otra espada ancha pavonada
-Otra espada ancha vieja
-Un estoque dorado viejo
-Otra espada vieja dorada
-Un estoque buido dorado
-Un estoque de tres esquinas
-Otra espada barniçada
-Otra espada ancha barniçada
-Una espada ancha de más de mano
-Otra espada vieja
-Otra espada vieja
-Otra espada vieja
-Otro estoque ancho [...]
-Una espada con punta de lança
-Un estoque viejo antiguo
-Otra espada vieja ancha
-Otra espada vieja
-Una espada pequeña dorada
-Otra espada vieja pomo dorado
-Otra espada vieja punta de lança
-Un estoque viejo casi dorado
(fol. 650 r)
-Una espada vieja sin puño ni pomo
-Otra espada vieja
-Otra espada vieja
-Otra espada vieja sin puño ni pomo
-Otra espada vieja sin vaina despuntada
-Otra espada vieja dorada
-Dos espadas viejas de torneos sin vainas
-Una lengua de pescado
-Dos espadas sin punta y sin vainas de torneo
-Otra espada de la misª forma
-Otra espada vieja
-Un alfanje viejo sin vaina
-Tres espadas viejas de esgrimir la una quebrada

-Dos espadas anchas sin guaniçión
-Un alfanje viejo sin guarnición
-Nueve vainas de tr°pelo de colores
-Un morrión viejo barreteado de hierro
-Otro morrión de hierro en parte dorado
-Otro morrión negro con orejeras
-Otro morrión con barretas doradas
-Otro morrión aforrado en tr°pelo con barretas doradas
-Otro morrión con puntas e los cantos dorados
-Otro morrión turquesco dorado
-Otro morrión grabado de figuras
-Quatro morriones de hierro blanco
-Dos sombreros de hierro guarneçidos de tr°pelo negro
-Un casco de hierro con unos garfios
-Dos rodelas de concha de galápago
(fol. 650 v)
-Quatro rodelas de a prueba
-Una rodela vieja pintada
-Otra rodela de madera dorada verde
-Una rodela açerada con una punta e fluecos azules
-Otra rodela açerada con un botón e fluecos morados
-Otra rodela açerada con una punta con fluecos naranjados
-Otra rodela açerada con un botón fluecos de seda blanca
-Otra rodela de madera pintada dorada con çintas de tr°pelo carmesí
-Otra rodela açerada de figuras dorada con fluecos de seda negra e plata
-Una rodela grabada açerada con punta y fluecos morados e amarillos e de plata
-Diez manojos de saetas turquescas
-Una piedra ymán de hasta media libra poco más o menos
-Una tablilla de piezas y un proceso contra Fr^co^ de Salinas
-Un retrato de una cabeça de León
-Seis alxabas turquescas con saetas
-Treze carcajes de cuero con saetas
-Honze carcajes de madera con saetas
-Seis cajas de madera negra con saetas
-Seis maças de hierro de armas
-Otra maça de madera negra
-Nueve arcabuzes chicos e grandes con cañones
-Cinco cajas de escopetas chicas quebradas
-Diez e siete lanças de sortijas las tres quebradas

(fol. 651 r)
-Yten diez e siete lanças y tres picas
-Çinco alabardas
-Dos partesanas
-Seis lanzones
-Otro lanzón con un hierro a manera de flor de lis
-Diez y seis arcos yngleses
-Dos arcos de azero
-Treçe arcos turquescos

-E se quedó comenzado el dho ymbenº siendo los t^{os} Antón Fer^{dez} E Miguel López vºs de esta vª

Ante mí Morales scrivº
(signed)

E después de lo suso dho en la dha villa de Cog^{do} a ocho días del mes de julio de mill e qºs nov^{ta} e quatro años el dho Pablo de Estrigana guardarropa de Su Sª ante el dho Do^{or} Molina el alcalde m^{or} est^{do} pres^{te} el dho Juº del Castillo depositaº prosiguiendo el dho ynventº de los bi^{es} de Su Sª ante mí el dho scriº mani^{tó} los siguientes

-Primeram^{te} diez e siete arneses de hombres de armas sin manoplas
-Otros tres arneses que les faltan las golas y al uno faltan grebas e quijotes e al otro las grebas e medio quijote- e a otro le faltan braçales quijotes e grebas
-Otro arnés de pieças negras con medio morrión está cumplido sin escarpes
-Otro harnés de Su Sª cumplido con dos çeladas y quatro guardabrazos sin una manopla
-Yten una caja de pino grande en que hay lo sig^{te}
-Veynte y una pieças grandes e chicas de armas de caballo buenas doradas
-Seis escudos de las armas de Su Sª
-Una çelada borgoñona dorada
(fol. 651 v)
-Otra çelada borgoñona dorada como la de arriba
-Una cofia e dos medios baberones dorados
-Otras tres pieças de armas de caballo
-Media visera y un guardabrazo e dos grebas
-Dos puntas de escarpes
-Dos sillas de armar con tres pieças cada una de tornillo

-Un manojo de saetas ynglesas
-Un cofre viejo en el qual hay lo sig^{te} sin cobertor
-Un cuerpo que se abre con dos escarçelas
-Un peto grabado y un espaldar
-Dos morriones e tres medios baberones
-Dos escarzelas
-Dos Brazales
-Dos guardabrazos
-Dos pares de quijotes
-Unas manoplas
-Una gola
-Yten otra arca llena de armas en la qual había e quedan quarenta e ocho pieças de piernas e quijotes y treinta e seis escarcelas
-Un taleguillo de hojuelas de hierro
-Otra caja en la qual hay un peto pavonado con dos escarcelas
-Un morrión pavonado
-Una çelada pavonada con clavazón dorada
-Un peto abierto con su gola e malla
-Un espaldar pavonado
-Dos petos pavonados
-Otro espaldar pavonado
-Una gola e dos guardabrazos pavonados
-Dos guardabrazos e brazales pavonados
(fol. 652 r)
-Una gola y dos escarçelas
-Otras doze piezas de armas y una bragueta de [...]
-Un taleguillo con hebillas de armar
-Yten una caja de madera en la qual hay lo sig^{te}
-Çiento e çinquenta e quatro pieças de armas chicas e grandes
-Dos faldones de tr^{o}pelo negro verde con tres tiras de malla
-Yten otra caja que tiene lo sig^{te}
-Çiento e noventa e dos pieças grandes e chicas de armas
-Otra caja que tiene quarenta piezas de armas chicas e grandes
-Dos çapatos aforrados en malla
-Dos puntas de escarpines con malla
-Una taleguilla con unas pezezuelas para adereçar las armas
-Más otra caxa que tiene sesenta e cinco pieças de armas y un rostro de caballo de hilo de aranbre
-Otra caja que tiene çinquenta e seis pieças chicas e grandes de armas

-Yten otra caja que tiene veinte pieças de armas grandes
-Otra caja que tiene treinta e siete pieças grandes e chicas de armas
-Otra caja que tiene çiento e çinquenta e dos pieças de armas y quinçe petos
-Otra caxa que tiene sesenta pieças de armas grandes e chicas e dos çapatos de malla y un talego de diez ristres de lança y doçe cosetes (sic)
(fol. 652 v)
-Una guirnaldilla de poner penacho a caballo
-Dos manoplas pavonadas
-Otra arca de pino con unos papeles viejos
-Yten treinta y una pieças de armas para caballos chicas e grandes
-Otra caja de madera con quatro frascos de almártiga quebrados vazíos e quatro orinales
-Una silla de armas de tr°pelo carmesí con tiras de oro e seda
-Otra silla de la brida de tr°pelo negro
-Otra silla de la misma suerte
-Otra silla de cuero vieja quebrada
-Otra silla quebrada de cuero
-Otra silla de cuero viejo de caballos todas quebradas
-Otra silla de gineta vieja sin recaudo
-Otra silla gineta sin recaudo
-Otra silla gineta turquesca con estribos
-Dos cogines de domar potros
-Seis rollos de colleras de caballos para coche
-Más unas correas viejas quebradas de tirar los coches
-Yten treinta y una piezas de caparazones e armas de caballos de cuero
-Yten diez esmeriles grandes y otro esmerilejo en un carretoncillo
-Un aranbre viejo de colores pequeño
-Una prensilla con un torno de hierro
-Un caballejo con una figura de bronze
-Una arquilla chica vieja sin llave
-Una linternilla vieja
(fol. 653 r)
-Tres candeleros de madera para hachas
-Un caldero y una caldera
-Tres regaderas las dos grandes y una pequeña
-Una silla de madera con asiento y espaldar de cuero viejo
-Otra silla rasa vieja
-Una caja llena de saetas
-Un manojo de cuerdas para arcos yngleses

-Una fuente de hoja de lata vieja quebrada
-Un alnafil (sic) de hierro bueno
-Un asiento de un coche de cuero
-Dos tablas de rima la una de nogal y otra de pino con sus pies de pino
-Un modelo de madera de una yglesia
-Una silla de madera de descanso quebrada
-Seis trozos de escaleras de palo
-Un banco con un cajón y una tabla con dos pies

E por hoy se quedó el dho ynventario començado siendo pr^{te} Juº del Castillo depositario e tºs Miguel López e Antón Fer^{dez} vºs de Cog^{do}

Ante mí Morales scriº
(signed)

Y después de lo suso dho en la dha villa de Cog^{do} a nueve días del dho mes de julio de mill e quiºs e nov^{ta} e quatro años ante el dho do^{or} Molina alcalde mayor e juez suso dho e el dho Pablo de Estrigana prosiguiendo el dho ynventario de más del de arriba declarados e manifestados manifº e declaró haber por bi^{es} de Su Sª del dho don Juº Luis de la Cerda duque (fol. 653 v) de Medinaçeli los bienes siguientes

-Quatro ruedas de coche viejas quebradas
-Dos caballos de madera para tiros de artillería
-Otros trastos viejos de madera
-Una estera vieja de esparto y dos de palma viejas
-Un peso de balanzas sin pesas
-Un talego con unas redes de pájaros
-Una varilla de hierro de las escaleras de palo
-Un asiento de servidor quebrado

Todos los quales dhos bienes arriba declarados el dho Pablo de Estrigana manifestó ante el dho alcalde ma^{or} declarados y espezificados en estas siete hojas con el comienço y este fin los quales están todos a su cargo y en su poder y que su cargo del dho gu^{to} no tiene notiçia que haya en esta viª otros ningunos bienes del dho duque don Juº Luis mi S^{or} que está en el çielo e que si a su noticia hubiere venido manifestará e declarará otros qualesquier que haya e que esto es verdad y el dho Pablo de Estrigana y Juº del Castillo depositº nombrado por el dho alcalde m^{or} que estará pres^{e} los dos juntam^{te} de mancomún a voz de uno e cada uno yn

solidun vereg^{do} por el todo renun^{do} como renun^{an} las leyes de la mancomunidad como en ellas se q^{ne} e se oblig^{on} de tener en depósito los dos bienes arriba declarados y espesificados en las dhas siete hojas de los quales se dieron por entregados y confesaron haberlos reçebido so requerimº de las leyes de la entrega prueba e paga e las demás que son en este caso hablan^{to} los quales dos bienes darán de manifiesto cada e q^{do} (fol. 654 r) que se les mande haber depósito e so la pena de los depositarios que no acuden con los depósitos para lo qual oblig^{on} sus personas e bienes habidos e por haber e dieron poder cumplido a la justª e juezes de estos re^{nos} e señoríos de su Mag^{t} [...] jurados se sometieron para que como por senª pasada e cosa juzg^{da} les compelan a ello por todo rigor de drº e renun^{on} su fuero e jur^{ón} e leyes de su favor e de general renun^{on} e los otorg^{on} en el dho día nueve de julio de mill y qu^{os} e nov^{ta} e quatro aºs siendo testigos Miguel López y Antón Fernández y Antón Ybáñez vºs y estantes en Cog^{do} y los dos otorgantes a los quales y el escriº di fee que conozco lo firmaron de sus nombres en este regº con el dho alcalde mayor que ynterpuso en ello su autoridad judicial e lo m^{dó} dar por testimº a los tºs dos
El D^{or} Molina Juº del Castillo Pablo de Estrigª Ante mí Morales escrivº.

Doc. 2.
1654, July 15.
Inventory of the Cogolludo palace (Partial due to the disappearance of half of the document).
Archivo Histórico Provincial de Guadalajara, Protocolo 3995.

(fol. r)
Tiendas
-Un pabellón de campaña de ~~peballares~~ [crossed out] colores blanco y colorado y açul con quatro ~~fajas~~ sogas en la rueda y diez y seis cordeles por abajo forrado en holandilla colorada
-Más una tienda conforme al pabellón y aforrada en holandilla colorada con siete piernas y dos medias y diez palos costillares con quinçe botones de madera
-Otra tienda de la misma manª que la de arriba con diez y ocho botones de madera
-Otra tienda con nueve listas de lienço açul
-Otra con diez listas blancas y con cabeceras açules
(fol. v)
-Otra con seis listas y una anesgada con ocho varas y listas açules
-Otra de la misma manª
-Otra con v^{te} y una listas blancas y v^{te} y dos varas aforradas en açul
-Otra de la misma manª
-Otra de la misma manª
-Otra con diez y nueve listas blancas y diez y nueve varas
-Otra tienda con v^{te} y una listas blancas y diez y nueve varas con listas açules
-Otra con diez listas blancas y nueve varas con listas açules
-Otra tienda con treçe listas blancas de lienço y dos anesgadas con diez y nueve listas angostas açules esto a cada p^{te} con almenillas de lienço açules y pajiças en las caydas y con listas açules y amarillas en medio con tres hierros en medio y tres cordeles y otros diez cordeles hechos patas de gallo
-Más un pabellón de campaña con listas blancas y açules y con rueda de madera y quatro cordeles grandes y diez y nueve cordeles en el vuelo de abajo hechos pies de gallo
-Otro pabellón de la misma suerte con quatro sogas por la rueda y por el vuelo diez y ocho cordeles hechos patas de gallo
-Más otra tienda juntos dos lienços blancos y pintados negros de labor de cantería con tres piernas cada uno
-Otra tienda de la misma suerte con doçe piernas la una p^{te} y diez la otra

-Otra tienda grande con quarenta listas blancas y quatro nesgadas al una lado y por el otro lado quarenta y nueve listas blancas y seis nesgadas y por abajo almenillas blancas y açules y por en medio de la misma suerte
-Más dos pedaços de lienço blanco (fol. r) y negro de labor de cantería con una lista açul al canto de hechura de puerta
-Otro pedaço de tienda muy viejo que es de pabellón con v^{te} listas blancas
-Quatro ruedas de tablas para dhas tiendas con sus lienços blancos y negros de labor de cantería
-Un maço de madera con dos cercos de hierro a las orillas
-Más v^{te} y una banderillas de lata para las tiendas
-Más diez y seis bolillos de madera para remate de las banderillas
-Más dos digo tres banderillas de lienço
-Una soga de cáñamo
-Otra tienda grande con cinq^{ta} y quatro listas blancas y diez y seis nesgadas con listas açules por una lado y por el otro quarenta y quatro listas blancas y quatro nezgadas con almenillas blancas y açules por abajo y por en medio unas listas blancas y açules están muchas piernas remendadas y están con treinta y tres cordeles hechos pies de gallo
-Otra tienda pequeña con diez listas blancas a cada lado y dos nezgadas a cada lado y por medio con unas listas amarillas y açules y siete cordeles y almenillas por abajo
-Diez y seis palos redondos para armar las tiendas con unos hierros al rem^{te} de arriba
-Diez maderos para otras tiendas algunos quadrados y otros de media caña
-Seis estacas de madera de hasta tres quartas de largo cada una
-Dos faroles de madera viejos
-Quatro argollas pequeñas de hierro de los palos de las tiendas
(fol. v)
-Más doçe botones de palo de otras tiendas
-Más catorce varas de las otras tiendas sueltas
-Más tres llaves del quarto de San Fran^{co} la una de la puerta principal las dos de adentro

En la vi^{a} de Cog^{do} a quinçe días del mes de Jullio de mil y sei^{o}s y çinquenta y quatro años pareció Bar^{mé} Felipe vz^{o} de esta vi^{a} Jardinero del jardín de Su Exª el duq de Medinaçeli mi S^{r} y dijo que se daba por contento y entregado a su voluntad y confesó haber recibido de Alonso Díaz Coronel mayordomo que fue en esta vi^{a} de todos los bienes contenidos en el ynventario y memorial de suso que son de su Exª y estaban a cargo del suso dho el qual está escrito en quatro hojas con esta y se obligó de los tener en su poder y dar de manifiesto y entregar siempre que por p^{te} de Su Exª le fuere mandado so pena de pagar los daños que se rescrecieren porque pueda ser ejecutado con sola la declaración de la p^{te} de Su Exª en que queda diferido y renº la exp^{ón} del engaño y leyes de la entrega y prueba de ella y para lo haber por firme obligó su persona y bienes habidos y por haber y dio poder cumplido a qualesquier Jueçes y Just^{as} a quien lo pueda y deba dar a cuya Justª se sometió para que le compelan como por sentª pasada en cosa juzgada y renº las leyes y der^{os} de su favor y la que prohíbe la general renunci^{ón} en testimº de lo qual lo otorgó ante mí el dho escrº siendo tesºs Diego Páez Juº de Isla y Luis de Sopeña vzºs de esta viª y el otorgante que yo el escrº conozco lo firmó de su n^{e}= t^{do} de peballares no valga=

Bar^{mé} Phelipe
Pasó ante mí Jerónimo Suárez de Ciruelas escrº
(signed).

Doc. 3.
1667, September 19.
Inventory of the Cogolludo Palace, including the armoury.
Archivo Ducal de Medinaceli, Cogolludo, Legajo 3, D. 44.

(fol. 1 r)
Cogolludo, 19 de Sept^{e} de 1567 (sic)
Ynventario de los muebles, Alhajas, y Armas que había en el día de la fecha en las Casas Palacio de su Exª en el día de la fecha.

(fol.3 r)
Luis de Bustares
En la villa de Cogolludo a diez y nueve días del mes de septiembre de mil y ss^{os} y sesenta y siete a^{s} el Sr L^{do} Don Manuel de Bedoya del Consejo de su Exª el duque de Medina Zeli mi señor y su contador mayor en virtud de un orden de Su Exª fue a sus casas palaçio de esta villa e hiço inventario de los bienes que su Exª tiene en ella y armas de su armería q están a cargo de Bar^{mé} Phelipe difunto para los entregar a Luis de Bustares q sucedió en el ofiçio de casero y jardinero tomando cuenta a sus albaçeas la qual dieron y entregaron los bienes de su cargo del tenor siguiente

=Primeramente en la primera pieça del dho Palaçio treçe armas de treçe cuerpos puestos en sus estantes sin manoplas
-Yten quatro lanças de en ristre una con un hierro de estruzo
(fol. 3 v)
-Yten otras quatro lanças de ristre las tres doradas
-Y la una de ellas está quebrada
-Yten diez picas
-Zinco chuzos las cuchillas de lanças
-Yten una lança pequeña
-Yten dos partesanas
-Yten siete mosquetes con sus cañones
-Yten un cañón de alcabuz (sic) suelto y la caja de por sí quebrada
-Un casco de silla morisca
-Veinte y nueve aljabas con flechas de diferentes hechuras
-Yten treçe arcos de disparar flechas el uno quebrado y otro sentido
-Otros catorçe arcos largos a forma de báculos algunos con sus fundas y algunos despuntados y unos mayores que otros
-Yten otra aljaba con flechas

-Una alabarda quebrada el asta

=Yten en otra pieza más adentro del dho Palaçio hubo lo siguiente
-Primeram^{te} treçe armas de tre (sic) cuerpos (fol. 4 r) en sus estantes sin manoplas
-Seis lanças de en ristre
-Dos alabardas
-Tres cajas de mosquetes
-dos atabales de cobre rotos los parches

En la villa de Cog^{do} a veinte días del mes de sep^{e} de mil y ss^{os} y sesenta y siete años el dho señor contador mayor prosiguiendo dho ynventario le hiço de los bienes q se hallaron en dho Palaçio del tenor siguiente

-Primeram^{te} zinquenta y tres manoplas q algunas les faltan los dedos y un braçalete
-Un peto y espaldar negro y es todo de una pieça
-Catorçe petos los quatro negros y los demás azerados
-Dos medias cabezas de caballos una pavonada negra
-Yten quarenta y dos pieças de armaduras diferentes
-Yten un brazalete

(fol. 4 v)
=En la tercera pieza de la armería de dho Palaçio hay lo siguiente
-Primeramente siete cuerpos armados puestos en sus estantes
-Otros seis cuerpos armados y en sus estantes pavonados negros
-Otro cuerpo de armas pavonado negro pequeño en su estante
-Un peto y espaldar gola morrión braçaletes y cañones de piernas pavonado negro
-Una armadura de caballo en su estante grabada de oro
-Una tabla de jugar a las damas compuesta de nácar
-Ocho rodelas açeradas
-Otra rodela grabada de oro con una batalla y un morrión de su hechura
-Dos rodelas de madera
-Dos adargas viejas
-Veinte y cinco espadas de diferentes hechuras una sin guarnición
-Una armadura de ancas de caballo de ante
(fol. 5 r)
-Veinte morriones de diferentes hechuras
-Un peto y espaldar pavonado negro con su morrión q el morrión está contado con los demás

-Otro peto y espaldar pavonado negro
-Ocho piezas de coraças carmesíes tachonadas doradas
-Tres piezas de malla
-Nueve maças de desarmar las dos sin hierros
-Tres armaduras de cabezas de caballo pavonadas negras
-Veinte y tres armaduras diferentes de caballos de diferentes hechuras
-Nueve arandelas de lanças las tres pavonadas negras
-Treinta y seis piezas diferentes de armaduras de silla

Bienes muebles
-Un quadro de la Madalena de vara y media con marco dorado
-Un retrato del Duque de Lerma de vara y media con marco dorado
(fol. 5 v)
-Quatro países de pastores con ganado y marcos dorados y negros
-Veinte y ocho cuadros de ermitaños del yermo de hasta vara y media de ancho con marcos dorados y negros
-Tres países menores con figuras pequeñas con marcos dorados y negros
-Dos quadros yguales y mayor es el uno de la ciudad de Sodoma y el otro de Moisen (sic) con marcos dorados y negros
-Otro cuadro de la Vía Sacra de Medinazeli
-Un quadro de San Fran^co^ de una vara con marco dorado y negro
-Otro cuadro de una vara de San Juan y el niño Jesús alcançando fruta de un árbol
Otros dos cuadros de una vara de San Ju^o^ Baptista con marco negro
-Un retrato de mi señora la Duquessa (fol. 6 r) Antonia q goça de Dios siendo niña

Escritorios y Bufetes
-Dos bufetes grandes de caoba
-Una mesa baja de nogal en que comía el Duque mi señor
-Otro bufete más pequeño en que comía el marqués mi S^r^ y señores
-Un retrato de Escandarbey con marco negro
-Quatro retratos yguales pequeños del rey nro Señor y señores infantes Don Carlos y Don Fernando y Conde Duque
-Un cuadro de san Jerónimo de tres quatras sin marco
-Un cuadro del Niño Jesús disputando con los dotores (sic) con marco dorado y negro
-Una mesa de nogal para la cama y un respaldo de nogal y vaqueta
-Dos bufetes de vaqueta viejos
-Una mesa larga de pino en que se sirve

(fol. 6 v)
-Una mesa de pino quadrada y grande q sirve en la escribanía baja
-Un estante de papeles q estaba en la secretaría baja y tres banquillos todos de pino

Sillas [...]
(fol. 7 r)
-Una mesa de trucos con paño verde y dos cajones para los tacos y su margolla y bolillo q está en el cuarto de mi Señora la marquesa de Astorga [...]
-Diez candeleros de madera que estaban colgados del techo alrededor de la mesa de trucos y el inventario antiguo diçe q el uno está quebrado y no se ha podido reconocer si están cabales por estar en el cuarto de mi Señora la marquesa de Astorga y no poderse entrar dentro
[...]
(fol. 7 v.)
-Dos faroles grandes de vidrio para las luzes
-Una cuna dorada de tablas con sus pies dorados q el inventario antiguo llama cama [...]
-Un lienço de pintura con dos cazadores
-Un caballo de madera para pintar
-Seis tablas yguales y una pequeña doradas y algunas solo los perfiles [...]
(fol. 8 r)
-Quatro pilares asiento y zielo dorado q es de la Custodia y seis tablollas de la misma color de los pilares
-Un juego de ajedrez de alabrastro con veinte piezas [...]
-Quatro bastidores grandes y pequeños los dos con pies [...]

(fol. 9 r)
Tiendas
-Un pabellón de campaña de colores blanco y encarnado y azul con quatro sogas en la rueda y diez y seis cordeles por bajo aforrado en holandilla colorada
-Una tienda con forme al pabellón encarnada con siete piernas y dos medias aforrada en holandilla y diez para los costillares con quinçe botones de madera
-Otra tienda de la misma manera q la de arriba con diez y ocho botones de madera
(fol. 9 v)
-Otra tienda con nueve piernas de lienço blanco y diez varas aforradas en lienzo azul
-Otra con diez piernas blancas y onçe varas aforradas en lienzo azul
-Otra con seis listas blancas y una [...] con ocho varas azules

-Otra de la misma manera
-Otra de la misma manera
-Otra con veinte y una piernas blancas y veinte y dos varas aforradas en açul
-Otra tienda con veinte y una listas blancas y diez y nueve varas aforradas en açul digo de la misma manera
-Otra de la misma manera
-Otra de la misma manera
-Otra con diez y nueve listas blancas con diez y nueve varas y aforradas en açul
-Otra tienda con veinte y una listas blancas y diez y nueve varas aforradas en açul
-Otra con diez listas blancas y nueve varas aforradas en azul
-Otra tienda con treze listas (fol. 10 r) blancas de lienzo y dos [...] con diez y nueve listas angostas azules en medio con almenillas de lienzo açules y pajizas en las caydas y con listas açules y amarillas en medio y con tres hierros en medio y tres cordeles hechos patas de gallo
-Yten un pabellón de campaña con listas blancas y açules y con rueda de madera en medio y quatro cordeles grandes y diez y nueve cordeles en el vuelto de abajo hechos pies de gallo
-Otro pabellón de la misma suerte con quatro sogas en la rueda y por el vuelto diez y ocho cordeles hecho patas de gallo
-Yten otra tienda juntos dos lienzos blancos y pintados negros de labor de cantería con tres piernas cada uno
-Otra tienda de la misma suerte con doze piernas por la una parte y diez por la otra
-Otra tienda grande con quarenta listas blancas y quatro rasgadas por el un lado y por el otro quarenta y nueve (fol. 10 v) listas blancas y seis rasgadas y por abajo almenillas blancas y açules y por el medio de la misma suerte
-Otros dos pedaços de lienzo blanco y negro de labor de cantería con una lista azul al canto de hechura de puerta
-Otro pedazo de pabellón muy bajo con veinte listas blancas
-Quatro ruedas de tallas para las tiendas son sus lienzos blancos y negros de labor de cantería un mazo de madera con dos arcos de hierro a las orillas
-Veinte y una banderillas de hasta para las tiendas
-Diez y seis bolillos de madera para remate de las banderillas
-Tres banderillas de lienzo muy viejas
-Una soga de cáñamo
-Otra tienda grande con zinquenta y quatro listas blancas y diez y seis nesgadas con listas azules por (fol. 11 r) blancas de lienço y dos anesgadas con diez y nue= digo un lado y por el otro quarenta y quatro listas blancas y quatro nesgadas con

almenillas blancas y azules por abajo y por medio unas listas blancas y azules y están con treinta y tres cordeles hechos pies de gallo
-Otra tienda pequeña con diez listas blancas a cada lado y dos nesgadas a cada lado la suya y por medio con unas listas amarillas y açules y siete cordeles y almenillas por abajo
-Diez y seis palos redondos para armar las tiendas con unos hierros al remate de arriba
-Diez maderos para dhas tiendas algunos cuadrados y otros de media caña
-Seis estacas de madera hasta tres quartas de largo cada una
-Dos faroles de madera viejos
-quatro argollas pequeñas de hierro de los palos de las tiendas
-Doze botones de palo de dhas tiendas
-Catorze varas sueltas de las dhas tiendas
-Otro lienzo pintado de negro de (fol.11 v) de treçe piernas por cada lado
-Otro lienzo blanco pintado de negro como abntería de dos anchos y el uno tiene doze piernas y el otro onze y este lienzo y el antecedente se hallaron de más de todo lo inventariado en el inventario q se hiço de los bienes q recibió Bar^{mé} Phelipe en quinçe de julio de mil y ss^{os} y çinquenta y quatro años
-Otra redoma de vidrio grande
-Una silla de mano aforrada en damasco carmesí con vidrieras de cristal para los lados en cada una tiene ocho vidrieras y los dos correones para los mozos de silla [...]
-En un cofre viejo sin tapa sesenta y ocho piezas chicas y grandes de armar de hombres y de caballos
-En otro cofre bajo sin tapa treinta y siete piezas de armaduras como las de arriba
-Unas piezas de armaduras de caballos de (fol. 12 r) cartón
-Una rueda de palo al parezer para limpiar las armas
-Una pieza de hoja de lata como pie de fuente
-Seis cañones de piezas de artillería pequeños y otro pequeño
-Media prensa
-Quatro cascos de sillas bajas de caballos
-Dos guarniziones bajas de mulas de coche
-Un pretal con sus campanillas
-Siete cañones grandes y pequeños de piezas de artillería

La qual dha cuenta y entrego de bienes referidos se hiço por el L^{do} Sánchez albaçea del dho Bar^{mé} Phelipe y pareció según el rezibo de dho difunto haber entregado todos los bienes de su cargo según los entregos q de ellos se le hiço el uno en quinçe de julio del año de mil y ss^{os} y çinquenta y quatro= y el otro de las armas en catorçe de abril de mil y ss$_{os}$ y sesenta y dos años= y demás de ellos entregó otros bienes q parezieron ser de Su Exª (fol. 12 v) que estaban por inventariar de q doy fee y su merzed lo firmó el L^{do} Don Manuel de Bedoya ante mí Jaçinto de Zeruelas s^{no}.

En la villa de Cog^{do} a veinte y un días del mes de septiembre de mil y ss^{os} y sesenta y siete años en presencia de mí el escribano y testigos parezieron Luis de Bustares vz^{o} de esta villa principal y el L^{do} Fran^{co} Duro presbítero y Manuel de el Mudo vz^{o} de ella como sus fiadores principales deudores pagadores y obligados todos juntos de mancomún a voz de uno y cada uno por sí y por el todo renunciando como renunziamos las leyes de la mancomunidad y el benefiçio de la división y excursión como en ellas se contiene dijeron y confesaron haber recibido de el L^{do} Don Manuel de Bedoya del Consejo de Su Exª el duque de Medina Zeli mi Señor y su contador mayor todos los bienes muebles y armas contenidos en el inventario de suso q va escrito en seis fojas de papel con este y por orden (fol. 13 r) de Su Exª se mandaron entregar al dho Luis de Bustares a quien Su Exª hiço merçed de los ofiçios de casero y jardinero de su Palaçio y jardín de esta v^{a} los quales se obligaron de tener en forma de depósito y los volver y entregar siempre q por Su Exª o su contaduría les fuere mandado so las penas de los depositarios q no cumplen con los depósitos demás de pagar lo que ansí dejaren de entregar con más los daños q se rezibieren según la declaración en la parte de Su Exª en q lo difiero porque puedan ser ejecutados sin otro instrumento ni averiguación alguna= y de todos dhos bienes se dieron por entregados por lo haber rezibido con efecto sobre que renunziaron la excep^{ón} del engaño y las demás leyes q en este caso hablan como en ellas se contiene

Otrosí se obligaron de q el dho Luis de Bustares cumplirá con las cargas y obligaciones q los demás jardineros han tenido y han sido a su cargo de cumplir so pena (fol. 13 v) q a ellos se han compelido por todo rigor de der^{o} demás de pagar los daños q se recrezieren porque puedan ser ejecutados con sola la declaración de la parte de Su Exª= las quales condiciones q tocan de cumplir al dho jardinero son del tenor sig^{te}

Primeram^{te} han de correr por su cuenta las quiebras q hiçiere la fuente q va al jardín desde la esquina de la plaça q era casa de Andrés de Arenas en q tiene su ofiçio Alonso Pérez es^{no} hasta la fuente principal del jardín, y desde allí a la fuente del patio bajo ansí [...]

(fol. 14 v)
Todas las quales dhas condiciones y demás referido en esta escritura cumplirán en la forma q quedan obligados y no lo haciendo pueda venir persona a su cobrança y ejecución la qual pagarán por cada un día de salario quatrozientos maravedís de venida estada y vuelta y renunziaron la premática q lo prohíbe y por ello se pueda ejecutar como por el principal con sola su declaración en q lo diferimos= y para lo haber por firme obligaron sus personas y bienes espirituales y temporales habidos y por haber y dieron poder cumplido a qualesquier justiçias a quien lo pueda dar a cuya jurisdizión nos sometemos para que nos compelan como por sentencia (fol. r) pasada en cosa juzgada y renunziamos las leyes y fueros y derechos de nro favor y la q prohíbe la general renunziaçión= […].

Doc. 4.
1685, October 26.
Inventory of the Cogolludo Palace, including the armoury.
Archivo Ducal de Medinaceli, Sección Cogolludo, Legajo 5, D. 1.

(fol. 1 r)
Cogolludo 26 de Oct^{e} de 1685
Ynventario de todos los bienes, y Alhaxas, y Armas que había en el Palacio de S. E. en dha V^{a}, q^{e} executó D^{n} Fran^{co} Gallo de Ledesma Cont^{or} de S. E.
Acompaña un traslado autorizado de Alonso Pérez; y una razón separada de ciertas alhajas, que se debían añadir en dho ynventario.

(fol. 3 r)
En la villa de Cogolludo a veinte y dos días del mes de octubre de mil y seiscientos y ochenta y cinco años el Señor D^{n} Fran^{co} Gallo de Ledesma contador del Duq^{e} mi S^{or} en su villa de Medina como tal y en virtud de orden particular de Su Exª= por ante mí el ess^{no} dijo pretende hazer inventario nuevo de todos los bienes y alhajas que han quedado en el Palazio de esta villa y en particular del nuevo herraje que se ha echado en el poco tiempo que Su Exª y familias han estado en él por haber costado más de siete mil Rs y así mismo de las Armas y hombres y otros adereços de guerra que hay en la Armería de Su Exª para lo qual fue llamado Luis de Bustares jardinero y guardaropa vezino de esta villa y para dar principio al dho inventario le fueron entregadas todas las llaves así nuevas como si hubo alguna vieja y con su asistencia se dio principio a dicho inventario en la forma siguiente [...]

(fol. 3 v)
-xUn tablón de la mesa de trucos que se deshizo quando Su Exª vino a su Palazio y sólo quedó dho tablón [...]
(fol. 4 r)
-Un caballo de madera para pintar [...]
-Un juego del aljedrez (sic) con v^{te} piezas de alabastro [...]
(fol. 4 v)
-Un tablero de damas embutido en nácar con su marco de pino. Falta
-Treinta marcos dorados y negros que son de las pinturas (fol. 5 r) q Su Exª mandó llevar a Madrid. Faltan
-Otros marcos quebrados reduzidos a dos líos [...]

Armería

-Primeram^te quarenta hombres armados enteros y otro medio cuerpo que ynvió Su Exª este año de ochenta y cinco que se compone de pecto espaldar, y morrión borgoñota brazeletes y manoplas con rodela

-xDiez y nueve pectos y espaldares y más otro mayor

-xVeinte y quatro morriones de diferente género

-xUn cuerpo entero de armas pavonado negro que se compone (fol. 5 v) de peto, espaldar, morrión, borgoñota, brazeletes y manoplas que era de Su Exª que goza de Dios

-xDiez rodelas de azero, y dos de palo

-xVeinte y ocho fronteras de cabezas de caballo de azero

-xQuarenta y quatro manoplas de azero

-xNoventa y dos piezas de azero sueltas para armar que son de diferentes hechuras

-xUn caballo armado que se compone de silla, freno, pechos, ancas, cabeza, con una punta de azero en la frente grabado y dorado

-Ocho mosquetes con sus cañones llaves y esparadores (sic). Faltan dos

-xTres cajas de mosquetes sin llave ni cañones

-Veinte y cinco espadas de a caballo anchas con sus vaynas excepto dos

-xNueve hachas de salmar (sic)

-Cinco cotas encarnadas y quatro faldillas entachonadas de lantejuelas doradas. Falta dos cotas y una faldilla

-xDos mallas y dos brazeros de azero colado

-(8) Nueve arcos de flechas

-xVeinte y nueve carcavos (sic) de flechas

-xOcho partesanas con sus astas de palo, tres alabardas la una quebrada el asta

-xVeinte y tres lanzas

(fol. 6 r)

-xUnas ancas de caballo de cartón doradas

-xUn casco de sillas

-xDos zimbales de campaña de cobre

-xDos adargas

-xUna tienda de Campaña encarnada blanca y azul con su pabellón

-xUn pedazo de tienda con onze varas de pino. Falta

-Otro pedazo de tienda con onze varas de pino

-xOtro pedazo de diez varas de lo dho

-Otro pedazo con veinte y dos varas de lo mismo. Se halló hecho pedazos

-Otro pedazo de veinte varas de lo mismo

-Otro con diez y nueve varas de lo mismo

-xOtro pedazo con diez y ocho varas de lo mismo

-xOtro pedazo con diez y seis varas de lo mismo
-xOtro pedazo con onze varas de lo mismo
-xOtro pedazo con ocho varas de lo mismo
-xOtra tienda grande de lienzo blanco y azul
-xOtra tienda grande de lienzo con listas azules
-xUn pabellón grande blanco con su rueda en medio para armalle
-xOtro pabellón grande con su rueda en medio para armalle
-xQuatro líos de madera de pino arqueados y dere(fol. 6 v)chos para las tiendas
-x Tres pedazos de tienda viejos con tiras azules al canto
-Catorze lenzones o pedazos de tiendas viejos
14-Diez palos para arcos de ballestas

Zerrajería [...]
(fol. 8 r)
-Más una zerradura que se hizo en la Armería que se quitó de otra puerta y de ella no tenía más que servirse la chapa y llave
-Más ocho tiros naranjeros incluso medio tiro de lo mismo y tiene cada uno ocho pies de largo y son todos lo hierro
-Más otro medio tiro más grueso menos largo con roscas alrededor dos en cada uno
-Más otros dos tiros más gruesos y más cortos con aldabones y roscas
-Más doze escudos de madera con las armas del duque mi Señor y unos florones viejos [...]
(fol. 8 v)

De los quales dhos bienes se constituyó por depositario Luis de Bustares jardinero y guardarropa de S. E. en esta villa de que se dio por entregado y renunçió las leyes de la entrega y del engaño y todas las demás que en este caso hablan y se obligó con su persona y bienes presentes y futuros a darlos de manifiesto cada y quando que por Su Exª o contadores de la contaduría de Medinazeli u otro juez competente le sea mandado so las penas de incurrir en las que caen e incurren los depositarios que no acuden con los depósitos que les son entregados y dio poder a todas las justicias y juezes del Rey nro S^or^ de qualesquier partes que sean para que a ello me compelan y apremien por todo rigor de dro y vía ejecutiva y lo recibió por sentencia pasada en autoridad de cosa juzgada renunció todas las leyes fueros y derechos de su favor y la general en forma y lo otorgó así ante el presente es^no^ y testigos en la villa de Cogolludo a veinte y seis días del mes de oct^re^ de mil y seiscientos y ochenta y cinco años (fol. 9 r) siendo testigos Antonio Pérez el Dr D^n^ Pedro Lázaro Vallejo y Juan de Lillo el mozo vecinos de esta villa y el otorg^{te} que yo el es^{no} doy fee conozco no firmó por no saber a su ruego lo firmó en testigo junto con el dho Señor D^{n} Fran^{co} Gallo de Ledesma= D^{n} Fran^{co} Gallo de Ledesma= Dr D^{n} Pedro Lázaro Vallejo= Ante mí Alonso Pérez= Concuerda con su original a que me remito= E Yo el dho Alonso Pérez es^{no} del rey nro S^{r} y del núm^{o} de esta villa de Cogolludo y su Marquesado fui presente a lo que dho es y en fee de ello lo signé y firmé día de su otorgamiento de pedimiento del S^{or} cont^{or} D^{n} Fran^{co} Gallo de Ledesma= en testimonio de verdad= Alonso Pérez

Concuerda este tras^{do} con su original que queda en la Cont^{ría} m^{or} de hacienza del Marqués Duque mi S^{or} (a que me remito) que para este efecto me fue entregado por los S^{res} Cont^{r} M^{or} y cont^{es} y en fe de ello yo Domingo Matheo Galán s^{no} por el Rey nro S^{r} pu^{co} del n^{o} de esta villa y mayor de rentas [...] en esta su villa de Medinaceli lo signé y firmé (fol. 9 v) en dha villa a veinte y nueve del mes de en^{o} de mil setecientos y onze años

En testimonio de Verdad
Domingo Matheo Galán
(signed).

Doc. 5.
1746, August 11.
Inventory of the Cogolludo Palace, including the armoury.
Archivo Ducal de Medinaceli, Sección Cogolludo, Legajo 5, D. 38.

(fol. 1 r)
Cogolludo 11 de Agosto de 1746
Ynventario authéntico de los bienes, y alhajas que existían en la capilla del Palacio, que S. E. tiene en la villa de Cogolludo con expres^{ón} de lo que había en dho Palazio, y Armería

(fol. 3 r)
Valga para el reinado del Sr Fernando Sexto
Ynventario de los bienes de la capilla
En la villa de Cogolludo a onze días del mes de Agosto de mill setez^{os} y quarenta y seis años el Sr D^{n} Carlos Galán, oficial mayor de la contaduría del estado de Medina Çeli, usando de la comisión q le está dada, por los señores contadores de aquel estado, requirió con ella, al Sr D^{n} Fran^{co} Xavier Cañamón, mayordomo de rentas, de ese Marquesado, para que en nombre del Padre Joseph Cañamón, su hermano, clérigo menor, capiller (sic) de esta capilla de S. E. (respecto hallarse accidentado en cama de un año a esta parte) diesse de manifiesto, todos los bienes y alhajas pertenecientes a dha capilla, que tiene en esta su villa de Cogolludo el Exmo Sr Duque de Medina Çeli &^{a}= mi S^{r} y que le fueron entregadas en veintidós de mayo, de mill setez^{os} y treinta y quatro como consta del ynventario, que se hizo por testimonio de Joseph de Frías Mazario ss^{no} de este Ayuntamiento y Número, de que salió fiador Man^{l} Cañamón su padre, ya difunto; y teniendo presente el referido ynventario se formó de nuevo en la manera siguiente [...]

(fol. 8 r)
Ynventario de los bienes y alhajas
De este Palacio de Cogolludo
En la villa de Cogolludo a doze días del mes de Agosto de mill setez^{os} y quarenta y seis años por ante mí el ss^{no} D^{n} Carlos Galán oficial mayor de la contaduraía del Exmo Sr Duque de Medina Çeli &^{a}= mi S^{r} requirió a D^{n} Fran^{co} Xavier Cañamón, mayordomo de rentas de S. E. de este Marquesado le dé y ponga de manifiesto los bienes y alhajas que en este Palacio y Armería pertenecen a la haçienda de S. E. para hacer ynventario de ellas, y poniendo p^{r} execution (sic) se dio principio a esta Dilix^{a} en la forma siguiente

Bienes de Palacio [...]
(fol. 8 v)

Armería
Bienes de la Armería
-Dos timbales de cobre buenos con sus tornillos y les falta (sic) los parches
-Siete varas con lanzas anchas
-Quatro con puntas a modo de rajones (sic)
-Otras diez varas sin hierros de diversos géneros
-Diez palos mediados, que se duda, para lo que pueden servir
-Ocho broqueles de Azero colado
-Dos de madera
-Quatro arcabuzes, el uno con llave y los tres sin ella
-Quarenta hombres con Armadura de Azero colado que aunque les faltan diversas piezas del vestido, están sueltas por el suelo
(fol. 9 r)
-Un armazón de Acero para caballo, bronceado que le falta (sic) muchas piezas
-Dos quotas (sic) de malla tachonadas
-Quarenta piezas de Azero sueltas para Armar, que son de diferentes hechuras
-Diez y ocho morriones
-Veinte y ocho fronteras de cabezas de caballo
-Una mesa de trucos grande
-Quarenta manoplas de hierro
-Diez y ocho Armaduras de sillas deshechas y viejas
-Un cañón grande como de cinco pies de largo y diez roscas, que se hace juicio pesar ocho arrobas
-Otro más pequeño, de una vara de largo, con tres roscas que pesará hasta cinco arrobas
-Otro más ancho y corto, con quatro asas, del peso de hasta ocho arrobas
-Otro de dos varas de largo ochavado, que pesará como diez arrobas
-Otros cinco lisos de nueve quartas de largo, más angostos que los antecedentes, que cada uno pesará tres arrobas
-Otro ancho de una vara de largo que pesará dos arrobas

De los quales dhos bienes se dio por entregado el expresado D^{n} Fran^{co} Xavier Cañamón, y se obligó a darlos de manifiesto siempre y quando que por el Duque mi S^{r} o su contaduría de Medina Çeli le fueren pedidos y no lo haciendo lo pagará de sus bienes y hacienda y lo firmaron cada una de las partes de que yo el ss^{no} doy fee= Carlos Galán= Fran^{co} Xabier Cañamón Segoviano= Ante mí Pedro Nolasco de Frías

(fol. 9 v)
Concuerdan los inventarios insertos con sus originales, que quedan en mi poder a q me remito, y porque conste de pedim^{to} de D^{n} Carlos Galán oficial mayor en la Contaduría que en su v^{a} de Medina Çeli tiene el Exmo Sr Duque de Medina Çeli &^{a} mi S^{r} doy el pres^{te} en esta v^{a} de Cogolludo, y Agosto veinte y tres de mill setez^{s} y quarenta y seis años y lo signo, y firmó

En testimonio de verdad
Pedro Nolasco de Frías
(signed)

(fol. 12 v)
Cogolludo, 11 de Agosto de 1746
Ynventario de los bienes y alhajas que hay existentes en la capilla que en este Palacio de la villa de Cogolludo tiene el Exmo Sor Duque de Medinazeli &^{a} mi S^{or} que han estado de cargo del Jph Cañamón, clérigo menor, su capiller.

Otro de las que al presente quedan en ser en este palacio de S. Exª y Armería que hay en él de que tiene dado R^{bo} Fran^{co} Xavier Cañamón May^{mo} de R^{tas} a cuyo cargo quedan
Por testim^{o} de Pedro Nolasco de Frías son en cuyo poder y oficio queda el original.

MONTEMAYOR, MARQUESES DE

Doc. 1.
1606, November 27.
Inventory of the castle of Montemayor.
Archivo Histórico de la Nobleza, Baena, C. 107, D. 98.

(fol. 1 r)
Montem^{or}
Ynventario de los bienes q^{e} había en la fortaleza de la villa de Montem^{or} propios del Sor D^{n} Juan de Silva y Xuera Marq^{s} de dho título; al tiempo q^{e} falleció Diego Ruiz de Sta María Gob^{or} q^{e} había sido de dha fortaleza y se entregaron p^{or} sus hixos de orden de dho Sor Marq^{s} a Pedro de Tórtoles Tostado; cuyo inventario se hizo en Montem^{or} a 27 de Nov^{e} de 1606 ante Manuel de Oviedo es^{no} de dha V^{a}

(fol. 2 r)
En la villa de Montemayor a v^{te} y siete días del mes de noviembre de myll y seysçientos y seys a^{o}s María de los Ángeles y Greg^{o} de Santa María su hermano hijos de Diego Ruyz de Santa María dif^{to} gobernador y Just^{a} mayor que fue en la dha villa dieron y entregaron a Pedro de Tórtoles Tostado escrib^{o} por orden y mandado del marqués my señor lo que hubo en la fortaleza y palaçio de Su S^{a} al qual ynventario estuvo pres^{te} Fr^{co} Nyeto Dorantes t^{e} de gobernador

-Primeramente se subió a la torre del homenaje y se halló lo sigui^{e}
-Seys cañones de arcabuz sin cajas
-Yten çinco arcabuçes antiguos con sus cajas y llaves de mecha
-Dos tiros de bronce de gordor de un braço el uno liso y el otro con sus molduras
-Quatro culatas de bronce con sus asas pequeñas
-Catorçe cascos buenos y malos de diferentes hechuras
-Treçe pieças de petos y espaldares pequeñas y grandes y algunos quebrados
-Yten hubo otros pedaços anejos a los petos y espaldares y a los cascos que por ser pedaços pequeños y de poca valía no se puso cada uno de por sí
(fol. 2 v)
-Dos frascos viejos de estaño de enfriar sin tapadores
-Un candelero de hoja de lata viejo
-Una cadena de hierro de eslabones delgada
-Unas armas de caballo de cuero doradas
-Un velador torneado pa una lámpara sin un pie
-Quatro frascos de pólvora viejos

-Un bastidor de cama sin lienço ni pies viejo
-Diez y nueve pieças de una reja de hierro desbaratadas machos y hembras
-Quatro bancos de mesas de estantes los dos quebrados y viejos
-Dos mesas de estantes de tarazea con sus dobladuras y bisagras y armas
-Yten otra vieja hecha pedaços
-Yten una tabla vieja con dos burrillos
-Una cruz de hierro con su varilla
-Una silla redonda de palo
-Otra silla vieja quebrada el respaldar
-Otras sillas de cadera antiguas de espaldar
-Dos ollas de cobre una mayor que otra
-En el patio un cajón que sirve de troje
(fol. 3 r)
-Quatro arcas grandes de pies viejas
-Dos coperos uno de pino bueno y otro más pequeño
-Un perfumador como arquilla
-Una tienda vieja desbaratada y rota con un faldón
-Unas gafas de hierro antiguas de armar ballestas con su tornillo
-Un veladorçillo pequeño
-Dos asadores grandes de cozina
-Diez tiros lameriles (sic) en sus carretones son de bronce y los dos sin culatas
-Un morillo de alumbre
-En el patio de palacio un tiro de hierro con dos culatas
-Y en las esparrillas un tiro de hierro grande

Todos los quales dhos bienes contenydos y declarados que este memorial Greg^{o} Ruyz de Santa María y su her^{na} dieron y entregaron al dho Pedro de Tórtoles Tostado por mandado del marqués mi señor y él los R^{o} y se dio por entregado en ellos como Su S^{a} manda a todo lo qual fue pres^{te} el dho Fr^{co} Ny^{o} t^{e} de gobernador que lo firmó de su nombre fho día mes e año suso dho siendo pres^{te} por t^{o}s Bernabé Gr^{a} y Diego de Fiadelo y Pedro Tostado alguaçil mayor v^{o}s de la dha villa y lo firmo el dho Fran^{co} Nyeto Dorantes tinyente de gober^{dor} y el dho Gregorio Ruyz de Santa María a los quales yo el scrib^{o} doy fee conozco y ansí mismo al dho Pedro de Tórtoles Tostado q los rescybyó los dhos bienes las quales firmas fueron en el registro como se sigue Fran^{co} Nyeto Dorantes Gregorio Ruyz de Santa María Pedro de Tórtoles Tostado pasó ante mí Manuel de Oviedo escribano e yo el dho Manuel de Oviedo escribano aprobado por el Rey nuestro señor y pú^{co} de rentas de la villa e jur^{ón} de Montemayor por n^{o} de Su S^{a} el marqués my señor presente fuy a lo que dho es en fee de ello fize aquí mi signo [...].

MONTERREY, CONDES DE

Doc. 1.
1550, November 22.
[Copy, 1840].
Objects linked to the entailed estate by the Count of Monterrey Alonso de Acebedo y Zúñiga.
Archivo Ducal de Alba, Casa de Alba, C. 158, nº 15.

(nº. 1)
Efectos y alhajas vinculados por Dn Alonso de Acebedo y Zúñiga, conde de Monterrey por escritura otorgada en Valladollid a 22 de noviembre de 1550 ante el escribano Simón de Cabezón de la qual presentó testimonio señalando con el nº 2º fº. 15 pieza corr^{te} de los autos sobre entrega de marcos de plata al curador ad liten del Sr D^{n} Duq^{e} D^{n} Carlos con su demanda de 1º de agosto de 1803 fº. 51 de la misma Pieza [...]
(nº. 24)
Cosas de guerra
-Unas sobrecubiertas de brocado de tres altos de caballo con unas bordaduras sobre terciopelo verde encima de tela de oro y plata con franjas anchas y angostas de oro y cerco verde q son once piezas con su guarnición de la misma color (con sus espuelas y estribos dorados) y un sayo de este mismo brocado con la misma bordadura [...]
(nº. 27)
-Cuatro paños de trompeta con sus cordones y con las armas de su señoría de tafetán amarillo, de una parte un escudo dorado con cinco estrellas y de la otra parte un spera (sic)
-Más cuarenta y cinco veletas pequeñas de hombres de armas de tafetán negro con unos escudos de oro con estrellas y franjas de cingo amarillo q hilan en las lanzas
-Más 45 cubiertas de caballo doradas llenas de estrellas
-Más 45 arneses de seguir
-Más otros tantos martillos y otros tantos estoques
-Más 45 collares de caballo de malla para sobre pescuezo de caballo
-Más 44 sayos de hombres de ar(nº. 28)mas de terciopelo negro y paño amarillo a cuartones con sus cruzes de tafetán carmesí
-Más 45 pares de estribos y 45 pares de espuelas y 45 frenos de la brida
-Nueve tiendas de campo con sus aderezo

-Dos arneses de Justa Real
-Cuatro arneses de Justa de guerra
-Cuarenta y cinco sillas de armas
-Ciento y diez coseletes de Ynfantería con sus guarniciones de brazos, golas y morriones
-Seiscientas picas de fresno con sus hierros de media naranja
-ciento y diez arcabuzes con sus frascos y frasquillos y moldes para pelotas y rascadores
-Diez y ocho partesanas
-catorce adargas [...].

Doc. 2.
1560, October.
Inventory of objects from the Count of Monterrey's chamber.
Archivo Ducal de Alba, Casa de Alba, C. 246, nº 41.

Memoria de las cosas q Jerónimo de Laguna trajo de Monterrey en el mes de octubre del año de mill y quiºs y sesenta años q estaban en la recámara de Monterrey como bienes del conde mi Sor q sea en gloria los quales el dho Gerónimo de Laguna entregó al señor don Alº de Fonseca heredero universal del conde my señor que sea en gloria los quales dhos bienes son los siguientes

-Primeramente un arca encorada castaño (sic) que tiene dentro una sábana vieja rota
-Más un estoque de armas y una daga el estoque tiene una vayna de terciopelo negro con unos cordoncillos de oro
-Más dos frenos de la brida dorados el uno tiene copas
-Más dos pares de estribos dorados de la brida los unos de ellos tienen açiones de terciopelo carmesí
-Más un hierro de lança de armas blanco y un ristre de arnés dorado y otro ristre y no está dorado
-Más unos çapatos de armas q tienen dentro unos tornillos para poner plumas son dos tornillos
-Más una guarnición corbata de terciopelo carmesy nueva q tiene grupera y cabeçada y pretal y riendas y açiones
-Más unas riendas de terciopelo carmesy con unas cadenas de hierro doradas
-Más unas riendas de terciopelo negro con una borla de oro y negro
-Más dos pares de espuelas de la brida doradas las unas de terciopelo negro y las otras de terciopelo carmesy
-Más los hierros de la silla del estandarte
-Más veynte y dos docenas de cordoncillos de hilo negro y amarillo de las cubiertas
(fol. v)
-Más quinçe dozenas de cordones menos quatro cordones q son de seda carmesy oro q sirven para las cubiertas de brocado
-Más nueve docenas de cordones de las mismas cubiertas de oro y negro faltan quatro cordones de las nueve dozenas
-Más tres doçenas de trenças de oro y negro y son tres çintas de estas coloradas para las mismas cubiertas

-Más seys doçenas de los mismos cordones y trenças de oro y verde y dos cordones menos son todas seys doçenas
-Más cinco cordones los quatro azules y el uno verde q son cinco para las mismas cubiertas
-Más tres correones de terciopelo carmesy q son para las cubiertas
-Más treinta y siete plumas amarillas y negras para las testeras de los caballos de armas
-Más treinta plumas amarillas y negras para los chapeos de los hombres de armas
-Más quarenta y ocho veletas de tafetán negro y franjas amarillas q son de las lanças de armas
-Más un pendón de damasco amarillo de trompeta [...]
-Más ocho escudos de armas de los atabales [...].

Doc. 3.
1843
Table of objects linked to the entailed estate of the different Houses attached to the Duchy of Alba.
Archivo Ducal de Alba, Casa de Alba, C. 158, nº. 15.

Resumen General de bienes y Alhajas vinculados [...]
1º. Alba [...] espadas, 02 [...]
6º. Monterrey [...] espadas, 01 mor^{ca} de oro [...]
It. Consta vinculado en este estado de Monterrey: veinte y nueve alfombras chicas y grandes de diferentes colores y medidas moriscas y turcas y cinco doseles= treinta y seis Almohadas de tela de oro brocado y terciopelo= Muchos jaezes, y adrezos ricos de caballo y otras cosas de guerra todo de mucho valor y se advierte que las 49 onzas y 4/8 de oro hacen seis marcos una onza y quatro ochavas= los dos mil y diez y seis marcos: seis onzas y media de plata labrada vinculada hacen quarenta arrobas, ocho libras, y seis onzas, y media. Contando a ocho onzas el marco. Y no entran en esta q^{ta} las dos fuentes de plata grandes, q^{e} regaló la ciu^{d} de Nápoles al Gran Duque D^{n} Fernando [...].

NÁJERA, DUQUES DE; MAQUEDA, DUQUES DE

Doc. 1.
1604, July 4.
Appraisal of the arms of the Duke of Nájera and Maqueda.
Archivo Histórico de Protocolos de Madrid, Protocolo 2619.

(fol. 1337 r)
Julio de 1604
Sepan quantos esta carta de poder vieren cómo nos don Jorxe de Cárdenas duque de Maqueda marqués de Elche usando de la venia y facultad real que de su magestad tengo para regir y administrar mis bienes y rentas de que yo el escribano doy fee y yo Juan de la Cotera escribano del número de la villa de Madrid como curador ad liten que soy de los señores don Jaime y don Juan de Cárdenas hermanos del dho señor duque don Bernardino de Cárdenas difunto y de mi señora doña Luisa Manrique de Lara duquesa de Maqueda y duquesa de Náxara cuya curaduría me fue disçirnida (sic) por el señor alcalde Silva de Torres corregidor de esta villa ante Françisco Testa escribano del número de ella [...]

(fol. 1343 r)
En la villa de M^{d} a quince días del mes de julio de mill y seiscientos e quatro a^{o}s ante mí el escrib^{o} e t^{o}s parecieron los ss^{es} B^{mé} Martínez de Quintana como curador que es [...] de la señora doña Marta e doña María de Cárdenas ~~hixas~~ y en nombre de la señora doña Luisa Manrique Manuel duquesa de Maqueda e Náxera= y en virtud del poder que de ella tiene y el l^{do} Di^{o} Nieto de Móxica abogado en la corte y consexos de su magestad y en esta v^{a} de M^{d} en nombre del señor don Jorxe de Cárdenas duque de Maqueda y de Juan de la Cotera como su curador ad liten y de don Xaime y don Juan sus hermanos y en virtud el poder que del dicho Sr Duque y del dho Ju^{o} de la Cotera tiene ambos por sí y en los dhos nombres dixeron q pa tasar los caballos y quartagos e demás cabalgaduras que haya y esclavos y esclavas ambos de conformidad nombraron a Diego Sánchez [...]

(fol. 1410 r)
En la villa de Madrid a veinte y siete días del mes de julio de mil y seiscientos e quatro a^{o}s los dhos B^{mé} Martínez de Quintana s^{o} de la dha señora duquesa de Maqueda y Náxara y licen^{do} Diego Nieto de Móxica abogado en los dhos nombres para lo tocante a la tasación nombraron a los tasadores siguientes

-A Juº León Camero para lo tocante a [...]
-A Herrera tapicero para tapicerías y alombras [...]
-A Pº de Allas entallador para bufetes sillas escritorios y arcas
-A Felipe armero para armas
(fol. 1410 v)
-A Baltasar López pintor para lo tocante a quadros y pinturas e imágenes y demás cosas del oratorio
-Para lo tocante a coches y literas a Briçuela
-A Berrielo librero para libros [...]

Las quales dhas personas nombraron los suso dhos para que cada uno tase las cosas que le tocan conforme al dho nombram^{to} y lo firmaron de sus n^{es} a quien yo el escribº doy fe que conozco

B^{mé} Mart^{z} de Quintana
El licen^{do} Diº Nieto de Móxica
Juº de Medina
Ante mí Fran^{co} Testa
(signed)

(fol. 1549 r)[1604, November 18]
E luego incontinenti en el dicho día mes y año dichos se prosiguió en la dha tasación (fol. 1549 v) por ante my el dicho escribano y Felipe de Namur armero vzº de esta villa el qual juró haçer esta tasación bien y fielmente e se hizo en la manera siguiente

Arca nº 29
-Primera un arca de pino en la qual va lo sig^{e}
-Un arnés grabado y dorado a listas de altobajo que es de seguir de Alemania que tiene çelada gola peto espaldar guardabrazos y braçales mandiletes grebas rodilleras y quijotes y zapatos con sus espuelas tasóse el dho arnés en treçientos reales
-Un morrión con trencillas de hierro sobredoradas aforrada la falda en terciopelo açul tasóse en tres ducados es de Calatayud
-Otro morrión con trençillas de ~~hierro~~ [crossed out] latón sobredoradas con unas tranzaderas blancas por dentro sin aforro se tasó en tres ducados
-Otro morrión con trençillas de ~~hierro~~ [crossed out] latón sobredoradas escritas de letra antigua sin aforro tasóse en doçe reales de Calatayud

-Dos morriones blancos (fol. 1550 r) lisos sin guarnición ninguna unas correas por dentro tasáronse en diez y seys reales cada uno
-Un estoque ancho con una cruz de guarnición dorada con un puño de plata viejo tasóse en quarenta y siete reales y medio
-Un estoque de tres esquinas a la tudesca con una vayna de hierro y çapa tasóse en tres ducados
-Otro estoque con lomo en medio guarnición de cruz barnizada tasóse en dos ducados
-Otra espada de a caballo ancha y corta con guarnición a la tudesca con vayna baya tasóse en diez y seys reales
-Otra espada corta con guarnición barnizada del moro de Zaragoza con vayna negra tasóse en un doblón
-Otra espada corta con una guarnición dorada toda de hierro con vayna de zapa baya con un cuchillo tasóse en treinta y tres Rs
(fol. 1550 v)
Con lo qual se çerró la dicha arca e se le puso el dicho número
Arca nº. 30
-Primeramente un arca de pino blanca en que va lo sig^e^
-Una espada ancha de a caballo con una guarnyçión de cruz dorada y baya de terciopelo viexa tasóse en diez y seys reales
-Otra espada corta de Sebastián Hernández con canal hasta la punta y una guarnición con cruz dorada y confitada y puño de plata viexo vayna de terciopelo viexo tasóse en veynte reales
-Otra espada de Sebastián Hernández con su garniçión (sic) de cruz dorada e confitada con su puño de plata vayna de terciopelo vieja y talabarte sin pretina tasóse en veynte reales
-Otra espada corta con su guarnición dorada de una cruz con su pon(fol. 1551 r)te (sic) en ella y un puño de plata viexo y una vayna de zapa y su cuchillo y punzón y un canuto de plata dentro la vayna tasóse en tres ducados
-Otra espada de a caballo corta de Tomás de Ayala con una guarnición avantanada (sic) y dorada con un puño de alquimia viexo y un talabarte viexo y su vayna baya sin pretina tasóse en tres ducados
-Una espada ancha turquesca con una guarnizión toda de hierro y la vayna de hierro dorada y un talabarte de ante pespuntado con seda verde tasóse en veynte reales
-Otra espada de a caballo del perrillo con una guarniçión ochavada de color de hierro y una vayna viexa tasóse en tres ducados
-Un alfanxe con guarnición barnizada vayna baya con cuchillo tasóse en diez y seis reales

(fol. 1551 v)
-Otra espada corta con guarnizión de color de hierro con una marca de mase Domingo con su vayna tasóse en tres ducados
-Una espada chiquita de niño con una guarnición de rey pavonada de Valençia la hoja con vayna de terciopelo con quatro reales (sic)
-Otra espada de niño la hoja de Valenzia con una cruz y su vayna viexa de terciopelo sin contera tasóse en dos reales
-Otra espada de Valençia la hoja con su guarnición de rey vayna de terziopelo viexa tasóse en quatro reales
-Otra espada corta de a caballo con guarnición barnizada vayna viexa de terciopelo tasóse en seys reales
-Otra espada ancha y corta la guarnición de dos (fol. 1552 r)[...] de color de hierro y la hoja de Françisco Ruyz el viexo con canal hasta la punta tasóse en cinco ducados con su vayna e cuchillo
-Otra espada ancha y corta del perrillo con una guarnición barnizada y una vayna viexa tasóse en dos ducados
-Otra espada ancha con una guarnición con una puente barnizada con su vayna tasóse en doçe reales
-Otra espada punta de alfanje con una guarnición barnizada corta la hoja turquesca tasóse en seys reales
-Otra espada corta con su daga que sirve de conpás (sic) guarnición de rey papavonada (sic) vaynas de zapa tasóse en veynte y quatro reales
-Una hoja sin guarnición que es estoque de armas con dos canales y un lomo hasta la punta en medio tasóse en tres escudos
(fol. 1552 v)
-Seys hojas las tres de un corte de Françisco Ruyz tasadas a veynte y quatro reales cada una
-Otra hoja con vayna y sin guarnición como cuchillo de monte de un filo de Françisco Ruyz tasóse en diez y ocho reales
-Otra espada corta sin guarniçión ni vayna tasóse en diez y seys reales
-Otra espada ancha sin guarniçión de a caballo con su vayna tasóse en doçe reales
-Otra hoja valenziana corta tasóse en diez y seys reales sin guarniçión
-Otra hoja de Sebastián Hernández ancha de a caballo tasóse en tres ducados
-Una espada de niño sin guarniçión con su vayna tasóse en quatro reales
-Un lajudo (sic) de alfanje (fol. 1553 r) con guarnición de hierro con su vayna tasóse en treinta reales
-Una espada corta y ancha sin guarnición adamascada tasóse en un doblón
-Un alfanxe viexo sin guarnición con vayna de zapa sin contera tasóse en quatro reales

-Otro alfanxe de la yndia con su vayna de madera y cuero barnizado tasóse en diez y seys reales
-Otro alfanxe con su guarnición de hierro dorada con dos cuchillos y un tenedor con vayna baya tasóse en veynte reales
-Otro cuchillo alfanxado con su guarniçión de hierro y cuchillo y punzón tasóse en veynte reales
-Un cuchillo de monte que corta por un cabo y por el otro es sierro (sic) con una vayna verde viexa tasóse en catorçe reales
-Otro cuchillo con guarnición (fol. 1553 v) dorada y un pomo de Brasil tasóse en diez y seys reales
-Otro cuchillo que corta por un cabo y por el otro es sierra con un cuchillo y punzón con vayna baya tasóse en diez y seys reales
-Otro cuchillo de monte con su cuchillo y punzón los mangos de cuerno de venado tasóse con su vayna verde en diez y ocho reales
-Otro cuchillo viexo con mango de cuerno de venado y vayna verde tasóse en quatro reales
-Otro cuchillo de monte quebrado el mango con vayna verde tasóse en dos reales
-Dos cuchillos en una caxa verde pequeños tasáronse en tres reales
-Un machete judío con el cabo de cuerno y vayna leonada y se tasó en tres reales
-Otro cuchillo judío con (fol. 1554 r) mango de cuerno negro y una cadenilla sortija y un brocal y su contera de latón labrado tasóse en seys reales
-una espada del moro de Zaragoza con una guarnición plateada con una cruz y una puente un puño de plata con vayna de zapa con su cuchillo el mango de marfil y un cañón de plata tasóse en tres ducados
-Una espada corta con una guarnición labrada a lo antiguo con cabezas de carnero y su daga talabarte pretina con vayna de zapa tasóse en veynte reales
-Un alfanje con una cruz y puño de granadino y vayna de sapa y un cuchillo tasóse en doçe reales
-Seys hojas alfanxadas con aguas como damasquinos tasáronse a escudo cada uno
-Una espada corta con una guarnizión dorada la hoja de Saagún (sic) con puño de plata viejo talabarte y pretina de terçiopelo viexo y vayna sin daga (fol. 1554 v) con su funda tasóse en treinta reales
-Otra espada de Sagún (sic) corta con una guarnición de almendrilla pavonada con su talabarte y pretina y daga tasóse en quarenta reales
-Otra espada ancha del perrillo con una guarnición de dos manos con su daga y talabarte sin pretina y un cuchillo tasadas en quatro ducados
-Otra espada corta ancha de Bilbao con guarnición barnizada daga y pretina sin talabarte un cuchillo en la vayna tasóse en treinta reales

-Otra espada corta de a caballo con una guarnizión de rey vayna quchillo y punzón y daga y talabarte y pretina de rizo tasóse en tres ducados
-Otra espada corta con una guarnición de habada y su vayna con cuchillo y punçón daga talabarte y pretina que la daga sirve de compás vaynas de zapa tasóse en quatro ducados
(fol. 1555 r)
-Otra espada con guarnición barnizada daga talabarte y pretina en su funda con cuchillo y punzón puño de seda viexo tasóse en treinta reales
-Una espada y daga con talabarte sin pretina cuchillo y punçón con puños de plata viexos la hoja con canal hasta la punta de Juanes de la Orta tasóse en seys ducados
-X Otra espada y daga con pretina sin talabarte la hoja de la espada de Pedro de Tor (sic) con una guarnición labrada cuchillo y punçón tasóse en quatro ducados es la guarnizión negra
-X Otra espada y daga de Saagún (sic) corta guarnición almendrina bernizada una vayna de terçiopelo con talabarte y pretina de terçiopelo viexo tasóse en quatro ducados en su funda
-Una espada ancha con una guarnición y puño de (fol. 1555 v) espada y daga de hierro dorado y negro labrado con su talabarte y pretina de terciopelo negro cayreles de seda negra en su funda tasóse en diez ducados
-Otra espada de Saagún (sic) el viexo con una guarniçión dorada de cordonçillo con su puño de plata y daga de la misma suerte con su talabarte y pretina de terciopelo blanco bordado de canutillo de oro y plata tasóse en doçe ducados
-X Otra espada y daga de Juanes de la Orta con guarniçiones doradas puños de plata dorados clavaçón dorada en talabarte y pretina tasóse en siete ducados
-Otra espada ancha y corta con su daga guarniçiones plateadas de cordonçillo talabartes y pretina de cuero bayo con trençilla con su funda tasóse en quarenta reales
-Otra espada y daga las guarniciones doradas (fol. 1556 r) de almedrillo con sus puños de plata la vayna y el talabarte y pretina de rizo en su funda tasóse en cinco ducados hoja de Saagún (sic)
-Otra espada ancha de a caballo con guarnición aventanada dorada con puño de plata viexo cuchillo y punçón sin talabarte ni pretina tasóse en tres ducados
-Otra espada de a caballo ancha de Bilbao con una guarnición aventanada con su daga y puños de plata vaynas baya (sic) con cuchillo y punçón tasóse en cinco du°s
-Otra espada corta de Saagún con una guarnición de almendrilla dorada con su daga vaynas de terçiopelo y pretina puños de plata con talabarte de terciopelo clavazón dorada tasóse en cinco ducados

-Otra espada (fol. 1556 v) y daga con guarnición de cordonçillo la hoja de la espada de Juanes de la Orta con daga talabarte y pretina de terciopelo con trencillas de oro los puños de hierro dorado vaynas de terziopelo tasóse en seys ducados
-Otra espada y daga guarniciones de almendrilla pavonado (sic) hoja del perrillo vaynas y talabartes y pretina de terçiopelo con cayreles de seda con daga en quarenta reales
-Otra espada corta con una guarnición plateada de Juanes de la Orta la guarnición de cordonçillo talabarte y pretina y vayna de terciopelo con trençillo de plata y puño de plata con cuchillo y punçón tasóse en quatro ducados
-Un alfanxe con el puño de barba de ballena broqual (¿) verde y funda de paño açul tasóse en dos ducados
(fol. 1557 r)
-Un talabarte y pretina con una daga de terciopelo bordado de canutillo de oro y plata con una daga de hidalgo guarnição y clavazón de cordonçillo plateado tasóse en diez y ocho reales
-Otra daga con talabarte y pretina bayo clavazón dorada tasóse en ocho reales
-Dos guarniciones viejas la una de nudo de çerezo en quatro reales
-Una daga buyda en tres reales con guarnizión de color de hierro
-Un tahalí negro de cuero tasóse en real y medio
-Otros dos tahalís de cuero bayos a quatro reales cada uno
-Una pretina çinto de cuero bayo clavazón dorada tasóse en dos reales
-Tres tiros de terciopelo negro son pretinas a lo antiguo clavazón pavonada tasáronse en dos reales cada uno
(fol. 1557 v)
-Un talabarte y pretina de terciopelo blanco con trençillos de oro y clavazón dorada tasóse en ocho reales
-Una pretina de terciopelo morado con trencillas y cayreles de oro clavazón dorada tasóse en doçe reales
-Otra pretina de cuero leonado con los hierros plateados y leonados tasóse en quatro reales [...]
-Un çinto de cuero de zapa aforrado en terciopelo negro tasóse en quatro reales
-Un tahalí de pistola de cuero bayo pespuntado tasóse en seys reales
-Otro tahalí pequeño de pistola pespuntado tasóse en dos reales
-Una pretina de terciopelo con clavazón dorada en dos Rs

(fol. 1558 r)
-Unas espuelas de la brida tasáronse en dos reales
-Tres boçinas de colmillo de elefante tasáronse en tres ducados

-Una caja vayna de zapa con tres cuchillos damasquinos puños de barba de ballena cabos de latón dorado tasáronse en cinco ducados
-Yten una caja de herramientas con diez piezas los cabos de marfil y dorados tasóse en siete ducados
-Una caja de pino blanca pequeña con diferentes herramientas tasóse en tres ducados
Con lo qual se çerró dha arca e se le puso el número treinta la qual dicha tasación hizo el dho Felipe de Namur armero en presencia de Juº de Montenegro Osorio el qual juró a Dios y en forma de drº a una cruz a tal como esta + que ha fecho la dha tasación bien (fol. 1558 v) y fielmente a lo que [...] e Dios le da a entender y lo firmó de su nombre el dho señor mayordomo y tasador [...]

Juº de Montenegro Osorio
Felipe de Namur
ante mí Fran^{co} Testa
(signed)

En la villa de Madrid a diez y ocho días del mes de mayo noviembre de myl y seiºs y quatro años prosiguiendo en la dha tasación e ynventario Frº de Puebla armero de alcabuces y ballestas por ante my el dho escribº se prosiguió en la manera sig^{e}

Arca nº 31
-Primeramente un arca de pino blanca con unas señales negras en que se puso nº treinta y uno iba en ella lo sig^{e}
-Un alcabuz de seys palmos la llave hecha en Granada de rueda con un cañón labrado de ataujía de oro embebido y plata con una caja de çerezo tasado en diez y ocho ducados digo veynte y dos ducados

-X Otro alcabuz de rueda con la llave hecha en Alemania (fol. 1559 r) con el cañón de cinco palmos y medio hecho de Alonso Rodríguez// con llave de armar en la dha llave con caja de çerezo tasado en catorçe ducados con funda de vaqueta
-Un alcabuz de rostro de Juan de Alemán de a quatro palmos y medio con llave de rueda hecha en Alemania de cubo çerrado caja de serbal con funda de vaqueta en ocho ducados
-Otro alcabuz con llave de rueda del ofiçial de la xarra con un cañón de cinco palmos y medio ochavado hecho en Bresa con su caja de çerezo tasóse en ocho ducados
-Un alcabuz de seys palmos con su llave de mecha y caja de serbal tasado en diez ducados

-Un alcabuz de rueda con el cañón de navajón hecho en Alemania de cinco palmos con su caja de serbal tasado en çien reales
-Un alcabuz que tiene por nombre retaco de quatro palmos y quatro dedos llave de rueda (fol. 1559 v) hecho de Alonso Rodríguez todo con su caja de çerezo munición de quinçe adarmes tasado en veynte escudos
-Un arcabuz de chispa todo de Vizcaya maltratado tasado en diez y ocho reales
-Otro alcabuz de rueda de seys palmos de diez y ocho adarmes de munyzión con su llave de rueda hecha en Granada el cañón de mase Simón con su caja de çerezo tasado en duçientos reales
-Un alcabuz de rueda curdo con el cañón de cinco palmos y medio de Al° Rodríguez caja de çerezo tasado en catorçe ducados
-Otro retaco que no es de a marca con su llave de rueda çurdo e hecho todo de Alonso Rodríguez con su caja de serbal con su molde tasado en doçe ducados
-Otro alcabuz de rueda con un cañón rayado hecho en Bresa con caja de nogal tasado en doçe ducados
-Un alcabuz de munição maltratado tasado en diez y ocho reales
-Treynta y nueve alcabuçes (fol. 1560 r) todos de rastillo con llaves que vuelven atrás con cañones de tres palmos y medio para de a caballo con sus cajas de çerezo tasados a ocho ducados cada uno
-Un cañón suelto de Alonso Rodríguez de munición de diez adarmes con su mira encubierta tasado en seys ducados
-Otro cañón suelto que fue cortado a uso de junco tasado e veynte y quatro reales
-Con lo qual se çerró la dha arca y se le puso el dicho número treinta e uno
Ojo
-Yten ocho cajas envueltas en unos anjeos y en la una de ellas dhas cajas hay tres esmerjones que por todos son veynte y quatro tasado cada uno a diez y seys ducados
N° 32 [...]
-Un arca tumbada de pino blanca en que va lo sig^e se le puso n° treinta y dos
-Un alfanxe de arzón con su ynpuñadura de ta(fol. 1560 r)raçea labrada de cuerno de venado con sus mascarones que es a modo de pistola la empuñadura en ocho ducados
-Una llave de alcabuz de rueda suelta nueva tasada en cinco du°s
-Otra del propio maestro de rueda suelta tasada en cinco du°s
-Un cañón digo dos sueltos de a quatro palmos munizión de nueve adarves hechos de Petrojo tasados a tres ducados
-Un molde y una llave de armar tasado en diez reales

-Dos pistolas de rueda juntas una caja (sic) de terçiopelo negro con los cañones de la piña (¿) hecho todo en Alemania con cajas de ébano de Portugal labrado de marfil en tramos de un maes (sic) tasado todo en diez y nueve ducados
(fol. 1561 r)
-Otro pistolete hecho en Alemania llave y cañón la llave de taraçea tasado en seys ducados
-Otro pistolete de rueda de jarra y piña hecho en Ytalia la caja cuajada de tarazea con su funda de vaqueta tasado en diez ducados
-Dos pistolas de rueda y las cajas de hierro embutidas de oro y plata con sus boçeles de plata de martillo con fundas de terciopelo bordado tasado en treinta e seys ducados
-Un pistolete de rueda hecho en Alemania caja de taracea tasado en treinta reales con funda de vaqueta
-Una pistola de dos palmos y medio con una llave de rueda labrada de ataujía con çiertos resaltos para ençender lumbre y el cañón de taujía (fol. 1561 v) labrados y dorado la caxa de Brasil labrada toda de montería con su funda de vaqueta con ella una caja con quatro guardafuegos tasado en diez y seys ducados
-Dos pistolas de jarra y piña y las llaves del propio maestro labradas las caxas de taujía con bolsa de vaqueta juntas en diez y seys ducados ambas
-Otra pistola con llave de rueda de cubo cerrada hecho en Alemania con caja labrada de ataujía funda de vaqueta tasado en seys ducados
-Una pistola de dos cañones con una llave de rueda que tiene dos fobones (sic) que hacen dos efectos al de las armas y los cañones de ataujía (fol. 1562 r) y la llave de lo propio la caja de taraçea labrada con funda de vaqueta tasado en doçe ducados
-Otra pistola de jarra y piña con llave de rueda con caja de granadillo y baqueta de marfil con caja de vaqueta tasado en diez ducados
-Otra pistola con llave de rueda de cubo la caja de taracea con su guardamonte funda de vaqueta tasado en veynte y ocho reales
-Otro pistolete hecho en Bresa con su llave de rueda la caja de boscaje labrada con su [...] de vaqueta tasado en quatro ducados
-Una llave de rueda suelta con dos perridos (sic) de enbo (sic) çerrado hecha en Aragón tasada en diez y seys rreales
-Una pistola de rueda hecho en Alemania todo con caja de hierro labrado de ataugía tasado en tres ducados
(fol. 1562 v)
-Otro pistolete hecho en Alemania con llave de rueda caja de rueda lisa que es de faldriquera tasado en tres du^os^

-Otro pistolete de hechura de daga con llave de rueda grabado y dorado con puño de zapa guarniçión a lo antiguo tasado en quatro du°s
-Otra pistola con llave de rueda la falta el medio perrillo maltratado tasado en ocho reales
-Un frasquillo de hierro que sirve de llave de armar y de polvorín tasado en ocho reales
-Otro frasquillo de hierro que sirve de llave de armar y de polvorín seys reales
-Otro frasquillo de hierro aovado (sic) labrado de oro que sirve de polvorín tasado en diez y seys reales
-Otro frasquillo (fol. 1563 r) aovado (sic) labrado de ataugía que sirve de polvorín tasado en ocho reales
-Un frasquillo que sirve de dos cargas de pólvora para llevar pelotas en él guarneçido de hierro sobredorado y viexo tasose en ocho reales
-Un frasco de cuerno de venado con un cordón de seda parda para llevar pólvora maltratado tasóse en ocho reales
-Un frasco de pólvora de tortuga guarneçido de hierro de tres adarmes la carga tasado en veynte y quatro reales
-Un molde de haçer perdigones enrramados con sus astiles tasado en diez y seys reales
-Un frasco de cuerno berberisco para llevar pólvora en él tasóse en quatro reales
-Una caja de hierro de hechura de morrión (fol. 1563 v) labrado de ataujía de oro con sus quatro piezas dentro diez y seys reales
-Un frasquillo de polvorín labrado de ataujía escudo y medio
-Otra caja a uso de morrión guarneçida de hierro tasóse en ocho reales
-Un frasquillo de cuerno berberisco con un cordón verde tasado en quatro reales
-Una bolsilla de cuero negro con ocho cargas de pistoletes tasado en tres reales
-Veynte y siete hierros que sirven de lavar los alcabuçes y otras cosas convinietes a ello veynte reales
-Un molde liso de pelota quatro reales
-Una turquesca grande que sirve de molde de haçer perdigones de todas suertes con sus astiles viejos con una ema (sic) en medio tasada ocho du°s
(fol. 1564 r)
-Una turquesa grande donde se haçen en ella diez y siete pelotas de una vez tasada en sesenta y ocho reales
-Una bolsa de alcabuz con su çinto y hierros de espada tasada en veynte reales
-Otra bolsa de alcabuz baya pespuntada con seda verde tasada en doze reales
-Una bolsa verde pespuntada de seda amarilla tasada en ocho reales
-Otras dos bolsas de cuero verde pespuntadas de amarillo en diez y seys reales

-Otra bolsa de cuero de benazon (sic) y cuero leonado pespuntada tasóse en diez reales
-Otra bolsa viexa de cuero negro y leonado tasóse en dos reales
(fol. 1564 v)
-Otra bolsa de cuero muy viexa que ha tenydo hilos de oro tasada en dos reales
-Otra bolsa de cuero verde nueva por guarneçer tasada en seys reales
-Dos bolsas de vaqueta para llevar cargas tasada en diez y seys reales [...]
-Una bolsa de enfriar agua rota tasada en doçe reales
-Un señuelo de cazador con un cordón de seda dorada y oro tasada en ocho reales
-Un hacha de armas guarneçida de zapa el astil con su martillo tasada en veynte e quatro reales
-Otra hacha de armas con su astil y remate al cabo tasada en veynte e quatro reales
-Un venablo sin guarneçer labrado de oro con guardamontes que salen de él tasado en seys du^{o}s
(fol. 1565 r)
-Una corneta dorada y en ella unas monterías con cordón de seda tasada en quatro ducados
-Un hierro que sirve de sacatrapos de pistolas tasóse en dos reales
-Un caja (sic) herramientas de cuero leonado de engerir árboles que tiene diez piezas en todas guarneçidas de ébano tasado en catorce ducados
-Con lo qual se çerró el dho arca tumbada e se le puso el dicho número treinta y dos
Arca nº. 33
-Un arca de pino blanca en que va lo siguiente e se le puso número treinta y tres
-Siete vergas sueltas de seys virotes y una de bodoques son de ballesta de Puebla de todas onzas a veynte y quatro reales cada verga
-yten ocho vergas sueltas (fol. 1565 v) de ballestas de virotes de todas onzas hechas de Luys Moreno el de la Puebla de Montalbán en treinta y ocho reales cada verga tasadas
-Dos vergas de ballesta de virotes sueltas una pequeña y otra mayor e dhas de Fran^{co} de Santa María el de Zamora a veynte reales cada una
-Otras verga de ballesta de virote (sic) hecha de Grajeras el viexo tasada en dos ducados
-Tres vergas sueltas las dos de ellas hechas del padre de Luys Moreno y la otra hecha en Badajoz tasadas en veynte reales cada una
-Diez y seys vergas de ballestas sueltas de virote de mano de Juan Blanco el viexo a tres ducados cada una entiéndese que son pequeñas
-Más treçe vergas (fol. 1566 r) del dho Juan Blanco que son más grandes sueltas de ballesta de virote tasada cada una a veynte y ocho reales

E con esto se dejó la dha tasaçión por ser hora de medio día para proseguir adelante en ella

Ante mí Fran^{co} Testa
(signed)

E luego yn continente en el dho día mes y año dho por la tarde se prosiguió en la manera siguiente en el dho ynventario por el dho Fr^{co} de Puebla pres^{te} Ju^{o} de Montenegro Osorio mayordomo
-Una ballesta de serbal labrada de ataujía de virote con sus gafas envuelto en una bayeta morada tasóse en seys ducados
-Una ballesta con verga de Juan Blanco la qureña (sic) de granadillo las gafas redondas de Zamora tasóse en catorce ducados
-Una ballesta de virote la verga de Luys Moreno la cureña de serbal con gafas redondas de Zamora tasóse en ocho ducados
(fol. 1566 v)
-Una ballesta de virote con verga de mase Myguel y tablero de alcar unas gafas quedradas ocho ducados
-Otra ballesta de virote pequeña con verga del de Zamora con unas gafitas quadradas sin encaxe tasóse en seys ducados
-Otra ballesta de virote guarnezida de latón con verga del sordo de Zamora y gafa redonda de Robles tasóse en çien reales
-Otra ballesta de virotes con verga de Luis Moreno y una gafa quadrada tasóse en ocho ducados
-Una ballesta de virotes la qureña de granadillo con una verga de Ju^{o} Blanco el viexo sin gafas tasóse en diez ducados
-Otra ballesta de virotes la qureña de ençina con verga de Luis Moreno sin gafas tasóse en çien reales
-Una ballesta de virotes con una cureña de alcar gafas quadradas verga de mano de Puebla tasada en ocho ducados
(fol. 1567 r)
-Una ballesta de virotes con verga de Juan Blanco el viexo y un tablero de alcar gafas quadradas tasóse en diez ducados
-Otra ballesta de virotes con verga de Pedro la Fuente cureña de enzina gafa quadrada tasada en seys ducados
-Otra ballesta de serbal de virotes con verga de mase Miguel gafas quadradas en quatro ducados
-Otra ballesta de virotes de serbal con verga de mase Myguel gafa quadrada negra tasada en quatro du^{os}

-Una ballesta de enzina con verga de Juan Blanco gafas quadradas cureña de enzina tasada en nueve ducados
-Una ballesta de virotes con cureña de alcar verga de Puebla de dos jaras con una gafa blanca de Zamora tasada en siete ducados
-Otra ballesta de ençina con verga de Luys Moreno (fol. 1567 v) una gafa redonda blanca tasada en diez ducados
-Otra ballesta de virote de granadillo con verga de Juan Blanco sin gafas tasada en ocho ducados
-Otra ballesta de virotes sin nuez guarneçida de latón con verga de Juan Blanco con gafas quadradas tasada en çinquenta reales
-Otra ballesta de virotes de serbal con verga de Juº Blanco unas gafillas quadradas tasada en quatro ducados
-Otra ballesta de virotes sin gafas con verga de Juan Blanco tasada en tres ducados
-Una ballesta de virotes con cureña de serbal una verga de Luys Moreno sin gafas tasada en ocho ducados
-Otra ballesta de virotes de serbal sobrepuesta con una verga de Juº Blanco tasada en çinquenta reales
-Otra ballesta de virotes cureña de serbal con una verga de Zamora tasada en seys duºs
(fol. 1568 r)
-Una ballesta viexa de virotes verga de Juan Blanco dos ducados es el tablero de serbal
-Tres pares de gafas de bodoques en siete reales todas tres
-Un tablero de serbal echada una canal tasado en quatro reales
-Una horquilla de aguardar espera tasada en seys reales
Con lo qual se çerró la dha arca e se le puso número treynta e quatro (sic)

Arca nº. 34
-Otra arca de pino blanca en que va lo siguiente e se le puso número treynta e quatro
-Primeramente diez y ocho arcos turquescos tasados en treinta e seys ducados que va a dos ducados uno con otro
-Yten quatro carcajes de cuero y uno de madera con duçientas jaras de pico de gorrión de pinavete (sic) tasadas en çien reales
-Yten tres carcajes de vaqueta con ochenta y cinco […] y pasadores y escoplillos (fol. 1568 v) tasadas en sesenta reales todas
-Diez virotes de armella çerrada de los once en diez reales tasados todos en un carcax de vaqueta leonado tasado en quatro reales que por todos son catorçe

-Un carcax verde viexo con dos arpones y tres escoplos tasados en diez reales todo
-Veynte y ocho xaras de ballesta para tirar con yerba tasados en veynte y ocho reales con un carcax leonado pespuntado amarillo el carcaj tasado en doçe que todo hace quarenta reales
-Seys rallones doce reales con un carcaj de vaqueta negra en ocho reales que son veynte
-Una burjaca (sic) de llevar querda para virotes de cuero de perro en ocho reales (fol. 1569 r)
-Una alxaba de benazon (sic) con sus remates de cuero torneados en que llevan jaras enarboladas tasado en diez y seys reales
-Un carcax verde de llevar virotes palomeros tasado en ocho reales
-Tres carcaxes de virotes pespuntados tasados en diez y ocho reales son de cuero leonado
-Una alxabilla verde para llevar dos jaras tasadas en seys reales
-Una alxaba de cuero leonado con cinco remates de cuerno negros torneados con seys virotes y sus cordones de seda verde tasado en treinta reales
-Otra alxaba con seys remates de marfil (fol. 1569 v) torneados y aforrada en raso amarillo guraneçida de terziopelo verde tasada en seys ducados
-Otra media funda berberisca de llevar arcos turquescos tasada en ducado y medio
-Otro carcax de jaras de cuero tasado en tres reales
Con lo qual por hoy se cerró la dha tasación y el dho Fr^{co} la Puebla juró a Dios y en forma de dr^{o} haberla fecho bien y fielmente estando pres^{tes} el dho mayordomo y lo firmó de su nombre

Ju^{o} de Montenegro
Fran^{co} de Puebla
Ju^{o} de Medina
Ante mí Fran^{co} Testa
(signed)

(fol. 1570 r)
Arca n° 35
En la villa de M^{d} a diez y nueve días del mes de noviembre de myll y sei^{o}s y quatro a^{o}s se ynventarió un arca que es de nogal que tiene tres repartimientos y cinco llaves en la qual se puso número treinta y cinco y se halló en ella lo sig^{e}
-Un saco de lienzo algo grande que tiene por señal una [...] ochenta y dos marcos y siete onzas y cinco ochavas de monedas estrangeras de plata y [...]

(fol. 1571 r)
En la dicha villa de Madrid en el dho día mes y año dhos el dho Bartolomé de Arramans prosiguiendo en el dho peso y tasación tasó y pesó lo sig^{te}
-Una espada corta morisca con el puño y toda la guarnición brocal y contera y otras piezas en la vayna de plata y tahalí de seda labrado de plata que tiene unas borlas de seda e [...] plata y hechura se tasó en çien escudos
-Otra espada morisca guarnizión y puño brocal y contera y otras piezas de plata en la vayna se tasó en sesenta escudos tiene por señal unas bandas atravesadas como escudos de armas
-Otra espada guarnición de plata blanca que llaman guarnición de hechura de rey con la contera de plata vayna de zapa cuchillo y punzón con cabos de [...] la espada de un corte tasada en quarenta ducados
-Un alfanxe con vayna (fol. 1571 v) de terciopelo carmesí guarnición puños y pomo contera y brocal de plata sobredorada la cuchilla damasquina la tasó todo (sic) en quinientos reales
-Una caja de tres cuchillos damasquinos con los cabos de marfil contera y brocal de plata sobredorados lo tasó en treinta ducados
La qual dha tasación se puso en juro a Dios y en forma de drº a haber fecho el dho Bar^{mé} de Escaramas bien y fielmente a lo qe alcanza que Dios le da a entender porque no se pudo pesar por no desguarnecer las dhas espadas a lo qual se halló presente el dho Juº de Montenegro Soria y lo firmaron ambos

Juº de Montenegro Soria
Br^{mé} Descarraman
Juº de Medina
Ante mí Fran^{co} Testa
(signed) [...]

(fol. 1637 v)
Arca nº 94
-Yten en una arquilla de pino blanco número noventa e quatro había lo sig^{te}
-Yten diez y siete arcos turquescos que se tasaron por el dho Francisco de Puebla a ocho reales uno con otro
-Yten dos carcaxes de vaqueta negra y en (fol. 1638 r) cada uno treinta y seis jaras que por el dho Francisco de Puebla se tasó cada carcax con sus jaras a veinte e seis reales
-Yten duçientas e noventa y siete jaras que están en siete atados que por el dho Françisco de Puebla se tasó cada una de las dichas jaras a medio real

-Yten noventa y zinco virotes sostrados de Armelea cerrada de Paones y de Renedo que se tasaron por el dho Francisco de Puebla a dos reales
-Yten tres viras de tirar albarrerones se tasó por el dho Francisco de Puebla en seis reales a dos cada una
-Yten doce arpones para tirar con yerba que se tasó por el dho Fran^{co} de Puebla a dos reales cada (fol. 1638 v) uno
-Yten tres escoplos que se tasaron por el dho Puebla a real y medio cada una (sic)
-Yten una xara en un real es de ballesta
-Y con esto quedó cerrada la dha arca del dho número noventa e quatro
Y por ser las doçe del medio día se quedó en este estado la dha tasación para la acabar e proseguir siendo necesario la qual se hiço con asistencia del dho Juan de Montenegro Soria y elos dhos tasadores joraron a Dios e a una cruz en forma haberla hecho bien y fielmente a su saber y entender e todos lo firmaron [...]

Ju^{o} de Montenegro Soria
Fran^{co} de Puebla
Al^{o} de Morales
Ju^{o} de Medina
Ante mí Fran^{co} Testa
(signed)

(fol. 1657 r)
En la villa de Madrid a seis días del mes de diciembre [...] del año de mil y seiscientos y quatro a^{o}s (fol. 1657 v) Jusepe González platero de oro por ante mí el dho escribano hiço la tasa del tenor sig^{te}

-Primeramente una espada corta con la guarnición a la morisca que el puño e pomo y guarnición y contera e brocal y dos pieças de que cuelga el tahaliz en el tahalí otras dos pieças con escudos de armas es de oro y un puñal que la guarnición contera vaina y brocal es ansimesmo de oro esmaltado de colores e la vaina de la espada es de cuero bayo labrada de oro tirado y el tahalí labrado de la misma suerte que la dicha espada y puñal oro y hechura se tasó por el dho Jusepe Gonçález en mil escudos de oro de a quatrocientos maravedís
-Yten una espada y daga que tiene la guarnición (fol. 1658 r) con unas cabecillas de carnero y otras labores de oro y hierro para escarçela que está con sus cuchillas tasóse por el dho Jusepe González en sesenta escudos de oro de a quatrocientos maravedís
-Y las dhas espadas se metieron en un cofre número veinte y tres

Y el dho Jusepe Gonçález juró a Dios en forma que la dha tasación la hecho (sic) bien e fielmente a todo su saber y entender sin fraude alguno la qual hiço con asistencia del dho Juan de Montenegro Soria y ambos lo firmaron

Ju° de Montenegro Soria
Jusepe Gonçález
Ju° de Medina
Ante mí Fran^co^ Testa
(signed).

OROPESA, CONDES DE

Doc. 1.
1573, January 31.
Arms inventory of the Oropesa Castle, of the Count of Oropesa. Followed by the inventory of arms that the residents of the town had.
Archivo Histórico de la Nobleza, Frías, C. 1277, Doc. 11.

[unfoliated, the digitized folio number is entered]
(nº. 1)
Oropesa= Averigüación de las armas en esta villa y demás de su estado
Memoria de las armas que están en casa del q^e my señor

-Treynta arneses enteros son toda armadura p^a hombre y caballo
-quatro coraçinas
-dos petos y dos espaldares negros
-Un arnés de Alemania con pieças dobles
-quatro tiros grandes de artillería
-Un trabuco grande todo de hierro
-Una culebrina de bronze
-seys paveses
-onze arcabuzes
-quinze ballestas
-Hay cantidad de coraçinas y cosoletes (sic) enteros
-quinze ballestas
-dos pistoletes
-una escopeta/ otra/ otra
-çinco lançones
-Un venablo
-una alabarda
-un dalle
-quatro rodelas
-Un broquel

(nº. 2)
Conçejo Justiçia y Regimiento de esta my villa de Oropesa sabed que la Mag^t del rey nro señor por justicia y respeto y por buenos motivos ha mandado averiguar las armas que hay en el reyno que puedan ser y sean útiles para la defensa de él y

exerçiçio de la guerra para cuyo efecto ha mandado prevenir a los señores para que en nras tierras hagamos la dha averig^{ón} e porque yo quiero en la mía se haga con mucha particularidad como se debe al cumplim^{to} de los mandamientos de su Mag^{t} y acatando al buen fyn para que es dirigido vos mando que luego como esta mi provisión os fuere mostrada hagáis averiguación de todas las armas arcabuzes y coseletes y otros qualesquier tiros de pólvora y qualesquier otras armaduras que haya demás de las ordinarias en poder de los v^{o}s de esa dha villa y cada uno de ella y hecha esta averigu^{ón} por ante escribano que de ello dé fee la ynviéis ante my por escripto originalmente y como pasare sin dexar traslado y lo cumplid dentro de tres días primeros siguientes porque así combi^{e} al servicio de su Mag^{t} fecho en Orop^{a} a 31 enero 1573 a^{o}s

El conde
(signed)

En las casas del ayuntami^{o} de la villa de Oropesa lunes día de ayuntami^{o} ordinario nueve días del mes de febrero de mill y qui^{o}s y setenta y tres años ante los muy magni^{cos} señores lic^{do} Gutierre de Castro juez de residencia en la dha villa e su tierra por el q^{e} mi Señor y P^{o} Guiral de Çerezuela y P^{o} de Montalbo y Al^{o} del Rincón regidores e Fran^{co} de Tapia pr^{co} general del q^{o} de esta dha villa, y en presencia de mí P^{o} de Solís escribano y t^{o}s par^{o} presente Al^{o} de Córdoba criado de Su Señoría del q^{e} mi Señor y en su nombre e presentó y leer hizo por mí el dho escribano la provisión de Su S^{a} de suso qu^{da} e pidió cumplimi^{o} de ella, e los (nº. 3) dhos señores estando en el ayuntami^{o} ordinario dixeron q obdedesçían e obedesçieron la dha provisión de Su Señoría con el acatamiento debido e que están prestos de cumplir e que la començarán a cumplir dende hoy e q no se alçará mano de ello hasta lo acabar de cumplir con toda la diligençia posible t^{o}s q son presentes Hernando de Orihuela e Alonso Myn Portero v^{o}s de la dha villa Di^{o} de Solís escrib^{o}
E después de lo suso dho en la dha villa de Oropesa a diez días del dho mes de febrero del dho año de mill e q^{o}s e setenta e tres a^{o}s los dhos señores juez de residencia y de [...] e Pedro Montalvo [...] dixeron en cumplimi^{o} de la dha provisión de Su S^{a} anduvieron por todas las casas de los v^{o}s de [...] on sus arrabales e calles [...] y en ellas se hizo la verificación y hallaron las armas q se siguen por ante my el dho escribano pú^{co} en esta manera

-En la casa del muy Yll^{e} señor don Fran^{co} de Toledo vicerey del Perú dos çeladas de ynfante y otra çelada de seguir la guerra y una testera de caballo y una maça e una silla alemana de caballo e unas grebas y dos pares de quixotes e unos boçetes (sic) de mangas de malla e una testera de arnés [...].

PASTRANA, DUQUES DE

Doc. 1.
1596, March 23.
Inventory of the Duke of Pastrana.
Archivo Histórico de Protocolos de Madrid, Protocolo 1617.

(fol. 1832 r)
Inventario de los bienes libres del duque de Pastrana

En la villa de Md a vte y tres días del mes de março de mill y quiºs e noventa y seis aºs ante el dotor Ortiz tenyente de corregidor en la dha villa y su tierra por Su Magd y ante mí el escribano público e tºs yusoescriptos la señora doña Ana de Portugal y Borxa princesa de Mélito Duquesa de Pastrana viuda mujer que fue del duque Don Rodrigo de Silva y de Mendoça defunto que sea en gloria como su albacea y testamentaria y tutora y administradora de las personas y bienes de Don Francisco de Portugal y Silva y Don Diego de Mendoca y doña Margarita de Borxa sus hixos y del dho Duque su marido y dixo que agora nuevamente ha venido a su noticia la muerte del dho Duque e para que siempre y en todo tiempo conste y se entienda los bienes libres que tenía y dexó y haya de ellos la luz y claridad que se requiere quiere hacer ynventario jurídico cierto y verdadero de todos ellos [...] el dho teniente le mande haçer por ante mí el presente escribº y que se le de por testimº e manera que haga fee y sobre todo pidió justicia

(fol. 1832 v)
En la villa de Md a veynte e nueve días del mes de abril de mill y quiºs e noventa y seis años por ante mí el escrº público la dha señora duquesa como tal testamentaria e tutora y administradora de los dhos sus hixos cumpliendo con la obligación que tiene dixo que quiere proseguir fenecer y acabar el dho ynventario (fol. 1833 r) dentro del término que es obligada e poniéndolo en efecto le fue prosiguiendo e prosiguió ante mí el presente escribano e puso en él los bienes siguientes

-Un coche con cuatro caballos y dos pares de guarniciones viexas e nuevas y sus frenos y demás adereço
-Quatro sillas de la brida y dos de jineta
-Cinco pares de estribos de la brida los dos pares plateados y dos dorados y unos barniçados
-Unos estribos de la jineta barniçados de negro
-Otros estribos de la jineta dorados de figuras

-Otros plateados y relevados
-Unas cabeçadas y pretal y espuelas de xineta con la clavaçón dorada y plateada
-Otras cabeçadas de la jineta barniçadas
-Dos pretales de cascabeles uno carmesí y el otro açul
-Dos frenos de la brida con sus apar^{s} doradas
-Otros dos frenos de xineta viexos
-Çinco cabeçadas de caballos de la brida
-Çinco guruperas viexas
Un pretal de terciopelo de la xineta
Çinco pretales de cuero viexos
(fol. 1833 v)
-Diez cinchas traídas
-Dos sillas de la brida viexas e rotas de teciopelo negro con pasamanos
-Tres maletas una de lienço y otra de jerga y otra de cuero con su pelo y un almofrex viexos
-Dos cabeçadas de terciopelo negro
-Un pretal y un par de riendas y una gurupera de lo mismo
-Un caparaçón de terciopelo negro con pasamanos e traído
-Una cuna de viento dorada con su pabellón de paño pardo
-Quatro sillas de terciopelo verde con franxas de oro e clavaçón dorada
-Otras quatro sillas de terciopelo carmesí con franxas de oro y clavaçón dorada y fundas de frisa colorada
-Tres sillas de terciopelo negro con franxas de oro y clavaçón dorada y la madera de la Yndia [...]
(fol. 1835 r)
Talabartes de torneo
-Dos talabartes y pretinas de terciopelo carmesí con clavaçón plateada de torneo
-Otro talabarte y pretina de cuero açul de torneo argentado de plata [...]
(fol. 1840 r)
-Dos escopetas viexas
-Dos pistoletes viexos
-Dos espadas con su dagas [...]

Pastrana
Yten dexó y declaró que en la villa de Pastrana hay ciertos bienes los quales puso por ynventario q son los siguientes

-Primeramente cinco rodelas tres negras y dos de la India pyntadas con oro grabadas

-Dos alxabas de cuero y terciopelo carmesí con un hueso de marfil
-Un broquel grande grande (sic) blanco con [...] de hierro
-Dos mapamundis con sus caxas de madera
-Cinco frascos de arcabuz los dos pequeños y dos de cuerno y uno de latón
-Dos arcos turquescos sin querdas
-Dos carcaxes con quarenta y cinco flechas turquescas
(fol. 1840 v)
-Siete cargas de cañas con cordones verdes y colorados de seda
-Un alfanxe con una vayna de Toledo negro
-Dos pistoletes labrados de marfil
-Un arcabuz pequeño de Mota con su funda
-Un arcabuz grande labrado de marfil la caxa
-Un mosquete con su horquilla
-Una hacha de armas
-Cinco espadas de jineta la una quebrada
-Tres adargas de Berbería blancas
-Una maleta de las adargas
-Un sillón de mujer de terciopelo negro viexo con sus cabeçadas de terciopelo y falsas riendas guarnecidas de plata
-Dos retratos de dos señores hixos del Duque de Medinasidonia
-Una silla de caballo de terciopelo negro bordada con canutillo de vidro (sic) y oro falso
-Unas guarniciones de caballo [...]
-Un adereço de coche viexo ençerado y frisas viexo las cubiertas de arriba con tres coxines de terciopelo amarillo viexo
-Tres badiles de lumbre y unas tenaças y un atiçador todo de hierro
-Siete vihuelas de arco la una grande en su caxa y las seis en un arca pintada aforrada en frisa
-Seis sillas de cuero pespuntadas de seda amarilla y franxas amarillas
-Quatro sillas viexas pespuntadas grandes de cuero leonado y dos taburetes de lo mismo
(fol. 1841 r)
-Cinco bancos el uno pequeño
-Un quitalumbre de bocací verde y palo teñido
-Una mesa grande de Alemania con dos pies
-Otra mesa de Alemania con unos pies de cadena teñida la madera de colorado
-Otra mesa negra viexa con sus pies de cadena
-Un atambor
-Dos bacines con sus caxas de madera

-Quatro candiles con sus bolas de açofar
-Unos morillos grandes de açofar
-Otros más pequeños de açofar
-Otros chicos de açofar
-Una mesa de cadera de pino con sus pies
-Once bufetes grandes y pequeños
-Un baçín grande de alambre
-Quatro caxas de madera viexas
-Un bufete grande de nogal
Lo que había en la contaduría
-Dos bancos de nogal
-Dos taburetes con cueros leonados el uno quebrado

Lo que había en un cofre
-Una almártaga (sic) de terciopelo negro con unos cordones de seda
-Unas fundas de terciopelo viexo
-Un adereço de caballo riendas y falsas e riendas y pretal y cabeçadas y gurupera bordado de canutillo de vidro (sic) y oro e plata que todas son siete pieças
-Dos pares de estribos viexos unos de xineta y otros de la brida
-Tres frenos viexos y un bocado
-Unas espuelas sin correas
-Un pergamino de la discreción de Mélito
-Un arca consagrada
-Una caxa de cuero aforrada de frisa del calentador de plata
(fol. 1841 v)
-Una tabla de San Diego

En la [...] de la capilla
-Una imagen de la Cena y otra de [...]= e otra de Nuestra Señora de la Quinta Angustia= e otra de un Eçe Homo y otra de un San Antonio= e otra imagen de la Anunçiaçión a los pastores= e otra imagen pequeña de Santa Catherina
-Un bufete de nogal en que estaba hecho el altar= y un caxón de sacristía vaçío
-yten puso por ynventario çiertos bienes que parecen estar en poder de Joan de Alcaraz que son los siguientes
-Unos açicates de hierro viexos
-Una caçoleta de derretir plomo
-Unos pies de escritorio
-y las mesas que hay de pino sobre que estaba la guardarropa
-y en este estado quedó el dho ynventario este dho día por ser noche= [...].

Doc. 2.
1627, January 9.
Post-mortem inventory of the Duke of Pastrana.
Archivo Histórico de Protocolos de Madrid, Protocolo 3361.

Ynventario de los bienes que quedaron por la fin y muerte del Exmo señor Ruy Gómez de Silva Prínçipe de Mélito y Duque de Pastrana

(fol. 647 v)
En la villa de Madrid a nueve días del mes de enero de mill y seiscientos y v^{te} y siete años se fue prosiguiendo en el dho ynventario de los bienes que quedaron de su Exª del dho señor Duq^{e} de Pastrana el Sor Don Gregorio de Borxa su caballeriço puso las cosas que estuvieron a su cargo tocantes al dicho oficio que son las siguientes

Sillas de manos [...]

(fol. 671 v)
En la villa de Madrid a Vte y seis días del mes de enero del dho año de seiscientos y v^{te} y siete se fue prosiguiendo en hazer ynventario de los bienes del dho Sr Duque de Pastrana y Prínçipe de Mélito y el dicho Miguel de la Peña puso por inventario las cosas de su cargo que son las siguientes

Armas espadas alcabuces ballestas rodelas y otras cosas de este género
Mosquete
-Primeramente un mosquete de rueda
Rodelas
-Seis rodelas dos doradas ~~con armas~~ [crossed out] y quatro negras
Un arco
-Un arco turco con su alxaba y saetas
Un hueso/ no se tasó por no ser de provecho
-Un hueso de marfil con algunas flechas
Alcabuces
-Quatro arcabuçes con sus fundas verdes que presentó el Duque de Florencia a su Exª
Una bolsa/ tasóse con los arcabuces
-Una bolsa con recado para hazer las pelotas de los quatro arcabuçes
Pistolas
-Seis pistoletes pequeños de puño que trajo su Exª de Nápoles

Pistolas
-Dos pistolas grandes de arçón sin fundas
Pistolas
-Dos pistolas con sus fundas que presentó don Luis de Portugal
Pistola
-Otra pistola con su funda
Arcos
-Tres arcos de mano de yndios
Alabardas
-Doçe alabardas con sus borlas
Pistolas
-Dos pistolas de arçón con sus fundas amarillas de vaqueta
frasquillos
-Dos frasquillos con sus cordones de seda para la pólvora
(fol. 672 r)
Frasquillos
-Tres frasquillos para pólvora los dos de cuerno y uno de bronce dorado
Adereço de espada
-Un adereço de espada y daga plateado con tiros y pretina
Espada/ojo
-Otro adereço de espada dorado con su funda
Bolsilla
-Una bolsilla con instrumentos de alcabuz
Frascos
-Tres frascos de búfano para pólvora
Frasco
-Otro frasco grande de búfano para pólvora
Cuchillo/ojo
-Una caxa de tres quchillos damasquinos caxa de çapa con guarniçión dorada
Cuchillo/ojo
-Otro cuchillo damasquino con vayna q parece de plata dorada con una cadenilla de plata
Pistolete
-Un pistolete con la caxa labrada pequeña
Guarnición/ojo
-Una guarnición de espada y daga con sus puños y dos conteras doradas
Guarnición
-otra guarnición y pomo pequeña de espada
Guarnición

-Una guarnición de daga negra lisa con pomo y contera
Espuelas
-Ocho pares de espuelas cinco negras y tres doradas
Guante
-Un guante de malla
Espadín
-Un espadín dorado
Espadín
-Otro espadín dorado y plateado
Espadín
-Otro espadín con las cabeças de león y dos veneras grandes por guarda con daga y demás guarniciones en su funda
Espada
-Otro adereço de espada y tiros y petrina negros
(fol. 672 v)
Adereço de espada
-Otro adereço de espada y daga con tiros y pretina negro enrejado
Pistolete
-Un pistolete pequeño negro
Una daga
-Una daga con guarniçión dorada y leonada
Un adereço negro
-Un adereço negro con tiros y daga en una hoja
Otro adereço
-Otro adereço de espada y daga dorado con tiros y pretina dorados çençilados (sic) de açul y oro
Un estoq[e]
-Un estoque largo negro sin adereço
Un adereço/este no se tasa que es del duq[e]
-Un adreço pequeño de espada con tiros y petrina con guarnición dorada
Un espadín
-Un espadín ancho dorado
Una espada
-Una espada de tornear negra ancha
Otro espadín
-Otro espadín dorado y labrado con su hoja ancha
Un alfanje
-Un alfanxe de la China guarnecida al pareçer de oro

Una daga
-Una daga de ganchos con guardamano
Una hoja
-Una hoja ancha de a caballo
Un cañón
-Un cañón de arcabuz/ corto de quatro palmos
Tiros y pretina
-Tiros y pretina sin hierros bordados viejos
Tiros y pretina
-Tiros y pretina con hierros negros de terciopelo riço prensado
Una pretina/ tasóse con un adereço
-Una pretina baya pespuntada de seda parda
Una vaina/ no vale nada
-Una vaina de a Yndia de un cuchillo
Cuchillos
-Seis cuchillos ordinarios con cabos de cuerno
(fol. 673 r)
Hojas de daga
-Dos hojas de daga pequeñas
Espinilleras
-Dos espinilleras de hierro
Caxa de piedras
-Una caxa de piedras de arcabuçes
Dos puntas
-Dos puntas de lanças doradas
Cuchillos
-Siete cuchillos sin cabos
Pedaço de çapa/ no se tasó
-Un pedaço de çapa blanca
Quchillos
-Una caxa con tres quchillos con sus cabos con tres cabeças de león guarneçida la caxa de plata dorada
Una almarada
-Una almarada de tres esquinas con caja de plata
Un cuchillo
-Un cuchillo con su caxa sin cabo la caxa dorada y labrada
Otro cuchillo
-Otro quchillo de campo con su caxa verde grande
Otro cuchillo

-Otro cuchillo de campo con una caxa blanca
Otro cuchillo
-Otro cuchillo con el cabo de piedra y la caxa verde guarneçida de plata
Guantes de malla/ tasáronse [...]
-Dos guantes de malla
Pistolas
-Quatro pistolas las tres medianas y la una pequeña
Frascos
-Dos frascos de hueso
Carcax
-Un carcax con virotes
Un jaco
-Un xaco suelto en una funda de bayeta colorada
Dos hojas
-Dos hojas de daga
Un adereço de espada
-Un adereço de espada pavonada con tiros llanos de vaqueta
Una daga
-Una daga suelta negra lisa
(fol. 673 v)
Unos tiros/ tasóse con las espadas
-Unos tiros de espada y pretina pespuntados
Otros tiros/ no se tasaron por no valer nada
-Otros tiros de espada sin pretina de terciopelo negro guarneçidos con molinillos de oro
Otros tiros/ no se tasaron por no ser de provecho
-Tres pares de tiros de espada llanos negros sin pretinas
Otros tiros/ lo mismo
-Otros tiros bayos
Guarnic[ón] de daga/ lo mismo
-Una guarnición de daga negra y pomo y contera de espada
Una guarnición
-Una guarnición de espada lisa
Una guarnición/ no se tasó por no valer nada
-Una guarnicioncilla de daga
Una pistola
-Una pistola de hierro de dos bocas y dos llaves
Un frasco
-Un frasco de cuerno de venado para pólvora

Cañones de arcabuçes
-Diez y ocho cañones de arcabuçes que trujo de Génova su Exª del Sr Duq^{e}
Rodelas
-Dos rodelas grandes de corcho con sus braçales
Ballestas
-Tres ballestas con sus gafas
Otra ballesta/ no se tasó
-Otra ballesta pequeña con sus gafas
Un carcax
-Un carcax grande con cantidad de virotes y unas gafas dentro con su funda de cuero verde
Otro carcax
-Otro carcax de cuero verde con virotes
Otro carcax
-Otro carcax viejo con virotes
Un hierro
-Un hierro de ballesta
Una pistola
-Una pistola de tres palmos rayada con su funda colorada y llave
Un arcabuz
-Un arcabuz de rueda de cinco palmos con su funda colorada
(fol. 674 r)
Un peto
-Un peto manopla y medio braçal
Un broquel
-Un broquel verde con su chapa de hierro en medio con çerco de hierro
Un peto
-Un peto y espaldar y faldones y manopla que se hizo en Nápoles con su funda colorada
Una ballesta
-Una ballesta de bodoques
Arcabuçes
-Cinco arcabuçes guarnecidos que vinieron de Roma largos a gatillo
Arcabuçes
-Otros dos arcabuçes a mencha (sic) que estaban en el retrete
Montante
-Un montante
Ballesta
-Otra ballesta de bodoques

Broqueles
-Dos broqueles viejos
Una funda
-Una funda de vaqueta para un pistolete
Otra funda
-Otra funda de pistolete […].

POZA, MARQUESES DE

Doc. 1.
1544, June 30.
Inventory of Sancho de Rojas, heir of the Marquis of Poza.
Biblioteca Francisco de Zabálburu, Altamira, 32, Grupo Documental 3, D. 12.

[unfoliated, the digitized folio number is entered]
(nº 1)
Memoria que se hizo último de junio de mill e quiºs y quarenta y quatro años de todo lo que está dentro de la armería ansí de armas como de otras cosas y ansí mismo en la recámara

Armas
-primeramente un arnés grabado que tiene las piezas siguientes
-una coraça con su volante
-otro peto de la mesma coraça pa justar de tarjeta
-más una gola con su çelada y baberón y su pieçeta
-más un braçal çerrado derecho
-más otro braçal derecho
-más otros dos braçales izquierdos
-más dos guardabraços con sus rodajas uno yzquierdo y otro derecho
-más otro guardabraço yzquierdo con su bufeta
-más dos quixotes enteros con sus grebas y escarpes
-más otras dos grebas con sus escarpes
-más tres medios quijotes uno yzquierdo y dos derechos
-más un escarçelón
-más otro medio escarçelón
-~~más un barbote~~ [crossed out]
-más otra celada de seguir grabada
-más otro medio barbote
-más dos escarçelas una yzquierda y otra derecha
-más dos medias grebas con sus alpartazes de malla
-más dos sobremanoplas una de justa de guerra y otra de tarjeta
-más una sobreguarda de braçal yzquierdo
-más otra sobreguarda pequeña de braçal
-más otras dos manoplas yzquierda y derecha
-más quatro mandiletes los dos izquierdos y los dos derechos

(nº. 2)
-más dos celadas encajadas
-más una bragueta
-más unos çapatos de malla con sus puntas grabadas
-más seys puntas de escarpes
-más una pieça de la gola que hizo Verdejo que no está grabada
-más una silla de ynfante
-más dos ristres
-más una testera
-más una tarjeta con su puerca y tres tornillos
-más una pieça de un guardabraço que está desclavada
-más dos escarçelas largas de torneo de a pie grabadas
-más dos launas de escarpes grabadas y otra launa grabada de una coraça del arnés grabado
-más una escofia grabada
-más tres arandelas
-más un guardabraço derecho grabado

Otro arnés
-más otro arnés que no está grabado que tiene las pieças siguientes
-una coraça
-más un peto
-más una çelada con su baberón y pieçeta
-más dos braçales uno yzquierdo y otro derecho
-más una pieça del braçal yzquierdo para justar de guerra
-más otra pieçeta pequeña
-más un guardabraço yzquierdo
-más otro guardabraço derecho con su rodaja
-más un medio escarçelón
-más otro guardabraço de la tarjeta sin ninguna laona
-más medio baberón
-más una sobre batyxa
~~-más una escarçela~~ [crossed out]
-más otra media escarçela
Ojo
-más una çelada de ynfante este arnés no tiene arnés de piernas porque lo dio don Sancho mi Señor a Estada, tiene un arnés de piernas grabado quijotes y grebas y escarpes

-más una pieça pequeña de silla de justa que se pone en el arzón delantero por de dentro

(nº. 3)
Otro arnés
-más otro arnés dorado con su coraça y ristre dorado
-más una celada con su baberón
-más un guardabraço yzquierdo con su bufa
-más otro guardabraço derecho
-más dos braçales uno yzquierdo y otro derecho
-más dos manoplas una derecha y otra yzquierda
-más dos escarçelas una derecha y otra yzquierda
-más un arnés de piernas con quijotes y grebas y escarpes
-más un mandilete
-más una gran bufa
-más una armadura de cabeça del mismo arnés de ynfante
-más otra armadura de cabeça de seguir
-más un morrión
-más una çelada borgoñona del Señor Don Pedro Sarmiento y un guardabraçillo, y una escofia. Esta çelada la llevó el Señor don Pedro
-más dos calvas de este arnés dorada la una está dorada por la una parte. Del S^{or} don P^{o}
~~-más una espada de armas de torneo~~ [crossed out]
Encaxe
-más un encaxe de plumaje del mismo arnés dorado
-más dos toneletes de hoja de Flandes
Espuelas
-más tres pares de espuelas de justa guarneçidas con sus correas blancas
Brida
-más una brida de lienço blanco con sus dos çintas de seda de armar
Malla
-más dos goçetes de malla
-más un alpartaz de la coraça
-más una cuera de malla y un peto pa con ella para sobre el estómago
-más unas mangas de la mesma cuera de malla enteras
-más dos guantes de malla
-más otros dos pedaços de plata malla (sic)

(nº. 4)
Sombreros de hierro
-más un sombrero de hierro desguarneçido
-más otros sombrero de hierro forrado en terciopelo negro de dentro y fuera
-más dos cascos el uno guarneçido de grana y el otro desguarneçido
-más un bonetillo de launas forrado en raso negro
-más un broquel de hierro
-más tres llaves y un martillo para armar
-más una puerca con su punta de diamante de tarjeta
-más un encaxe de penacho todo de hierro
-más una bigornia y una desclavadera y un punçón y un martillo
~~-más un morlete hecho de hilo de hierro~~ [crossed out]
-más dos espuelas de tornillo de justa que se ponen y se quitan en las grebas
-más otra llave para armar la tarjeta
-más tres cañones con sus tornillos pa poner los plumajes de las çeladas
Espadas de torneo
-más dos espadas de torneo con sus guarniciones doradas y con sus vainas de terciopelo encarnado
-más otra espada de torneo con una guarnición dorada vieja con su vaina de terciopelo negro
-más dos espadas de torneo con su guarnición estañadas
-más otras dos espadas de torneo con sus dos cruzes sin otras guardas con vainas de cuero blanco
-más una espada de armas y una daga y un estoque con sus cruzes y pomos y conteras doradas y vainas de terciopelo colorado
(nº. 5)
-más otra espada de armas con su guarnición barniçada y vaina de terciopelo negro que se guarneçió para la guerra y una daga con ella
-más un estoque de los de Florençia con su guarnición barniçada
-más otras dos espadas con sus guarniciones barniçadas y la una tiene vaina de terciopelo negro y la otra tapetada
-más otra espada con su guarnición plateada y puño de plata y vaina de terciopelo negro
-más otra espada con su guarnición dorada y la labor dorada y lo de abaxo pavonado, o morado con su vaina de terciopelo negro y su puño de oro y morado con su daga de la misma manera
-más otra espada dorada de las de Florençia con su puño de oro de unas puntas y su daga de la mesma manera con sus vainas de terciopelo negro

-más otras dos espadas con sus guarniçiones doradas la una grabada con su puño de plata y oro y la otra con un puño de oro tirado
-más una hoja de una espada nueva que la dio el marqués de Denia y una cochilla de una daga q compró su md con una medalla y unas labores en el recaço sacadas de buril- esta [...] q tiene una guarniçión dorada y plateada con unas medallas y hojas relevadas y un puño de oro y plata que es esta mesma que aquí
-más una guarnición de espada y una daga dorada y plateada nueva
Ojo
-más dos espadas doradas con sus vainas de terciopelo negro chiquitas del Señor don Luys y del Señor don Di°
(n°. 6)
-más un cochillo (sic) de monte don sus dos vainas la una verde y la otra negra y dorada y dos cochillos (sic) en la una vaina
-más dos pedaços de vocaçí verde q pueden ser tres varas con que están cubiertas las espadas
-más tres cuchillos y un punçón en una caxa de unos cabos colorados
-más unas dos fundas de espadas de camino
-más una funda de cuero pa una espada de su md pa de camino
Sillas de justa
-más tres sillas de justa con sus hierros y tornillos las dos nuevas grabados los hierros y la otra de conteras todas tres con sus fundas de frisa
-más otra silla de armas pa guerra con sus hierros y tornillos barniçados y hechas unas labores en el barniz con su funda de lienço ençerado y de cuero ençima
Lanças
-más una partesana
-más tres lanças de la gineta con sus hierros
-más dos lanças de pino pa torear con sus hierros
-más una lança de justa
-más un hierro de un venablo dorado
Rodelas
-más dos rodelas de las de Módena desguarneçidas sin las embraçaduras
Penachos
-más un penacho de las Yndias que tiene dos plumas verdes muy grandes
-más otro penacho de las Yndias pequeño que tiene unas plumas verdes y otras encarnadas con unas mariposas y su argentería
-más quatro penachos de las Yndias fhos a manera de ventalles con sus colgaderas viejos pelados
-más otro penacho de plumas blancas verde y naranjado
-más otro penacho de pluma blanco y negro y encarnado con dos colgaderas

-más otro penacho pequeño blanco y negro y encarnado para la testera
(nº. 7)
-más otro penacho azul pequeño pa la testera
-más un penacho grande de negro de pluma con dos colgaderas
-~~más una pluma blanca~~ [crossed out]
-más seys plumas blancas viejas
-más un penacho negro de seda grande con tres colgaderas
-otro penacho pequeño negro pa la testera
-más otro penacho grande de seda amarillo
-más otros dos pedaçicos de penachos azules y encarnados
-más en otra caxa de madera un penacho de papel encarnado y de hoja de lata
-y más papeles teñidos de colorados y amarillos que están en la caxa y muchas puntas de la misma hoja de lata en el suelo de la mesma caxa
-más dos cruzes y dos pomos de espadas de torneo
-más estaba dentro del armario tres pares de çapatos blancos de armar y unos pedaços de cuero colorado
-todo esto arriba dicho está dentro del almario (sic)
-más una vaina de terciopelo negro cubierta de tafetán blanco, de espada de torneo
-más están en dos guantes de cuero colorado viejos xxi tornillos de los arneses

En los cajones del armario está lo siguiente
-más dos faldamentos de raso morado y blanco y debaxo de hoja de lata que se hiçieron para un torneo para sobre los toneletes
-más dos chapeos el uno de raso falso blanco y el otro morado
-más una cuera de tafetán encarnado acuchillado y debaxo de las cuchilladas de hoja de lata
-más un coplón y petral de raso blanco y tafetán azul con mucha hoja de lata acuchillado
-y una capa de la mesma manera q se hizo para correr sortija
-más dos sayetes de sobre armas de angeo

(nº. 8)
-más otro sayete con mangas de bocací amarillo
-más otro sayete de raso falso morado
-más un coplón y petral y sayo de armas de terciopelo encarnado con unas fajas del mesmo terciopelo acuchilladas con sus borlas
-más un coplón y cuello y petral de terciopelo negro guarneçido con sus franjas de sirgo negro y sus borlas y un sayete de terciopelo negro guarneçido de la

misma manera con una cruz colorada de raso carmesí q se hizo para la guerra de Perpiñán
-más otros cuerpos de terciopelo negro
-más dos adereços de caballo de justa de terciopelo azul bordado con unas hojas relevadas de raso azul con mucha argentería que son dos coplones y dos petrales y dos cuellos y dos testeras y un florón q se ponía en medio del coplón y un faldamento del mesmo terciopelo azul y bordado de la mesma manera todo esto guarnezido con franjas y borlas de plata
-más unas sobrecubiertas de terciopelo negro q son un coplón y un sobrelomo y dos costeras y baticola y un cuello q son onze piezas todas estas guarneçidas con sus franjas y por lo baxo una flocadura larga
-más treinta y seys cordones de seda negra pa las cubiertas

(nº 10)
-más otros treinta y nueve cordones de hiladillo negros clavados para las cubiertas
-más cinco cordones de seda naranjada y morada clavados para armar. Más xiiijº borlas de seda encarnadas q fueron de un adereço de un torneo
-más veynte çintas nuevas de armar clavadas de sedas de colores
-más hasta dos ~~quatro~~ varas de raso falso amarillo ~~y morado~~ [crossed out]
-más seys pedaços de a quarta de tafetán encarnado y otros seys negros que se compraron para atar la testera del caballo
-más otros seys azules q se compraron pa lo mismo. ~~Más 5 pedaços de tafetán negro pequeños pa atar la testera~~ [crossed out]
-más un talabarte de terciopelo negro que se hizo para colgar la espada del arçón
-más un çincho de terciopelo azul pa sobre las armas con sus hierros dorados
-más otro çincho colorado de terciopelo pa sobre armas sin hierros ningunos. ~~Más un talabarte de terciopº negro con sus hierros barnizados que se hizo para colgar la espada del arzón pa un torneo~~ [crossed out]
-más otro çincho de terciopelo negro con sus hierros negros
-más otro çincho de cuero blanco pa sobre armas
-más una darga (sic) grande que se truxo de Granada de ençerado con su funda
-y otras dos dargas más pequeñas con sus fundas de cuero
-más un caparaçón de terciopelo leonado guarneçido de dos tiras de raso blanco falso
-más una coraça de terciopelo negro forrada en fustán guarneçida con un pasamano de sirgo negro pa silla de la gineta
-más un jaez de seda negro q tiene unas encaladas y dos sementales y quatro borlas del petral
-más un petral negro labrado de hilo de oro con sus caxas y hebillas y puntas doradas y labradas de negro

(nº. 11)
-más otro jaez negro q tiene una cuerda y dos sementales
-y un petral negro labrado de hilo de oro con sus caxas y cabos y hebillas doradas y labradas digo q le falta un cabo
-más otro jaez de seda azul que tiene encaladas y una cuerda azul y dos sementales y quatro borlas del petral
-y un petral labrado de hilo de oro sobre cuero colorado con sus caxas y cabos y hebillas doradas y una mochila de terciopelo azul con sus rapacejos de seda azul
-más dos cabezadas de la gineta de cuero colorado con su clavaçón dorada buenas tiradas y labrado de oro y negro y las otras cabezadas con su clavaçón dorada y plateada nuevas q se truxo de Portugal la clavaçón
-más unas espuelas de la gineta doradas nuevas con una clavaçón dorada y plateado q se truxo la clavaçón de Portugal con la clavaçón de las cabezadas arriba dhas
-más otros dos pares de espuelas doradas con su clavaçón dorada
-más tres pares de estriberas traídas de la gineta doradas y plateadas las unas nuevas traídas los dos pares con sus fundas
Espada
-más una espada de la gineta con su vayna de cuero colorado labrada de hilo de oro con su guarnición de plata que es pomo y puño y cruz y en la vayna tres cajas y una contera todo de plata y un texillo de oro y morado con sus cabos y sortijas de plata q está asido en la dicha espada
(nº. 12)
-más una almártaga de hierro dorada con un cordón de hiladillo azul
-más un petral de cascabeles que tiene quarenta y seys cascabeles
-más tres reatas de seda la una blanca y encarnada y la otra azul y naranjada y la otra toda negra
-más otra reata de hiladillo blanco y colorado y azul. Más unos çaraguelles de tafetán amarillo con sus cordones de sirgo amarillo
-más dos pares de çaraguelles de tafetán encarnado y leonado con sus cordones
-más una toca de tafetán blanco con sus rapacejos de oro falso largos
-más quatro veletas la una de damasco encarnado guarneçida de unas franjas de oro de Luca alrededor y la otra de damasco amarillo/ y la otra de tafetán azul/ y la otra de tafetán negro con sus borlas de oro y cordones está borlas y cordones tiene la de damasco amarillo. Más otra veleta de tafetán amarillo con sus franjas y cordones de sargo amarillos
-más un pedaço de hasta dos terçias de frocadura (sic) de oro de Luca larga y una borla de oro y hasta cinco varas de franja angosta y ocho alamares todo esto de

oro de Luca. Más un albornoz y una marlota y un capirote de estameña blanca. Más v varas y mª de franjas de oro de Luca y una borla de oro
-más otros tres pedaços de frocadura (sic) larga de oro y plata de Luca q terná todo ello hasta dos varas
-más otros dos pedaços de frocadura (sic) larga de plata que terná hasta dos varas y media y una borla y tres alamares de plata de Luca
-más dos caperuzas la una de terciopelo encarnado y la otra de terciopelo morado. Más otra caperuza de terciopelo amarillo
-más dos máscaras [...]
(nº. 13)
-más unos borceguíes naranjados con un laço verde
-y otros borceguíes negros
-y unos çapatos de la gineta bayos
-más un penacho de garçota
-más tres pares de açiones de los de Granada de la gineta colorados. Más un albornoz de damasco amarillo guarnecido de franjas de plata de Luca con sus alamares
-más otros dos pares (un par) de açiones de Granada de la gineta
-más quatro pares de riendas de la gineta colorados de Granada. Más treze alamares de plata de Luca
-más otros dos pares (un par) de riendas vayas de las de Granada. Más viiiº alamares de oro de Luca
-más tres çinchas de Granada la una verde y la otra de muchas colores blanco y negro y amarillo y morado/ y la otra de blanco y colorado y verde y azul y encarnado todas tres con sus látigos esto hasta aquí estaba en los cajones de la armería
-más otra çincha de la gineta toda de cáñamo q se hizo pa el caballo gallego
Sillas de gineta
-más una silla de la gineta nueva
-más dos sillas viejas de la gineta la una de ellas con sus estriberas nuevas y su petral de cuero con sus caxas y hebillas barnizadas
Sillas de estadiota
-más un silla de la estadiota de terçiopelo morado con sus franjas de oro y morado y su funda de friseta verde
-más otra silla de terciopelo negro de las de Nápoles con los fustes dorados y pintados con su funda de cuero
-más otra silla de paño negro con los fustes dorados vieja
(nº. 14)
-más un fuste dorado pintado nuevo

-más dos sillas françesas nuevas de faxas y cojín de terciopelo negro la una pa el señor Don Luys
-más otra silla francesa vieja que tiene un cojín de terciopelo viejo
-más una silla de caballo vieja
-más tres sillas de mula buenas con sus çinchas todas tres y las dos tienen estribos y açiones
Estribos
-más unos estribos nuevos de justa estañados con sus açiones blancos
-más otros estribos de justa q solían tener unas varillas con sus açiones blancas
Çinchas
-más tres çinchas de cáñamo pa el caballo gallego
-más dos pares de çinchas nuevas guarneçidas de terciopelo negro de a tres çinchas cada par
-más otras tres çinchas de cáñamo guarneçidas de cuero blanco
-más otras tres çinchas qbradas q son tres guarneçidas de terciopelo
Cadenas
-más dos cadenas una de caballo y otra pa mula allende de otras tres que tienen los caballos en la caballeriça
Açiones
-más dos pares de açiones de cuero negras
Estribos dorados
-más unos estribos dorados buenos
-más otros estribos de caballo barniçados con sus açiones de cuero
Guarniçiones de caballos
-más una guarniçión de terciopelo morado de Ramales guarneçida de franjas de oro el pretal y morado con su clavaçón dorada (nº. 15) y tiene petral y grupera y cabezadas y riendas y açiones de terciopelo morado
-más otra guarnición corbata con costeras de terciopelo amarillo y tiene petral y grupera y cabezadas y riendas
-más otra guarnición corbata sin costeras de terciopelo encarnado q tiene sus cabezadas y petral y grupera y riendas
-más otra guarnición de terciopelo negro de Ramales con su clavaçón dorada que tiene sus cabeçadas y riendas y petral y grupera
-más otra guarnición de Ramales de terciopelo negro con ribetes de cuero pardo con su clavaçón dorada que tiene sus cabeçadas y riendas y petral y grupera y açiones de la mesma manera
-más otra guarnición de terciopelo negro con ribetes de cuero naranjado con su clavaçón dorada q se hizo para el caballo del señor Don Luys q tiene sus cabezadas y riendas y petral y grupera de Ramales y açiones de la mesma manera

-más otra guarnición corbata con costeras de terciopelo negro q tiene sus cabezadas y riendas y petral y grupera esta es nueva
-más otra guarnición corbata sin costeras de terciopelo negro q tiene sus cabezadas y riendas y petral y grupera
-más unas cabezadas de caballo de terciopelo negro ~~que se hiçieron pa la guerra~~
-más unas medias riendas de terciopelo negro q se hiçieron para la guerra pa poner en ellas riendas de hierro
(nº. 16)
-y más unas cabeçadas de terciopelo azul con sus riendas de lo mismo
-más un petral de cuero y una grupera que se hiço para un torneo para debaxo de las cubiertas
-más una guarnición de cuero tapetada corbata con costeras vieja
-más una almártaga de cuero colorado
Guarniçiones de mulas
-más una guarnición nueva de mula de terciopelo negro entera con sus cabeçadas y falsas riendas y petral y grupera
-más otra guarnición vieja de paño de mula entera
-más tres guarniçiones de cuero todas tres enteras con sus cabeçadas y petrales y falsas riendas y gruperas eçeto que a la una le falta una grupera
-más unas cabeçadas y falsas riendas y grupera de terciopelo negro viejas q están cortadas por de dentro y hechas labores
-más seys gualdrapas tres nuevas y tres viejas
Frenos
-más tres frenos de mula con sus riendas y el uno con sus copas
-más tres frenos nuevos de la brida que se hicieron pa el caballo gallego el uno con sus copas doradas
-más otro freno nuevo de la brida con sus copas doradas que es el que truxieron de Monçón con el caballo alaçanejo
-más dos pares de copas doradas allende de los dos pares q están puesta (sic) en estos dos frenos arriba dhos
-más otro freno nuevo de la brida que es un cañón
(nº. 17)
-más diez y seys frenos de la brida
-más otros dos frenos de la brida el uno con sus copas enbarnizadas
-más un freno de la gineta del caballo morçillejo
-más otros dos frenos de la gineta que les faltan las varillas y tornillos
-más otros dos bocados de la gineta sin barbadas
Espuelas
-más unas espuelas doradas con sus correas de terciopelo

-más otras espuelas negras de la hechura de las doradas con sus correas de terciopelo
-más otras espuelas negras pa de camino de correa entera

En un arca encorada de un cuero blanco y castaño está lo siguiente [...]
(nº. 18)
-Más un cochillo (sic) de monte con su vaina verde pequeño con un cochillo y un punçón [...]
-más un caparaçón de paño pardo para monte para sobre la silla del caballo con su rodaxa de cuerno y su bolsa de cuero para colgar del arçón pa llevar la ballesta
-más una reata de cuero para sobre el dho caparaçón
-más unas cabezadas y un petral y un çinto de cuero de tigre para monte q está yapelado (sic) parte de ello
-más unas riendas bayas de la gineta pa monte
-más una almártaga de cuero negro con su cordón de hilo verde y pardo para monte
-más unas alforjuelas de lana pa monte
-más una çebadera de angeo para monte [...]
-más un alxabón de cuero negro con su candado y llave que tiene dentro una alxaba grande de cuero azul con diez y seys xaras en ella
-más otra alxabilla de cuero negro pequeña con dos xaras y tres cuerdas nuevas de ballesta y un ovillo de hilo para hazer cuerdas
-y una taleguilla con eslabón y pedernal y yesca
-más un carcax de cuero negro sin virote ninguno
Ballestas
-más tres ballestas con sus fundas de cuero negro y con sus gafas y la una gafa con su çapatilla de cuero negro
-más otras dos gafas
-más una curueña (sic) de ballesta
(nº. 19)
-más otra media curueña de ballesta
-más quatro vergas de ballesta
-más tres horquillas de madera
-más quatro palos de serbal pa hazer curueñas de ballesta
-más están en una arquilla de la ballestería quarenta y quatro jaras
-más están cinco virotes de los muy buenos y el uno de una madera leonada
-más están otros nueve virotes y un rallón
-más están tresçientos y treinta casquillos de xaras
-más quatro cuernos de yerba

-más una sierra y un çepillo de hierro que está ya des soldado
-y una anguijuela pa labrar xaras y una raspa de hierro
-otro hierro con un cabo de madera pa igualar las plumas de los virotes
-más un pedaço de hilo de conexo
-más unas trébedes de hierro chequitas (sic) para poner crisol pa hazer las esmeraldas
-más un libro de pinturas
-más una piedra de jaspe grande llana y dos chequitas pa moler colores para pintar
-más una tablilla de nogal pa el dedo pa tener las colores para pintar
-más una varilla de fresno pa sostener el braço pa pintar
-más siete vasillos con sus tapaderos pa tener las colores los seys de hoja de Flandes y el otro de vidrio
-más un compás de hierro también pa la pintura
-más una piedra negra qbrada que la qbró el conde de Nieva q es de un pintor q está encaxada en una tabla de nogal
(nº. 20)
-más un juego de ajedrez y de tablas labrado de ataraçea (sic) de marfil y de otras colores que tiene una sola pieçesica (sic) desencolada y está dentro de este mesmo tablero con su funda de cuero no tiene piezas de ajedrez ni tablas dizen q están en la cámara de mi señora
-más dos tiquetaques (sic) con todas sus tablas y dados y el uno de ellos tiene un juego de ajedrez ençima y tiene sus piezas de juego de axedrez
-más una traza de una casa en un cuero de pargamino (sic)[…]
-más un sello de hierro y el cabo de marfil con las armas de Rojas […]
-más una caxa grande que se hizo para un penacho q está cubierta de cuero negro con su tapador
-más otra caxa de madera de tres esquinas q fue de çiertos penachos
-mas dos manojos de cordeles para liar dos cofres
-más quatro barbas de ballena dos grandes y dos pequeñas […]
(nº. 28)
-más están en la recámara tres caballos de madera pa poner las sillas de caballo
Digo yo Andrés de Pedrosa q el día q yo fuy despedido q fue a veinte y seis del mes de junio de mil y quinientos y quarenta y quatro años q qdó en la recámara de don Sancho de Rojas mi señor todo lo q en estas treze hojas y esta media plana está escrito lo qual se puso todo por memoria estando presente Diº de Soto

Andrés de Pedrosa
(signed).

Doc. 2.
1605, January 19.
Inventory and appraisal of Marquis of Poza, Francisco de Rojas.
Archivo Histórico de Protocolos de Madrid, Protocolo 2176.

(fol. 204 r-205 r)
Enero en^{o} 1605
Ynventario de los bienes que quedaron por fin e muerte del Sr Don Fran^{co} de Roxas marqués de Poça y tas^{ón} de ellos
Sacada un treslado (sic)
Sr Tamayo

(fol. 205 r)
Doña Françisca Enrriq^{ez} de Cabrera marquessa de Poza viuda mujer que fui del Sr Don Fran^{co} de Rojas marqués de Poça mi S^{or} del Consejo de Estado y Guerra de Su Mag^{d} digo que por fin y muerte de su S^{a} quedaron en esta villa de Madrid bienes muebles de que haçer ynventario y en la çiu^{d} de Valladolid quedaron algunos bienes y para cumplir con lo que soy obligada pido a Vm, mande dar liçençia para que se haga el dho ynventario y pido Justiçia

La marquesa de Poça
Doña Fran^{ca} Enríquez
(signed)

Yo el Liçenciado Ruiz como albaçea testamentario del marqués de Poça mi señor pido lo mismo

El Liz^{do} Ruiz
(signed)

(fol. 205 v)
En la villa de Madrid a diez y siete días del mes de enero de mil y sei^{o}s y cinco a^{o}s ante el S^{r} licen^{do} Orive de Verg^{a} ten^{te} de corr^{or} en esta vi^{a} de Madrid y su tierra y ante mí el escrib^{o} la parte de la S^{a} Doña Fran^{ca} Enrríquez de Cabrera marquesa de Poça viuda mujer que fue del Sr Don Fr^{co} de Roxas marqués de Poza difunto presente estuvieron e pidió Jus^{a}

Aquí la petición

E por su mrd vista dio licenª para q por ante escribº y en forma se haga ynventario de los bienes que quedaron por la fra^{te} muerte del marqués de Poza y ansí lo nº

Ante mí Diego Román
(signed)

En la villa de M^{d} a diez y nueve días del mes de en^{ro} de mil y seisçientos y cinco años ante mí Diº Román escribº del rey nro S^{or} y del número de esta villa de M^{d} y de los testigos uso (sic) escriptos pareció la S^{ra} Doña Françisca Enrríquez marquesa de Poza viuda mujer que fue del Sor Don Françisco de Rojas marqués de Poza y en cumplimiento del auto antes de este y para cumplimº con lo que tiene obligación puso por ynventario las casas y cosas y bienes sig^{tes}

-Primeramente unas casas principales en esta dha villa en la parroquia de St Mrtn en la calle que llaman del marqués de Poza con sus jardines corrales caballeriças y ofiçinas que está todo yn[...] (fol. 206 r) en las dhas casas y alindan con otras en que vive la condesa de Valençia que compró de los herederos de Don Juan de Minchaca y están enfrente del Colegio de St Agustín que llaman de Doña María de Aragón y la plaça y campillo de las dhas casas y monasterio
-Yten las casas que eran de Don Juan de Minchaca que compró la condesa de Valençia de sus herederos linde de las dhas casas principales del dho marqués y por abajo casas de Juan Díaz de Vega que son y quedaron por bienes del dho S^{or} marqués [...]

(fol. 234 r)
Armería
-Yten se pone por ynventario dos arneses de Alemania blancos grabados y dorados el uno de justa y el otro de seguir de caballo con todas sus pieças de lo uno y de lo otro que se tasaron en dusientos ducados__xxvU
-Yten un arnés grabado y dorado con la labor de unas cruçes que es hecho en Fuentidueña y tiene un peto y espaldar de ynfante que se tasó en ochoçientos Reales__xxvyUcc
-Yten un arnés grabado y dorado de Milán de siete listas que sirve de seguir y de torneo de a caballo que se tasó en quarenta ducados__xvU
-Yten veynte y nueve arneses grabados los dos de estos de Alemania y los demás de Milán con todas sus pieças y silla y testera de caballo y morriones y çapatos que se tasó cada arnés en treçientos reales ccxcvUdcccº

-Yten diez y siete arneses lisos de Milán q sirven de seguir la guerra y para torneo de a caballo con todas las dichas pieças que se tasó cada uno a veynte ducados_ _cxxvyUd

-Yten çinquenta arneses blancos de seguir de Milán y algunos de ellos sirven para tornear a caballo y a otros les falta (sic) algunas pieças que se tasó cada uno a veynte ducados__ccclxxvU

-Yten se pone por ynventario treinta y seys pares de escarselas (sic) de ellas forradas y de ellas por forrar que se tasó cada par a quatro reales__iiijºUdcccºxcvi

-Yten diez pares de quixotes lisos de Milán que se tasó cada par a seis Reales__ yUxl

-Yten treinta açeros de sillas lisos todos con sus tornillos que se tasó cada uno a veinte Reales__xxvUccccº

(fol. 234 v)

-Yten cantidad de piezas de armas sueltas que se tasaron a montón por ser todas piezas perdidas en cien Reales__iiiUcccº

-Yten una cubierta de caballo de azero con sus dos crines y petral y testera q se tasaron en veynte ducados__vyUd

-Yten cinquenta y quatro maças de hierro con sus cordones de colores de media seda que se tasó cada una a seis Reales__xiUxvi

-Yten cinquenta y quatro bocados de frenos de hechuras diferentes que se tasó cada uno a doze Reales__xxyUxxxy

-Yten sesenta y cinco cadenillas de falsas riendas de hierro estañado que se tasó cada una a tres Reales__vYUdcl

-Yten çinquenta pares de estribos plateados hechos en Aljofrín finos que se tasó cada par a diez y seys Reales__xxvYUcc

-Yten onze pares de espuelas sin guarneçer plateadas que se tasó cada par a quatro Reales__viUdccccºxcv. Ojo

-Yten tres alabardas cubiertas las astas de terciopelo carmesí con clavos y tachuelas doradas q se tasó cada una a quatro ducados__iiijºUd

-Yten una alabarda cubierta el asta de terciopelo verde con clavos dorados que se tasó en quatro ducados__lUd

-Yten quatro alabardas las astas de palo sin guarnición y con clavos y tachuelas doradas q se tasó cada una a treinta reales__iiijºUlxxxxº

-Yten un venablo guarnecida el asta de él con franjón de oro y seda azul y el hierro todo dorado que se tasó en diez ducados__iijUdccl

(fol. 235 r)

-Yten otras dos alabardas doradas con los palos llanos y en cada uno tres guarniciones de un franjón enredado de oro que se tasó cada una a tres ducados_ _yUccl

-Yten dos martillos por la una parte una punta y por la otra tres puntas y las astas llanas que se tasó cada uno a tres ducados__yUccl
-Yten dos guadañas doradas y grabadas con los palos llanos con dos guarniciones de un franjón de oro y seda enredado q se tasó cada uno (sic) a treinta y seis Reales__yUcccc°xlviiij°
-Yten dos partesanas doradas que tienen las astas llanas y con dos guarniciones de franjón de oro y seda cada una que se tasaron a quarenta reales cada una__ yUdcxx
-Yten dos chuços dorados con tres astas llanas guarnecidas con dos franjones de oro y seda cada uno que se tasaron a dos ducados__iUd
-Yten un martillo con su asta blanca de fresno que se tasó en diez y ocho Reales_ _Udcxy
-Yten dos hachas de Milán las astas llanas q se tasaron a dos ducados cada una_ _iUd
-Yten una arma de tres puntas que se tasó en tres ducados__iUcxxv
-Yten una jineta pequeña que se tasó en doze reales__Ucccc°viij°
-Yten una maça de armas partida el asta y con su gozne que se tasó en doze Reales__Ucccc°viij°
-Yten una partesana vieja con el asta blanca que se tasó en diez y ocho Reales_ _Udcxy
-Yten un venablo de Vizcaya que se tasó en ocho reales__Ucclxxy
-Yten dos lanças jinetas finas con astas de frexno (sic) y sus hierros q se tasó cada una en catorce reales__Udccc°ly
(fol. 235 v)
-Yten otro venablo de Vizcaya con el asta blanca que se tasó en diez Reals__ Ucccxl
-Yten un lançón de Vizcaya que tiene su asta blanca q se tasó en ocho Reales__ Ucclxxy
-Yten tres lanças de hombres de armas con banderillas de tafetán blanco y negro y hierros de punta de diamante que se tasó cada una a ocho Reales__Udccc°xvi
-Yten quarenta y ocho lanzas con banderillas de tafetán carmesí con cruz de Jesusalén q se tasó cada una a diez y seis Reales con sus hierros de punta de diamante__xxviUcxy
-Yten sesenta y una lanças de pino de hombres de armas con sus hierros de punta de diamante que se tasó cada una a diez Reales__xxUdcxl
-Yten tres lanças para sortija de pino y pintadas con sus casquillos que se tasó cada una a diez Reales__lUxx
-Yten treinta morriones grabados de Milán q se tasó cada uno en diez y seis Reales__xviUcccxx

-Yten veynte y ocho morriones blancos de Milán q se tasó cada uno en diez Reales__ixUdxx
-Yten ciento y quinze estançias de arneses de pino de Cuenca q se tasó cada uno a doze Reales__xlviUdccc^{o}xx

En la villa de M^{d} a v^{te} y siete días del mes de febr^{o} de mill y seisçientos y cinco a^{o}s ante mí el pres^{te} scrib^{o} y testigos paresçieron Joan de Bruselas y Ju^{o} de Çuaço armeros de su Mag^{d} y dijeron que ellos han visto y tasado las pieças de armas contenidas en esta tasaçión las quales han tasado en los pre^{o}s que en ella van declarados y está bien y fielmente fecho y ansí lo juraron a Dios en forma siendo t^{o}s Andrés de Arazo y P^{o} de Chaves y lo firmaron y por Ju^{o} de Çuaço un t^{o}s

Ju^{o} de Bruselas
por t^{o} Andrés Arazo
Ante mí, Fran^{co} Román
(signed)

(fol. 236 r)
Espadas
-Yten se ponen por ynventario çinquenta y quatro espadas de a caballo de hombres de armas las hojas de Bilbao con guarnición de una puente con sus tiros y petrina q se tasó cada una a catorze Reales__xxvUdciiij^{o}
-Yten treinta y ocho dagas con guarniçiones chatas que se tasó cada una a seys Reales__vyUdcly
-Yten un montante de el Cornadillo que el pomo y cabos de los gavilanes de la guarniçión son de plata y la contera y se tasó en seis ducados__yUccl
-yten una espada hecha en Cuenca con guarniçión aventanada que se tasó en quatro ducados__lUd
-Yten otra espada hecha en Galiçia de a caballo como la de arriba con guarniçión aventanada con una vayna baya q se tasó en diez y seys Reales__Udxliiij^{o}
-Yten una espada de un corte hecha en Alemania con guarniçión aventanada que se tasó con su vayna negra en tres ducados__lUcxxv
-Yten una espada de Bilbao con la guarniçión barniçada y vayna de terçiopelo negro vieja que se tasó en dos ducados__Udcl
-Yten una espada de Toledo labrada a lo antiguo con guarniçión pavonada acanalada la labor de dicha guarniçión con vayna baya que se tasó en quatro ducados__lUd
-Yten otra espada vieja de un corte con guarnición barniçada q se tasó en ocho Reales__Ucclxxy

-Yten un alfanje berberisco con guarniçión pavonada y dorada q se tasó en tres ducados__lUcxxv
(fol. 236 v)
-Yten otro alfanje morisco con guarniçión de plata labrada a lo turquesco con un tahalí de tela morisco que se tasó en diez ducados con plata__iijUdcl
-Yten otro alfanje turquesco corbado que se tasó en quatro ducados__lUd
-Yten otro alfanje turquesco con la guarnición y contera y pommo de plata que se tasó todo en diez ducados__iijUdcl
-Yten otro alfanje derecho con guarniçión de plata contera y brocal y remate de puño que se tasó en diez ducados__iiiUdcl
-Yten un cuchillo de monte con un brocal de plata y contera y remate de puño que se tasó en quatro ducados__lUd
-Yten una rodela de Milán grabada guarneçida de terçiopelo con sus franjas con su punta y roseta en el medio de la dicha rodela que se tasó en diez ducados__ iijUdcl
-Yten otra rodela negra de yguera (sic) con su guarnición y cerco de hierro que se tasó en quarenta reales__lUccclx
-Yten un broquel grande con su copa de hierro pavonado y forrado de cuero amarillo q se tasó en treinta y seis Reales__lUccxxiiij^{o}
-Yten un estoque de armas bydo de tres esquinas que se tasó en tres ducados__ lUcxxv
-Yten dos estoques de armas con las guarniciones labradas y doradas q se tasó cada una (sic) a dos ducados__lUd
En la villa de M^{d} a v^{te} y siete de febr^{o} de mill y sesizientos y zinco a^{o}s ante mí el pres^{te} escrib^{o} y t^{o}s pares^{o} Andrés Díaz espadero v^{o}z de esta v^{a} y con juram^{to} que hizo dijo que él ha fecho la tasación de espadas y armas tocantes a su of^{o} que es lo antes de esto contenido la qual está bien y fielmente fecha y lo firmó t^{o}s Ju^{o} de Çuaço y Chaves

Andrés Díaz
Ante mí Diego Román
(signed)

(fol. 237 r)
Arcabuçería y tiros
-Yten se pone por ynventario tres ballestas las dos con sus gafas y la otra sin ellas tasada cada una a ciento y cinq^{ta} Reales__cvUccc
-Yten se pone por ynventario sesenta arcabuces de Aragón de rueda con sus llaves de armas que se tasó cada uno en cinco ducados__cxyUd

-Yten se pone por ynventario dos mosquetes de seys palmos de rueda que se tasó cada uno en seis ducados__iiijºUd
-Yten otro mosquete de Vizcaya de mecha que se tasó en tres ducados__lUcxxv
-Yten quatro arcabuces de mecha de Vizcaya que tiene cada uno su frasco y frasquillo que se tasó cada uno a dos ducados__iijU
-Yten un arcabuz de dos cañones y dos llaves que se tasó en ocho ducados__IIjU
-Yten un arcabuz quebrada la caxa y sin baqueta de rastrillo que se tasó en diez y seis reales__Udxliiijº
-Yten un mosquete de rueda de quatro palmos que se tasó en seis ducados__ yUccl
-Yten un pistolete de Alemania de quatro palmos que se tasó en tres ducados_ _lUcxxv
-Yten un arcabuz de mecha de Milán de ataujía que se tasó en tres ducados__ lUcxxv
-Yten seys tiros de bronze con sus ruedas encabalgados que se tasó cada uno en quatroçientos Reales__lcccºUdccvi
En la villa de M^{d} a v^{te} y siete de febrº de mill y seisçientos y (fol. 237 v) çinco aºs ante el pres^{te} escribº y tºs paresº maestre Simón arcabucero vzº de esta vª y dijo que él ha visto y tasado los arcabuçes y tiros ante de esto en las cantidades de ms que van declarados y está bien fecha y lo juró a Dios siendo tºs Andrés Díaz y Juº de Çuaço y lo firmó

Simón Marquarte
Ante mí Diego Román
(signed)

-Yten se pone por ynventario diez guarniciones blancas de caballos de armas y a la una guarnición le falta grupera y açiones tasadas a diez y seis Reales cada guarnición__vUcccºxl
-Yten se pone por ynventario ocho sillas de armas con sus azeros que se tasó cada una a seis ducados__xviijºU
-Yten se pone por ynventario catorze sillas de armas muy viexas sin azeros que se tasó cada una a tres ducados__xvUdcl
-Yten se pone por ynventario veynte y ocho pares de riendas para los caballos de los hombres de armas de las dichas sillas de arriba q se tasó cada par a dos Reales__lUdcccºiiijº
-Yten veynte y ocho pares de gruperas de las dichas riendas tasadas al dicho preçio y más veynte pares que son por todas quarenta y ocho__iijUcclxiiijº

-Yten quarenta y seis cabezadas para las dichas guarniçiones arriba tasadas a tres Reales cada una__iiij^{o}Udcxxy
-Yten más quarenta y cinco petrales para las dichas guarniçiones de arriba tasados a dos Reales cada uno__iijUlx
-Yten diez y seis sobrecinchas que son con todo lo demás anexo a la dicha armería tasadas a tres Reales cada una__lUdcxxxy

En la villa de M^{d} a v^{te} y siete días del mes de febr^{o} de mill y seisçientos y cinco a^{o}s ante mí el pres^{te} escrib^{o} y t^{o}s pares^{o} pres^{te} P^{o} de Chaves guardicionero vz^{o} de esta v^{a} y dijo que él ha visto y tasado las cosas tocantes a su of^{o} que es la [...] contenidas y la tasación está bien y fielm^{te} fecha y ansí lo juró a Dios y lo firmó t^{o}s Andrés Díaz y Ju^{o} de Çuaço

P^{o} de chaves
Ante mí Diego Román
(signed)

(fol. 238 r)
Más armería
-Yten se pone por ynventario quarenta casacas de saya entrapada que son para las armas de la dicha armería guarnecidas con una faxa de terciopelo carmesí y sobre la dicha faxa un flueco de terciopelo blanco sobre el cual va asentado un pasamano dorado y rosa seca con alamares y botones de las dichas colores forradas en lienzo morado que se tasó cada una quinientos ducados__ccxxvU
-Yten se pone por ynventario quarenta y seis sombreros de fieltro colorado forrados por dentro en tafetán encarnado con trencillas y cayreles de plata fina que se tasó cada uno en catorze Reales__xxiUdccc^{o} ccv

Vestidos de hombre [...]

(fol. 242 r)
Espadas
-Yten se pone por ynventario una espada hecha en Toledo de Ju^{o} Martínez que tiene la guarnición dorada manquilla con su daga de la misma suerte con vainas y talabarte de terciopelo negro y puños de plata que se tasó en ocho ducados__iijU
-Yten una espada de Bilbao y daga con guarniçiones doradas y ochavadas con sus puños de plata que se tasó en seis ducados__yUccl

-Yten una espada y daga dorada las guarniçiones doradas y torneadas con vainas y talabartes de terciopelo que se tasó en seis ducados__yUccl
-Yten espada y daga las guarniçiones de claraboyas con vainas y talabarte de terçiopelo y vainas y puños de seda q se tasó en tres ducados__iUcxxvi
-Yten una espada y daga de Sahagún con guarniçiones rayadillas y barniçadas con su talabarte y vainas de terçiopelo negro que se tasó en seis ducados__yUccl
-Yten una espada de Pedro de Belmonte que tiene la guarnición torçida con su daga con vainas de cuero que se tasó en tres ducados__iUcxxvi
-Yten una espada de Sahagún con guarniçión de claraboyas pavonada con talabarte y vaina de terçiopelo que se tasó en siete ducados__yUdcxxviii^{o}

En la vi^{a} de M^{d} a seis días del mes de abril de mill y sei^{o}s y cinco años ante mí el escrib^{o} y t^{o}s Gaspar Sánchez vz^{o} de esta vi^{a} dixo que él ha visto e tasado las espadas e dagas referidas en las par^{das} de arriba y con juram^{to} declaró que está bien fha a su saber y entender sin hacer agravio a ning^{a} de las p^{tes} y lo firmó t^{o}s Fr^{co} de […] e Fran^{co} López e Ju^{o} de […] v^{o}s de M^{d}

Gaspar Sánchez
Ante mí Diego Román
(signed)[…]

(fol. 245 r)
Caballeriza […].

PRIEGO, CONDES DE

Doc. 1.
1579, March 28.
Inventory of the Count of Priego, Fernando Carrillo de Mendoza.
Archivo Histórico de Protocolos de Madrid, Protocolo 759.

(fol. 385 r)
Ynventario que la Illma. Señora doña Jhuana Carrillo condesa de Pliego hace de los bienes que dexó al tiempo de su fin e muerte el Illmo. Señor don Hernando Carrillo y Mendoça mayor^{mo} de Su Mag.^{d} y conde de Pliego (sic) su marido ya defunto que esté en gloria

(fol. 386 r)
Ynventario q hace la muy Ill^{e} señora doña Jhoana Carrillo condesa de Pliego viuda mujer que fue del muy Ill^{e} señor don Hernando Carrillo y Mendoça conde de Pliego ya defunto que esté en gloria

1579
En la villa de Madrid a veinte e ocho días del mes de marzo de mill e quinientos y setenta y nueve años la muy Ill^{e} señora doña Juana Carrillo viuda mujer q fue del muy Ill^{e} señor don Her^{do} Carrillo y Mendoça conde de Pliego (sic) y mayor^{mo} de su mag^{t} que esté en gloria comenzó a haçer ynbentario de los bienes que el dho señor conde dexó en cumplimiento de lo proveido de su pedimiento por la Just^{a} de esta dha villa el qual dho ynbentario la dha señora condesa hizo en la forma e manera siguiente [...]

(fol. 388 r)
Camarín
En la villa de Madrid a seis días del mes de abril de myll y qui^{o}s y setenta e nueve a^{o}s estando dentro de una pieça q llaman camarín de los despojos de la batalla de la liga la mem^{a} de ellas es lo sig^{e}

Lienzos
-dos lienzos grandes al olio guarnesçidos de madera blanca el uno es de quando se vieron las dos armadas y se yban zercando a pelear y el otro de quando estaban ya enbestidas y pelearon
quadro

-Un quadro de madera blanca barniçada es de tafetán carmesí estanpado de la batalla como está puesta y pintada en Roma en la sala grande del Papa donde asysten los actos prinz^{s}.

Retrato del Papa
-Un retrato del papa Pío quinto al olio con una guarnición de madera blanca barniçada

Retrato S^{or} don Ju^{o}
-Otro retrato del S^{or} don Ju^{o} de Austria al olio con la mysma guarnición de madera blanca

S^{or}. S. fran^{co}
-Una imagen de Sor San fran^{co} con el capucho puesto con una cortina de tafetán azul guarneçida con un pasamano de oro- es al olio-

ymagen de nro S^{r} en la coluna
de Michael Ángel
-Una imagen de Xpo en la coluna al olio que es de Michael Ángel guarneçida de ébano y con la cubierta de ébano y pasa de plata

Sta Elena
-Una imagen de Sta Elena al olio guarneçida de ébano

ymagen
-Una imagen pequeña del naçimi^{o} pintada al olio

anus dey
-Un anus dey grande aovado (sic) guarnezido de plata con unos cordones de seda azul y plata yluminado- tiene la [...] de una pte y por otra el cordero y las armas de Pío quinto- con su tafetán azul

(fol. 388 v)
anus dey
-Otro anus dey grande rredondo yluminado que es adoraçión de los reyes y por otra pte el cordero guarneçido de raíz de oliva con una presilla de plata

caxa de anus dey
a my S^{a} la condesa

~~-una caxa rredonda anus deyes~~ [crossed out]

-dos retratos de la suzesión de los emperadores turcos- de la suzesión de Selín que fue a quien se dio la batalla están pintados al olio todos en una chapa de bronze y al natural guarneçidos con sus caxas de nogal

Retrato de Barbarroja
-Un retrato de la misma guarnición q los de arriba q es de Barbarroja

Retrato prínçipe Doria
-Otro retrato de la misma guarnición q es del príncipe Doria
-Una carta guarneçida con sus palos de todo el mapa donde se ve el lugar donde fue la batalla

corona
-Una corona o guirnalda q se presentó al S^{or} don Ju^{o} puesta en papel de versos en lor del S^{or} don Ju^{o}- y en él una çelebraçión de canto de órgano

túnica del papa
Pío quinto
-Una túnica que traya vestida el Papa Pío quinto- blanca

bonete colorado
del Papa
-Un bonete colorado carmesí de los que traya el papa Pío quinto

otro bonete del
Papa
-Otro bonete de raso blanco aforrado e tafetán de los que traya el Papa Pío quinto

çapatos del papa
-Unos çapatos de terciopelo carmesy con unas cruzes de unos listones de oro que traya Su Santidad Pío quinto

horas del Papa
-Unas horas en q reçaba el Papa quinto (sic) el rosario de nra Señora

cortinas pequeñas
de tafetán

-dos cortinillas de tafetán azul q están debajo de estos despojos

(fol. 389 r)
q^tas de perdones
-Unas quentas de prs blanco y pardo engastadas en alquimia- con una medalla es de muchos perdones el rosario-

corales
-Una corona de corales ensartados engazados y con estremos de oro con su medalla que dizen es de muchos perdones

Rosario
-Un rosario de diez quentas de madera blanca y parda de perdones con una calavera al cabo y una medalla

Rosario
-Otro rosario de honze quentas de ébano grueso con un cordón de seda negra q dizen es de perdones

Rosario
-Otro rosario de onze quentas de ébano con su cordón de seda negra q diçen es de Perdones

çimitarra
-Una cimitarra con la guarnizión de plata es damasquina tiene la vayna de çapa

zimitarra
-Otra cimitarra con la guarnición de hierro dorado y la vayna de çapa con sus guaniçiones de hierro dorados es damasquina

espada
-Una espada con otra manera de guarnición que pareze a las que traen los moros de hierro o cobre tallado y es maltado con su çinto de seda y oro texido y sus pieças abaxo baçiadas y talladas y su vayna de terciopelo pardo

alfanxe
-Un alfanxe con el puño y pomo labrado de hueso y los encajes de tauxía de oro la guarnición de la vayna y del mismo alfanxe tallada y dorada y esmaltada con su vayna de cuero colorado

cuchillos
-Unos cuchillos q son quatro en una vayna de çapa guarneçida muy bien de plata de modo turquesco con tres vueltas (fol. 389 v) de una cadena de plata y guarneçidos los quatro cuchillos los puños de madera y los cabos de diente de pescado

cuchillos
-Tres cuchillos en una vayna de çapa con la guarnición de ella de plata dorada y oro baxo labrado a modo turquesco con las [...] y cabos de diente de pescado con una cadena de plata de muchas vueltas

cuchillos
-Otros tres cuchillos su otra vayna de çapa guarnezida de plata y los cuchillos los puños de madera y los cabos de diente de pescado

cuchillo
-Un cuchillo a manera de navaja que está en un cabo de diente de pescado está guarnezido de oro. Es damasquina la cuchilla muy fina

cuchillo
-Otro cuchillo con la vayna de çapa guarnezida de plata lisa y el cuchillo damasquino muy fino con el puño de tartaruga y el cavo de diente de pescado

cuchillo
-Otro cuchillo con la vayna de çapa guarnezido de plata tallada es damasquino el cuchillo tiene el manguillo de tartaruga y el cabo de diente de pescado

cuchillo
-Otro cuchillo con una vayna de çapa q está guarnezida de plata lisa está rota
cuchillos
-dos cuchillos y un punzón de plata en una caxa pequeña de tartaruga guarnezida de plata el uno de los cuchillos el puño de tartaruga y el cabo de diente de pescado y el otro el cavo de cuerno y el puño de marfil

cuchillo
-Un cuchillo en vayna de çapa guarnezida de plata lisa el puño y cavo de cuerno

(fol. 390 r)
corales

-cient cuentas de corales encarnados con otra quenta gruesa y el rremate que eran en las que reçaba el baxá

sustinete
-Un sustinete para tener en el suelo el turbante labrado de cuerno de hueso con el cartofiçio (sic)

arco
-Un arco turquesco por de dentro negro y por de fuera labrado de blanco y dorado y negro fino sobre el blanco

arco
-Otro arco turquesco por de dentro negro y por de fuera labrado y pintado muy bien del oro fino

arco
-Otro arco turquesco por de dentro de negro y por de fuera de otro color jaspeado
sortijas de hueso
-Tres sortijas de hueso y la una más fina con que tiran los turcos con los arcos

carcax
-Un carcax de cuero leonado labrado por de fuera al modo turquesco en que hay veinte e seis flechas. Está con su correa

carcax
-Otro carcax negro con su correa labrado por de fuera al modo turquesco y dentro dél treinta y una flechas

(fol. 390 v)
flechas
-Un atadillo con seis flechas

flechas
-Un manojo con diez y seis flechas
flechas
-Veynte y ocho flechas en un ençerado puestas con su cubierta de los mismo como los turcos lo acostumbran traer

cuchilla

-Una cuchilla con su martillete con que los turcos suelen pelear

martillo del Papa
-Otro martillo y hacheta guarneçida con un cordón de seda carmesí y oro que tenía el Papa Pío quinto en su cámara por despojo de los turcos

turbante
-Un turbante sobre un bonete azul

penacho
-Un penacho de los que suelen traer los matasietes que llaman de muchas garcostas (sic)

penacho
-Otro penacho de plumas de gruas (sic) guarneçido por debajo con una trenzilla de seda y oro

cuerda de arco
-Una cuerda de arco

(fol. 391 r)
bonetillo
-Un bonetillo turquesco de pellejos de lobo

çintura de los turcos
-Una çintura con que se çiñen los turcos de texillo de seda colorado y azul

libros turquescos
-Dos libros turquescos el uno con las cubiertas azules lisas y el otro con las cubiertas doradas sobre el cuero labradas de una labor menuda

abano de la Yndia de Portugal
-Un abano de los que traen de la Yndia de Portugal que también los alcançan los turcos que vienen de la China grande con una manga grande lacreado y él todo de colorado
abano
-Otro de la misma manera sino que es más pequeño dorado y colorado

plato
-Un plato de madera turquesco pintado y dorado

espuelas
-Unas espuelas turquescas de metal doradas guarneçidas en correas de cuero coloradas

(fol. 391 v)
Esponja
-Una esponja con una correhuela de cuero que traen los turcos dentro de las bolsas para sacar agua de las fuentes y ríos y beber con ella

bolsa para agua
-Una bolsa turquesca de cuero granujado (sic) con capa de las pequeñas con unas guarniciones labradas del mismo cuero

bolsa agua
-Una bolsa turca grande turca (sic) para beber agua con un cordón de seda colorada

bolsa agua
-Otra bolsa turca para beber agua con su correa de cuero con una guarnición de cuero negro bien labrada es grande.

bolsa agua
-Otra bolsa de cuero nueva algo más pequeña con la misma guarnición

çintos
-Dos çintos de texillo de colores el uno de azul y colorado y el otro de amarillo y azul

cabeçada
-Una cabeçada de caballo con su freno y riendas que se tomó a un moro de los que llaman matasietes estando sobre Nabazin le tomó un criado del conde de Pliego habiéndole muerto de un arcabuzazo

doselejo
-Un doselejo de terciopelo carmesí con una faja alrededor de tela de oro rasa con una flocadura de oro y seda carmesí sobre q estaban los anus deyes y ymágines

de suso referidas y con esta se acabó de inventariar todo lo que estaba en el dho camarín a donde lo dho estaba &^{a}

(fol. 392 r)
En otra pieça de la sala de abajo estaba lo sig^{e}/ quera el escritorio del conde

Caxeta
-Una caxeta de alemanya grande de las que se usaba antigua embutida está lo sig^{e} [...]

Un libro
-Un libro de trajes de los turcos [...]

Relox
-Un relox dorado de muelles y un pereco

-un martillo de hierro dorado nuevo con su tenaza de los mismo [...]

(fol. 400 v)
En la villa de Madrid a diez días del dho mes de abril del dho año de myl e qui^{o}s e setenta e nueve años prosiguiendo en el dho ynbentar^{o} se ynbent^{o} dentro del dho escriptorio en la dha sala baxa q está dho las cosas que allí faltaban de inventariar q son las siguientes [...]

(fol. 404 v)
arma de los yndios
-Un arma de los indios de çierta madera de haya del Perú negra que pareze ébano con unas labores abaxo [...]

(fol. 405 r)
carcax
-Un carcax dentro del cual va el aparejo del [...] arriba dho con su zerradura [...]
Con la dha linterna q es la última partida de suso se acabó de poner e ynbentariar todo lo q se halló e hubo en el dho escriptorio e camarilla del dho conde q está dentro de una sala en el patio de la dha casa en estrando en ella a mano derecha

Recámara
Después de lo ql entrando dentro de la recámara del dho conde se ynventarió lo siguiente [...]

(fol. 410 r)
Espadas talabartes dagas cuchillos y bolsa de agua y otras cosas
cota
-Una cota de malla
-Dos mangas de la dha malla

(fol. 410 v)
guantes
-unos guantes de la misma malla

malla
-Otras nueve pieças pequeñas de malla

espada
-Una espada ancha dorada una guarnición dorada de las antiguas- ytaliana con su cuchillo dorado sin punzón con su vayna de terciopelo pardo

espada
-Otra espada ancha vieja con guarnición de dos puentes con su vayna de cuero colorado

estoque
-Un estoque de armas con vaina de terciopelo carmesí

espadas
-Dos espadas de tornear la una con vayna de terciopelo blanco y la otra de carmesí

alfanxe
-Un alfanxe con un león de marfil por puño con vayna de terçiopelo verde

maça de armas
-Una maça de armas dorada con un cordón de seda parda

Espada que tráela
el S^or don Antonio
a lo que se dize
-Una espada de çinta que su señoría traya con su vayna de cuero negro y guarnición barnyçada. Es la vayna de çapa

daga
-Una daga dorada con su vayna de terciopelo blanco

talabarte
-Un talabarte de terciopelo negro con sus tiros

talabarte
-Otro talabarte de camino con su pretal de vaqueta

talabarte
El S^{r} don Antonio
-Otro talabarte de vaqueta con su petrina que el conde dizen traya

guarnición de espada
-Una guarnición dorada labrada de relieve. Está en una bolsa de cuero

zeñidor
-Un zeñidor de almaizar morisco grande

cuchillo
-Un cuchillo de monte con dos cuchillos y un punzón y tres cargas de arcabuz

puñal
-Un puñal dorado con una vayna baya

(fol. 411 r)
cuchillo
-Un cuchillo con el cabo de marfil es turquesco con la vayna guarneçida de plata con su tenedor y una amoladura guarneçida de plata y su cadenilla de plata asida en la mesma vayna

bolsa pa agua
-Una bolsa de quero turquesca con las cantoneras de plata metida en una funda de terciopelo amarillo con unos pasamanillos de plata y una cuerda larga de seda carmesí y plata para virilla del çinto

cuero turquesco
- Un quero largo turquesco q es la mesa adonde dicen comen los turcos sirve dizen para dormir la siesta

cordobanes
-Tres cueros de cordobán colorado turquescos para capotes de halcones

sobrevayna
-Una sobre vayna de vaqueta

vayna
-Una vayna de terciopelo negro con su contera

vaynas
-Seys vaynas biejas de espadas

cuchillos
-Diez y seis cuchillos de mesa con sus thenedores

Jaezes
-Jaeçes de la gineta guarniciones de caballos frenos ginetes y de brida adreços de monte de la gineta caparazones y corazas estribos de la brida gualdrapas de terciopelo y de paño y çinchas de la gineta y de la brida y cogines de la posta con sus estribos y pretales de cascabeles [...]

(fol. 416 v)
Arcabuzes Ballestas y otras cosas para la caza
arcabuz
-Un arcabuz grande con el punto de plata hecho en Flandes

ballesta
-Una ballesta con la berga de la Puebla y la culeña (sic) de serbal con su gafa de encaxe hecha en Almozaziz de Çurita con su funda

ballesta
-Una ballesta grande la cureña de serbal hecha en Madrid. Está sin gafas tiene su funda

ballesta
-Otra ballesta con la cureña de enzina negra tiene su gafa y funda

(fol. 417 r)
ballesta

-Otra ballesta de las de dos enpulgueras hecha en Aragón. Tiene su funda y su gafa

ballesta
-Una ballesta de bodoque de las de Nápoles de hombro y de armar con la mano

carcax
-Un carcax de cuero bayo que es también aljaba para llevar un par de saetas tiene un par de virotes dentro

carcax
-Otro carcax de cuero negro con una bolsilla en el mismo carcax y çerradores

aljaba
-Una aljaba morisca con quatro cachuchos para yerba con tres jaras y su correa y bolsa

aljaba
-Una aljaba de arcabuz con una cubierta de cuero ençerado con seis saetas para tirar con arcabuz y con sus senos para tener las cargas de pólvora y sus medidas de ellas de una parte y de otra y sus apartados para tener pólvora y perdigones y una sartenilla en la bolsa para derretir plomo para haçer perdigones y pelotas y una lima y ella con sus correas para asylla del caballo

bolsa
-Una bolsa de arcabuz redonda con tres bolsas por de fuera para llevar lo necesario en ellas para el arcabuz

bolsa
-Una bolsa de caza blanca de las de Nápoles con su hierro con un guante de caça dentro

Sinuelo
-Un sinuelo que es frasco y sinuelo

trayllas
-Çinco trayllas de çerdas para los perros blancas y leonadas y negras unas gruesas y otras delgadas son cinco líos de ellas.

(fol. 417 v)
cadenillas pa perros
-Seys cadenillas cada una con sus dos colleras de cuero de ante para llevar perros atados son doze colleras y sus cadenas como está dho

corneta
-Una corneta de marfil hecha de los indios

corneta
-Otra corneta del mismo marfil labrada de los mesmos indios ambas a dos don labradas de relieve

cornetas
-Dos cornetas negras de cuerno de bufano
-Una hamaca de algodón de las q usan los yndios para caminar

Adargas Rodelas tableros de juego y otras cosas que hay en la dha recámara
adarga
-Una adarga de cuero blanco

Rodela
-Una rodela labrada de turquesas de Yndia

tablero
-Un tablero para jugar a las tablas q se çierra con su llave y tiene por ençima juego de ajedrez tiene veynte tablas blancas y quinze negras tiene en una bolsilla dentro tres dados cinco gitones de bronze y dos de hueso para las daotas (sic) y tantos de caracoles y habas

Truque
-Un truque que llaman de madama. Es un juego q juegan en Flandes es de nogal

Laúd
-Un laúd con su llave

parasol
-un parasol labrado de hueso y de ébano curiosamente con su palo de otra madera estraña cubierto de damasco carmesí y tiene su funda de lienzo y su caxa de haya.

(fol. 418 r)
dientes
-Dos dientes de alefante (sic) el uno mayor que el otro

jaula
-Una jaula de caña [...]

Retrato
-Una estatua del retrato del negrillo de Julio Céssar hecho de papelón y fingido de bronce

Estatua
-Otra estatua de barro fingida de bronze de Laoconte [...]

bastones
-Un legajo de bastones algo reçios q son catorze

bastones
-Otro legajo de bastones delgados en que hay quarenta y un bastones diferençiados

Medallas de plomo y alquimia
y otras suertes
-Una caxa con çiento noventa y una medallas grandes y pequeñas y otras historias de plomo y otras beynte y tres historias de medallas de alquimia

más medallas
-Y más ansimismo en la dha caxa doçientas y veinte y siete medallas Antiguas y modernas de bronce- y ansimismo hay más de veinte e ocho medallas y historias guarneçidas de madera y de cuerno y de ellas de bronce y de ellas de cuerno y más quatro guarniciones de madera y tres de cuerno sin medallas cinco bezotes de metal de los que traen las mujeres yndias- Una culebrilla y un lagarto de plomo sacados al natural- Unas yerbas y un tlaber baçiadas y sacadas y al natural cinco bolillas del trueco de madama de hueso. Un cuerneçito pequeñito de tabali un pedazo de colmillo de pescado tres guijarros escriptos todo lo qual está en la caxa sobre dha- Un kanselete (sic) de yndias a manera de un monillo. Está ansimismo en la caxa con las demás cosas dhas

(fol. 418 v)
çapatos
-Unos çapatos de los que traen los yndios [...]

torno
-Un torno de hierro para labrar de torno con sus porguezuelas que son quatro y sus anillos y otras pequeñas

torno
-Otro torno de palo para poner en una mesa

torno
-Otro torno de madera para labrar con él puesto en una mesa con sus dos porguezuelas y otras dos piezas

cartabón
-Un cartabón y otro para yr señalando con sus rayas y son de madera [...]

ynsignia
-Una ynsignia militar de los yndios [...]

piezezuelas
-Tres piezezuelas para labrar de torno

punzón
-Un punzonzillo para hazer agujeros [...] para el torno

(fol. 419 r)
marfil
-Un remate de salero de marfil que era de Montezuma

cuchillejo
-Un cuchillo de entallador

Escofina
-Una escofina con su mango de box guarnezida de cuerno

guirnalda
-Una guirnalda de yndios labrado de mosaico de turquesas en tres o quatro pedazos

canulla
-Una atadillo de unas canullas con que sacrificaban los yndios

quixada
-Una quixada de un tigre guarnecida de las mismas turquesas […]

zapato
-Un zapato de los yndios. Es solo uno del cuero labrado que ellos traen

guantes
-Unos guantes del mismo cuero labrado

faldón
-Un faldón de pluma de los que traen los yndios de guerra

ventalle
-un ventalle de pluma larga de los que traen los yndios […]

(fol. 425 r)
coco
-Un coco de mucho valor que es de man[…] Guarneçido de plata que por ser de mucho valor no se […] para saber el peso de la plata y por agora no se le quyda su valor y es a manera de barco con su cubierta y hecho un caño por donde se bebe con mucha curiosidad- si esta pieça estuviere asentada otra vez esta partida y la otra es una cosa.

aXXii de abril
lo que vino en un arca de Pliego
En veinte y ocho días del mes de abril del dho año de mil e quinietos e setenta e nueve años se prosiguió el dho ynventario de cosas que dizen se truxeron de la villa de Pliego en una arca grande que es de pino estaban las cosas sig^{tes} digo en veinte e dos días del mes de abril

-Dos caxas ahovadas de madera con agnus deyes
-Una salva de ébano negra
-tres zurrunzillos de yerba
-Un reliquiario de azofar con su cubierta
-dos bandexicas chiquitas de la Yndia la una negra la otra laceada
-otra cosa redonda de la Yndia pareze tapador […].

SALINAS DEL RÍO PISUERGA, MARQUESES DE

Doc. 1.
1617, September 20.
Inventory and appraisal of Marquis of Salinas, Luis de Velasco.
Archivo Histórico de Protocolos de Madrid, Protocolo 2320.

(fol. 1 r)
Ynventario y tasación de la plata y demás bienes muebles que quedaron por fin y muerte del Sor Don Luis de V^{co} marqués de Salinas y Presidente del Real Consejo de Yndias q está en el çielo hecho por parte de sus testamentarios

(fol. 9 r)
Alonso de la Fuente testamentario del S^{r} marqués de Salinas don Luis de Velasco difunto Presidente que fue del Rl Cons^{o} de Yndias por mí y en n^{e} de los ss^{es} Condesa de Santiago doña María de Velasco don Luis de Velasco e Ybarra marqués de Salinas Liçen^{do} don Ju^{o} de Villela del dho Cons^{o} y Padre Diego de Ybarra de la Comp^{a} de Jhs de esta villa ansímismo testamentarios de quien tengo poder p^{a} este efecto- digo que por escusar costas y dilación conviene que como se vayan inventariando los bienes que quedaron del dho S^{r} marq^{s} difunto se vayan tasando por pers^{as} q lo sepan y entiendan e yo nombré las sig^{tes} […]

-a Esteban de Fuentes espadero p^{a} las espadas y alfanjes […]
-a Ju^{o} de Çuaço armero de Su Mag^{d} para las armas […]

(fol. 9 v)
-a Simón de Vall^{d} guarnicionero para las guarniciones de la caballeriça

A Vmd sup^{co} los haya por nombrados y m^{de} se les notifique lo acepten y hagan la dicha tas^{ón} con juramento pido just^{a} &^{a}

Al^{o} de la Fuente
(signed)

Q se haga el ynvent^{o} y tas^{ón} según y como por esta pet^{ón} se pide y sean por nombrados para haçerla a los en ella cont^{os} y notifíqueseles luego lo açepten e juren de haçer la dha tas^{ón} bien y fielmente ante X^{o}ptobal Vaçán escrib^{o} de Su Mag^{d} a quien se comete en M^{d} a veynte días del mes de sep^{e} año de mill y sei^{o}s y diez y siete tomando el S^{or} L^{do} don Fr^{co} de Roxas y Oñate ten^{e} de correg^{or} de esta v^{a}

Ante mí Di^{o} Ruiz de Tapia escrib^{o}
(signed)

N^{ón}
En la dha vi^{a} de Madrid el dho día v^{te} (fol. 10 r) de septiembre del dho año de myll y seisçientos y diez y siete yo Xpbal Vaçán escrib^{o} de Su Mag^{d} leí e not^{é} la pet^{ón} y auto de esta otra p^{te} a Andrés de Pedrera maestro platero marcador y contraste de esta vi^{a} en su pres^{a}= e aceptó el off^{o} de tasador= y dijo está presto de haçer el jur^{to} y tasación de lo que a él toca cada y qu^{do} q se le pida y esto respondió de que doy fee

Xpobal Vaçán
(signed)[...]

(fol. 10 v)
Espadero
-En la dha vi^{a} de M^{d} el dho día mes y año yo el escrib^{o} n^{é} la petición y auto a Esteban de Fuentes espadero v^{o}z de esta v^{a} en su pres^{a} y dijo estaba presto de cumplir lo que por el dho auto se le nada y esto dio por su resp^{ta} de q doy fee

Xpobal Vaçán
(signed)[...]

(fol. 11 v)
Armero
-En la dha v^{a} de M^{d} dho día v^{te} de sep^{e} del dho año yo el escrib^{o} hiçe otra tal not^{ón} como las de arriba a Ju^{o} de Çuaço armero de Su Mag^{d} en su pres^{a} y dijo estaba presto de cumplir con lo q por el dho auto se le m^{da} y en fee de ello lo firmé

Xpobal Vaçán
(signed)[...]

(fol. 57 v)
Alfanjes
Las hojas a 5 du^{o}s cada uno
En F^{co} de Rosales en 100 Rs como las hojas
En la dha villa de M^{d} dho día mes y año dicho el dho Alonso de la Fuente puso por ynvent^{o} las guarniciones de cobre [between lines:](que están en los alfanjes) ensacadas (sic) en oro que tienen el de número ~~de oro~~ [crossed out] uno de v^{te}

quilates y se apreçia el oro de él en çiento y quarenta y dos reales= y el otro número dos tiene oro de los dhos vte quilates aprecióse en çiento y treinta y ocho reales= q todo mta duçientos y ochenta reales y el dho apreçio y tasón hiço el dho Andrés Pedrera debajo de su jurto y ambos lo firmon tdo de oro- q tiene [...] en los alfanjes

Aloº de la Fuente
Andrés Pedrera
Ante mí Xpobal Vaçán
(signed)[...]

(fol. 80 v)
Espadas y alfanjes
En la dha villa de Madrid el dho día primero de octubre del dho año de myll y seisçientos y diez y siete el dho Alonso de la Fuente mayormo y testamentº suso dho (fol. 81 r) puso por ynventº como bienes del dho Sr marqués de Salinas las espadas alfanjes y adreços que de Su Sa quedaron q son las que adelte se dirán y se tasaron por Esteban de Fuentes espadero vºz de esta via debajo de juramto q primero hiço en fora de drº el dho ynventº y tasación se hiço en la manea sige

-Un adreço de espada y daga sogueado con su pretina y la hoja de lomo de Domingo Aguirre de Toledo tasóse en diez ducados__3U750
-Dos adreços de espadas y dagas çiçavados (sic) con hojas de Frco Ruiz cada uno tasáronse a cinco ducados cada uno__3U750. En Frco de Rosales en 100 Rs
-Dos hojas de alfanjes que pareçen ser hechos en la tierra del Japón tasáronse en cinco ducados cada uno q son diez ambos__3U750

Los quales dhos alfanjes espadas y adreços quedaron en poder de Luis de Velasco de q yo el escribº doy fee y que los recibió y el dho Esteban de Fuentes= dijo haber fho la dha tasón bien y fielmte a su saber y entender sin fraude algº y no lo firmó por no saber firmólo el dho Alonso de la Fuente

Alº de la Fuente
Ante mí Xpobal Vaçán
(signed)[...]

(fol. 88 v)
Armas
En la dha villa de Madrid el dho día cinco de octubre del dho año de mill y seisçientos y diez y siete ante mí el escribano el dho Alonso de la Fuente puso

por ynventº las armas que quedaron por muerte del dho S^{r} marqués de Salinas q son las que adelante yrán declaradas y se tasaron por Joan de Çuaço armero de Su Mag^{d} q vive enfrente de sus caballeriças Rles debajo de jur^{to} q hiço en forª de drº en la forma y manera sig^{e}

-Primeram^{te} un peto de açero dorado y grabado con sus carcelas (sic) y ristre espaldar gola çelada braçales con guardabraços dos grebas con sus quijotes dos manoplas testera de caballo açeros de silla y un peto q sirve de tornero (sic) con su gola y çelada de encaxe y guardabraço para torneo y una bufa pª torneo de a caballo y un morrión de infantería y un faldón de malla y dos çapatos de malla con puntas todo de açero tasáronse en ochoçientos reales estas armas__27U200
-Pusiéronse por ynventario otras armas fuertes de açero açules q son dos petos dos espaldares el uno con escarçelas y el otro sin ellas dos golas dos guardabraços tres manoplones largos un morrión fuerte y una çelada (fol. 89 r) tasáronse todas estas armas fuertes en zinquenta duºs__18U750
-Pusiéronse por ynventº otras armas fuertes con listas doradas q son de açero peto y espaldar con su gola dos guardabraços un morrión con sus orejeras otro viejo tasáronse estas armas en cien reales__3U400
-Pusiéronse por inventº otras placas de armas sueltas viejas milanesas q son una rodela una testera de caballo y un copón de lo mismo dos guardabraços dos grebas enteras dos escarçelas una arandela una faldilla de peto tasóse todo en ochenta reales__2U720
-Seis armas de la China con todas sus pieças que cada arma es peto y espaldar murrión y sus faldillas braçaletes y manoplones y sus grebas y los morriones con sus mascarillas los morriones tasáronse a v^{te} ducados cada una de las dhas seis armas que monta çiento y v^{te} ducados__45U
-Púsose por ynventario una ballesta con un rétulo (sic) q diçe de Santa Mª y con sus gafas tasóse en quatro ducados__1U500
-Una escopeta con cañón dorado y la caja guarneçida de cuero dorado con sus frasco y frasquillo para pólvora con sus cor(fol. 89 v)dones guarneçidos de oro tasóse en diez ducados__3U750
-Otro arcabuz largo ordinario tasóse en ocho ducados__3U

Todo lo qual se tasó por el dho Juº de Çuaço en la forª dha debajo del dho juram^{to} con el dho Alº de la Fuente y todo quedó en poder del dho Luis de Velasco de q yo el escribº doy fee y de ello fueron tºs Blas Hernández Miguel Serrano y Fr^{co} Choque res^{tes} en esta q^{te} y porque el dho Luis de Velasco dijo no saber firmar a su ruego o firmó un tº-

Alº de la Fuente
Juº de Çuaço
Por tº Miguel Serrano
Ante mí Xpobal Vaçán
(signed)[…]

(fol. 91 r)
Guarniçiones
En la dha villa de Madrid el dho día seis de otu^{e} del dho año de myll y seiºs y diez y siete ante mí el escribº el dho Alonso de la Fuente como tal mayor^{mo} y testamentº puso por ynventº las cosas de guarniciones y gualdrapas de caballos que quedaron del dho S^{r} marqués q son las que adel^{te} yrán declaradas y se tasaron por Simón de Valli^{d} guarnicionero vzº de esta viª debajo de juram^{to} q hiço en forma de derecho en la manera sig^{e}

-Quatro guarniciones de caballos de coche enteras las dos y las otras dos medias con silla de vaqueta clavaçón negra tasóse en duçientos reales__6U800
-Una silla gineta de cordobán colorado con su caparaçón de lo propio y otro negro de cordobán y todo su recado y estribos y freno tasóse en siete ducados__2U625
-Un caparaçón de cuero de color con la labor de enzima de cuero verde y su petral y cabeçada y una reata tasóse en ochenta reales__2U720
-dos guarniciones de mula viejas q servían pª los caballos de vaqueta tasáronse en zien rales__3U400
-Una gualdrapa de trºpelo negro guarneçida con dos fluecos de seda desviado uno de otro aforrado en su fustán y una guarnición del mismo trºpelo con sus pasamanos y cordonería (fol. 91 v) que es petral cabeçadas guruperas açiones riendas y estribos dorados y la clavaçón de toda ella dorada tasóse todo en çinquenta duºs__18U750
-Una doçena de riendas de la gineta de Yndias y seis pares de açiones tasóse en tres ducados__1U125
-otros tres pares de riendas açules de gineta y brida tasáronse en quatro reales y mº__U153
-diez y siete pieças de guarniciones nuevas de brida de la Yndia petrales cabeçadas guruperas açiones riendas y almártigas tasáronse todas en tres duºs__1U125

Todo lo qual tasó en esta forª el dho Simón de Valli^{d} y dijo haber fho la dha tas^{ón} a su saber y entender sin fraude algº debajo del dho jurº y no lo firmó por no saber firmólo e dho Alº de la Fuente y el dho Luis de V^{co} lo recibió en mi presª y de los tºs yuso escriptos de q doy fee y porq dijo no saber firmar a su ruego lo firmó un tº siéndolo Fr^{co} Choque Bernardo Ochoa de Salinas y Juº Álvarez Res^{tes} en esta q^{te}- t^{do} quince su-

Alº de la Fuente
Por tº Ber^{do} Ochoa
Ante mí Xpobal Vaçán
(signed)[…]

(fol. 92 r)
Biobos (sic) y retasa de las cosas de madera del Japón= y de las Yndias […].

SANTA CRUZ, MARQUESES DE

Doc. 1.
1537, October 18.
1549, January 8.
Foundation of the "*mayorazgo*" or entailed estate of Álvaro de Bazán. Inventory of artillery in Gibraltar.
Archivo Histórico de la Nobleza, Santa Cruz, C. 32, D. 1.

(fol. 17 v)
Item que el dicho señor don Álvaro de Baçán y doña Ana de Guzmán su mujer por Razón del dicho desposorio y casamiento que se concierta que se ha de hazer entre el dicho señor don Álvaro de Baçán su hijo maior y la dicha señora doña Juana de Baçán hija maior de los dichos señores Conde y Condesa de Miranda. Y porque haia efeto y se effetue han de hazer maioradgo de todos los bienes Raízes que ellos al presente tienen así villas lugares, fortalezas, vassallos, juros, Rentas e diezmos e de otras qualesquier Rentas e bienes (fol. 18 r) Raízes y así mesmo de todo lo que su magestad debe al dicho señor don Álvaro de Baçán del pleito del soto de Roma de Asquerosa y de Moçatea así de los fructos corridos como de lo principal. Y también han de fazer el dicho maiorazgo de todo lo que su magestad debe al dicho señor don Álvaro de Baçán el padre de lo de las galeras del tiempo que fue Capitán general de ellas. Para que luego como se cobraren todas las dichas deudas se empleen los dineros en bienes Raízes para el dicho maioradgo que así han de façer. Y que así mesmo han de hazer maioradgo de toda la artillería que el dicho señor don Álvaro el padre tiene y así mesmo de la munición de ella. Y entiéndese que es la dicha artillería y munición de ella de que así se ha de hazer maioradgo toda la contenida en la dicha escriptura y memoria que de esto tiene hecha el dicho señor don Álvaro de Baçán que pasó y se otorguó (sic) en la ciudad de Gibraltar en diez y ocho días del mes de Otubre de Mil y quinientos y treinta y siete Años por ante Juan Fructuoso escribano público y del número de la dicha cibdad de Gibraltar que es la que Rescibió Gómez de Balboa alcaide de la fortaleza de Gibraltar según que más largo en la dicha escriptura se contiene el tenor de la qual es este que se sigue

En la noble e más leal cibdad de Gibraltar estando en la fortaleza de ella a diez e ocho días del mes de Otubre Año del nasçimiento de nro salvador Jhu Crhisto (sic) de Mil e quinientos e treinta e siete años en presencia de mí el escribano público e de los testigos de iuso escritos paresció (fol. 18 v) presente el señor

Gómez de Balboa alcaide e capitán de esta dicha cibdad por el yllustre y muy magnífico señor don Álvaro de Baçán alcaide e capitán de esta dicha cibdad por sus majestades e dixo que por quanto Pero González de Cimancas (sic) aio del señor don Álvaro de Baçán le ha dado e entregado las cosas de artillería e muniçión e otras cosas de iuso escritas que son del dicho señor don Álvaro de Baçán e él las ha Recebido que por tanto se dava e dio por entregado en ellas e las cosas que ha Recebido e el dicho Pero González de Simancas le dio e entregó de que el dicho señor Alcaide se dio por entregado e contento con las siguientes.

Inventario del artillería (sic) que está en la fortaleza de Gibraltar e otras partes de ella q es del señor don Álvaro de Baçán.

-Un falconete de cuenta de Siete Quintales e una arroba e seis libras es ochavado.
-Medio Cañón Redondo de la fundición de Málaga que tiene de quenta deciséis (sic) Quintales y tres arrobas y deciocho (sic) libras.
-Más un cañón Redondo de la fundición de Málaga de cuenta de veinte Quintales e tres arrobas e seis libras.
-Un cañón pedrero de quenta de treze quintales e veinte libras tiene por armas las Columnas de Hércules con una corona imperial encima en una venera por baxo de las dichas columnas.
-Otro medio cañón Redondo de la fundición de Málaga cincuenta.
-Otro medio cañón de la dicha fundición de quenta de Diez e seis Quintales e dos arrobas e diez y siete libras.
-Más otro falconete ochavado de quenta de Ocho quintales e una arroba e catorze libras.
-Más un sacre doble con unas columnas de Hércules con una Corona imperial en (fol. 19 r) cima de ellas e una Venera debaxo sin quenta de peso.
-Más otro falconete morisco que tiene una señal de pelotazo a par de los bunñones (sic) con un letrero de lestras moriscas par del fogón e otro par de la boca.
-Más un sacre que tiene de quenta Quinze quintales e dos arrobas e veinte libras.
-Otro sacre ochavado sin quenta con una salamandra ençima del fogón y en el fogón una gegodica (sic).
-Otro medio sacre Ochavado de quenta de siete Quintales y tres arrobas e onze libras e unas Columnas de Hércules delante de los buñones.
-Un cañón pedrero que tiene de quenta Doze Quintales e dos arrobas e ocho libras con las Colunnas (sic) de Hércules e una venera por baxo de ellas.
-Otro medio cañón pedrero de quenta de nueve Quintales y tres arrobas e diez y siete libras con las columnas de Hércules e una venera por baxo de ellas e baxo de los buñoles (sic) una herradura.

-Otro medio cañón Redondo que dixeron es de la fundición de Málaga tiene de quenta Diez e seis Quintales y tres arrobas e tres libras.
-Otro medio Cañón que dixeron es de la fundición de Málaga sin quenta.
-Medio cañón pedrero Redondo tiene de quenta Doze Quintales e cinco libras.
-Otro Cañón pedrero que tiene Qatorze (sic) Quintales y veinte y quatro libras.
-Otro Cañón pedrero tiene de quenta treze Quintales y tres arrobas y veinte libras.
(fol. 19 v)
-Otro cañón pedrero que tiene de quenta Doze Quintales e tres arrobas y Ocho libras con las Columnas de Hércules.
-Otro Cañón pedrero de quenta de Onçe quintales y Ocho libras con las columnas de Hércules y una venera por baxo de ellas y en el fogón una herradura.
-Un falconete cámara de quenta de Ocho quintales y tres arrobas y una libra.
-Otro cañón pedrero de quenta Onze quintales e Ocho libras.
-Otro falconete de cámara de quenta Ocho quintales e tres arrobas e Diez e nueve libras.
-Un Rebodoquín Redondo de quenta tres quintales e quatro libras.
-Un falconete de los buñoles atrás Ochavado con las columnas de Hércules con un molinete de hierro en los buñoles tiene de quenta Siete quintales e una arroba e diez libras.
-Media culebrina encabalgada y entorchada las Ruedas desguarneçidas la cureña guarnecida. Y tiene la culebrina la imagen de Sant Andrés y las armas imperiales tiene de quenta por letras del guarismo tres mil e Quinientas e treinta de suma.
-Más otro sacre de la fundición de Flandes con las armas imperiales con una cabeça de león en la culata con un caxcavel de metal en la boca que tiene de suma en letra de guarismo Mil e Setecientas e quarenta e dos de suma.
-Una culebrina Ochavada con dos escudos adelante del fogón tiene de quenta Sesenta y quatro quintales y una arroba.
-Otro (sic) culebrina Redonda.
-Un cañón pedrero Redondo que tiene de (fol. 20 r) quenta treinta e dos quintales e una arroba y veinte libras.
-Más un medio cañón que tiene de quenta Diez y seis quintales y una arroba e Diez libras.
-Más otro medio cañón que tiene de quenta Diez e seis quintales e diez arrobas e siete libras.
-Otro medio cañón pedrero que tiene de quenta Doze quintales e tres arrobas e Quinze libras.
-Un sacre ochavado quebrado.
-Medio cañón pedrero Redondo.

En la plaça de esta cibdad.
-Media culebrina entorchada con una imagen de San Pablo y las armas imperiales y tiene de quenta en suma de letra de guarismo tres mil y Quatroçientos y setenta y ocho.
-Otra media culebrina entorchada con una imagen de Santo Thomás con las armas imperiales tiene de quenta en letra guarismo tres mil e Quinientos e Sesenta y seis de suma.
-Otra media culebrina entorchada con las armas imperiales e una imagen del apóstol San Felipe tiene de quenta en guarismo tres mil e seiscientos e Quatro de suma.
-Otra media culebrina con una mira de hierro y una salamandra y una gegodica (sic) en el fogón sin quenta de los bullones adelante entorchada e de los bullones a la culata Redonda sin quenta de peso.
-Otra media culebrina Redonda que tiene de peso veinte e un quintales y una arroba y quinze libras.
-Un cañón Redondo que tiene de quenta veinte Quintales y tres arrobas y (fol. 20 v) diez e seis libras.
-Medio cañón pedrero Redondo tiene de quenta treze quintales y una arroba.
-Otro medio cañón pedrero tiene de quenta treze quintales e tres arrobas e seis libras.
-Un sacre ochavado de los bullones a la culata con una cabeça de león en la culata en la boca un caxcavel y tiene de suma en guarismo mil e seisçientas e dos en suma.
-Un falconete ochavado tiene de quenta siete quintales e tres arrobas e una libra.
-Otro falconete ochavado que tiene de quenta ocho quintales.
-Otro falconete ochavado que tiene de quenta ocho quintales e treze libras.
-Otro falconete Redondo con una letra en el fogón de esta manera G (?) sin quenta de peso.
-Otro falconete ochavado tiene de quenta ocho quintales e dos arrobas e e (sic) Ocho libras.
-Otro falconete ochavado con unos verdugos tiene de quenta Siete quintales una arroba y diez y ocho libras.
-Otro falconete ochavado de los bullones a la culata con las columnas de Hércules y va venera abaxo de ellas tiene de quenta Siete quintales y una arroba e veinte e una libras.

De las quales dichas cosas inventariadas de artillería e de todo lo demás contenido en el dicho ynventario el dicho señor alcaide se dio por contento y entregado a toda su voluntad e en Razón de ello Renunció qualesquier leies que sean en su favor e se obligó a su persona e bienes Raízes e muebles habidos e por haber de dar e entregar toda la dicha artillería e otras cosas contenidas (fol. 21 r) en el dicho inventario al dicho señor don Álvaro de Baçán o a quien su poder hubiere cada que le sea pedida e demandada o pagar su valor e sin pleito e sin contienda alguna e para lo ansí cumplir e mantener dio poder a las justicias de sus majestades para que lo constringan (sic) a ello por todo Rigor de justicia como por cosa pasada contra él por sentencia difinitiva dada e pasada en cosa juzgada e dada a entregar cerca de ello Renunçió todas e qualesquier leies de que se pudiese aiudar e aprovechar en general para que le non valan y en especial Renunció la ley e Regla del derecho que diz que general Renunciaçión de leies non vala. En testimonio de ello otorgó lo suso dicho ante mí el escribano público e testigos de iuso escritos en el dicho día mes e año susodichos testigos Juan de Figueroa e Pero Hernández Albani vecinos de esta dicha cibdad e Alonso de Balboa criado del dicho señor alcaide Gómez de Balboa estante en esta dicha cibdad e el otorgante lo firmó de su nombre en el Registro de esta carta pasó ante my Juan Frutuoso escribano público. Gómez de Balboa. E io Juan Frutuoso escribano de sus majestades en la su corte e en todos los sus Reinos e señoríos e su escribano público uno de los del número de esta dicha cibdad presente fui en uno con los dichos testigos al otorgamiento de esta carta e la fize escrebir según que ante mí pasó. E por ende fize aquí este mío signo a tal como este. En testimonio de verdad. Juan Frutuoso escribano público.

E así mesmo han de hazer el dicho maioradgo el dicho don Álvaro de Baçán y la dicha doña Ana su mujer de más y allende de la artillería y (fol. 21 v) muniçión contenida en la dicha escriptura que de suso va incorporada de Dos mil y quinientas y siete pelotas de hierro colado de culebrina y de cañón de las que están en el primer cargo de los alcades (sic) y de tres mil e quinientas pelotas de hierro colado de la galera quinq Remi y de trecientos y Setenta quintales de metal de cinco cañones gruesos de la bastarda que está todo en la dicha fortaleza y es del dicho señor don Álvaro y que ansí mesmo han de façer maioradgo el dicho don Álvaro de Baçán y doña Ana de Guzmán su mujer de la mitad de todos los bienes que de aquí adelante adquirieren y la otra mitad de los que adquirieren ha de quedar para que los dichos señores don Álvaro de Baçán y doña Ana de Guzmán puedan disponer de ellos como bienes propios y libres suios de todos los quales dichos bienes de suso declarados han de hazer maioradgo los dichos don Álvaro de Baçán y doña Ana de Guzmán su mujer en el dio don Álvaro de Baçán su hijo mayor varón legítimo para que case con la dicha señora doña Joana de Baçán su esposa e mujer que será plaziendo a Dios y en sus descendientes varones del dicho don Álvaro y de la dicha doña Juana de Baçán y a falta de ellos en las otras personas que los dichos don Álvaro e doña Ana de Guzmán quisieren llamar al dicho maioradgo [...]

(fol. 24 r)
Item que en lo que toca al artillería y munición de ella que los dichos señores don Álvaro e doña Ana de Guzmán han de meter en el dicho su maioradgo como arriba se dixo el dicho señor don Álvaro el padre la pueda vender en su vida si quisiese con que si la vendiere en su vida lo que por (fol. 24 v) por ella y por la dicha munición le dieren se compre de bienes Raízes para el dicho maioradgo y se meta en él con tanto que si el dicho don Álvaro el padre en su vida antes que venda la dicha artillería hiziere alguna jornada por la mar y en ella se perdiere parte de la dicha artillería que lo que así se perdiere de la dicha artillería no sea obligado el dicho señor don Álvaro a pagar ni a meter en el dicho maioradgo en lugar de lo que ansí se perdiere de la dicha artillería bienes algunos. Y que después de sus días el dicho señor don Álvaro el moço su hijo maior varón si el dicho señor don Álvaro su padre no hubiere vendido la dicha artillería y muniçión la pueda también vender con que el precio porque se vendiere y lo que por ella se diere se emplée en bienes Raízes que se subroguen en el dicho maioradgo en lugar de la dicha artillería y munición y que todo esto vaia así expresado y declarado dispuesto y mandado en el dicho maioradgo que ansía han de hazer los dichos señores don Álvaro e doña Ana de Guzmán en el dicho don Álvaro su hijo maior y en sus descendientes varones y en las otras personas que llamaren al dicho maioradgo como de suso dicho es.

Item que los maravedís porque ansí se vendiese la dicha artillería y munición de ella y así mesmo todas las otras quantías de maravedís que se cobraren de su magestad q debe al dicho señor don Álvaro el padre su magestad del dicho pleito del soto de Roma de Asquerosa y de Mocateaali (sic) del principal como de los fructos corridos y también de lo que debe su magestad (fol. 25 r) al dicho señor don Álvaro de lo de las galeras. Todas estas dichas quantías de maravedís no puedan entrar ni entren si al tiempo que se cobraren fuere muerto el dicho señor don Álvaro de Baçán el padre en poder del dicho señor don Álvaro de Baçán el moço ni de ningún suççessor del dicho maioradgo sino que se haian de depositar y depositen en un monasterio de la Orden de San Gerónimo en un arca que esté en el depósito del tal monasterio con dos llaves la una que tenga el prior del dicho monasterio y la otra el suççessor que a la sazón fuere del dicho maioradgo, para que se allí se paguen a los que vendieren bienes Raízes para meter y surogar en el dicho maioradgo con todas las fuerças e firmeças dél, y q de otra manera para otra cosa no se puedan sacar de la dicha arca ni del dicho depósito los dichos dineros.[...]

(fol. 29 r)
Ante mí Juan de Santistevan es(fol. 29 v)cribano público y del número de esta noble villa de Valladolid e la firmaron de sus nombres en este registro la dicha Señora condesa y el dicho señor don Álvaro el padre y el dicho señor don Álvaro el hijo y la dicha Señora doña Juana de Baçán siendo presentes por testigos a todo lo que dho es que vieron otorgar esta dicha escriptura a la dicha señora condesa de Miranda y a los dichos señores don Álvaro de Baçán el padre y don Álvaro de Baçán el hijo y a la dicha señora doña Juana de Baçán a los quales y a cada uno de ellos yo el escribano doy fee que conozco y que son los mismos que la otorgaron siendo para todo ello llamados y rogados Alonso Serrano e Diego de Villafelis e Francisco de Castro criados de los dichos señores conde e condesa de Miranda e el bachiller Juanes de Caraes vezino y estante en la dicha villa de Valladolid. Que fue fecha y otorgada en esta dicha villa de Valladolid en presencia de mí el dicho escribano e de los dichos testigos, a ocho días del mes de henero, año del nascimiento de nuestro Señor salvador Ihesuchristo de mill y quinientos y quarenta y nueve años [...]

(fol. 39 r)
E ansimismo meto e yncorporo en la dicha mejora e mayoradgo toda el artillería que yo tengo y ansí mismo la munición de ella y entiéndose que es la dicha artillería y munición de ella que yo así yncorporo e meto en el dicho mayoradgo e mejora toda la contenida en la dicha escriptura y memoria que de esto yo tengo hecha que pasó y se otorgó en la ciudad de Gibraltar en diez y ocho días del mes de otubre de mill e quinientos e treinta y siete años por ante Juan Frutuoso escribano público y del número de la dha ciudad de Gibraltar que es la que rescibió Gómez de Balboa alcayde de la dicha fortaleza de Gibraltar según que más largo en la dicha escritura se contiene el tenor de la qual de suso va incorporado en la dicha escriptura de capitulación y demás y allende de la dicha artillería y muni(fol. 39 v)ción contenida en la dicha escriptura que de suso como dicho es va incorporada meto e incorporo en esta dicha mejora e mayoradgo dos mil e quinientas e siete pelotas de hierro colado de culebrina y de cañón de las que están en el primer cargo de los alcaydes y de tres mil e quinientas pelotas de hierro colado de la mi galera quinque Remi y de trecientos y setenta quintales de metal de cinco cañones gruesos de la bastarda que está todo en la dicha fortaleza de Gibraltar y es mío propio, pero quiero y es mi voluntad que yo el dicho don Álvaro en my vida pueda vender si quisiere la dicha artillería y

munición de ella con que si la vendiere en mi vida lo que por ella y por la dicha munición me dieren yo sea obligado de empleallo en bienes Rayzes para esta dicha mejora o mayoradgo e para que los dichos bienes que así se compraren del preçio porque se vendiere la dicha artillería e munición de ella se subroguen en su lugar en esta dicha mejora e maioradgo e yo desde agora para en tal caso los he por subrogados e puestos en él con tanto que si yo el dicho don Álvaro en mi vida antes se venda la dicha artillería hiziere alguna jornada por la mar y en ella perdiere la dicha artillería, o parte de ella que lo que así se perdiere de la dicha artillería no sea obligado yo el dicho don Álvaro a paguar (sic) ny meter en esta dicha mejora e mayoradgo en lugar de lo que ansí se perdiere de la dicha artillería bienes algunos. E que si yo el dicho don Álvaro en mi vida no hubiere vendido la dicha artillería y munición que el dicho don Álvaro my hijo mayor ligítimo la pueda tan bien vender con que el preçio porque la vendiere y lo que por ella se le diere se emplée en bienes Raízes que se subrogue en esta dicha mejora y mayoradgo en lugar de la dicha artillería y munición y con que los maravedíes por que (fol. 40 r) así se vendiere la dicha artillería y munición de ella si al tiempo que se cobraren yo fuere muerto no puedan entrar ni entren en poder del dicho don Álvaro de Baçán mi hijo mayor ny de ningún successor de esta dicha mejora sino que se mande depositar […]

(fol. 63 v)
En firmeza de lo qual yo el dicho don Álvaro de Baçán por mí e en nombre de la dicha doña Ana de Guzmán mi mujer e por virtud del dicho poder que de ella tengo otorgué esta carta de mayorazgo e mejoría ante el escribano público e testigos de yuso escritos e lo firmé de mi nombre que fue fecha e otorgada en esta noble villa de Valladolid estando en ella la corte y consejo Real de sus majestades e asimismo la su chançillería e presidente e oidores en diez e seys días de el mes de abril de mill e quinientos e quarenta e nueve años testigos que fueron presentes para ello llamados e Rogados Alonso Serrano secretario de la ylistrísima señora condesa de Miranda e Diego de la Serna su maestresala e el bachiller Juanes de Garaiz e Diego de Villafeliz maestresala de la dicha señora condesa e Christóval de Carrión criado de mí el presente escribano vecinos y estantes en la dicha villa de Valladolid. Y el dicho ilustrísimo señor don Álvaro de Baçán otorgante de esta carta al qual doy fee que conozco e que es el mesmo que lo otorga lo firmó de su nombre en el registro de esta carta en presencia de los dichos testigos.

Doc. 2.
1584, November 15.
Incorporation of goods and increase of the "*mayorazgo*" of Álvaro de Bazán, Marquis of Santa Cruz.
Archivo Histórico de la Nobleza, Santa Cruz, C. 32, D. 1.

(fol. 65 r)
Conocida cosa es y aprobada por grandes exemplos y confirmada con la espiriencia (sic) los muchos grandes provechos que se siguen de las instituciones de los mayoradgos y grandes casas porque con ellas se consiguen tres cosas notables y señaladas la primera es del servicio de Dios nro señor [...] Por tanto yo don Álvaro de Baçán marqués de Santa Cruz señor de las villas del Viso y Valdepeñas, comendador mayor de León Capitán General del mar océano y reynos de Portugal, considerando todas estas cosas (fol. 65 v) y la grande nobleça claridad y antigüedad de mi linaje y de mis predecesores y que don Álvaro de Baçán mi hijo mayor ha de suçeder después de mis días en el mayoradgo antiguo que los dhos mis predecesores instituyeron y ordenaron digo q aviendo sido mis padres y abuelos capitanes generales del invictísimo emperador Carlo quinto máximo y de los reyes sus antecesores de gloriosa memoria y haviendo yo seguido la disciplina militar y procurando imitarles en sus obras y servido con mi persona de más de al dicho emperador a la magestad del rey don Philipe nuestro señor que hoy reyna en las Españas heredando a mis pasados en la prosecución y nombre de Capitán General, y porque habiéndome dado Nuestro Señor muchas vitorias en el discurso de mi vida y tiempo que he servido a los dichos emperador y rey mis señores y demás de haberme hecho Su Magestad mercedes por mis servicios principalmente de la encomienda mayor de León que vale diez mill ducados de renta cada año y del cargo de Capitán General del mar océano y reynos de Portugal con doze mil ducados de sueldo al año y a don Álvaro de Baçán mi hijo mayor de la encomienda de Solana y Alhambra que vale seis mill ducados de renta y a mis hijos don Francisco de Baçán y don Pedro de los hábitos de Calatrava y Alcántara con promesa de hazerles merced y demás de lo suso dicho he ganado muchas quantías de maravedís en las batallas que he vencido tierras e yslas que he conquistado y ganado, así en Berbería como en las yslas de los Açores y otras partes donde he hecho jornadas que más largamente se hallarán en otras partes escriptas y con lo en ellas he habido e fabricado el palacio y casa de mi villa del Viso y comprado la villa de Valdepeñas y otras muchas rentas que tengo por previllegios de Su Magestad y de particulares y considerando quanto justo y conviniente es hazer de los bienes que así Dios me ha dado y adqri(fol. 66 r)do aumento de mayoradgo en el dicho don Álvaro de

Baçán mi hijo mayor dando y adjudicando asimismo de ellos a los demás mis hijos e hijas las partes que han sido necesarias y comvinientes, yo y la ilustrísima marquesa doña María Manuel mi ligítima mujer ya difunta hicimos escritura de partición entre nuestros hijos e hijas y acrecentamiento de mayoradgo en la ciudad de Badajoz por ante Pero Vázquez año de mill y quinientos y ochenta y uno y su magestad confirmando la dicha escritura me dio facultad para hazer el dicho acrecentamiento y aumento de mayoradgo como más largamente por la dicha facultad parece que es la que se sigue.

DON PHILIPPE por la gracia de Dios rey de Castilla de León de Aragón [...] por quanto por parte de vos don Álvaro de Baçán marqués de Santa Cruz nos ha sido hecha relación que vos y la marquesa doña María Manuel vuestra mujer ya difunta pocos días antes que falleciese hizistes quenta y abanco de todos los bienes que vos y ella teníades así de los dotales de la dicha marqsa como del capital y gananciales vuestros y averiguada y hecha la dicha quenta y sacadas las deudas qe se habían contraído durante vuestro matrimonio restaron por bienes de ambos ciento y dieziséis mill y veinte y seis ducados de los quales y de toda la demás hazienda que adelante tuviésedes hasta el tiempo de vuestra fin y muerte hizis(fol. 66 v)tes donación y vínculo en don Álvaro de Bazán vuestro hijo mayor ligítimo e inmediato sucesor en vuestro estado y mayoradgo antiguo al qual acrecentastes y agregastes la dicha donación y nuevo vínculo con las cláusulas y llamamientos prohibiciones y sostituciones del dicho vuestro mayoradgo y vículo antiguo y mandastes a los demás vuestros hijos e hijas cierta cantidad de maravedís demás de las legítimas porciones q les podían pertenecer de los dichos bienes y especialmente mandastes a don Francisco de Bazán vuestro hijo segundo veinte mill ducados y se los señalastes en ciertos bienes entre los quales fueron los cortijos de Asquerosa y Daragedí que son en la vega de Granada con todas sus tierras riego y secano montes y alameda casas y molino y lo demás a ello annexo (sic) y perteneciente los quales dichos cortijos son bienes del mayoradgo antiguo de vos el dicho marqués según más largo se contiene y declara en la escritura que sobre ello vos y la dicha marquesa otorgastes en la ciudad de Badajoz a diez y siete días del mes de marҫo del año pasado de mill y quinientos y ochenta y uno ante Pero Vázquez escribano del número de ella y que después de hecha y otorgada la dicha escriptura tuvistes otra hija más y la dicha marqsa falleció y pasó de esta presente vida y por otra escriptura que vos otorgastes en la villa de Madrid a diez y ocho días del mes de mayo de este presente año de mill y quinientos y ochenta y quatro incorporastes en el dicho vuestro estado y mayoradgo la villa de Valdepeñas con todos sus vasallos términos y juridición cevil (sic) y criminal alta baxa mero misto imperio y con todas sus rentas pechos y derechos y todo lo demás a ellas annexo y perteneciente que comprastes de nos y ansímismo incorporastes en el dicho vuestro mayorazgo otros bienes y señalastes a la dicha vuestra hija última la parte que avía de aver por sus ligítimas como se (fol. 67 r) contiene más largo en la escriptura q sobrello otorgastes ante Pedro de Velasco nuestro escribano el tenor de las quales dichas escripturas es como se sigue

In dei nomine amen sepan los que esta escriptura vieren como nos don Álvaro de Bazán Capitán General de las galeras de su Mag[t] e doña María Manuel su mujer marqueses de Santa Cruz estantes en la ciudad de Badajoz e yo la dicha doña María Manuel pido licencia al dicho marqués mi señor y marido para otorgar esta escriptura el qual me la concedió en forma de que yo el escribano doy fee e usando de ella marido e mujer de un acuerdo y voluntad dezimos que por quanto yo el dicho marqués de ordinario ando en servicio de Su Magestad por mar como tal capitán general de Su Magestad de las dichas sus galeras y armadas trayendo mi vida a riesgo y peligro y ansí raramente nos juntamos marido e mujer y no sabemos quándo otra vez nos veremos juntos ni quándo Dios Nuestro Señor será servido de llevarnos de esta presente vida atento a esto e a que tenemos muchos hijos para su quietud y sosiego y el nuestro queremos dar orden y modo de vivir e gobierno e alimentos necesarios a los dichos nuestros hijos para que correspondan a sus antepasados sirvan a Dios Nuestro Señor y al invictíssimo nuestro señor rey don Phelipe segundo de este nombre e a sus sucesores como deben de los bienes y sustancia y patrimonio que Dios nuestro señor fue servido darnos y para que haya efeto lo sobre dicho hazemos donación gracia pura y perfeta e irrevocable que es dicha entre vivos valedera de la forma e con las condiciones e gravámenes infraescriptos a don Álvaro de Baçán nuestro hijo ligítimo y mayor e sucesor en nuestro estado casa y mayoradgo e a otro qualquiera de nros hijos legítimos q a falta del dicho don Álvaro de Baçán sucediere en el dicho nuestro estado y mayoradgo conforme a las cláusulas e firmezas de sus fundaciones […](fol. 67 v) para que en todo ello el dicho don Álvaro de Baçán nuestro hijo mayor suceda y lo tenga y posea por suyo e sus sucerores y los que después dél sucedieren en el dicho nuestro estado casa y mayoradgo por vía de mayoradgo sin que el susodicho y sus sucesores los puedan vender ni enagenar en manera alguna sino que pasen en ellos como tales vienes vinculados y de mayoradgo por vía de aumento y acrecentamiento de la dicha nuestra casa y estado y mayor perpetuación y con que a de ser obligado el dicho don Álvaro nuestro hijo mayor y el que a su falta hobiere de cumplir lo contenido en esta escriptura a dar y pagar y hazer de su parte todo lo infraescripto […]

Memoria de los bienes libres que tenemos (fol. 68 r) el marqués y marquesa de Santa Cruz hoy diez y siete de março de mil y quinientos y ochenta y un años […]

(fol. 68 v)
Memorial de las deudas que se deben según relación q tenemos oy diez y seis de março mill y quinientos y ochenta y un años […]

(fol. 80 r)
Y nos suplicastes que porque vos deseáis que se cumpla y guarde todo lo contenido en las dichas escripturas de suso incorporadas por ser en tanta utilidad y acrecentamiento del dicho vuestro estado y mayoradgo y sucesores en él fuésemos servido de las aprobar y confirmar particularizando que podáis sacar del dicho vuestro mayoradgo antiguo los dichos cortijos […] (fol. 80 v) confirmamos loamos y aprobamos las dichas escripturas de suso incorporadas en todo y por todo como en ellas se contiene y suplimos qualesquier defetos obstáculos impedimientos assí de hecho como de derecho […] para que valgan y sean firmes y valederas en todo y por todo como en ellas y en cada una de ellas se contiene para agora y para siempre jamás y mandamos que contra ellas no se vaya ni pase en tiempo alguno ni por alguna manera y damos licencia y facultad a vos el dicho marqués de Santa Cruz para que de nuevo podáis hazer la dicha incorporación y acrecentamiento en el dicho vuestro mayoradgo antiguo de los bienes contenidos en las dichas escripturas y para que también la podáis hazer de otros qualesquier bienes que al (fol. 81 r) presente tenéis y adelante tuviéredes con los vínculos y condiciones gravámenes llamamientos y sostituciones del dicho vuestro mayoradgo antiguo y para que atento al beneficio que le hazéis con la dicha incorporación y acrecentamiento podáis poner gravamen a los sucesores en el dicho vuestro mayoradgo antiguo e incorporación de que no se puedan casar sino con mujer noble hijadalgo y de limpia generación libre de toda mácula y raza […] y atenta la incorporación que habéis hecho en el dicho vuestro mayoradgo antiguo de la dicha villa de Valdepeñas y otros bienes y a la que también habéis de hazer de los demás que queréis incorporar os damos asimismo licencia y facultad para que podáis sacar y dismembrar (sic) de ellos dichos cortijos […](fol. 81 v) que nos de nuestra cierta ciencia y propio motu y poderío real absoluto de que en esta parte queremos usar y usamos lo aprobamos y damos por bueno firme y estable y valedero y desde agora lo habremos por puesto en esta nra carta como si de palabra a palabra aquí fuese inserto e incorporado y lo confirmamos loamos y aprobamos según y como y con las condiciones vínculos e firmezas cláusulas posturas derogaciones submisiones penas y restituciones que en los dichos maioradgos por vos hechos y otorgados

fueren y serán puestos y contenidos y suplimos todos y qualesquier defetos [...] (fol. 82 v) y encargamos al serenísimo príncipe don Philipe mi muy caro y muy amado hijo y mandamos a los infantes, perlados duques marqueses condes ricos hombres priores [...] que guarden y cumplan y hagan guardar y cumplir a vos el dicho marqués de Santa Cruz y a los dichos vuestros hijos y descendientes y personas que sucediesen en los mayoradgos e incorporación esta nuestra licencia y facultad poder y autoridad que nos os damos para hazerlos y todo lo que por virtud y conforme a ella hiziéredes y ordenáredes en todo según que en esta nuestra carta se contiene y que en ello ni en parte de ello impedimiento alguno os no pongan ni consientan poner y si vos o ellos quisiéredes o quisieren de esta nra carta y de lo que por virtud de ella hiziéredes e instituyéredes previllegio y confirmación mandamos a nuestros concertadores y escrivanos mayores de los previllegios [...] que os la den libren pasen y sellen la más fuerte y firme y bastante que les pidiéredes y menester hubiéredes dada en san lorenço el real a tres días del mes de julio de mill y quinientos y ochenta y quatro años va escripto entre renglones de, que, y multiplicado y que adquiriésemos, mis, hijas vezes vala (fol. 83 r) yo el Rey yo Juan Vázquez de Salazar secretario de su Católica Magestad la fize escrevir por su mandado registrada Jorge de Olaalde Vergara chanciller mayor Jorge de Olaalde Vergara e licenciado Juan Tomás el licenciado Chumacero de Sotomayor.

En cumplimiento de la qual dicha facultad y en virtud de ella y del poder a mí dado por la dicha marquesa por lo que a ella le toca y en aquella vía y forma que mejor de derecho lugar haya hago e instituyo en favor del dicho don Álvaro de Baçán mi hijo mayor el aumento de mejora y mayoradgo de los dichos mis bienes adqueridos y ganados en la guerra que son los siguientes

Primeramente incorporo e junto en el dicho mayoradgo dos arneses de mi persona labrados en Alemania y dorados con que he seguido la guerra y todas las demás armas artillería piezas y arcabuzes que yo tengo en la casa de mi villa del Viso y el estandarte real de damasco carmesí con las armas reales y la figura del apóstol Santiago que yo yo (sic) llevaba en la popa del galeón San Martín en que iba mi persona quando vencí el armada francesa sobre la isla de San Miguel día de Señora Santa Ana el año de mill y quinientos y ochenta e dos siendo Capitán General del armada y exército de Su Magestad con veinte y cinco naos que llevaba en la qual dicha armada francesa venía Philippi Stroci gran marichal de Francia por capitán general con sesenta e tres naos de alto borde y seis mil soldados sin la gente de mar y aventureros y por almirante de la dicha armada el conde de Brisaque.

Ansí mesmo junto e incorporo las armas y rodela fuerte del dicho Phelippe Stroci y el estandarte real que traya de seda e tafetán blanco que se le dio para la dicha armada el Rey Enrique tercero de Francia

Ansí mesmo junto e incorporo en el dicho mayoradgo las armas de Mosiur de Chatres hermano del duque de Giosa (sic) y cuñado del dicho Rey de Francia que era su capitán general de la infante(fol. 83 v)ría y exército francés y portugués que estaba en guarda de la isla Tercera y de las demás circunvezinas que llaman de los Açores en favor y ayuda de don Antonio Prior de Ocrato de la orden de San Juan que se intitulaba rey de Portugal a los quales venzí en el campo allanando las dichas yslas y reduciéndolas al servicio de Su Magestad.

Así mesmo junto e incorporo sesenta banderas de infantería de tafetanes de colores que trayan los dichos capitanes franceses y portugueses con el estandarte de la caballería que es de damasco blanco con la cruz de Christus.

Ansí mismo meto e incorporo dos venablos que el uno era del dicho capitán general monsiur de Chatres y el otro de mosiur de Garambaques maese de campo general del exército.

Ansí mismo meto e incorporo en el dicho mayoradgo el bastón de capitán general que tenía Manuel de Silva conde de Torres Bedras que por el dicho don Antonio era gobernador y capitán general de todas las yslas de la corona de Portugal.

Ansí mesmo meto e incorporo quatro fanales es a saber el del Rey de Francia que traya su nao capitana y el otro la capitana de Portugal quando vencí su armada en el río de Lisboa y el otro de Açam baxá nieto de Barbaroxa capitán general que fue de Sultán Solimán gran turco y fue este fanal el q le presentó la señoría de Venecia al dicho Barbaroxa el qual tomé en la galera de Mahamete Bay nieto de el dicho Barbaroja peleando con ella sobre la ysla de la Sapiença a vista de las dos armadas de la Liga y del turco. Y el otro de Acam Cheveri que yo gané en su galera el año antes quando vencimos la armada del turco sobre las yslas de Lepanto.

Así mesmo meto e incorporo en el dicho mayorazgo las caxas de atanbores y pífanos de la dha infantería ducientos mosquetes ducientos arcabuces y ducientas picas que yo escoxí en las armas que me rindieron después que los vencí en campo e me entregaron las armas que trayan.

Yten meto e incorporo en el dicho mayoradgo (fol. 84 r) una espada gineta antigua que era de mi abuelo guarnecida de esmaltes de oro con que sirvió a los reyes cathólicos en la guerra de Granada.

Y por quanto en el sitio de la casa del mayoradgo en la villa del Viso yo la he reparado reedificado y labrado y héchola una casa muy principal y no embargante que lo que se reedifica y labra en bienes de mayoradgo es visto ser en más acrecentamiento de él para más claridad declaro que meto e incorporo en

el dicho mayoradgo todo lo que he labrado mejorado y acrecentado en la dicha casa como más largo consta y parece por la escriptura que yo e la dicha marqsa mi mujer doña María Manuel otorgamos en la ciudad de Badajoz.
Ansí mesmo meto e incorporo la villa de Valdepeñas con su jurisdicción cevil (sic) y criminal alta baxa mero mixto imperio y con todo lo a ello annexo (sic) y perteneciente en qualquier manera e con todas sus rentas términos montes bosques abrevaderos fuentes corrientes y manantes (sic) y todo lo demás según y de la manera que me pertenece conforme a la venta que su magestad me hizo y a las averiguaciones que por su real comisión hizo Alonso de Rosales contador de su magestad.[…]
(fol. 84 v)
Ansí mesmo con facultad que yo en mis días pueda sacar del dicho mayoradgo qualesquiera destos bienes trocallos y cambiarlos y enaxenarlos darlos y donarlos y hazer de ellos como bienes nuestros libres fuera de mayoradgo con q después de mis días todos los juros e bienes de suso declarados o lo que de ellos hobiere quedado qden metidos e incorporados en el dicho mayoradgo con las mesmas cláusulas vínculos condiciones gravámenes restituciones que están los otros bienes del dicho mayoradgo antiguo y con las q de nuevo yrán puestas por mí el dicho marqués en esta escriptura. […]
(fol. 85 r)
Ansí mesmo meto e incorporo en el dicho mayoradgo dos mesones que tengo en la villa del Viso el uno llaman el mesón nuevo que hizieron los maestros italianos que vinieron de Génova a fabricar la dicha casa del Viso que está al cabo de la calle mayor de la dicha villa saliendo camino de Granada a la mano derecha y el otro mesón es el que era casa de Balbuena que linda con las casas que eran de Juan Rodríguez de Pisa. […]
Ansí mesmo junto e incorporo en este mayoradgo quatro fuentes de plata y quatro aguamaniles todo ello dorado por de dentro y por de fuera labrados de una labor antigua con las armas de Baçán las fuentes en medio esmaltadas que todo ello pesa ciento y diez y ocho marcos cinco onzas quatro ochavas.
(fol. 85 v)
Ansí mesmo meto e incorporo en el dicho mayoradgo doze paños de tapicería muy antiguos grandes y pequeños de la historia de Hércules.
Ansí mesmo junto e incorporo cinco paños muy antiguos de la historia de Alexandro Magno de seda y oro.
Así mesmo catorce paños de tapicería amarillos grandes y pequeños que eran de mi abuelo.
Yten junto e incorporo en el dicho mayoradgo ocho reposteros de tapicería finos hechos en Flandes los quatro con sola las armas de Baçán en medio y los otros

quatro con las armas de Baçán y Guzmán con trofeos e insignias de guerra por orla que las hizo mi padre por memoria de la armada francesa que venció en el cabo de Finisterra reyno de Galicia de que era general mosiur de Sunna
Así mesmo meto e incorporo en el dicho mayoradgo una cama pequeña de campo cielo tuncado de brocado turquesco rico que hizo de la ropa que hera de Allí Baxá general de la armada de el turco la qual se tomó en la batalla de Lepanto q vencimos siendo general de la Liga el señor don Juan de Austria y tiene las corredizas de damasco amarillo y carmesí y el cielo es de la dicha ropa con goteras de otro brocado turquesco y madera dorada.
Ansí mesmo meto e incorporo en el dicho mayoradgo otra cama de campo grande de terciopelo pardo labrado con goteras de tela de oro y unos escudos de tela de plata y oro con las armas de Baçán en el cielo cabezera e paño de atrás y corredizas de damasco pardo y la madera dorada.
Así mesmo meto e incorporo un dosel de brocado rico de tres altos con canefa (sic) de terciopelo morado bordado de oro e plata y en medio las Armas de Baçán.
Ansí mesmo meto e incorporo otro dosel de tela de oro enrizada amarilla con una bordadura sobre terciopelo carmesí de trofeos de guerra.
(fol. 86 r)
Ansí mesmo incorporo otro dosel de terciopelo verde con una canefa (sic) a la redonda de tela de oro frisada e las Armas de Baçán en medio.
Ansí mesmo meto e incorporo en el dicho mayoradgo otro sitial de terciopelo verde con çanefa (sic) de tela de oro y dos almohadas de tela de oro.
Yten dos mesas de piedra que están en la casa del Viso labradas en Roma de matices como jaspeado
Yten meto e incorporo en el dicho mayoradgo dos sillas a la castellana de sentar de brocado de tres altos antiguo y labradas de taracea.
Ansí mesmo meto e incorporo en el dicho mayoradgo seys sillas de asiento hechas en Nápoles las tres de tela de oro y otras tres de terciopelo carmesí con flocaduras de oro.

Todos los quales dichos bienes e cosas contenidas e declaradas en esta dicha escriptura yo el dicho don Álvaro de Baçán marqués de Santa Cruz meto e incorporo en el dicho mi mayoradgo antiguo y quiero y es mi voluntad que después de los días de mi vida para siempre jamás anden juntos unidos e incorporados en el dicho mi mayoradgo antiguo con los mismos llamamientos condiciones gravámenes vedamientos restituciones y según y de la manera que están los otros bienes del dicho mi mayoradgo antiguo y demás de las condiciones e gravámenes del dicho mi mayoradgo usando como uso de la dicha facultad real de suso incorporada teniendo consideración a que este dicho acrecentamiento que ansí hago en el dicho mayoradgo son a fin de hazer mayor mi estado así en rentas e vasallos como en calidad para que se conserve siendo Dios servido perpetuamente en limpieza e claridad de generación quiero y es mi voluntad que el que hubiere y heredare el dicho mi mayoradgo antiguo y bienes y cosas que ansí por este le acreciento que siempre ha de andar incorporado en el dicho mayoradgo se haya (fol. 86 v) de casar e case con mujer noble hijadalgo e de limpia generación libre de toda mácula e rraza de confeso, moro o judío e de tal calidad e suerte que sus hijos e decendientes (sic) sean capaces de tener qualquier de los tres hábitos de las órdenes militares de Santiago del espada Calatrava y Alcántara sin que para tenerlos sea necesario valerse de dispensación de Su santidad ni de otra gracia remedio ni refugio alguno de los reyes e administradores perpetuos, o maestres de las dichas órdenes sino que los hayan e alcancen por pura justicia y limpieza de sangre con pena que le pongo al heredero y sucesor del dicho mi mayoradgo que no cumpliendo en todo y por todo esta cláusula y gravamen e no casando con mujer noble hijadalgo de sangre e limpia generación libre de toda mácula e rraza de suerte que sus hijos sean capaces para tener qualquier de los tres hábitos militares como dicho es por el mismo caso desde luego le expelo privo y desheredo del dicho mayoradgo antiguo y acrecentamiento como a indigno y quiero y nombro al siguiente en grado por legítimo sucesor concurriendo en él las dichas calidades q dho tengo de suerte que sus hijos puedan ser capaces e gozar de los dichos hábitos o qualquiera de ellos y si el siguiente [...]

(fol. 87 v)
Otrosí mando y quiero usando de la dicha facultad real y leyes destos reynos que por la conservación e perpetuación del dicho mayoradgo antiguo y de este dicho acrecentamiento e bienes de él quiero y es mi voluntad que ninguna persona q en él sucediere y tuviere derecho de suceder e ser llamado para lo poseer o que lo posea no pueda enagenar ni vender ni trocar ni dividir ni empeñar ni hipotecar ni intentar enagenar ni dividir ni hipotecar los bienes del dicho mayoradgo y de este dicho acrecentamiento en todo ni en parte ni cosa alguna de ellos por vía de venta ni donación ni permutación ni por otra especie de enagenación aunque sea con notoria e grande mejoría ni por transación ni pacto ni prenda ni hipoteca causa ni razón alguna ni por otra manera de enagenación entre vivos ni en la muerte ni por título honoroso (sic) lucrativo ni necesario ni por causa de dote ni arras ni por obra pía ni en otra manera ni por alimentos de ninguna persona por propinqua que sea del poseedor aunque el derecho naural (sic) e cevilla obligue ni para redención de cautivos (fol. 88 r) aunque sea el mesmo poseedor el que se hobiere de redimir ni por otra causa ni razón alguna ygual ni mayor de estas aunque para ello intervenga y haya licencia e facultad real y aunque por

qualquiera de los dichos títulos o sin ellos hayan poseydo los dichos bienes por luengo o longuísimo tiempo que baste de derecho para prescribir qualesquier bienes y aunque hayan seydo poseídos por tanto tiempo de cuyo principio no haya memoria de hombres [...] e quiero e mando que qualquiera que enagenare qualquiera de los bienes contenidos en el dicho mi mayoradgo antiguo y en este dicho acrecentamiento de él o parte alguna de ellos de hecho o de derecho por las vías e maneras suso dichas o por qualquiera de ellas o por otra qualquiera pensada o no pensada que el q la hiziere o intentare de hazer pierda y haya perdido el dicho mi mayoradgo antiguo y esta dicha mejora y acrecentamiento de él y todo el derecho q tuviere a ello y en ello así a la propiedad como a la posesión y uso fruto y desde agora le privo de él e quiero y mando que venga e suceda en el dicho mayoradgo e acrecentamiento el siguiente en grado [...](fol. 88 v) porque mi intención e voluntad es que todos los dichos bienes se conserven juntos para siempre jamás en este dicho mayoradgo e no se dividan disminuyan aparten antes anden siempre unidos e incorporados e lo mismo quiero y mando que se guarde y cumpla si el poseedor del dicho mi mayoradgo antiguo e este dicho acrecentamiento o llamado a él pidiere o ganare facultad real para hazer qualquiera enagenación de los bienes en el dicho mayoradgo antiguo y en este dicho acrecentamiento contenidos e qualquier parte de ellos [...]

(fol. 89 v)
Lo qual todo que dicho es y lo contenido en esta dicha escriptura de acrecentamiento dispongo ordeno quiero y mando que se guarde y cumpla a la letra según que en esta dicha escriptura se contiene para que para siempre jamás haya cumplido efeto lo susodicho y pido y suplico a su magestad y a qualesquier sus justicias manden executar y executen las penas en esta dicha escriptura contenidas en los que contra ello fueren o pasaren en qualquier manera en firmeza de lo qual otorgué esta escriptura de acrecentamiento y mayoradgo ante el escribano público y testigos de yuso escriptos que fue hecha y otorgada en la villa de Madrid estando en ella la corte y consejo real de su magestad a quinze días del mes de noviembre de mill y quinientos y ochenta y quatro años testigos que fueron presentes a lo q dicho es e vieron otorgar esta escriptura al dicho señor marqués de Santa Cruz y firmar su nombre en este registro al qual yo el presente escribano doy fee que conozco el capitán Ynacio (sic) de Alderete y Christóval de Sotomayor e Gonçalo de la Concha e Alonso de Herrera estantes en esta corte don Álvaro de Baçán pasó ante mí Pedro de Velasco.

Doc. 3.
1748, 1819, 1839.
Inventories of Marquis of Santa Cruz armoury.
Archivo Histórico de la Nobleza, Toledo, Santa Cruz, C. 35, D. 9–16.

(fol. 1 r)
Sta Cruz
Nota de las armaduras existentes en el Palacio de la Villa de El Viso, remitida en 1819 por Miguel del Campo Alcayde del referido Palacio.

(fol. 2 r)
Sta Cruz
Razón Genealógica de la Ill^{e} Familia Bazán sacada de las inscripciones del Palacio del Viso, y una Nota de las armaduras existentes en el mismo.

[D. 9]
(fol. 3 r)
D^{n}. Fernando Casero y Bravo en carta de 10 de Mayo de 1839 dio aviso a S. E. de que con el carretero Fran^{co} Guzmán, vecino de la villa de Santa Cruz de Mudela remitía los efectos siguientes:
-Seis cascos de morrión
-Veinte piezas sueltas que parece correspondieron a morriones
-Cuatro espaldares
-Medio peto
-Una pieza que parece perteneció a otro peto
-Cuarenta y cuatro piezas sueltas de brazaletes y rodilleras
-Tres adargas muy destruidas, o rodelas de cuero

Los efectos espresados anteriormente los entregó Francisco Guzmán en 14 de Mayo de 1839.

[D. 10]
(fol. 1 r)
El Adm^{or}. Central del Estado Marquesado de Santa Cruz, D^{n}. Fernando Casero y Bravo, en carta núm. 38 dice, que con el arriero de Valdepeñas Francisco Morales, remite a S. E. el fanal de la Galera Capitana del Sr. D^{n}. Álvaro Bazán, que se custodiaba sobre la puerta del Oratorio del palacio de la Villa del Viso, y además en tres serillas, nueve petos y espaldares, siete morriones, una rodela de

hierro y once piezas sueltas de brazaletes, que se hallaban en el referido palacio del Viso, y posteriormente fueron trasladados a la Villa de Santa Cruz de Mudela En 8 de Mayo de 1839 entregó Fran^{co} Morales el cajón que contenía el fanal de que queda hecho mérito anteriorm^{te}; una rodela de hierro, y los tres serijos en los cuales estaban colocadas las piezas siguientes:
4 Petos
4 Espaldares
7 Morriones
11 Piezas sueltas de brazaletes

[D. 11]
(fol. 1 r)
Ynventario de los Trofeos de Guerra, bienes y algunos despojos que hay en el Palacio del Exmo. Sr. Marqués de Santa Cruz en la Villa del Viso
(fol. 2 r)
Ynventario de los Tropheos de Guerra, y bienes, y algunos Despojos que hay en el Palazio del Exmo. Señor Marqués de Santa Cruz y de Bayona Conde de Monte Santo &^{a} mi señor en la Villa del Viso son los siguientes

Armería
-Primeram^{te} dos cañones pedreros de bronze de dos varas de largo
-Otro cañón pedrero de lo mismo más pequeño con su cureña; está con una rueda
-Dos bodoques de hierro
-Treze vestidos de hierro con quinze morriones
-Otro vestido de hierro grande con su morrión que dizen traía Su Exª
-Dos casquetes de hierro
-Dos rodelas de hierro grandes forradas en badana
-Zinquenta pedazos de hombros, y rodilleras de hierro
-Veinte adargas de vaqueta de Moscobia
-Seis adargas (de moros) de madera forradas en cuero
(fol. 2 v)
Bienes
-Un barrón de hierro pequeño con una veleta de cobre de una de las torres
-Otro barrón de hierro grande de otra thorre con su cruz de lo mismo, bola, y veleta de cobre
-Otros dos barrones como el antecedente el uno con su crus, bola, y veleta de cobre
-Otro barronzico de hierro de una vara de largo
-Una rexa de hierro muy grande de uno de los quartos bajos

-Una zerradura de hierro grande
-Siete pestillos de hierro
-Tres bolas de cobre grandes en seis mitades de las torres
-Diez y siete pedazos de canelones de hoja de lata
-Seis armarios de ofizina
-Tres puertas de madera viejas
-Otra puerta con su bastidor
-Otras dos puertas grandes de la cochera
-Seis puertas ventanas
-Dos umbrales de toza de la cochera
-Seis pedazos de toza
-Una quartiza
-La caja de una cureña de un tiro grande
(fol. 3 r)
-Dos faroles de hoja de lata grandes sin vidrieras
-Onze cajones p^{ra} papeles del archivo con sus pies
-Dos bastidores de país
-Un cofre grande con barras de hierro
-Ocho baraustres, y ocho varas de piedra jaspe blanca que eran del brocal del aljibe
-Quatro bolas de piedra jaspe grandes para remates de edificio
-Otra bola de lo mismo más pequeña
-Catorze ladrillejos de piedra Jaspe blanca quadrados
-Veinte y tres baldosas negras ochavadas
-Una porción de pizarras, y ladrillos de fragmentos de las torres
-Ocho fanales, de los q se traían antiguam^{te} en las Galeras, que están alrededor de los corredores
Orat^{o}
-Un lienzo pintura del Deszendim^{to} de la Cruz puesto en bastidor embutido en la pares de más de tres varas de alto, y dos de ancho que haze altar ma^{or} en el Oratorio
-Un frontal de droguete
-Una sabanilla
-Un velo de tafetán encarnado
-Tres lienzos en bastidor que están enfrente del Altar Mayor. En el uno están pintados el Padre Santo, D^{n}. Juan de Austria, y un Cardenal (fol. 3 v) y en los otros dos están pintados seis Grales (sic)
-Dos payses grandes, y dos pequeños en que está pintada la Batalla de Naval (sic)

Cuios bienes están entregados a Pedro Fran^{co} del Campo Alcayde de dho Palazio. Viso y Septiembre 24 de 1748

Ygnazio Uzquiano y Murga
(signed)

[D. 12]
(fol. 1 r)
Nota de las Armaduras que existían en el Palacio de la Villa del Viso en 24 de Setiembre de 1748, de las que quedaron en el año de 1819, y de las que se encontraron en Octubre de 1838, según las relaciones remitidas por D^{n} Ygnacio Uzquiano y Murga, Miguel del Campo, y D^{n} Juan Pedro de Córdova.

Existían en 24 de Septiembre de 1748	Quedaron en 1819	Faltaron desde 1748 hasta 1819	Quedaron en Octubre de 1838	Faltaron desde 1819 hasta 1838	Faltaron desde 1748 hasta 1838
2 Cañones pedreros de bronce de dos varas de largo		2 Cañones pedreros de bronce de dos varas de largo			2 Cañones pedreros de bronce de dos varas de largo
1 Cañón pedrero de lo mismo más pequeño en su cureña, está con una rueda		1 Cañón pedrero de lo mismo más pequeño en su cureña, está con su rueda			1 cañón pedrero de lo mismo más pequeño en su cureña, esta con una rueda
2 Bodoques de hierro 13 Vestidos de id 15 Morriones	8 petos y 8 espaldares 15 Morriones	2 Bodoques de hierro 05 Vestidos de id	5 espaldares y ocho petos 12 Morriones incompletos	Algunas piezas para completar los otros 3 espald^{s} 3 Morriones incompletos	2 Bodoques de hierro 5 vestidos de hierro, y algunas piezas p^{a} el completo de 3 espaldares 3 Morriones uncompletos

1 Vestido de hierro grande con su morrión, que se decía ser el que gastaba S. E.	El peto y morrión de D[n] Álvaro de Bazán	Todo el resto del vestido de D[n] Álvaro de Bazán, excepto el peto		El peto y morrión de D[n] Álvaro de Bazán	El vestido de hierro y morrión de D[n] Álvaro de Bazán
2 Casquetes de hierro	Dos sombreros, uno con un balazo de mosquete		Dos sombreros, uno con un balazo de mosquete		Dos sombreros, uno con un balazo
2 Rodelas de hierro grandes, forradas en badana	1 rodela de hierro	1 Rodela de hierro	1 Rodela de hierro		1 rodela de hierro
50 pedazos de hombros y rodilleras de hierro	Varias piezas de brazaletes	Se ignora si faltarían algunas piezas de los 50 pedazos de hombros &	65 piezas sueltas de cascos, petos, espaldares y brazaletes		
20 Adargas de vaqueta de Moscobia	10 Rodelas de cuero, con la cifra de José esclavo de María	10 Adargas de bqueta de Moscobia	3 Rodelas de cuero	7 Rodelas de cuero	17 Rodelas de vaqueta de Moscobia
6 id (de Moros) de madera, forradas en cuero		6 id (de Moros) de madera forradas en cuero			6 id (de Moros) de madera forradas en cuero

La falta de piezas que se observó en el año de 1819 se atribuye a estravío, con motivo de la guerra de la independencia.

La falta que apareció en 1838, pudo dimanar de haber entrado en aquel año los facciosos en la Villa del Viso

[D. 13]
(fol. 1 r)
Noticia de las Armaduras que existían en el Palacio del Viso y han sido conducidas de orden de S. E. a Santa Cruz

-Una rodela de hierro
-Tres id de cuero
-Doce morriones incompletos
-Ocho petos
-Cinco espaldares
-Sesenta y cinco piezas sueltas de cascos, petos, espaldares y brazaletes

(fol. 1 v)
Esta nota la entregó D^{n} Pedro María Laguna, en la Contaduría del Exmo. Sr. Marqués de Santa Cruz, en 15 de Octubre de 1838

[D. 14.]
(fol. 1 r)
Razón q^{e} yo Miguel del Campo Alcaide del Palacio del Exmo Sr. Marq^{s} de Sta Cruz (mi Sr.) doy de las armas q^{e} existen en dho a saber

-Dos sombreros el uno con un balazo de mosquete recibido en la acción de la toma de Lisboa
-Del vestido de D^{n} Álvaro de Bazán solo ha quedado peto y morrión sin espaldar, y una rodela de hierro de dos q^{e} había
-Quince Morriones descompuestos con varias piezas de brazaletes
-Ocho espaldares
-Ocho petos
-Diez rodelas de cuero con la cifra de Joseph esclavo leal de Manuela

Miguel del Campo (rub.)

(fol. 1 v)
Año de 1819
Nota de las Armaduras q^{e} existen en el Palacio del Viso

[D. 15]
(fol. 1 r)
Esta Nota fue entregada por D^{n} Pedro María Laguna, en la Contaduría del Exmo Sr. Marqués de Santa Cruz, el día 15 de Octubre de 1838.

(fol. 2 r)
Nota de las Ynscripciones que hay en el Palacio del Viso, en los sitios que ocupaban seis fanales que de orden de S. E. se han trasladado a Santa Cruz=

Fanal de D^{n}. Álvaro Bazán
Sobre la puerta del Oratorio
Del Marqués de Santa Cruz. Los capitanes antiguamente acostumbraban cuando dejaban de navegar, poner los fanales que habían traído en las Galeras Capitanas donde sus personas andaban, en Yglesias o partes decentes, y siguiendo esta costumbre se puso el que traía el primer Marqués de Santa Cruz en esta capilla, acompañándole de los otros Fanales de las Armadas que venció, que están puestos en los corredores de este Palacio.

Fanal de Hacán
Sobre la primera puerta de la derecha
Demetrio victorioso en su Lámina, no ganó tan honroso nombre contra Ptolomeo, como en Lepanto el primer Marqués de Santa Cruz rindiendo a Hacán Chirivi, que traía este Fanal en su Capitana como se verá en la Historia de la Batalla Naval en este Palacio.

Fanal de Messina
Este se hallará sobre la puerta figurada que sigue a la anterior, sin otra inscripción que la de, Mesina.

Fanal de Navario
Sobre la puerta *figurada a la derecha del Salón* grande del Balcón con honrosa gloria fue ganado este Fanal del primer Marqués de Santa Cruz, e la Galera que rindió a Mahamet Vey, hijo de Hacán Vajá Rey de Argel, como se verá en su Historia en este Palacio.

(fol. 2 v)
Fanal de Francia
Sobre la puerta también figurada de la izquierda del Salón

No triunfó con más gloria Luctancio contra Carthago, que la que consiguió el primer Marqués de Santa Cruz ganando este Fanal a la Armada de Francia, a favor de D^{n} Antonio, como se verá en su Historia en este Palacio.

Fanal de Portugal
Sobre la primera Puerta de la izquierda, subida la escalera de la derecha
En el Comicio, lugar en Roma dedicado a las Victoriosas Estatuas, merecía estar este Fanal conquistado por el primer Marqués de Santa Cruz, libertando al Reyno Lusitano de la Tiranía de D^{n}. Antonio, como se verá en su Historia en este Palacio.

[D. 16]
(fol. 1 r)
Destrozos ocasionados por los facciosos de Oregita en este Palacio el 17 de Abril de 1838.
-Destruidas dos de las cuatro piedras labradas que contenía el Brocal del Argibe pequeño.
-Uno de los caños roto por la mitad, y los 3 restantes bollados por varias partes.
-La hoja derecha de la ventana del segundo descanso de la escalera a la derecha, arrancada y hecha mil pedazos.
-La puerta del cuarto llamado de las Monjas, tiene destrozado un tablero, y a la ventana de la derecha le quitaron todos los quadrillos o tableros.
-La puerta del Salón de los Linajes, q^{e} sale a los corredores solo le ha quedado la armadura.
-La seg^{da} ventana de dho salón, tiene dos cuadrillos quitados.
-Las cerrajas de ambas puertas de dho Salón, así como la del cuarto que le sigue llamado el de Proserpina, fueron robadas: A la ventana de este le faltan dos tableros o cuadritos.
-La puerta q^{e} hay en dho cuarto, saliendo p^{a} el salón grande, se halla destrozada casi en su totalidad, con pérdida de la cerradura.
-A las dos ventanas del Cuarto de la Torre, q^{e} hace esquina contigua a la Casa Cochera les han roto y destruido 11 tableros.
-Las dos puertas del Cuarto Despacho, conocido p^{r} el de D^{n} Pedro María, solo contienen las cruces, o armaduras, por haberles destrozado los 8 tablones de que ambas se componían.
-A la puerta de la cocina q^{e} usa D^{n}. Pedro María (fol. 1 v) le falta un cuadrito.
-La puerta del Oratorio tiene hecho pedazos la mitad del tablero grande, con pérdida de la cerradura.
-En el piso bajo, ni en el de los Camarachones, nada ha sufrido este edificio.

-Las armaduras q[e] se conservan en el Oratorio, han sido rotas y extraviadas. De estas he recogido algunas, así como 5 cerraduras: Para el paradero de las restantes estoy haciendo averiguaciones.

Viso 23 de Abril de 1838

Córdova
(signed)

VALLE DE CERRATO, MARQUESES DEL

Doc. 1.
1616, January, 6.
Inventory of Juan de Acuña, Marquis of Valle de Cerrato.
Archivo Histórico de Protocolos de Madrid, Protocolo 2661.

(fol. 1 r)
Ynventario de los bienes que quedaron y fincaron del Iltr^{mo} Sor Don Juan de Acuña marqués del Valle pres^{te} que fue de Castilla
Va en diez y nueve legajos y una q^{ta} con el Sor Suero de Vega

(fol. 687 r)
Pinturas
En la villa de Madrid a seis días del mes de enero de mill y sei^{o}s y diez y seis a^{o}s ante los Ss^{es} Licen^{do} Ju^{o} de Aguilera del q^{o} de su mag^{d} alcalde de su casa y Corte y don Pedro de Guzmán S^{r} de la villa de Valtierra com^{or} de Sagra y Cenete de la Orden de Sant^{o} gentilhom^{e} de la boca de su mag^{d} y su correg^{or} de Gran^{a} de [...]

(fol. 832 r)
Arcabuçes
Y para los arcabuzes se llamó a Simón Marquarte arcabucero de Su Mag^{d} quien debajo de juramento q hiço en quatro de março de mill y seiscientos y diez y seis lo tasó y se ynventarió en la manera sig^{te}
Escopeta
-Primeram^{te} una escopeta larga con su llave de rueda y caja de siete palmos de largo poco más o menos con su punto dorado que se tasó en diez seys ducados_ _U176
Otra
-más otra escopeta larga menor que la dha con su llave de rueda de seis palmos de largo con su baqueta y marca del año de myll y qui^{o}s y ochenta que se tasó en doçe ducados__U132
Otra
-más otra escopeta de cinco palmos de largo mitad ochavada y mitad redonda con un cañón punto y mira sin llave en su caja con marca del año de myll y qui^{o}s y cinquenta que se tasó en seys ducados__U066

Arcabuz
-más otro arcabuz de cinco palmos (fol. 832 v) de largo con su llave de rueda y caja tiene por marca una pynnao (?) y tiene un pedaço ynjerido (sic) a la boca de una quarta de largo con que se alarga y acorta el dho arcabuz con su tornillo y medio cañón en la mira que se tasó en siete ducados__U077
Arcabuz
-más otro arcabuz de quatro palmos y medio de largo con un letrero de Gaspar de Taloçón (?) [...] su llave de rueda en su caja que se tasó en ocho ducados__ U088
Otro
-más otro arcabuz de quatro palmos y medio del toledano con su llave de rueda y caja de nogal y vaqueta que se tasó en cien Rs__U100
Otro
-más otro arcabuz de quatro palmos de largo poco más con las armas de Acuñas doradas y con tres hoçes por marca con su llave de rueda y caja de vaqueta que se tasó en doçe ducados__U132
Otro
-más otro arcabuz de quatro palmos de largo redondo todo de Al° Rodríguez con su llave de rueda y caja que se tasó en siete ducados__U077
(fol. 833 r)
Otro
-más otro arcabuz pequeño de tres palmos y medio mitad ochavado y mitad redondo con su llave de rueda y caja y vaqueta y un cañón y medio por mira que se tasó en quatro ducados__U044
Pistolete
-más un pistolete reventado con su caja de hierro todo grabado sin llave con su funda de vaqueta que se tasó en quatro Rs__UC04
Frasco
-más un frasco de cuerno para pólvora guarnecido de hierro el brocal su nede medida [...] y un frasquillo menor que el dho guarnecido de la misma manera que se tasaron en ocho reales entrambos__U008
Frasco
-más otro frasco de hierro grabado con un armador dorado el frasco es todo grabado que se tasó en ocho Rs con su llave__U008
Frasco
-más otro frasco de cuerno negro antiguo a modo de concha es verde que se tasó en un real__U001

Otro
-más otro frasquillo de hierro (fol. 833 v) polvorín y armador que se tasó en ocho Rs__U008
Frascos
-más otros dos frascos de cuerno de cuerno (sic) largos que se tasaron en quatro reales__U004
Bolsa
-más una bolsa de cordobán verde para muniçión de arcabuz con tres cargas de hoja de lata y un sacatrapos lavador sacapelotas y botador (sic) que se tasó en seis reales__U006
Bolsa
-más otra bolsilla de cuero bayo con dos cargas de box y dos sacatrapos y un tornillo para armar y desarmar la llave del arcabuz que se tasó en doce Rs-_U012
Bolsilla
-más una bolsilla vieja antigua no se tasó q no vale nada
Trayllas
-más dos trayllas de cuero berberisco verdes para los perros de caça se tasaron en seis Rs__U006
(fol. 834 r)
Caracol
-más un caracol grande guarnecido con dos puntas de plata para bramar los venados que se tasó en doce Rs__U012
Cañón pa bramar
-Un cañón negro para bramar de cuero que se tasó e dos Rs__U002
Puntas de cuerno
-más tres pares de puntas de cuernos del croço (sic) se tasaron en tres reales__U003
Llaves
-más tres llaves de arcabuz todas de ruedas grandes que se tasaron en tres ducados__U033
Armadores
-más dos armadores el uno pavonado con una vuelta y su cordón de seda parda y oro y el otro de color de hierro que se tasaron en quatro Rs__U004
Estribos
-Unos estribos de monte grandes de madera verdes para el monte con sus açiones cerrados por delante que se tasaron en doçe reales__U012
Bolsilla de cordobán
-Más una bolsilla de cordobán con su brocal de box para llevar pólvora en dos Rs__U002

Fundas
-más ocho fundas de vaqueta para las escopetas en quatro du^{o}s__U044
(fol. 834 v)
Turquesas
-Más una turquesa de perdigones grande de hierro que se tasó en doçe reales_ _U012
Turqsa
-más otra turquesa de hierro pequeña para pelotas y postas que se tasó en tres Rs__U003
Turqsa
-más otra turquesa de latón de pieças para pelotas postas (sic) que se tasó en tres Rs__U003

Los quales dhos arcabuçes y escopetas que de suso van declaradas suman y montan myll y noventa y un Rs y el dho apreçiador debajo de dho juram^{to} declaró haberlo apresçiado por su justo presçio y valor y lo firmó de su n^{e} t^{o}s Domingo Muñoz y Gr^{rio} R^{o}z estantes en M^{d} pre^{te} el dho don Fran^{co} Ruyz vot^{ce} lo tasó en los pre[…] para armar y desarmas las llaves

Simón Marquart
Ante mí Agustín Galván
(signed)

(fol. 835 r)
Todos estos arcabuçes m^{dó} el S^{or} marqués por su testamento al S^{or} don Diego marqués seg^{do} del Valle su hijo y así se le quedan

Galván
(signed)

(fol. 836 r)
Ynvent^{o} de las ballestas y cuchillos
[…]andolos el S^{or} marqués su hijo
Y para los cuchillos se llamó para apreciarlos a Juan Grande cuchillero y la de las ballestas fueron tasadas por Juan de Azcotia (sic) ballestero de Su Mag^{d} en quatro de março del dho año debajo de juramento que hiço en la manera siguiente

Ballesta
-Primeramente una ballesta de virotes de mano de P^{o} de la Fuente de tablero de serbal y la verga de Juan Blanco con su funda blanca y el cavo de la llave con pelo

con unas gafas ~~quadradas~~ [crossed out] redondas y çapatillas verdes que se tasó en diez ducados- es de enjerto __U110
Otra
-más otra ballesta de virotes el tablero de serbal de Pedro de la Fuente con su verga de Juan Blanco con su funda blanca y una gafa de Pedro de la Fuente (fol. 836 v) con çapatilla verde y es de ynjerto que se tasó en doçe ducados__U132
Otra
-más otra ballesta de virotes el tablero de ençina negra de Pº de la Fuente y la verga de Puebla y el rótulo de don Juº de Acuña con funda de vaqueta negra con una gafa blanca que tiene un escudo de armas y çapatilla verde que se tasó en doçe ducados__U132
Otra
-más otra ballesta de virotes con tablero de serbal de Ortega con tres canales para virote y jaras y dos nueçes para las jaras y con verga de Luis Moreno con su funda de vaqueta negra con su gafa blanca redonda con çapatilla verde en su funda que se tasó en doce ducados__U132
Otra
-más otra ballesta de virotes con tablero de serbal de mano de Juan Criado y verga de Pedro de la Fuente con su funda de (fol. 837) vaqueta negra tiene un ynjerto postiço con su faga (sic) blanca redonda y çapatilla baya que se tasó en seys ducados__U066
Otra
-más otra ballesta de virotes con tablero de serbal sin funda que tiene verga de Luis Moreno y letrero de don Juº de Acuña con una gafa negra redonda y çapatilla negra digo verde que se tasó en seys ducados__U066
Otra
-más otra ballesta de virotes de mano de Hernández tablero de ~~serbal~~ ençina sin funda con verga que diçe Santa María con su gafa negra redonda çapatilla negra que se tasó en cinco ducados__U055
Otra
-más otra ballesta de virotes de Pº de la Fuente muerta tablero de serval y verga de Juan Blanco y rótulo que dize ynternatus con su funda de vaqueta no tiene gafas porque con las gafas arriba dhas se armaba esta que se tasó en seys ducados__U066
Tablero
-más un tablero de serbal (fol. 837 v) y una verga con chapas de latón y nuez y llave que se tasó en diez y seys Rs__U016

Verga
-más una verga suelta de Puebla con rótulo de don Juº de Acuña que se tasó en diez y seys Rs- es de virotes__U016
Otra
-otra verga suelta de virotes de Juan Blanco con un rótulo que diçe façile non es atinjere médium que se tasó en dos ducados__U022
Palos
-más cinco palos de serbal para hazer ballestas que se tasaron a seys reales cada uno q m^{ta} treinta Rs__U030
Ballesta
-otra ballesta de bodoques con una verga viscayna y tablero de serbal con una gafa quadrada con çapatilla baya que se tasó en tres ducados__U033
Otra
-más otra ballesta de bodoques tablero de serbal que dize Jesuxpo la verga vizcaína con gafa redonda negra que se tasó en quatro ducados- tiene la gafa funda__U044
Otra
-más una ballesta de bodoques arqueada de las de Florencia (fol. 838 r) con su mira querda verga puntería y llave que se tasó en quarenta y quatro reales__ U044
Virotes
-más quarenta virotes que se tasaron a real y quartillo cada virote que montan cinquenta reales__U050
Carcaje
-más un carcaje de cordobán verde en que están los dhos virotes con su correón que se tasó en diez y ocho reales__U018
Jaras
-más çinquenta y dos jaras blancas con sus casquillos todas emplumadas a tres cuartillos cada una monta treinta y nueve reales__U039
Tragaçetes
-más cinco tragaçetes a dos reales cada uno diez Rs__U010
Carcaje
-más un carcaje pequeño viejo de cordobán verde que se tasó en seys reales__ U006
Otro
-más otro carcaje blanco de cuero de venado que se tasó en seys reales__U006

(fol. 838 v)
Aljabilla
-más otro carcaje digo una aljabilla para llevar jaras que se tasó en seys Rs y es con su cincha sin hierros__U006
Bolsa
-una bolsa de cuero blanco de venado para cuerdas y recado de caça que se tasó en dos reales__U002
Çinto
-más un çinto de cuero blanco que se tasó en dos Rs__U002
Empulguera
-más una enpulguera de cuero bayo se tasó en dos Rs__U002
Bolsa
-más una bolsa de cuero bayo para llevar bodoques en quatro reales__U004
Casquillos
-más ochenta y seys casquillos diferentes para jaras y saetas todos de hierro que se tasaron en doze reales__U012
(fol. 839 r)
Cuchillos
Ynvent° de los cuchillos de monte y mesa que se tasaron por Ju° Grande cuchillero en quatro de marzo de myll y sei°s y diez y seis a°s debajo de juram^{to} que hiço en forma
Puñal
-primeramente un puñal con una marca en forma de cetro y su cabo puño de piedra de leche con su cuchillo chico con medio cabo de cuerno negro y su vayna de cuero negro que se tasó en ses^{ta} y quatro Rs el puño en cinquenta y lo demás en catorze Rs__U064
Cuchillo
-más otro cuchillo de monte mayor que el dho con una flor de lis y a la dha p^{te} una cruz ençima con otro cuchillo pequeño con medio cabo de cuerno de çiervo y un martillico que sirve de aguja punçón y lima guarnecido el cuchillo grande con dos medios cabos de cuerno de ciervo con su vayna de cuero verde que se tasó en veynte Rs__U020
Cuchillo
-más otro cuchillo de monte ma(fol. 839 v)yor sincelado de una p^{te} y otra de corte a sierra guarneçido el cabo con su cabo de cuerno (sic) con un remate de un grifo y una estrella en el brocal con vayna verde con aguja lima y otro cuchillo pequeño y tenaça y tenedor sinçelado y todo bueno y se tasó en siete ducados__U077

Cuchillo
-más otro cuchillo de monte menor que tiene el cabo de atauxía dorado con su vayna verde con brocal y contera de ataujía dorado con otro cuchillo pequeño y punçón pequeño de hierro lisos y una pluma de plata larga de más de una quarta que se tasó todo en catorce Rs__U014
Otro
-más otro cuchillo de monte mediano con dos bocas sinçelado por entrambas partes con sus tenaçuelas de raspa lima y aguja y su vayna verde que se tasó en doçe reales__U012
Otro
-más otro cuchillo de monte antiguo guarneçido el cabo de cuerno negro con dos orejas de hierro que se tasó en quatro reales__U004
Otro
-más otro cuchillo pequeño en vayna de çapa con un tenedor guarneçi (fol. 840 r)dos entrambos de cuerno de bufano y unas forcas al cabo de la vayna que se tasó en seys Rs__U006
Puñal
-más otro puñal largo de tres quartas con su cuchillo pequeño y punçón que se tasó en veynte y quatro reales__U024
Y está en vayna verde
Alfanje
-más un alfanje turquesco con su vayna negra que se tasó en treinta reales__U030
Machete
-más un machete con su punta alfanjeada y su cuchillo chico sin punta con cabo de cuerno negro en su vayna de cuero verde que se tasó a veynte y ocho Rs__U028
Puñal
-más un puñal con su cruz guarnecido de cuerno de çiervo con su cuchillo digo raspa lima y punçón y aguja guarneçido de cuerno de ciervo y su brocal para la çinta de hierro que se tasó en diez y ocho reales__U018
Puñal
-más otro puñal con su cuchillo y punçón guarneçido de cuero de çiervo con su brocal de hierro que se tasó en veynte reales__U020
Caja
-más una caja de cuchillos de diez (fol. 840 v) y seys pieças guarneçido de cuerno negro todos con las armas de Acuña grabados que se tasó en çinquenta y dos reales__U052

Cuchillos
-más doçe cuchillos y dos tenedores de mesa guarneçido de cuerno de vaca negro que se tasaron en catorze reales__U014
Por manera que suma y monta el ynventario de las ballestas y cuchillos en los presçios que van hechos por los dhos Ju° Grande y Ju° de Azcoyta (sic) myll y quatroçientos y sesenta reales y el dho Juan de Azcoyta y lo firmó de su n^{e} siendo t°s Domyngo Muñoz y Gr^{rio} R°s estanes en M^{d} pres^{te} el dho don Fran^{co} Ruiz Var^{es} quadrada y bor serbal que/ quatro ençina

Juan de Azcoytia
Ante mí Agustín Galván
(signed)

(fol. 841 r)
Y la tasa de los dhos cuchillos fue hecha por el dho Ju° Grande en los dhos apresçios

Juan Grande
Galván
(signed)

Todas estas ballestas y cuchillos m^{dó} el S^{or} marqués por su testamento al S^{or} don Diego marqués segundo del Valle su hijo y así se le quedan

Galván
(signed)[...]

(fol. 864 r)
Caballeriça
Ynvent° de los coches literas y carros y otras cosas tocantes a la caballeriça que se tasaron por Andrés G°s cochero para literas y carros y a P° Salgado albéitar para cabalgaduras y a Gr^{mo} de Rivera para cosas de cordonería y a guarniciones sillas gualdrapas por Ju° Al. sillero los quales debajo de juram^{to} lo apreciaron en lo sig^{te}

Coche grande
-Primeramente un coche grande de pesebrón aforrado de terciopelo carmesí con su encerado verde aforrado en paño y frisa carmesí que agora está cubierto de luto con quatro ruedas y lança clavijas y balancines y lo demás necesario que

se tasó en ducientos ducados y está con sus respaldos quatro almohadas de terciopelo y damasco carmesí y sus banquillos__2U200
Litera
-más una litera grande aforrada en terciopelo açul con su ençerado verde aforrado en frisa açul con sus estribos en forma de coche y sus varas de encaje alto y tiene varillas y cortinas y se tasó en myll y ducientos reales__1U200
Litera
-otra litera menor que la dha guarnecida de cordobanes bayos con su ençerado blanco aforrado en paño (fol. 364 v) verde con sus varas lisas y cortinas y varillas que se tasó en ochoci°s Rs__U800
Carro [...]
(fol. 865 r)
Guarniciones
-más quatro guarniciones enteras de los dhos caballos de coche con sus frenos y cinchas que se tasaron en çiento y cinquenta reales y es todo de vaqueta negro_ _U150
Silla
-más una silla con sus estribos que sirve a un caballo de los quatro del coche que se tasó en dos ducados__U022
Caballo [...]
(fol. 866 r)
Silla
-más una silla de brida con estribos açiones cabeçadas riendas pretal y guruperas que anda sirviendo al presente y se tasó en por su justo valor en treynta reales_ _U030
Silla
-más otra silla del machuelo chico con todas sus guarniciones y estribos que se tasó en tres ducados__U033 [...]
Silla
-más otra silla con sus guarniciones y estribos de la dha jaca castaña (fol. 866 v) que se tasó en veynte y dos reales__U022 [...]
Tornillo
-más un tornillo de tres pies para levantar el coche y carro en alto para mudar las ruedas que se tasó en diez reales__U010 [...]
Gualdrapa
-más una gualdrapa de terciopelo liso negro guarnecido con su flueco de seda que se tasó en ciento y cinquenta reales__U150

Guarniciones
-más unas guarniciones de terciopelo negro que tienen cabeçadas pretal sobrelomo açiones riendas almártaga (fol. 867 r) de la dha gualdrapa que se tasaron en doze ducados son çiento y treinta y dos Rs__U132
Antojos
-más unos antojos de cuero que se tasaron en dos reales__U002
Gualdrapa
-más otra gualdrapa de paño de luto que se tasó en cinquenta reales__U050
Cabeçadas
-más unas cabeçadas riendas y pretal sobrelomo açiones de luto para con la dha gualdrapa que se tasaron con su almártaga en treinta y tres Rs__U033
Estribos
-más unos estribos pavonados que se tasaron en diez y seys reales__U016
Hierros
-más quatro hierros antiguos que sirven para alçar las entrepuertas del coche de camino que se tasaron a dos Rs cada uno son ocho Rs__U008
(fol. 867 v)
Silla
-más otra silla buena negra de cordobán con el correón dorada de brida con sus tres cinchas que se tasó en tres ducados q son treinta y tres Rs__U033
Gualdrapa
-más una gualdrapa de cuero negro corta para de camino que se tasó en diez y seys reales__U016
Sillones
-más dos sillones de machos de litera viejos con todos sus adereços que se tasaron en çien reales__U100
Guarniciones
-más dos ~~medias~~ guarniciones de caballos de coche viejas que se tasaron son enteras her^{nas} de las de abajo quarenta Rs__U040
Guarniciones
-más otras dos guarniciones enteras de caballos de coche viejas que se tasaron en çinquenta reales__U050
Çinchas
-más tres cinchas de caballo de brida buenas tejidas con seda y cáñamo que se tasaron en ocho reales__U008
(fol. 868 r)
Tirantes
-más quatro tirantes de cuero de vaqueta con sus garabatos de hierro raçonables para el coche que se tasaron son del Sor Suero de Vega y no se tasan

Sillas
-más siete sillas viejas hechas pedaços que se tasaron a quatro Rs cada una son v^{te} y ocho Rs__U028
Caparaçones
-más tres caparaçones amarillo y negros viejos que se tasaron los dos negros a ocho Rs y el amarillo en dos son diez y ocho reales__U018
Frenos
-más nueve frenos y bocados de caballos de coche y de [...] quebrados y viejos que se tasaron ~~en~~, un bocado de gineta y brida con copas ocho Rs y dos cañones de formar con copas seis Rs, y uno largo de aformar q^{tro} Rs monta v^{te} y q^{tro} rs lo demás qda por hierro viejo__U024
Barras
-más nueve barras de sillas viejas, quatro almohadas viejas, tres arcos, un estribo, dos barros, un çerco, una varilla, una asa, todo hierro viejo que se tasó y diose por hierro viejo
(fol. 868 v)
Cabeçadas
-más se pone por ynventario y tasa siete cabeçadas de caballos y machos diferentes viejas y desbaratadas que se tasaron a real y medio cada una montan diez Rs y medio__U010 ½
Çincha
-más una çincha de gineta de seda guarneçida y aforrada y con [...] que se tasó en ocho reales__U008
Riendas
-más unas riendas blancas de cuero berberisco y dos pares de correas de estrivos que se tasaron a siete Rs cada cosa es v^{te} y un Rs__U021
Guarniciones
-más unas guarniciones de terciopelo negro de mula antiguas que tienen pretal sobrelomo falsa riendas açiones riendas y otras pieças diferentes que se tasó en ocho Rs__U008
Pretal
-más un pretal de cuero bajo labrado a lo morisco con su clavaçón dorada y labrada a lo turquesco que se tasó en seys reales__U006
(fol. 869 r)
Almártaga
-más una almártaga de terciopelo verde con su hierro dorado antigua que se tasó en dos reales__U002

Guarniciones
-más unas guarniciones de terciopelo carmesí antiguas angostas con sus borlas de seda encarnada que tienen açiones pretal sobrelomo cabeçadas y riendas con clavaçón ordinaria que se tasó en diez y seys reales__U016
Otras
-más otras guarniciones de terciopelo negro diferentes que tienen sobrelomo açiones falsa rienda y riendas y pretal y otro sobrelomo, y otra pieça que no sé de qué sirve se tasó en diez reales__U010
Cuerdas
-más quatro cuerdas con quatro borlas blancas y adereço de caballo antiguas que se tasó en dos reales__U002
Otra
-más una cuerda de seda (fol. 869 v) amarilla con tres borlas para caballo antigua se tasó en dos reales__U002
Cuerda
-más una cuerda de seda carmesí para aderezo de gineta con tres borlas y otros aderentes berberiscos, que se tasó en ocho reales__U008
Otra
-más otra cuerda con unas borlas ~~grandes~~ [crossed out] pequeña con pataletas de cuero carmesíes y otras cosas berberiscas para gineta que se tasaron en ocho reales__U008
Otras
-más otras quatro borlas de seda carmesí grandes con sus cordones que se tasaron en doze reales digo v^{te} Rs__U020
Otra
-más otra cuerda berberisca de seda carmesí con su guarnición dorada que tiene dos borlas grandes y una pequeña en medio que se tasó en ocho Rs__U008
(fol. 870 r)
Petoral
-más un petoral (sic) de gineta labrado a lo turquesco con un escudo en medio redondo esmaltado y otro hierro triangulado encima dorado con una borla grande y dos pequeñas a los lados carmesíes que se tasó en veynte y seys reales__U026
Alpargatas
-más unas alpargatas de hilo blanco y las suelas de cáñamo que se tasaron en tres reales__U003
Copas
-más unas copas doradas lisas de macho antiguas que se tasaron en ocho reales__U008

Copas
-más otras copas de alatón doradas de molduras antiguas que se tasaron en ocho reales__U008
Almártaga
-más una almártaga de cuero negro vieja que se tasó en dos reales__U002
(fol. 870 v)
Espuelas
-Unas espuelas barniçadas lisas para de camino con sus correas negras que se tasaron en seys Rs__U006
-más otra espuela dorada para con çapato que tiene una venera por remate en un cabo que se tasó en seys Rs__006
-más otra espuela açicate dorado liso para gineta con sus correas de terciopelo negro con su cascajo dorado que se tasó en seys reales__U006
-más otro par de espuelas de brida plateadas con sus correas de terciopelo verde que se tasaron en quatro reales__U004
-más otro par de espuelas doradas sin correas con rosetas diferente que las dhas que se tasaron en tres reales__U003
-más otro par de espuelas antiguas doradas y largas de roseta arqueadas para los tobillos (fol. 871 r) con sus correas de terciopelo negro que se tasaron en seys reales__U006
-otro par de espuelas de gineta a modo de açicates color de hierro con sus correas y sus coscojas de hierro dorado y pavonadas que se tasaron en ocho reales__U008
-más otro par de espuelas açicates para gineta doradas sin correas que se tasaron en diez reales__U010
-más otro par de espuelas de gineta doradas y esmaltado los talones con sus correones bajos y anchos que se tasaron en doze reales__U012
-más otra espuela pequeña de talón ancha para con çapato que se tasó en dos reales__U002
-más otro par de espuelas largas y barniçadas para con çapato que se tasaron, digo son de color de hierro en tres Rs__U003
(fol. 871 v)
-más una espuela chiquita para con armas agujerada todo el talón que se tasó en un real__U001
-más otro par de espuelas pavonadas para con çapato que se tasaron, no parecieron y así no se tasaron
-más otro par de espuelas de pico de gorrión antiguas mal doradas con sus correas negras y viejas en quatro reales__U004

-más seys frenos de cañón estañados grandes el uno con copas de alatón a seys Rs cada uno de los sin copas y el que tiene copas doce Rs quarenta y dos Rs__ U040
-más dos guarniciones de cuero blanco de compás que tienen dos cabeçadas dos riendas dos pretales dos guruperas con sus açiones que se tasaron en diez y seys Rs__U016
Estribos
-más un par de estribos con açiones de cuero blanco de brida ocho Rs__U008
(fol. 872 r)
-más otro estribo de brida bajo con su açión de cuero blanco dos Rs__U002
Guarniciones
-otros dos pares de guarniciones de cuero blanco de compás como las de arriba a ocho Rs__U016
Espuelas
-quatro espuelas de brida con correones de cuero blanco de compás dos Rs__ U002
Estribos
-más otro par de estribos de gineta antiguos de hierro con unas rosas abiertas en los suelos labrados de ataujía y plateados con sus fundas en quarenta reales_ _U040
Estribos
-otro par de estribos de rua dorados altos de buena hechura con sus fundas en quatro Rs__U004
Estribos
-más otro par de estribos dorados de rua de buena hechura con quatro barras por los suelos con sus fundas en seys Rs__U006
(fol. 872 v)
Estribos
-más otro par de estribos dorados de dos arcos dorados tres barras en el suelo bajos y anchos en quatro Rs__U004
Guarnición
-más una guarnición de terciopelo negro viejo que tiene pretal y cabeçada riendas y açiones y gurupera que se tasó en seys reales__U006
Espuelicas
-unas espuelicas de talón redondas que se tasaron en dos Rs__U002
Tiros y pretinas
-más dos tiros y dos pretinas de terciopelo blanco antiguos que se tasaron en quatro reales__U004

(fol. 873 r)
Espadas tiros y vaynas
Espada
-Una espada de Toledo de Gonçalo Simón un dedo menos de marca con su guarnición de una puente un dedo menos de marca con su guarnición de una puente y un laçico y una guarda al pomo con vayna de çapa con cuchillo y punçón de alquimia viejo apreçiada en quatro ducados__U044
Otra
-más otra espada de Bilbao de las viejas corta de marca un dedo con guarnición como la de arriba de una puente y braço y una guarda pomo con vayna de çapa algo maltratada con un punçón y puño roto de seda en dos ducados__U022
Otra
-más otra espada de Bilbao en triángulo rabiangosta con la guarnición de una puente y un bre[…] y una guarda al pomo un puño de hierro largo vayna de çapa muy rota se tasó en ocho Rs__U008
Estoque
-más un estoque en triángulo buida de largo de vara y ochavo con un puño de cuero y con la guarnicion(fol. 873 v)çilla baladí con una cruz con vayna de bezerro rota sin contera que se tasó en ocho Rs__U008
Vayna
-más una vayna de terciopelo negro traída con unos tiros y pretina clavaçón lisa pavonada que se tasó en quatro Rs__U004
Otra
-más otra vayna de terciopelo viejo negro con una conterilla abollada en un real_ _U001
Guarnición
-más una guarnicioncita aventanada y labrada de ataujía tiene una puente dos braçitos y una guarda al pomo con puño y punçón y no tiene contera que se tasó en seys reales__U006
Tiros
-más unos tiros de terciopelo con una vayna y su pretina del mismo terciopelo negro todo pespuntado de negro clavaçón pavonada lisa que se tasó en ocho reales__U008
Guarnición
-más otra guarniçioncita de una puente y dos laçicos labradilla (fol. 874 r) que se tasó en dos reales- con un terçio de espada__U002
Tiros
-más unos tiros y pretina de vaqueta bayos con clavaçón dorada que se tasaron en quatro Rs__U004

Tiros
-más otros tiros françeses de terciopelo negro de quatro anyllos con clavaçón de ataujía que se tasaron en dos Rs__U002
Almártaga
-más una almártaga de cuero negro que se tasó en dos Rs__U002
Pretina
-más una pretina de llave de cuero doradilla que se tasó en un real__U001
Guardasol
-Más un guardasol de cordobán bayo aforrado en tafetán carmesí con su flueco de seda que está quebrado el un palo que se tasó en dos ducados__U022
Guardasol
-más otro guardasol de bocací verde grande maltratado con varillas de barba de ballena que se tasó en quatro reales__U004
(fol. 874 v)
Borceguíes
-más unos borceguíes de cordobán blancos sin laço que se tasaron en seys reales__U006
Borceguí
-un borceguí con su rodillera de cordobán en dos Rs__U002
Espadita
-más una espadita pequeña de tres quartas de largo con guarnición dorada y çiçelada (sic) puño de plata todo maltratado con su daga de lo propio tiros y pretina de terciopelo con sus trencillas de oro que se tasó en quatro ducados__U004
Otra
-más otra espadica del mismo tamaño con su daga guarnición plateada puños de alquimia tiros y vaynas de cordobán açul con trencillos de plata todo muy viejo que se tasó en diez y seys reales__U016
Vayna
-más otra vayna de espada de terciopelo negro con su contera y cuchillo con el cabo de ataujía dorado que se tasó en quatro Rs__U004

(fol. 875 r)
Ynventario de jacos petos cascos y cosas de arnés por Gabriel Mont° armero de Su Mg[d] en 15 de março

Jaco
-Primeramente un jaco de chapas de açero cubierto de terciopelo negro con el hábito de Santiago claveteado de chachullas (sic) de latón sin mangas con faldillas cortas que se tasó en cinquenta Rs__U050

Jaco
-otro jaco como el dudo dho algo mayor con una encomienda de Santiago que se tasó en cinquenta reales__U050
Jirones
-más quatro puntas o jirones de los mismos jacos de açero cubiertas de terciopelo y clavaçón va tasado con los jacos de arriba y entra en la tasa
Peto
-más un peto fuerte de azero con sus tres agujeros para el riçe (sic) que se tasó en dos ducados__U022
Cascos
-más tres cascos de hierro el uno con orejas y el otro sin ellas y el (fol. 875 v) uno redondo con una escofia vieja de raso carmesí aforrado se tasó de orejas en doce Rs y el sin ellas v^{te} reales, y el redondo en doce Rs es quarenta y quatro rs__U044
Rodela
-yten una rodela grande de corcho vieja que se tasó en veynte y seys Rs__U026
Rodela
-otra rodela más pequeña y quebrada con unas listas doradas que se tasó en doce Rs__U012
Çapatos
-más unos çapatos de malla guarneçidos de terciopelo negro que se tasaron en ocho Rs__U008
Martillo
-más un martillo de armas con su puño y cordón que se tasa en dos ducados__U022
Hierros
-más dos hierros de lança de [...] harce (sic) uno con las armas de los Acuñas y el otro con dos escudos a los lados dorados con sus bolsas que se tasó en diez y seys reales entrambos__U016
Vanablo
-más un venablo con su hierro sin punta granujado viejo y roñoso en quatro Rs__U004
(fol. 876 r)
Faldones
-más siete faldones de paño amarillo guarneçidos de paño negro los dos muy viejos y los otros raçonables para hombres de armas que se tasaron a tres reales cada uno monta v^{te} y un Rs__U021
Caparaçón
-más un caparaçón de terciopelo carmesí guarneçido con rapacejos y franjas ~~de oro~~ y seda carmesí muy viejo y antiguo que se tasó en doçe Rs__U012

Otro
-más otro caparaçón de terciopelo açul guarneçido con franjas y rapacejos de oro y seda açul antiguo viejo que se tasó en veynte y seys reales__U026
Otro
-más otro caparaçón o cubierta de caballo de terciopelo encarnado guarnecido con fluecos de plata aforrado todo en fustán blanco acuchillado que se tasó en veynte reales__U020
Reatas
-más dos reatas de silla de gineta la una de seda, y la otra de hilo que se tasó la de seda en seys Rs y la de hilo en dos Rs__U008

(fol. 876 v)
La qual dha tasa de jacos petos cascos rodela çapatos martillos hierro venablo faldones caparaçones fue hecha por el dho Grabiel (sic) Montero que monta doçiºs y ses^{ta} y q^{tro} Rs y debajo del dho juram^{to} dijo haberla hecho jurídicam^{te} y lo firmó tºs Juº de Salzedo y Gr^{mo} Rºs estantes en M^{d} pres^{te} el dho don Fran^{co} de Luque tr^{do} de oro

Gabriel Montero
Agustin Galván
(signed)

Y lo que toca a cordonería que de suso va inventariado lo tasó Gr^{mo} de Ribera cordonero e los dhos apreçios debajo de dho juram^{to} y monta los apreçios en que van apreçiados y lo firmó tºs los dhos de suso

Gr^{mo} de Ribera
Agustín Galván
(signed)

(fol. 877 r)
Y los bienes tocantes a cosas de guarniciones sillas gualdrapas espuelas frenos jaquinas mantas que van declaradas fueron tasadas en los dhos apresçios por Juan Alº sillero de Su Mag^{d} debajo de juramento que en forma de derecho hizo tºs Juº de Salzedo y Jerónimo Rºs estantes en esta corte

Juan Alonso
Ante mí Agustín Galván
(signed)[...].

VELADA, MARQUESES DE; ASTORGA, MARQUESES DE

Doc. 1.
1616, July 27.
Inventory of Marquis of Velada.
Biblioteca Francisco de Zabálburu, Altamira, 198, D. 1.

[unfoliated, the digitized folio number is entered]
(nº. 71)
Ynventario general
De los bienes muebles que quedaron por fin y muerte del marqués de Velada mi Sr así en su casa como en los lugares de su estado en 29 de julio 1616 q Dios llevo a Su Exª

En la villa de Madrid a dos de setiembre de mil y seiscientos y diez y seis aºs don Antº Sancho Dávila Marqs de Velada gentilhombre de la Cámara de Su Magd mi Sor como hijo y heredero universal y testamentario de don Gómez Dávila Marqs de Velada mi Sor del Conssº de Estado y maymo Mayor de Su Magd comendador de Mançanares de la Orden de Calatrava difunto dixo q para cumplir con lo que dispone en las leyes de estos reinos quería hazer ynventrio de todos los bienes muebles que quedaron por fin y muerte del dho Sor Marqs su padre en vte y siete de julio del dho año de seiscientos y diez y seis q Dios llevó a Su Exª y así me mdo a mí Juº Simón de Ravarena su contor como perssª q tiene la razón de la hazda de Su Exª hiziese el dho ynvvrio el qul yo hiçe con la mayor puntualidad que se pudo habiendo visto las cosas e en él irán declaradas y cotejando los inventarios particulares de las personas q a su cargo tenían las cosas q en él van puestas el qual se hiço en esta manera &ª

Cosas que hay en Velada a cargo de Franco Hernández casero de palaçio [...]
(nº. 73)
Gafas
-Unas gafas
Alxafas
-Dos Alxafas (sic)
Horquilla
-Una horquilla para tirar con ballesta
Bolsa
-Una bolsa de caça de vaqueta

Hierros
-Unos hierros de guarni^{ón} de cav^{o} en otra bolsa
Colmillo de elep^{e}
-Un pedaço de colmillo de elefante
Banco
-Un banco de madera para torno […]
Mosquete
-Un mosquete grande con su caxa que llaman esmuil […]
Palas
-Dos palas de jugar a la pelota
Arandelas
-Tres arandelas de lança de Ristre
Hierros
-Quatro hierros para jugar a la pelota
Axedrez
-Una tabla de axedrez ordinaria
Otra
-Otra tabla de nácar con una funda de tafetán verde […]
Cueros de gamo
-Tres queros de gamo uno blanco y dos negros
Cofre
-Un cofre viejo encorado con unas armas dentro que es un arnés entero blanco con fundas de frisa colorada
Espadas
-Seis espadas y tres montantes […]
(nº. 76)
Clavaçón
-Una clavaçón de adreço de caballo […]
Armas
-Unas armas antiguas cerradas q son piernas un braçalete çelada una pieça de guarnición de silla
Silla
-Una silla verde de mujer con su gualdrapa […]
Silla
-Una silla de brida vieja para niños de terciopelo negro […]
Rallo
-Un rallo de hierro para hazer perdigones […]
Moldes
-Dos moldes de perdigones

Pelotas
-Otro de pelotas
Caño
-Un caño de hierro
Sierra
-Una sierra
(nº. 77)
Tenaças
-Unas tenaças y quatro barsillas de hierro agujeradas para tirar hierro […]
Hierros
-Quatro hierros antiguos que diçe son de jugar a la pelota […]

En la Recam^{ra} de mi S^{ra}
Retrato
-Un Retrato de mi S^{ra} la Duquesa de Medinaçeli quando niña arrollado en una caxilla de pino […]
Retratos
-Seis retratos de pinçel de mujeres extranjeras […]
Báculo
-Un báculo de may^{mo} m^{or} de nácar con su funda
Coco
-Un coco […]

En el camarín de los açulejos
Libros
-Treinta y seis libros diferentes grandes y pequeños en un estante de madera […]
(nº. 78)
Hierro
-Un hierro para marcar los corchos de las colmenas de Su Exª
Escopeta
-Una escopeta larga de seis palmos sin llave q estaba depositado por la Justª de lo q hurtaron
Ballestas
-Dos ballestas que también son de las q habían hurtado
Gafas
-Yten dos gafas de las dhas ballestas hurtadas q todas estas tres partidas estaban depositadas e poderes de Pº de Badaxoz alcalde […]

(nº. 83)
Cosas de caballeriça en M^{d} a cargo de Rodrigo de León
Coche
-Primeram^{te} un coche negro con todo su recaudo quatro almohadas de cordobán negro colchado (sic) y dos cortinas de chamelote negro
Otro
-Otro coche para criadas viexo
Varas
-Unas varas de litera con sus correones
Sillones
-Dos sillones para los machos de litera cubiertos de ençerados
Carro
-Un carro largo con sus ruedas clavijas y balançines
Litera
-Una litera de cuero bayo con ençerado verde y todo su recaudo
Ruedas
-Quatro ruedas viejas q están en la cochera
(nº. 84)
Herradas
-Quatro herradas para servir en las caballeriças
Sillas
-Ocho sillas de gineta que sirve a las del campo con sus caparaçones y frenos y estribos faltan unos estribos y dos bocados
Caparaçones
-Siete caparaçones de cordobán negros buenos y malos
Otro
-Un caparaçón de terciopelo negro con siete pass^{os} darda (sic)
Guarniçiones
-Tres guarniciones de terciopelo negros falta de a una gurupera petral y açiones
Almártigas
-Dos almártigas de terciopelo negro
Banda
-Una banda de tafetán negro
Gualdrapa
-Una gualdrapa de terciopelo negro
Arquilla
-Una arquilla para guardar estribos
Almártiga
-Una almártiga de Yndias labrada

Otras
-Tres almártigas de Yndias labrada no es nada
Otras
-Tres almártigas de cuero negras
Muserolas
-Quatro muserolas bayas y negras
Reatas
-Cinco reatas negras
Cabeçadas
-Tres cabeçadas y dos riendas de cuero blancas
Cabeçadas
-Siete cabeçadas de brida buenas y malas todas usadas
Cabeçadas
-Çinco cabeçadas de jineta viejas
Adreço
-Un adreço de la sortija que es de cordobán plateado
Guarnición
-Una guarnición de mula vieja
(nº. 85)
Pretales
-Dos pretales negros de jineta viejos […]
(nº. 99)
Picas
-Seis picas con sus hierros
Lanças
-Dos lanças largas sin hierros para atajar quadros […]
(nº. 103)
Montante
-Un montante negro
Banderilla
-Una banderilla de paxa de Valençia […]
(nº. 104)
Guarnición
-Una guarnición lisa dorada para espada y el puño de plata y unos tirillos y vaina de terciopelo amarillo […]
Mosquete
-Un mosquete grande

Bolsilla
-Una bolsilla de paño verde con dos turquesas para pelotas y quatro hierros para limpiar arcabuçes y un sacatrapos [...]
Sillas
-Doze sillas brida viejas [...]
Caçoleta
-Una caçoleta de cobre con su paleta para colores [...]
(nº. 105)
Requetas (sic)
-Dos requetas (sic) de madera para jugar a la pelota
Lança
-Una lança y una adarga [...]

De lo que solía tener a su cargo D. Mª de Lira [...]
(nº. 107)
Arca
-Una arca grande cerrada que no se abrió núm^{ro} 25 en que ha de haber lo sigu^{te} [...]
Ballesta
-Una ballesta tablero de nogal y la verga de Juº Blanco con sus gafas en funda negra
Otra
-Otra que dize la llave de Ortega y en la verga Jesús María y se entiende era del marq^{s} de Poça y pesa la verga 33 onças y las gafas no tienen nombre funda negra
(nº. 108)
Ballesta
-Otra el tablero de madera negra y la verga dize Jesús María y en tres partes dize 7º y las gafas dizen en Pº de la Fuente y pesa la verga 33 onças funda de paño verde
Otra
-Otra pequeña tablero de madera negra sin nombre y tiene en la verga señal a manera de flor de lis no tiene gafas
Otra
-Otra q dize la verga Don Fadrique de T^{do} y pessa 26 onças y tiene gafas y no tienen señal con funda negra y la verga dize 7º en dos partes

(nº. 120)
Lo q tiene Fran^{co} G^{ez} de las Fraguas may^{mo} del haz^{da} de su Exª en el castillo de vª Toro

Pieças
-Prim^{ra}mente veinte y una pieças de piernas y muslos de armas chica y grandes partidas
Pieças
-Yten otras çinquenta y quatro pieças de armas chicas y grandes de armar desde la çintura para arriba hasta el percueço e rostro sin peto ni espaldar q no le hubo
Pieças
-Yten otras çinq^{ta} y dos pieças de armar para braços manos en pieças chicas y grandes medias y enteras

(nº. 121)
Las cosas q hay en M^{d} a cargo de Juan de Alba guardarropa de su Exª [...]
(nº. 127)
Retrato
-Más un retrato de mi S^{ra} Doña Vitoria abuela del S^{or} don Antº con moldura de ébano en su caxa
Otro
-Más otro retrato de una mujer vestida de negro con la mano en el pecho y en ella tres sortijas con su marco de madera
Sillas
-Yten ocho sillas de armas blancas con cinco pares de estribos riendas cabeçadas çinchas y guruperas
Caballos
-Dos caballos de madera en que están estas sillas
Arneses
-Catorçe arneses blancos que tiene cada uno peto y ristre y dos escarçelas y espaldar dos braçales dos manoplas dos quixotes para los muslos y su çelada y siete azeros de silla de caballo que cada azero son cinco piezas y falta una çelada porq no hay sino treze y están en cinco cofres y dos cajas
Escarçelas
-Yten diez y nueve escarçelas q sobran de las dhas armas
Quixotes
-Yten diez quixotes que también sobran [...]
Mosq^{tes}
-Dos mosquetes grandes con sus caxas [...]
(nº. 129)
Martillo
-Un martillo de pico de desarmar negro [...]

(nº. 130)
Ruedas
-Dos ruedas pequeñas de una pieça de artillería [...]
Carcax
-Un carcax turco con veinte y tres flechas
Otro
-Otro sin flechas
Arco turco
-Un arco turco amarillo [...]
Espada
-Una espada barniçada sin tiros [...]
(nº. 131)
Espada
-Una espada ancha a lo antiguo con los seis roeles que se halló en Ávila en un sepulcro metida en una caxa de cuero negro [...]
Arcabuz
-Un arcabuz que se compró en Cuenca [...]
Caxa
-Una caxa de pino grande baxa con bisagras y su llave para meter arcabuçes [...]
(nº. 132)
Morriones
-Yten tres morriones antiguos [...]
(nº. 133)
Adereço
-Un adreço de cuero bayo cabeçadas pretal estribera açiones esmaltadas de blanco azul y verde y un caparaçón de terciopelo carmesí con dos faxas de brocado con vuelta y molinillos de oro q fue del Sor Don Fer^do^ de Toledo [...]
Vaina de espada
-Una vaina de espada de terciopelo negro con contera [...]
Hoja de espada
-Una hoja de espada con vaina de çapa baya contera dorada puño de hilo de oro cuchillo y punçón
Espada
-Una espada con guarniçión dorada cuchillo y punçón y daga con puños de oro tiros petrina y vaina de terciopelo
Daga
-Una daga con guarn^on^ dorada con puño de oro tiros y petrina baya
Otra
-Una daga con gu^on^ negra pavonada con petrina negra y unos tiros como ella

Otra
-Una daga con guarnición enbarniçada
Tiros
-Más unos tiros y petrina de riço negro con unos molinillos y una vaina para daga
Espuelas
-Unas espuelas plateadas y otras negras para la brida [...]
Virotes
-Çinco virotes grandes y seis saetillas
Gu^{on} de espada
-Una guarnición de espada con pomo y contera dorados
Otra
-Una guarnición dorada para espada y otra negra
Tirillos
-Unos tirillos de ter^{o}pelo con clavaçón dorada
Daga
-Una daga con guarn^{on} dorada con puño de oro vaina con tiros petrina con dos caireles de oro
Tiros
-Unos tiros de cuero naranjados con clavaçón dorada
Hierros
-En un papel unos hierros para tiros y petrina guar^{on} dorada cuchillo y punçón todo dorado
Otros
-Los hierros para tiros y petrina con guar^{on} de daga y pomo para espada todo negro
Otros
-En un papel doçe hierros dorados como aldabillas para ventanas
Clavos
-Más nueve clavos pequeños con las cabeças doradas [...]
Hebillas
-Quatro hebillas rodadas en un papel
Cascos
-Dos cascos [...]
Coraça
-Más una coraça sobre terciopelo negro [...]
(nº. 135)
Espada
-Una espada pavonada con su daga taelí (sic) de cuero y tiros

Otra
-Otra espada con la guarn^{on} dorada digo labrada color de hierro con su daga tiros y petrina de cuero negro [...]
Un pabellón
-Un pabellón de Holanda que tiene veinte piernas con unas listas labradas de seda colorada y capp^{an} en tres pieças y está roto
(nº. 137)
Un coco
-Un coco guarneçido de plata [...]
Cocos
-Dos cocos sin labrar [...]
Cofre
-Otro cofre viejo núm^{ro} siete en q hay lo sig^{te} [...]
(nº. 138)
Espada
-Una espada dorada pequeña con vaina de terciopelo verde y funda tiros y petrina de lo mismo [...]
(nº. 140)
Cuchillo
-Un cuchillo damasquino con vaina de çapa [...]
Cuchillo
-Otro cuchillo damasquino con cavo de piedra y vaina de çapa [...]
Un alfanje
-Un alfanje turco con vaina baya y puño de plata
Un modo de alf^{xe}
-Un modo de alfanje de latón con vaina negra y en el puño un artificio [...]
Arnés
-Un arnés grabado de oro tomado de herrumbre peto espaldar gola manoplas braçales escarçelas sin çel^{da} en un cofre viejo
(nº. 141)
Otro
-Otro arnés grabado muy bueno q fue del duque de Medinaçeli q es çelada peto espaldar testera de cav^{o} piernas quixotes manoplas gola braçales y todas las pieças sin que falte nada
Testeras
-Dos testeras de caballo una çelada y un morrión dos piernas una arandela de lança dos quixotes dos escarçelas dos hombros y un braçal de ristre y un peto y diez y siete pieças chicas y grandes y unos tornillos en una bolsa y unos toneletes de tafetán verde y vei^{te} y una fundas de frissa colorada en dos cofres

Una Req^{ta}
-Una Requeta (sic) de jugar a la pelota [...]
Tornillos
-En un lienço muchos tornillos de hierro pequeños
Vergas
-Dos vergas de ballesta de hierro
Arco
-Un arco de hierro con el asidero de terçiopelo açul [...]
Cofres
-Ocho cofres viejos hechos pedaços con sus q están los arneses (sic) [...]
Fundas
-Quatro fundas de badana p^{ra} espadas
Badil
-Un badil y unas tenaças de hierro
Turquesa
-Una turq^{sa} de hierro
(nº. 142)
Piedra
-Una piedra que parece piedra ymán
Collares
-Tres collares de perros de cuero leonado
Reglas
-Dos reglas de nogal
Baq^{ta}
-Una baqueta de arcabuz
Frasco
-Un frasco de cuerno
Alicates
-Unos alicates de hierro [...]
Un arnés
-Otro arnés dorado y grabado que es peto espaldar gola çelada braçales y manoplas con siete fundas de frissa colorada [...]
Un martillo
-Un martillo de hierro [...]
(. 144)
Un bastón
-Un bastón de may^{mo} m^{or} guarn^{do} de oro con su funda [...]
Una pieça
-una pieça de pelo de cabra [...]

(nº. 146)
Dos cocos
-Dos cocos de ébano a modo de frascos con sus pies de marfil [...]
Un frasco
-Un frasco de arcabuz de ébano con un cordón de seda parda [...]
(nº. 147)
Una llave
-Yten una llave de armar arcabuçes [...]
(nº. 148)
Ballestas
-Ocho ballestas y quatro gafas la una con funda de terciopelo verde [...]
(nº. 149)
Petos
-Yten tres petos de armas sin espaldar [...]
(nº. 151)
Un frasco
-Un frasco de cuerno con su llave pavonada y llave de arcabuz pendiente de un cordón de seda cabellado
Un estuche
-Un estuche con muchas herramientas q entran unas en otras y la caxa negra que parece de çapa [...]
(nº. 155)
1 Talabarte
-Yten un talabarte petrinas vainas de espada y daga de terçiopelo negro [...]
(nº. 156)
1 espada
-Yten una espada con vaina de terciopelo y tiros
Otra
-Yten una espada con vaina de cuero negro [...]
(nº. 166)
1 báculo
-Más un báculo de may^{mo} de madera negra y blanca con remates de marfil
1 frasco
-Un frasco grande cubierto de cuero negro para pólvora
1 carcax
-Un carcax de lienço con virotes para palomas [...]
(nº. 172)
Libros q tenía a su cargo Negrete son los sig^{tes} [...].

Doc. 2.
1678, November 26.
Inventory of arms in the cabinet and in the armoury of the palace of the Marquis of Velada and Astorga, in the town of Velada.
Instituto Valencia de Don Juan (Archivo y Biblioteca), Manuscritos, 25-V-13.

(fol. 122 r)
Inventario de las armas de fuego espadas y otras cosas que hay en el camarín del Palazio de Velada hecho de orden del Exmo Sr Marqués de Velada y Astorga mi señor por Don Gabriel Dirsen y del Vaus tesorero y gentilhombre de la cámara del marqués mi señor y por ante mí el presente ess^{no} en la manera s^{e}

-Primeramente treze pistolas las ocho con llaves de rueda dos de chispa franzesa una de chispa castellana y las otras dos sin llaves
-Una daga dorada
-Una carabina de rueda
-Otra carabina de piedra de tres quartas de largo
-Un arcabuz de seys quartas de largo que no tiene llave ni baqueta
-Una hoja de espada de dos cortes fina del maestro Domingo Corte
-Otra hoja de espada de lomo de dos cortes
-Dos cajuelas de tres dedos de alto media vara de largo y una quarta de ancho con ynstrumentos matemáticos en que faltan algunas piezas
-Quatro estantes de a tres cajones de madera dorados azul y blanco y con sus remates y armas de la casa de Su Exª en que están los libros
-Seys retratos de poco más de quarta de alto yguales de mujeres estranjeras con diferentes trajes
(fol. 122 v)
-Tres mapas grandes arrollados el uno roto y otro maltratado
-Otro mapa pequeño de la planta de la villa de Gueldes
-Dos barros yguales el uno sin pie

Y en la manera dha se hizo dicho ynventario por dicho don Gabri^{l} Dirsen y del Vaus que le firmó Velada y nobiembre veynte y seys de mill y seyszientos y setenta y ocho años= Don Gabriel Dirsen y del Vaus ante mí= Ju^{o} Resino= E yo el dho Ju^{o} Resino ess^{no} Público del número y ayuntamiento de esta villa de Velada por md y nombramiento del Exmo Sr Marqués de Velada y Astorga mi S^{r} que presente fuy con el dho Sr Don Gabriel Dirsen y del Vaus a este ynventario saqué de él este traslado que va zierto y verdadero y concuerda con su original que queda en mi poder en fee de ello lo signé y firmé

En testimº de Verdad
Juº Resino
(signed)

Digo yo Don Pº Garzés Alcalde mayor y mayor^{mo} de rentas del Exmo S^{r} marqués de Velada y Astorga mi S^{r} en el partido de esta su vª de Velada que rezibí [...] me doy por entregado de las armas de fuego espadas y lo demás contenido en este ynventario para tener [...] de pronto para quando se me pida por Su Exª [...] sona que tenga su poder y lo firmé Velada y dizi[...]bre veynte y dos de mill y ss^{os} y setenta y ocho años

D. Pedro Garçés
(signed)[...]

(fol. 124 r)
Inventario de la armería que el Exmo Sr Marqués de Velada y Astorga mi S^{or} tiene en el Palazio de esta su villa de Velada hecho de orden de Su Exª por el Sr Don Gabriel Dirsen y del Vau tesorero y gentilhombre de la cámara del Marqués mi S^{r} y ante mí el presente escribano en la manera s^{e}

-Primeramente diez y nueve morriones con sus viseras
-Siete morriones sin ellas algunos labrados
-Yten veynte y seys petos con sus espaldares
-Otro peto sin espaldar con media cabeza
-Veynte y un brazos enteros con sus hombros
-Onze brazos sin hombros
-Otros dos hombros solos
-Veynte y dos bolillas
-Diez y siete pares de guantes
-Veynte y quatro pares de rodilleras con sus medios muslos
-Otro par grande de rodilleras de goznes
-Zinco piernas las dos con pies y las demás sin ellos
-Otra pierna y medio muslo con su pie
-Quatro testas de caballos
-Treynta y siete faldillas de los petos y sueltas
-Dos medios brazos con sus manoplas
-Zinco medios brazos de la muñeca al codo
-Sesenta y dos pedazos que se reconocen ser de las armas y de diferentes géneros
(fol. 124 v)

-Un atado de faldillas angostas para las armas
-Treze ballestas de virotes maltratadas
-seys gafas para armarlas
-Tres montantes
-Tres cañones de mosquetes el mayor como de dos varas y media de largo y los dos poco menores con sus cajas sin llaves
-Un cañón de mosquete sin caja
-Dos cajas rotas de arcabuzes con sus llaves de rueda y maltratadas
-Una caja de mosquete rota con su llave de rueda
-Una caja de pistola con su llave de rueda maltratada
-Dos llaves de rueda que parezen de mosquetes
-Una lanzilla del Japón con vayna
-Un cañón de hoja de lata para papeles
-Un artifizio de hierro pequeño quadrado con pie de madera
-Un zepo para armar lobos
-Tres mangas de malla

Todas las quales dhas armas están en la torre que está en dho palazio al lado del parque puestas en el suelo y abiertas las ventanas muy maltratadas todas y orimientas (sic) y en esta forma se hizo dho ynventario por el dho señor Don Gabriel Dirsen y del Vaus que lo firmó en la vª de Velada en zinco de noviembre de mill y ss^{os} y setenta y ocho años= Don Gabriel Dirsen y del Vaus= ante mí Juº Resino= E yo el dho Juº Resino ess^{no} Pú^{co} del número y ayuntamiento de esta vª de velada por md y nombramiento del Exmo Sr Marqués de Velada y Astorga mi S^{r} que presente fuy a este ynven(fol. 125 r)tario con el Sr Don Gabriel Dirsen y del Vaus tesorero y gentilhombre de la cámara del marqués mi S^{r} saqué este traslado que va sierto y verdadero y concuerda con su original que queda en mi poder en papel del quarto sello en diez y nueve de noviembre de mill y ss^{os} y setenta y ocho años en dee de ello lo signé y firmé

En testimº de Verdad
Juº Resino
(signed)

Digo yo Don Pº Garzés Alcalde mayor y administrador de la hazienda del Exmo S^{r} marqués de Velada y Astorga mi S^{r} en el partido de esta su vª de Velada que recibí y me doy por entregado de todas las armas contenidas en este ynventario y que las tendré de manifiesto para cada y quando que se me pida por Su Exª o la persona que tenga su poder y lo firmé Velada y diciembre v^{te} y dos de mill y ss^{os} y setenta y ocho años

D. Pedro Garçés
(signed)

(fol. 205 r)
En la villa de Velada en seys días del mes de junio de mill y seyszientos y ochenta y nueve años ante mí el escribano y testigos pareció presente el Sr Don Juan de Villarroel Maldonado alcalde mayor y mayordomo de las rentas del Exmo Sr Marqués de Velada Astorga y Ayamonte conde de Trastámara mi S^{r} en esta dha villa y su pp^{do} y confesó haber rezibido de Don Pedro Garzés mayordomo que fue de dhas rentas por mano del Sr Don Domingo Virto del Rincón teniente cura de la Parrochial de esta villa todos los libros que consta por el ynventario original que queda en mi poder hay en la librería que los Ex^{mos} Señores Marqueses de Velada tienen en esta villa arrimada al convento de ella y los de la librería del camarín= y en las armas de fuego espadas y otras cosas que contiene otro ynventario de las cosas que están en dho camarín y todo lo que contiene otro ynventario de pinturas y diferentes cosas que están debajo de las llaves de la librería principal y en los papeles contenidos en el ynventario de papeles que están en diez y seys cajones chicos y grandes y de diferentes hechuras en el archivo que está arrimado a la dha librería principal y en las armas que se menzionan en el ynventario de la armería y en todo lo contenido en otro ynventario que se intitula de diferentes cosas que hay en el palazio jardines y huerta de la Palma y de herramientas de cozina en que se entregó dho señor Don Juan de Villarroel excepto en los libros y otras alhajas que han faltado en dhos inventarios y no se han entregado que son las siguientes
En la librería principal de junto al convento faltó [...]

(fol. 205 v)
Armería
-En el ynventario de la armería faltó dos petos= tres brazos sin hombros= y una caja de arcabuz con llave de rueda [...]

(fol. 206 r)
En todos los quales dhos libros bienes y alhajas contenidas en dichos inventarios excepto los que aquí se declara haber faltado el dho señor Don Juan de Villarroel se dio por entregado renunzió las leyes de la entrega y se obligó a tenerlos en su poder y de manifiesto para dar quenta de ellos y volverlos a entregar cada que le

sean pedidos por Su Exª el marqués mi Sʳ u otra persona en su nombre o quien lo pueda hazer legítimamente a que se obligó en forma con su persona y bienes muebles y rayzes habidos y por haber y para la execuzión y cumplimiento de ello dio su poder a todos y qualesquier juezes y justizias de su real Magestad de qualesquier partes que sean para que a ello le compelan y apremien por todo rigor de derecho y vía executiva como por sentencia pasada en cosa juzgada renunzió las leyes de su favor la general y sus derechos en forma (fol. 206 v) y otorgó este rezibo y entrega de dhos bienes siendo testigos el liz^{do} Don Diego Fernández de Agüero Gabriel González y Gabriel Resino todos vecinos de esta dha villa y el dho Sʳ otorgante a quien yo el escribano doy fee conozco lo firmó= Don Juº de Vªrroel Maldonado= ante mí Juº Resino= Testado= otro libro= otro libro= e yo el dho Juan Resino escribano Público del numero y ayuntamiento de esta vª de Velada por md del Exmo Sʳ Marqués de Velada Astorga y Ayamonte conde de Trastámara mi Sʳ que presente fuy a lo que dho es con los dhos testigos y otorgante saqué este traslado día de su otorgación en este pliego del quarto sello por no lo haber en el estanco de esta vª del segundo ni otro ningún sello, de su original que queda en mi poder en el mesmo sello en fee de ello lo signé y firmé

En testimonio de Verdad
Juan Resino
(signed)

(fol. 207 r)
Inventario de las armas de fuego espadas y otras cosas que hay en el camarín del palazio de Velada hecho de orden del Exmo señor marqués de Velada y Astorga mi Sʳ por Don Gabriel Dirsen y del Vau tesorero y gentilhombre de la cámara del Marqués mi Sʳ y por ante mí el presente escribano en la manera siguiente

-Primeramente treze pistolas las ocho con llaves de rueda dos de chispa francesa= una de chispa castellana y las otras dos sin llaves
-Una daga dorada
-Una carabina de rueda
-Otra carabina de piedra de tres quartas de largo
-Un arcabuz de seys cuartas de largo que no tiene llave ni baqueta
-Una hoja de espada de dos cortes fina del maestro Domingo Corte
-Otra hoja de espada de lomo de dos cortes
-Dos cajuelas de tres dedos de alto media vara de largo y una cuarta de ancho con instrumentos matemáticos en que faltan algunas piezas

-Quatro estantes de a tres cajones de madera dorados azul y blanco y con sus remates y armas de la casa de Su Exª en que están los libros
(fol. 207 v)
-Seys retratos de pocos más de cuarta de alto yguales de mujeres extranjeras con diferentes trajes
-Tres mapas grandes arrollados el uno roto y otro maltratado
-Otro más pequeño de la planta de la villa de Gueldes
-Dos barros yguales el uno sin pie

Y en la manera dha se hizo dho ynventario por dho Don Gabriel Diesen (sic) y del Vaus que le firmó Velada y noviembre veynte y seys de mill y seyszientos y setenta y ocho años D. Gabriel Dirsen y del Vau= ante mí= Juº resino= e yo el dho Juº Resino ess^{no} Pú^{co} del número y ayuntamiento de esta vª de Velada por md del Exmo Sr Marqués de Velada Astorga y Ayamonte conde de Trastámara mi S^{r} que presente fuy a lo que dho es saqué este traslado en pliego del segundo sello de su original que queda en mi poder en papel del quarto sello en veynte de mayo de mill y ss^{os} y ochenta y nueve años en fee de ello lo signé y firmé

En testimº de verdad
Juº Resino
(signed)

(fol. 215 r)
Inventario de la armería que el Exmo Sr Marqués de Velada y Astorga mi S^{r} tiene y está en la torre del palazio de esta su villa de Velada al lado de solano hecho de orden de Su Exª por el Sr D^{n} Gabriel Dirsen y del Vau tesorero y gentilhombre de la cámara del Marqués mi S^{r} y ante mí el presente escribano en la manera siguiente

-Primeramente diez y nueve morriones con sus viseras
-Siete morriones sin ellas y algunos labrados
-Yten veynte y seys petos con sus espaldares
-Otro peto y espaldar enterizo
-Otro peto sin espaldar con media cabeza
-Veynte y un brazos enteros con sus hombros
-Onze brazos sin hombros
-Otros dos hombros solos
-Veynte y dos colillas

-Diez y siete pares de guantes
-Veynte y quatro pares de rodilleras con sus medios muslos
-Otro par grande de rodilleras de goznes
-Zinco piernas las dos con pies y las demás sin ellos
-Otra pierna y medio muslo con su pie
-Quatro testas de caballos
-Treynta y siete faldillas de los petos y sueltas
-Dos medios brazos con sus manoplas
-Zinco medios brazos de la muñeca al codo
-Sesenta y dos pedazos que se reconocen ser de las armas y de diferentes géneros
-Un atado de faldillas angostas para las armas
(fol. 215 v)
-Treze ballestas de virotes maltratadas
-Seys gafas para armarlas
-Tres montantes
-Tres cañones de mosquetes el mayor como de dos varas y media de largo y los dos poco menores con sus cajas sin llaves
-Un cañón de mosquete sin caja
-Dos cajas rotas de arcabuzes con sus llaves de rueda y maltratadas
-Una caja de mosquete rota con su llave de rueda
-Una caja de pistola con su llave de rueda maltratada
-Dos llaves de rueda que parezen de mosquetes
-Una turquesa para hazer balas de mosquetes
-Una lanzilla del Japón con vayna
-Un cañón de hoja de lata para papeles
-Un artificio de hierro pequeño quadrado con pie de madera
-Un zepo para armar lobos
-Tres mangas de malla

Todas las quales dhas armas están en la torre que está en dho palazio al lado del parque en el suelo y abiertas las ventanas muy maltratadas todas y orimientas (sic) y en esta forma se hizo dho ynventario por el dho Sr Don Gabriel Dirsen y del Vaus que lo firmó en la villa de Velada en zinco de noviembre de mill y seyszientos y setenta y ocho años= Don Gabriel Dirsen y del Vaus= ante mí= Ju^o^ Resino= E yo el dho Ju^o^ Resino ss^no^ Pú^co^ del número y ayuntamiento de esta v^a^ de velada por md del Exmo Sr Marqués de velada Astorga y Ayamonte conde de Trastámara mi S^r^ que presente fuy a lo que dho es saqué este traslado en pliego del segundo sello de su original que queda en mi poder en el quarto sello en veynte y ocho de mayo de mill y ss[os] y ochenta y nueve años y en fee de ello lo signé y firmé= entre R[s]= de la armería=

En testimonio de verdad
Ju[o] Resino
(signed).

Doc. 3.
1690, 1698, 1737.
Inventory of arms in the cabinet and in the armoury of the palace of the Marquis of Velada and Astorga, in the town of Velada.
Instituto Valencia de Don Juan (Archivo y Biblioteca), Manuscritos, 25-V-14.

(fol. 303 r)
Y en la forma referida se hizo y acabó este inventario (fol. 303 v) de pinturas papeles y otras alhajas contenidas en él con asistenzia del dho S^{or}, D^{n}, Joseph de Yanguas y Morales y dho S^{or}, D^{n}, Domingo Virto del Rincón y de mí el ess^{no} que le hizieron bien y fielme^{te} y firmaron de que doy fee= D^{n}, Joseph de Yanguas y Morales= D^{n} Carlos de Villarroel= D^{n}, Domingo Virto del Rincón= Ante mí Gabriel Fran^{co} Resino

Armas de fuego y otras cosas
Continuando en dhos ynventarios hizo su Merzed de las armas de fuego espadas y otras cosas que hay en el camarín del Palazio de esta dha villa ante mí el presente escribano en la forma siguiente

-Primerame^{te} treze pistolas las ocho con llaves de rueda y dos de chispa unas y otras franzesas una de chispa castellana y las dos sin llaves
-Una carabina de rueda
-Otra carabina de piedra de tres quartas de largo
-Dos caxas de tres dedos de alto media vara de largo y una quarta de ancho con instrume^{tos} matemáticos en que faltan algunas piezas
-Quatro estantes de madera dorados de azul y blanco de a tres cajones con las armas del Marqués mi S^{or}, en que están los libros
-Otros dos cajones de ébano con redes y llaves en que están los libros que se inviaron a s^{n}, Ju^{o} de Villarroel el año pass^{do} de nove^{ta} y siete para que se colocasen en dho camarín
(fol. 304 r)
-Seis retratos de poco más de quarta de alto yguales de mujeres estranjeras que están fijadas en el techo del dho camarín
-Tres mapas grandes arrollados el uno roto y el otro maltratado
-Otro más pequeño de la villa de gueldres y su planta

Y en la manera dha se hizo y acabó este ynventario con asistenzia de los referidos y de mí el ess^{no}, y firmaron= D^{n} Joseph de Yanguas y Morales= D^{n} Carlos de Villarroel= D^{n}, Dom^{o} Virto del Rincón= a^{te} mí Gabriel Fran^{co}, Resino=

Alhaxas que tiene el Marqués mi S^{or} en el palazio de esta villa y los xardines
Ynventario de las alhajas que el Marqués mi Sor de Velada y Astorga tiene en el palazio de esta villa y otras varias cosas en los jardines y huerta de la Palma

Armería

En uno de los desvanes del palazio hay diferentes piezas de armas ballestas y otras que serán poco más o menos en todas ziento y znquenta pero ya tan pasadas y mohosas del tiempo que no pueden servir de nada [...]

(fol. 306 r)
En la villa de Velada en zinco días del mes de octubre de mil seis^{tos} y nove^{ta} y ocho años su mer^{d} el S^{or}, D^{n}, Joseph de Yanguas y Morales contador de resultas de su Mag^{d} y secretario y contador de los estados del Exmo S^{or}, Marqués de Velada y Astorga mi S^{or}, y juez de este ynventario e virtud de orden del Sor Gobernador de ellos habiendo visto los autos y ynventarios prezedentes mandó se le dé traslado de él a la parte de Dª Ysabel de Villarroel y Ábalos madre tutriz tutora y curadora de los hijos menores de D^{n} Juan de Villarroel difunto a cuyo cargo (fol. 306 v) la administrazión y cobranza de las rentas del Marqués mi S^{or}, para que dentro de segundo día responda si tuviere que y con lo que dijere o no en vista de ello proveer justizia así lo proveyó mandó y firmó= D^{n} Joseph de Yanguas y Morales= Ante mí Gabriel Fran^{co} Resino [...]

(fol. 308 r)
Relazión de las alhajas que faltan
En cumplimiento del auto antezedente se hizo relazión de las faltas de las alhajas y libros en esta forma

(fol. 309 r)
Más alhaxas
-Dos hojas de espada una de dos cortes fina del Mro Domingo Cortes y otra de lomo de dos cortes.
Y en esta forma referida se hizo esta rela^{ón} (fol. 309 v) de faltas con la asistencia del dho liz^{do} [...]

(fol. 319 r)
Velada
Año de 1737

Testimonio con inserçión del ynventario de las executorias, papeles, librerías, ganados, colmenas, y demás alhajas que en esta su villa de Velada tiene la Exma Señora Marquesa de Astorga mi Señora y rezivo de todo dado por el S^{r}, D^{or}. D^{n} Mathías Antonio del Vado cura propio de la Yglesia parrochial de esta dha villa y el S^{r}, Liz^{do}. D^{n}. Ag^{n}. Fran^{co} Thirado abogado de los R^{s} Consejos vezino de la de Talavera, ambos apoderados de Su Exª para la administración, beneficio y cobranza de sus rentas; fecho dho ynventario en este año de 1737= por el Sr Joseph Gómez alcalde ordin^{o} de esta dha villa y por ante Fernando Pérez de Aguilera ss^{no} de S. M. y del núm^{o} y ayuntamiento de ella por graçia y merçed de Su Excelençia (fol. 319 v) D^{n}. Agustín Françisco Thirado= Ante mí= Fernando Pérez de Aguilera [...]

(fol. 340 v)
Armería
Y continuando su md en dho inventario pasó con dhos apoderados a uno de los desvanes que tiene dhas casas palaçio en donde se halló diferentes piezas de pistolas (fol. 341 r) escopetas, ballestas, petos, y morriones de açero todo muy oriniento y maltratado por el trascurso del tiempo p^{r} no haberse ruido que es lo mismo, que el ynventario antiguo expresa estar de esta calidad y sus piezas ser de armería antigua [...].

VILLAFRANCA, MARQUESES DE

Doc. 1.
1650, March 30.
Inventory of the Marquise of Villafranca.
Archivo Histórico de Protocolos de Madrid, Protocolo 4499.

30 de março de 1650
Inventario hecho por la S[a] D. María de Mendoça Marq[sa] de Villafranca viuda del Exmo Sor D. G[a] de Toledo Osorio marq[s] de Villafranca Duque de Ferrandina su marido

Aquí está la tas[on] de joias [...]

(fol. 97 r)
Espadines
-Dos espadines negros franceses
-Un espadín ancho culebreado con una concha
-Más otra espada con un adereço blanco q remata en memorias
-Más otro adereço culebreado las hojas de espada y daga que remata en memorias
-Un espadín de açero de Ynsplue (sic)
-Otro espadín de a caballo como sierra
-Otra espada con una adereço çinçelado
-Otra espada que llaman de las batallas
-Otra espada y daga con un adereço bruñido
-Otro espadín la guarniçión de hierro embutido de plata
(fol. 97 v)
-Otro espadín de plata de dos veneras
-Otro espadín angosto la guarnición de p[ta]
-Otro espadín la guarnición de plata
-Otro espadín de p[ta] con sus veneras
-Otro espadín francés con la guarnición plateada
-Otro espadín de plata con sus veneras
-Otro de plata con otras seis veneras y mescollo por pomo
-Otro espadín de açero de Ynsplue (sic)
-Son de la misma parte
-Otra espada con adereço de memorias con su daga
-Otros dos espadines de açero

-Otro espadín con una conchilla
-Otra espada y daga con adereço negro
-Otra espada con un adereço de dos conchas [...]
(fol. 99 r)
-Una pistola con la caxa de madera ordinaria
-Un frasquillo [...]
(fol. 100 r)
-Una arca de pino y en ella las pistolas sig^{tes}
-Dos pistolas yguales con sus fundas las cubiertas de ellas de terciopelo
-Otra pistola forrada en frisa blanca digo la funda de ella
-Otra pistola con su funda de frisa colodada
-Un çafión
-Otra pistola de faltriquera con llave de rueda [...]
-Otra pistola del mismo tamaño que es de chispa y el cañón labrado
-Otra del mismo tamaño en funda açul y el cañón rebordado
-Otra pistola del mismo tamaño llave de chispa caja blanca y funda de paño pardo
-Una daga pistola con su vayna y funda blanca
-Otra pistola de tres cañones llave chispa y funda parda
-Otra de dos bocas en funda de vaqueta negra
-Otra pistola de faltriquera con llave de rueda caja de nogal y funda negra
-Otra pistola de faltriquera llave de chispa caxa y funda blanca
(fol. 100 v)
Otra pistola de rueda con su caja de nogal y funda colchada
-Otra pistola de chispa caja blanca y funda blanca
-Dos cajas de pistolas con llaves y sin cañones
-Una pistola de arçón de chispa funda blanca
-Otra pequeña de chispa con la caja de nogal
-Otra de faltriquera llana con funda colorada
-Otra pistola catalana algo larga con cañón labrado y caja labrada de bronce y funda colorada
-Otra que es compañera de la anteçedente
-Otra pistola de cinta con boca de mosquete y una almareda y llave de rueda con su funda parda
-Otra pistola llana de chispa con una funda blanca
-Otra pistola de llave de rueda caxa labrada de p^{ta} y el cañón pareçe hecho por la Caracominas
-Otra pistola de la misma forma en funfas frailescas
-Un carabinón con su coz de hierro atrás que se dobla

-Otra pistola de faltriquera llana de chispa con funda colorada
-Otra pistola del mismo tamaño labrado el guardam^{s} en una funda morada
(fol. 101 r)
-Dos pistolas de arçón yguales de chispa labrados los cañones y caja de ébano con funda de vaqueta colorada y flueco verde
-Otras dos q las llaves son de rueda y sin cañones la caxa blanca en una funda de vaqueta e terciopelo roxo con fluecos pardos
-Otras dos lisas con llaves de chispa y figuras de bronce dorado en los remates con cajas de vaqueta (sic) parda
-Otras dos de rueda y atauxía e figuras [...] en caxas de vaqueta con terciopelo verde
-Tres llaves de montar un martillo y quatro turquesas
-Una bolsa de martas con un frasco [...]

(fol. 116 r)
Armas
En la villa de Madrid a dieciocho días del mes de m^{ço} año de mill y sei^{o}s y çinquenta Juan Sánchez maestro armero con junramento que hiço a Dios y a una cruz en forma de derecho hiço la tasaçión de las armas q le fueron mostradas y quedaron por muerte del Sor Marq^{s} Don Garçía de Toledo, Osorio, la qual hace en la manera sig^{te}

Juan Sánchez mro armero
-Primeramente tasó un par de pistolas de a caballo con llave de chispa en çinquenta Rs__U050
-Otro par del mismo género en lo mismo__U050
-Un par de pistolas catalas (sic) largas tasadas en ocho du^{os}__U088
-Otro par de pistolas de rueda de munição y de la Caricominas guarneçidas de plata se tasaron en diez y seis ducados
-Otro par de pistolas de rueda con cajas de ébano (fol. 116 v) se tasaron en ocho ducados__U088
-Otro par de pistolas de pretina con llaves de chispa se tasaron en diez ducados_ _U110
-Otro par de pistolas catalanas cajas labradas de latón de çinta tasadas en doce ducados__U132
-Otro par de pistolas françesas guardamontes y guarnición labradas se tasaron en once du^{os}__U121
-Otro par de pistolas llanas en diez ducados__U110
-Una pistola de rueda de Granada de munición se tasó en doce ducados__U132

-Otra pistola de chispa de çinta con cañón rayado en çien Reales__U100
-Otra pistola de zinta con llave de chispa con rastrillo de bisagra en diez ducados_ _U110
-Otra pistola de çinta de llave francesa cañón grabado se tasó en ocho ducados_ _U088
-Otra pistola de rueda pequeña se tasó en seis ducados__U066
-Otra pistola de rueda del mismo tamaño en lo mismo__U066
-Otra pistola de rueda en quarenta Rs__U040
-Otra pistola francesa también pequeña tasada en zinco ducados_U055
-Otra pistola de rueda de a caballo tasada en tres ducados__U033
-Otra pistola larga catalana tas^{da} en tr^{a} y seis Rs__U036
-Otra de lo mismo en lo mismo__U036
-Otra pistola de çinta de dos cañones en doce du^{os}__U132
(fol. 117 r)
-Una pistola dentro de una daga en cinco ducados__U055
-Una pistola francesa de dos cañones y dos llaves en doce ducados__U132
-Otras dos pistolas de faltriquera con caxas de bronçe se tasaron en doçe ducados ambas__U132
-Otra pistola de una llave con armas en ocho du^{os}__U088
-Otra en una llave más pequeña en çinq^{ta} Rs__U050
-Un arcabuz pequeño tasado en sesenta Rs__U060
-Un arcabuz de llave de Simón en treçientos Rs__U300
-Un arcabuz viexo tas^{do} en zinco ducados__U055
-Otro arcabuz maltratado en sesenta Rs tas^{do}_U060

La qual dha tasaçión el dho tasador declaró haber hecho bien y fielmente a todos su saber y entender y ansí lo juró en forma de derecho y lo firmó testigos Don Gregorio V^{a}mañe Pedro Xil del Valle y Françisco Sanctos criados de la dha S^{ra} marquesa doy fee= y dijo que es de edad de tr^{a} y q^{tro} años poco más o menos

Juan Sánchez
Ante mí Joan de Béjar
(signed)

(fol. 117 v)
Espadero
En la villa de Madrid a dieciocho días del mes de março del dho año de mil y sei^{o}s y cinquenta Julián de Urande maestro espadero con juramento que hiço a Dios y a una cruz en forma de derecho tasó las cosas tocantes a su ofiçio q

quedaron por fin y muerte del Exmo Sor Don García de Toledo Osorio marqués de Villafranca y de Ferra^{na}, la qual dicha tasaçión hiço en esta manera

Ju^{o} de Urande mro espadero
-Primeramente tasó una espada de Sahagún con guarnición de memorias conchas llanas en diez ducados__U110
-Una espada de Alemania con guaniçión de garapiña en tres ducados__U033
-Una hoja vieja maltratada de Seb^{án} Hernández con guarnición pavonada en çinq^{ta} Rs__U050
-Una espada gineta con guanición de dos órdenes de mem^{as} en tres ducados_ U033
-Otra espada de Toledo mala guarnición grabada en quarenta Rs__U040
-Otra espada y daga con guarniziones de vueltas y gavilanes retorçidos en seis du^{os}__U066
-Un espadín de a caballo culebreado en cinco du^{os}__U055
-Otro espadín françés de a cav^{o} en çinq^{ta} Rs__U050
-Otro espadín alemán concha en redondo tasado en quatro ducados__U044
-Otro espadín de Alemania de lomo conchas en redondo pavonado en diez y seis Rs__U016
(fol. 118 r)
-Otro espadín como el mismo en lo propio__U016
-Otro espadinillo de una concha ocho Rs__U008
-Dos espadinicos pequeños tasados en diez y seis reales__U016
-Otro espadín con guanición de plata en quinientos reales de concha y media_ _U500
-Otro espadín guarnición plateada con seis conchas se tasó en çien reales__U100
-Otro espadín con guarnición de plata en duçientos y ochenta Rs sin conchas tas^{do}__U280
-Otro espadín de concha y media guarnición de hierro embutida en plata en çiento y çinq^{ta} Rs__U150
-Otro espadín de plata de una concha en doçe du^{os}__U132
-Otro espadín guarnición de azero en diez ducados tasado__U110
-Otro espadín de açero guarnición de plata de concha y media tasado en veinte ducados__U220
-Un cuchillo de monte dorado con vaina de terciopelo se tasó en ocho ducados_ _U088
-Otro cuchillo de m^{te} cabo de venado en catorce Rs__U014
-Dos puñales de çinta a seis Rs__U012
-Un cuchillo damasquino tas^{do} en diez y seis Rs__U016

La qual dha tasación el dho tasador declaró haberla hecho a todo su saber y entender sin fraude algu^{o} presente dho contador s^{do} testigos D. Gregorio Villamañe P^{o} Xil del Valle y Fran^{co} Santos residentes en esta corte= y que es de edad de cinq^{ta} y tres años poco más o menos= y no firmó porque no supo

Ante mí Joan de Béjar
(signed).

VILLANUEVA DEL FRESNO, MARQUESES DE

Doc. 1.
1634, September 9.
Inventory of Marquis of Villanueva del Fresno, Antonio Moscoso.
Archivo Histórico de Protocolos de Madrid, Protocolo 5201.

(fol. 1 r)
Doña Fran^{ca} Fernández de Portocarrero marquesa de Villanueba del Fresno y Balcarrota (sic), señora de Moguer, viuda de don Antonio Moscosso Marq^{s} de Villanueba del Fresno mi señor y marido= digo que el suso dho ha fallecido y pasado de esta presente vida en la jornada del serenísimo señor Ynfante don Fer^{do} haciendo of^{o} de enbaxador de su Mag^{d}= y p^{a} que en todo tiempo conste de los bienes que han quedado por su muerte es neçess^{o} haçer ynvent^{o}= Por tanto a Vm^{d} supp^{co} le mande haçer por ante escribano Pido Justiçia ett^{a}=

La marquesa de Villanueba
(signed)

(fol. 1 r)
Auto
Que se haga el ynvent^{o} de los bienes que por la pet^{o}n se pide [...] por en [...] En M^{d} a nueve de sep^{e} de myll seis^{o} y tr^{a} y quatro

Ante mí Nicolás Gómez
(signed)

En la villa de Madrid a nueve de sep^{e} de mill y seisçientos y tr^{a} y quatro años la S^{ra} marquesa de Villanueba del Fresno doña Fran^{ca} Fernández de Portocarrero por ante (fol. 2 r) el escribano en cumplim^{o} del auto del señor teniente estando en las casas de su morada començó hacer ynvent^{o} de los bienes q han quedado por m^{te} del señor don Ant^{o} Moscosso su marido en la forma y manera siguiente [...]

(fol. 27 r)
-Quatro sillas de caballos
-Quatro guaniçiones de mula de coche

Y en este estado quedó el dho ynventº y declaró su ssª no tiene noticia de otros ningunos bienes con protestación de que si tuviese de otros los pondrá por ynventario y lo firmó de que doy fee=
La marquesa de Villanueba

Ante mí Nicolás Gómez
(signed)[...]

(fol. 62 r)
Tasa^{ón} de espadas
En la villa de madrid a v^{te} seis de otbre de mil y seiºs y trª y cinco años Pº Palín espadero tasador nombrado pª las cosas de su oficio habiendo jurado en forma de drº fue haciendo la tasación de los bienes sig^{es}

Pedro Palín
Espadero
-Primeram^{te} tasó un adereço de espada y daga plateado con un estoque de Belmonte de Toledo y la hoja de la daga vizcaína con sus tiros y pletina en ciento y cinquenta rs- 150
-Otro adereço de espada dorado con sus tiros y pletina q la hoja es de un corte de Fran^{co} Gómez de Toledo en çien reales- 100
-Otro adereço con la guarnición bruñida con dos veneras la hoja de Tomás de Ayala entre fallada con tiros y pletina en cien rs- 100
-Otro adereço negro de Granada digo espada daga tiros pletina que la hoja es de Toledo muy antigua que parece es de Pedro de Toro y la daga vizcaína en çiento y cinquenta reales- 150
-Otro adereço ochavado negro liso espada y daga tiro y pletina que la hoja es de Fran^{co} Ruiz de las viexas y la daga vizcaína en ciento y cinquenta rs- 150

(fol. 62 v)
-Yten tasó un espadín con la hoja de Alemania menor de marca con mascarillas negras y doradas en la guarnición con una venera por guardamano en tres ducados- 033
-Otro espadín la guarnición color de hierro la hoja menos de marca extranjera en tres ducados- 033
-Yten una espada ancha de a caballo la guarnición manquilla y la hoja extranjera en quatro ducados- 044
-Otra espada de a caballo con la g^{ón} manquilla lisa con la hoja extranjera tasó en treinta rs- 033

-Un cañón de un alcabuz sin llave ni caxa tasó en tres ducados- 033
-Dos pistolas de rueda de alarçón de a caballo las caxas enbutidas de marfil y acero en quarenta ducados- 044
-Unas espuelas negras aventanadas diez reales- 010
-Otras espuelas doradas en diez rs- 010
-Otras espuelas plateadas en seis rs- 006

Y en la manera que dho es el dho Pº Palín hizo la dha tasación debajo de dho juramento y [...]

(fol. 69 r)

Ynformación de cómo murió abentestato el señor don Antº de Moscosso y el día de su muerte=

Doña Fran^{ca} Portocarrero Marquesa de Villanueba del Fresno= Viuda de Don Antonio Portocarrero y Moscoso= difunto= en el pleito de partición a los bienes del suso dho= Digo que estando en servicio del S^{mo} Car^{l}. Ynfante don Fernando falleció en la ciu^{d} de Rotamburgº en la provincia de Alemania abintestato y sin haçer testamento en veinte y nueve de julio del año pasado de treinta y quatro= a Vm^{d} pido y spp^{co} que para que conste en la dha partición de lo suso dho para los efectos que fueren necesarios mande recebir ynformaçión q ofrezco de lo suso dho con çitación del curador ad liten de don Juan Domingo Portocarrero mi hijo y del dho marq^{és} pido justicia y para ello ettª

La marquesa de Villanueva
(signed)[...]

(fol. 70 r)
Ynformazión
Tº
En la vª de Madrid a v^{te} y tres días del mes de abril de mill y seiszientos y treinta y siete años de presentación de la señora doña Fran^{ca} de Portocarrero viuda marquesa de Villanueva del Fresno juró para esta información Juan de Villena gentilhombre de la dha señora marquesa= y siendo preguntado al tenor del pedim^{to} para esto presentado dixo sabe que el señor don Antonio Portocarrero y Moscosso marqués que fue de Villanueva del Fresno y marido de la dha señora marquessa fallezió en la ziu^{d} de Rotenberg en la provincia de Alemania y país del Tirol abintestato sin hazer testamento estando en servizio del serenísimo Cardenal Ynfante don Fernando en v^{te} y nueve días del mes de julio del año pasado de mill y seiszientos y treinta y quatro a cossa de las zinco de la mañana

poco más o menos= y esto lo sabe este testigo porque a la sazón estaba en su serviçio y hazía ofizio de su mayordomo por haber quedado enfermo el que lo era en la ciu^{d} de Milán y haberse hallado presente a su muerte y haberle visto muerto naturalmente porque murió en los brazos de este testigo y en esto es lo que sabe y la verdad para el juramento que tiene hecho y público y notorio en que se afirmó y ratificó y lo firmó de su nombre dixo ser de edad de quarenta y zinco años poco más o menos=

Juan de Villena
Ante mí [...] Criado
(signed)

(fol. 70 v)
T^{o}
Juro sobre lo suso dho Fran^{co} de Mendoza mozo de silla que al presente es de la dha S^{ra} marquesa y lacayo que fue del dho señor marqués el qual hizo cumplidamente y por medio de decir verdad y siendo preguntado al tenor del dho procedimiento= dixo que como tiene dho este testigo fue lacayo del dho señor marqués y pasó con su ss^{a} a los estados de Alemania y sabe que estando al servicio del Sereníssimo Ynfante Cardenal don Fernando= le dió una enfermedad en la ciu^{d} de Ynsprug Corte del archiduque Leopoldo y vino con ella a la ziu^{d} de Rotemberg donde se echó en una cama y se le quitó luego la habla de forma que no pudo hazer testamento y murió abintestato en v^{te} y nueve días del mes de julio del año pasado de mil y sei^{o}s y tr^{a} y quatro a cosa e las cinco de la mañana poco más o menos y esto lo sabe este testigo porque como tal su criado asistió a su enfermedad juntamente con Juan de Villena que hazía el oficio de su mayordomo y otros criados y lo vió muerto naturalmente y él y los demás criados lo trujeron desde los dos estados de Alemania hasta la dha v^{a} de Villanueba del Fresno donde está sepultado en su entierro en el convento de San Fran^{co} de la dha v^{a} y esto es lo que sabe y la verdad para el juramento que tiene ff^{o} en que se afirmó e rratificó y no firmó porque dixo no saber y dixo ser de edad de treinta y siete años poco más o menos [...]

(fol. 71 r)
T^{o}
En la dha villa a v^{te} y quatro días de esta dho mes y año juró sobre lo suso dho don Antonio Portocarrero caballero de la horden de Calatrava residente en la dha villa= dixo que al tiempo y cuando el dho señor don Antonio Portocarrero y Moscoso marqués de Villanueba del Fresno pasó a los estados de Flandes a

servir a su alteza el Sereníssimo Cardenal Ynfante este testigo pasó arrimado a la persona de Su Sª y por su camarada y anduvo con él todo el tiempo que estuvo en los dhos estados y en los de Alemania= y sabe que en la ziu^{d} de Esprug le dio una enfermedad y vino con ella hasta la ziu^{d} de Rotemberg donde se echó en una cama y luego se le quitó la habla de forma que no pudo hazer testamento ni disponer de sus cossas y murió abintestato en v^{te} y nueve diías de mes de julio del año pasado de mill y seiszientos y treinta y quatro= y este testigo lo vió muerto naturalmente y como deudo y camarada suyo se halló a su muerte y vino por cabeza de la gente que traya el cuerpo de su Sª hasta la villa de Villanueba del Fresno y meterle en su entierro= y sabe ansimismo murió el dho día veinte y nueve de julio del dho año a las zinco de la mañana poco más o menos= y esto es la verdad púbº y notorio pública voz y fama y lo que sabe para el Juramento que tiene hecho en que se afirmó ratificó y lo firmó de su mº dixo ser de hedad de treinta y zinco años poco más o menos [...]

Don Antº Portocarrero y Monrroy
(signed) [...]

(fol. 89 r)
Relaçión de los bienes y haz^{da} que el Sr D. Antº Moscosso Portocarrero entró por su capitán al tiempo que casó con la s^{ra} D. Francisca Portocarrero marquesa de Villanueba del Fresno presentada por mí Antonio de Castro curador aliten del Sr Don Juº Gaspar Domingo Portocarrero hixo de los dos S^{res} para las particiones que están haciendo [...]

(fol. 91 v)
Espadas
-Un adreço de espada y daga plateado con un estoque de Belm^{te} de Toledo
-Otro adreço de espada dorado con la hoxa de un corte de Fr^{co} Gómez de Toledo
-Otro adreço con la guarnición bruñida y la hoja de Tomás de Ayala
-Otro adreço ochavado negro liso que la hoja es de Fr^{co} Ruiz de las viexas
(fol. 92 r)
-Una espada de a caballo la guarnición manquilla y hoja estrangera
-Otra espada de a caballo con la guarnición manquilla lisa la hoja estranxera
Espuelas
-Unas espuelas negras aventanadas
-Otras espuelas aventanadas
-Otras espuelas plateadas

Carroças
-Una carroça de media vuelta de baq[ta] negra y damasco carmesí con cortinas de verano y ebierno (sic) con sus vidrieras y cristales a los lados y adelante
-Yten quatro mulas de este coche que se consumieron durante el matrimonio
-Quatro guarniciones viexas
-Quatro sillas de caballos [...].

VILLANUEVA DEL RÍO, MARQUESES DE

Doc. 1.
1599, July 23.
Inventory and appraisal of Marquis of Villanueva del Río.
Archivo Histórico de Protocolos de Madrid, Protocolo 1418.

(fol. 1663 r)
Doña María Manrique Marquesa de Villanueva del Río viuda mujer q fuy del marqués don Fr^{do} Enríquez dí firmado por mí y como tutora y curadora de las personas y bienes de mis hijos y myos digo q sea hecho ynventario solene (sic) de los bienes que quedaron por fin y muerte del dho marqués mi marido, y hay muchos muebles axuar y preseas de casa vestidos plata y otras cosas las quales conviene se apreçien sus valores ansí la partición que se haya de hazer como por lo que hubieren de ellos menester los dhos menores y pa hazer almoneda o pa otro qualquier efeto a Vm pido y sup^{co} mande se aprecien los dhos bienes y pa ello nombrar personas q con juram^{to} los hagan bien y fielmente pa lo qual

Doña Mª Marq^{sa}
(signed)

(fol. 1663 v)
En la villa de Madrid a veynte y tres días del mes de julio de mil e qui^{o}s e noventa e nueve a^{o}s ante el señor liçen^{do} Antonio Rodríguez tiniente de correg^{dor} en la dha villa de M^{d} e su tr^{a} por el rey nro señor por parte de la marquesa de Villanueva del Río se presentó esta petición e por su merced vista mandó que se haga la tasación y apreçio de los bienes del dho marqués de Villanueva del Río don Fernando Enríquez la qual se haga por ante Françisco de Quintana escribano de su Mag^{d} e nombró para la tasación del apreçio de los dhos bienes a Antonio Juberto y Roque de Santa Cruz plateros para las cosas de oro y a Juan Sánchez e Francisco de Fresnillo plateros para las cosas de plata y a Çamora sastre para las cosas de vestidos y a Juan de Sola calcetero prara las cosas de calças y a Carrión tapiçero para la tapiçería y a León Camero para las colgaduras y camas y a Simón alcabuzero para los arcabuçes y ballestas y cosas semejantes y a Andrés Díaz espadero para las espadas y a Beatriz López para la ropa blanca y a Juan Pérez entallador p^{a} las cosas de madera y a Çinbranos pintor p^{a} las pinturas y a Briçuela cochero para los coches y literas y para los vidrios y barros y porcelanas a Manuel

Loma e pª las demás menudencias a Melchor de Tapia criado de la dha marquesa y ansí lo proveyó mandó e firmó

El Licº Rodríguez
Pasó ante mí Grabriel de Rejas
(signed)[…]

(fol. 1691 r)
En la dha villa de Madrid a treze días del dho mes y año se continuó la dha tas^{ón} y pareció Andrés Díaz espadero y dixo que ha sido nombrado para tasar las espadas y dagas del dho S^{or} marqués por tanto q haçía y hizo la dha tas^{n} en la forma sig^{te}

-Una espada de ataujía de oro con sus tiros y petrina de tropº con sus trencillas de oro y daga de lo mismo- digo q la guarnición es de ataujía y la hoja de Sebastián Hernández el viejo y puños de oro y hierro en veinte duºs
(fol. 1691 v)
-Otra espada dorada acanalada guarneçida de tiros y petrina de cuero y daga de lo mismo con puño de plata dorado la hoja de la espada de Juº Mtnez de Toledo en diez y seis duºs
-Otra espada labrada de ataujía negra y puño dorado con tiros y petrina de trpº digo la guarnición es de Milán manquilla labrada de relieve y la hoja de Domingo Rodríguez de Toledo y daga de la misma guarnición quinze duºs
-Una espada la guarnición plateada y de negro conteras y petrina de trºpelo negro birdados de plata y daga de lo mismo digo q la guarnición es de cadenilla nielada y puños de plata blanca y la hoja de Ortuño en veinte y quatro duºs
-Otra espada dorada acanalada con tiros y petrina de cuero sin daga puños de plata y la hoja de Bilbao quatro duºs
-Otra espada el pomo acabalado con tiros y petrina de cuero quatro duºs sin daga- hoja de Ortuño
-Otra espada la guarnición labrada de negro con tiros y petrina de cuero la guarnición ençestada y hoja de Sahagún- quatro duºs
-Otra espada y daga la guarnición de ataujía de Milán de oro y negro (fol. 1692 r) con tiros y petrina de negro y sin contera- es la guarnición manquilla puños de plata hoja de Milán tres duºs- digo quatro duºs
-Otra espada pavonada de la guarnición lisa sin ningún adereço puño de plata y seda cinquenta reales- hoja de Sahagún
-Un estoque de más de la marca la guaniçión de color de hierro- hoja de Sebastián Hernández cinco duºs

-Quatro dagas con sus guarniciones doradas las tres labradas y la otra lisa- la una q es la lisa es de caracolillo y puño de plata en doze Rs y otra labrada de claraboyas y el puño de plata en diez y seis Rs y la otra de ramos y coronas y puño de plata con tiros y petrina de la misma obra bordado de plata y oro tres du^{o}s y la otra con un adereço todo dorado y labrado de cuerdas y frutaje sin contera de espada ni brocal de daga ni clavaçón de talabarte en seis du^{o}s
-Un adereço dorado de espada y daga y talabarte con unos laços todo el pomo quajado y los (fol. 1692 r) botones de la guarnición en v^{te} du^{o}s
-Otra daga la guarnición barniçada seis Rs
-Una hoja de daga suelta dos Rs
-Otras dos hojas de daga a dos Rs
-Un cuchillo de monte dorado la guarnición quatro du^{o}s
-Un alfanxe la guarnición negra de cuerno y la vaina de paño amarilla en seis Rs
-Un cuchillo damasquino con una vaina de çapa guarneçido de plata v^{te} y quatro du^{o}s
-Otro cuchillo de monte con dos cuchillos y un punçón guarneçido de verde hecho en Guadalaxara en doze Rs
-Otro cuchillo de monte con un escudo con una vaina verde vieja quatro Rs
-Otro cuchillo de monte con un cuchillo de Palermo y una vaina vde ocho Rs
-Dos hojas viejas anchas de a caballo la una del perrillo y la otra hecha en Galicia la del perrillo dos du^{o}s y la otra seis du^{o}s
-Una hoja de Toledo vieja en doze du^{o}s
-Dos hierros de lança quatro du^{o}s

(fol. 1693 r)
Y juró en forma de dr^{o} q la dha tas^{on} la ha hecho bien y fielm^{te} en sus justos y moderados preçios a todo su saber y entender sin hacer agravio alg^{o} y lo firmó t^{o}s P^{o} Bravo y Diego de Ysla y Ant^{o} de Cuéllar estantes en esta corte

Andrés Díaz
Ante mí Fr^{co} de Qui^{na} es^{o}
(signed)[…]

(fol. 1709 r)
En Madrid este dho día [23]- Simón Marquart arcabucero dixo que ha sido nombrado para tasar de los bienes que quedaron por fallecimiento del dho marqués de V^{a}n^{a} los bienes de suso conthenidos- por tanto que tasaba y tasó los dhos bienes […] y forma sig^{te}

(fol. 1709 v)
Arcabuzes y ballestas
-Un arcabuz de la jarrilla con la llave de Simón q no tiene la marca le tasó en doze ducados
-Otro arcabuz de quatro palmos y mº el cañón y llave del dho Simón hecho en duçientos Rs tasado
-Otro cañón y llave del dho Simón hecho en dhos duçientos Rs tasado
-Otro pistolete con dos llaves la caja guarneçida de taracea con su herramienta y funda en ciento y zinq^{ta} Rs
-Otro pistolete viejo en veinte Rs
-Una ballesta des[...]ita con su gaxa (sic) de lo mismo ducientos Rs
-Otra ballesta de Pº de la Fuente el tablero y gafas en ciento y zinq^{ta} Rs
-Otra ballesta de Azqueta sin gafa çien Rs
-Otra que tiene el tablero de ençina negra de Uzeda q en la verga diçe el conde de Osorno sin gafa en seis ducados
-Otra ballesta el tablero de Reina (¿) y la verga de Luis Moreno con una gafa vieja ocho duºs
-Dos carcaxes de cuero el uno bayo y el otro con cubierta de cuero de tigre en tres duºs
-Otro carcax pequeño bayo para virotes redondo quatro Rs
(fol. 1710 r)
-Otro carcax chico para tres jaras en seis Rs
-Otro carcax redondo con veinte y siete jaras con su arco bajo en tres ducados
-Tres dezenas de virotes y catorce jaras tasado a tres [...] uno con ocho que montan treinta y siete Rs y mº
-Dos horquillas grandes y tres chicas las grandes a quatro Rs y las chicas a tres Rs
-dos çintos a quatro Rs cada uno
-Una llave de armar arcabuzes en quatro Rs
-Un cañón de arcabuz en quatro Duºs es rayado por de dentro
-Un pistolete de hierro que está en casa de Simón tasado por él dos ducados
Y juró en forma de drº que ha tasado los dhos bienes en sus justos precios a todo su saber y entender sin hacer agravio algº y lo firmó siendo tºs Juº [...] y Hernando de Casanova estantes en esta corte

Simón Marquarte
Ante mí Fr^{co} de Qui^{na} escº
(signed)[...].

ONOMASTIC APPENDIX[1]

---

1 Not all the names of people cited in the text of the book are reflected in the onomastic appendix, especially in what has to do with the documentary appendix. The names of people related to arms and armour, and armouries, have been collected (owners, manufacturers, personnel related to their custody, as well as culturally and politically relevant figures). But to avoid confusion and excess size of the appendix, the names of notaries, witnesses in documents, and people who have no direct relationship with the subject of this study have been omitted.

www.ingramcontent.com/pod-product-compliance
Ingram Content Group UK Ltd.
Pitfield, Milton Keynes, MK11 3LW, UK
UKHW040857081026
15587UKWH00016B/496

9 783631 915578